T0277010

A HISTORY OF MAGIC AND EXPERIMENTAL SCIENCE

VOLUMES VII AND VIII
THE SEVENTEENTH CENTURY

VOLUME VII

A HISTORY OF MAGIC AND EXPERIMENTAL SCIENCE

VOLUMES VII AND VIII
THE SEVENTEENTH CENTURY

By LYNN THORNDIKE

VOLUME VII

NEW YORK

COLUMBIA UNIVERSITY PRESS

In the publication of this volume the author has received assistance from the fund for the encouragement of historical studies bequeathed to Columbia University by Professor William A. Dunning.

Clothbound editions of Columbia University Press books are Smyth-sewn and printed on permanent and durable acid-free paper.

PUBLISHED 1958, COLUMBIA UNIVERSITY PRESS

ISBN: 978-0-231-08800-8

LIBRARY OF CONGRESS CATALOG CARD NUMBER: 23-2984

PREFACE

With these two volumes on the seventeenth century *A History of Magic and Experimental Science,* upon which I began work over fifty years ago, draws to its close.

The process of publication has been slow and, because of the heavy cost of printing, it has usually not been feasible to take note of books and articles which have appeared since the typescript went to press. A bibliography of books on magic in MS 500 at Besançon, seen in the summer of 1956, dates the printing of the dissertation of Dorscheus, *De . . . Satanae obsessione,* at Leipzig in quarto in 1646, rather than posthumously at Rostock in 1666, as quoted from Hauber at VII, 368-69.

It may be useful to list here a number of subsequent articles which supplement or revise certain portions of the first four volumes of this work. "Albumasar in Sadan," *Isis,* 45 (1954), 22-32, bears this relation to I, 651. "Traditional Medieval Tracts Concerning Engraved Astrological Images," *Mélanges Auguste Pelzer,* Louvain, 1947, pp. 217-74, revises I, 340, 664-66; II, 223-27, 234-35, 257-58, 280, 389-91, and 399-400. "More Manuscripts of the *Dragmaticon* and *Philosophia* of William of Conches," *Speculum,* XX (1945), 84-87, adds to II, 63-65. The long chapter on Peter of Abano with its eight appendices at II, 874-947, has been supplemented by "Relations of the Inquisition to Peter of Abano and Cecco d'Ascoli," *Speculum,* I (1926), 338-43; "Peter of Abano and the Inquisition," *Speculum,* XI (1936), 132-33; "Translations of Works of Galen from the Greek by Peter of Abano," *Isis,* 33 (1942), 649-53; "Manuscripts of the Writings of Peter of Abano," *Bulletin of the History of Medicine,* XV (1944), 201-19; "The Latin Translations of the Astrological Tracts of Abraham Avenezra," *Isis,* 35 (1944), 293-302; "Henri Bate on the Occult and Spiritualism," *Archives internationales d'histoire des sciences,* 27 (1954), 133-40; "Peter of Abano and Another Commentary on the Problems of Aristotle," *Bulletin of the History of Medicine,* 29 (1955), 517-23. The chapter on Cecco d'Ascoli which follows that on Peter has been further supplemented by "More Light on Cecco d'Ascoli," *The Romanic Review,* 37 (1946), 293-306. To Chapters 62 and 63 of this second volume should now be added: "Further Consideration of the *Experimenta, Speculum astronomiae,* and *De secretis mulierum* ascribed to Albertus Magnus," *Speculum,* 30 (1955), 413-43.

"Pliny and *Liber de presagiis tempestatum,*" *Isis,* 34 (1942), corrects III, 273 and 707-14. "Milan Manuscripts of Giovanni de' Dondi's Astronomical Clock and of Jacopo de' Dondi's Discussion of Tides," *Archeion,* 18 (1936), 308-15, is supplementary to Chapter 24 in my third volume. Concerning Oresme I have added: "Coelestinus's Summary of Oresme on Marvels," *Osiris,* I (1936), 629-35, and "Oresme and Fourteenth Century Commentaries on the *Meteorologica,*" *Isis,* 45 (1954), 145-52. Besides the two chapters on Heingarter in the fourth volume, there is: "Conrad Heingarter in Zurich Manuscripts, especially his Medical Advice to the Duchess of Bourbon," *Bulletin of the Institute of the History of Medicine,* IV (1936), 81-87.

I am indebted to many libraries and librarians but especially to Columbia University for providing a work-place for me in Low Memorial Library. Dr. C. Doris Hellman has been very helpful in the verification of references, other research assistance, and the early and last stages of proof-reading. Grants from the William A. Dunning Fund have aided in meeting the expense of printing.

LYNN THORNDIKE

Columbia University
October 11, 1956

CONTENTS

ABBREVIATIONS

AE	Acta eruditorum
Alegambe	Bibliotheca scriptorum societatis Iesu, Antwerp, 1643; Rome, 1676
BEC	Bibliothèque de l'Ecole des Chartes: revue d'erudition consacrée spécialement à l'étude du Moyen Age, Paris, 1839 to date
BL	Bodleian Library, Oxford
BM	British Museum, London
BMsl	British Museum, London, collection of Sloane manuscripts
BN	Bibliothèque Nationale, Paris
c	century
c.	circa
Col	Columbia University Library, New York
comm.	commentary or commentator
Cornell	Cornell University Library, Ithaca, N.Y.
Correspondance	Correspondance de P. Marin Mersenne, publiée par Mme. Paul Tannery, ed. Cornelis de Waard, 4 vols., 1932, 1936, 1946, (1955)
dedic	dedication, dedicated to, etc.
DNB	Dictionary of National Biography, London, 1885-1901, 63 vols.
Duveen	Denis I. Duveen, Bibliotheca Alchemica et Chemica, London, 1948
EB	Encyclopedia Britannica
ed	edited by, edition, editor
Ferg or Ferguson	John Ferguson, Bibliotheca Chemica: a catalogue of rare alchemical, chemical and pharmaceutical books, manuscripts and tracts... in the collection of J. Young, Glasgow, 1906
fr	français or French
Graesse	J. G. T. Graesse, Trésor de livres rares et précieux, ou, Nouveau dictionnaire bibliographique, Dresden, 1859-1869, 7 vols.
GS	George Sarton, Introduction to the History of Science, 3 vols. in 5, Baltimore, 1927, 1931, 1947
Hoefer	J. C. F. Hoefer, Histoire de la chimie..., Paris, 1842-1843, 2 vols.
Jöcher	Ch. G. Jöcher, Allgemeines Gelehrten Lexicon, Leipzig, 1750-1751, 4 vols.
JS	Journal des Sçavans
LC	Library of Congress
LR	Lindenius Renovatus (J. A. van der Linden, De scriptis medicis, revised edition by G. A. Mercklin, Nürnberg, 1686)

MS, MSS	Manuscript, Manuscripts
n.	note or footnote
n.d.	no date of publication
NH	Natural History
NYAM	New York Academy of Medicine
NYP	New York Public Library
p. pp.	page, pages
Poggendorff	J. C. Poggendorff, Biographisch-Literarisches Handwörterbuch
pr	printed, printer
Pritzel	G. A. Pritzel, Thesaurus literaturae botanicae
pt.	part
PT	Philosophical Transactions
Sbaralea	Supplementum et castigatio ad Scriptores trium ordinum S. Francisci a Waddingo aliisque descriptos, Rome, 1806, 2 vols.; revised edition, 1909, 1921
s.l.	sine loco (without place of publication)
Sudhoff (1902)	Karl Sudhoff, "Iatromathematiker vornehmlich im 15. und 16. Jahrhundert," Abhandlungen zur Geschichte der Medizin, Heft II (1902), Breslau, viii, 92 pp.
T	Lynn Thorndike, A History of Magic and Experimental Science, New York, 1923, 1934, 1941, 6 vols.
tr	translated, translation, translator
VA	Vatican, and Vatican Latin manuscript
VAb	Vatican, Barberini Latin manuscript
vol.	volume
Wadding	Luke Wadding (1588-1657), Scriptores ordinis minorum..., editio novissima, Rome, 1906
Will	Georg Andreas Will, Nürnbergisches Gelehrten-Lexicon, 1755-1758, 4 vols.
Zedler	J. Zedler, Grosses Vollständiges Universal Lexicon aller Wissenschaften und Künste, Halle and Leipzig, 1732-1750, 64 vols.
Zetzner	Lazarus Zetzner, Theatrum chemicum, Strasburg, 1659-1661, 6 vols.

A HISTORY OF MAGIC
AND EXPERIMENTAL SCIENCE

THE SEVENTEENTH CENTURY

CHAPTER I

BACKWARD GLANCES

Le Gendre—Brucker—Facciolati—Witchcraft Delusion—Method of our investigation—Nature and Magic—Arcana—New and Old.

Bei einer Reihe von Männern wie Nicolaus von Cusa, Bernardinus Telesius, Franciscus Patritius, Thomas Campanella und Giordano Bruno verklingen nur die alten pythagoreischen und platonischen Naturphantasien. Bei Anderen verbanden sich dieselben mit den wilden und verworrenen Träumen der Cabbala . . .

Schwärmerische Geister wie Paracelsus, Robert Fludd, van Helmont, Jakob Böhme ergriffen diese Phantasien . . .

So verband sich mit dieser Philosophie der Natur allmälig aller Aberglaube der Astrologie, Alchemie, Zauberkunst und Dämonenlehre.

—APELT

In 1733 Gilbert Charles le Gendre published his Treatise Concerning Opinion or Memoires to aid the history of the human mind.[1] The point of view from which he wrote was professedly sceptical, and he described his subject matter as the opinions that had reigned in the past in the profane sciences. He believed that he had found a new way to instruct the mind, namely, by experience and by its own history, something which, he affirmed, no author had done hitherto. A chief aim of his book was to tear the veil of obscurity from the abstract sciences and render them intelligible. And furthermore to show that beneath their veil of mystery they had often sheltered the most deceptive arts and the most idle

---

Apelt, *Die Reformation der Sternkunde,* Jena, 1852, pp. 253, 254, 255.

[1] *Traité de l'Opinion ou Memoires pour servir à l'Histoire de l'Esprit Humain,* Paris, 1733, 6 vols. in-12. BN Z. 24425-24430. Second edition, 1733-1735, BN Z. 24431-24436. Third edition, 1741.

studies, and thereby had seduced weak and credulous minds.

He was writing in the age of reason, when Voltaire and Montesquieu were in their prime, and he granted that the errors to which he referred had recently been abandoned. But he nevertheless intended to consider them in order to demonstrate what excesses the human mind was capable of, and to make it impossible that they should ever delude mankind again. Otherwise he feared that they might do so, for men had already begun to avoid profound learning and science, and to think that it was enough to be bright and witty, an attitude which might lead to ignorance and error again.

Beginning with literature, then history and chronology, Le Gendre eventually comes to the belief in demons, souls, magic, cabala, number, oracles, sibyls, augurs, dreams, fortune and destiny. He further considers mathematics, physics, chemistry, astronomy, medicine, astrology, so-called natural divination, naturalists, the arts, metaphysics, sense and imagination, politics and political thought, morals, laws, customs, sorrow and death.

Our viewpoint and plan in our previous volumes have been not unlike those of Le Gendre, and in this latest instalment on the seventeenth century we carry our survey up to the very brink of that period when Le Gendre felt that errors had been abandoned. He admitted, however, as well he might, that he was unable to present natural history with that precision of discernment which would separate what was true from that which was false and fabulous. He proceeded to tell of men with twelve feet, and that in 1535 the Portuguese had found in the East Indies a man who was 395 years old, whose teeth and beard—which last was now once more black—had grown again three times, and who had a son aged ninety. Le Gendre had another failing of which he was less cognizant and which he shared with seventeenth century sceptics and critics of superstition and occult arts, namely, historical inaccuracy. Thus he repeated the old canard that Stoeffler predicted a flood for the year 1524.[2] I shall try to be more accurate in this respect.

Le Gendre's work had run through three editions in less than a decade before the appearance of Jacob Brucker's Critical History

[2] T V, 181.

of Philosophy, in which, too, much space was given to the occult philosophies of the past.[3]

Eleven years after Le Gendre's book had reached its third edition, Facciolati published a volume on the past of the University of Padua,[4] where Peter of Abano and Pomponazzi had once taught, and William Harvey had studied under Fabricius ab Aquapendente and Cremonini. To illustrate the cases and controversies which were referred from all quarters of Europe to the legal faculty of Padua for its opinion or decision, he chose thirty-two questions which were submitted to it in 1601 by Maximilian I, duke of Bavaria, as to the application of torture in cases of accusations of witchcraft. Facciolati speaks of such "magic superstition" as an atrocious, execrable and detestable crime, yet sometimes so ambiguous, because of the secrecy in which it is veiled, that some deem it poisoning covered up with vanities; others think it the fraud of a crafty inclination toward gain and error; while some would have it considered the fault of an unsound mind, to say nothing of those who refer every magic artifice to the sleight-of-hand of mountebanks. How it should be punished is stated in the laws; how it is to be recognized and detected is less clear. After this non-committal introduction, he states the case itself.[5]

Paul Papponerius with his wife Maevia and three sons, Gumpert aged 22, Jacob 20, and Cyprian 11, were arrested, found guilty of many crimes, especially magic superstition, and burned. But Cyprian, whom his mother had consecrated to the devil while he was yet in her womb, was first baptized in prison. When about to die, they indicated certain accomplices, among them Johann Clasius, a weaver, and Anna with her daughter Ursula, who, duly condemned, suffered the same penalty. Three other women were accused and thrown into prison, named Barbara, Sempronia and Maria, and the whole questionnaire was with regard to them. Barbara was first named by Paul's son Jacob, who said that he had seen her twice at a nocturnal sabbat of witches, then by Clasius the weaver,

[3] Jacob Brucker, *Historia critica philosophiae*, 1742-1744, 4 vols. in 5.

[4] *Jacobi Facciolati De gymnasio Patavino syntagmata XII ex eiusdem gymnasii fastis excerpta*, Patavii, 1752, in-8, 239 pp.

[5] *Ibid.*, Syntagma IX, pp. 109-15.

Ursula, the daughter of Anna, and Anna herself who added the further detail, that Barbara, when she had two children, a boy in 1577 and a girl two years later, had slain both with poisons (or, sorceries) and devotions, and that during their first year, as is the custom of witches. The same Jacobus and Gumpert his brother indicated Sempronia, asserting that they had seen her at nocturnal sabbats of witches. Their mother Maevia confirmed this and added that seven years before she had by her magical incantations aroused a storm which blighted the crops, and had killed a pregnant woman who suffered abortion from her sorceries. Finally Jacobus, Anna and Ursula said that they had seen Maria celebrating nocturnal solemnities at gatherings of witches.

The messengers were sent to Padua to inquire, first whether these charges were serious enough to put the women in prison and whether they should be detained until they purged themselves from the accusation. And whether, in the case of a crime which theologians and jurists classed as excepted, women who were defamed by no suspicion could be tortured on the accusation of several persons, who, however, did not remember the time or place? And, because these women repelled the accusations with many exceptions and demanded advocates, the 32 other questions had arisen which the Paduan faculty was asked to answer one by one. Facciolati, however, does not give the replies to each but only the questions and the general tenor of the response, and it will here suffice to repeat a few of the questions:

1. Whether the information of several persons without other reasons is enough to torture those who are indicated?
2. Whether enmity on the part of the informer, to infirm his testimony, ought to be proved by two witnesses . . .?
3. Whether those condemned to death, when they inform on others, ought to have given signs of penitence, lest they be thought to have informed from hatred or levity; although penitence of witches is very rare?
5. Whether in excepted crimes, such as witchcraft, an infamous person should be believed who accuses a person of good repute? What if the infamous who accuse are several?
7. Whether a person may be interrogated under torture, if several

witnesses assert that he was at sabbats, although he has harmed no one?

16. Whether penitence and revocation by the informer made *extra judicium* so weaken his judicial delation, that the accused may not be tortured further?
17. If no magic instruments are found on the accused, whether faith in the informers is so weakened, that no place is left for torture?
21. Whether in more atrocious crimes the accused may be more severely tortured than in others, and, since witches are often very tough (*pervivaces*) and obdurately silent through sorcery, what is the most suitable kind of torture in this crime?
22. How often and how long the accused may be tortured in each case?
25. What is felt as to dripping cold water on the back of an accused who is being tortured by the rope?
26. Since witches are wont to be tough and obstinate, whether and when torture may be repeated; not indeed in consequence of new testimony, for as to that there is no controversy, but that from which they already seem to have cleared themselves by the first questioning under torture?
27. Whether the common notion, that no one may be questioned under torture more than three times, is based on law and right; and whether this triple questioning should be held on one day or several? What if one is charged with so many crimes, that his examination cannot be completed in one day?
31. Whether in more atrocious crimes the accused should be allowed counsel, by confidence in whom he may become bolder in denials?

Such was the grim retrospect of the witchcraft delusion which had prevailed all through the seventeenth century and still hung like a pall over the early years of the eighteenth. But from such backward glances, which suggest, nonetheless, how much a matter of concern the subject of our investigation still was to thinking men in the middle of the eighteenth century, let us plunge into the thought of the seventeenth century itself. Our treatment, as in previous volumes, will be partly by men and partly by subjects, now topical and now chronological. It will of necessity be selective and illustrative rather than all-inclusive and exhaustive. Two or three preliminary generalizations may be offered.

There is some difficulty in attaining chronological order. The frequency of posthumous publication raises the question as to when the work was actually finished or if it ever was, to say nothing of whether the author intended it for publication or would have preferred to have it remain unpublished, and how much it has been emended since his death. Again, it is sometimes difficult to obtain access to the first edition, while later editions may have undergone substantial changes either at the hands of the original author or of subsequent editors.

Manuscript material as compared to printed books and periodical publications becomes of less relative importance than it was in the age of incunabula or even in the sixteenth century. It was in general less widely read and less influential in the history of thought. Still, it cannot be entirely neglected even in the case of an investigation like our own which aims only to select examples more or less at random from the multitudinous phenomena of past intellectual history. Newton left nothing in print on the subject of alchemy, but alchemy is a leading interest of over a million words left in his own handwriting. And he was far from being the only one to lay claim to priority in an idea or a discovery, although he had failed to be the first in print with it. But while in the case of experimental science there may be disagreement as to by whom first the new was tried, in the case of magic, at least in the seventeenth century, the question is rather who was last to lay the old aside.

Nature itself was repeatedly thought of and spoken of through the century in a way favorable to magic. It was personified even by scientists of the calibre of Galileo, Harvey, Leibniz and Newton. It was glorified again and again. Not only its arcana and secrets, and mysteries and secret archives, but its marvels and miracles were matters of incessant remark. Nature was called a beneficent mother, or wise and provident, sentient by Campanella, stupendous by Munting, purposive by Harvey. Boyle was unusual in speaking of its irregularity. Helmont had held that it was not subject to mathematics and, on the other hand, knew no contraries. Le Grand contended that it was not restricted by time. Sturm became alarmed at the chorus of adulation and at the concept itself, and wrote a book in 1692 in which he argued that an idol had been made of nature that derogated from God and that should be expunged from the mind and from writings, and the very word not used any more.[6] Five years later Schelhammer published his Vindication of Nature for itself and for physicians, or a bipartite book on nature, in which it is not only accurately inquired what it is, but also all its force and power is placed most clearly in view and it is circumscribed by its own limits.[7]

Use of the word, *arcana,* like the use of the words, occult and secret, is almost always a relic and sign of a magical and unscientific attitude. One of the treatises in the alchemical collection of Zetzner was entitled, The Ark of the Most Artful Arcanum of the Supreme Mysteries of Nature. And when the first of three works by Hermann Grube cited by C.F. Garmann in his thick tome on Miracles of the Dead,[8] itself a very promising title from the standpoint of magic, is on the arcana of medicine,[9] we are not surprised to find that the other two are on the transplantation or magic transfer of disease[10] and on the bite of the tarantula,[11] a poison supposed to operate on anniversaries of the original puncturing and to be relieved only by dancing to certain music. An exception to the rule is seen in Leeuwenhoek's quite appropriate use of the title, *Arcana naturae,* for his letters to the Royal Society announcing his truly scientific researches with the microscope. However, it also attests the cur-

[6] Joh. Chris. Sturm, *Idolum naturae,* Altorf, 1692.

[7] *Guntheri Christophori Schelhammeri Natura sibi et medicis vindicata, sive de Natura liber bipartus, in quo non modo quid illa sit accurate exquiritur sed etiam omnis eius vis et potestas clarissime ante oculos ponitur suisque limitibus circumscribitur. Simul etiam patet quid medici per eam intelligant, ab eaque debeant expectare. Denique artis medicae existentia ac certitudo solide demonstratur, et methodus medendi eiusque principia et fundamenta e tenebris suis eruta exhibentur,* Kiliae, J. S. Riechelius. 1697, in-4, 335 pp. and Index. BN 4° T19.29.

[8] *De miraculis mortuorum,* Dresden and Leipzig, 1709.

[9] *De arcanis medicorum non arcanis commentatio,* Hafniae, 1673: BN 8° Te7.24.

[10] *De transplantatione morborum,* Hamburgi, 1674: BN 8° Td10.13.

[11] *De ictu tarantulae,* Francofurti, 1679: BN 8°Te64.6.

rency and popularity of the word arcana. Paschius still employed it frequently in his *De novis inventis* of 1700. For example, he said: "Since many arcana which we possess are due to invention, curiosity demands that we look into the history of inventions." [12]

Kepler said that, although much of Arabic astrology was nonsense, there were arcana of nature in it which were not nonsense and should not be discarded with the nonsense.[13]

As the reader may note for himself as we go along, a striking feature of both scientific and pseudo-scientific works of the seventeenth century is the frequency with which such words as "new" and "unheard-of" appear in their titles.[14] Frommann, writing on fascination in 1675, spoke of "this prurient world which is not satisfied with anything except what is new, rare, curious and fresh."[15] On the other hand, in the closing year of the century, Thomas Baker declared that many recent opinions, which had little but their novelty to recommend them, really lacked that too, "and might be easily shown to be only the spawn of the ancient philosophers."[16] A good example is the *Discursus astronomicus novissimus of* Pietro Cortesio[17], which turns out to be a mere repetition of the *Sphere* of Sacrobosco of the early thirteenth century.

We shall indeed encounter plenty of stale ideas and outworn beliefs in the seventeenth century. Moreover, many volumes, including even some of those that laid claim to novelty, or professed to be experimental, were made up largely of quotations from previous writings.

[12] *Op. cit.*, pp. 3-4.

[13] *Opera*, I, 419; quoted by N. Herz, *Keplers Astrologie*, 1895, p. 28.

[14] See my "Newness and Craving for Novelty in Seventeenth-Century Science and Medicine," *Journal of the History of Ideas*, XII (1951), 584-98.

[15] *De fascinatione*, 1675, Praefatio ad lectorem.

[16] *Reflections upon Learning*, 2nd edition, London, 1700, p. 79.

[17] Palermo, 1642, 114 pp.: BM 531. k.17 (5.).

CHAPTER II

KEPLER AND GALILEO

Kepler's attitude in 1606—Its alteration—His *New Astronomy*—Not so easy to shake off Ptolemy—Altobelli—Kepler's attitude towards astrology—New Star of 1604—Three chief avenues of celestial influence—Harmony and Sympathy—Letter to Wallenstein—Other presages—Animistic and occult interpretation—Conception of gravity—Kepler and Galileo on Comets—Galileo on Tides—Other Errors—Declining Interest in Astrology—Assertion of novelty—Atonism—Advent of modern experimental method—The tower of Pisa—Magical remnant—Survival of old ideas—Citation of ancients, neglect of medievals—Slow diffusion of *Two New Sciences*—The Bible and Science.

Not mine own fears, nor the prophetic soul
Of the wide world, dreaming on things to come
—SHAKESPEARE

Perchè la natura non si diletta delle sceniche poesie
—GALILEO

Kepler held the erroneous view, but one all too common then and since, that the world had been asleep for a thousand years after the fall of Rome and the barbarian invasions, plunged in barbarism and ignorance, but that from the year 1450 on civilization had revived. Peace and order and communication by post existed in Germany; even the Turks had made great progress in civilization; the voyages of discovery had brought incredible increments to European trade; artillery and printing had been invented; learning flourished once more; there was a new theology and a new jurisprudence; the Paracelsists had renovated medicine, and the Copernicans, astronomy.[1]

Therefore, writing of the new star of 1604–1605 and the events

[1] *De stella nova in pede Serpentarii,* cap. 29: *Opera,* II (1859), 730-32. I chiefly cite this old edition by Ch. Frisch of Kepler's works, but a new edition by Max Caspar is in process. Between 1937 and 1951 vols. 1-4, 6, 13-15 had appeared. In 1953, vols. 5, 7; in 1954, vol. 16.

which might be expected to follow it, Kepler declared that the astrologers were stupid if they thought that more and greater things were to be looked for in the next two hundred years than had happened in the one hundred and fifty years since 1450. "Unless perchance they think," he continued sarcastically, "that some new orb will be discovered and an art of flying by which we may go to the moon."

Writing this in 1606, Kepler did not realize that, before even that first decade of the seventeenth century closed, the telescope would have been invented, that Galileo in his *Nuncius sidereus* would have announced the discovery of mountains on the moon and of the satellites of Jupiter, and that, even before this, he himself in his work on the movements of Mars in 1609 would promulgate the first two of his planetary laws. That in the same year weekly newspapers would begin in Germany, that the first professorship of chemistry in Europe would be founded at Marburg, and the first chair of political science at Upsala. Kepler was probably not even aware that already in that decade Shakespeare had presented *Hamlet* and Cervantes published the first part of *Don Quixote*. He foresaw none of the subsequent scientific advance of the century from Harvey's announcement of the discovery of the circulation of the blood in *De motu cordis* of 1628 to Newton's *Principia* in 1687.

It may be added that the slur on the period before 1450 came with especially bad grace from Kepler, since he had made the thirteenth century work of Witelo on optics the foundation and starting point for his own *Ad Vitellionem paralipomena quibus astronomiae pars optica traditur*, printed only two years before. Or consider the association of the spheres of the planets with the five regular solids in Kepler's *Mysterium cosmographicum* of 1596. Modern historians of science have regarded it as a fantastic aberration on his part, although Maestlin asserted that all astronomy would be reformed in consequence of it, while Kepler himself represented it as a divine revelation[2] such as he had never read in the book of any philosopher,[3] and that he would not renounce the

[2] *Opera*, I (1858), 108, "Divinitus id mihi obtigisse arbitrabar."

[3] *Ibid.*, p. 3, "inventum hoc quippe in nullius philosophi libro talia umquam legeram."

glory of its discovery "for the whole electorate of Saxony."[4] But if Kepler had turned to the commentary on the *Sphere* of Sacrobosco (early thirteenth century) which Prosdocimo de'Beldomandi completed in 1418, and which was printed in the collection, *Sphaerae tractatus,* of 1531 (Venice; L. A. Junta), he could have read that Campanus of Novara in the thirteenth century in his commentary on Euclid's *Elements,* penultimate conclusion of Book 13, told that certain disciples of Plato said that the sky or whole mass of the heavens and each of the elements was angular and not spherical, and that the number of essences corresponded to the number of regular solids: the pyramid to fire, hexahedron or cube to earth, icosahedron to water, octahedron to air, and duodecahedron to the fifth essence, "as may be inferred in conclusions 13, 14, 15, 16, 17, of the 13th book of Euclid's *Elements,* and more clearly from Campanus in his comment on the 17th conclusion of the said 13th book."[5] Perhaps Kepler had read the passage and it subconsciously suggested his own theory to him. In any case, he had no license to scorn medieval science before 1450 or before Copernicus. And let us not do so in our survey of the seventeenth century.

Kepler's attitude underwent a marked change with the appearance of the *Nuncius sidereus* of Galileo. The discovery of four new planets had made a great sensation, and Kepler realized that it seemed to overthrow his pet theory in the *Mysterium cosmographicum,* unless the four Medicean stars could be regarded as moons of Jupiter—as they proved to be. If so, Kepler suggested that there should be six or eight about Saturn, two for Mars, and one each for Venus and Mercury. Once he had thought that the observations of Tycho Brahe would never be superseded but now he recognized that they were no longer the last word in astronomy. Kepler's own previous conjectures had been in some cases disproved, in others confirmed by Galileo's recent observations. Now he eats his sarcastic words of 1606 and suggests that both our moon and Jupiter may be inhabited and that we may some day fly to them. For what good will it do to have four moons coursing about Jupiter, if there is no one on that planet to watch them? Which shows that for

[4] H. W. Tyler and R. P. Bigelow, *A Short History of Science,* 1939, p. 242.

[5] *Sphaerae tractatus,* 1531, fol. 17v.

Kepler man is still the measure and center of all things; he is still more astrologically than astronomically minded. He goes on to say that some will now think that his terrestrial astrology and doctrine of aspects are groundless. But he holds that the new stars are so small and so close to Jupiter, that they may be identified with its influence so far as the earth and its inhabitants are concerned, and that their particular influence will be upon the inhabitants of Jupiter. But the possibility of dwellers on other planets at last raises in his mind the question whether we are the noblest of rational creatures and whether all is made for man.[6] It is further noteworthy that Kepler repeatedly cites, not only other astronomers such as Maestlin, but Giordano Bruno as to an infinite number of worlds or earths and the fixed stars being suns and the planets moons or earths.[7] He also quotes at length from the *Magia naturalis* of Giovanni Battista Porta on lenses. Thus his astronomy is mingled with mere speculation and with natural magic, as well as with astrology.

Let us not, however, forget or discount Kepler's own New Astronomy[8] which, after years of tedious calculation and frequent discouragement, he had finally triumphantly completed just on the eve of the discoveries with the telescope. This was the greatest single work in astronomical theory between the *De revolutionibus* of Copernicus in 1543 and Newton's *Principia* in 1687.[9] In it we

[6] *Dissertatio cum nuncio sidereo nuper ad mortales misso a Galilaeo Galilaeo mathematico Patavino,* Pragae, 1610: *Opera,* II (1859), 485-506; also in Galileo, *Opere,* III, i (1892), 105-25.

[7] *Opera,* II, 490, "infinitos alios mundos (vel ut Brunus Terras) huius nostri similes esse"; 500, "ut Bruni verbis utar, illas esse Soles, hos Lunas seu Tellures?"

[8] *Astronomia nova* Αιτιολογητος *seu Physica Coelestis tradita commentariis de motibus stellae Martis ex observationibus G.V. Tychonis Brahe iussu et sumptibus Rudolphi II Romanorum imperatoris plurium annorum pertinaci studio elaborata Pragae a . . . Joanne Keplero,* 1609, in-fol., 337 pp. No index. The work is found in III (1937) of Caspar's edition.

[9] Since writing this sentence I find that Apelt a century ago made the stronger statement that Kepler's work was the chief contribution to astronomical theory since the *Almagest* of Ptolemy:

"Geschichtlich giebt es nur zwei Hauptquellen der theoretischen Astronomie: die eine ist der Almagest des Ptolemäus, die andere Keppler's Commentar über den Stern Mars. Das Werk des Kopernikus enthält keine neue Theorie der Planetenbewegung, sondern nur eine Umformung der ptolemäischen in Sinne der heliocentrischen Hypothese." E. F. Apelt, *Die Reformation der Sternkunde. Ein Beitrag zur deutschen Culturgeschichte,* Jena, 1852, p. 269.

see Kepler at his best and at the height of his career. Although based largely upon the observations made by Tycho Brahe,[10] it erected a new structure upon these which was as ruinous to the Tychonic as to the Ptolemaic system. Of especial interest to the historian of thought is the fact that Kepler tells how he came to develop the subject and presents his results in their historical sequence and not merely by geometrical demonstration, and also keeps an eye upon physical causes. And we can understand his pessimistic tone back in 1606, when at one point he says to the reader, If you are bored by this laborious method, I can certainly sympathize with you, since I have had to go over it at least seventy times with great loss of time.[11] But he triumphed at last and demonstrated that Mars moved in an elliptical orbit not about the earth but about the sun, and that its speed increased as it approached the sun and decreased as it moved away from the sun. Furthermore, that the earth and other planets described similar orbits about the sun. This shattered forever the long-standing Aristotelean notion that the motion of the celestial bodies was circular and so perfect, a doctrine which Ptolemy had done his best to preserve by the devices of epicycles and eccentrics, devices which even Copernicus and Tycho still employed to a more limited extent. It abandoned mere hypotheses for physical reality based upon careful observation and mathematical calculation. It should be added, however, that Kepler's laws do not stand out in his *Astronomia nova* as they do in modern histories of science, and that Schoock, writing in 1663, still emphasized the irregularity and difficulty of the motion of Mars rather than Kepler's solution thereof.

Kepler not merely advocated the Copernican theory but based his observations upon it. Thus, if the earth was immobile at the center of the universe, it would afford the astronomer only a single fixed viewpoint, whereas on a moving earth the star-gazer could observe the other planets from varying angles and relate their motions and positions to the sun as a fixed point instead of the

[10] In chapter XV, after giving ten observations by Tycho and two of his own in 1602, Kepler says (p. 87), "Sed quia observationes a morte Tychonis rariores a nobis (p. 88) sunt habitae nec continuatis diebus v, lubet securitatis causa consulere etiam illas observationes quas David Fabricius in Frisia Orientali . . . mecum communicavit."

[11] *Ibid.*, p. 96.

earth. Kepler, says Apelt, found the key to the secrets of astronomy in measuring the distance of the planets from different positions of a moving earth.[12]

It was not so easy, however, to shake off Ptolemy. Just as Copernicus had depended upon his observations, so Kepler, investigating the movement of the apogee and nodes, said, "This investigation will be as certain as are the Ptolemaic observations, or rather, traditions."[13] And later on he remarked:

> Since we are without suitable observations of antiquity, this state of affairs forces us to leave this disputation concerning the motion of the nodes, like many other matters, to posterity, if it please God to grant the human race time enough . . .[14]

Again, he did not foresee that a great time and space saving instrument, the telescope, was about to come into use.

Even while Kepler was toiling away, Ilario Altobelli, addressing in 1607 to one of the cardinals an astrological prediction as to the fate of the Venetian republic, was complaining that Mars by its irregular motion had eluded all astronomers since the world began, that he could not trust the Alfonsine or Copernican or Prutenic Tables which erred enormously as to the position of this "unobservable erratic." While he recognized that no one could succeed at astrology without a thorough knowledge of astronomy, in this case he found it necessary to use astrology to help astronomy out. The disastrous defeat of Venice by Genoa and Pisa in 1250 must have come from the direction of the sun to the left tetragon of Mars, which would place Mars at that time in Libra 20°52′. Then by combining the observations of Tycho Brahe with the position of Mars given by the Alfonsine Tables for the date of the foundation of Venice on March 25, 421, he arrived at the result of Libra 20°50′, almost equivalent to the other.[15]

[12] E. F. Apelt, *Die Reformation der Sternkunde,* Jena, 1852, pp. 215-16.

[13] *Astronomia nova,* cap. xvii, p. 106.

[14] *Ibid.,* p. 323.

[15] BN Latin MS 7452, De proxima reipublicae Venetae inclinatione ex astris rita solidaque coniectatio multiplex, 45 fols. See especially fols. 10r-11r. The date for the foundation of Venice is based on a *Thema coeli* which, he says (fols. 4v, 9r-v), was found in the archives of Padua before the Palatium there was burned down and which was attested by the names of six nobles delegated by the Senate.

With this close association of astrology and astronomy in mind, we turn from Kepler's great accomplishment in explaining the motion of Mars to the problem of his interest in astrology.[16] There is no doubt that he composed and issued numerous astrological predictions.[17] But, because of two or three passages in his works and letters, which have been quoted over and over again, it has been maintained that he was not a willing and sincere worker on such predictions, but composed them because he needed the money, whereas his heart was only in astronomical observation and theory, and he disbelieved in astrology. Several times, in his treatise of 1606 on the new star of 1604,[18] in *Tertius Interveniens* of 1610,[19] in a letter of 1617 to Bernegger,[20] and in the preface to the Rudolfine Tables of 1627,[21] he referred to astrology as the stupid or meretricious daughter of astronomy, who, however, depended on that daughter for support, as Kepler himself did at times.[22] In the work of 1606 he also affirmed that he had been hired by the emperor not to be a public predictor but to continue the astronomical work of Tycho Brahe.[23] Undoubtedly Kepler was rather fond, not to say,

[16] Norbert Herz, *Kepler's Astrologie,* Vienna, 1895, was a beginning in the right direction, but tended to quote more of Kepler's German than Latin, which apparently gave Herz some difficulty. Also he chiefly cited the first volume of Frisch's edition, although all eight volumes had appeared by 1870, twenty-five years before his own work.

A later work is Heinz Artur Strauss und Sigrid Strauss-Kloebe, *Die Astrologie des Johannes Kepler. Eine Auswahl aus seinen Schriften,* 1926.

Kepler's belief in astrology is recognized by W. C. Rufus and E. H. Johnson, (*Johann Kepler, 1571-1630*). —*A Tercentenary Commemoration of his Life and Work,* 1931, pp. 35-38, 69-76; and by N. T. Bobrovnikoff in his review of that volume in *Isis,* 18 (1932), 197-200.

Dermul, "Tycho Brahe et Kepler croyaient-ils à l'astrologie?" *Gazette astronomique,* 33 (1951), 64-67, is trifling and incompetent.

[17] Besides those in Frisch's edition of Kepler's *Opera* see: Walther von Dyck, *Zwei wieder aufgefundene Prognostica von Johann Kepler,* München, 1910, in-4; Max Caspar und von Dyck, *Prognostikum auf das Jahr 1620,* in *Abhandlungen d. Bayer. Akad. d. Wiss., Math.-nat. Kl.,* Neue Folge, Heft 17 (1933), 58 pp.

[18] *Opera,* II, 657.

[19] *Opera,* I, 560-61.

[20] *Opera,* I, 660.

[21] *Opera,* VI, 670; also p. 666, "tum deinde per somnia et nugas praedictionum genethliacarum educata paulatim adolevit . . ." etc.

[22] *Opera,* VIII, 705: "Quod nativitates et Calendaria interdum scribo, ea me Christe molestissima mihi servitus est sed necessaria . . ."

[23] *Opera,* II, 748.

proud, of the mother and daughter simile, and pretty certainly at times he found the composing of annual predictions and the like quite boring and regarded his *Mysterium cosmographicum* as far preferable.[24] That he nonetheless composed them shows not merely that he needed the money, but that such predictions still exerted a strong hold upon society and that they were expected, or at least hoped for, even from astronomers of the highest rank. This was quite possibly a matter of more importance than Kepler's own personal attitude and sincerity or insincerity. These may be revealed in letters in which he unbosomed himself to a single correspondent. Far more significant for the history of thought are the works which he offered to the reading public and the world of science. But let us see what his attitude towards astrology and the influence of the stars really was.

Kepler set forth his position with regard to astrology and the influence of the heavens several times very fully, clearly and explicitly: in Latin in 1602 in his treatise on the more certain foundations of astrology in 75 theses,[25] and in 1606 in his work on the new star of 1604,[26] in German in his *Tertius interveniens* of 1610,[27] so called because he intervened in the controversy between Schärer, a pastor who had issued prognostications and defended astrology, and Feselius, physician to the Markgraf of Baden, who had attacked it. The full title continues "That is, a warning to sundry theologians, medical men and philosophers, especially Philippus Feselius, that they, while very properly overthrowing star-gazing superstition, do not chuck out the baby along with the bath-water and thereby unwittingly injure their profession."[28] Then in 1619 in the fourth book of *Harmonice Mundi*[29] Kepler

[24] *Opera*, I, 97.

[25] *De fundamentis astrologiae certioribus: Opera*, I, 417-438.

[26] *De stella nova in pede Serpentarii . . . libellus astronomicis, physicis, metaphysicis, meteorologicis et astrologicis disputationibus . . . plenus: Opera*, II, 611-750, especially caps. 24-30, pp. 699-750.

[27] *Opera*, I, 547-651, with 140 theses.

[28] *Tertius Interveniens. Das ist, Warnung an etliche Theologos, Medicos und Philosophos, sonderlich D. Philippum Feselium, dass sie bey billicher Verwerffung der Sternguckerischen Aberglauben, nicht das Kindt mit dem Badt ausschütten und hiermit ihrer Profession unwissendt zuwider handlen.*

[29] *Opera*, V, 211-67: *Harmonices Mundi Liber IV*. De configurationi-

once again discussed the matter in Latin. He also often treated of astrology in his letters.[30]

Therefore, while at times Kepler may have been reluctant and unwilling to compose predictions, he seems never to have tired of the problem to what extent the traditional rules and technique of the art of astrology might be retained or reformed, and the other problem how and to what extent the heavenly bodies exerted influence upon this earth and its inhabitants.

It is true that he condemned the general run of vulgar astrological or supposedly astrological predictions. Only one was correct against a hundred that were wrong.[31] Kepler had read the book of Pico della Mirandola against astrology and agreed with much of it.[32] The names of the signs of the zodiac had been bestowed arbitrarily; even its division into twelve signs was a human figment.[33] Relating triplicities to the four elements was also arbitrary, as was the division into twelve astrological houses.[34] In a conjunction of planets, their number and closeness to one another counted for more than the triplicity in which the conjunction occurred.[35] Making annual predictions from the entry of the sun into Aries was idle.[36] Future contingents depended upon human free will and could not be foreknown.[37] Political and religious change could not be predicted.[38]

But Kepler did not go all the way with Pico's attack upon the past technique of astrology.[39] He would keep the old names and divisions as a matter of convenience.[40] He not merely retained

---

bus harmonicis radiorum sideralium in terra earumque effectu in ciendis meteoris aliisque naturalibus.

[30] *Opera,* I, 295-384: Literae Kepleri aliorumque mutuae de rebus astrologicis.

[31] *Opera,* I, 420-21; also V, 234: "Sic cum prognosticum aliquod millies errat, neglegitur tamen hoc; et si semel scopum attingit, hoc memoria dignum censetur, hoc omnium sermonibus celebratur."

[32] He cites it repeatedly in both *De stella nova* and *Harmonice mundi.*

[33] *Opera,* II, 629; I, 139, 581; II, 626. Strauss, *op. cit.,* p. 23.

[34] *Opera,* II, 632; I, 431 (*De fundamentis,* Thesis 49).

[35] *Opera,* II, 728-29 (*De stella nova,* cap. 29).

[36] *Opera,* I, 431, 581.

[37] *Opera,* I, 595 (*Tertius interveniens,* Thesis 55).

[38] *Opera,* II, 728.

[39] *Opera,* II, 637 (*De stella nova,* cap. 8).

[40] *Opera,* II, 625 (*De stella nova,* cap. 3).

planetary aspects[41] but increased their number from five to eight, adding quintile, biquintile and sesquadrum to conjunction and opposition, sextile, quadrate and trine.[42] He maintained that experience showed that all sorts of meteors were seen when the planets were configured in aspects, whereas the air was undisturbed, when they were not.[43] In 1619 he wrote that for twenty years past he had observed the relation of the weather to planetary aspects.[44] He further retained the theory of the significance of great conjunctions of the planets, although he agreed with Pico that it was inept and superstitious to set periods to religions and empires from them.[45] He also held that the planets impressed sublunars by their colors as well as by their meetings and configurations.[46] In fine, many great secrets of nature were hidden in astrology, and study of the sound and unsuperstitious variety was as little forbidden by the Bible as was the study of anatomy.[47]

In Kepler's *Harmonice mundi* of 1619 were discussed the harmony of rays from the heavenly bodies descending to earth, their effects on sublunar nature and the human soul, and the relation of planetary aspects to musical consonance.[48] There was an efficacious configuration when the rays of two planets made such an angle as

[41] *Opera,* II, 624. Frisch erred in saying (*Ibid.,* II, 578), "Keplerus igitur rejicit illam quam astrologi somniabant vim aspectuum . . ." and later admitted (III, 816, n. 13), "Sententiam hanc de vi astrologica aspectuum in 'naturam sublunarem' firmiter tenuisse Keplerum posteriori quoque tempore . . ."

[42] *Opera,* II, 642 *(De stella nova,* cap. 9). In the *Epitome astronomiae Copernicanae* of 1618, *Opera,* VI, 490-91, he suggested yet others, such as decile for 36 degrees and octile for 45 degrees.

[43] See further *Opera,* I, 586 (*Tertius interveniens,* Thesis 46). And at p. 650, in Thesis 138, he indicates how the weather from day to day in the winter of 1608-1609 conformed with the planetary aspects.

[44] *Opera,* V, 251 (*Harmonice mundi,* IV, 7.).

[45] *Opera,* II, 635, 641 (*De stella nova,* caps. 7, 8); and V, 259 (*Harmonice mundi,* IV, 7).

[46] *Opera,* II, 638-39 (*De stella nova,* cap. 8): ". . . colores planetarum . . . et ipsos eorum congressus et configurationem naturis seu facultatibus rerum sublunarium imprimi et his objectis illas permoveri, cum ad formandum, tum ad movendum corpus cui movendo praesident."

[47] *Opera,* I, 561, 565 (*Tertius interveniens,* Theses 8, 16).

[48] On the title page the fourth book is described as: "Quartus Metaphysicus, Psychologicus et Astrologicus, De harmoniarum mentali essentia earumque generibus in mundo; praesertim de harmonia radiorum ex corporibus coelestibus in Terram descendentibus, eiusque effectu in natura seu anima sublunari et humana," *Opera,* V, 75.

was apt to stimulate sublunar nature and the faculties of inferior animate beings to more effective operation at the time of the configuration. The first degree of efficacy and the strongest was that of conjunction and opposition. In his treatise on the new star Kepler declared that "geometry in the rays of the stars affects sublunar nature."[49] In *De fundamentis* he again affirmed that the earth was stimulated by a geometric concourse of rays.[50]

In 1601 Kepler had composed a special treatise on directions, in which he asserted that they were the noblest part of astrology and strongly confirmed by experience.[51] Despite the fact that both his own first son and the son of Maestlin, for whom at their birth he had predicted a favorable future, died in their first year,[52] Kepler retained faith in nativities and in the importance of the position of the planets at the moment of birth. "In general there is no expedite and happy genesis, unless the rays and qualities of the planets meet in apt, and indeed geometric, agreement." He even thought that sons, especially the first-born, often were born under similar horoscopes to their parents. "And these things cannot be done except by the impression of the whole character of the heavenly position on the very faculties of man, generative, altering, formative, sensitive, animal." And "these species of celestial things are imprinted within by some occult way of perception."[53] Those born on December 9-10, new style, had the sun in conjunction with the new star, and all have a tendency to innovation.[54] Even in the midst of his great work on the movements of Mars, Kepler tells how his mother noted the positions of the stars, and that she herself was born in a configuration of Saturn, Jupiter, Mars, Venus and Mercury in sextile and trine aspects, and strove to have her children, "especially me the first-born," delivered under similar configurations.[55] In a passage in his diary he says that mid-sky then could signify no one but mother and her pitiable state.[56]

[49] *Opera*, II, 645 (cap. 9).

[50] Thesis 43.

[51] *Opera*, VIII, 295; also I, 316, and other passages listed in the Index under "Directiones." Also Strauss, *op. cit.*, p. 25.

[52] Herz, *Keplers Astrologie*, 1895, p. 7.

[53] *De stella nova*, cap. 10, *Opera*, II, 646-47.

[54] *Ibid.*, cap. 28, *Opera*, II, 725-26.

[55] *De motibus stellae Martis*, Pars III, cap. 39, *Opera*, III, 319; noted by Strauss, *op. cit.*, pp. 16, 24.

[56] Herz, p. 17.

In her old age Kepler's mother was accused of practicing magic arts by a woman whom she had offended by her plain-speaking, and after some years was arrested. Kepler then procured her acquittal, but he now characterized her as "delira et garrula."[57]

Coming back to nativities, we may note that in another passage Kepler states that those who are born when there are many aspects between the planets are apt to be hardworking and industrious, whether in money-making, public life or scholarship, and cites his own geniture as an example.[58] He held that the babe began to live only with birth, and that its geniture was impressed upon its subconscious memory.[59] A clear proof of this was the relationship between the genitures of parents and children.

For when the foetus is ripe, the formative faculty of the soul presiding over generation girds up its loins most potently to thrust out the foetus and by that extrusion to kindle the new vital faculty of the soul, when the stars, returned to the seats of maternal and paternal genesis or to the same configurations, remind the soul of itself and its celestial character.[60]

Kepler composed nativities of Wallenstein, the Emperor Rudolf, and others, and it would seem that he took more interest in drawing up the horoscopes of individuals and had less objection to it, than he voiced against calendars and annual prognostications.

The Strausses, without citing any particular passages, represent Kepler as not regarding solar and lunar eclipses as especially active and influential.[61] Herz cited one passage to show how little influence Kepler ascribed to eclipses—to me, however, the passage seemed to attribute considerable influence to them—but went on to say that Kepler did not completely deny such influence.[62] Indeed, he spoke of them as very ominous, stating that, unless the earth were endowed with a soul-like faculty which was vehemently disturbed by the shutting off of light, or unless, rejecting all phy-

[57] *Opera,* VIII, 359-562, Judicium matris Kepleri.

[58] *Harmonice mundi,* IV, 7, *Opera,* V, 261.

[59] *Ibid.,* p. 265.

[60] *Idem.*

[61] *Op. cit.,* p. 18.

[62] Herz, *Keplers Astrologie,* 1895, p. 25: "So gering auch der Einfluss ist den Kepler hier den Finsternissen zuschreibt, so ist nicht zu verkennen dass er ihn nicht vollständig leugnet."

sical causes, this ordinary work of nature was ascribed to divine providence, "you cannot explain why eclipses are so ominous."[63] On the other hand, in private correspondence he remarked that not every solar eclipse brought drought,[64] and expressed doubt as to any reason for the common practice of estimating the duration of the effects of an eclipse by as many months in the case of lunar, and years in the case of solar eclipses, as the obscuration had lasted hours.[65]

Duhamel in 1660 wrote that Kepler held that comets were nothing but celestial *aura,* condensed by some occult influence, which collected the sun's rays and tinged them with certain colors.[66] Yet in Kepler's estimation comets were signs of the future, and this in a triple mode: natural, sympathetic, and purely significative—the last from God. Comets had preceded the births of Alexander the Great, Mithridates and Mohammed, while those of the present century frequently indicated religious disturbances. Those of 1618 would be especially ominous to the New World, and, since all three were retrograde, would result in the greatest confusion of empire.[67] Thus political and religious change, which Kepler said could not be predicted from the stars, were indicated by comets.

Kepler affirmed that the new star of 1604 was the result, not of chance, nor necessity of nature, but the certain plan and special providence of God.[68] It was not produced, as some astrologers held, by the great conjunction of the planets with which it was coincident, but had been miraculously produced by divine omnipotence at that time in order to conform with the rules of astrology and to attract the attention of astrologers, just as the Star of Bethlehem, which was similarly coincident with a great conjunction, had drawn the attention of the three Magi, and—in both cases—to promote human salvation.[69] Kepler was inclined to think that the natural effects of the new star, if any, would be confined to the weather,[70] although he asserted that there is nothing in the visible heaven which does not in some occult way reach earth and affect all the faculties of

[63] *De fundamentis,* Thesis 46, *Opera,* I, 430.

[64] *Opera,* I, 354.

[65] *Opera,* I, 320.

[66] *Astronomia physica,* p. 72.

[67] *De cometis,* III, De significationibus, *Opera,* VII, 119-37.

[68] *Opera,* II, 734.

[69] *Opera,* II, 708-11, 717, 744.

[70] Cap. 28, *Opera,* II, 719-21.

natural things, as the sky itself is affected.[71] He was more outspoken as to its significations: varied rumors, varied forebodings, consternation and stupefaction of the populace, many impediments to many undertakings, many new occupations, and much writing and printing concerning it—a favorite joke of Kepler.[72] But even those who had not heard of the new star would be by some occult instinct prone to innovations, and there would be movements through Hungary, Austria, Moravia, and Brittany; more atrocious ones in Livonia, Muscovy, Turkey, Persia, India, and nearer home in Brunswick, France, and most of all in Italy over the Venetian affair, while movements already under foot in Poland, Lithuania and Russia would be greatly stimulated by this new star.[73]

The new star appeared coincidently with a conjunction which had not occurred more than seven times since the foundation of the world at the beginning of the fiery triplicity, a thing most celebrated among astrologers for many years.[74] So, despite what we have heard Kepler say against the doctrine of triplicities, he finally hazards the following predictions "from the natural effect of the fiery" triplicity (and of the star). The kings of Europe will fight for power; there will be new factions and new opinions; no mean changes will occur; and the Turks might be overthrown. But Kepler sees no natural cause for the conversion of Islam.[75] In the last chapter, however, he concludes astrologically that there will be a Christian victory over Turkey, some general conversion of Jews to Christianity, a new sect and great contention, followed by quiet. Ecclesiastical discipline will be restored by a public council, preachers will no longer be permitted to issue prognostications, the Church will be reformed; freedom of speech for the young will be reduced; popular fury checked; the aristocracy of the gilds (*collegiorum*) will prevail; and the pomp, luxury and pride of the monarchists be overcome.[76] And because on the sixtieth day after the appearance of the Nova, Saturn came to it, these changes will occur after sixty years. And because then the sun too was present with them, "this ratification will have great solem-

[71] *Ibid.*, p. 719.
[72] *Ibid.*, p. 723.
[73] *Ibid.*, p. 725.
[74] *Ibid.*, p. 705.
[75] *Ibid.*, pp. 732-33.
[76] Cap. 30, *Opera*, II, 744-45, 747.

nity and will be hidden from the eyes of the vulgar and treated by men of learning, not in any public meeting... but by exchange of letters."[77]

Thus, despite Kepler's assertion that political and religious change cannot be predicted, we have heard him declare that both were indicated by recent comets and by the *Nova* of 1604. It is true that he ascribed the new star and the significations of the comets directly to God. Nevertheless they observed astrological conventions and were intended to attract the attention of astrologers. Nor were they created out of nothing, for Kepler belived comets and *novae* to be of watery origin and nature.

Kepler distinguished three chief physical causes by which the heavenly bodies acted upon and influenced the sublunar world and upon which prediction could be based. First and most potent was the access and recess of the sun; second, the moon, with which experience proved that all humor waxed and waned, and which influenced the crises in diseases; third, the varied natures of the other planets, revealed by their colors. Kepler thought Saturn cold and wet rather than cold and dry. The planets influenced more while in the sign Cancer, because then they were longer above the earth. They also had more virtue when in the north and when stationary, especially Mercury because it was normally the swiftest, whereas Saturn when stationary effected least because it had little motion to lose.[78] The sun moved the other planets,[79] and the earth moved the moon, although this might well be a reflected virtue from the sun.[80] Kepler attributed little influence to the fixed stars because of their great distance and regarded as a questionable innovation the practice of some astrologers of noting the aspects of the planets with them.[81]

But the mere physical action of the heavenly bodies is for Kepler

[77] *Ibid.*, p. 748.

[78] *De fundamentis*, theses v, xv, xix, xxxiii-iv: *Opera*, I, 421-27. On the moon and crises, *Tertius Interveniens*, Thesis 70, *Opera*, I, 608-11. N. Herz, *Keplers Astrologie*, 1895, p. 30.

[79] *De motibus stellae Martis*, cap. 33: *Opera*, III, 300.

[80] *Opera*, II, 8, Caspar ed. XIV, 123 (letter no. 166); (from a letter to Ferdinand, archduke of Austria, July, 1600): "In Terra igitur inest virtus quae Lunam ciet. Antea vero primarius eius fons in Sole erat...," etc.

[81] *Tertius Interveniens*, Thesis 43, *Opera*, I, 584.

only a part of the relationship between superiors and inferiors which forms the foundation for astrological prediction. There are also the relationships of harmony and sympathy, on which we have already touched to some extent. We heard Kepler refer to his terrestrial astrology. Copernican though he was, this was based not upon the movement of the earth but upon the attribution of a soul to the earth.[82] Indeed, in his *Mysterium cosmographicum* of the previous century he had declared the whole world full of soul and that there was a peculiar soul in each planet.[83] In the *Harmonice mundi* of 1619 he reiterated that there was "a soul of the whole universe, set over the movements of the stars, the generation of the elements, conservation of animals and plants, and finally the mutual sympathy of superiors and inferiors. He thought it not unlikely that this soul of the universe resided at the center of the world, "which for me is the sun," and was propagated thence by rays of light which corresponded to the spirits in the body of an animal.[84] But for his terrestrial astrology the important point is the existence of an animal-like faculty by which the earth and men on it sense occult changes in the sky, "sympathies bound with imagination in the globe of earth," and "a sympathetic consensus of human nature with the stars."[85] Similarly the effect of a conjunction is not the work of the planets in conjunction, from which there is merely heat and light, but is the work of sublunar nature.[86] The essence of Kepler's "terrestrial astrology" then, is that Earth is the chief cause (*principatus causae in Terra sedeat*). Earth, like animals, has its circuits of humors and quasi diseases.[87] And it has a twin faculty of attracting sea waters into secret seats of concoction, and of expelling the vapors which have been thus concocted. By its

[82] *De fundamentis*, Thesis 42, *Opera*, I, 429. *Harmonice mundi*, IV, 7, *Opera*, V, 254, 258-59.

[83] Cap. 22, *Opera*, I, 183.

[84] *Harmonice mundi*, IV, 7, *Opera*, V, 251.

[85] *Opera*, VII, 7-8, 129, 130.

[86] *De stella nova*, cap. 8, *Opera*, II, 637: "opus quod superioribus adscribimus junctis quod non competit separatis id nequaquam planetarum ipsorum est (praeter nudam illuminationem et calefactionem), sed ipsius naturae sublunaris." Morin, *Gallia astrologica*, 1661, XXI, i, 4, p. 501b, quotes inexactly, "Opus conjunctionis non esse opus planetarum conjunctorum, a quibus tantum est illuminatio et calefactio, sed est opus ipsius naturae sublunaris."

[87] *De fundamentis*, Thesis 47, *Opera*, I, 430.

perception of the celestial aspects it is stimulated and excited to excrete these vapors with a pleasure akin to that which an animal feels in the ejaculation of its semen.[88] Man, too, is not merely a rational being but is endowed with a natural faculty like that of the earth of sensing celestial configurations, "without discourse, without learning, without progress, without even being aware of it."[89]

Such was Kepler's occult explanation of the relations of cause and effect between heavens and earth, such his "terrestrial astrology and doctrine of aspects," based upon an animistic and non-mechanical interpretation of nature. This was too much for even an astrologer like Morin (1583–1656), who objected that for Kepler the planets did not move sublunar nature as natural agents but as foreign objects affect the senses, and that Kepler attributed sense perception and even intelligence and free will to plants, minerals and earth.[90] To tell the truth, Morin concluded, Kepler was quite ignorant of astrology.[91]

So much for theory. How Kepler dealt with astrology in practice may be illustrated by his letter of Easter, 1611, to Wallenstein.[92] He protests his German honesty and loyalty to the Emperor (Rudolf II); the Bohemians and Austrians have failed to corrupt him. The Emperor is credulous; should he hear of the prognostication of that Frenchman, he would put too much faith in it. Vulgar astrology can easily be induced to say pleasing things to both sides. When Kepler is questioned by those whom he knows to be hostile to the Emperor, he tells them that the stars favor Caesar, but he doesn't tell the Emperor himself so, lest he become over-confident and negligent. He is about to tell Wallenstein in confidence what he has never told Matthias and the Bohemians, namely, that since 1606 both directions and revolutions, hostile before, have been favorable to them, whereas the Emperor has adverse directions. Kepler has written this with the intention that Wallenstein should

[88] *De stella nova,* cap. 28, *Opera,* II, 721.

[89] *Ibid.,* p. 722.

[90] Kepler is quoted and criticized in two passages of the *Astrologia Gallica* (1661), XXI, i, 4 (pp. 501-2) and XXII, iii, 2 (pp. 561-62).

[91] "In rei veritate Keplerus fuit astrologiae valde ignarus": *Ibid.,* p. 562b.

[92] *Opera,* VIII, 343-45.

see how little reliance is to be put in the Frenchman's Prognostication.

In short, I think that astrology ought to depart not only from the senate but also from the minds of those who today wish to give the Emperor the best advice, and hence should be kept out of his sight entirely.[93]

Kepler felt that there were worse forms of curiosity than even superstitious astrology,—geomancy for instance. On the other hand, he did not limit presages of the future to the aspects of the planets. He interpreted monstrous births as ominous. In January, 1606, at Strasburg, were born female twins with distinct abdomens, two livers, with veins to arms and feet, and a twin spine and cerebellum carrying motor nerves to the double arms and feet, and two pairs of ears, but one face, nose and mouth and a single pair of eyes, one thorax, stomach, throat, heart and lungs. After making the suggestion that this might forecast religious union, Kepler added, "unless perchance this monster was a prelude to the expedition of the Hanseatic cities to raise the siege of Brunswick, which more accords with ancient examples."[94] Kepler again took note of a monstrous birth in his prediction for the year 1620 dedicated to the Estates of Carinthia.[95]

Kepler's animistic and occult interpretation is seen in yet other fields than astrology and presages of the future. He chided Feselius for dismissing the doctrine of signatures as sheer fancy.[96] He accepted the great force of the imagination of a pregnant woman upon the foetus.[97] He believed that the physiognomy of the body was one of three things on which human fortune depended, and that another was the tutelar genius or guardian angel.[98] Like Bodin in the previous century, he dwelt upon the political significance of numbers.[99] He explained gravity, like Gilbert, as the action

[93] *Ibid.*, 345.

[94] *De stella nova*, cap. 28, *Opera*, II, 724.

[95] Published by Caspar and von Dyck in *Abhandlungen d. Bayer. Akad. d. Wiss., Math.-Nat. Kl.*, Neue Folge 17 (1933), 58 pp. See p. 37.

[96] *Tertius Interveniens*, Thesis 126, *Opera*, I, 639.

[97] *Opera*, II, 726; V, 263.

[98] *Harmonice mundi*, IV, 7, *Opera*, V, 263, 265.

[99] *Ibid.*, III, 16, De tribus medietatibus digressio politica, *Opera*, V, 195-210.

of the earth, which was a great magnet, by the means of immaterial effluvia.[100] But Kepler here further introduced the action of the soul of the earth.[101]

Kepler's notions of the earth and gravity were, indeed, subject to variation, as Fahie has already observed.[102] In his *Astronomia nova* of 1609 he correctly described gravity as follows:[103]

> Gravity is a mutual physical affection between related bodies towards union or conjunction (the magnetic faculty is of this order), so that the Earth draws a stone much more than the stone seeks the Earth.

He proceeds to say that whether the Earth is immobile at the center of the universe or is placed elsewhere or transported wheresoever by its animal faculty, heavy objects will always fall towards it. If the earth were not a sphere, heavy bodies would not fall straight down towards its center but to different points. Moreover,

> If the Moon and Earth were not retained, each in its circuit, by animal force or something equivalent, the Earth would rise toward the Moon by a fifty-fourth part of the interval between them, the Moon would descend toward the Earth by about fifty-three parts of the interval, and there they would unite, assuming that they are of the same density. If the Earth ceased to attract its waters to itself, all the water of the sea would be lifted and flow into the body of the Moon.

Kepler goes on to explain the tides as produced by the movement of the moon.

A century ago Apelt said that Kepler still thought of gravity as specific terrestrial force, and something essentially distinct from

[100] Copernicus had defined gravity as "nothing else than a certain natural inclination imparted to parts by the divine providence of the Maker of the universe that they may come together into their unity and integrity in the form of a globe"; *De revolutionibus* (1543), p. 7, quoted in Mersenne, *Correspondance,* II, 471. "Gilbert, concevant la Terre comme un aimant qui agitait par le moyen d'*effluvia,* imagina qu'elle pouvait transmettre son attraction par des fils magnétiques (fibres ou filamenta); les *effluvia* étaient à son avis immateriels": *Ibid.,* p. 472.

[101] Letter of 1605 to Herwart, "Et cum terra ut magnes attrahat gravia per effluxum immateriatum, terra per suam animam translata transfertur etiam effluxus . . .": *Opera,* II, 87.

[102] In Charles Singer, *Studies in the History and Method of Science,* II, 254.

[103] Introductio: *Opera,* III, 151.

the cause of the movements of the heavenly bodies. It worked in a straight line; those movements were circular. There was a bond of gravitation between earth and moon, but it did not disturb the movement of either. The moon only caused the tides by drawing the waters of earth upward vertically. The course of the planets about the sun was regulated by forces entirely different from gravity.[104]

Similarly Koyré has recently contended that this Keplerian gravitation or mutual attraction was limited to cognate or related bodies, such as stones and water, earth and moon, moon and water, and was not extended by him to celestial bodies beyond the moon. The planets did not gravitate towards the sun but were moved by it by means of immaterial species which were diffused from it through space but only in the plane of the ecliptic. Only when the qualitative and astrological distinction between terrestrial and celestial bodies had been completely dispelled, and the conception adopted of one "perfectly and absolutely homogeneous matter," could a theory of universal gravitation become possible.[105] To me this does not seem quite fair to Kepler. He not only states that the planets move faster as they approach the sun and slower as they recede from it, but that, were the system of Tycho followed, we would have to say the same of the earth, but that it is absurd to hold that the sun, which is so much greater than the earth, is moved by the earth. Thus a relation between two bodies which are not

[104] E. F. Apelt, *Die Reformation der Sternkunde,* Jena, 1852, p. 248: "Was man auch dagegen gesagt hat, so scheint es doch gewiss, dass sich Newton zuerst zur Idee der allgemeinen Schwere erhoben habe. Keppler dachte sich die Schwere noch als specifisch terrestrische Kraft. Sie ist ihm wesentlich verschieden von der Ursache der Himmelsbewegungen. Denn sie wirkt geradlinig, diese Bewegungen aber sind Umläufe. Die Erde steht zwar nach ihm noch mit dem Monde in einem Gravitationsnexus, aber die Schwere übt weder auf die Bewegung des Mondes noch auf die der Erde einen störenden Einfluss. Der Mond bringt nur die Ebbe und Fluth hervor, indem er die Gewässer der Erde senkrecht in die Höhe zieht. Der Umlauf der Planeten um die Sonne wird aber durch ganz andere von der Schwere gänzlich verschiedene Kräfte regiert."

[105] Alexandre Koyré, "La gravitation universelle de Kepler à Newton," *Actes du VIe Congrès International d'Histoire des Sciences,* Amsterdam, 14-21 Août, 1950, I, 196-211. Kepler introduced the term, *inertia,* but not in its present sense; was still obsessed by the simplicity and naturalness of circular motion.

cognate is made a matter of bulk. Also Kepler says that two stones, *anywhere in the universe*, "if outside the sphere of influence of a related body," would come together at an immediate place determined by their comparative mass (*moles*).[106] But whatever his momentary approaches toward a correct theory of gravitation, he also kept suggesting other explanations. Indeed, at the same time that he represents the planets as moved by immaterial species diffused from the rotating sun, he speaks of the earth as moved by animal or rather magnetic motion.[107]

In 1619 he carried such animism still further[108] and said that the Earth sometimes seemed sluggish and contumacious, at another time exacerbated. The Earth was an animal, not like a dog quick to respond to every stimulus, but, like a cow or elephant, slow to anger but so much the more violent when enraged. Terrestrial phenomena were so analogous to animal physiology that Kepler was convinced that the Earth too had a soul, and that like other animals it ate and drank, digested metals and the like in its maw, or formed and brought forth like a pregnant woman. Earth's sucking in the waters of the ocean explained why the sea never increased from the inflow of so many rivers. Storm-breeding springs confirmed Kepler's belief, but sometimes, instead of tempests of rain, Earth emitted mere winds or sulphurous exhalations and pestilential sweats. Such terrestrial respiration especially resembled that of fish which take in water through the mouth and expel it from the gills, and might serve to explain the tides rather than the movements of the moon. One day at Antwerp there were no tides, although the moon kept on its course as usual.

What Kepler has to say of comets is not of much astronomical value. He had observed very few and even for these combined the observations of others with his own.[109] He thought that comets were formed from the denser parts of the ether and were gradually dissolved by the rays of the sun.[110] He asserted that their trajectory was a straight line,[111] partly because this seemed to fit in better

[106] See pp. of his *Introductio* with signatures (***) 3 r, and (***) 4 r.

[107] *Ibid.* (***) 3 v; (***) 6 r.

[108] In his *Harmonices mundi libri quinque*, IV, 7; *Opera*, V, 254-55; Caspar's recent edition, VI, 268-70.

[109] Caspar and von Dyck, *op. cit.*, p. 39.

[110] *Opera*, VII, 5.

[111] *Opera*, VII, 3, 5.

with the Copernican system. He said that his treatise on comets of 1619 was not well received in Italy on this account, the prohibition of the Copernican theory having been recently issued there in 1616.[112] Sarsius, in a work of 1619 to be mentioned presently, stated that Kepler's having comets move in a straight line required the circular movement of the earth which Catholics were forbidden to hold.[113]

Horatius Grassius, a Jesuit who taught mathematics at Genoa and then at Rome, published at the latter place in 1619 An Astronomical Disputation concerning the three comets of the year 1618 held in the Collegium Romanum. Affirming that no part of the sky now escapes our gaze since the discoveries of Galileo and Kepler, Grassius says that a rumor in August which spread in Italy as to a comet was confirmed by letters from Germany, and that the comet was observed on August 29 between two stars of Ursa Maior. On November 18 was first seen another comet near the constellation called the Bowl which covered nearly 24° in twelve days. Some said that the third comet appeared on November 1, others not until November 29. It had a tail which was 40° long. Observations of December 13 at Cologne and Rome showed that it had practically no parallax. Its tail always opposed the sun, it moved in a great circle like the planets. Grassius computed its size as 490,871,150 cubic miles and its distance from the earth as 572,728 miles, somewhere between the moon and sun.[114]

This initiated a controversy in which a disciple of Galileo, Mario Guiducci, replied to Grassius,[115] who answered under the pseudonym of Lotharius Sarsius in a work printed at Perugia,[116] to which

[112] *Opera*, VII, 21: "... in Italia vix vendendum, quia salvo apparentias cometae per eius trajectum aequabilem in aethere et per motum Telluris circa Solem."

[113] *Opera*, VII, 151.

[114] Kepler, *Opera*, VII, 149-50.

[115] *Discorso delle Comete*, Florence, 1619: *Opere*, VI (1896), 37-108, with a MS version and indication of the parts really written by Galileo.

[116] *Libra astronomica et philosophica qua Galilaei opiniones examinantur*, Perusii, 1619. *Opere*, VI (1896), 109-80, "con postille di Galileo."

According to Cesi (Bernardus Caesius, *Mineralogia*, 1636, p. 32a) "Horatius Grassius in libra astronomica, examine 3, propositione 2," argued against Galileo that *corpora attrita* were not necessarily diminished. He hammered copper (*aes*) until it became too hot to touch, yet its weight remained the same.

Galileo replied in *Il Saggiatore* of 1623,[117] while Sarsius published yet another response in 1627.[118]

We will not follow these later publications in which, after the tiresome fashion of sixteenth and seventeenth century controversial writing, the opponents reply to each other paragraph by paragraph and at great length. We limit our account to a brief summary of Galileo's views as to comets, as set forth in the *Discorso.* The treatise begins, as was the fashion with Latin treatises on comets back in the thirteenth and fourteenth centuries,[119] with a review of the various explanations of comets given in antiquity, such as that they are planets or conjunctions of planets. Incidentally it is argued that a comet does not move like a planet, that it is large at first and soon grows smaller or disappears entirely, and that the brevity of its appearance, space traversed, and length of its occultation would require a huge epicycle for it.

Coming to Aristotle's opinion, that a comet is a terrestrial exhalation which takes fire in the upper air near the sphere of fire, it is said that it seems probable but is really as unreliable as those it pretends to confute. The figure and movement of a comet are too regular for a tumultuous, wandering fire, as well as the fact that its tail or beard is always diametrically opposite to the sun. And Aristotle's holding that comets were sublunar is repugnant to the smallness of its parallax, "observed with exquisite diligence by so many excellent astronomers."

Galileo inclines to the view of the ancient Pythagoreans that a comet is not a real visible object but a refraction of our vision of the sun. The method of parallax applies in the case of actual and permanent objects but not for mere apparitions such as haloes and rainbows. Venus and Jupiter seen in the daytime are not a hundredth

[117] *Il Saggiatore nel qualè con bilancia esquisita e giusta si ponerano le cose contenute nella Libra astronomica e filosofica di Lotario Sarsi Sigensano,* Rome, 1623. *Opere,* VI (1896), 197-372.

[118] *Ratio ponderum librae et symbellae, in qua quid e Lotharii Sarsii Libra astronomica quidque e Galilaei symbellatore contra Libram edito de cometis statuendum sit, collatis utriusque rationum momentis philosophorum arbitrio proponitur,* Paris and Naples, 1627. *Opere,* VI (1896), 373-500, "con postille di Galileo."

[119] See my *Latin Treatises on Comets Between 1238 and 1368 A.D.,* The University of Chicago Press, 1950.

of their apparent size by night. With the naked eye we see the radiation, through the telescope the actual body of a star.

Reverting to the question of the movement of comets, Galileo argues against their having a sphere of their own, pointing out the differences in course, speed and direction of the comets of 1577 and of 1618. Although to a person at the center of a sphere motion in a great circle seems to be in a straight line, the converse is not true—that what seems in a straight line is really in a great circle. Tycho Brahe argued from the movement of the comet of 1577 to a sphere for comets about the sun outside of Mercury and Venus. Galileo or Guiducci asserts that the path for the present comet indicated by Grassius would traverse the four elements and the inferno (at the center of the earth). Galileo himself argues for a quite simple and equable motion in a straight line from the earth's surface towards the sky. He rejects the distinction between the elements and the heavens. However, the atmosphere about the earth is not pure air but, at least to a certain height, mixed with gross fumes and vapors which render it much denser than the upper ether which expands pure and limpid through immense space.

Galileo was equally in error with regard to the tides[120] which he attempted to explain by the movement of the earth, the container of the sea, and that without making any reference to the correct view that they follow the phases of the moon. He held that the combination of the earth's diurnal and annual movements retarded the tides at the point nearest to the sun and accelerated them at the point farthest from the sun.[121] Thus in his anxiety to supply another indication of the truth of the Copernican system and movement of the earth, Galileo fell into a false hypothesis and tried to support truth by error and error by truth. His treatise, addressed to Cardinal Orsini, was not printed because of the ecclesiastical decree of 1616 against the movement of the earth and holding the sun to be stationary, but fourteen manuscripts of it are extant, and Galileo's explanation of the tides was known to and mentioned by other writers of the century. He concluded it by saying that his hypothesis was based only upon reason and philosophy and astronomical ob-

[120] *Discorso del flusso e reflusso del mare*, *Opere*, V (1895), 378-95.

[121] *Ibid.*, pp. 382-83.

servations, but that it has been declared false and erroneous by virtue of more eminent (i.e., theological) knowledge, which makes what he has written vain. So let them find the true explanation, or conclude that God wished it to transcend the power of the human intellect, or drop such idle curiosity, which might better be spent on more salubrious studies.

Comets and tides were not the only subjects where Galileo went astray. He thought that a chain hung from two nails driven into the wall at the same level would take the form of a parabola,[122] but actually the curve is a catenary.[123] He repeatedly stated that the air offered increasing resistance to the increased acceleration of a falling body, which would not go on increasing indefinitely because of this resistance, which "finally reduces its speed to a constant value which is thereafter maintained."[124] In one passage he even stated that the air offered greater resistance to a body in rapid motion than to the same body when in slow motion, but soon contradicted this by saying that the resistance of the air did not affect motions of high velocity more than those of low velocity, "contrary to the opinion hitherto generally entertained."[125] He also still retained in large measure the late medieval doctrine of impetus (*virtù impressa*).[126] Nor had he abandoned the notion that nature abhors a vacuum,[127] although he suggested the possibility of innumerable very minute vacua in bodies, and although he affirmed that various experiments indicated that a vacuum might be produced by violence.[128]

Galileo was not so addicted to astrology as Kepler but he had

[122] *Dialogues Concerning Two New Sciences by Galileo Galilei,* translated . . . by Henry Crew and Alfonso de Salvio, New York (1914), p. 149. I have sometimes altered the wording of their translations.

[123] Jacques Bernouilli first gave its equation in 1691.

[124] *Dialogues* (1914), pp. 74, 255-56, 93-94.

[125] *Ibid.*, pp. 253, 255; *Le Opere* (ed. naz.), VIII (1898), 276: "l'altra è nel contrastar più alla velocità maggiore che alla minore dell' istesso mobile"; 277, "E questa osservazione ci assicura congiuntamente delle 2 pro- (278) posizioni, cioè che le massime e le minime vibrazioni si fanno tutte a una a una sotto tempi eguali, e che l'impedimento e ritardamento dell'aria non opera più ne i moti velocissimi che ne i tardissimi; contro a quello che pur dianzi pareva che noi ancora comunemente giudicassimo."

[126] *Dialogues*, pp. 165-66.

[127] *Ibid.*, p. 11 *et seq.*

[128] *Ibid.*, p. 67.

read such astrological writings as Porphyry's Introduction to the *Tetrabiblos* of Ptolemy and Hermes on the revolutions of nativities. He also composed many genitures, but the last extant is in 1624.[129] He left an autograph note as to the exact time of his birth which served as the basis for astrological calculation,[130] although modern historians have utilized it to show that he was not born on the same day that Michelangelo died. By 1633, however, Galileo would seem to have had little faith left in astrology, for on January 15th he wrote to Elia Diodati at Paris that he marveled at the great esteem in which Morin held judicial astrology and at his attempt to establish its certitude by "his conjectures, which seem to me uncertain enough, not to say most uncertain," and to locate it "in the supreme seat of human sciences." Galileo expressed great curiosity to see "si maravigliosa novità."[131]

But it is time to turn to Galileo's great service as an exponent of experimental science, which is especially manifested in his *Dialogues Concerning Two New Sciences.*

As might be expected from its title, the *Dialogues Concerning Two New Sciences* are full of passages laying claims to extreme novelty. "I am at your service," says Salviati, "if only I can call to mind what I learned from our Academician (i.e., Galileo), who had thought much upon this subject and according to his custom had demonstrated everything by geometrical methods, so that one might fairly call this a new science."[132] The interlocutors prefer ideas which are somewhat startling from their novelty to "dead books which raise many doubts but remove none."[133] The Peripatetics would regard these ideas "as mostly new."[134] When Salviati explains condensation and rarefaction by the "contraction of an infinite number of infinitely small parts without the interpenetration

[129] A. Wolynski, *Nuovi documenti inediti del processo di Galileo,* Florence, 1878, p. 164; Antonio Favaro, "Galileo astrologo secondo documenti editi ed inediti," *Mente e Cuore,* VIII (1881), 99-108.

[130] Favaro, "Il 'Racconto Istorico della Vita di Galileo,'" *Archivio storico italiano,* 74-2 (1916), 129: "e che del resto contribui a fornire i dati necessari ad alcuni temi astrologici di varia fonte."

[131] *Correspondance du P. Marin Mersenne,* III, 368.

[132] *Dialogues* (1914), p. 6.

[133] *Ibid.,* p. 26.

[134] *Ibid.,* p. 48.

or overlapping of finite parts," and the "expansion of an infinite number of indivisible parts by the interposition of indivisible vacua," Sagredo finds the idea "new and strange," while Simplicio "finds difficulty in following either path, especially this new one."[135] The notion that even a great difference in weight does not affect the speed of falling bodies is "so new" that it requires most convincing proofs.[136] But Sagredo is "exceedingly fond of choice and uncommon propositions."[137] Or we are assured that "the questions pertain to natural science and have not been treated by other philosophers."[138] The treatment of local motion is "a very new science concerning a very old subject."[139] How acceleration varies with slope is a new feature and opens the door to a new vista.[140] Galileo's many new discoveries have already brought odium upon him, but, after many thousands of hours of speculation, he has finally reached conclusions "which are far removed from our earlier ideas and . . . remarkable for their novelty."[141] Discussion of musical intervals and consonance is described as "these novelties," and an argument with reference to the motion of projectiles in a parabola is termed "new, ingenious and conclusive."[142]

Galileo repeatedly asserts that bodies are composed of an infinite number of very minute corpuscles or atoms. The extremely fine particles of fire are able to penetrate the slender pores of metals, although these are too small to admit even the finest particles of air or of many liquids, while a vast number of tiny vacant spaces bind together the least particles of a metal.[143] Bodies are composed of an infinite rather than a finite number of atoms.[144] Similarly a continuous quantity contains an infinite number of indivisibles.[145] Gold and silver may be reduced "into their ultimate, indivisible, and infinitely small components."[146] Or again, many examples show

[135] *Ibid.*, pp. 51-52.

[136] *Ibid.*, p. 83.

[137] *Ibid.*, p. 57. A rather free translation of "Ed io, che sento tanto diletto in certe proposizioni e dimostrazioni scelte e non triviali . . ."

[138] *Ibid.*, p. 92.

[139] *Ibid.*, p. 153, but for the translator's "I have discovered by experiment," the original text has only "comperio."

[140] *Ibid.*, p. 243.

[141] *Ibid.*, pp. 262, 271.

[142] *Ibid.*, pp. 107, 250. I have modified the latter translation a little to conform better to the original.

[143] *Ibid.*, pp. 19-20.

[144] *Ibid.*, p. 25.

[145] *Ibid.*, p. 34.

[146] *Ibid.*, p. 41.

that *materie fisiche* are made up "of infinitely small indivisible particles."[147] Thus Galileo favored atomism before or contemporaneously with Gassendi and Descartes, and foreran the "corpuscular philosophy" of Robert Boyle.

Although abounding also in mathematical demonstrations, the *Two New Sciences* of Galileo perhaps more than any other single book, marks the advent of modern experimental method. There are experiments with a wooden rod fitted into a wall at right angles, with weights supported by two nails of different size driven into a wall, with a weight attached to a cylinder hanging vertically.[148] The binding effect of spirals is illustrated by a device used in sliding down a rope.[149] The experiment with two smooth flat surfaces which will slide over each other readily but are difficult to lift apart, is still attributed to avoidance of a vacuum and not to air pressure,[150] and apparatus is devised to measure the *forza del vacuo*.[151] There are experiments to find to what length cylinders of metal, stone, wood, glass etc. of any diameter can be elongated without breaking of their own weight, and experiments with molten gold, silver and glass.[152] A rather crude experiment is described to determine whether the propagation of light is instantaneous,[153] but it was a first step towards the Morley-Michelson experiment.

There are experiments with falling bodies,[154] including one from a tower two hundred cubits high,[155] but those to show that a vacuum might be produced by violence "would here occupy too much time"[156] and so are omitted. Two bodies which differed very little in speed falling through the air, in water sank with a speed ten times as great for one as for the other.[157] Sagredo "often tried with the utmost patience to add grains of sand to a ball of wax until it should acquire the same specific gravity as water" and maintain equilibrium therein, but with all his care was never able to accomplish this.[158] When physicians perform a similar experiment in

[147] *Ibid.*, p. 55.
[148] *Ibid.*, pp. 3-4, 6-7.
[149] *Ibid.*, pp. 9-10.
[150] *Ibid.*, pp. 11, 18.
[151] *Ibid.*, pp. 14-15.
[152] *Ibid.*, pp. 17-18.
[153] *Ibid.*, pp. 42-43.
[154] *Ibid.*, pp. 62-67, 71-74, 82.
[155] *Ibid.*, p. 75. (Ed. naz., VIII (1898), 120.)
[156] *Ibid.*, p. 67.
[157] *Ibid.*, p. 68.
[158] *Ibid.*, pp. 68-69.

testing waters, the addition of two grains of salt to six pounds of water is enough to bring the ball from the bottom to the surface.[159] Evidently others than Galileo were performing exact and meticulous experiments. If a globe full of water with a very narrow mouth—"about the same diameter as a straw"—is inverted, the water will not run out nor the air enter, but if a vessel of red wine is applied to the opening, the water and wine will very slowly interchange places.[160]

There follow experiments with compressed air and an air-pump, weighing water against air, experiments with pendulums, showing that the time of descent is the same along all arcs, and "some easy and tangible experiments" concerning marvelous accidents in the matter of sound,[161]—all these before even the first of the four days of the dialogues is over. Those with sound included the scraping of a brass plate with an iron chisel which produced rows of fine parallel marks when it made a noise. When the tones were higher, the marks were closer together and when deeper, farther apart.[162] An experiment noted in the dialogue of the second day was producing a parabola by rolling a perfectly spherical brass ball on a metallic mirror held somewhat inclined.[163]

The experiments of the remaining two days are concerned with motion, and Galileo claims to have discovered notable properties of motion not hitherto observed or demonstrated, such as that "the distances traversed during equal intervals of time by a body falling from rest stand to one another in the same ratio as the odd numbers beginning with unity,"[164] and that the path of projectiles is a parabola. Experiments are made with a ball rolling in a channel down inclined planes and the time measured by a water clock.[165] We are told on the one hand that principles which have once been established by intelligent experiments become the foundation of the entire superstructure,[166] but on the other hand that understanding why a thing happens far outweighs the mere information received from others or even gained by repeated experiment, and

159 *Ibid.*, pp. 69-70.
160 *Ibid.*, p. 71.
161 *Ibid.*, pp. 78-80, 84 *et seq.*, 95 *et seq.*
162 *Ibid.*, pp. 101-2.
163 *Ibid.*, pp. 148-49.
164 *Ibid.*, p. 153.
165 *Ibid.*, pp. 170-79.
166 *Ibid.*, p. 178.

that the knowledge of a single fact by discovering its causes enables one to understand and ascertain other facts without need of recourse to experimentation,[167] and further that "the same experiment which at first glance seemed to show one thing, when more carefully examined, assures us of the contrary."[168] Thus Galileo recognized the existence of certain perils in experimentation and set limitations to the use of the experimental method, preferring rigid geometrical demonstration where that was possible.

Professor Lane Cooper in *Aristotle, Galileo and the Tower of Pisa* (1935), noted that the story of Galileo's employing the campanile or leaning tower of Pisa in experiments with falling bodies,[169] first appeared in Vincenzio Viviani, *Racconto istorico de la Vita del Signor Galileo Galilei,* which was composed in 1654 but not published until 1717, and that this original statement had been greatly enlarged upon and embroidered by subsequent imaginative writers. In a passage cited above concerning experiments with falling bodies, Galileo spoke of a tower 200 cubits or 300 feet high[170] whereas the leaning tower is only 181 feet. The figure of 200 may have been selected for purposes of easy computation, but it seems clear that Galileo had often experimented with dropping bodies from heights and that towers would provide the most convenient and sheer drops. In his *De motu,* discussing why less heavy bodies at first fall faster than heavier bodies, he asserts that wood at first drops faster than lead, but after a little the motion of the lead is so accelerated that it leaves the wood behind, "and, if they are released from a tall tower, precedes it by a great space; and of this I have often made trial."[171]

Despite the rigid mathematical demonstration and experimental proof which are said to characterize the two new sciences, an

[167] *Ibid.*, p. 276.

[168] *Ibid.*, p. 164.

[169] It rather was first so employed by an opponent of Galileo's principles of motion and defender of "the statement of Aristotle, in the first book of *De Caelo,* that the larger body of the same material moves more swiftly than the smaller, and in proportion as the weight increases so does the velocity." Translated by Lane Cooper (1935), 29, from Coresio, *Operetta intorno al Galleggiare de corpi solidi,* Florence, 1612; reprinted in *Opere,* IV, 242. We speak in a later chapter of Riccioli's experiments from the Asinelli tower at Bologna, which is 312 feet high.

[170] Lane Cooper does not include this in his "Passages for Reference."

[171] *Le Opere,* ed. naz., I, 333.

element of the marvelous, not to say magical, is not entirely lacking in them, or at least in their exposition. We are told that "sometimes a wonder is diminished by a miracle."[172] When Sagredo, speaking of burning glasses, remarks, "Such effects as these render credible to me the marvels accomplished by the mirrors of Archimedes," Salviati retorts that it was Archimedes' "own books (which I had already read and studied with infinite astonishment) that rendered credible to me all the miracles described" by other writers.[173] Or we have explained "a phenomenon upon which the common people always look with wonder,"[174] while even Salviati's "surprise is increased," when he sees every day enormous expansions occurring almost instantaneously, as in the explosion of gunpowder.[175] Later he alludes to the supernatural violence with which projectiles are launched from fire-arms.[176] Even secret mysteries are still the order of the day.

> The fact that one can take the origin of motion either at the inmost center or at the very top of the sphere

leads Simplicio to think

> that there may be some great mystery hidden in these true and wonderful results, a mystery related to the creation of the universe (which is said to be spherical in shape), and related also to the seat of the First Cause.

And Salviati agrees with him.[177]

Writing in 1606 of his own experiments with the magnet, Galileo said that he had made it sustain twice its own weight and that he had in mind (*nella fantasia*) "some other artifices to render this yet more marvelous."[178] Twenty years later he wrote that for three months he had been busy with an admirable device of multiplying artificially and extremely the virtue of the magnet in sustaining iron. A piece weighing six ounces which by its natural force would not

[172] *Dialogues* (1914), p. 27.
[173] *Ibid.*, p. 41.
[174] *Ibid.*, p. 56.
[175] *Ibid.*, p. 60.
[176] *Ibid.*, p. 255.
[177] *Ibid.*, pp. 193-94.
[178] Antonio Favaro, "Di alcuni inesattezze nel 'Racconto Istorico della Vita di Galileo' dettato da Vincenzio Viviani," *Archivio Storico Italiano*, 74-2 (1916), 127-50, p. 131, citing *Opere*, X, 207.

sustain more than an ounce of iron, by art sustains 150 ounces. At first he had been thrilled to support forty times more than its innate strength would, but now he is not content with 150 times it.[179] Thus experimental science has not yet freed itself from the appeal of the surprising, the unexpected, the marvelous, the secret and the mysterious, which had always been characteristic of magic.

Experiment is also still employed for purposes of deception and entertainment, as well as causing astonishment. Sagredo fooled some friends into thinking that he could make a ball of wax remain in equilibrium in fresh water by secretly putting salt water, upon which the ball floated, at the bottom of the vessel and fresh water above this. Then, whether the ball was pushed to the bottom of the vessel or lifted to the surface, it would come back to the middle.[180]

Salviati, discussing the standing out and up of large drops of water on cabbage leaves, and the fact that water will not flow out from nor air enter an inverted glass globe with a very small mouth, says that he thinks there must be a "great dissension" between air and water or "una disconvenienza . . . occulta a me." Simplicio interrupts to say that he has to smile to see the great antipathy that Salviati has for antipathy, which he avoids naming, and Salviati accepts it, for Simplicio's sake, as the solution of the matter.[181] But Galileo would not seem to have had much sympathy with the doctrine of sympathy and antipathy.

In such passages Galileo, despite his new science, is in large measure repeating old and familiar ideas. One or two further examples may be offered of his clinging to old clichés: Man is represented as surpassed by the animals in a thousand operations, and the fallacy of many popular beliefs is dwelt upon.[182] On the other hand, it was a workman called in to repair a pump who told Sagredo that no pump could lift water above eighteen cubits, and gunners were the first to inform him that the maximum range for artillery was obtained from an elevation of 45 degrees.[183] Or re-

179 *Ibid.*, p. 132.

180 *Dialogues*, p. 69.

181 Ed. naz., VIII, 116, "Or sia questa, in grazia del Sig. Simplicio, la soluzione del nostro dubbio."

182 *Dialogues*, pp. 69, 168-69.

183 *Ibid.*, pp. 16, 276.

course is had to the method used by those "who are skilled in drawing gold wire."[184] Nature is still personified in an allusion to "the habit and custom of Nature herself."[185]

Although Galileo thought that he had invented two new sciences, he recognized to some extent the work of predecessors. He was a great admirer of Archimedes.[186] And while he chided Aristotle for such views as that a stone weighing ten times another would fall ten times as far in the same time,[187] he cited the *Mechanics*, which was then attributed to Aristotle, a number of times with respect.[188] He accepted Aristotle's assertion that all elements except fire have weight, and his experiment showing that a leather bottle weighs more when inflated than when collapsed.[189]

The difficulty with Galileo's position was that he assumed that there had been no progress in mathematical or experimental physics since Aristotle and Archimedes, except for one or two very recent works, such as that on the center of gravity of solids by Luca Valerio,[190] "the Archimedes of our age," or that of the Jesuit, Buonaventura Cavalieri, on burning glasses.[191] Thus the impression is given that one should begin where Aristotle left off and correct his *Physics*, whereas its doctrines, especially as to motion, had been repeatedly criticized in the later Latin middle ages. The sole medieval Latin author mentioned in *Two New Sciences* is Sacrobosco.[192] Yet the experiment with two flat plates or surfaces goes back at least to Henry of Hesse in the fourteenth century,[193] while that with the inverted water jar went back at least to Adelard of

[184] *Ibid.*, p. 52.

[185] *Ibid.*, pp. 160-61.

[186] *Ibid.*, pp. 41, 110, 144, 147, 242, 251.

[187] *Ibid.*, p. 62.

[188] *Ibid.*, pp. 20, 110, 125-26, 135, 291.

[189] *Ibid.*, pp. 77-78.

[190] *De centro gravitatis solidorum libri tres*, Rome, 1604, in-4. Cited in *Two New Sciences* at pp. 30, 148, 294.

[191] *Lo specchio ustorio . . .*, Bologna, 1632 and 1650. Cited in *Two New Sciences*, at p. 41.

[192] *Op cit.*, p. 57: "I remember with particular pleasure having seen this demonstration when I was studying the *Sphere* of Sacrobosco with the aid of a learned commentary." Some passages in *Trattato della sfera*, written before Galileo became an adherent of the Copernican theory, seem reminiscent of Sacrobosco: *Opere*, II, 215, "che il cielo sia sferico e si muova circolamente"; 220, "Che la terra sia constituita nel centro della sfera celeste"; 223, "che la terra stia immobile."

[193] T III, 480.

Bath in the twelfth century.[194] The "wonderful phenomenon . . . that a vibrating string will set another string in motion,"[195] had long been remarked. And the *Calculationes* of Richard Suiseth are immediately brought to mind by such a proposition as this:

> The time in which any space is traversed by a body starting from rest and uniformly accelerated is equal to the time in which that same space would be traversed by the same body moving at a uniform speed whose degree of velocity is half the highest and last degree of the aforesaid uniformly accelerated motion.[196]

Most modern physicists do not repeat Galileo's mistake of studying only ancient science and neglecting medieval activity in physics, for the simple reason that they do not pay any attention to the history of science. But is not this doubling his error?

Galileo's *Dialogues Concerning Two New Sciences* were not as widely and rapidly circulated and read as one might think. A letter of Mariotte to Huygens of February 1, 1668, shows that he had only just read them, thirty years after their publication, and three years later Roberval remarked that Galileo "a multis non recipitur." Huygens' father, however, had been engaged in negotiations with Galileo, and Christiaan knew the Dialogues well by 1652.[197]

The Bible was still a factor to be reckoned with, in the history of science in the seventeenth century. Not only was it cited against the Copernican theory and for creation, and accepted as authoritative by many in scientific matters. It also to some extent stimulated or guided scientific curiosity. Men still wondered at what time of year creation occurred, what the carnivorous animals lived on in Noah's ark, and directed their zoological and mineralogical investigations to the animals mentioned in Scripture or the gems found

194 T II, 38-39.

195 *Dialogues*, p. 98.

196 *Ibid.*, p. 173. I have revised the English translation of the closing clause, "cuius velocitatis gradus subduplus sit ad summum et ultimum gradum velocitatis prioris motus uniformiter accelerati," from Crew and De Salvio's, "whose value is the mean of the highest speed and the speed just before acceleration began." There was no speed then, since the body was at rest, and the revised translation makes the resemblance to Suiseth the more apparent. For Suiseth, see T III, 370-85.

197 Huygens, *Oeuvres complètes*, XXII, 639, 670, 433, 458.

in the breastplate of the high priest, or their medical speculations to the wounded side of Christ.

On the other hand, ever since the twelfth century, if not before, there had been opposition to taking the Bible literally as a scientific authority. While temporarily weakened by Protestant Biblicism and Catholic reaction, this tendency increased with the growing independence of natural science and the development of distinct mathematical and experimental methods.

Galileo gave voice to it, as well he might, in a letter addressed to Christine of Lorraine, Grand-Duchess of Tuscany, at about the time of the decree of 1616 against the teaching of the Copernican system. It was first printed at Treves in 1636,[198] with a camouflaged title which puts Galileo's contentions in the mouths of the Church Fathers and theologians.[199] A Prussian, Robertus Robertinus, writing on January 6, 1635, from Danzig, sent to Matthias Bernegger, translator into Latin of Galileo's dialogues on the two systems of the world, a manuscript copy which he had brought back from Italy fifteen years before, and speaks of the letter itself as written eighteen or twenty years ago. It arrived too late to appear with Bernegger's translation,[200] as Robert had hoped, and was printed separately, as we have seen. It is said, in the Libri Catalogue of 1861, to have been "so rigidly suppressed that only a few copies" are extant.

Galileo says that, if one always followed the sacred text verbatim, one would not only fall into error but attribute to it contradictory statements and even false, heretical and blasphemous ones. When the Bible attributes hands and eyes to God, this is of course not

[198] *Nov-Antiqua sanctissimorum patrum et probatorum theologorum doctrina de sacrae scripturae testimoniis in conclusionibus mere naturalibus quae sensata experientia evinci possunt in gratiam Christinae Lotharingae*, Augustae Trebocorum, 1636, in-4, 60 pp. BM T. 727 (9.).

[199] In an Italian MS copy at Oxford, All Souls College 193, 47 fols., it is more accurately described as "De S. Scripturae testimoniis in conclusionibus mere naturalibus temere non usurpandi," with the titulus, "Alla serenissima madama la Grandduchessa Madre Galileo Galilei." But Coxe's Catalogue calls her Archduchess of Austria and dates the MS as 16th century.

[200] *Systema cosmicum . . . in quo quatuor dialogis de duobus maximis mundi systematibus, Ptolemaico et Copernicano . . . disseritur*, Strasburg, D. Hauttus for Elzevir, 1635.

meant literally but is in order to make things plain to vulgar minds. Nature is the word of God, too, and in disputing natural problems one should by no means start from the authority of passages of Holy Scripture, but from sense experiences and necessary demonstrations. Augustine said that he hadn't time to discuss whether the sky moved or stood still. Galileo quotes a high cleric that the Holy Spirit teaches us how to get to heaven, not how the heavens move. To condemn the books of Copernicus, after they have been accepted for so many years and increasingly proved true by many new observations, is to slam the door in the face of truth and forbid the whole science of astronomy. Some hold that where Holy Scripture always speaks in the same way concerning natural phenomena, there it is to be followed literally, and that this is the case as to the sun moving and earth standing still. But Galileo contends that in matters where the soul's salvation is not at stake, Scripture conformed not only to the capacity of the common man but to views then prevalent, and came closer to received usage than to the true essence of things. He quotes Jerome that many things in the Scriptures are said according to the opinion of the time when they occurred and not according to absolute truth. It does not follow, because all the Church Fathers admitted the immobility of the earth, that this is an article of Faith. But the Fathers are not all in agreement.

CHAPTER III

THEIR CONTEMPORARIES

Bayer—Aguillon—Geraldinus—Biancani—Mulerius—Tanner—Cavalieri—Fortunio Liceto—Fromondus—Metius—C. Borri—Oregius—Deusing—Boulliau—Cabeo—Henao.

When He prepared the heavens, I was there
—PROVERBS 8.27

We may parallel our account of Kepler and Galileo by noting the views and achievements of some of their contemporaries in related fields. Johann Bayer's *Uranometria*, published in 1603, has been called the first complete celestial atlas.

Aguillon, a Jesuit of Brussels who died in 1617, in 1613 published six books of Optics,[1] which have been called a classic in the history of that subject. The fourth book dealt with optical illusions; the sixth book, with three kinds of projection, orthographic, stereographic and scenographic.[2] Aguillon distinguished between celestial and elementary qualities. Examples of the former were the heat of the sun, the dominance of the moon over all humid things, and the occult virtues of the other stars. Very similar to the last were the qualities in sublunar bodies called specific and found in the magnet, nephritic stone, rhubarb and six hundred other simples. Inquiring why flames are more lively in winter and bitter cold, Aguillon says that the surrounding cold prevents expiration and forces the inflammable spirits together, so that by the unity of its material the fire is made sharper. This also explains why cold bodies do not shed cold as hot bodies radiate heat, for cold closes up the body while heat expands it and throws off hot corpuscles.[3]

Almost a century later Garmann repeated a prodigious personal

[1] F. Aguilonius, *Opticorum libri sex*, Antwerp, Plantin, 1613, 24 ff., 684 pp., 22 ff. BN V. 1652.

[2] *Ibid.*, pp. 504-, 575-, and 643-84 respectively.

[3] *Ibid.*, pp. 358-59.

experience which Aguillon related of himself.[4] While he was asleep, his natural spirits were so kindled that they would have burst forth in flame, had he not waked up just in time. His arteries were marvelously distended, his pulse violent, breathing very frequent but almost stifled, a very sharp ringing in the ears, and the whole state of the body perturbed, as when buildings are on fire. A bright light seemed diffused in the eyes and disappeared only after some time. It was of such sort that no external objects could be seen but blotted them out like a lucid cloud. Hence it was nowhere else than in the eyes and came from the spirit with which the entire head and the eyes themselves were distended. For Aguillon agrees with the most experienced medical men that this spirit glows with perpetual fire.

Johannes Geraldinus included comets in his treatise on meteors of 1613, holding that they were of elementary, not celestial, nature and located in the supreme region of the air. He believed that they announced winds, tempests, sterility, bad weather, pestilence, earthquakes and floods, but he was doubtful whether they signalled the death of leading men.[5] Geraldinus noted that three common opinions as to the origin of rivers were all included in Aristotle's *Meteorology* and that he rejected them in favor of the explanation that air was turned into water in the caverns of mountains.[6] But Geraldinus prefers the theory that they come from the sea, which he believes is favored by *Ecclesiastes* 1,7.[7]

In the year 1615 Giuseppe Biancani (1566–1624) of the Society of Jesus published two works relating to mathematics.[8] The longer was one of those sterile and futile compilations at which members of his Order delighted to spend their time and which required little or no thought, and least of all any independence of thought. It

[4] *Ibid.*, p. 14. "Accidit et nobis aliquando in somnis hunc spiritum praeter consuetum morem vehementius accendi, sic ut impetu quodam eruptura flamma, ac simul vita etiam ipsa videretur, ni soluto somno vigilantia succurrisset, quemadmodum in Ephialte usuvenit." Garmann, *De miraculis mortuorum*, 1709, p. 683.

[5] *De meteoris tractatus . . . in quinque partes distinctus*, Paris, 1613, in-8, 300 pp., BN R. 12816, pp. 122-129.

[6] *Ibid.*, p. 188.

[7] *Ibid.*, p. 193.

[8] *Aristotelis loca mathematica ex universis ipsius operibus collecta et explicata . . . Accessere de natura mathematicarum scientiarum tractatio, atque clarorum mathematicorum chronologia*. Bologna, 1615, 4to. Copy used: BN V. 7470.

ran, however, to only 283 pages, for the mathematical passages in the works of Aristotle are not very extensive or numerous. With it was bound a shorter treatise on the nature of the mathematical sciences together with a chronology of famous mathematicians. The latter shows that, however good a mathematician Biancani may have been, his chronological and historical information was very faulty. He puts Campanus of Novara in the eleventh century with Alhazen and Arzachel, but goes on to say that he wrote on computus in the year 1200, "as he himself says." The first translation of the *Almagest* from the Arabic is ascribed to Frederick II; Thebit ben Corat is placed in the thirteenth century; Roger Bacon and Marco Polo, in the fourteenth; Leonard of Pisa, in the fifteenth; Giovanni Bianchini in the sixteenth.[9]

Biancani, however, has to yield the palm for a useless compilation to Samuel Reyher who, over half a century later, produced at Kiel a *Mathesis Mosaica,* or mathematical passages of the Pentateuch explained mathematically, with an appendix of other mathematical passages of Scripture.[10]

In response to repeated requests from his students, Biancani also published a book on the Sphere.[11] A new presentation was needed, he said, in order to summarize the great recent progress in astronomy from discoveries made with the telescope. Furthermore he aimed to suit the general reader by omitting subtleties, trigonometry, and sines, tangents and secants. He still, however, cited Pliny[12] and he still put the earth at the center of the universe and denied that it moved, stating that the Copernican system was forbidden by the Church.[13] He still distinguished between the celestial spheres and

[9] *Op. cit.,* pp. 57-60. Hugo Sempilius and Andr. Stirbius are said to have composed similar works to the *Mathematicorum chronologia,* and J. B. Riccioli to have written a *Chronicon astronomorum sive astrologorum, cosmographorum ac polyhistorum.* Probably his *Chronologia reformata* etc., Bologna, 1669, is meant.

[10] *Mathesis Mosaica sive loca Pentateuchi mathematica mathematice explicata cum Appendice aliorum S. Scripturae locorum mathematicorum,* Kilonii, 1679, in-4.

[11] *Sphaera mundi seu cosmographia demonstrativa ac facili methodo tradita in qua totius mundi fabrica una cum novis Tychonis Kepleri Galilaei . . . adinventis continetur,* Bologna, 1620, 11 fols., 447 pp., in-4. Copy used: BN V. 7469, 2nd ed., Modena, 1635.

[12] *Ibid.,* pp. 55, 94.

[13] *Ibid.,* pp. 72-75.

the inferior, sublunar, elementary world, although he left it to physicists to determine whether there were three or four elements.[14]

Biancani ascribes the tides to the action of the moon and discusses the height of the lunar mountains.[15] He states that recent geographers have estimated the surface of water on our globe as about equal to that of the dry land.[16] He attributes the invention of measuring the depth of the sea by releasing a float as the weight hits the bottom and timing its return to the surface, to Leo Battista Alberti, unmindful of the earlier *Metrologum de pisce cane et volucre* of Giovanni da Fontana.[17] He repeats the time-honored problem, how far would a stone fall, if dropped into a hole running through the center of the earth to its opposite side.[18] He cites Tycho Brahe and Kepler with respect and often refers the reader to them for further information. He discusses the satellites of Jupiter and Saturn, sunspots, and the new stars of 1572, 1600 and 1604.[19] His observations of the recent comet of 1618 agreed with those at Parma, Rome and Antwerp that it had no parallax and so was in the highest heaven.[20] He suggests that comets and new stars may be only new-appearing, moving towards and away from us. He retains the ninth sphere "to save all the movements of the fixed stars."[21] He goes back farther than that, and treats of the rising and setting of the signs and stars, first according to the poets, repeating some of Sacrobosco's poetical quotations, and then "according to the astronomers."[22] On the other hand, he gives the declination of the compass as six degrees, whereas Crescentio had given it for Rome as 11°15′ and Mersenne in 1625 as 13°8′.[23] The book concludes with a section on measuring echoes.

And so, cannily or unconsciously, our author has to move on a little. He may have started with the mathematics of Aristotle, the astronomy of Sacrobosco, and the history of past mathematicians. But if he wants to pose as an astronomer and a teacher of that sub-

[14] *Ibid.*, p. 67.

[15] *Ibid.*, pp. 100, 161.

[16] *Ibid.*, p. 109.

[17] *Ibid.*, p. 108. For Fontana, T IV, 172 *et seq.*

[18] It is one of 35 problems at pp. 118-27.

[19] *Sphaera mundi*, pp. 250, 287, 292, 344.

[20] *Ibid.*, p. 305.

[21] *Ibid.*, p. 352.

[22] *Ibid.*, p. 340, "Aliter de ortu et occasu secundum astronomos."

[23] *Correspondance du P. Marin Mersenne,* I (1932), 202, citing Bartolomeo Crescentio, *Nautica Mediterranea,* 1601.

ject, and to satisfy his students and his readers, he has to make observations with the telescope; he has to pay lip-service at least not merely to fellow Catholics such as Tycho and Galileo, but even to a Protestant astronomer like Kepler. And he makes one fruitful suggestion: comets may return. He died in 1624, but new editions of his book appeared at Modena in 1635 and 1654, so that its influence was more than momentary.

Nicolaus Mulerius, M.D., was ordinary professor of medicine and mathematics at the University of Groningen, where in 1616 he published an elementary textbook of astronomy which was reminiscent of the *Sphere* of Sacrobosco, but with additions concerning geography and navigation.[24] In the preface Mulerius says that astronomers today divide into two sects of Peripatetics and Pythagoreans. The former, who hold that the earth is at rest at the center of the universe, and that sun and stars go round it with a double motion, diurnal and annual, are supported in this view by Aristotle, Hipparchus, Ptolemy, Alfonso the Wise, Tycho Brahe, "and with them an infinite multitude of the learned." But some eminent astronomers put the sun at rest at the center, and the fixed stars motionless in highest heaven, and only the earth and other planets in movement. The contention between the two schools is erudite and keen but without ill-feeling—a statement which of course applies to the period before 1616 and the prohibition by the Church of the Copernican system. As for Mulerius, after twenty-five years' examination of both views, he cannot find sufficient reason to abandon the former. It may involve absurdities, but they are mere hypotheses, and Copernicus, while avoiding them, runs into others that are equally bad.

> For he is forced willy-nilly to attribute to the universe a huge and enormous center, including the three bodies, sun, moon and earth, and so capacious that it could embrace over a billion and a half billion of earths.

Mulerius would be better pleased if Copernicus had left the earth at the center of the universe but had it revolve on its axis. On the

[24] *Institutionum astronomicarum libri duo, quibus etiam continentur geographiae principia, necnon pleraque ad artem navigandi facientia,* 8vo, 176 small pages: copy used, BN V. 20949 (1). There was another edition in 1649: BN V. 20950.

other hand, Mulerius accuses the Peripatetic school of holding that the heavenly spheres are solid, whereas the present age believes that there is no solid sphere between the earth and the fixed stars. Pena first represented sky and air as one and the same matter. Tycho Brahe regarded the heavens as most liquid and rarefied, which he proved from the wanderings of comets in the highest ether and by optical refraction.[25] But discussion whether stars are born and die, and whether all the fixed stars are at the same distance from the earth Mulerius postpones to another time.[26] He believes that the sun keeps our vital spirits going,[27] and he connects the tides with the moon.[28]

Despite the foregoing criticism of the Copernican system, Mulerius supplied the *Notae breves* which accompanied the edition of *De revolutionibus* issued the next year, 1617, at Amsterdam.[29]

In 1621 Adam Tanner, a Jesuit, discussed the heavens from a theological and Peripatetic angle.[30] In trying to explain away the new phenomena recently observed through the telescope, he kept citing Augustine, Pliny and others, and asked in one place if Tycho Brahe was to be followed against all antiquity. The answer of course is, although Tanner did not give it, Yes, in so far as his observations were confirmed after his death by the telescope.

Another Jesuit, Buonaventura Cavalieri, who was professor of mathematics at Bologna, where he lectured on Euclid and Ptolemy, besides mathematical treatises published in 1632 a work on the burning glass,[31] which we heard Galileo mention favorably, and in 1639 an astrological treatise of which we treat in another chapter.

Fortunio Liceto of Genoa, ordinary professor at Padua, published six books on new stars and comets in 1623, in which he held that there were both elementary or sublunar comets on the one hand, and new stars and celestial comets on the other, and that both had

[25] *Ibid.*, pp. 46-47.

[26] *Ibid.*, p. 56.

[27] *Ibid.*, p. 64.

[28] *Ibid.*, p. 164.

[29] *Nicolai Copernici . . . Astronomia instaurata libris sex comprehensa qui de revolutionibus orbium coelestium inscribuntur. Nunc . . . restituta notisque illustrata opera et studio D. Nicolai Mulerii*, Amsterdam, 1617, in-4, 487 pp.

[30] *Dissertatio peripatetico-theologica de coelis . . .*, Ingolstadt, 1621, in-4. Copy used: BM 1016.1.12.

[31] *Lo specchio ustorio overo trattato della settioni coniche et alcune loro mirabili effetti . . .*, Bologna, 1632, in-4, 224 pp. BN V. 6138 (1).

been envisioned by Aristotle and were supported by his writings and authority. He listed twenty-one other opinions as to comets. There had been previous instances of new stars and celestial comets before those of modern times. They were made by condensation and rarefaction. There were many little stars in the heavens which were invisible to us, and the Intelligences that moved the spheres might bring some of them together.[32]

Libertus Fromondus (or, Froidmont), chief professor of philosophy in the College of the Falcon at Louvain, born in 1587, in 1627 published six books of meteorology. There were further editions in 1631, at Antwerp; in 1639, at Oxford; in 1646; in 1656, at London; and in 1670.[33] The six books are on meteors in general, on ignited meteors, comets, winds—under which earthquakes are considered, watery meteors which include the sea, fountains and rivers, and apparent meteors such as the rainbow. Aristotle is usually preferred to other authorities, but not in the case of comets, although a sop is thrown to him by saying that some of them are sublunar. But the minimum parallax of others proves them celestial. Three regions of the air are accepted, and the common question is put whether the supreme region follows the circular movement of the heavens. Against Bodin it is held that demons are not the cause of thunder

[32] *De novis astris et cometis libri sex,* Venetiis, Apud Io. Guerilium, 1623, 410 pp. with 51 lines of small type per page: BN R.2897. Consult our index for references to some of Liceto's other works which covered a great variety of topics. The title of one, *Ulysses apud Circen sive de quadruplici transformatione deque varie transformatis hominibus* (Utini, ex typibus N. Schiratti, 1636, in-4, vi, 55 pp. BN R.2926), attests the interest which the question whether human beings could be transformed into beasts had then, but the text is a fanciful dialogue in the course of which the Cumaean sibyl quotes Cesalpino on demons, cap. 9, at length. Bound with it, in the copy I examined, was a similar dialogue addressed to a physician of Urban VIII and canon of the Vatican, *Mulctra sive de duplici calore corporum naturalium dialogus physico-medicus ad Cl.V. Thad. Colicolam S.D.N. Urbani VIII archiatrum et canonicum Vaticanum* (Utini, 1636, 164 pp. BN R.2928), and a third work of the same year addressed to Gaffarel, *Athos perfossus sive rudens eruditus in Criomixi Quaestiones de alimento, ad Cl. Virum Iacobum Gaffarellum D. Aegidii Priorem* (Padua, 1636, 185 pp. BN R.2927).

[33] I have used a copy of the Oxford, 1639 edition in the NYP: *Liberti Fromondi S.Th.L., Collegii Falconis in Academia Lovanensi philosophiae professoris primarii Meterologicorum libri sex,* Oxoniae, 1639, in-16, 505 pp. The dedication is dated in 1627 For other works by him see *Bibl. Curiosa* (1676), pp. 240-41.

storms, nor angels of winds.[34] The question is raised whether earth and sea have the same center, but that of the origin of fountains is passed over. The moon is recognized as the chief cause of the tides, the sun as a contributory cause. The various wonders told of waters are viewed somewhat sceptically, and Fromondus soon turns away to the living waters which Christ offered to the woman of Samaria. This religious interest is characteristic. Discussing whether apparitions of horsemen and battles in the sky can be explained naturally, Fromondus grants that clouds may sometimes take the form of horses or men. But no one in his right mind can account for all such apparitions recorded in histories as the results of natural chance. They are unmistakeable evidence of a directing divine intelligence, which is further indicated by their signifying future events. He rejects the attempt of atheists to account for them as reflections of such happenings in the clouds.[35] Fear of God is the final cause of earthquakes.[36]

In general Fromondus is not averse to signs of the future, treating of presages of storms and prodigious rains,[37] and, while denying that comets are causes of calamities, accepting them as signs divinely instituted. Kepler in his book on the comets of 1607 and 1618 makes them the cause of sublunar disturbances by the sympathy which the elementary world has with the celestial. Thus the comet of 1577 made Sebastian, king of Portugal, cross to Africa to his destruction. The faculty of the sublunar world or of the earth is thrown into consternation, exudes vapors, and produces rains and floods. But Fromondus regards this Keplerian doctrine of an animated earth as a blot upon his otherwise great science.[38]

Fromondus's own science is not above reproach. He says that winds are not made of air but of vapors, and by their gravity they flow downwards to earth like torrents and rivers. He queries whether a ship could circumnavigate the globe with the same wind.[39]

Adriaen Metius (1571–1635) was born at Alkmaar in Holland. In 1598 he became professor of mathematics at Franeker and held this

[34] *Ibid.*, pp. 57, 186.
[35] *Ibid.*, pp. 499-501.
[36] *Ibid.*, p. 260.
[37] *Ibid.*, pp. 90-93, 399.
[38] *Ibid.*, p. 155.
[39] *Ibid.*, pp. 177-79, 191, 207.

position for the rest of his life. He published various works in mathematics and astronomy, in 1625 acquired the M.D. degree, and is said to have spent a fortune on alchemy. The one book by him in which we are now interested is the *Primum mobile,* published at Amsterdam, 1630-31.[40] In it Metius praises Gemma Frisius; laments the recent untimely death of Mulerius on September 5, 1630; alludes to Maestlin's observations of the situation of stars or comets without the aid of instruments;[41] and tells how the whole face of the starry heavens can be described as the result of new observations.[42] But he remains conservative both in his astronomy and his astrology. Despite Copernicus, he regards the earth as immobile at the center of the universe,[43] although he grants that it forms one globe with the element water.[44] He repeats Sacrobosco as to the rising of the signs and stars.[45] He discusses astrological houses, the figure for the horoscope, revolutions and directions, *significator* and *promissor.*[46]

Adriaen's brother, Jacob Metius, is often credited with having invented the telescope before Galileo in 1608 and even with having discovered the moons of Jupiter which Galileo first announced. It has been suggested that the Metius brothers did not publish their results because, when another professor at Franeker who had opposed Aristotle was asked to resign in 1609, Adriaen's salary was reduced by fifty guilders.[47]

Christophorus Borri of Milan became a Jesuit in 1601 and served as a missionary in the Orient for many years, then taught mathematics at Coimbra and Lisbon. In 1631 there was published a book by him on the three heavens, aerial, sidereal and empyrean.[48] It was in six parts. The first, on ancient astronomy and its confutation, rejected the Copernican as well as the Ptolemaic hypothesis, as false and vain, although admitting that all the phenomena were not badly saved by it. The second part was on new phenomena, "observed in

[40] Copy used: BN V. 7678 (1-4).

[41] *Ibid.,* I, 147. Gemma and Mulerius are mentioned either in the preface or first pages of the text.

[42] *Ibid.,* I, 148.

[43] *Ibid.,* I, 11.

[44] *Ibid.,* I, 15.

[45] *Ibid.,* I, 84-85.

[46] *Ibid.,* I, 102, 107, 113, 115.

[47] G. Tierie, *Cornelis Drebbel,* Amsterdam, 1932, pp. 22-24.

[48] *Collecta astronomica ex doctrina P. Christophori Borri Mediolanensis ex societate Iesu de tribus caelis aereo sydereo empyreo issu (iussu?) et studio domini Gregorii de Castelbranco . . .,* Lisbon, 1631. BM 531.g.16.

our time:" the comet of 1577 and new star of 1572; the appearance of Mars sometimes above and sometimes below the sun—a fact, Borri says, accepted by all mathematicians; digressions concerning parallax and telescope; the comet of 1618, and the fact that comets properly so called are not found in the third region of the air; new phenomena in sun and moon seen through the telescope; also in Mercury and Venus, as that Venus is horned and has phases like the moon; and the satellites of Jupiter and Saturn.

The third part is on the tenuity of the heavens and the movement of the planets in the ethereal atmosphere. The first conclusion reached is that there is one heaven of the planets and that it is fluid, not solid. The second conclusion is that angels are the cause moving the planets. Borri rejects the theory of Kepler, in his work on the planet Mars, that the sun like a magnet draws the other planets. Borri, who regards the sun as moving, says, "after it," but this misrepresents the view of Kepler, who held the heliocentric theory and said "about itself." Borri accepts rather the hypothesis of Tycho Brahe, adding to it a spiral motion.

The fourth part on the number of the heavens denies that there are any mobile heavens beyond that of the fixed stars. Whether the latter is solid or tenuous is a matter of indifference to Borri's hypothesis, but it more probably is not solid, and the stars in it are moved, like the planets, by intelligences or angels. The question is raised whether an angel can extend the scope of his activity to any distance. The answer is that he cannot do so in three dimensions, which immensity is characteristic of God alone. But he can do so in two dimensions by making the area of his activity narrower and longer. This is graphically illustrated by a figure of a seated angel which fills up most of a square, and one of the same angel stretched out at great length within an oblique parallelogram having the same base and altitude, and consequently the same area as the square. Scriptural passages implying solid heavens are explained away, and Borri concludes that the heavens are three in number: aerial of the planets, sidereal of the fixed stars, and the empyrean beyond it, although Cajetan held that there was no scriptural authority for the last, and that God and the blest were in the *primum mobile.* The empyrean heaven may be fluid but more likely is solid; it may be four-square but more likely is round. Certainly of all bodies it is the simplest, most excellent, farthest removed from corruption and radiant in marvelous splendor.

From these celestial reveries we come back in the fifth part to physical questions according to the new astronomy. The questions themselves, however, are not new. It is asked whether the heaven other than the empyrean is corruptible, whether its matter is a mixture of the elements or is one of them, whether there is a sphere of fire in the concave of the sphere of the moon. The last is answered in the negative, nor is fire the matter of the heavens. Nor is water, Borri prefering air as the material of the heavens. It is very probable that both heavens and stars are subject to corruption, though to a less degree than sublunar bodies. The stars are solid bodies, not perfectly spherical, and differ from the heaven as earth does from air. Nay more, that the stars differ from one another is suggested by their diversity of color and of influence, and their substantial form differs from all the forms of sublunar bodies.

Borri makes no approach towards any universal law of gravitation, holding that the matter of each star tends towards its own center and has no relation either of gravity or levity with respect to the earth or center of the world. He denies the earth's motion of nutation. New stars are very likely formed from condensed ether. Angels might do this or bring some of the smallest stars close together or unite them. But there would be nothing miraculous about either action. Following Cysatus, Borri similarly explains comets as new aggregations; he has no idea of their coming from afar and returning.

The sixth and last part is on the creation of the heavens, including the empyrean and following the account in the Book of Genesis. The firmament was formed from water, and Borri accepts the existence of waters both above and below it, although all this may seem inconsistent with his previously expressed preference for air as the material of the heavens. When God said, "Let there be light," that light was not the sun but something like a very bright cloud, and an angel placed that lucid body about the earth.

In conclusion Borri states that he has won others over to his views not only in Europe but almost the world over in his journeyings, although for various reasons he has put off publication until now.[49] His book shows how variously the new astronomical discoveries might be interpreted according to one's preconceived ideas and interests, and how curiously it might be combined with such notions as the belief in four elements, in Aristotelian moving Intelligences transformed into Christian angels, in a material heaven of the blest, and in the Biblical story of creation. The association seems to us today incongruous, but at the time it may have aided the new astronomy more than it injured it.

How the heavens were represented on the basis of the biblical account of creation by a theologian contemporary with Kepler and Galileo may be further seen from a work of 1632 by Augustinus Oregius, a canon of the Vatican, adviser of the Holy Inquisition, and private secretary to pope Urban VIII. After two treatises on God, one and triune, and a third on angels of 216 pages, Oregius turned to creation and polished off that of the heavens in about twenty pages, devoting most of his 172 pages to questions concerning Adam and the immortality of the soul, the last 72 pages being an explanation of passages from Aristotle, "which prove and confirm the immortality of the human intellect." Indeed, Aristotle's attitude towards the immortality of the soul had been discussed for fifty pages preceding.[50]

Oregius understands the word abyss to apply to all that pellucid

[49] According to de Backer and Sommervogel, I, 1821-22, citing p. 470, the book was printed at Lisbon in 1629. A Vatican MS cited by Mai, *Scriptorum veterum nova collectio*, IV (1831), ix, contains a compendium of the work, which was translated from Latin into Persian at Goa in 1624, and an Italian version of 1631. And according to a none too reliable French *Relation* made in Portugal in 1627, which incorrectly represents Borri as having formerly taught in Italy, he had to do penance at Rome for his astronomical views. Possibly the sixth book was added to satisfy the ecclesiastical authorities. Borri left, or was expelled by, the Jesuits shortly before his death in 1632.

The work by Cysatus (Johann Baptist Cysat) which Borri cites is *Mathemata astronomica de loco motu magnitudine et causis cometae qui sub finem anni 1618 et initium anni 1619 in caelo fulsit,* 80 pp. in-4, Ingolstadt 1619. BM 8561.c.30.

[50] *De opere sex dierum,* Rome, 1632, in-4. In the copy which I used, BN D.9075 (2), it is preceded by his *Tractatus tertius de angelis,* of which we treat in Chapter 38.

and diaphanous body which extends from the earth's surface to the empyrean heaven. Probably it is further indicated when it is said that the spirit of God moved over the face of the waters. The firmament is the heaven of the fixed stars, which divided the abyss so that there were waters above and below it. This raises some difficulty, not only as to the waters above the firmament but those below, since there are no waters directly below the heaven of the fixed stars but, according to one view, other heavens which are as solid and firm as the firmament, and then, below these, spheres of fire and air before any water is reached. The word, ether, is now used for all celestial body within the firmament, and sometimes the word, firmament, is employed to include the spheres of the planets or ether as well as the heaven of the fixed stars. Some have suggested that the waters above the firmament are for the blest above to look at. Or three, instead of ten, heavens are distinguished: one of fluid ether, rather than solid spheres, in which the planets move as birds do in air or fish in water; second, the firmament; and third, the empyrean heaven with the waters above the firmament in which its splendor is reflected. As for the waters above the firmament, since it is very difficult to philosophize about them, "and far be it from us to do so"—a weak withdrawal, in view of his frequent citation of Aristotle—no one can affirm anything with certainty except what is revealed in the Bible. And what it does not reveal, requires the observations of astronomers and mathematicians rather than the contemplation of theologians. Oregius then proceeds to interpret the expression, luminaries, in the account of creation to include the planets and fixed stars and not apply merely to sun and moon.[51]

Anton Deusing (1612–1666) in his earlier years composed a Catholic Cosmography and Astronomy According to the Ptolemaic Hypothesis.[52] The next year he published a Dissertation on the True System of the World, in which the Copernican system is reformed and the well nigh infinite circles of the Ptolemaic system,

[51] *Ibid.*, pp. 6-7, 10-11, 13-15, 19.

[52] *Cosmographia catholica et astronomia secundum hypotheses Ptolemaei*, Amsterdam, 1642, in-8, BN V. 20948.

by which the mind of man is distraught, are removed.[53] This work, however, is less favorable to the Copernican hypothesis than one might infer from its title, since the earth remains fixed at the center of the universe, and the only movement accepted for it is one of rotation upon its axis, while the sun is represented as moving and is treated along with the three superior planets. Frequent reference is made to the system of Tycho Brahe and to the recent *Philolaus* of Boulliau, of which we are about to speak, but Kepler is not cited for the movement of Mars, only for that of Mercury in his Epitome of the Copernican astronomy.[54] Later we shall find Deusing opposing the sympathetic powder and a similar superstition called Man-Schlacht. He also took note of Boyle's experiments with the elasticity of air.[55]

Ismael Boulliau or Bullialdus (1605–1694), who in 1638 published a treatise on the nature of light, the next year issued his *Philolai sive Dissertationis de vero systemate mundi libri IV*.[56] After reviewing the ancient system of the world and that of Tycho Brahe, he investigated the true system and came to the conclusion that the three superior planets moved with one motion and in one circle, and that Venus and Mercury also had one movement and one circle, and moved about the sun. He asserted that he had proved by geometrical demonstrations that the earth moved, and that the diurnal movement of the earth necessarily followed from the annual movement.[57] He would, however, restore the annual anomalies to the circle in which the sun appears to move.

In his *Astronomia Philolaica* of 1645, Boulliau, like Biancani, dated the Latin translation of the *Almagest* under Frederick II, further dated Henri Bate of Malines about 1350, and made Dominicus Maria Novara the teacher alike of Peurbach and Regiomontanus, Walther, Werner and Copernicus. However, he praised Kepler's hypotheses, except that rejection of real mean motion

[53] *De vero systemate mundi dissertatio mathematica qua Copernici systema mundi reformatur sublatis interim infinitis pene orbibus quibus in systemate Ptolemaico humana mens distrahitur*, Amstelodami, Apud Ludovicum Elzevirium, 1643, 173 pp. I own a copy. BN V.7736 (1).

[54] *Ibid.*, p. 120.

[55] *Considerationes circa experimenta de vi aeris elastica*, 1662. BN R. 14562.

[56] Copy used: BN V.7736 (2).

[57] *Ibid.*, p. 131.

offended him. He also accepted the Rudolfine Tables for the most part but had showed by most certain observations that the orbit of Mercury was more contracted. He recognized that the heavens were not solid, that comets had proved that generation and corruption went on in the heavens, and that sunspots showed that the sun revolved about its axis in twenty-six days. Elliptical orbits satisfied all the phenomena so far as eccentricity was concerned, but, whereas Kepler had represented the planets as of themselves inert and moved by the force[58] of the sun, Boulliau held that they were moved by their own form.[59]

The Jesuit Cabeo, in his huge four volume commentary on the *Meteorology* of Aristotle, examined Galileo's views concerning the motion of projectiles in connection with the doctrine of impetus,[60] and accused him of "unbearable boastfulness" for asserting that the proportion in which the velocity of falling bodies increased had been unknown to all philosophers since Adam and first demonstrated by himself. Cabeo questions whether Galileo had read all the previous literature on the subject, and adds that in the year in which Galileo's *Dialogues* appeared

> while I was at Genoa, Giovanni Battista Baliani told me that he had demonstrated the increment of velocity many years before he heard anything of Galileo, as he affirmed in print years later.[61]

Cabeo rejected Galileo's explanation of the tides, and correctly attributed them to the moon, but incorrectly to its peculiar virtue of exciting sulphurous and "salnitrous" spirits from the bottom of the sea.[62]

The empyrean heaven continued to receive attention in the second half of the century. In 1652 Gabriel de Henao, a Jesuit from Valladolid and professor of theology in the Jesuit College at Salamanca, published there two folio volumes on Empyreology or Christian

[58] Kepler used the word "Kraft."

[59] Prolegomena to *Astronomia philolaica*. Copy used: Col 520 B66 Q.

[60] Nicolaus Cabeus, *Commentarius in Meteorologica Aristotelis*, Rome, 1646, I, 88.

[61] *Ibid.*, I, 423-24. The work referred to is Baliani's *De motu naturali gravium solidorum et liquidorum*, Genoa, M. Farroni, 1646, in-4, 174 pp.

[62] *Comm. in Meteor.*, II, 42-62, 63-70.

Philosophy concerning the empyrean heaven.[63] On the title pages of both volumes the book was called "A New Work, very necessary to philosophers, theologians, scholastics and mystics, interpreters of sacred scripture and preachers of the divine word." The first volume divided into four books dealing with the name, existence, production and antiquity of the empyrean heaven; with its nature, or matter and form, with a ninth exercise on it as the beginning of motion and rest; with its three chief properties, namely, light, immobility and incorruptibility; and with its other accidents and qualities. The second volume, also divided into four books, dealt briefly with the problem of its influence, and more fully with its inhabitants, discussing such points as souls existing there before the general resurrection of bodies, and other than human bodies to be found there, such as the eucharist and cross of Christ. The questions were also raised whether the angels would assume bodies and whether the blest would wear clothes. The final book was on their external and internal material actions. As to the influence of the empyrean heaven, various authors are cited who conceded its influence upon the other celestial bodies and upon sublunar bodies. But it is noted that Thomas Aquinas, after diligent consideration of the matter, changed his opinion and denied such influence. It is further added that Aquinas has written variously as to the dependence of sublunar bodies upon the motion of the celestial bodies.[64]

[63] *Empyreologia sive Philosophia Christiana de caelo empyreo*, Salamancae, 1652, in-fol. Also at Lyons, 1652, 2 vols., 324, 336 pp. and Index, for which I have used BM 473.d.8.

[64] *Ibid.*, II, 4b, 6b.

CHAPTER IV

FRANCIS BACON

His character—Publication of his works—Mode of composition—Claim to novelty and superiority—Mersenne's criticism—Bacon's influence and merits—Attitude towards past science—Method—Experiments—Division of Natural History—Nature—Heat and sound—Astronomy—Tides—Astrology—Biology—Spirits—Medicine—Prolongation of life—Witchcraft, Fascination, Imagination—Natural magic—Faulty logic—Beeckman's interests and outlook compared with Bacon's—His reactions to Bacon's works—Bacon weighed and found wanting.

If parts allure thee, think how Bacon shined,—
The wisest, brightest, meanest of mankind!

—Pope

Without whole-heartedly subscribing to this violent antithesis, it may be said that, as Bacon was impeached as Lord Chancellor for accepting gifts of money from suitors, while their cases were still pending, so there has been a recent tendency among historians of science to censure him, as a professed natural philosopher and reformer of learning, for taking what did not belong to him in that sphere also, and failing to own his debt to predecessors such as Roger Bacon in the thirteenth century. This note of censure has replaced a former chorus of adulation based upon ignorance and misapprehension of the middle ages and typified by the following quotation:

It took more than twelve centuries for a Bacon to rescue the principle of scientific causality from a world which had become enveloped in medievalism.

However, even an early Victorian like Whewell wrote over a century ago that Francis Bacon's precepts as to scientific method "are now practically useless." And back in 1861 Draper, after condemning Bacon for rejecting the Copernican system, and after stating that "his chief admirers have been persons of a literary turn," that he never accomplished any great practical discovery, and that "few scientific pretenders have made more mistakes" than he, concluded:

It is time that the sacred name of philosophy should be severed from its long connexion with that of one who was a pretender in science, a time-serving politician, an insidious lawyer, a corrupt judge, a treacherous friend, a bad man.[1]

We should perhaps remind ourselves that the *Advancement of Learning* appeared in English in 1605, and *Novum Organum* in Latin in 1620, the year before Bacon's impeachment and retirement from public life. Other scientific or pseudo-scientific treatises were published during his last years or posthumously, as was the case with *Sylva sylvarum*, a collection of a thousand observations and experiments arranged in ten centuries, in 1627; *Descriptio globi intellectualis* and other treatises, in 1653; *Historia densi et rari*, in 1658; yet others only in 1679, 1688 and 1734.

Francis Bacon resembled his thirteenth century namesake Roger not only in some of his leading ideas but in his plan and method of composition. As Roger planned a comprehensive work of philosophy which he never completed, but of which some of his extant writings were probably intended to serve as sections, so Francis planned an *Instauratio Magna*, and his philosophical works consist in large part of "works published, or designed for publication, as parts of the *Instauratio Magna*."[2] And as Roger left various versions of the same treatise or used the same material over again in different treatises, so another chief group of Francis's works were "originally designed for parts of the *Instauratio Magna*, but superseded or abandoned,"[3] and are further described by their editors as writings which Bacon himself would not have cared to preserve and which contain

but little matter of which the substance may not be found in one part or another of the preceding volumes, reduced to the shape in which he thought it would be most effective.[4]

[1] *Intellectual Development of Europe*, edition of London, 1902, II, 260.

[2] These occupy vols. I-IV and part of V in the seven volumes of *Philosophical Works* as edited by Spedding, Ellis and Heath.

[3] *Ibid.*, V, 417-56, and VI and VII.

[4] *Ibid.*, V, 419.

In the dedication to King James I of the Great Instauration, Bacon represented his position as completely novel, saying, "It is at least new, even in its very nature," and that the

only wonderful circumstance in it is that the first conception of the matter and so deep suspicions of prevalent notions should ever have entered into any person's mind.[5]

By making a virtue of not citing authorities he avoided any mention of the sources of his ideas. This was bad enough, but he made it worse by assailing some of the greatest names in the past. Thus, after having spoken depreciatingly of Aristotle, Plato and Ramus, he continued:

Let us now proceed to physicians. I see Galen, a man of the narrowest mind, a forsaker of experience, and a most vain pretender.

No statement could be more unjust and erroneous. Of all extant ancient writers Galen approaches most closely to the conception of experimental method,[6] and was often cited by medieval authors for the recognition of experience as a criterion of truth. There is a trifle more of verisimilitude in Bacon's scornful characterization of Hippocrates as one who

sheltered by brevity . . . does nothing but either deliver certain sophistications in sentences abrupt and suspended, thus withdrawing them from confutation; or invest with stateliness the observations of rustics.[7]

But when we come to consider Bacon's own medicine, I feel sure that the reader will prefer that of Galen and Hippocrates. And when Bacon exhorts to attend to things themselves, he is only repeating what Galen—and many others—had said long ago.

Bacon's defects in the rôle of a protagonist of experimental science did not escape the eyes of his contemporaries. Mersenne

[5] "Sunt certe prorsus nova etiam toto genere . . . Illud enim in eo solummodo mirabile est initia rei et tantas de iis quae invaluerunt suspiciones alicui in mentem venire potuisse."

[6] T I, 151-62.

[7] These passages are from the Interpretation of Nature. In the *Sylva sylvarum* both Aristotle and Hippocrates are cited approvingly.

in 1625 made three criticisms of him: first, that he should have consulted the savants of different countries before laying down rules which were either practiced already or of no use; second, that he often proposed experiments which had already been performed; third, that he introduced innovations in terminology which would retard scientific progress.[8]

Yet the title of Lord Verulam under which his works appeared commanded universal respect in England and abroad. He possessed undoubted ability, breadth of view, and intellectual insight. He was not a scientist by profession or training: few men as yet were. But in a sense there was something to be said for having an outsider and man of good general education—rather than an astronomer or astrologer or chymist or alchemist or physician or pharmacist or mathematician or mechanic—consider the general problem of natural and experimental science. Bacon was not bound by any university curriculum or professional limits. His wide and restless curiosity kept him out of ruts and beaten paths, though it did not keep him from trespassing on others' preserves as if he had an intellectual right of way or of eminent domain. He marked, in the British Isles at least, that amateurish interest of the upper classes in natural science and experimentation which led to the founding of the Royal Society. As an outsider he naturally joined to the credulity of the layman a certain amount of lay contemptuous scepticism for the niceties and the traditions of academic and professional science. He also tended to conjectural speculation on the one hand and to accept generalizations then in the air on the other hand, and occasionally to express himself in Elizabethan terms of metaphorical fancy rather than of literal science.

Bacon called for "an absolute regeneration of science," a fresh start—as Descartes was to do again presently, "an entirely different way from any known to our predecessors."[9] For one thing, he wished to keep natural philosophy unadulterated, and to mark off natural science as a distinct field and discipline. He complained that Aristotle had mixed it with logic; Plato, with natural theology; and the Neo-Platonists, with mathematics.[10] This suggests what

[8] Mersenne, *Correspondance,* I (1932), 172.

[9] Preface to *Instauratio magna.*

[10] *Novum organum,* I, 96.

from the standpoint of modern science was his chief defect, his total disregard of mathematical method. He spoke of pure mathematics as, like the game of tennis, of no use in itself but as good exercise to cure intellectual defects.[11] He further held that attention to final causes and the argument from design belong to metaphysics, but serve only to obstruct the search for immediate physical causes. He was therefore critical of such statements as that the clouds are to water the earth, the leaves to protect the fruit, the bones to support the frame, the skins of animals to protect them from heat or cold, and the eyelids to protect the sight.[12] But such criticism perhaps accords none too well with the doctrine of evolution and survival of the fittest.

Bacon would study nature as a whole and all at once. He wants a broad collection of particular facts "capable of informing the mind,"[13] something on the order, it would seem, of a revision and reformation of Pliny's Natural History. He speaks slightingly of the experimental specialization of Gilbert with the magnet, and of that of the alchemists.[14] Bacon's attitude was the not uncommon one of condemnation of alchemists for their endless efforts, perpetual hope deferred, and waste of time and money, along with admission that they had made not a few discoveries and useful inventions.[15] And of course he had his own little recipes for making gold.[16]

Like Aldrovandi, Bacon would try to exclude fables and marvels, curiosities and traditions. He would furthermore not merely collect observations and experiences like ants or the Empirics—as a

[11] *Advancement of Learning*, Bk. II; *Works*, VI (1863), 227.

[12] *De augmentis scientiarum*, III, iv; *Works*, II (1861), 294.

[13] *Novum Organum*, I, 98.

[14] *Ibid.*, I, 70: "Quod si magis serio et constanter ac laboriose ad experimenta se accingant, tamen in uno aliquo experimento eruendo operam collocant; quemadmodum Gilbertus in magnete, chymici in auro." Also *Advancement of Learning*, Bk. I; *Works*, VI (1863), 132: "So have the alchemists made a philosophy out of a few experiments of the furnace; and Gilbertus, our countryman, hath made a philosophy out of the observations of the loadstone."

[15] *Novum organum*, I, 85.

[16] Joshua C. Gregory, "Chemistry and Alchemy in the Natural Philosophy of Francis Bacon," *Ambix*, II (1938), 93-111, says at p. 106: "Bacon's recipe for making gold impresses the modern chemist no more than the projected elixir and alchemical over-firing impressed Bacon himself."

matter of fact, that was often as far as he got—but would work them over by the inductive method as bees make honey, and ascend from facts and experiments to laws of nature. Mere observation, as in Aristotle's History of Animals, is not enough; purposive artifical experimentation is further required.[17] Bacon realized the need of apparatus and expenditure—as indeed did Gilbert and the chemists—and urged original research and endowment thereof, new inventions and things out of the common track. I must confess that I fail to appreciate his criticizing the naturalists of his time for observing the differences between various animals, plants and minerals instead of noting their resemblances. If differences are noted, the residue will be resemblances and *vice versa.* What is the difference? Or, one might argue, if you note differences, you are assuming or approaching some norm from which they differ; while, if you note resemblances and analogies, you appear to be picking these out of a chaotic sea of differences. One might argue either way.

However, experiments are the fundamental elements for science, as the letters of the alphabet are for language. Bacon supposes the existence of prerogative and crucial experiments which are more decisive and more convincing than others. His prerogative instances subdivide into solitary, migrating, conspicuous, clandestine, constitutive or collective, similar or proportionate, singular—for example, the elephant among quadrupeds, deviating—i.e., errors of nature and monsters, to which we shall find men of the seventeenth century giving much attention, instances of power, accompanying and hostile, subjunctive, of alliance or union, crucial or decisive, of the lamp or immediately informative and assisting the senses, of the door—for example, the microscope, citing or invoking, of the road or itinerant, of refuge or supplementary and substitutive, lancing or twitching—i.e., surprising, of the rod or rule—having to do with time, of the course—having to do with space, quantitative, wrestling—under which head nineteen varieties of motion are distinguished, suggestive, generally useful, and finally magical[18]—of which more anon. This odd assortment shows the quaintness of Bacon's thought and method. He admitted that his *Sylva sylvarum* was an undigested heap of particulars, and that many of the "experiments" were "vulgar

[17] *Novum organum,* I, 95.

[18] *Novum organum,* II, 22-52.

and trivial, mean and sordid, curious and fruitless." But he regarded it as such a natural history as might be fundamental to the future erection of a building of true philosophy. He had avoided any exact method of arrangement, though there was a secret order—another remnant of magic. In it his measurement of time is by so many pulse beats, a rather personal and subjective method. More scientific was his tabulation of the relative weights of the same quantity of different substances from gold, quicksilver, lead, down.[19]

Bacon's use of the word, experiment, is still rather loose. There are natural as well as artifical experiments. For example, in mentioning wells that produce storms, if you cast a stone into them, he adds that the analogy of volcanoes is an indirect experiment for this.[20]

> Our experiments we take care to be, as we have often said, either *experimenta fructifera* or *lucifera:* either of use, or of discovery: for we hate impostures and despise curiosities.[21]

Yet he asserts elsewhere that experimental science has hitherto been a failure because men have "sought out experiments for the sake of gain and not of knowledge."[22] In the *Advancement of Learning* he writes:

> As a man's disposition is never well known till he be crossed, nor Proteus ever changed shapes till he was straitened and held fast; so the passages and variations of nature cannot appear so fully in the liberty of nature as in the trials and vexations of art.[23]

A noiseless gunpowder "is a dangerous experiment, if it should be true, for it may cause secret murders."[24] In the *Sylva* he includes a solitary experiment in gold-making.[25] He also proved by testing it that a vessel full of ashes would not hold as much water as it did when empty.[26] In testing another belief, that pearls, coral and

[19] *Phaenomena universi; Works* (1863), VII, 237-9.

[20] *Natural History of Winds*, p. 14.

[21] *Sylva*, Century VI, opening passage before Item 501.

[22] *Phaenomena universi, Works* (1863), VII, 230.

[23] *Works*, VI (1863), 188.

[24] *Sylva*, 120.

[25] *Sylva*, 326-27.

[26] *Sylva*, 34.

turquoises which had lost their colors would recover these, if buried in the ground, Bacon did not display much patience, persistence or perseverance, for, when he dug up the stones after they had been buried for six weeks and found no change, he discontinued the experiment.[27]

In the *Novum Organum* Bacon made a threefold division of natural history: the first being concerned with species, the second with monsters, and the third with artificial products; or the history of generation, pretergeneration and arts. This emphasis upon monsters or errors and freaks of nature was to remain characteristic of the science of the rest of the century. Bacon held, however, that monsters which were merely prodigious and natural might be considered with the history of generation, leaving unnatural monstrosities and the superstitious history of miracles to a separate treatment. The history of generation he divided into five parts dealing with the ethereal and celestial, meteors and regions of the air, land and sea, the four elements more particularly considered, and particular species of things. For him the mechanical and experimental went with the history of the arts.

Considering that a fresh start had to be made in science, and that "nothing is rightly inquired into or verified, noted, weighed or measured in natural history,"[28] Bacon indulged in some surprisingly sweeping statements as to Dame Nature herself. In his Description of the Intelligible Globe he said that Nature was "accustomed to alternate fine gradations and distinct transits in her processes."[29] And in the opening passage of his Aphorisms he affirmed:

> Nature is placed in three situations and subject to a threefold government. For she is either free and left to unfold herself in regular course, or she is driven from her position by the obstinacy and resistance of matter and the violence of obstacles, or she is constrained and moulded by human art and labor.

He declared that heat and cold "are Nature's two hands whereby she chiefly worketh," and he was certain that heat was the chief of all the powers in nature.[30] In another passage he personified time

[27] *Sylva*, 380.

[28] *Novum Organum*, I, 98.

[29] Cap. 6.

[30] *Sylva*, 68, 99.

and heat as "fellows in many effects."[31] Or he asserts that sound is one of the most hidden portions of nature, and is a virtue incorporeal and immaterial, of which there are very few in nature.[32] The transmutation of one plant into another Bacon ranks among the mighty works (*magnalia*) of nature. Common philosophy pronounces the transmutation of species impossible, but there are manifest instances of it, so that it should be investigated further.[33]

When Bacon suggested weighing vegetation "that sprouts out of the ground" to discover whether the air had contributed to its growth,[34] he went a step farther than Nicholas of Cusa, who had suggested a similar experiment but had no idea that plants received sustenance from the air.[35] But neither Bacon nor Nicholas thought of the soil itself receiving such sustenance, although perhaps a plowman could have told them so. Bacon may have been influenced by van Helmont's prolonged experiment with a tree, from which he concluded that it took all its nourishment from water and none from the soil, rather than have taken the suggestion from Nicholas of Cusa.

A good illustration of Bacon's employment of the inductive method is his investigation of the nature of heat. He first listed "all the known instances of heat which agreed in the same nature," or "a table of existence and presence." Next he drew up "a table of deviation" or of absence in proximity. Third, he noted cases of the presence of heat in a greater or less degree. Then, upon an individual review of all the instances recorded, motion was found always present when heat was, always absent when heat was absent, and to increase and decrease in degree with it. But Bacon did not stop there. Just as the scholastics, after arriving at a conclusion based upon reason, experience and authority, confirmed it by answering in detail all the arguments and authorities to the contrary, so Bacon by use of his tables excludes or rejects other things as not being the form of heat, and so finally rests at the conclusion that motion must be the form of heat. All this has taken ten double-columned quarto pages and reminded me not a little of the method

[31] *Sylva*, 294.
[32] *Sylva*, 290.
[33] *Sylva*, 525.
[34] *Sylva*, 29.
[35] T IV, 389.

employed by Peter of Abano in his *Conciliator* of the differences of philosophers and especially physicians. But we are not quite through yet. "We must lastly consider the true differences which limit motion and render it a form of heat," which Bacon does for two more pages.[36]

Bacon engaged in a long series of experiments with sound intended to test its relation to the movement of the air, and including experiments with sounds under water and as to the different sounds emitted by iron when hot and cold. He put such questions as why the voice changes at puberty and why cock birds are always the better songsters.[37] Some of the ancients thought that the humming of bees did not "come forth at their mouth," but was made internally. Bacon suggests that it may be produced by the movement of their wings, since it is only heard when they are on the go.[38] As for the relation of the air to sound, Bacon writes that it is certain that sound is not produced without some movement of the air.

> But you must attentively distinguish between the local motion of the air which is but . . . a carrier of the sounds and the sounds themselves conveyed in the air. And it is the more probable that sound is without any local motion of the air, because, as it differeth from the sight in that it needeth a local motion of the air at first, so it paralleleth in so many other things with the sight and the radiation of things visible, which without all question induce no local motion in the air.[39]

We have heard Mersenne criticize Bacon, but in 1636 he proposed much the same questions as to sound as Bacon had put earlier.

It was to Bacon's credit that he believed that astronomy would be advanced by employing the methods of physics, and that such phenomena as expansion and contraction prevailed in the heavens as well as on earth. He also graciously granted that much had already been accomplished in astronomy, since he would add to the treasury of Ptolemy and Copernicus the observations of moderns. Even the questions which he suggested for further investigation were not new: for example, Is there a system of the universe?

[36] *Novum Organum,* II, 11-12. He alludes in the same work to a thermometer by the designation *vitrum calendare,* but the word, thermometer, is said first to occur four years later in Leurechon, *Recréation mathématique,* 1624.

[37] *Sylva,* 180, 239.

[38] *Sylva,* 175.

[39] *Sylva,* 125.

What is its center, depth, connection and distribution of parts? Is the substance of the heavenly bodies different from that of the earth? Are the interstellar spaces void or full of substance? Are the stars kept alive by due sustentation? Are they spherical? How far off are they and of what dimensions? Are they produced and decomposed through long periods of time?[40]

To the moon Bacon still attributed such influences as eduction of heat and induction of putrefaction, increase of moisture, exciting the spirits in the human body—of which lunatics are a crucial instance,[41] and the effect of new and full moons upon winds and weather.[42]

New moons presage the dispositions of the air, but especially the fourth rising of it, as if it were a confirmed new moon. The full moons likewise do presage more than the days which come after.

He added that there might be other secret effects of the moon not yet brought to light,[43] and further promised to discuss the general question of astrological influence more fully later. Earlier in the same work he had expressed doubt as to "a sympathy between the sun, moon and some principal stars and certain herbs and plants."[44] In his Natural History of Winds he affirmed that winds both preceded and followed planetary conjunctions, unless the conjunction was with the sun, in which case there would be fair weather.[45]

At the rising of the Pleiades and Hyades come showers of rain, but gentle ones; after the rising of Arcturus and Orion, tempests. Returning and shooting stars . . . signify winds to come from that place whence they run or are shot. But if they fly from several or contrary parts, it is a sign of great approaching storms of wind and rain.[46]

Bacon followed Pliny in locating rich soil at the ends of the rainbow.[47]

In his treatise on tides Bacon favored the incorrect theory that

[40] *Intell. Globe,* caps. 5-7.

[41] *Sylva,* 889-895.

[42] *Natural History of Winds,* Artic. 16.

[43] *Sylva,* 896.

[44] *Sylva,* 493.

[45] *Op. cit.,* Artic. 31.

[46] *Ibid.,* 32-33.

[47] *Sylva,* 665. Pliny, Nat. Hist., XVII, 3.

they were produced by the American continent interfering with the general tendency of the ocean to move from east to west. This view had been advanced by Pandulpho Sfondrato in a work addressed to Gregory XIV in 1590. Sfondrato held that the Straits of Magellan were the sole passage from the Atlantic to the Pacific and that the tides were caused by a large part of the Atlantic ocean not being able to get through and so rebounding. He did not know whether more water entered or went out through the Straits of Gibraltar, and in general his treatise seemed inferior.[48] Ellis has noted that in the *Novum Organum* Bacon cited Acosta that the time of high tide roughly corresponded on the coasts of Europe and America, whereas what Acosta really said was that the tides flowed together on both the Atlantic and Pacific coasts of South America, meeting in the Straits of Magellan about seventy leagues from the Atlantic and thirty from the Pacific, which Bacon in his treatise on the tides had admitted would be fatal to his theory.[49]

Astrology was discussed at some length by Bacon in *De augmentis scientiarum,*[50] where he declared that it was full of superstition but should be expurgated rather than utterly rejected. He discarded the reign of the planets in turn over the hours of the day, although he admitted that they got their names in this way. He would also drop horoscopes, astrological houses, and emphasis upon the hour of birth or initiating an undertaking or making an inquiry. In other words, he opposed nativities, elections and interrogations. Towards revolutions he was more favorable. He granted that the celestial bodies exerted other influences than those of heat and light. He had no doubt that the moon in Leo had more power over terrestrial bodies than when in Pisces, or that a planet was more active when in its apogee, and more communicative when in its perigee. Prediction of comets, the weather, epidemics, wars, schisms, and folk migrations he thought possible, and his final word was that even elections were not altogether to be rejected. With astrological ceremonial he had little sympathy. Describing the process of

[48] *Causa aestus maris,* Ferrariae apud Benedictum Mammarellum, 1590, 44 fols. Copy used: BN R.3301.

[49] R. L. Ellis, Preface to *De fluxu et refluxu maris, Works,* V (1862), 224. Ellis also mentions the treatise of Sfondrato: *Ibid.*, p. 239.

[50] III, iv; *Works,* II (1861), 272-80.

weapon ointment in his Natural History, he said that one thing he liked about it was that there was no observation of any certain constellation in confecting the ointment.[51] In discussing the length of human life according to the time of birth, he "omitted for the present" horoscopes and other astrological data.[52]

We turn from Bacon's physics, astronomy and astrology to some instances of his biology. In the *Advancement of Learning* he waxes eloquent on the theme of animal inventions and sagacity. The art of inventing arts is deficient, and logic does not pretend to invent sciences or even axioms.

Men are rather beholden to a wild goat for surgery, or to a nightingale for music, or to the ibis for some part of physic . . . It was no marvel, the manner of antiquity being to consecrate inventors, that the Egyptians had so few human idols in their temples, but almost all brute . . . Who taught the raven in a drought to throw pebbles into a hollow tree, where she espied water, that the water might rise so that she might come to it? Who taught the bee to sail through such a vast sea of air and to find the way from a field in flower, a great way off, to her hive? Who taught the ant to bite every grain of corn that she burieth in her hill, lest it should take root and grow? [53]

In the *Sylva* are some generalizations concerning animals, such as that no birds have teeth, that hard-shelled animals lack bones within, and horned animals lack upper teeth, while any beast with upper teeth had teeth in the lower jaw as well.[54] Bacon also indulges in such fanciful explanations as that, because birds have no means of urinating, all their excess moisture goes into their feathers.[55] He still accepted spontaneous generation and "living creatures that come of putrefaction."[56] If it is true that buried oak branches put forth wild vines, this is not because the oak turns into a vine but because its rotting "qualifieth the earth to put forth a vine of itself."[57]

During a recent outbreak of the plague many toads were seen in the vicinity of London with tails two or three inches long, "whereas

[51] *Sylva*, 998.

[52] *Hist. vitae et mortis*, Topica particularia.

[53] *Advancement of Learning*, Bk. II: *Works*, 1863, VI, 263-64.

[54] *Sylva*, 747, 753.

[55] *Sylva*, 680.

[56] *Sylva*, 525.

[57] *Sylva*, 522.

toads usually have no tails at all." This was a sign of putrefaction in the air and soil. On the other hand, carrots and parsnips were said to be sweeter and more luscious in infectious years.[58] Speaking of toads, Bacon thought it odd that venomous beasts appeared to be fond of sweet-smelling and wholesome herbs. Snakes liked fennel; toads sat under sage; "frogs will be in cinque-foil." Bacon suggested that they were attracted by the shade rather than the virtue of the herbs.[59] Apparently it did not occur to him that the toad might be lying in wait for insects. Yet he was aware that a chameleon would eat flies as well as air, although he regarded the latter as its chief food.[60] He asserted thrice that bears waxed fat during their winter sleep.[61] He doubted whether the flesh of deer and snakes would greatly prolong life,[62] but conceded that there might be some truth in the notion that application of the guts or the skin of a wolf would cure colic, for the reason that the wolf was an animal "of great edacity and digestion,"[63]—a close approach to a magical association of ideas.

Bacon scorned sympathy as the explanation why some plants grew best in close proximity, the real explanation being that they required different nourishment from the soil. In this connection he further remarked that most experiments concerned with sympathy and antipathy forsook the true indication of causes.[64] Yet he repeated as a creditable report the statement that earth from the Nile valley, although removed to a great distance, would increase in weight during the inundation of that river.[65] He very properly condemned as a foolish bit of magic, the burning of a chameleon on a house-top in order to cause a storm, the theory being, "according to their vain dreams of sympathies," that because the chameleon lived on air, its carcass and ashes should exert great virtue in affecting the air.[66] But in other passages he was himself guilty of similar magic logic and argument from mere association, physical or mental, as we have already noticed in one case and of which we shall presently adduce further examples.

[58] *Sylva*, 691.
[59] *Sylva*, 674.
[60] *Sylva*, 360.
[61] *Sylva*, 57, 746, 899.
[62] *Historia vitae et mortis, Intentiones.*
[63] *Sylva*, 972.
[64] *Sylva*, 479.
[65] *Sylva*, 743.
[66] *Sylva*, 360.

Other questions that attracted Bacon were why the feathers of birds had livelier colors than the hairs of beasts, why the sweat of Alexander the Great had a sweet odor, why parts of rhubarb purged, and other parts of it were constipating.[67] Such queries continued a long line of tradition from the Problems of Aristotle down the centuries.

So it is observed by some that there is a virtuous bezoar and another without virtue, which appear to the show alike. But the virtuous is taken from the beast that feedeth upon the mountains where there are theriacal herbs, and that without virtue from those that feed in the valleys where no such herbs are.[68]

And if Bacon cites Galen for the cure of scirrhosis of the liver by drinking the milk of a cow which eats only certain herbs,[69] let us remember that recent news items inform us that it has been dis-discovered that an extract made from the liver of a pregnant cow relieves stiffness of the joints, and that arthritis is cured by a chemical extracted from ox bile.

We pass on to Bacon's favorite explanation of such phenomena, which is that in all tangible bodies there are very fine, rarefied, subtle and invisible spirits, which are neither heat nor vacuum, air or fire, but differ from one another as much as tangible bodies do. They are almost never at rest and are easily dissipated, evaporate, infuse and boil away. They govern nature principally, and in animate bodies vital spirits are added to those found in inanimate bodies.[70] Gems have in them fine spirits, as their splendor shows, and they may work upon the spirits of men to comfort and exhilarate them, but no credit is to be given to their particular properties.[71] Yet we presently find Bacon not only suggesting that, if wearing the bloodstone really checks nosebleed, it is doubtless by astriction and cooling of the spirits, but further asking if the stone taken out of the head of a toad is not of like virtue, "for the toad loveth shade and coolness,"[72]—than which there could hardly be a better example of far-fetched magical association and logic.

[67] *Sylva*, 5, 8, 19.

[68] *Sylva*, 499.

[69] *Idem.*

[70] *Sylva*, 17-20, 98; *Historia vitae et mortis*, Canon IV.

[71] *Sylva*, 960.

[72] *Sylva*, 967.

But to continue with examples of the efficacy of sprits. The leaf of burrage has "an excellent spirit to repress the fuliginous vapor of dusky melancholy and so to cure madness.[73] The force of explosives proves that a small amount of spirits in the brain and sinews will suffice to move the whole body.[74] The emotions affect the body largely through the spirits, and an intoxicated person thinks that the room is going round because his spirits are whirling.[75] Infectious diseases are more in the spirits than in the humors, and putrefaction is caused by the spirits trying to get out of the body.[76] The reason why blows and bruises induce swellings is that the spirits rush to relieve that part of the body and draw the humors with them.[77] The upper parts of the body sweat more than the lower, because they are more replenished with spirits.[78] But in worms, flies and eels the spirits are diffused all over the body, and therefore, if they are cut to pieces, the pieces continue to move for some time.[79] Cats and owls could not see by night, were there not still a little light, sufficient for their visual spirits, and the reason why we see better with one eye shut is that "the spirits visual unite themselves more."[80] Nitre, though cold, cleans clothes because it has a subtle spirit, and quicksilver is the coldest of the metals because it is the fullest of spirit.[81] But heat refines the spirits, and makes the cock song bird excel the hen.[82] The spirits are affected more immediately through hearing than any other sense except perhaps that of smell, hence the effect of music upon manners.[83] Tears are caused by a contraction of the spirits of the brain, which further leads to wringing of the hands, "for wringing is a gesture of expression of moisture"[84]—another example of a magical association and way of thinking. Somewhat similarly, Adelard of Bath in the twelfth century had explained his nephew's weeping for joy at his return on the theory that his excessive delight overheated his brain and

[73] *Sylva*, 18.
[74] *Sylva*, 30.
[75] *Sylva*, 713, 725.
[76] *Sylva*, 297, 328.
[77] *Sylva*, 862.
[78] *Sylva*, 708.
[79] *Sylva*, 400.
[80] *Sylva*, 866, 868.
[81] *Sylva*, 362, 73. J. C. Gregory in *Ambix*, II (1938), 108, notes that, according to *Historia vitae et mortis*, cold things are usually poor in spirits.
[82] *Sylva*, 851.
[83] *Sylva*, 114.
[84] *Sylva*, 714.

distilled moisture thence.[85] That children and some birds learn to speak so easily, and that in darkness as well as by light, makes Bacon even wonder if there is not some transmission of spirits from the teacher to the pupil which predisposes the latter to imitation of the sounds.[86] In short, Bacon uses material spirits within bodies to explain anything and everything. But we have already seen this tendency in Telesio and other authors of the sixteenth century, and we shall find it continued after Bacon in the seventeenth.

Bacon illustrates the tendency in seventeenth century England for every lord and lady to be his or her own physician, dabbling in medical recipes and perhaps chemical experiments.[87] Indeed, one French lady of quality made such a collection for the poor. The *Journal des Sçavans,* reviewing in 1678 "this last" and much enlarged edition, said:

> One finds nut oil sovereign for stomach-ache, fevers, pest, dropsy and many other ills; imperial water admirable against poison, melancholy, headache, toothache, etc. Golden liquor for the squeamish, for insomnia, for indigestion, for women in travail; oil of balm to assuage the pains of sciatica.[88]

Another earlier work of this sort was *The Ladies Dispensatory* by Leonard Sowerby in 1652. More professional were the *Secrets et Remedes eprouvez* of l'Abbé Rousseau, based upon experiments which he had made at the Louvre by royal command and which were published in 1697 after his death. They included the right way to concoct the Water of the Queen of Hungary, essence of viper, the elixir of propriety, laudanum, Minerva's Lily, tranquil balm, and perfect essence of manna.[89]

Bacon believed that the wise physician should diligently search for medicinal simples with extremely subtle parts, such as elder-flowers for the stone, fumitory for the spleen, dwarf-pine for jaundice, and hartshorn for agues and infections. Since putrefaction is

[85] T II, 34. But Adelard's explanation is more mechanical and less magical.

[86] *Sylva,* 236.

[87] See the MSS listed at T II, 806.

[88] JS VI (1678), 318.

[89] JS XXV, 75-76. John Locke on May 6, 1676, gave the price in France of Queen of Hungary's water as forty shillings per pound, but on February 13, 1677, paid £5-9-4 for two pint bottles. *Travels in France,* 1953, pp. 91, 124.

the subtlest of all motions in the parts of bodies, the putrefied parts of plants and animals make excellent medicines. So do "creatures bred of putrefaction, though they be somewhat loathsome to take," such as earthworms, timber-sows and snails.

> And since we cannot take down the lives of living creatures, which—some of the Paracelsians say—if they could be taken down, would make us immortal; the next is for subtility of operation to take bodies putrefied such as may be safely taken.[90]

Bacon noted that medicinal earths were few in number, but he still listed *terra Lemnia, terra sigillata communis,* and *bolus Armenus.*[91] The wife of the English ambassador at Paris cured his warts by rubbing them with a bacon rind and nailing it up with the fat side towards the sun. Within five weeks every wart had disappeared.[92] Instead of the usual account of replacing noses by plastic surgery, Bacon tells of men with large and ugly noses who have cut off the excess flesh and then healed the wounds by making a gash in their arms and holding their noses there for a time, "which, if it be true, shows plainly the consent of flesh and flesh."[93]

Bacon seems to have been more interested in the prolongation of life and health than in the cure of disease. He thought that purges were more conducive to a long life than exercise and sweats, arguing that perspiration drove out not only noxious humors but also good juices and spirits. On the other hand, frequent blood-letting might be beneficial by renewing the fluids of the body.[94] He held that persons with long legs were likely to live longer than those with long trunks.[95] For adults he recommended "an opiate diet" every year about the end of May. He knew a great man who attained a long life and whose custom it was to have a fresh sod of earth brought to him every morning while in bed and he would hold his head over it for some time.[96] Unicorn horn was rather out of favor when Bacon wrote, but the bezoar stone, gold and powdered pearl, emerald or jacinth were still regarded highly as medicines promoting longevity. Among Bacon's own favorites were "Grains of Youth" and "Methusalem water." The former comprised

[90] *Sylva,* 692.
[91] *Sylva,* 701.
[92] *Sylva,* 997.
[93] *Historia vitae et mortis,* IX, 27.
[94] *Historia vitae et mortis,* X, 3; Medicinae, 6.
[95] *Ibid.,* in homine, 38.
[96] *Sylva,* 928.

four parts of nitre, three of ambergris, two of orris-powder, one-quarter of white poppy seed, one-half of saffron, with water of orange blossoms and a little tragacanth. These ingredients were to be made into four small grains which were to be taken at four o'clock or upon going to bed. The latter was the product of repeated washing, steeping, drying and powdering of shells, the tops of rosemary, pearl, ginger, white poppy seed, saffron, nitre, ambergris, cucumbers sliced in milk and stewed in wine, vinegar, spirits of wine, and so forth.

Of thirty-two extracts from a book on the prolongation of life for his own use we may note a few. Take Mithridate thrice a year. Before retiring for the night eat a bit of bread dipped in scented wine with syrup of roses and a little amber. Never keep the body in the same posture for more than half an hour at a time. Break off custom, shake off spirits ill disposed, meditate upon youth, and do nothing contrary to your personal equation. At supper time take one drink of wine in which gold has been quenched.[97] In the *Sylva* Bacon remarked that among the ancient Greeks and Romans and modern Turks bathing was as usual as eating or sleeping, but "with us" it was used only for medicinal purposes.[98] In the History of Life and Death he recommended bathing of the feet at least once a week in a bath made of

> lye with bay-salt, and a little sage, camomile, fennel, sweet-marjoram, and pepper-wort, with the leaves of angelica green.[99]

"Barbarossa in his extreme old age"—he was not yet seventy when he died on the third crusade—by advice of his Jewish physician applied young boys to his abdomen to warm and comfort it, and other old men "lay whelps (creatures of the hottest kind) close to their stomachs every night."[100] Bacon died at sixty-five.

[97] Medical Remains; *Works*, VII (1863), 432.

[98] *Sylva*, 740.

[99] *Historia vitae et mortis*, V, 41; *Works* (1862), III, 454.

[100] *Ibid.*, IX, 26. Those desiring a fuller and sometimes more favorable presentation of Bacon's medicine may turn to G. W. Steeves, "Medical Allusions in the Writings of Francis Bacon," *Proceedings of the Royal Society*, Section of the History of Medicine, II (1913), 76-96; or Max Neuburger, "Lord Bacon's Relations to Medicine," *Medical Life* (1926), 149-69; or Helmut Minkowski, "Einordnung Wesen und Aufgaben der Heilkunst in dem philosophisch-Naturwissenschaftlichen System des Francis Bacon," *Janus*, 37 (1933), 325-53.

Bacon had little faith either in witchcraft, fascination or the power of imagination over other bodies, especially at great distances. Neither the confessions of witches nor the evidence against them were to be rashly accepted, since they were imaginative themselves and other people were credulous.[101] Paracelsus and "the disciples of pretended natural magic" had grossly overestimated the power of imagination in fascination, and had justified ceremonial magic as strengthening the imagination rather than being indicative of a pact with the devil.[102] But for Bacon

> The experiments which may certainly demonstrate the power of imagination upon other bodies are few or none: for the experiments of witchcraft are no clear proofs; for they may be by a tacit operation of malign spirits.[103]

He further was opposed to ceremonies, characters and charms *per se*,[104] and regarded resort to occult virtues as slothful.[105] He classed astrology, natural magic, and alchemy together in the *Advancement of Learning* as "sciences . . . which have had better intelligence and confederacy with the imagination of man than with his reason."[106]

But in the *Advancement of Learning* of 1605 he was speaking of a degenerate natural magic

> whereof now there is mention in books, containing credulous and superstitious conceits and observations of sympathies and antipathies and hidden properties, and some frivolous experiments . . .[107]

He distinguished between it and

> the true natural magic which is that great liberty and latitude of operation which dependeth upon the knowledge of forms.[108]

He found it, too, deficient at that date but endeavored to fill the gap with the *Sylva sylvarum*, which he described as "not natural

[101] *Sylva*, 903; also 26, 859.

[102] *Advancement of Learning*, Bk. II; *Works* (1863), VI, 256-57.

[103] *Sylva*, 950.

[104] See note 102.

[105] *Sylva*, 36.

[106] Bk. I, *Works* (1863), VI, 127; Bk. II, *Ibid.*, 229. Also *De augmentis scientiarum*, Bk. I; *Works*, II (1861), 133.

[107] Bk. II; *Works* (1863), VI, 229.

[108] *Ibid.*, p. 230.

history, but a high kind of natural magic."[109] The last of his twenty-seven types of prerogative instances in the *Novum Organum* were magical, since he argued that superstition and magic were not to be entirely omitted, but should be investigated for some natural operation concealed under their cover.[110] Indeed, he had already written in *Advancement of Learning:*

Neither am I of opinion, in this history of marvels, that superstitious narrations of sorceries, witchcrafts, dreams, divinations, and the like, where there is an assurance and clear evidence of the fact, be altogether excluded. For it is not yet known to what cases, and how far, effects attributed to superstition do participate of natural causes; and therefore howsoever the practice of such things is to be condemned, yet from the speculation and consideration of them light may be taken, not only for the discerning of the offences, but for the further disclosing of nature.[111]

Bacon also resorted to the old excuse for including matter of dubious authenticity, that previous writers had repeated it, and that there might be something profitable in it after all.[112] He gave the old theory of natural divination without either accepting or rejecting it.[113] He said that physiognomy and the interpretation of natural dreams had a solid foundation in nature and were useful in daily life. But he soon added that chiromancy was an imposture, and that at present the interpretation of natural dreams was full of ineptitudes.[114]

Often Bacon qualifies the beliefs and traditions which he repeats, with some such expression as, "if it be true." For instance, if it be true that the salamander lives in fire and can extinguish fire, it must have "a very close skin" and further, "some extreme cold and quenching virtue" in its body.[115] Here the tradition itself is self-contradictory enough, for why should an animal that spends all its life in fire wish to quench it or be able to quench it, and how could there be any fire for it to live in after it had quenched it? Bacon mildly questions the facts but swallows the contradiction hook, line and sinker, and even heightens it. Yet surely a native of

[109] *Sylva,* 93.

[110] *Novum Organum,* II, 31, near close, and 51.

[111] Bk. II; *Works* (1863), VI, 185.

[112] *Sylva,* 500.

[113] *Advancement of Learning,* Bk. II; *Works* (1863), VI, 256.

[114] *De augmentis scientiarum,* IV, i; *Works,* II (1861), 315-16.

[115] *Sylva,* 860.

the element fire should be hot and fiery, not thick-skinned and possessed of extreme cold and destructive to fire. Otherwise, fish should be very dry in order to live in water and birds very heavy in order to fly in air. A little more logic and less use of traditional and magical experiments would not have done Francis Verulam any harm.

We have heard Mersenne make three general critictisms of Francis Bacon. It may be worth while to give further a criticism of details in his works by a contemporary, Isaac Beeckman, in his Journal for the years 1604-1634,[116] in which he from time to time jotted down reflections upon his reading, or on natural questions which had been long debated—such as whether fountains originated from the sea or from rain, why the sea was salt, why the thoughts of a pregnant woman affected the foetus, why round wounds take longest to heal, why the stars twinkle, why quenching hot iron in cold water hardened it, why a varied diet is more agreeable, why it is hotter at the tropics than at the equator, whether cold is something real or mere privation of heat, whether the stars can be seen by day, and so on. He also discussed recent inventions, like telescopes, pumps, air thermometers, and Drebbel's devices. He participated in 1626 in the foundation of a Mechanical Academy at Rotterdam, but he thought that the mechanical instruments of the people could hardly be improved upon, or new ones be much better than those already invented.[117] Like Bacon, he treated of such general subjects as motion and sound, also light and color, and was especially full concerning music and medicine. He came back again and again to the experiment of the candle going out when covered with a glass,[118] and to that of oil in the lamp climbing the wick to the flame.[119] Top-spinning fascinated him.[120]

Along with such adumbrations of modern science,[121] Beeckman's Journal shows a considerable resemblance still to the collections of secrets and experiments that we noted in the thirteenth century

[116] *Journal tenu par Isaac Beeckman de 1604 à 1634*, ed. par. C. de Waard, La Haye, 4 vols., 1939, 1942, 1945, 1953.

[117] *Ibid.*, II, 429: III, 15, 306.

[118] *Ibid.*, I, 38; II, 144, 195, 228, 327, 382; III, 64.

[119] *Ibid.*, I, 102; II, 48.

[120] *Ibid.*, I, 30-32, 242, etc.

[121] Others are noted in paragraphs on Beeckman in our chapters on Sennert and Descartes.

manuscripts. A ball held between the index and middle fingers will seem two.[122] Garlic will glue a broken vase together.[123] Perpetual motion[124] and perpetual clocks[125] are considered again and again. Speaking secretly through a tube goes back to Hippolytus in the early third century or his still earlier sources,[126] but to it is added secret writing through telescopes. Hearing at a distance is from Albertus Magnus.[127] Writing with the left hand to be read in a mirror is nothing new,[128] nor is the camera obscura,[129] nor a burning glass of immense virtue,[130] nor amusing tricks with an artificial wind.[131]

Beeckman also was not entirely inattentive to the occult. He discussed the Lullian art and the Jewish Cabala.[132] Although in one passage he spoke of himself as abhorring divinations, in others he inquired why persons with headaches or aching bones predicted storms, and why imagination was sometimes more exact in sleep and sickness.[133] He doubted tales of witchcraft and explained *incubi* upon a physical basis, but, while holding that devils change nothing in nature, discussed their influence upon our thoughts and soul.[134] He objected to Gilbert's describing magnetic force as incorporeal and seeming to attribute intelligence to the earth, "which is unworthy of a philosopher."[135] But he repeated the story that the heart of Zwingli would not burn and told of a son who, though absent, was affected at the hour when his father died.[136] He usually avoided resort to occult qualities, but once, influenced by Galen, spoke of "force according to the whole substance."[137]

Beeckman did not appeal to material spirits in bodies for his explanation of natural phenomena as often as Bacon did, but he said that pain was caused when they contracted, and that there was a close connection between the nerves and the animal spirits.[138]

122 *Journal*, I, 28.

123 *Ibid.*, I, 37.

124 *Ibid.*, I, 39, 67; II, 199-200, 202, 344, 352, 353, 355, 359; III, 228.

125 *Ibid.*, III, 203-4, 302, 358.

126 *Ibid.*, I, 46; T I, 468.

127 *Journal*, I, 83.

128 *Ibid.*, I, 193.

129 *Ibid.*, II, 12.

130 *Ibid.*, II, 371-72.

131 *Ibid.*, III, 23.

132 *Journal*, I, 294; III, 5.

133 *Ibid.*, III, 214-15; I, 126, 270-71. His attitude towards astrology is treated in our chapter on Descartes.

134 *Ibid.*, III, 288; I, 281; II, 241-42.

135 *Ibid.*, III, 18.

136 *Ibid.*, I, 227; III, 122.

137 *Ibid.*, II, 118.

138 *Journal*, I, 125, 136.

Some of his observations have a faint Baconian flavor, as when he counted 2100 beats of his pulse in half an hour, as he lay in bed on April 12, 1614, or planned in 1628 to observe weather changes from a tower which the government of Dordrecht was building for him.[139] He was intent on such problems as why the front wheels of a cart are smaller than the rear wheels, why a needle would float on water, why the bottom of the pot remained cold while the water boiled, whether caves are hotter in winter, why men are not sometimes generated spontaneously and why they never generate beasts, why birds were smaller than quadrupeds and these than fish, why there were no animals of immense size.[140] He also made bold and sometimes erroneous assertions such as that fish die in winter when the ice prevents the air from reaching them, or that water is hotter just before boiling than when it boils, or that heavy bodies fall faster.[141] He affirmed that the cause of frigidity in the air was greater or less density; that the nitre of the urine of a healthy person was good for stones in the kidneys; and that thoughts were impressed on the membranes of the brain as images were on the retina of the eye.[142] He declared that all diseases arise from the four first qualities—hot, cold, dry and moist.[143] In one place he said that water passed through cracks more readily than air, light, smoke or fire.[144] But years later we find him explaining to his brother why snow goes through shoes more than water does, the explanation being that snow is drops of water dissolved into minuter particles.[145]

In view of the general resemblance—rough rather than close—between the two men in their outlook upon nature, particular interests, and mode of approach, it is interesting to note Beeckman's reactions upon reading several of Bacon's works.

When Beeckman read the *Novum Organum* in 1623, his comments in the Journal included the following criticisms. Bacon argued that the heavens revolved daily from the same phenomena that would appear if the earth did. He perhaps erred in saying

[139] *Ibid.*, I, 34; III, 85.

[140] *Ibid.*, I, 59, 233, 345; II, 342; III, 59; II, 69, 300; III, 71.

[141] *Ibid.*, I, 158, 174-75.

[142] *Ibid.*, III, 110, 203, 199.

[143] *Ibid.*, II, 304.

[144] *Ibid.*, I, 81.

[145] *Ibid.*, III, 144.

that weight corresponded to mass (*copia materiae*). He held that missiles would not make so strong a percussion at the start as a little further on. Like Keckermann, he believed that water contracted on freezing. And he did not believe that a vacuum was intermixed with things.[146]

On December twelfth of the same year Beeckman criticized Bacon's History of Winds for the statement that water turns into air and occupies a space one hundred times greater than before. Also for saying that smaller ships move faster because they can carry more sail in proportion to their size. This, Beeckman points out, disregards the facts that a heavy body once in motion persists in it longer, and that air and water offer more resistance proportionally to the smaller ship. Bacon further thought that the wind in the upper sails moved the ship more than that in the lower sails on the ground that it exerted a longer leverage. Finally, he thought that air had no weight.[147]

In the case of Bacon's History of Life and Death, Beeckman was not satisfied with the statement that aging bodies inclined to dry up and contract as in the case of parchment held near the fire.[148] He held that there could be no inclination in an irrational body like parchment, and that its shrivelling was produced by the fire entering the portion nearest to it, mingling with its humor, and carrying it away with it. In the case of the parts farther from the fire, it is able only to dilate but not carry off some humors, and their dilation protrudes fibres towards the portion which is already destitute of humor. Beeckman also rejected Bacon's explanation of a candle's going out when a glass was placed over it, that the air within the glass, dilated by heat, over-crowded the flame and extinguished it by pressure.[149] He pointed out that, if this were the case, water would not rise into the glass when it was imposed

[146] *Journal,* II, 251-54.

[147] *Journal,* II, 276-77. See III, 297, 331, 336, for experiments proving that water cannot be turned into air. The editors of *Correspondance du P. Marin Mersenne,* I (1932), 299, have pointed out that Gorlée, Basson, d'Espagnet, de Caus and Helmont also denied that water by rarefaction changed to air, but that Mersenne adhered to the common opinion which had come down from Aristotle.

[148] *Journal,* II, 327, citing p. 64 of the 1623 edition (*Works,* III [1862], 353).

[149] *Ibid.,* citing p. 373 of the 1623 edition (*Works,* III, 470).

over water. But his own explanation was little better, that the attenuated air drew the fire with it through the glass. Earlier in his Journal, however, he had held that only the fire and not the air could pass through the pores of the glass.[150] Still earlier he had suggested that the flame had consumed the air and so had no more pabulum.[151]

In conclusion there is not much that one can say for Francis Bacon. He was a crooked chancellor in a moral sense and a crooked naturalist in an intellectual and scientific sense. He did not think straight. Or put it in this way, if you prefer. Even a Lord High Chancellor, even a Francis Bacon, could not think straight when he thought as a naturalist and tried to amass "experiments" on the one hand and to grapple with magical tradition and superstition on the other hand. The path of magic and experimental science was no straight and narrow one; it was not true, and its course did not run smooth. It was a relatively easy thing to criticize the past and present state of learning, and to advocate a new program including "experimental science." Roger Bacon had done it three and a half centuries before. But when it came to getting down off one's high horse of generalities and putting one's shoulder to the problem of particular phenomena of nature and dealing with specific facts and beliefs and traditions and errors, Francis Bacon was as helpless as Pliny had been in antiquity or as any one else was in the early seventeenth century. The best that one can say for him is that he really tried.

It must be admitted, however, that he was much cited and admired by many writers of his century, Mersenne being something of an exception. And his tendency to explain natural phenomena by the action of corporeal spirits became widespread and general.

[150] *Journal,* II, 227-28.

[151] *Journal,* II, 195.

CHAPTER V

ASTROLOGY TO 1650

Introductory—England: John Chamber, Sir Christopher Heydon, George Carleton, Wright, Dekker, Cotta, Davenport, Weigel—Macrocosm and Microcosm: Guibelet, Nancelius, Bourdin, Fludd—Potentates and Astrology: Chavigny, Senelles, Urban VIII and Father Morandi—French opposition to astrology: Cauvigny, Heurtevyn, Bulenger, Pithoys, Saumaise—French interest in astrology: Morisot, Octoul, Alleaume—Italian discussion of climacteric years and critical days: Florido (the Tides), Claudinus, Silvaticus, Codronchi, Columbus—Of astrology in general: Altobelli, Baranzani, Ferrante de Septem, Bartolini, Giuffi, Alexander de Vicentinis—Bologna: Bonhombra, Roffeni, Ghiradelli, Cavalieri, Pandolfo—Padua: Tomasini, Cremonini, Argolus—Rodriguez de Castro on meteors of the microcosm—Spigelius on the influence of the moon—Merenda against astrology—Carena and the attitude of the Inquisition towards astrology—Najera—Sempilius of Madrid—German astrological medicine: Ampsing, Etzler, Pleier, Bicker—Other Germans on Astrology: Aquaviva, Kirchner, Goclenius, Combach, Dieterich, Avianus, Origanus or Dost, Eichstadt, Tidicaeus, Linemann—Budowez—Northern astrology: Caspar Bartholinus, Forsius, Heldvad, Lomborg, Beckher, Franckenius, N. Malmenius, Rhodius, A. A. Malmenius.

I can see no justification whatever for the attitude which refuses on purely a priori grounds to accept action at a distance... Such an attitude bespeaks an unimaginativeness, a mental obtuseness and obstinacy.

—P. W. BRIDGMAN

Attacks upon astrology were numerous in the seventeenth century. We have already described several of them in our sixth volume and need not here repeat what was said there concerning George of Ragusa, Alexander de Angelis, and Giannini.[1] On the other hand, we have already seen that the papal bulls against astrology of 1586 and 1631 had only a limited effect, and that the subject continued to be taught at the University of Bologna into the seventeenth, and at Salamanca into the eighteenth century.[2] We shall now examine

P. W. Bridgman, *The Logic of Modern Physics*, 1949, pp. 46-47.

[1] T VI, 198-206.

[2] T V, 247-51; VI, 164-78.

further into its status and the books written for and against it during the first half of the seventeenth century in various regions of Europe: England, France, Italy, Portugal and Spain, Germany, and northern Europe. We shall not attempt to cover annual astrological predictions or works elicited by particular comets, eclipses and planetary conjunctions. But even the authors of such judgments might assert that they were free from all superstition.[3]

i. ENGLAND

In England, in the opening year of the century, John Chamber published *A Treatise against Judicial Astrologie.*[4] Chamber remarked that astrological superstition had been long tolerated, but held that it was inconsistent with Christianity. He further objected that the number of stars was not known nor the exact time of birth, and that casters of nativities hesitated whether to rely on the latter or the hour of conception. He disapproved of applying astrological rules and prediction to man alone, taking no cognizance of the belief which was almost universal then that man is a microcosm, whereas other animals are not. He passed on to such criticisms as the uncertainty of astrologers' predictions, the impossibility of foretelling events dependent upon chance or free will, the uselessness of predicting events which occurred necessarily and so could not be avoided, and the diversity of twins. Both philosophers and emperors had opposed the art in the past, and God had reserved to Himself knowledge of the future. He charged the astrologers with wresting a passage of Aristotle to favor their art, discussed the attitude of the ancient Greeks towards astrology, and compared it with other arts. He argued that elections or the selection of favorable moments for action were inconsistent with astrological pre-

[3] Paul Nagel, *Explicatio oder Auszwicklung der himmlichen Kräffte aus . . . Grunde der astrolog. Kunst ohn alle Superstition, gerichtet auff das Jahr so uns zeiget das Wort Judicium,* Leipzig, 1613, in-4.

[4] With a dedication to Sir Thomas Egerton. BM 719.e.12 is an interleaved copy of 132 printed pages, with writing on many of the otherwise blank leaves and on the margins of some of the printed pages. There follows *Astronomiae Encomium a Ioanne Chambero ante annos 27 peroratum quo tempore Ptolemaei Almagestum in alma universitate Oxonien. publice enarravit,* London, by the same printer, John Harison, 1601. With an English translation it fills 39 pp.

diction, and concluded with an attack upon belief in climacteric years and critical days.[5]

The defense of astrology was assumed by Sir Christopher Heydon, a member of Parliament, who rebutted Chamber's arguments in a series of parallel chapters with both longer headings and longer text.[6] Heydon had studied at Cambridge and his book was issued by John Legat, "Printer to the Universitie of Cambridge."

Chamber composed a reply to Heydon which was not printed but is preserved in a manuscript at the Bodleian, dedicated to King James I and entitled: "A Confutation of Astrologicall Daemonologie or the divells schole, in defence of a treatise intituled against Iudiciarie Astrologie & oppugned in the name of Syr Christopher Heydon, Knight."[7]

Bound with the British Museum's copy of Heydon's work is a reply to it by George Carleton, an Oxford master of arts and fellow of Merton College who became bishop of Llandaff and then of Chichester. It was not printed until 1624,[8] but the dedicatory epistle by Thomas Vicars explains that almost twenty years have passed since its composition, and that numerous requests to print it have finally prevailed. Carleton held that astrology had been invented by the devil and spread by Zoroaster; that it and magic were

[5] The numbering of chapters and pages becomes confused towards the close of Chamber's treatise. In the text there are two chapters numbered XIX and two pages numbered 117, while what was Chapter XXIII in the Table of Contents precedes XXI and XXII.

[6] *A Defence of Juridicall Astrologie, in answer to a treatise lately published by M. John Chamber, wherein all those places of Scripture, Councells, Fathers, Schoolemen, later Divines, Philosophers, Histories, Lawes, Constitutions, and reasons drawn out of Sixtus Empericus* (sic), *Picus, Pererius . . . and others against this Arte, are particularly examined, and the lawfulness thereof by equivalent proofes warranted,* Cambridge, 1603, 551 pp., not including preface, table of contents, errata, and index. BM 718.e.14.

[7] Savile 42, dated February 2, 1603-4, ending at fol. 230r. From the table of contents at fol. 4r-v, under Chapter 2 may be quoted: "Thirdly you have a large yet necessary digression in the commendation of woman to cleare that sex of a rash and false imputation of witchcraft and sorcery imposed upon them by the adversary, as if they were fit for no good arte or studie but only for witchery, and such like divellish lewdnesse."

[8] *Astrologomania, the Madnesse of Astrologers: or, an Examination of Sir Christopher Heydon's Booke, entituled, A Defense of Judiciarie Astrologie,* London, 1624, in-4, 123 pp. In ten chapters.

inseparable in practice; and that magic was no part of natural philosophy.[9]

Another justification of astrology by Heydon, who had died in 1623, together with his Judgment from the great conjunction of 1603, was printed posthumously in 1650,[10] while a third astrological "Recital of the Caelestiall Apparitions of this present Trygon" was never published.[11]

Astrological medicine met with milder opposition. Thomas Wright of Oxford, who in 1601 had published a moral discourse on Passions of the Mind,[12] thought not unfit to be inserted in its last book, *A Succinct Philosophicall Declaration of the Nature of Clymactericall Yeeres, occasioned by the death of Queen Elizabeth*[13] in 1603 at the age of seventy.[14] Physicians by long experience had found that "men of lusty constitution" usually lived an even score of years, dying at 40, 60, 80, 100 or 120. Moses was an example of the last; Ecclesiastes XVIII, 8, for one hundred; and Psalm 89, 10, for eighty. But the most dangerous years were 49, 63, 70, and 81.

> Those humors which alter the bodie and dispose it to sicknesse and death, the same bend the soule to take inordinate affections and passions.[15]

Some physicians give an astrological explanation for climacteric years. Others say that God created all things in number, measure and weight.

> These . . . I will not confute. For, albeit I do think them both in some things most true, yet they are too general and remote.[16]

Wright notes other bodily periods. Man grows in height until 21

[9] *Ibid.*, chapters 7 and 9.

[10] *An Astrological Discourse with mathematical demonstrations, proving the powerful and harmonical influence of the planets and fixed stars upon elementary bodies, in justification of the validity of astrology, together with an astrological judgment upon the great conjunction of Saturn and Jupiter, 1603*. . . now published by Nicholas Fiske, Cornhil, 1650, in-12, xviii, 111 pp.

[11] DNB.

[12] BM c. 70.aa.26.

[13] London, 1604, 17 pp. BM 1141. a.43.

[14] *Ibid.*, p. 2, "I think it not unfit to be inserted in the last book of the Passions of the Minde; because the same temper of body and propension to death which is the base of Clymactericall yeres; the very same conferres much either to moove Passions or hinder the opperations of the soule."

[15] *Ibid.*, p. 4.

[16] *Ibid.*, pp. 5-6.

or 25 years of age; in thickness from 25 to 40; from then on declines. Infancy, boyhood, adolescence and youth (*iuventus*) extend to the seventh, fourteenth, twenty-first and twenty-eighth year; manhood from 28 to 49; old age, from 49 to 63; decrepitude, from 63 on. After a digression concerning an attack of ague which he had at Como, and citing Vallesius, *De sacra philosophia,* as to critical days, Wright argues that noxious humors accumulate during a period of six or eight years. If insufficient to do harm at seven, they go on multiplying until nine; if they fail then, to fourteen; then to eighteen; and so on. This is the reasons why doctors advise purging in spring and autumn, although their patients may not be conscious of the accumulated humors.

> I myself have known a man almost with half his lungs rotten with a consumption, and yet boldly avouch that he was strong, for *Ab assuetis non fit passio.*[17]

Although a man cannot exceed his clymacteric period, there are many ways in which he can shorten it.

Such is Wright's explanation, scarcely more satisfying than the astrological doctrine. His final word is that he "would give any physician most hearty thanks who in few words would teach me a better way."[18]

Thomas Dekker parodied astrological predictions in *The Raven's Almanacke* of 1609. John Cotta, who also wrote against witches,[19] included astrological medicine in an attack upon quack medical practitioners which had three editions between 1612 and 1619.[20] Of ten compositions by Franciscus a S. Clara or Christopher Davenport (1598–1680), a Roman Catholic, which are listed in the *Dictionary of National Biography,* all are on religious subjects except a treatise against judicial astrology, and very likely his opposition to it was based largely upon religious grounds.[21]

[17] *Ibid.,* p. 15.

[18] *Ibid.,* p. 17.

[19] *The Triall of Witch-craft,* 1616; *The Infallible, True and Assured Witch,* 1624.

[20] *A Short Discoverie of the Vnobserved Dangers of Severall Sorts of Ignorant and Vnconsiderate Practisers of Physicke in England,* 1612.

[21] The article in DNB gives no indication whether it was printed, and, if so, when and where. It does not appear in the BM and BN printed catalogues, but somewhere I have seen a reference to an *Epistolium de judiciis astrologicis,* Duaci, 1626, by him. And it may be contained in his *Opera,* Duaci, 1665-1667, in-fol.

A different point of view appears in the *Astrologie Theologized*[22] of Valentin Weigel, who grants the stars vast powers and who would theologize astrology by laboring for six days and sanctifying the seventh and "by the benefit of regeneration in the exercise of the Sabbath."[23] He puts under astrology

> all orders, states and degrees of men, distinctions of persons, dignities, gifts, offices, and every kind of life as well naturally ordained by God himself as thought of and invented by humane wit . . . All these are the fruits of the Starrs.[24]

Astrology is synonymous with philosophy or universal knowledge "of all the wonderful and secret things of God."[25] There is much talk of macrocosm and microcosm, and a chapter on the seven governors of both these worlds.[26] But observance of the Sabbath day seems the chief concern of the author.

ii. FRANCE

The conception of macrocosm and microcosm, that man is a little world and corresponds member for member and faculty for faculty with the universe, or, more particularly, with the earth on the one hand and the heavens on the other, is evidently closely connected with the belief that inferiors are ruled by superiors and that man is related to and governed by the stars. It is not merely a fitting foundation for astrology, but really part and parcel of astrology in the broad sense of that word.

All this is well illustrated by one of three philosophic discourses which Jourdain Guibelet, a physician of Evreux, published there in 1603, and which is entitled, *De la comparaison de l'homme avec le monde*.[27] He compares the rational soul with God, human faculties with the Intelligences that move the heavens, the head with the heavens, the heart with the sun, and the liver with the moon. The

[22] *Wherein is set forth what Astrologie and the light of nature is; what influence the stars naturally have on man, and how the same may be diverted and avoided*, London, 1649, in-4. BM E.562.(14.). The book was reprinted in 1886: BM 8610.ee.10.

[23] *Ibid.*, p. 26.

[24] *Ibid.*, pp. 2-3.

[25] *Ibid.*, p. 4

[26] *Ibid*, chapters 4 and 7; pp. 19, 22, 31.

[27] *Trois discours philosophiques*, Evreux, Antoine le Marie, 1630, in-8: copy used, BN R. 37931.

liver presides over human infancy, as the first age of other animals is under the government of the moon. To Jupiter corresponds the brain; to Venus, the generative organs; to Mercury, the tongue;[28] to Saturn and Mars, the gall and spleen.[29] Guibelet relates the hair to the fixed stars and other parts of the human body to the signs of the zodiac, but he adds that some give the eyes to sun and moon, the ears to Mars and Venus, the nostrils to Jupiter and Saturn, and the mouth to Mercury.[30] Man further comprehends the elements, meteors and minerals, plants and animals.[31] And in the little world as in the great there is republic, aristocracy and monarchy, and cities with all sorts of artisans and instruments to ply each trade.[32] But the world is now in its decrepit old age, and all that heaven and earth engender in their senility is but as mere excrement in comparison with previous periods.[33]

Despite the close connection between human faculties and members and the heavens and stars, which Guibelet made in this first *Discours*, in the second on the principle of human generation the influence of the stars is not mentioned, while in the third on melancholy he declares that the predictions of astrologers seem to him as ill-founded as those of augurs, and that they often turn to magic or demons for assistance.[34] He also now notes that the astrologers assign an excess of melancholic humor to the influence of Mars and Saturn, but that we see many melancholics who are not under those planets, and many persons who are under those planets who are not melancholy.[35] Thus he tacitly accepts planetary influence on men, although denying the truth of astrological prediction. This apparent discrepancy shows that the house of astrology is being divided against itself, and that, as the seventeenth century opens, a man may condemn prediction, although he accepts doctrines upon which it is based. It further indicates that these doctrines are being disassociated from astrology, although they may seem logically to go with it.

A much more exhaustive and exhausting treatment of the analogy

[28] *Ibid.*, fol. 61r.
[29] *Ibid.*, fol. 64v.
[30] *Ibid.*, fol. 67r.
[31] *Ibid.*, cap. xvii, fols. 78v-97v.
[32] *Ibid.*, cap. xviii, fols. 97v-105r.
[33] *Ibid.*, fols. 26v, 28v.
[34] *Ibid.*, fol. 279v.
[35] *Ibid.*, fol. 255v.

of the microcosm to the macrocosm was turned out by Nicolas Nancelius of Noyon, physician to Leonore Bourbon, abbess of Fontevrault, in 1611.[36] It stretches to thirteen books and 2232 columns in folio, with quotations galore from the classics and church fathers marked by large capital letters across the column and set off by leaving a blank space above and below. It is not worth while to try to pick out his own views, if any, from the mass of citations, quotations and indirect quotations, and most of the text has little or nothing to do with the analogy of microcosm and macrocosm, which merely serves as a springboard for a dive into a sea of quotations and opinions. The subject is said to be treated theologically, physically, medically, historically and mathematically. After a Proemium of forty-eight columns on God, the first book deals with the analogy of man with God, of the soul with the ether, the head with the sky, and "the seven conjugations of nerves" with the planets. Book two has more concerning the spirits of the human body in particular and "the miracles of air and fire." The third book is devoted to the earth and the analogy of parts of the human body with it, while a few columns are devoted to the theme of sleep and waking. Book four proceeds from esophagus to diaphragm, and by Book seven we reach the sexual organs with discussion of various problems of generation, such as whether the eighth month's child will live, and which lead finally to remarks concerning the Gregorian calendar. Book VIII on the arms and hands, in treating of the arts of chiromancy and physiognomy, seems to accept the *Physiognomy* ascribed to Aristotle as a genuine work, yet condemns those arts as false, inane, ridiculous, and full of tricks and impostures. It is absurd to predict one's whole fate from one little part of the body. Nancelius wonders that such grave men as Conciliator, Cardan and Albertus Magnus could waste time over such matters—although he himself devotes considerable space thereto[37]—while he has no use whatever for such writers as Corvo, Tricasso, John of Indagine and Cocles.[38] Later on we find him pointing out the analogy of the

[36] Nicolaus Nancelius Trachyenus Noviodunensis Leonorae Borboniae rev. abbatissae Fontebraldensis medicus, *Analogia microcosmi ad macrocosmum*, Paris, Claude Morellus, 1611, in-fol. Copy used: BN R. 1057.

[37] *Ibid.*, cols. 1311-18.

[38] For them see T VI, Index.

humors of the human body with earth's waters and with the four elements, quoting from John of Sacrobosco and Euclid, writing of the origin of fountains and rivers and their marvels, discussing why the sea, especialy the Dead Sea, is salt, and such other favorite and time honored questions, as whether the semen comes from the brain or the whole body, whether heart or brain is superior, and whether the world will have an end. He thereby illustrates the narrow range of ideas and problems that then occupied and beset men's minds, even when, like Nancelius, they took plenty of space in which to express themselves.

The Jesuit, Pierre Bourdin (1595–1653) of Moulin, who taught rhetoric for seven, and mathematics for twenty-two years at La Flèche and Paris, besides a number of works in mathematics and related subjects,[39] published together in 1646 a work on the sun as flame and aphorisms on the analogy of microcosm and macrocosm.[40] In the former he not only held that the sun was flame but nourished by vapors from our globe of earth and water, which were impregnated by the influence of the planets. The three chief fluids in the microcosm were chyle, venal blood and arterial blood; in the macrocosm, water, air and fire. The eight founts of fluids in the small world were the mouth, stomach, mesentery, spleen, liver, right sinus of the heart, lungs, and left sinus of the heart. Those of the great world were Saturn, Jupiter, Mars, Venus, Mercury, the moon, our terraqueous globe, and the heart of the world whence the flame of the sun bursts forth like vital spirit from the heart of man. Solar spirits retarded the movement of the superior planets westward. These solar spirits were changed into celestial, which were distributed through the world. Bourdin held that the earth was at rest and did not move about the sun. Vital spirits corresponded to solar; animal, to celestial. Air was the equivalent of the empyrean; and skin, of the firmament. Brain, arms and thighs paralleled the starry spaces; above the diaphragm, corresponded to planetary space; below it, to the moon and earth.

[39] The third edition of his *Le cours de mathematique,* in which pages of figures alternated with pages of text, appeared in 1661, 186 pp. BM 529.d.5.

[40] *Sol flamma... eiusque pabulo... Aphorismi analogici parvi mundi ad magnum, magni ad parvum,* Paris, 1646, in-8. BM 534.c.35 (2.).

Robert Fludd (1574–1637) in 1617 published the first part on the macrocosm of a work on macrocosm and microcosm.[41]. At the close of the Appendix to his *Harmonice mundi* of 1619 Kepler compared the two works as to subjects covered and added qualitative distinctions. Fludd drew from old authorities; Kepler, from the nature of things by observation and experience. Fludd's affinities were with alchemists, Hermetics and Paracelsans; Kepler's, with astronomers and mathematicians. Fludd interpreted harmony in terms of light and darkness; Kepler, in terms of motion. Fludd was arbitrary, mystical and obscure; Kepler, geometrical and natural. Fludd dealt in enigmas, symbols and analogies; Kepler, in demonstrated measurements.[42]

As in the days of the Roman emperors, the attitude of monarchs and governments to astrologers was largely swayed, not by the validity or vanity of the art of astrology, but by the favorableness or unfavorableness of the prediction.[43] Jean Aimes de Chavigny flattered Henri IV by a collection, under the title of *Pleiades*,[44] of seven past predictions which, he insisted, all foretold the happy advent of that monarch. The first, composed by Cataldus, bishop of Trent,[45] over a thousand years ago, was a forecast of future ills of Italy which was brought to light only just before the French invasion by Charles VIII. The second was the vaticination of the Erythraean sibyl; the third, an anonymous tract given to Chavigny twenty years ago by Jaques Gohorry;[46] the fourth, by Lorenzo Bonincontri di San Miniato. Even these first four predictions, according to Chavigny, "make authentic mention of Your Majesty," and at the close of the fourth he has taken occasion to "discourse on some points of your happy birth." The remaining items are the celebrated prediction

[41] *Utriusque cosmi maioris scilicet et minoris metaphysica physica atque technica historia . . .*, Oppenhemii, 1617, in-fol.

[42] Johannes Kepler, *Gesammelte Werke*, Bd. VI (1940), 373-77. For the resulting controversy with Fludd, *Ibid.*, 513-17.

[43] On the legal standing of astrology in France, as illustrated by the 1615 and 1671 editions of Bouchel, *La bibliothèque . . . du droit français*, see T VI, 170-71.

[44] His preface to the king is dated from Lyon, 15 Avril 1603.

[45] This was corrected later in the margin in handwriting to archbishop of Taranto.

[46] Spelled "Iaques Gohorri" by Chavigny. He wrote alchemical tracts under the pseudonym, Leo Suavius: see T V, 636-40.

of Antonio Torquato[47] to Matthias of Hungary in 1480; a translation into French of a previous translation by a German into Latin from Turkish; and a prayer extracted from Hippolytus. Chavigny has annotated them and compared them with the prognostications of Nostradamus.[48] He cites Cyprian Leowitz[49] and prefers his connecting mutations of kingdoms and empires with planetary conjunctions to Bodin's ascribing them to "the force and virtue of numbers."[50] He also cites such astrological authors as Junctinus and Cardan.[51]

But when the physician Senelles predicted from the horoscope of Louis XIII the death of the king in September, 1631, he was accused of lèse-Majesté together with Duval, another royal physician, condemned to the galleys, and his property confiscated.[52]

Similar circumstances and considerations moved Urban VIII, pope from 1623 to 1644, to issue a new bull against astrology. Father Morandi, as a result of astrological calculations and the fact that the pope would be in his sixty-third year or grand climacteric then, came to the conclusion that Urban would die in 1630. He submitted his reckonings to three friends, abbot Luigi Gherardi of Padua, Francesco Lamponi and Father Raffaelo Visconti, for verification or correction. The first two agreed with his conclusion, but Visconti thought that, if the pope did not leave Rome, he would live until 1643 or 1644, and on February 21, 1630 composed *Un discorso sulla vita di Urbana VIII*, which was communicated to many cardinals, prelates and diplomats. But the view of Morandi prevailed and drew various foreign cardinals to Rome in expectation of a conclave to elect Urban's successor. Morandi, despite his reputation

[47] For Arquato or Torquato, T IV, 467 *et seq*.

[48] At p. 59 he cites the prognostication of Nostradamus for 1564; at p. 85, that for 1563, etc.

[49] *Pleiades*, 1603, pp. 29, 98.

[50] *Ibid.*, p. 31. For Bodin's theory, T VI, 465.

[51] *Pleiades*, pp. 113, 121.

[52] Gui Patin wrote on October 28, 1631: "Le médecin Senelles qui estoit dans la Bastille pour l'horoscope du Roy, où il se promettoit que le Roy mourroit au mois de septembre, est condamné à perpetuité et ses biens confisquez au Roy: sa charge de médecin par quartier, donnée à un de nos compagnons nommé M. Baralis": *Lettres* (1907), 23. Senelles' real offence seems to have been that he was implicated in a plot against Richelieu and brought from Lorraine letters from an exiled lady-in-waiting of the queen, Anne of Austria. In 1643 his condemnation was commuted to exile, but he died soon after.

for personal piety and high standing in his Order, was imprisoned on July 13, 1630, and died of fever in November, whereupon all the other accused were set at liberty, except that Visconti was rusticated to Viterbo. Urban suspended the process against them on March 15, 1631, but on April first he issued the new bull against astrologers. And on April 22, 1635, for astrological predictions of the pope's death together with incantations, necromancy and sorcery aimed at his life, Giacinto Cantini, nephew of cardinal Felice Cantini, was decapitated, fra Cherubino da Foligno of the Order of Zocolanti and Fra Bernardino, called il Romito, were hanged and afterwards burned in the Campo di Fiori at Rome, while five other friars were condemned to various terms in the galleys.[53]

François de Cauvigny, who was related to Malherbe, published a refutation of judicial astrology in 1614, and the Sorbonne forbade its practice on May 22, 1619.[54] In the same year appeared at Paris *L'incertitude et tromperie des astrologues judiciares* of B. Heurtevyn.[55] After chapters on the dates of creation, of the end of the world, of the deluge, of the birth and death of Jesus, and on religious changes, with the aim to show that astrologers have disagreed or been wrong as to these, comes another series of chapters on incorrect past predictions by them. A long chapter on the failure of astrologers to foresee their own deaths is then followed by a shorter one arguing that the devil is the author of judicial astrology, after which the volume terminates in a series of chapters criticizing the Copernican astronomy and astrological technique. The book is not so well arranged or expressed as this brief summary might seem to suggest, and Heurtevyn matches the incorrect predictions of the astrologers by historical errors and exaggerations of his own, such as the assertion that all the astrologers of Asia, Africa and Europe, and Stoeffler in particular, had predicted a universal flood for 1524,

[53] Arturo Wolynski, *Nuovi documenti inediti del processo di Galileo*.., Florence, 1878, pp. 157, 160-63. The Franciscan, Candido Brognolo, in his *Alexicacon*, Venice, 1668, I, 36, ¶ 134, tells of a sorcery plot against Urban VIII which was revealed by one of the plotters, "qui omnes poenas tanto sceleri debitas Romae publice solverunt."

[54] *Correspondance du P. Marin Mersenne*, by Mme. Paul Tannery, Cornelis de Waard, and René Pintard, I (1932), 42.

[55] Copy used: BN V. 21820, xv, 182 pp.

whereas there was such a drought that it withered all the fruit.[56] We shall encounter other instances of misuse of history by opponents of astrology, indicating that they were special pleaders and not strictly accurate or judicial in their attitude.

The *Opuscula* of Julius Caesar Bulenger (1558–1628), a member of the Society of Jesus, printed at Lyons in 1621, include a work on all kinds of divination,[57] of which the main feature is an attack upon astrology. However, he grants that the weather, price of crops, and diseases may be predicted without superstition. He contends that Aristotle did not recognize other influences of the heavens than by their motion and light, but he admits that bodies which are placed under the sky so depend upon the celestial bodies that they cannot long persevere without them. Also he accepts the existence of occult virtues in inferiors, such as the softening of adamant by the blood of a goat, the lion's terror of the cock, Thessalians fascinating by laudation, the animal *catoblephes*[58] killing men from afar by its glance. He further admits that these occult virtues may be produced by the stars.[59] But he regards genitures as fallacious and forbidden, astrological elections as frequently fraudulent, and astrological images as forbidden and magical. The astrologers associate six religions with the relations of Jupiter to the other six planets, but there are actually over a hundred religions. They cannot predict contingent events, and it is ridiculous to say that the stars incline me to play or read or walk or drink at this or that time. But he does not say what does determine his choice in these cases. He makes use of previous authors a great deal: Sixtus ab Hemminga, Augustine and Favorinus, Cicero and John of Salisbury, Origen, Gregory of Nyssa and Eusebius. He denies that matter is prepared for forms by the heavens alone, or even that critical days are due to the moon. Astrologers could not have learned the forces of the stars from experience.[60] He further attacks various particulars of astrological technique, such as the *monomoeriae* of the Egyptians, *antiscia*,

[56] *Ibid.*, p. 10. For correction of this misstatement see T V, 181-82, 231-2.

[57] *Opuscula,* I, *de tota ratione divinationis.*

[58] *Catoblepas* in Pliny, NH, VIII, 32.

[59] *Opuscula,* I, 111, 144-45, 147.

[60] *Ibid.*, 111, 133, 135, 137, 141-42, 170, 175, 145, 148, 158.

conjunctions, and horoscopes for the founding of cities.[61] But there is little logical order, plan or structure to his arrangement and argument. He dismisses astrology as a vain and infidel art, puffed up with lies and day-dreams.[62]

Bulenger was born at Loudun and died at Cahors. He left the Order in 1594 to supervise the education of his brothers and nephews; taught at Paris, Toulouse and Pisa; then re-entered the Order in 1614. He was a doctor of theology and also wrote upon classical antiquities.[63]

Judicial astrology was condemned in French in a little book of 272 pages by Claude Pithoys (1596–1676), printed at Sedan in 1641[64] and reflecting a Huguenot point-of-view, as Pithoys had declared himself a Protestant in 1632, and taught philosophy at Sedan. The author is described on the title page as a theologian and professor of philosophy and law, and advocate *consultant* at Sedan. He lists a number of -mancies or forms of divination besides astrology, or "astromantie des genethliaques," and regards them all as relics of pagan darkness. He holds that judicial astrology is a magic art, condemned alike by God, canon law, the Fathers and theologians, civil law, philosophy, medicine and astronomy. Kepler and Tycho Brahe are represented as among its foes. It is pernicious to its practitioners, their employers, and the public at large. It attempts the impossible, for the stars cannot act upon the rational soul; the rules and methods of astrology are absurd and ridiculous; the art is not justified by experience, and its predictions are often false. Pithoys distinguishes five kinds of prediction: moral, political, natural, divine, and diabolical. His book was reprinted in 1646.[65]

[61] *Ibid.*, 139-40, and again at 164, 161, 168.

[62] *Ibid.*, 131. Therewith the first of his five books on divination ends, but he renews the attack upon astrology in *Liber II, Adversus genethliacos et mathematicos*, at pp. 132-90.

[63] Augustin et Alois de Backer, *Bibliothèque des écrivains de la compagnie de Jésus*, I (1869), 945-48.

[64] *Traitté curieux de l'astrologie iudiciaire ou Preseruatif contre l'Astromantie des Genethliaques.* Par C. Pithoys Theologien, & Professeur en Philosophie, & en Droit, & Aduocat Consultant à Sedan. A Sedan par Pierre Iannon, Imprimeur de l'Academie. The Columbia University Library copy (156.4 P682), then has the date MDCLXI on its title page, but this seems an error for MDCXLI.

[65] BM 718.d.8 (1.). 718.d.8 (2.) is the 1641 edition.

Another attack upon astrology from an authoritative quarter was made by Claude Saumaise or Salmasius, the great French classical scholar, in his *De annis climactericis et antiqua astrologia diatribae*, a long work published at Leyden in 1648. Gui Patin, in a letter to Spon of May 8, 1648, says that a bookseller in Paris who received twenty copies sold them all within four days.[66] The preface, which occupies most of the preliminary 64 leaves, is devoted to an onslaught upon astrology, which, however, continues to be the object of occasional criticism in the 844 numbered pages of the text proper.

The vanity and unreliability of astrology, Saumaise maintains, are shown by the disagreement between astrological authors, by the fact that the present art differs from the ancient, which is today unknown, and by the gross errors made by Arabic translators from the Greek and by medieval Latin translators from the Arabic. Moreover, the signs of the zodiac are mere creations of human imagination. The Chaldeans recognized only eleven signs, having no Libra, and Scorpio covering sixty degrees. Our zodiac is late and based upon Greek mythology, and much astrological detail is drawn from the fables of the poets. Tartars, Hindus and Chinese have other zodiacs. There probably are other stars, invisible to us, in the vast intervals which separate the named constellations, and so there is nothing solid and true about present celestial configurations. Salmasius also criticizes the division of the signs into decans. He notes that astrology was condemned by the Christian emperors and by many classical authors, and asks why it continues, when other forms of divination have disappeared.

If the stars are inanimate, they can produce only physical effects, not grammarians or rhetoricians or medical men or musicians or smiths or astrologers. How can they indicate future good or evil, when what is good for one man is evil for another? If their influence alters with their changing positions, they are evidently not gods, and Saumaise puts the dilemma: Either the stars are gods, or there is no astrology. Moreover, the telescope has shown us more than seven planets, and if the Copernican system is true, astrology is outlawed along with the Ptolemaic astronomy. Ptolemy makes the absurd statement that the planet Saturn is cold because so far from

[66] *Lettres* (1907), p. 592.

the sun, and is dry because it so far removed from the vapors which rise from earth, as if such vapors would reach any planet. After questioning whether the nativity should take into consideration the moment of birth or of conception or when the mother first feels the embryo moving, Saumaise asks why not begin to predict anew from each new period of human life, such as childhood, adolescence and youth. Or, if twins have different horoscopes, why should there not be a different nativity for the child who emerges head first, from that whose feet are first to appear?

In the body of the text, Saumaise holds that climacteric years depend upon the horoscope and so stand or fall with it, that sixty-three is not necessarily the grand climacteric, and that even twins have different climacteric years, as they do horoscopes. He argues that the art or science of physiognomy is possible without astrology, and that the tract on astrological medicine according to the position of the moon in the signs should not be ascribed to Hippocrates. He criticizes the use and meaning of technical terms like *hyleg* and *aphetes* at considerable length, and displays a fairly wide acquaintance with medieval Latin and Arabic authors, as well as with classical and more recent writers, upon astrology.

The arguments of Saumaise against astrology were, on the one hand, fresher and more original than were those of most opponents of that art, and, on the other hand, more historical and scholarly, attacking astrology from a factual and linguistic, instead of a primarily rational or religious, standpoint.

Saumaise might attack astrology, but he still believed that there were "marvelous secrets" of chemistry and medicine, and praised his friend, Johann Elichman of Silesia, for his knowledge of them.[67]

Of the chief French defender of astrology, Morin, we shall speak in a later chapter. But we may here adduce two or three instances that faith in it still prevailed widely towards the middle of the century, despite the attacks upon it which have just been noted.

The astrological point of view is still prevalent in the large folio history of the maritime world and events by Claude Barthélemy Morisot of Dijon, where the book was printed in 1643 with a dedi-

[67] Mersenne, *Correspondance*, III, 462.

cation to Louis XIII.[68] He affirms that many are called by the stars to a nautical life and naval victories, and that sailors are born rather than made. Among the Chaldeans Berosus, among the Greeks Eudoxus of Cnidus, Aratus, Aristotle and Empedocles, among the Egyptians Ptolemy, among the Latins Julius Firmicus and Marcus Manilius state that persons born under Pisces or the Dolphin are excellent sailors and divers. The lack of spleen is a great advantage to swimmers and divers. Under Aries are born sailors, towmen, and shipbuilders. Morisot even notes that among the seals of Solomon is a stone with the image of a ship under full sail, carved when the sun was in Leo, with Saturn and Mars to the south. One wearing it becomes a good sailor and fortunate in navigation. He who has Pisces in his horoscope will win naval battles, seek out new worlds, and be a wonderful shipbuilder, pilot, and forecaster of winds and tempests.

Father Octoul, who was a Minime, published *Inventa astronomica* at Avignon in 1643, with diffuse dedicatory epistles to the Virgin and Louis XIV, then a child of five.[69] The brief book is primarily chronological and states that the Church puts the birth of Christ 5199 years and nine months after creation. But it contains some astronomy and astrology: a catalogue of astronomical observations with the Tables of Lansberg, the construction of a *thema caelestis* for the observation of two sun-spots, and a discussion of the restitution of the celestial houses,[70] which is its chief, if not sole, astrological feature.

Jacques Alleaume was a pupil of Vieta and, although a Huguenot, prominent at Paris in mathematical and scientific circles. Snellius in his *Eratosthenes Batavus,* 1617, p. 103, said that he owed the mechanical division of the circle by compasses to "our most illustrious and ingenious friend, Jacobus Alealmus." Peiresc wrote with admiration of his burning glasses, his machine for shaping parabolic

[68] C. B. Morisot, *Orbis maritimi sive rerum in mari et littoribus gestarum generalis historia in qua inventiones navium, earundem partes, armamenta . . . urbes et coloniae maritimae . . . leges navales . . . venti . . . etc.,* Divione, 1643.

[69] *Rev. P. Stephani Octoul Minimi Inventa astronomica. Primae mundi epochae a priori constructae eodem tempore,* Avenione, J. Bramereau, 1643, in-4, 22, 95 pp.

[70] This last occupies pp. 56-63.

lenses, and other scientific apparatus. He drew up tables of longitudes and in 1624 is spoken of as royal engineer. He was a friend, too, of Paolo Sarpi, to whom he sent a manuscript copy of his work on perspective, which, however, was not printed until 1643, sixteen years after his death in 1627. Yet he also was interested in astrology and translated into French the sixteenth century book of Rantzovius or Rantzau on nativities, which had five Latin editions between 1597 and 1615. Alleaume's French version did not appear until 1657, but this shows that the interest in astrology still continued at that date.[71]

iii. ITALY

Turning now from France to Italy, we first note discussion of climacteric years and critical days. The work of Magini, professor of mathematics at Bologna, in 1607, on the astrological basis and use of critical days and on astrological medicine, has been already treated in a previous volume.[72]

Codronchi composed a defense of climacteric years in 1609, but deferred publishing it until 1620 in order to get others' criticisms of it first. Meantime at Padua in 1612 appeared the treatise of Ambrose Floridus, on climacteric years and critical days, "in which a marvelous doctrine, taken from sources astrological and philosophical, is revealed; how the whole course of human life, regulated by groups of seven years, is at diverse times seriously disturbed and distorted according to varied conjunctions of the planets."[73] The book is dedicated to cardinal Boniface Cajetan, and closes with the statement that, if anything has been said which does not conform to the edicts

[71] For the facts in this paragraph see *Correspondance de P. Marin Mersenne*, II (1936), Index, where there are over a score of page references to Alleaume. The earliest edition of the French translation listed there is 1657.

[72] T V, 250-51.

[73] Ambrosius Floridus Patavinus Augustinensis, *Tractatus de annis climactericis ac diebus criticis, dialogistico contextus sermone, in quo mira doctrina ex fontibus astrologorum excerpta et philosophorum panditur quo pacto totius vitae humanae cursus per septennarium annorum numerum regulatus secundum varias planetarum coniunctiones in diversis temporibus graviter exagitetur et torqueatur* . . . , Patavii, apud Matthaeum de Meniis, 1612, 4 fols., 43 pp., in-4. Copy used BN 4°.Te30.3. Also in BM 718.e.37 (7.).

Sudhoff (1902), 74, characterized the treatise as "völlig schematisch und abstrus."

of the Holy Roman Catholic Church, "that we completely reject and reprove as false." Floridus abandons any resort to Pythagorean theory of number in favor of a purely astrological explanation. The fifty-fourth year of one's life is very perilous because of the lordship of Mars, and the fifty-sixth year because of the rule of Saturn. Astrologers say that humidity reaches its height in one's twenty-first year, heat at forty-two, dryness at sixty-three, and cold at eighty-four. The last part of old age is governed by Venus whose mild and placid influence preserves men of that age in marvelous wise, so that they rarely die then, but the sixty-sixth year is dangerous for the phlegmatic, the sixty-eighth for the choleric, sixty-ninth for the sanguine, and seventieth for all temperaments. Then decrepitude or the last age of man sets in, which is under the rule of the sun to seventy-seven, and of Mars to eighty-four. After dealing with critical days, which are governed by the moon, Floridus asks what aspects and conjunctions of the planets are more fatal in the whole regimen of human life; why Saturn, when in the terrestrial triplicity, always portends some calamity, especially in one's sixty-third year, the grand climacteric, if in its own house and dignity. And if death comes during that year, one ought to render immortal thanks to God for His kindness in prolonging one's life that far.

In the next year Florido published a brief work on the sea and tides in the form of a dialogue in which a philosopher and a *Philonauticus* were the interlocutors,[74] and the tides were attributed to the influence of the sun as well as to that of the moon. This was likewise the contention of two other treatises, which were bound together with that by Florido in the copy that I consulted. One, by Marcus Antonius de Dominis, archbishop of Spalato, was addressed to Cardinal Barberini and printed at Rome in 1624;[75] the other, in Italian rather than Latin, by Sempronio Lancione, a Roman doctor of philosophy and theology, was addressed to the archbishop

[74] Ambrosio Florido, *Dialogismus de natura universa maris ac eius genesi et de causa fluxus et refluxus ejusdem atque de aliis accidentibus quae ejus naturam comitantur,* Padua, 1613, in-4, xvi, 43 pp. Copy seen: BN R. 3298.

[75] *Euripus seu de fluxu et refluxu maris sententia Marci Antonii de Dominis archiepiscopi Spalatensis ad illustrissimum principem Franciscum Barberinum S.R.E. Card. amplissimum,* Romae apud Andream Phaeum, 1624. Dedication and 72 pp. BN R.3297.

of Salzburg and apostolic legate, and printed at Verona in 1629.[76] De Dominis in his treatise speaks favorably of the aspects of the planets.

Giulio Cesare Claudini, who taught the practice of medicine at Bologna from 1578 to 1618, had questioned the astrological explanation of critical days in 1612, but Hyppolitus Obicius, in the appendix to his *Iatroastronomicòn* of 1618, held that Claudini had not rightly understood Galen. The work of Claudini, however, was printed again, this time at Basel, in 1620.[77]

The work which Silvaticus published in 1615 against the doctrine of climacteric years[78] strikes one as labored and inferior to that of 1605 on the unicorn, bezoar stone, emerald and pearls, of which we treat elsewhere. There is excessive citation of Hippocrates, Galen and other ancients, while his list of recent writers on the subject[79] does not include the treatise by Ambrose Floridus of 1612, or even that by de Rossi of Sulmona back in 1585.[80] The most recent book cited by him is that of Federigo Bonaventura on the eighth month's child from the previous century.[81] Sometimes it is adduced in favor of Silvaticus's own contentions. Thus he says that Federicus Bonaventura, a most erudite and learned writer in the fiftieth chapter of *De octimestri partu* ridiculed and confuted those who placed the cause of climacteric years in numbers and boasted that he had found an evident natural cause.[82] In another passage Silvaticus says that he will not go into the astrological argument for climacteric years

[76] *Trattato sferico nel quale con dimostrative ragioni si discorre del flusso e riflusso del mare,* di Sempronio Lancione . . . etc., Verona, 1629, 34 pp. BN R.3299. Also bound in the same volume (BN R.3300, 3301, 3302) are treatises on the tides of the sixteenth century: Nicolo Sagri, *Ragionamenti sopra le varietà de i flussi et riflussi del mare oceano occidentale,* Venice, 1574, 105 pp., in Italian; Pandulfo Sfondrato, *Causa aestus maris,* Ferrara, 1590, to Gregory XIII, 44 fols.; *Dialogo . . . d'Alseforo Talascopio,* Lucca, 1561, about 67 pp.

[77] Other works on critical days were: Edm. Hollyng, *De crisibus et diebus criticis,* 1606; Pietro Castelli, *De abusu circa dierum criticorum enumerationem,* Messina, 1642; Pietro Cortesio, *De diebus decretoriis,* Palermo, 1642, in-8.

[78] Jo. Bapt. Silvaticus, *De anno climacterico,* Pavia, 1615, 94 pp. Copy used: BM 784.d.5 (3.).

[79] *Ibid.,* p. 18.

[80] T VI, 139-40.

[81] *De octomestris partus natura adversus vulgatam opinionem Peripatetica disputatio,* Urbino, 1596 in-fol.; Francof., 1601; Venice, 1602.

[82] Silvaticus, *De anno climacterico,* 1615, p. 9.

because Pico delle Mirandola and many since have condemned it, and Bonaventura has demonstrated more particularly that Mercury is not their cause.[83] But in Bonaventura's work itself we find him defending Galen on critical days against Pico, Fracastoro and others, while the evident natural cause which he had found was the stars.[84] And Silvaticus says towards the close of his treatise: "I say against Bonaventura . . . that, just as critical days do not mark diseases by reason of number as number, as has been demonstrated, so neither have numbers any force in climacteric years."[85]

Baptista Codronchi, whose credulous book on witchcraft and cures therefor had appeared in 1595,[86] finally published his treatise on climacteric years and how to avoid their dangers at Bologna in 1620. It was also printed at Cologne in 1623,[87] and at Ulm in 1651. For several pages he gives lists of the names of men who have died at such ages as 94, 77, 63, 49 and 42.[88] But one reason for his writing the book is his belief that climacteric years are not merely dangerous but may mark a change for the better in one's health.[89] He argues that Hippocrates believed in climacteric years, treats of their causes from astrologers, whose doctrine he defends, and then from philosophers and medical men, and finally answers a celebrated recent writer who had attacked them, possibly Silvaticus. The second part of his book then deals with avoidance of their perils.[90] Codronchi included even the patriarchs of the Old Testament among his examples, reckoning their grand climacteric as 910, or seven times 130, instead of 63, which is seven times nine.

Septimius Columbus, a member of the Academy of *Insensati,* in 1625 addressed to Cardinal Francesco Barberini a brief tract on

[83] *Ibid.,* pp. 19-20.

[84] *De octomestri partu,* VI, 6-46; cited by Joh. Ant. Magini, *De astrologica ratione ac usu dierum criticorum . . . ,* Venice, 1607.

[85] *De anno climacterico,* p. 90. The treatise ends at p. 94 with the words, ". . . Quantam itaque vanitatem habeant anni climacterici dicti ex his arbitror constat manifeste."

[86] T VI, 544-47.

[87] Baptista Codronchius Imolensis, *De annis climactericis necnon de ratione vitandi eorum pericula itemque de modis vitam producendi,* Coloniae sumptibus Matthaei Smitz, 1623. This is the edition I have used: BM 1038. e.17.

[88] *Ibid.,* 12-20.

[89] *Ibid.,* Preface, also pp. 29-30.

[90] *Ibid.,* 116-68. On Codronchi see Dr. Giuseppe Mazzini, *Di Battista Codronchi, medico e filosofo Imolese* (1547-1628), Terni, 1924, pp. 3-26.

climacteric years.[91] He contended that they were not superstitious, but were supported by both authorities[92] and experience, and by analogy with critical days in disease and with the division of man's life into seven-year periods, marked by teething at seven, puberty at fourteen, and so on. The most reverend bishop (of Volturara), Simon Maiolus, in his work on dog-days had already listed six hundred instances of death in climacteric years.[93] There was considerable difference of opinion as to which climacteric year—49, 63 or 81—was most critical and perilous, but Columbus regarded 63 as the most crucial. He also, although writing as a philosopher and physician rather than astrologer, and suggesting other possible causes for climacteric years, tended to select the stars as the chief cause, the order of the planets reverting every seventh year to Saturn. But the fact of their existence was enough for him.

We turn from the subject of climacteric years to other writings for or against astrology by Italians, and shall treat of those from the same city together.

Ilario Altobelli of Montecchio in Piceno received the doctorate in 1591, was made historian (*chronologus*) of the Franciscan Order in 1617, and died in 1628.[94] He also was interested in astronomy.

[91] MS VAb 284. On the illuminated title page: "Perbrevis tractatus de annis climacteris illustrissimo ac reverendissimo D. Francisco Barberino S.R.E. Cardinali amplissimo a Septimio Columbo accademico insensato compilatus."

Below in a tiny hand: "Franciscus Pallantes Cappellanus Triremis S. Sebastiani scrib. et miniab. anno domini 1625." After a dedicatory paragraph on fol. 1r, the text begins on fol. 2r and ends at fol. 14r. Although the MS is a large paper octavo, the handwriting is so large that there are only 17 lines to a page.

[92] He lists a large number at fols. 6v-7r.

[93] Simone Majoli lived from 1520 to 1597. His *Dies Caniculares, seu colloquia tria et viginti quibus pleraque naturae admiranda quae aut in aethere sunt aut in Europa Asia atque Africa, quin etiam in ipso orbe novo et apud omnes Antipodes sunt . . .*, first published at Rome in 1597, was a very miscellaneous work on nature, as the title just quoted shows. Probably on this account it was all the more popular, being reprinted at Ursel in 1600, Mainz in 1607 in one volume of 780 pp., 1610-1612 in three volumes, and 1614 in one folio volume, Paris in 1610 in French translation, and at Mainz and Frankfurt, 1615-1619, in 7 vols.

I examined the 1607 edition, which has a slightly different title, *Dies caniculares, hoc est, Colloquia tria et viginta physica . . . etc.*, on September 12, 1928, but by inadvertence failed to give any account of the work in my fifth and sixth vols.

[94] Wadding.

He wrote on the new star of 1604,[95] in 1610 addressed a letter from Ancona to Galileo on the satellites of Saturn,[96] and in 1615 published a treatise on the occultation of Mars.[97] A letter in Italian by him, in which he opposed the Aristotelian doctrine of comets, was printed at Venice in 1627.[98] In the last year of his life appeared his Tables for dividing the heavens into the twelve signs,[99] and the year following a Demonstration that Regiomontanus's method of directions and determining astrological houses did not agree with that of Ptolemy.[100] The last two works also indicate an interest in astrology, and a prediction from the stars by him in 1607 as to the destiny of the republic of Venice is preserved in a manuscript at Paris,[101] and has already been mentioned in connection with the motion of Mars in our chapter on Kepler. The use of the words *inclinatio* and *conjectura* in its title show a desire to avoid any appearance of attributing fatal necessity to the influence of the stars. At the same time horoscopes or *themata coeli* are given for the foundation of Venice about noon on March 25, 421; for the great conjunction of Saturn and Jupiter at 10.32 P.M., December 19, 1603—which is compared with that of March 2, 411, which preceded the founding of Venice; for the first appearance of the new star at Verona at 5.35 P.M., October 9, 1604; for the solar eclipse at 2.36 P.M., October 12, 1605; and for Leonardo Donato, the present doge of

[95] Filippo Vecchietti, *Biblioteca Picena, I* (1790), p. 89, lists it as printed; Riccardi, *Bibliotheca matematica italiana*, says it was not. All of Altobelli's works are very rare, not being listed in the printed catalogues of either BM or BN.

[96] Houzeau et Lancaster, *Bibliographie générale de l'astronomie*, Brussels, 1882-1889, II, 1443.

[97] *De occultatione stellae Martis*, 1615. Wadding, Vecchietti, Riccardi and Mazzuchelli give no further details as to place of publication etc.

[98] *Nova doctrina contra opinionem Aristotelis circa generationem cometarum epistola*, italice, Venet., typis Jacobi Sarzinae, 1627. Noted by Wadding only.

[99] *Tabulae regiae divisionum duodecim partium caeli et syderum obviationum ad mentem Ptolemaei*, Maceratae, typis Io. Baptistae Bonomi, 1628, in-4, with a full page portrait of Altobelli at the close. Riccardi notes a copy at Siena; it was also recently offered for sale bound with the tract mentioned in the following note.

[100] *Demonstratio ostendens artem dirigendi et domificandi Ioannis de Monteregio non concordare cum doctrina Ptolomaei*, Foligno, apud Augustum Allerium, 1629, in-4, 12 fols.

[101] BN 7452, fols. 1-45r: "De proxima reipublicae Venetae inclinatione ex astris rita solidaque coniectatio multiplex. Annum Domini MDCVII de mense Ianuarii."

Venice, and it is predicted that he will meet with a violent death either in 1611 at the age of 75 or in 1614 at 78. Four unfavorable astrological directions are also noted between 1607 and 1612, and it is held that two very strong movements, never before made in the heavens since the origin of Venice, are about to threaten its destruction. God is angry at the violation of religion by Venice, but if she amends her ways, she will be renewed like the eagle; "alioquin ad extremum." It is noteworthy that the work is addressed to one of the cardinals.[102]

Father Redento Baranzani was a Barnabite from Vercelli. His *Uranoscopia* was described in the long Latin title as "a new work, necessary, pleasing and useful to natural philosophers, astrologers, medical men and all professors of good arts,"[103] and his students seem to have set great store by it. A letter by John Baptista Murator, dated at Annecy on February 20, 1617, states that, because of war, he had left his native place for Annecy, Savoy, where there was a school of the Paulist fathers. But when his teacher in philosophy there, Baranzani, came down with fever, he feared that hope of a path-breaking, methodical work of philosophy, and especially astrology, by the intervention of his genius, was gone. But Baranzani recovered and dictated solely from memory his very rare and out-of-the-way (*peregrinae*) opinions, seldom heard in courses of philosophy, although Murator realizes that the space of only two months cannot attain the heights of his teacher's archetype, and he has had to omit most of the citations.[104] A few pages later is given another letter from Ludovicus des Hayes of Paris,[105] who is also editing the work. And when the second part begins with a new pagination, there is another prefatory letter by both Murator and Hayes, dated at Annecy on March 28, 1617, in which they explain that Baranzani followed a twofold way in saving all the celestial

[102] *Ibid.*, fol. 2r, "Illustrissimo et reverendissimo D. D. Alfonso Vicecomiti S. R. E. Cardinali Agri Piceni legato . . ." signed at fol. 4r, "Humillimus et obsequentissimus servus et subditus f. Hilarius Altobellus."

[103] *Uranoscopia seu de coelo in qua universa coelorum doctrina . . . Opus novum philosophis naturalibus astrologis medicis et omnibus bonarum artium professoribus necessarium iucundum et utile.* Authore R. P. D. Redempto Baranzano Vercellensi. Coloniae Allobrogum (i.e., Geneva) Apud Petrum et Iacobum Chouet, 1617.

[104] *Ibid.*, I, 4-6.

[105] *Ibid.*, I, 13-14.

phenomena, following Copernicus in some respects and Aristotle in others.[106] Two years later, Baranzani himself, in his New Opinions in Physics, advises reading his *Uranoscopia* which has been printed again in an enlarged and revised edition at Paris by his dearest disciple of subtle genius, Ludovicus des Hayes.[107]

The book of Baranzani is more astrological than either philosophical or astronomical. An astronomical foundation is laid, but then astrological definitions and rules are given, the substantial influx of the celestial bodies is set forth, and such questions are discussed as whether metals, herbs and plants are produced by the heavens, whether celestial influences are impressed in an instant, how long they last, whether the sky is the cause of some fortuitous events. Furthermore, whether celestial form is nobler and more perfect than any other, whether the heavens are the cause of animals born of putridity, whether there is any occult force from the heavens, whether astrologers can divine human actions and how, whether uncertain knowledge is prohibited by recent canon law? In the second part, with a new pagination and beginning with the empyrean heaven, it is soon asked whether in the primum mobile there are triplicities, houses, exaltations, and other dignities of the planets; and, with regard to the heaven of libration or tenth sphere, whether it is visible, influences sublunars, and what that influence is. Soon the question is raised how to draw up a horoscope, and it is noted that Firmicus, Avenezra, Campanus, Alcabitius and Ptolemy differ as to this. After some discussion of direction, *significator* and *pro-*

[106] *Ibid.*, II, 2-3. For Copernicus see I, 102, "Dubitatio 10 de ordine coelestium spherarum. Membrum i, Quid sentiat Nicolaus Copernicus" (Baranzani says that it is difficult to tell); p. 106, "Membrum ii, Quibus fundamentis innitatur Nicolaus Copernicus"; p. 107, "Membrum iii. Quomodo solvantur ea quae proponuntur contra Copernicum"; pp. 112-13, "Membrum iv, Quomodo solvantur argumenta Copernici" (very unconvincing). Also II, 119, Tables of longitude and latitude to 1620 according to Tycho and Copernicus; II, 152, "Quaenam sit Copernicea sphaerarum revolutio?"

Tiraboschi, *Storia della letteratura italiana*, VIII (1824), 346, following Mazzuchelli, says that, when Baranzani learned that his presentation of the Copernican system was displeasing to the pope, he added in closing a refutation of it. I did not see this in the edition of Geneva, 1617; perhaps it was added in the Paris revision by des Hayes.

[107] *Novae opiniones physicae*, Lyon, 1619, p. 145.

missor, pages 66-122 and, after discussion of the planetary spheres, pages 153-176 are devoted to Tables, some of which are astrological, such as the regions subject to each sign of the zodiac, and of diseases under each planet. Then comes what is called "a last question," whether the fixed stars exert more influence than the planets, the superior planets than the inferior, and the moon than any other planet except the sun. But it is followed by one more question, whether all the planets make critical days. Appendix I then deals with climacteric years, and Appendix II inquires whether the sun is the center for Saturn, Jupiter, Mars, Venus and Mercury. Here Tycho's hypothesis is called more improbable, because that of Copernicus is simpler. This is immediately followed by the theory of the planets according to Copernicus and Magini reduced to Tables. A *Prooemiolum* then says:

> While Tychonic Tables are being worked out, Keplerian already happily initiated from the movement of Mars are being perfected, Marianae against Copernicus are awaited, others of Magini with a view to restricting Venus, Mercury and the sun to a single sphere are being prepared, and many others are being composed by the most learned mathematicians of this astrological age, try these (of mine) too, though they may not agree with the proximity of the perigee of Mars to the earth.

There follow tables of the effects of the moon on agriculture, of prognostics from thunder attributed to Bede, of signs of rain, and of fair weather; a golden booklet instructing how to draw up an Almanach and predict crops and sterility, and yet other appendices.

A seventeenth century manuscript at the Vatican contains a defense of judicial astrology by a Ferrante de Septem,[108] which seems to have been composed soon after the opening of the century. Astrology rests upon the assured basis of the influence of the heavens over inferior bodies. "It is certain that the sun by heating, the moon by moistening, Saturn by chilling, and Mars by drying, are natural

[108] Barberini Latin 231, fols. 75r-77r, opening, "Frustra astrologiae rationes defendendas susciperem . . ." and closing, ". . . nedum Christiano vero indigna sed cuique optimo ac religiosissimo convenire." Septem. may be an abbrevation for a longer place name.

causes." Consequently an astrologer may predict as to length of life and the constitution (*complexio*) of the human body. "For those effects neither happen fortuitously nor are dependent upon the air." If they were, persons born in July would be of a hot nature, and those born in January would be very cold. But popes Urban VII and Innocent IX came into the world while the sun was in Leo, yet were very cold by nature, while Sixtus (V) and Clement (VIII), whose horoscopes were in winter while the sun was in Capricorn, were of warm and moist physical constitutions. Astrologers cannot predict purely rational effects and can note only inclination in those mixed events dependent on both soul and body. It appears difficult for them to forecast a violent death or the hour of death, yet Ptolemy treats at length of violent death and was borne out in the cases of Pier Luigi Farnese (natural son of Paul III who was killed on September 10, 1547) and Sebastian, king of Portugal (slain in the battle of Alcazarquivir, August 4, 1578). Many writers, like Pico della Mirandola, Sixtus ab Hemminga, and Benedictus Pererius (1535–1610), do astrologers an injustice by accusing them of what they do not teach or of what they even reject. One should observe astrological conditions in all one's actions. The Bull of Sixtus V against astrology is no deterrent to Ferrante. Ptolemy may seem to predict with certainty events which are dependent upon human free will, but he says to begin with that the stars only incline and do not necessitate. Ferrante's closing words are that not merely is astrology not unworthy of a Christian but that it is fitting for every good and pious man. His defense is followed in our manuscript by astrological figures for such years as 1552 and 1536, or, more recently, 1603 and 1604, 1606 and 1607. These are sometimes accompanied by daily statements of the weather.

An astrological prediction for the year 1618, which was addressed by Gioanni Bartolini to the Cardinal of Santa Susanna, Scipione Cobelluzzi, is preserved in another manuscript at the Vatican.[109] It is written in Italian and calculated for the meridian of Rome according to the observations of Tycho Brahe. The four seasons of the year are taken up in turn, beginning with winter, which is to

[109] Vatican Latin MS 6304, a little paper pamphlet of 44 leaves.

open on December 22, 1617, at 22.27 and 23 seconds, and with an astrological figure for the moment when each season opens. Prediction is limited to the weather, which is noted for each month, adding changes dependent upon the fixed stars,[110] and to agriculture, crops and vintages, navigation—including days unfavorable for sailing,[111] sickness and medicine, and such astronomical occurrences as eclipses. Bartolini thus keeps roughly within the limitations fixed by the Bull against astrology of Sixtus V in 1586.

Giovanni Antonio Giuffi of Palermo dedicated "these my astrological lucubrations concerning eclipses" in March, 1621.[112] They appeared in print at Naples in 1623.[113] The work purports, according to its full title, to discuss what should be considered in the prognostication of eclipses, when they begin to produce their effects, how long they will last, and when their influence will be at its height. How to find the lord or *dispositor* of the eclipse, in what lands and provinces its effects will occur, and in what sort of things, and if it will be good or bad? And what it signifies in each sign and house, "and some other points worthy of consideration." Ptolemy and Haly are much cited.[114] Each planet is taken up for each sign of the zodiac and what it signifies when lord thereof.[115] Many ills follow, when there is both a solar and a lunar eclipse in one month, especially in those places for which they have especial significance.[116] The work is accompanied by Tables. At its close Giuffi refers to the Bull of Sixtus V against astrology and explains that he is writing only for physicians, sailors and farmers and so is not violating it.

Alexander de Vicentinis, in a treatise on heat and the influence of the heavens, printed at Verona in 1634,[117] "never departing from the principles of Aristotle," was not ready to admit that heat was "of the substance of the heaven." He concluded that motion was the

[110] *Ibid.*, fol. 35r, "Delle Mutationi de Tempi che d'pendono dalle stelle fisse piu insigni et verticali."

[111] *Ibid.*, fol. 40v, "Giorni Cattivi pro Navigare."

[112] "12 Idus (*sic*) Martii."

[113] *Tractatus de eclipsibus*, Neapoli, Oct. Beltrani, 1623, in-4, 174 pp. Copy used: BM 531.k.9 (3.).

[114] *Ibid.*, pp. 57-61 especially.

[115] *Ibid.*, pp. 94-100.

[116] *Ibid.*, p. 2.

[117] *De calore per motum excitato et de coeli influxu in sublunaria corpora*, Verona, 1634, in-4: BM 549.e. 13 (1.).

cause of heat and that no one, unless rash and demented, would deny that the heavens by light and especially by motion exerted great influence upon these inferiors. But he held that the heavens influenced only by light and motion, denied the existence of occult qualities, and rejected the details of astrological technique, of which he thought that Pico della Mirandola had sufficiently exposed the vanity.[118] He held that the will was free, the mind divine, education potent, and that hence astrological predictions concerning individuals were undependable.[119] He denied the contention of astrologers that dreams were caused by the stars, and the opinion of Albertus Magnus—and Dante—that a continuous effluvium from a celestial form affected the imagination of the dreamer, and the view that the Intelligences which moved the heavenly spheres were responsible for divining dreams. He seemed to think, however, that some divination from dreams was possible.[120] In his final chapter, devoted to an argument that the fact that different things were produced in different places was not to be ascribed to occult virtue, he noted that Aristotle attributed spontaneous generation to the force of the heavens. But he denies that southerners are timid and short-lived because the planet Saturn rules over them, and northerners bellicose and long-lived because Mars governs them. The southerners are short-lived because the necessities of life are wanting there. "Also Saturn rules in India, and yet there they are long-lived." With which double-faced talk he terminates the treatise.[121]

A prediction for the year, 1607, published by Lodovico Bonhombra at Bologna, takes up the four seasons of the year in turn, then in a closing paragraph states that the inclinations of the stars over this year are subject to answer to prayer by Divine Majesty, also to human prudence, and in all things subject to the Holy Mother Roman Catholic Church.[122] Giovanni Antonio Roffeni of Bologna did not die until 1643 and had issued annual predictions for some

[118] *Ibid.*, pp. 6, 66, 70, 83, 91, 122-23.

[119] *Ibid.*, pp. 171-72, 174-75, 179-81.

[120] *Ibid.*, cap. viii, "Quomodo ex insomniis divinatio contingit," pp. 183-202.

[121] *Ibid.*, pp. 218, 229.

[122] *Discorso astrologico delle mutationi de' tempi e de' più notabili accidenti sopra l'anno 1607*. Copy used: BM 4051.c.5 (10.).

thirty years before.[123] Orlandi said that his explanation of meteorological matters extended to the year 1660. In 1614 he had published a work praising true astrology and against its calumniators.[124] He corresponded with Kepler and Galileo, and defended the *Sydereus Nuntius* of the latter against the Bohemian, Martin Hork.[125] Nor did astrological prediction at Bologna cease with Roffeni's death, since nine different *Practica's* for the year 1648 alone were issued there.[126]

Cornelio Ghiradelli of Bologna, a Franciscan and a member of the Academy of *Vespertini*,[127] published *Discorsi Astrologici* from 1617 on for a number of years,[128] also Considerations on the Solar Eclipse of 21 May, 1621, Astrological Observations on the weather in 1622, a weather prediction for the year 1623, and a tract on leap-year[129] of 1624, while a prediction for 1634 is found in the Bibliothèque Nationale, Paris.[130]

Buonaventura Cavalieri, professor of mathematics at Bologna

[123] P. A. Orlandi, *Notizie degli scrittori Bolognesi e dell'opere loro stampate e manoscritte* (1714), 150. P. Riccardi, *Biblioteca matematica Italiana* (1893), 387, specifies those for the years 1612, 1617, 1618, 1619, 1621, 1624, 1630, 1641, 1643 and 1644 (1642?), and had seen those for 1621 and 1642: *Discorso astrologico delle mutationi de' tempi e altri notabili accidenti dell' anno MDCXXI*, Bologna, Bart. Cochi, 1621, 4to; *Discorso astrologico delle mutationi de' tempi e d'altri accidenti dell' anno 1642 di Gio. Antonio Roffeni*, etc., Bologna, presso Gio. Battista Ferrari, 1641, 4to. BM 718.f.38 (2) is that for 1619: *Discorso astrologico delle mutationi de' Tempi e d'altri accidenti dell' anno 1619*, del dottore G. A. Roffeni, Bologna, 1618, 4to.

[124] *De laudibus verae astrologiae et adversus eiusdem calumniatores*, Bologna, 1614. BN Rés. V. 1614.

[125] *Epistola apologetica contra coecam cuiusdam Martini Horchii peregrinationem circa Sydereum Nuntium excellentissimi Galilaei Galilaei*, Bologna, H. Rossi, 1611, 4to.

[126] G. Hellmann, *Neudrucke von Schriften und Karten über Meteorologie und Erdmagnetismus*, XII (1899), 28, n. 6.

[127] He so styles himself in his *Cefalogia fisonomica*, 1630.

[128] Orlandi says for about twenty years, but the dedication by the printers of *Cefalogia fisonomica*, dated 10 Nov. 1630, in which they speak of printing the work lest oblivion triumph over the virtuous labors of Ghiradelli and of dedicating it to the hereditary chamberlain of the archduchy of Austria in order to assure Ghiradelli's literary immortality, seems to indicate that he had died before that date. However, the prediction for 1634 disproves this.

[129] Orlandi, *Notizie*, 94.

[130] *Partimento delle quattro stagioni del presente anno 1634 astrologicamente dedotto dalle cause celesti*, Bologna, C. Ferroni, 1634, 4to. 31 pp. BN Rés. V. 1240.

and who was praised by Galileo, together with a Hundred Varied Problems to Illustrate the Use of Logarithms, published a New Practice of Astrology at Bologna in 1639.[131] The license of his ecclesiastical superiors is dated July 31, 1636, and the dedication on April 1, 1637. The book deals especially with astrological directions and was composed for some of his students who, having visited his *Direttorio Uranometrico,* wished to use logarithms in finding directions. There are chapters on finding declinations, right ascensions, *significator* and *promissor* as well as directions.

> The *significator* among astrologers is called that point, place or star in the celestial sphere which carries the lordship and signification of anything. Just as the *promissor* is that which promises any accident when it reaches the site of the *significator.*

Five customary *significatores* are the sun, moon, ascendent, zenith and *Pars fortunae,* adding the governing planet. *Promissores* may be the planets, their aspects, *termini, antiscii, contra antiscii,* fixed stars, beginnings of houses, and so on.[132]

Alphonsus Pandulphus, bishop of Comacchio, had died in October, 1648,[133] but his Disputations as to the End of the World were not printed until ten years later at Bologna.[134] Since Raphael Aversa is cited in them, they would seem to have been written after 1625. Of these eight disputations, in which the views of philosophers were refuted and evangelical and prophetic doctrine alone received, only

[131] *Nuova Prattica astrologica . . . con una centuriata di varii problemi e con il compendio delle regole de' triangoli*, Bologna, 1639, in-8. BN V. 18405. The latter work has a new title page and pagination–526 pp. He states that he employs a different kind of logarithms than Kepler.

[132] *Ibid.,* p. 25. The *Nuova Prattica* of 133 tiny pages is followed by an *Appendice della Nuova Prattica astrologica . . . un Exemplare di fare le Direttione secondo la Via Rationale,* Bologna, 1640, with a new pagination of 108 pp. Then follows *Centurie di varii problemi per dimostrare l'uso e la facilità de' Logaritmi nella Gnomonia, Astronomia, Geografia, Altimetria, Pianimetria, Stereometria et Aritmetica Prattica . . . ,* Bologna, 1639, 526 pp.

Bouché-Leclercq, *L'astrologie grecque,* 1899, employs the neuter form, *antiscia,* but most seventeenth century writers seem to prefer *antiscii.*

[133] Eubel, *Hierarchia,* IV (1935), 157.

[134] *Disputationes de fine mundi . . . Opus posthumum,* Bonon., 1658, a double-columned folio of 383 pages: copy used, BM 526.m.9 (1.).

that on astrology will concern us, omitting the Pythagorean, Platonic, Aristotelian, Stoic and astronomical discourses which precede it, and the scriptural and theological sections which follow it. In a *Disputatio Prooemialis* Pandulphus lays claim to novelty of matter and treatment, but there is nothing very new in his arguments against astrology. He denies occult qualities to the heavens or that they are the cause of metals. The empyrean heaven does not act upon the inferior world; the heavens influence the intellect only indirectly; and critical days do not depend upon the moon. No change of kingdoms and empires may be inferred from the movement of stars from one sign of the zodiac into another, and astrological houses are rejected.[135]

At Padua, Giacomo Filippo Tomasini (1595–1655?) both praised and practiced astrology. The first work listed under his name by Vedova[136] is on the Revolution of the Year for 1614, 1615 and 1616. In his Eulogies of Illustrious Paduans, first published in 1629, Tomasini praised past devotees of astrology from Peter of Abano on, or affirmed, in the case of opponents like George of Ragusa, who died in 1622 at the age of only forty-three,[137] that their deaths had been accurately forecast by that art, as has been occasionally noted in our previous volumes. Another instance, not previously noticed, was that of Hieronymus Capivaceus, who taught medicine at Padua from 1552 to 1589. When he was an old man, an astrologer advised him not to undertake any journey, but he persisted in going to Mantua to give medical attendance to its reigning prince. On his return he was suddenly taken ill and died.[138]

The famous Aristotelian philosopher, Cesare Cremonini (1550–1631) taught at Padua from 1590 to 1631. There is a manuscript in the library of St. Mark's at Venice dated 1628 of a treatise or lectures

[135] *Ibid.*, pp. 220, 234, 223, 225, 236, 292, 296.

[136] Giuseppe Vedova, *Biografia degli Scrittori Padovani*, Padua, 1832-36, II, 355.

[137] T VI, 201-2.

[138] Tomasini, *Elogia illustrium virorum . . .*, 1630, p. 90. According to *Correspondance*, III, 531, n. 1, Jacopo-Filippo Tomasini was born at Padua in 1597 and died at Città-Nuova in 1654; "son évêché lui avait été donné par Urbain VIII," who, despite his bull of 1631 against astrology, seems to have overlooked or condoned Tomasini's attachment to that art.

Other spellings of Capivaceus are Capivaccius and Capovacceus: in Italian, Girolamo Capo di Vacca.

on the influence of the heavens by him which opens with the assertion that it is a tenet of Aristotle that sublunars are governed by the heavens.[139] The manuscript is so poorly written, with a number of passages crossed out and marginal summaries or substitutions, that it is difficult reading.[140] But it asserts that the first and last efficient cause is no other than the heavens, by which the elements are constituted, and inquires how motion heats, and especially how the motion of the sun acquires the power of heating by motion.[141] Earlier Jean de Jandun came in for considerable criticism.[142] Other works by Cremonini display a similar attitude as to the influence of the heavens. He maintained the view of Aristotle that they were a fifth substance, distinct from the elements,[143] and that there were movers of the heavens.[144] In a treatise on these Intelligences in a manuscript now at Florence he says that the treatment of separate substances is difficult because they are not perceptible to the senses. After setting forth the Peripatetic doctrine, he adds that the true opinion concerning all separate substances is to be had from the theologians.[145]

Cremonini is said in *Naudaeana* to have lived in a magnificent palace at Padua with a maître d'hotel, valets de chambre and other servants, two coaches and six fine horses. When he died, he left four hundred scholars and two thousand crowns of securities.[146] The Inquisition more than once took exception to the teaching of Cremonini, but he insisted that he was merely setting forth the

[139] MS VI, 176, a. 1628, 58 fols. "Celo sublunaria gubernari est Aristotelis propositio . . ." The Meteorology and *De celo* are cited in confirmation.

[140] It appears to divide into three parts or sections of five, six and six chapters respectively.

[141] *Ibid.*, fols. 28v, 38v, 45v, 47v.

[142] *Ibid.*, fols. 24v, 25v, 31v.

[143] *Apologia dictorum Aristotelis de quinta coeli substantia adversus Xenarcum, Joannem Grammaticum et alios*, Venetiis, 1616, in-4.

[144] *Disputatio de coelo in tres partes divisa: De natura coeli, De motu coeli, De motoribus coeli abstractis*, 2 vols., Venice, 1613.

[145] Florence, Bibl. Naz. Palat. 879, 17c, fols. 3r-133v: "Tractatio de substantiis abstractis est difficilis quia est tractatio de ente quod est ultra sensum . . ./. . . debent accipi tamquam ex principiis Peripateticis; veram sententiam omnium istorum a sacris theologis accipite."

[146] *Naudaeana et Patiniana, ou Singularités remarquables prises des conversations de Messieurs Naudé et Patin.* By Ant. Lancelot, published by P. Bayle, Paris, 1701, in-8, p. 45.

philosophy of Aristotle, as he was hired to do, and, in the case of objection to his *Apologia* of 1616, that the book had already been approved by the Doge and Senate of Venice, and could not be altered.[147]

The years of the birth and death of Andreas Argolus or Andrea Argoli are given by Zedler as 1570–1651,[148] by Sudhoff[149] as 1568–1657. He was born at Taglіacozzo in the Abruzzi, was a student with Magini, and taught Wallenstein and his astrologer Giambattista Zenno at the University of Padua. Perhaps his earliest extant or recorded printed work was Tables of the Primum Mobile with the particular purpose of more easily determining astrological directions.[150] Besides other Tables based upon the hypotheses of Tycho Brahe,[151] Ephemerides for the years, 1631–1700,[152] a dissertation on the comet of 1652–1653, and a *Pandosion sphaericum,*[153] he composed a work on critical days in two books, of which Sudhoff has already given some account, and which we shall further consider here as an example of the continued prevalence of astrology in the seventeenth century.[154] The title, *De diebus criticis etc.*, somewhat obscures the real character and content of the work, which is concerned chiefly with astrology in general and astrolo-

[147] See D. Berti, *Di Cesare Cremonini e della sua controversia coll' Inquisitione di Padova e di Roma, Mem. de l'Acad. dei Lincei, Scienze morali etc.*, III (1877-78), tom. ii. L. Mabilleau, *Cesare Cremonini,* Paris, 1881. A. Favaro, *Cesare Cremonini e lo studio di Padova,* Venice, 1883. J. R. Charbonnel, *La pensée italienne au XVIe siècle et le courant libertin,* Paris, 1919. Ernest Renan, *Averroes et l'Averroïsme,* Appendice XI, printed the letter of the inquisitor of Padua of 1619 and Cremonini's reply.

[148] Michaud alters this to 1570-1653, but as he says that Argoli died when 81, 1653 appears to be a misprint.

[149] Sudhoff (1902), 79.

[150] *Tabulae Primi Mobilis quibus veterum rejectis prolixitatibus directiones facillime componuntur,* Rome, 1610. Michaud lists a two volume edition of Padua, 1644, with a portrait of the author.

[151] *Secundorum mobilium juxta hypotheses Tychonistas et coelo eductas observationes Tabulae,* Padua, 1634. There was another edition at Padua in 1650, *Exactissimae secundorum mobilium tabulae,* etc.

[152] Printed at Venice, 1638; Padua, 1648; Lyons, 1659.

[153] *Pandosion sphaericum in quo singula in elementaribus regionibus atque aetherea mathematice pertractantur,* Padua, 1644, and again in 1653.

[154] *De diebus criticis et aegrorum decubitu libri duo,* Padua, 1639: copy used, BN V.8365. Sudhoff used the edition of 1652 (BN V.8366).

gical medicine in particular. Having in his early chapters asserted the influence of the stars, Argolus devotes the ninth chapter of the first book to instruction how to predict from one's nativity the coming train of events of the human body. The tenth considers the subjection of the external and internal parts of the body to the planets and signs of the zodiac, and the diseases which are attributed to particular signs and planets. The eleventh chapter maintains that the outcome of illness may be more rationally and evidently investigated by astrological method than by the medical art; the twelfth instructs how to forecast the nature and time of sickness from those superior causes. The thirteenth deals with determination of good or ill health from the revolution of the year, and it is only with chapter 14 that we at last come to critical days of which the discussion continues to chapter 21, where the first book ends. Even then the discussion continues to be primarily astrological, and we are assured that the crises in diseases do not follow a numerical order or proportion,[155] and that a horoscope should be drawn up at the beginning of the illness.[156]

The first six chapter headings of Book II are all astrological: whether the disease is curable, short or long, signs of death, signs of convalescence, relation to the course of the moon, and precepts to be observed in medicine such as that one's nails are to be cut when the moon is waxing in Aries, Taurus, Leo or Libra with Venus or the sun in friendly aspect.[157] After reprinting the *Iatromathematica* of Hermes Trismegistus to Ammon of Egypt, Argolus regales us with the horoscopes of the nativities and the falling sick of four recent popes, Sixtus V, Clement VIII, Paul V, and Gregory XV. Of the illnes of the last-named, who died in 1623, Argolus says that he was present almost daily with other physicians throughout the entire course of the disease. The new star of 1604 is said to have announced the election of Paul V. Argolus then gives similar data for Henries II, III and IV of France, connects a comet with the horoscope and death in battle on August 4, 1578, of king Sebastian of

155 *Op cit.*, cap. xvii, p. 70, "Quod non per rithmum numerorum crises morborum inoriantur."

156 Cap. xix, p. 78, "De figura coelesti erigenda in morbi initio."

157 *Op cit.*, Book II, p. 15.

Portugal, then passes on to Gustavus Adolphus[158] and the nativities of various other princes, cardinals and the like. In the later edition used by Sudhoff, Gregory XIII and Urban VIII (1623–1644) were added to the four popes above mentioned, showing in what light, with what qualifications, and within what limits we should interpret the bulls of Sixtus V in 1586 and Urban VIII in 1631 against astrology. We further see that astrology went on at the University of Padua in the seventeenth century as well as at Bologna and at Salamanca.[159]

The wide currency of the conception, man the microcosm, is further attested by a work of Stephanus Rodericus Castrensis, or Estevan Rodrigues de Castro,[160] of Portugal, first professor of medicine at the University of Pisa, on the Meteors of the Microcosm, in four books, dedicated to the Grand Duke of Tuscany, and printed at Florence in 1621.[161] The idea suggested by this title goes back at least to Severinus, *Idea medicinae philosophicae,* 1571, where it is said that fevers, epilepsy, dropsy, catarrhs, and so forth correspond to meteors in the great world. The representatives of the Inquisition, in their approbation of the work, note that Rodericus revives atomism and regards water as the principle of things, but only philosophically. The first book, after demonstrating the conformity between the world and man, and that there is a world soul—to which the inquisitors do not seem to have objected, and pointing out the similarity between it and the human soul, attacks Aristotle's arguments for four elements and argues for atoms. The second book inquires whether the origin of the microcosm has been correctly stated by others, considers the human anatomy, asks whether the blood and spirits are alive, and treats of the internal signature of the microcosm. With the third book we at length come to the meteors of the microcosm, and various kinds of fevers are dealt

[158] *Ibid.*, pp. 39-40; the remainder of the book to p. 148 is devoted to other nativities and times of falling sick.

[159] T V, chap. 12; T VI, pp. 165-166.

[160] BM catalogues his works under Rodrigues; BN, under Castro.

[161] *Stephani Roderici Castrensis Lusitani medici ac philosophi praestantissimi et in Pisana schola medicinam primo loco docentis De meteoris microcosmi libri quatuor cum indice rerum et verborum,* Florentiae apud Iunctas, 1621, in-fol., 224 double-columned pages. Copy used: BM 548.m.7.

with, which are compared to the fiery impressions in the universe. Asking whether fever is a quality or a substance, Rodericus concludes that all diseases are substances.

The fourth and last book is miscellaneous in content. First it takes up the cure of fevers, then whence *alexipharmaca* or antidotes for poison derive their virtue. Antidotes from gold and pearls are given as well as from animals, plants and minerals in general. "Fires of Vulcan and other fiery meteors" are represented by such diseases as elephantiasis. Epilepsy, apoplexy and paralysis are the lightning and thunderbolts of the microcosm. Menstrual blood is called poisonous. After discussing the nature and causes of winds in the great world, we turn to wind in the microcosm. Sweat is the inundation of the microcosm; catarrh, its rain; and finally we consider its stones.

The same conception appeared again in the Anatomy of the Microcosm by J. S. Kozak, printed at Bremen in 1636, in which are chapters on meteors of the macrocosm, and both salutary and morbid meteors of the microcosm.[162] Fortunio Liceto narrowed the comparison to lightning and fevers in particular.[162a] Paul Virdung and Genathius had applied it to winds.[163]

Adrian Spigelius or Spiegel (1578–1625) in his anatomical work published posthumously at Venice in 1627, accepted the observation of Falloppia that in the case of deep wounds of the head the brain swelled at the full of the moon and subsided as the moon waned. Furthermore Spigelius held that from this it followed that greater inconveniences would result from a blow on the head when the moon was waning. For then veins were more apt to be severed because of the interval left between the cranium and the hard membrane. Epilepsy was apt to come on at the time of the new

[162] *Anatomia vitalis microcosmi, in qua naturae humanae proprietates quas homo cum rebus extra se sitis communes habet,* Bremae, 1636, in-4, 209 pp., 29 caps.: BM 548.g.6, caps. 18-20, pp. 133-69.

Kozak also published *Septimana horologii microcosmi,* 1640, which I have not seen: not in BM and BN or LC catalogues.

[162a] *Pyronarchia sive de fulminum natura deque febrium origine libri duo,* Padua, 1634, in-4, 126 pp. BN R.2919, R.2920.

[163] Paul Virdung. *Discursus medicus de ventis in microcosmo,* 1619. J. J. Genathius, *Decas disp. medic.: De vento in microcosmo seu spiritu flatulento,* 1618: BM 1179.g.3.

moon, because then the humors, denied mixture with humidity, grew sharp and lacerated the brain, especially in the case of melancholy.[164]

Like many, if not most critics of astrology, Antonio Merenda was not an astronomer. A professor of civil law at Pavia, he published his work against astrology in Italian in order to convince those who could not read Latin of what theologians and philosophers had often demonstrated in Latin, the falsity of that art.[165] His argumentation, however, is so quibbling, involved and difficult to follow that it may be doubted whether it would convince any average reader. His attack is directed especially against prediction as to particular individuals, which he stigmatizes as a diabolical art of divination, and contends that no superstitious art is more fitted to forward the aims of the devil than the astrology of Ptolemy.[166] He objects that the influence of the stars at the time of conception or birth is not immutable but subject to change from subsequent celestial influences, and that the astrologers should stick to either the moment of conception or that of birth.[167] He denies that the Bible supports astrology, declares that papal bulls detest it, and that the church fathers were unanimously against it. He repeats Augustine's argument from twins against astrology, and is careful

[164] *De humani corporis fabrica libri decem*, Venice, 1627, in-fol., 328 pp., Index, 96 Plates. Neither *epilepsia* nor *luna* appears in the index. The passage cited is at p. 316 (lib. X, cap. 4). In Col Medical Library which also has the edition of Frankfurt, 1632.

Some further titles identifying human anatomy with microcosm are: Geo. Graseccius (or, Graseck), *Microcosmicum theatron in quo . . . fabrica humani corporis masculum . . .*, Strasburg, 1605, in-8. Fabrizio Bartoletti, *Anatomica humani microcosmi descriptio*, Bologna, 1619, in-fol. W. Bosschaert, *De microcosmo sive humani corporis fabrica, 1626;* not in BM or BN. G. Rolfinck, *Anatome microcosmi*, Jena, 1631; not in BM or BN.

Other microcosmic titles are: Adam Weinserer, *Praecognita majoris et minoris mundi. Item secreta naturalia*, in-12: no BM. J. G. Schenck, *Lithogenesia, sive de microcosmi membris petrefactis*, Frankfurt, 1608, in-4. J. Remmelin, *Catoptrum microcosmicum*, 1613, 1619. Fortunio Liceto, *Analogia hominis et mundi*, Udine, 1635: BN R.2911; R.2918. J. C. Dürr, *Disputatio de analogia corporum coelestium et sublunarium*, Jena, 1649: no BM or BN.

[165] *La destrutione de' fondamenti dell' astrologia giudiciaria, a gl' huomini particolari predicente dignitá, ricchezze, sanità ouero malattie del corpo, et altri successi accidentali*, Pavia, 1640, 68 pp. with 57 paragraphs numbered in the margins.

[166] *Ibid.*, pp. 41, 55.

[167] *Ibid.*, pp. 9-16.

never to cite Aquinas as favoring it. Yet a page which he finally quotes from him turns out to be his usual half-favorable attitude, saving free will but stressing the influence of the stars upon natural inclination. Merenda, on the other hand, had previously represented Aquinas (*Secunda secundae,* quaest. 95, artic. 5) as agreeing with Augustine (*Genesis ad literam,* cap. 17) that the devil, God permitting, interested himself in the predictions of astrologers.[168]

Merenda grants that it is easier to predict the weather or the health of a community than the fortune of an individual. But astrologers do not know all the stars and so cannot foresee even their general influence. They have not yet had sufficient experience of the effects of a great variety of constellations to formulate rules as to these. Ptolemy rejected previous experience and method as too particular and in his *Quadripartitum* laid down general rules largely on the basis of reason and probability. And after him Arabic astrologers were divided into discordant sects.[169] Why should a brief eclipse exert great influence, when daily we are deprived of sunlight all night long, and all day long of that of the moon? Merenda further attacks the doctrine that some one planet is lord of the year.[170]

A birth may be delayed or hastened and so miss its proper and natural constellation. Moreover, few clocks keep exact time, and clocks are found chiefly only in such places as fortresses, monasteries, convents, hospitals and colleges, so that it is difficult to tell the precise moment of birth.[171] Merenda further asks, perhaps sarcastically, why astrologers do not take the nativities of the father, brothers and sisters into consideration as the ancients did.[172] But he above all objects to predictions of violent death, or a rich marriage, or obtaining an office, as violations of freedom of the will.[173] He is much offended that astrologers promise even a cardinalate or the papacy to their clients.[174] Merenda professes originality in his discussion and denies that his method of treatment will be found in other authors, but it has resemblances to that of John Chamber,

[168] *Ibid.,* pp. 44-45, 42.

[169] *Ibid.,* pp. 23, 25, 37-39, 49-51, 53.

[170] *Ibid.,* pp. 125-26.

[171] *Ibid.,* pp. 20-22.

[172] *Ibid.,* pp. 28-29.

[173] *Ibid.,* pp. 32-34, 48-49.

[174] *Ibid.,* p. 56.

while he admits that he has not read them all.[175] He closes with a satirical parody, giving nine crafty rules for practical procedure on the part of predicting astrologers.[176]

Urban VIII, who issued a bull against judicial astrology in 1631, appointed Caesar Carena an inquisitor. The latter has a good deal to say concerning astrology in his Treatise on the Office of the Most Holy Inquisition, printed at Cremona in 1636 and 1641,[177] later at Bologna in 1668 and Lyons in 1669.[178] He states that natural astrology, which conjectures what naturally happens from the aspect of the stars, and, if it considers the nativity, infers from it human temperaments and propensities, is licit. So say Suarez and Sanchez.[179] But one should not consider the horoscope of Christ despite d'Ailly's doing so, nor predict concerning the pope and the church, which is forbidden by the papal bull of Urban VIII of April 1, 1631. An astrologer from the nativity may predict the child's temperament and future infirmities and when they will occur. "This conclusion is certain, nor can there be any doubt about it, because the stars act directly upon the body and its humors," as Aquinas well proved and many others of the Fathers of Coimbra. But actual prediction is an uncertain matter, and Carena does not agree with Hurtado that astrologers can know for certain a time beyond which a person cannot live.[180]

It is licit to construct a *figura coeli* of the beginning of a disease and for the moment of taking to one's bed, of which Maginus treats in his work on the legitimate use of astrology in medicine. But such a figure alone by itself is unreliable, and Campanella in his medical works says that it is insufficient unless it agrees with the horoscope

[175] *Ibid.*, p. 64.

[176] *Ibid.*, pp. 66-67.

[177] *Tractatus de modo procedendi in causis S. Officii*, Cremonae, apud M. A. Belpierum, 1636, 387 pp. in-4. BN E. 2290.

De officio sanctissimae inquisitionis et modo procedendi in causis fidei tractatus, Cremona, 1641, in-fol. Listed *Bibl. curiosa*, 1676, p. 411.

[178] I have used the last named edition which contains a letter to Carena dated June 1, 1641 and approbations of 1639, 1649 and 1667. Col 933.1 C18. *Tractatus de officio sanctissimae inquisitionis.*

[179] *Ibid.*, p. 178, citing Suarez *De relig.* lib. 2, *de superstit.* cap. 11, num. 8 *et seq.;* Sanch. *in Decalog.* tom. I, lib. 2, cap. 38, num. 28.

[180] *Ibid.*, p. 179, citing P. Hurtado *de caelo et mundo,* disp. 2, sect. 6, ¶ 70, num. 3, impressionis sextae.

taken at birth. When an astrological direction occurs opposed to the person's temperament, the astrologer is justified in saying that illness threatens. But Carena disapproves of erecting figures for the parents from that of the son, although Cardan, Campanella and Regiomontanus in one of his problems do so. It is more difficult to predict inclinations than physical temperament. One cannot predict, for example, that a person will be a sodomite. And future contingents may have little or no connection with inclinations. Raphael de la Torre holds that predicting human inclinations is not superstitious nor forbidden by the bull of Sixtus V, but Carena insists that prediction of inclinations and of future contingencies are both bad and are forbidden by both papal bulls. The division of the zodiac into astrological houses is uncertain, and astrologers disagree among themselves as to directions and other points, so that there is no sure foundation for judgments, although Carena grants that the planets exert influence not only *per se* but according to their positions in the signs. But one cannot predict future contingent events for individuals, much less political changes. Campanella in his judicial astrology claims to have purged it from all superstition, but actually whatever is superstitious in all astrology is found in that book, which is only a compendium of Cardan's commentary on the *Quadripartitum* of Ptolemy.

iv. PORTUGAL AND SPAIN

Antonio de Najera, whose *Discursos sobre o cometa* were printed at Lisbon in 1619, also published there, in 1632, a *Summa astrologia* which was devoted chiefly to the art of weather prediction and the judgment of eclipses, the revolution of the year, and more particularly the conjunctions, oppositions and quartile aspects of sun and moon.

Hugo Sempilius was a Scot who entered the Society of Jesus in 1615 at the age of twenty-one.[181] On January 1, 1635, from the Jesuit Royal College in Madrid, where he taught mathematics, he dedicated to Philip IV of Spain a work in twelve books on mathe-

[181] Zedler, 36, 1795.

matical disciplines.[182] This was a rather general and brief treatment, devoting only a few pages each to geometry and arithmetic, optics, statics, music, and cosmography,[183] and concluding with topical bibliographies at pages 262-310. Astrology was treated in a separate book from astronomy, but the properties of the planets, including such astrological matters as houses, triplicities and decans, had already been considered under Astronomy[184] before Book XI on Astrology began.[185]

Asking what astrology is licit and what is illicit, Sempilius gives the bull of Sixtus V against astrologers but adds that many points are raised by theologians affecting the interpretation of this bull, and that the first point is whether from the aspect of the stars not merely storms and the sterility or fecundity of the soil may be predicted, but also from consideration of the time of birth the temperaments and propensities of men. He affirms that they may be, because experience has taught by the clearest and most trustworthy testimonies that they are subject to the science (of astrology). In favor of his view he cites a number of authorities,[186] but notes that of these Suarez and Sanchez wisely warn that such divination as to temperament and inclinations cannot exceed the bounds of conjecture and surmise without imprudent rashness and mendacity. All this reminds us of Carena, who may have borowed from Sempilius.

Sempilius accordingly recognizes a threefold astrology which may be practiced with great utility and pleasure: first, weather prediction; second, genitures within natural limits; third, obser-

[182] Hugonis Sempilii Craigbaitaei Scoti S. J. *De mathematicis disciplinis libri XII ad Philippum IV*, Antwerp, 1635, folio: copy used BN R.764.

[183] I, de dignitate matheseos; II, de utilitate scientiarum mathematicarum; III, de geometria et arithmetica, pp. 54-61; IV, de optica, 61-86; V, de statica, 87-101; VI, de musica, 102-115; VII, de cosmographia (i.e. the Sphere), 115-20. VIII, de geographia, is longer, 121-79, with cap. 4, Asia, 165-; cap. 5. America, 172-; cap. 6, on the polar regions, 179a-b.

[184] *Ibid.*, pp. 224a-226b.

[185] *Ibid.*, pp. 229-242.

[186] "Ita tenent Abbas cap. 2, n. 7, *de Sortilegiis*, Caietanus 2.2.q.95.a.5 *in fine*. Suarez Tom. I *de Religione*, lib. 2 *de Superstitione*, cap. xi a num. 8 ad 11. Sales 1.2, quaest. 9, art. 5, tract. 5, disp. 2, sect. 8, n. 79. Manuel I Tomo *Summae*, cap. 7, num. 3. Thomas Sanchez lib. 2 *Summae*, cap. 38, num. 28."

vance of the times, days and places necessary for medical men, agriculturists and mariners. All judgments beyond these he holds suspect.[187] After remarking that Pico della Mirandola's criticisms of astrology have been sufficiently rebutted by Cirvelo,[188] Sempilius specifies after Cirvelo what each planet signifies as lord of an eclipse.[189] Albumasar's book on great conjunctions was condemned by the theologians of Paris for subjecting religious change to these, but this does not prevent Sempilius from following Origanus, Cirvelo and Cardan in choosing as the lord of a great conjunction the planet which is higher in eccentric and epicycle.[190] He adds a list of regions and cities which are under each sign of the zodiac, and instructs the reader who wishes further information to turn to Garcaeus and Gauricus.[191] Astrological elections are not permissible for religious rites such as baptism, but are licit for flebotomy, bathing, scarification, and when to take medicine, shear sheep, sow crops, wean children, or begin a journey. But it is superstitious to consult the stars as to the best time to meet a magistrate or teacher, and such matters as riches and the fate of kinsmen should not be predicted from one's nativity.

Sempilius either did not know of Urban VIII's new bull against astrology in 1631 or chose to ignore it. Thus in Spain at the Jesuit College of Madrid as well as at the University of Salamanca astrology continued to find favor in the seventeenth century.[192]

Comets were discussed by Sempilius at the close of his ninth book on hydrography, the air, atmosphere, twilight, meteors, fire and comets. He recognized the importance of parallax, but noted that there was disagreement among philosophers and astronomers whether comets were in the uppermost region of air, or in the heavens, or both above and below the moon. Another dispute was whether they were bodies temporarily illuminated, of matter condensable and dissipable, yet illuminated by the sun, as the turning

[187] *Ibid.*, 237a.

[188] Concerning Pedro Cirvelo and the other astrological authorities mentioned in this paragraph consult the Index to T V and VI.

[189] Sempilius, 237b-238a.

[190] *Ibid.*, 238a-b. The condemnation referred to is that of books owned by Simon de Phares in 1494. See T IV, 549.

[191] *Ibid.*, 239a-240a.

[192] For Urban's bull see T VI, 171; for astrology at Salamanca into the eighteenth century, T VI, 165-66.

of the tail away from the sun indicates. Kepler made them exhalations from the ethereal globes, but Sempilius holds that they are made from sunspots and are eternal bodies.[193]

v. GERMANY

In Germany on comets there had appeared at Wittenberg, in 1602 by Abraham Rockenbach, what was described as "a new methodical treatise, in which not only the causes of comets are expounded by the method of simple question but also their effects are noted.[194] Rockenbach still held the Aristotelian theory of comets and does not even mention the observations of Tycho Brahe and other astronomers to the contrary. But he was chiefly intent on the effects of comets. For those of the thirteenth, fourteenth and fifteenth centuries he seldom cites a source.[195] His treatise was reprinted with that of Philipp Müller on the comet of 1618 as A Treatise on Comets with an enumeration of comets from the foundation of the world even to this day.[196] Rockenbach was a doctor of laws and ordinary professor of mathematics and Greek, and extraordinary of law at Frankfurt on the Oder.

The hold that astrological medicine maintained in northern Germany through the first third of the seventeenth century is well illustrated by the *Dissertatio Iatromathematica* of Johannes Assverus Ampsing (1558–1642). *Lindenius Renovatus* lists three editions of it, all at Rostock, in 1602, 1618 and 1629,[197] while I have examined it in an "Editio secunda" of Rostock, 1630.[198]

[193] *De mathematicis disciplinis,* Antwerp, 1635, pp. 193a-194b.

[194] *De cometis tractatus novus methodicus in quo non tantum causae cometarum per methodum simplicis quaestionis exponuntur sed etiam effectus eorum . . . annotuntur,* Witebergae, 1602, in-8, 238 pp. BN V. 29270.

[195] An exception, at p. 195, is *Peucerus in Chronic.* for the defeat of Bajazet by Tamerlane following the comet of 1391.

[196] *De cometa anni 1618 commentatio physico-mathematica . . . Philippi Mulleri . . . Accessit A.R. Tractatus de cometis cum enumeratione cometarum a mundi conditu usque ad hunc diem,* Lipsiae, n.d., in-8. BN V.21122.

[197] LR 522-23. Sudhoff (1902), p. 73, had not seen Ampsing's work and depended on LR. There is a copy of the 1618 edition in BM 1141.b.15. The editions of 1602 and 1618 were in-4; that of 1629, in-8.

[198] *Johannis Assueri Ampsingii Trans-*

In the preface Ampsing holds that chronic diseases follow the movement of the sun; acute diseases, that of the moon. Later he defends the dependence of critical days upon the moon, and testifies from his own experience in Zeeland that patients die with the ebbing tide. Eclipses, too, are followed by tragic events. The times of the equinoxes and solstices are not safe for medication. The heliotrope is a proof of the occult influence of the sun.[199]

After opening chapters on the preeminence of medicine and the dignity of astronomy, the third treats of the marriage of medicine and astronomy.[200] To a large extent the book is a marshalling of authorities. Ampsing contends that Hippocrates, Galen and Pliny favored astrological medicine, and that Regiomontanus, Garcaeus and Cardan even went a little too far in the direction of Chaldaean superstition.[201] Levinus Lemnius and Fernel are great favorites.[202] There even appears to be some unjustifiable citation in favor of astrology of authors who really opposed it, for instance, Piccolomini and Vicomercati.[203]

Discussing the influence of the stars upon pest and popular epidemics, Ampsing argues that the primary cause of pestilence is not elementary but divine and celestial, although the pest may be more fatal in bodies which are already in a bad condition. Mercurialis in his treatise on the plague[204] confesses that there is something divine in it and an occult poisonous force brought into the air by the stars. When archduke Matthias consulted the medical college of Vienna

---

isulani Med. D. et Professoris publici et physici ordinarii in Academia et Repub. Rostochiensi Dissertatio Iatromathematica in qua de medicinae et astronomiae praestantia deque utriusque indissolubili coniugio disseritur tum vero ipsa etiam (this word is not in the title as given in LR) *astrologia quae pars est astronomiae (quatinus quidem arti Medicae inserviens et rationibus physicis et gravissimorum hominum observationibus procul omni superstitione nititur) a contemptu quorundam vindicatur.* Editio secunda, Rostochi, 1630. 311 pp. The text proper ends at p. 284 and a *Confutatio* of Frischlin begins at p. 285. This last is broken off in order that Ampsing may refute Jacobus Dürfeldius of Osnabrück, who has again written against him.

[199] *Ibid.*, pp. 6, 55 and 245 *et seq.*, 59, 62, 74, 95, for the foregoing passages.

[200] *Ibid.*, pp. 20-, 44-, 72-.

[201] *Ibid.*, pp. 62, 119.

[202] *Ibid.*, pp. 119-27, 133, 200, 203, 208-10, 212-13, 231-32.

[203] *Ibid.*, pp. 127-28; cf. T VI, 180, 368.

[204] *Ibid.*, p. 205, "Hier. Mercurialis *lib. de peste,* cap. 5, 6, 7."

as to the cause of a plague among the cattle in 1598–99, they designated as the remote astrological cause a comet in Taurus in 1597 and many conjunctions in the ruminating signs; as the near physical cause, humidity.[205]

The Confutation at the close of Ampsing's volume is of Frischlin's attack upon astrology printed back in 1586.[206] But Ampsing had already repeatedly criticized Frischlin's views in the earlier course of the volume.[207] Ampsing had intended to add in an appendix to his volume a clear and easy way of drawing up horoscopes (*themata coeli*) derived from Regiomontanus, Gemma Frisius, Metius and Rantzovius, but the necessity of refuting an attack upon himself by Jacobus Dürfeld prevented his carrying out this plan.[208] Despite his belief in astrological medicine, Ampsing had no liking for the followers of Paracelsus, whom he characterized as "a ridiculous kind of philosophers."[209]

Another German exponent of astrological medicine was August Etzler or Ezler. According to Sudhoff[210] he first published at Bautzen in 1610 a work which I have seen in the later edition of 1631 at Strasburg.[211] As its long Latin title shows, it emphasizes the doctrine of signatures in plants and animals, and relations of sympathy and antipathy, as well as the rule of superior celestial over inferior terrestrial bodies. A chapter is devoted to each planet and the plants, animals, human types, places, metals and diseases related to it.[212] Such statements are made as that, if the genitals are touched with the juice of aconite, men die, because the ruler of the genitals is

205 *Diss. Iatromath.*, pp. 202-7, and similarly on to 219.

206 T VI, 191-92.

207 *Diss. Iatromath.*, pp. 81, 85, 87, 245, 255, 260.

208 *Ibid.*, p. 308. LR does not include the name of Dürfeld. The BN catalogue lists one different work of 1642 by him.

209 *Diss. Iatromath.*, p. 139.

210 Sudhoff (1902), p. 77.

211 *Isagoge physico-magico-medica in qua signaturae non paucorum vegetabilium et animalium tam internae quam externae accurate depinguntur ex quibus mundi superioris astralis cum inferiori elementali mundo concordantia et influentia mirabilisque et occulta sympathia et antipathia rerum . . . elucescunt,* Argentinae, 1631, in-8, 176 pp. Copy used: BN R.12721; BM 718.b.19. Borel (1654), 265, lists an edition of Strasburg, 1651, but it may be a misprint for 1631.

212 Some of these are omitted in the later chapters. The first chapter on Saturn, which covers fifty pages, is the longest.

Venus, with whom Saturn wages extreme enmity.[213] Etzler issued another medical work in 1613[214] which contains a great deal of astrology but which Sudhoff did not mention, and in 1622 an introduction to astrological medicine.[215]

Yet another volume of astrological medicine was published by Cornelius Pleier of Coburg at Nürnberg in 1627.[216] In the preface to Christian, margrave of Brandenburg, Pleier attacks the defamers of alchemy and astrology, and praises astrological medicine. A treatise on critical days concludes with ways to determine them exactly according to the astronomers and Tables.[217] The pseudo-Hippocratic and pseudo-Galenic tracts on astrological medicine are then given in parallel columns, after which a briefer second book is devoted to the election of favorable times for dosing and bleeding according to the motion of the moon and configuration of the stars.[218]

An oration on climacteric years which Andreas Stechanus delivered publicly "in the school of the Arnstadters," was printed at Erfurt in 1633.[219]

The *Hermes Redivivus* of Joannes Bickerus was printed at Giessen in 1612[220] and dedicated on September 1, 1611, to Johann Georg, duke of Saxony.[221] A preface to the reader says that the work attempts to reconcile Galenic with Hermetic medicine. The first chapter takes up the history of medicine and represents ancient Egypt as a paradise of most secret arts with huge libraries "and men excelling in knowledge of all the arts who cultivated this me-

213 *Ibid.*, p. 10.

214 *Brevis tractatus fundamentum medicinae aeternum explanans et ad quintuplicis entis morbifici cognitionem viam sternens*, Halae Saxonum, 1613. Copy used: BM 1141.b.12 (5.).

215 *Introd. Iatro-math.*, 1622, in-8. LR 97; not in BM and BN catalogues.

216 *Medicus-criticus astrologus ex veteribus iatromathematicis productus*, Noribergae, sumptibus et typis Simonis Halbmayeri, 1627, in-12, 237 ff. Copy used: BM 1170.a.20. The work was described by Sudhoff (1902), p. 78, but by a misprint was dated 1527.

217 *Ibid.*, ff. 67-110.

218 *Ibid.*, ff. 111-199, ff. 202-237.

219 *Oratio de annis climactericis publice habita in schola Arnstatiensium*, Erfurt, 1633, in-4.

220 *Hermes redivivus declarans hygieinum de sanitate vel bona valetudine hominis conservanda*, Giessae, 1612, in-8, 480 pp.: BM 1039.d.11.

221 Later sections of the volume are dedicated to Augustus, duke of Saxony, Johann Philipp, duke of Saxony, and so on.

dicine of Hermes under the title of magic." The second chapter, on the balsamic temperament, cites the Hermetics as saying that the "dust of the ground" out of which Adam was created was the quintessence of all parts of the earth. The Hermetics also recognize only two elements in inferiors, earth and water, and place fire and air in the ethereal globe. Herewith we are launched upon a comparison of macrocosm and microcosm. The vessels of the heart parallel the rivers of Paradise: the aorta corresponds to Pison; the *vena cava* to Gihon or the Nile; the *vena arteriosa* to the Euphrates; the *arteria venosa* to the Tigris. Astral man and the balsam of life are like the sun and in sympathy with the stars. The diversity of color in living beings depends on diverse constitution of balsam. To the seven planets correspond seven chief members or organs of the human body, and Bickerus would replace the triple action of the soul—natural, vital and animal—by a sevenfold activity related to the planets.

"The ens of incantation, proceeding from the force of malign imagination, affects our imagination and thence strives to corrupt the harmony of our life."[222] Bickerus, however, disapproves of astrological images or of making a homunculus to attract and avert all sorcery by magnetic force, and instead advises prayer to God for aid against incantations. Later on he tells how the Brahmans of India, Pythagoras and the Magi, to provide for balsam and avoid the tyranny of hostile Saturn, clad themselves in white vestments and daily used jovial and solar songs and smoothest sounds and delights of Venus. So martial temperaments may avoid wrath and strife by use of Venus and thus change Mars to Venus. Or, *vice-versa,* those whom Venus agitates should not scorn the gifts of Mars and Saturn.[223] In another passage, discussing whether the stars affect mind and soul, he says he does not so esteem amulets and pentacles that he thinks all mental perturbations can be tranquilized by wearing these. But one won't go far wrong in making such pentacles, if one awaits the benign influences of the stars. Some persons are gay in bad weather and sad in fair weather. Not only ancient but recent philosophers attribute mental effects to gems, especially if carved with astrological figures—a statement

[222] *Ibid.*, cap. xxv, p. 167.

[223] *Ibid.*, II, i, 2, p. 269.

inconsistent with his previous rejection of astrological images—and also ascribe marvelous virtues to herbs.[224]

If star or sky is adverse, moderate it by application of solar, jovial or Venerial spirits, or by exaltation of your star against it. Porphyry in the life of Plotinus tells how an Egyptian magician and astrologer tried in vain to fascinate Plotinus by use of images. Bickerus advises medicaments which abound in solar and jovial virtue, compounded at the time when the moon hastens from conjunction with Venus to sextile aspect with the sun. Or compose the medicine when a constellation, which has before been found favorable, returns.[225] The halcyon bird is a fine medicine with which to renew one's youth, because it yearly sheds its wings and assumes new ones, and breathes forth a fragrant odor.[226] Such is the magic and astrology which are found scattered through *Hermes Redivivus.* Bickerus believes that drugs have signatures, and that there are signs of pest from the superior globe, but incubus is for him merely imagination of suffocation in sleep.[227] Furthermore, he has hygienic convictions of his own, advising against studying before dawn and recommending that one sleep until the second or third hour after sunrise.[228]

Air is the vehicle of life or aether. If one is without it for even a quarter of an hour, one is sure to die. "Respiration alone never ceases without peril of life." Both the inferior elements in the body are restored and nourished by the most subtle spirit of the blood and draughts of purer air. Air is diffused through the entire body like balsam or our aether and sustains the parts of the body, as it does the sphere of earth in the macrocosm, which Hippocrates had in mind in calling air the vehicle of earth. No one feels the weight of his members because of internal aethereal air. Another function of air is that balsam or star may live in it. As fire burns more freely when fanned with air, so the balsam of the microcosm is stronger and more vigorous, when the air is clarified. "And since there is no vacuum in nature, who will assert that there is any internal cavity . . . without air?"[229]

224 *Ibid.*, pp. 358-59.
225 *Ibid.*, pp. 396-97.
226 *Ibid.*, p. 404.
227 *Ibid.*, pp. 5, 410-11, 459.
228 *Ibid.*, p. 119.
229 Ibid., pp. 40-42.

Andreas Matthaeus Aquaviva, duke of Atri, published at Frankfurt-am-Main in 1609 a volume with a pretentious title which opens, Four Books of Illustrious and Most Exquisite Disputations, in which the Arcana of all Divine and human Wisdom, particularly of Music, moderator of the soul, and Astrology...[230] Actually the volume contains: first, the Greek text of Plutarch's Precepts concerning Moral Virtue; second, Aquaviva's Latin translation thereof; third, his commentary on it in four books. He asserts that Pythagoras investigated the secrets of nature by music and through music penetrated to the arcana of the universe and of human nature. He also discusses the matter of the seventh, eighth and ninth month's child.[231] This is about as close as the work comes to astrology, although the second book treats of the complicated movements of the planets, including the sun, and illustrates them with novel diagrams. The third and fourth books are moral.

Hermann Kirchner, orating at the university of Marburg in 1609 on the recent decline of higher education, recalled that almost all astrologers had predicted from a solar eclipse in 1598 the ruin of arts and letters and divine wrath and punishment for all ranks of *literati.* Evil Saturn was lurking in the worst angle with hostile aspect, and envenomed rays, made sharper by conjunction with Arcturus, poured down on the heads of learned men. He well remembered how the learned ridiculed this prediction as superstitious and to be classed with the past vanities and dreams of Stiffelius and Eustathius Posselius, of whom the former put the end of the world in 1598, and the latter, in 1606. But although Kirchner thinks that no fatal necessity should be placed in the courses of the stars,

[230] *Libri quatuor illustrium et exquisitissimarum disputationum quibus omnis divinae atque humanae sapientiae praesertim animi moderatricis musicae atque astrologiae arcana in Plutarchi Chaeronei de virtute morali praeceptionibus . . .*, Helenopoli apud Joh. Theob. Schonwetterum, 1609, in-4.

[231] Some later titles dealing with this subject are: Joh. Bergius, *Disputatio de partu cuiusdam infantulae Agennensis, An sit septimestris an novem mensium?* Extat Parte VI Operum Jacobi Sylvii, 1630, in-fol. Paul Strectes, *Consilium de partu nonimestri*, Pisa, 1651, in-fol. Joh. Cont. Axtius, *Dialogus de partu septimestri, an nempe ille sit perfectus vegetus et per consequens legitimus?* Jena, 1679, in-12. And see Index.

yet subsequent events have demonstrated that it was no idle signification of the solar eclipse that was then predicted.[232]

It was also at Marburg that Rudolphus Goclenius the Younger (1572–1621) in 1611 delivered.

An Apologetic Discourse *pro astromantia* against present scourgers of the same, written in the form of an oration and presented in the University of Marburg at a public gathering and distinguished assembly by Rudolphus Goclenius, M. D. and ordinary professor of physics, in which divination from the stars is defended, its certitude and utility is demonstrated, and the objections which are wont to be offered to the contrary are solidly and clearly refuted.[233]

In a preface of May 20, 1611, to the consuls and senate of Bremen, which opened the work as printed, Goclenius complained that astrology, like all noble sciences, had been fouled and distorted by the vulgar crowd. But divine goodness gave one Ptolemy to offset so many thousand sycophants, and insane attacks upon astrology and its followers are not to be tolerated in well ordered institutions of learning. In the text Goclenius speaks slightingly of Arabic astrology,—so many Albumasars, Abenragels, Alchabitii, Albubaters, Zahels, Messahalas.[234] But he does not think much of the attack on astrology by Pico della Mirandola who, although very erudite otherwise, did not understand that art and who revamped time-worn arguments which Bellantius and others have refuted. Yet recent opponents of astrology employ the same arguments and display the same ignorance of astrology itself.[235] Goclenius holds that universal influences of the heavens upon men at large do not override the particular genitures of individuals. Thus when the constellations produce wholesome air—as during the

[232] Hermann Kirchner, *De fatalibus Academiarum dissipationibus ac ruinis oratio*, Marburg, 1610. Michael Stifelius, at least in the first instance, had predicted that the world would end on October 3, 1533. See T V, 393.

[233] *Apologeticus pro astromantia discursus contra eiusdem huius aevi mastiges orationis forma scriptus ac habitus in Academia Marpurgensi publico conventu et honoratissimo consessu a Rod. Goclenio med. D. et professore physic. ordinar. in quo divinatio ex astris defenditur, certitudo et utilitas demonstratur, obiectionesque quae in contrarium afferri solent solide et perspicue refutantur*, Marpurgi Hessorum, 1611, 132 pp. Copy used: BM 718.f.26 (3.).

[234] *Ibid.*, p. 10.

[235] *Ibid.*, pp. 105-6.

past month—a person whose own nativity indicates serious illness at that time will not escape it, although its seriousness may be somewhat abated by the fine weather. Or, on the other hand, an individual may escape from an epidemic of the pest or a shipwreck because his own nativity does not decree death then.[236] Goclenius connects critical days with the moon, and says that lunatics evidently experience the force of that star.[237] He remarks that Saturn is now far distant from the earth and by reason of its eccentric and epicycles is moving through the farthest parts of Sagittarius. So it is of less efficacy than when it is in Gemini as in the year 1441 under Frederick III, when its nearness to the earth produced a winter so cruelly cold and long that it is still remembered.[238]

In 1612 Goclenius became professor of mathematics as well as physics at Marburg, and astrological treatises continued to come from his pen: *Urania divinatrix* or *Astrologia generalis* in 1614; Urania with twin daughters, that is, astronomy and astrology,[239] in 1615; in 1618, *Acroteleution astrologicum* in which he again attempted to distinguish "against new criminations" false astrology from true by reasons, examples and experiments.[240]

In the first of these three works, which accompanied a reprinting of the sixteenth-century treatises of Nifo on auguries and on critical days,[241] Goclenius prefaced to all three a dedicatory epistle, dated August 13, 1614, to the estates of Groningen and Ommelanden,[242] in which he noted that his date of publication was the same as that of the founding and opening of the new University of Groningen. In it he asked how anyone could oppose astrology. The wonders of the sky had not been placed by God before our eyes idly and without force. He admitted that many wild fancies of superstitious

[236] *Ibid.*, pp. 108-9.

[237] *Ibid.*, p. 30.

[238] *Ibid.*, p. 45. Goclenius again alludes to this winter of 1441 in his book of 1614, p. 111.

[239] *Urania cum geminis filiabus, hic est, Astronomia et Astrologica, speciali nunc primo in lucem emigrans*, Francofurti, 1615: BM 718.e.18 (1.).

[240] *Acroteleution astrologicum triplex hominum genus circa divinationem ex astris in scenam producens falsamque astrologiam a vera rationibus exemplis et experimentis distinguens contra novas criminationes*, Marburg, 1618, 78 pp.

[241] Copy used: BM 718.f.26. Nifo's treatises occupy pp. 1-68 and 69-143; then comes Goclenius, *Astrologia generalis*, with a new pagination, 1-150.

[242] "Groningae et Omlandiae ordinibus."

men had crept into astrology, and that it had been so confusedly handed down by the ancient astrologers that no one yet had found the right method to save youth a great expenditure of time and labor, so that many were deterred from the divine science. This he hoped to remedy. In the work proper he now cited Arabic astrologers, such as Alkindi and Haly Abenragel, Messahala and Albumasar, although he also noted a point on which Ptolemy and the Arabs disagreed.[243] He further cited such Latin astrological writers of the twelfth, thirteenth and fourteenth centuries as John of Seville, "Lincolniensis" on the weather in 1255, and John of Eschenden.[244] He discussed such matters of astrological technique as to what triangle or square of signs each people or nation was subject, and tables of dignities.[245] With reference to the latter, however, he held that those of essentials and accidentals handed down from the ancient astrologers were too laborious, and that it would suffice to observe merely the general and principal dignities in the table. But, despite his promised remedy in the preface, he now states that he does not propose to found new rules and canons from his own judgment but to collect in one repertory the opinions of the old astrologers as to varied weather changes.[246]

In his work of 1618 he carried this principle so far as to copy extensively and *verbatim* from Rantzovius without acknowledgement, as has been noted in our discussion of Rantzovius in previous volumes. He also, however, cited a number of past authorities by name as favorable to astrology. In his preface he shows the usual pessimism as to his own age, saying, "I do not think that there ever was any age so rich in contumelies and envy, so sterile in piety and sincerity, as ours." The text again maintains that the study of astrology must be maintained at all cost in well ordered universities as a public service.[247]

Even the use of astrological images for operative purposes was defended by Goclenius in a work defending himself from an attack by the Jesuit Roberti.[248] Goclenius protests that he has no faith

[243] *Ibid.*, 105, 117, 63.

[244] *Ibid.*, 71, 116, 29. The reference is to Grosseteste's *De aeris impressionibus*.

[245] *Ibid.*, 51, 71.

[246] *Ibid.*, 101.

[247] *Acroteleution*, p. 7.

[248] *Synarthrosis Magnetica opposita*

in the images of Ragael, Chael, Thetel, Hermes and the pseudo-Solomon.[249] He wonders whether his papist opponent is ignorant of the fact that a past papal physician, Arnald of Villanova, published a work on images.[250] Goclenius proceeds to quote at length its twelve seals for the twelve signs of the zodiac, and declares them too superstitious as involving prayers, carving of the names of angels, and the like. Goclenius's images are purely astrological and derive their virtue not from art or figure but from celestial radiation. An astral spirit flows into them in a purely natural way without any adjuration, consecration, and invocation of demons.[251] For Goclenius maintains that the sun, moon and other planets affect inferiors not merely by their light, heat and motion, but also by an occult and magnetic virtue,[252] and that occult properties come not from the elements but from the heavens.[253] Every thing or individual, when it comes under a determined constellation, receives a marvelous power of operating or suffering apart from that which it has as a member of a species.[254] Some thirty pages of the work are also devoted to another "Assertion of true astrology."[255] Frommann, writing in 1675, held that Goclenius had defended the employment of characters in the former treatise which Roberti had criticized,[256] and had altered his attitude in his reply to Roberti.[257]

Johannes Combach[258] (1585–1651), whose printed volumes on a great variety of subjects, including works of Roger Bacon,[259] number at least 174, stated his views with regard to astrology and the

---

infaustae Anatomiae Joh. Roberti D. Theologi et Jesuitae pro defensione tractatus de magnetica vulnerum curatione, Marpurgi apud Jonam Saurium impensis Petri Musculi, 1617; copy used, BN V.20949, where it is bound with the *Institutiones astronomicae* of Nicolaus Mulerius.

[249] *Ibid.*, 22.

[250] *Ibid.*, 87.

[251] *Ibid.*, 101-2.

[252] *Ibid.*, 103-5.

[253] *Ibid.*, 192.

[254] *Ibid.*, 193.

[255] *Ibid.*, 156-188, "Assertio verae astrologiae."

[256] Johann Christian Frommann, *Tractatus de fascinatione,* Nürnberg, 1675, p. 280, citing *Theatrum Sympatheticum,* p. 193.

[257] *Ibid.*, p. 293.

[258] On Combach see Fr. W. Strieder, *Grundlage zu einer hessischen gelehrten und schriftsteller Geschichte seit der Reformation,* Göttingen, 1781-1868, 21 vols. in 15, II, 244-62; III, 540-41; VIII, 507.

[259] *Perspectiva* and *Specula mathematica,* both at Frankfurt, 1614.

influence of the stars in a book printed in 1620.[260] Combach expressed himself unfavorably towards "the dreams and idle predictions" of the casters of nativities, but he did not lean in the opposite direction so far as to restrict the varying aspects of the stars merely to universal operations.[261] The fourteenth chapter of his *De homine* is on the relation of man to the heavens, whose influence upon man is accepted. The *Paramirum* of Paracelsus is quoted to the effect that planets, stars and all the firmament constitute no part of our body, do not act on us with respect to color, beauty, mores, virtues or peculiarities. If Saturn had never existed, there would nevertheless be men of Saturnine disposition. However, men and the rest of lower creation cannot exist without the stars and firmament, although they do not exist by them,

> because cold, heat and digestion of those things which we use and enjoy come from the stars. Man alone is not from them. And the stars are useful to us only insofar as we cannot get along without cold, heat, food, drink and air.

Combach, however, finds it difficult to accept this statement,[262] and he elsewhere notes that Paracelsus called man a microcosm,[263] and that chemists see in man another heaven, planets and stars, in accordance with whose motion and influence that microcosm is ruled.[264]

In the disputation on divination Combach allows astrology, physiognomy, chiromancy, and divination from dreams, if natural. Nature is rich and its exhaustless forces are shown by experiment today to exceed the bounds that many once set for them.[265] The force of putrefaction is so great that, if not the cause, it is at least the vehicle for seeds. Some think that some seed is left in the decaying animal which vegetates and generates again. But often generation occurs where there was no such animal before. Such

[260] *Liber de homine . . . Appendicis loco addita est disquisitio duplex, prior de caseo, altera de divinationibus*, Marburg, 1620. In the text itself the last item is headed, "Disputatio posterior de divinatione et astrologia judiciaria."

[261] *Disquisitio prior de caseo*, pp. 14-15. He again seems to condemn nativities at p. 26 *et seq*. of the other Disputation.

[262] *De homine*, pp. 194-95.

[263] *Ibid.*, pp. 158-59.

[264] *Ibid.*, p. 190.

[265] *Disp. posterior*, p. 23.

spontaneous generation Combach attributes to the stars, for the sky is full of forms in the sense that it produces them in inferiors. He does not deny, however, that worms are generated from the scattering of the seed of flies in decaying matter. But he tells a story of Albertus Magnus's saving a cowherd from being burned at the stake for sodomy together with a cow which had given birth to a human being by testifying that this might have been caused by the influence of the stars.[266]

Thus the attitude of Combach, like that of many of his contemporaries, is somewhat double-faced, pro and con, come and go, with reference to astrology. He rejects the excesses of the *genethliaci,* but will not limit prediction to generalities. He quotes Paracelsus against direct influence of the stars upon man, but prefers to regard man as a microcosm. He not merely attributes the spontaneous generation of lower forms of life to the heavens, but even suggests that a human being may be formed by their influence in a cow.

Helvig Dieterich, in his *Elogium planetarum caelestium et terrestrium, macrocosmi et microcosmi,* published at Strasburg in 1627 and dedicated to the landgrave of Hesse, held that the influence of the heavens was manifest in human sympathies and antipathies, astral diseases and remedies, Saturnine herbs, animals, stones, regions and cities. The planets should be observed in collecting herbs and compounding love philters. The hermaphroditic hyena was susceptible to lunar incantations.

Wilhelm Avianus of Thuringia published at Leipzig in 1629 a catalogue of stars for the benefit of astrology both genethliacal and meteorological.[267] Bound with it in the copy at the British Museum is a commentary on the Tables of Directions of Regiomontanus by Avianus which appeared much later.[268]

Daniel Beckher entitled *Medicus microcosmus* a work which first appeared at Rostock in 1622 and dealt with remedies drawn either

[266] *Ibid.*, pp. 11-13, 15.

[267] *Catalogi stellarum illustriorum 101 ex progymnasmatis nobilissimi Tychonis Brahe desumptorum ... in gratiam astrologiae tam genethliacae quam meteorologicae ad annum Christi 1633 diem 1 Maii supputati,* Lipsiae, 1629. BM 718.g.10. (1.).

[268] *Directorium universale,* with a table of contents of five parts, but only the first in found here, Lipsiae, 1665. BM 718.g.10.(2.).

from the living human body or the cadaver. The book has astrological features, such as collecting *usnea,* or the moss from the cranium of a man who has died a violent death, preferably on the scaffold, when the moon is waxing and in a favorable house. But it is more concerned with magnetic medicine and sympathetic magic than it is with macrocosm and microcosm or the influence of the stars, and so will be considered further in Chapter 34 on Medicine and Physiology.

David Origanus or Dost of Glatz, professor of mathematics at Frankfurt on the Oder, of whose Ephemerides for the years 1595–1630, printed in 1599, mention was made in our sixth volume,[269] had died in 1628. But it was not until 1645 that his effort to replace the vanity, superstition and impiety of past judicial astrology by a natural astrology of the effects of the stars, was published, and then it was printed at Marseilles[270] far away from Frankfort on the Oder, seat of the north German university and place of publication of his Ephemerides.

The Prooemium states that judicial astrology is of moderate or mediocre certainty, although not to be compared in this respect with astronomy. The influence of the sky is generally recognized, but whether the future can be predicted is disputed. The text proper is divided into four *Membra.* The first *Membrum,* on principles and fundamentals, outlines the properties of planets, fixed stars, and signs of the zodiac, the familiarity and dignity of the planets in the signs, astrological houses, and accidental dignities of the planets.

The second *Membrum* is on prognostication of general events, which may be based either on rare and long lasting causes or on anniversaries. Origanus holds that it is not impious to inquire into the celestial causes of political changes, and regards the time of the founding of a city or state as significant. Weather is predictable from the lord of the year, and particular forecasts are given for

[269] T VI, 60-61.

[270] David Origanus, *Astrologia naturalis sive tractatus de effectibus astrorum absolutissimus in quo omni astrologiae ut vocant judiciariae vanitate superstitione ac impietate . . . eversa vera physica coelestis . . . astruitur,* Massiliae, 1645, in-fol., 454 pp. BM 718.h.8.

single days. Crops, disease and pest, wars, conflagrations, earthquakes and floods are other matters of astrological prediction. Saturn is the significator of Jews; Jupiter and the sun, of Christians; Mars, of Turks and Tartars; Venus and the moon, of Ethiopians; Mercury, of learned men; Venus, of women. The status of magistrates is known from the sun; that of subjects, from the moon. Forecasts for journeys are taken from the third and ninth houses, also from the moon and Mercury which are general *significatores* for travel. The significance of a comet is judged from its place, position with reference to the sun, motion, figure and duration. A chapter on elections terminates the second *Membrum.*

The third is devoted to genethlialogy or nativities of individuals, including such matters as riches, brothers and sisters, parents, enemies and violent death. It is in vain that we inquire of the stars as to the saints of God, but ethnic religions not instituted divinely are without doubt subject to the heavens.[271] The quality of dreams depends on the planets which are *significatores.*

The fourth *Membrum* on special genethlialogical judgment treats of directions, annual profections, revolutions, and transits. An appendix then deals with the *trutina* of Hermes, *animodar,* correction of the nativity by the accidents of the individual, and the calculation of directions.

Lorenz Eichstädt (1596–1660) received the doctorate at Wittenberg in 1621, served as municipal physician at Stettin from 1624 to 1645, and then went to Danzig as professor of medicine and mathematics or astronomy. In 1624 he issued a treatise on theriac and Mithridatic. In 1625 he published a discussion of a new portent of five parhelia on May 25 at Alt Stettin in Pomerania. Various Ephemerides by him are dated in 1634, 1637[272] and 1639; and Harmonic Tables of the celestial motions in 1644.[273] Gassendi spoke of him in the preface to his Life of Tycho Brahe as "ille optimus Laurentius Eichstadius, Ephemeridum scriptor." Abdias Trew in 1663 cited his *Paedia astrologica, Introductio in parte Ephemeridum secunda,* sect. 3, for the influence of Mars upon acute diseases, but

[271] *Ibid.,* p. 299.
[272] BM 532.g.29.
[273] BN V.1814.

added that his arguments against astrologers in his *De crisibus*[274] were unworthy of an answer and of so great a man.[275] Morin, in the introduction to his *Astrologia Gallica,* said that he might have examined the hundred aphorisms which Eichstadt in the third part of his Ephemerides had taken from Pico della Mirandola, Argolus, and especially Kepler, who was totally ignorant of astrology. A very few of them were true, most of them false, and many contradictory. But he had no time for such nonsense.[276] Other pharmaceutical writings by Eichstadt than that on theriac were on Alchermes in 1634[277] and camphor in 1650.[278]

Franciscus Tidicaeus, born at Danzig in 1554, Ph.D. and M.D., was municipal physician and professor of medicine at Thorn. In 1607 he had published a book on theriac, containing the Greek poem of Andromachus on theriac with two Latin translations of it, other descriptions of theriac from classical authors, and a long commentary with a digression on *mumia* and weapon ointment.[279] He died in 1617, and a long work by him on the microcosm[280] contains a letter addressed to him by Bartholomew Keckermann from Danzig dated January 12, 1608, a *Privilegium* of 1613, and a dedicatory epistle by Tidicaeus dated January 1, 1615, from Thorn. But the volume does not seem to have been printed until 1638 at Leipzig. A preface to the reader by the pastor of the church at Thorn speaks of the book as "long in press and now at last published" and of "the grateful memory of the name of Tidicaeus." Beckher,

[274] Trew perhaps had reference to the work by Eichstädt which LR 731, describes as, "De Diebus Criticis, Libellus. Extat pag. 321 Partis alterius Ephemeridum, Stetini, apud Georg. Rhetium, 1639, in 4."

[275] Abdias Trew, *Astrologia medica quatuor disputationibus comprehensa,* Disp. III, sect. 32.

[276] *Op cit.,* 1661, p. xvi.

[277] *De confectione Alkermes*... Stetini, 1634: BM 546.d.22.(1.).

[278] *An camphora Hippocrati Aristoteli Theophrasto... fuerit incognita, et quid de eius ortu, natura... et usu recentiores medici prodiderunt?* BN S.3782.

[279] *De theriaca,* 1607, 320 pp. and Index. Copy used: BM 1038.i.29.

[280] *Microcosmus, hoc est, descriptio hominis et mundi...,* Lipsiae, 1615, in-4, 762 numbered and at least a hundred more unnumbered pages. The date, 1615, is given by the British Museum printed catalogue for the volume with the shelf mark, 549.f.2, but I did not see it as the date of publication in the volume itself. For the full Latin title see Vander Linden (1662), 183; LR 301, where the volume is dated, "Lipsiae apud Nerlichios, 1615, 1638."

Medicus microcosmus, 1633, cites Tidicaeus, *De theriaca,* of 1607 but not his *Microcosmus,* which had therefore probably not yet appeared in 1633.

The work of Tidicaeus, although not quite so long as that of Nancelius, is like it filled with quotations, often in Greek or poetry. After two prolegomena and a general introduction, the relation of man to God is considered, then the fact that man is the image of the whole world. Then we are told how the heavenly bodies with their spheres are expressed in man, how he exhibits the four elements and four qualities, how meteors and inanimate things, plants and animals are all delineated in and represented by man. The sun puts before our eyes the nature of science; the erratic stars, the variety of opinions. Diseases in plants and men correspond. We hear of the geometry of bees, the lion and his generosity, examples of gratitude in brutes. But man alone regards the heavens.[281] Even works of art are held to be contained in man, and he is compared to a house, and his cranium to a roof. States of life, too, are depicted in him, political, ecclesiastical, economic. Even then Tidicaeus cannot stop. There is an Appendix on man himself being in man himself—a marvel and yet proved by experience. There is an *Appendicula* on the approaching end of both macrocosm and microcosm. And finally an Epilogue of 120 pages on how man by consideration of the relations between the two worlds may progress to God.

The *Microcosmus* of Tidicaeus was called "an evidently new work" (*opus plane novum*) in the aforesaid *Privilegium* of 1613, although Guibelet had published his book on the subject in 1603, and Nancelius his tremendous tome in 1611.

We have seen in previous volumes that it was a not uncommon practice for astrologers to accompany their annual predictions by a more general introduction in which they discussed some astrological theory or other scientific or pseudo-scientific question. This custom continued in the seventeenth century, as may be illustrated by the case of Albert Linemann who from 1636 to 1654

[281] For the six passages immediately preceding: *Ibid.,* 277, 443, 487, 490-91, 493, 529. Otherwise my description follows the order of Tidicaeus's ten tractates.

accompanied his annual prognostication and calendar by sometimes as many as twelve questions. These discussions were brought together by Anne Linemann under the caption of secrets of nature in a single volume published at Königsberg in 1654,[282] and dedicated to the astronomer Hevelius.

A few of the questions may be noted as illustrative of the popular scientific interest of that period. The first query from the "Prognostico des 1636" is whether air is colder than earth and water. Others are why clouds do not fall, why days are longer than stated in the Calendar, why there is greater heat in mid-spring than in summer, when dog-days begin and end, and whether one should observe the planets "in der Metall-Arbeit." As to whether leapyear is a good time to set out trees and plant a lot of cabbage, Linemann answers that it is no different from any other year in this respect. He also answers in the negative the first question of the prediction for 1637, namely, Can the weather for the entire year be predicted from its twelfth day? The second problem is how long it would take a mill-stone dropped from the stars to reach the earth. Queries five and six are whether all the stars get their light from the sun, and whether the slight light of the eclipsed moon is its own or from the sun. Among the problems for 1638 are why astrologers are often at fault in weather prediction; whether it would be possible, if the moon, Venus and Mars did not exist, to find the distance of the sun from the earth—which is answered affirmatively; and why the powder tower at Königsberg was struck by lightning in 1636. In 1651 there were ten questions; in 1652, eleven, of which we note only the first two: namely, how far the shadow of the earth stretches, and whether more trees can grow on a mountain than in a plain of the same size as the base or surface on which the mountain stands. Among the twelve queries for 1653 are the old one why the sun and moon seem larger near the horizon; why the sun was blood-red at Königsberg on February 25, 1652; why head-wounds made in full light are far more dangerous than those made by moonlight, for which an astrological explanation is given; and why the bricks in

[282] *Deliciae Calendario-Graphicae . . . Geheimnüsse der Physic, Astronomi . . . aus den jährlichen Calendar des A.L. . . . zusammen getragen*, Königsberg, 1654, in-4. BM 8562 bb.44.

a building gradually disintegrate, while the mortar holds firm. Finally in 1654 it is asked why a good year for wine is a bad year for vegetables and vice versa; why plain glass windows frost more than stained glass in wintry weather; and why horses are less subject to plague than cattle.

A fleeting glance in the direction of Czecho-Slovakia shows that Wenzel Budowez, a counselor of Ferdinand II who was beheaded by that monarch's order together with other rebels at Prague in 1621, had published at Hanover in 1616 a Circle of the Solar and Lunar Clock, that is, a brief synopsis, historical, figurative and mystical, illustrated by various figures and emblems, representing from Old and New Testament a continuous series of the chief changes in church and state.[283] Events before Christ are grouped under twelve hours of the moon, and those since under twelve hours of the sun. The two closing chapters are "De cacochymicis" and "De tempestatibus mundi." But the work is primarily religious.

vi. NORTHERN EUROPE

Caspar Bartholinus (1585–1629) had seven distinguished sons. One of them, Albert, in *De scriptis Danorum,* a work published posthumously by *his* son, Thomas, in 1666, and then reprinted with revision and additions by Johann Moller in his *Bibliotheca septentrionis eruditi,* 1699, gave a full list of his father's works in the varied fields of rhetoric, logic, metaphysics, theology, medicine, anatomy, physics, elements, waters, minerals, amulets, occult qualities, and pygmies. In the work which now concerns us Caspar speaks of himself as *adolescens* in both dedication and preface. The dedication is dated in 1607 at Wittenberg, but I have used the seventh edition of 1624.[284] This rapid republication would seem a sign of

[283] For the Latin title, which I have translated, see Johannes Hallervordius, *Bibliotheca curiosa,* 1676, p. 402. For a copy of the book itself: BN G.3743 (1).

[284] *Astrologia seu de stellarum natura affectionibus et effectionibus exercitatio qua difficultates praecipuae de stellarum definitione, causis, ordine, divisione, quantitate, coloribus, luce et lumine, motu ingenito, distantia a terra, scintillatione, de calore coelesti, influentiis, praedictionibus astrologicis, eclipsibus, de maculis lunae, via lactea, de stella Magis exhibita, de novis nostri seculi stellis etc. succincte et nervose expediuntur.* Cum gratia et privileg. Elect. Sax. Cum indice quaestionum. Editio septima correctior mendosa tertia et melior. Apud

popularity. As the long Latin title just quoted in a footnote shows, the work is a rapid survey of matters both astronomical and astrological. Its 260 tiny pages cite almost that number of authors. The stars are said to act on these inferiors by a certain peculiar virtue transmitted with light, and to have other actions on inferiors than that of heat. The question is discussed whether any natural effect with respect to the sky happens by chance, and the influence of the stars is limited to material phenomena, while the will is left free except as it may be influenced by accident through material creation. It is granted that astrologers differ greatly in their methods of predicting. But is is held that nothing in astrology is necessarily contrary to Scripture; rather the two are often in wonderful agreement. The solar eclipse at the time of the Passion was not natural, and the star which appeared to the Magi was not a new creation, since God had created a perfect world to begin with, but was miraculously produced in the air from sublunar matter. Bartholinus goes on to say that it was shown to the Magi as astrologers, as if they would have mistaken a mere aerial apparition for a star. Yet he accepts the existence of new stars other than comets in the ethereal region above the moon, and inclines to restrict the term, comet, to natural or preternatural phenomena generated in the air. It is most certain that such celestial apparitions warn us of divine vengeance and are announcers of the future. Some astrologers go too far, but no discipline or science is so maligned as astrology. Astrological prediction is either certain, as in the cases of forecasting eclipses; or probable, as in the case of the weather and of disease; or false, as in the case of the superstitious Chaldeans and their followers.

In 1609, Sigfridus Forsius, professor of mathematics at the University of Upsala, published at Stockholm in Swedish an astrological prediction for that year and another for the years 1611–1620 with a long preface addressed "to the senators" of the kingdom of Sweden. Two years before he had dedicated to king Charles a discussion of

---

Casparum Heiden Bibliop. Anno MDCXXIV. Copy at New York Academy of Medicine. I have since examined the fifth edition of 1612 with the same title: BM 718.a.19, 264 pp.

comets in general and of that of 1607 in particular, while in 1608 he had treated of celestial apparitions with especial attention to parhelia.[285]

Annual predictions for 1608 and 1622 and other astrological writings by Niels Hansen Heldvad (1564–1634) are listed in Nielsen's bibliography of Danish mathematicians.[286] In the years 1621, 1622, 1624 and 1625, C. S. Lomborg (1562–1647) successively issued four disputations on astrology. The first maintained against its adversaries that it possessed some certitude, and that the stars influenced by their motion and light. The second disputation dealt more specifically with their influence on sublunar things. The third discussed the disposition of sublunar matter to receive this influence, and the fourth, the effect of solar and lunar eclipses.[287]

Johannes Franckenius (1591–1661), professor of medicine at Upsala, published there in 1626 a discussion of the influence, force and efficacy of the celestial stars upon sublunar bodies, and Nicolaus Malmenius, who was to die seven years after Franckenius, issued a similar work there, likewise seven years after the other, in 1633.[288]

[285] All four works were in Swedish, but their titles are given in Latin by Johannes Mollerus in his *Bibliotheca septentrionis eruditi*, part 2, *Joh. Schefferi Suecia Literata, Hypomnematis Historico-Criticis ab eodem J. Mollero illustrata*, 1698, pp. 68-69.

[286] Niels Nielsen, *Matematiker i Danmark, 1528-1800*, Copenhagen, 1912, pp. 92-93.

[287] *Ibid.*, p. 133, for the full titles of the four disputations.

[288] For the Latin titles see Moller, *Suecia literata*, pp. 123, 331. Neither work appears in the BM and BN printed catalogues. Another Malmenius, Andreas A., published *De veritate astrologica* in 1674 at Dorpat in Livonia. Heinrich Eckstorm, *Historiae eclipsium, cometarum et pareliorum . . . collectae cum eventuum quos portenderunt narrationibus succinctis . . .*, Helmaestadii, 1621, may also be noted.

CHAPTER VI

ALCHEMY AND IATRO-CHEMISTRY TO 1650

Alchemical scribbling—Collections of past literature—Basil Valentine—Works of the first years of the century—Bongars—Sendivogius—Ruland—Reinneccer—Tanckius—Dienheim—Michel Potier—P. Müller—pseudo-Flamel—Roussel—Rosicrucians—Angelo Sala—Potable gold: Anthony and Gwynne—Timothy Willis—Michael Maier—Besard—Scheunemann—Poppius—Hildebrandt—Finck—Mylius—Bartoletti — Cornacchini — Goclenius — Nuisement — Thornborough — Burggrav — d' Espagnet—Censure by the Sorbonne—Chandoux—Solombrino—Dieudonné—1630 edition of Norton—Arthur Dee—Jean Saignier misdated—Billich—de Clave —Rhumelius—P. J. Fabre—Zaccagnini—Locatelli—*Non-Entia Chymica*—Glauber —Conring—Zobell.

For preparing this (tincture of coral) legitimately, they list as many processes as there are flies in Armenia or poets in Germany

—*Non-Entia Chymica*

Some writer of the early seventeenth century whose name has escaped me said that the alchemists of his day spent more time in scribbling than they did in experimenting. This tendency is illustrated by various manuscript collections such as that of Agnolo della Casa of Florence who, between 1592 and 1618, filled some eighteen volumes of from 100 to 900 leaves each with matter that was mainly alchemical.[1] And much of this scribbling, and even of what was printed, consisted of quotations *ad nauseam* from earlier writers. But this failing also too often characterizes books of that period in other fields. Pierre Borel, in his mid-century bibliography, professed to list 4000 chemical authors (or titles?) past and present,[2]

[1] Florence, National Library, Palatine 867, tomes I-XVII, XIX.

[2] *Bibliotheca Chimica seu catalogus librorum philosophicorum hermeticorum . . . usque ad annum 1643*, Paris, 1654; Heidelberg, 1656. Copy at NYAM. There are not that many authors or even titles in its 272 tiny pages, as many are mere names mentioned in other books, while other items are repetitions and cross references.

but still omitted many according to Borrichius.[3] Of the vast output of alchemical literature in the seventeenth century only a few specimens and samples will be here presented.

That the writings of past adepts and authorities in alchemy still commanded a wide circle of readers as well as scribblers, is seen from the currency of Zetzner's *Theatrum Chemicum*, an *omnium gatherum* of such literature. First issued at Ursel in 1602 in four volumes, it was reprinted at Strasburg in 1613. Many of the treatises were from hitherto unpublished medieval manuscripts. A fifth volume was added in 1622. From 1659 to 1661, also at Strasburg, appeared the final edition in six volumes which is today usually cited, each containing a score or more of treatises.[4]

Meanwhile Nicolas Barnaud, under the title, *De occulta philosophia*, Leyden, 1601, printed several past alchemical works which were to re-appear in Zetzner. Others, ascribed to Roger Bacon, were published in *Sanioris medicinae* . . . , at Frankfurt in 1603. Joachim Tanckius edited *Opuscula Chemica* in German at Leipzig, 1605, and others in *Promptuarium alchemiae*, in two volumes of 1610 and 1614. Borel lists as published at Frankfurt in 1605, Six Very Old Writings of Chemical Philosophy, and, Very Old Writings of Chemical Philosophy latinized from Arabic from the Bodleian Library.[5] In 1608 Benedictus Figulus edited two alchemical collections: *Thesaurinella Olympica aurea tripartita*, at Frankfurt, dedicated to Rudolf II, and containing works of Paracelsus, Bernard, Koffsky and Raymond Lull;[6] and *Pandora magnalium naturalium*,

[3] Olaus Borrichius, *De ortu et progressu chemiae*, Copenhagen, 1668, p. 143.

[4] The first five volumes were practically identical in content with the four of 1613 and fifth of 1622, except that in the fourth volume were added Avicenna, *De congelatione et conglutinatione lapidum* (at pp. 883-87), and the work of Guilhelmus Tecensis, *Lilium de spinis evulsum*, with a colophon, no doubt reproduced from the MS used for the text, by a Hungarian bachelor of arts who copied the treatise in 1557 for Rheticus (see p. 911).

[5] Borel (1654), 60: *Chemicae philosophiae sex vetustissima scripta*, Francof., apud Io. Berner, in-8; *Philosophiae chimicae vetustissima scripta ex Arabice sermone latina facta*, Francof., in-8, ex Bibliotheca Bodleiana. The latter, of which a more exact title is *Philosophiae chymicae IV vetustissima scripta*, comprised four tracts already printed in 1566: T V, 622; Duveen 472.

[6] I have used an edition of 1682 at NYAM. For the 1608 ed.: BM 1033. h. 9 (2.).

at Strasburg, in which are *Liber Apocalypseos Hermetis* and various treatises by Alexander à Suchten.[7] But most of the medieval alchemical works which J. D. Mylius printed in the second book of his *Philosophia reformata* of 1622, under the caption, "De authoritatibus philosophorum," had already been included in the first volume of *Artis auriferae quam chemiam vocant* in 1597. Its second volume appeared in 1610.

At Frankfurt in 1625 H. Condeesyanus had published his *Harmonia inperscrutabilis chymico-philosophica* in two volumes, each of which contained a decade of treatises, for the most part by medieval authors. In 1647 Ludwig Combach edited such alchemical works as those of Ferrarius, John Belye, Edward Kelley and John Isaac Hollandus; and in 1649 the *Opera omnia* of George Ripley of the fifteenth century. In mid-century Zetzner's title and collection were imitated by Elias Ashmole in his *Theatrum Chemicum Britannicum*, published in 1652 at London, of which, however, only the first part in one volume appeared. Two years later there came out at Paris the *Bibliotheca Chemica ad annum 1653* of Pierre Borel, a bibliography already mentioned. Also in 1653 there was issued at Geneva the *Bibliotheca Chemica Contracta* of Nathan Albineus which comprised the Emerald Tablet of Hermes, *Chrysopoeia* of Augurellus,[8] another treatise which had already been printed by Zetzner, and works by Sendivogius and d'Espagnet which were here represented as anonymous.[9] In the preface Albineus advised further reading of the opuscula of Dionysius Zacharias, the Testament of Raymond Lull, Dialogue of Aegidius de Vadis, Ficino (pseudo) *de arte chemica* with the fourteen questions of the necromancer Illardus, the Secret of Wisdom of Jodocus Greverus, Khunrath's *Amphitheatrum*,[10] the poem and hieroglyphic figures of Nico-

[7] Concerning whom see T V, 641. Figulus wrote or edited other alchemical treatises in 1600, 1609, etc.

[8] T V, 534-35.

[9] *Cosmopolitae Novum lumen chemicum*, second tract or part at pp. 89-175, (otherwise all the items have each its own pagination) opening, "Sulphur non est postremum inter principia . . ."; *Anon. Galli Enchiridion Physicae restitutae*, 178 pp., opening, "Postquam nuper a publicis curis . . ."; and *Arcanum Hermeticae philosophiae opus*, 80 pp., with the incipit, "Divinae huius scientie principium est . . ." See the treatment of Sendivogius and d'Espagnet later in this chapter.

[10] Treated in our chapter 10 on Natural Magic.

las Flamel, and, for the connection between geometry and nature, the *Monas Hieroglyphica* of John Dee. In 1696 the *Conspectus scriptorum chemicorum,* a posthumous work of Borrichius, appeared at Copenhagen, and in 1702 most of Zetzner's content was reprinted in J. J. Manget's *Bibliotheca chemica curiosa,* in two large folio volumes.

There are no manuscripts of Basil Valentine earlier than the middle of the seventeenth century, and the name first appears in 1599.[11] Works of alchemy under his name, which were really composed by Johann Thölde, with such titles as occult philosophy, the microcosm, and the triumphal chariot of antimony, were first printed in German in the early years of the seventeenth century,[12] and did not appear in Latin translation until the forties.[13] Meantime, however, the German text was quoted by writers like Rudolph Goclenius of Marburg in 1625[14] and Zacharias Brendel of Jena in 1630.[15] Basil Valentine was supposed to have been a monk of the fifteenth century and precursor of Paracelsus, who in point of fact had genuine forerunners in the alchemists of the fourteenth century, when too John of Rupescissa had sung the praises of the fifth essence of antimony, centuries before the *Triumph-Wagen Antimonii* was published.

Oswald Croll (1580–1609) and his *Basilica chimica* of 1609 have been treated in a previous volume,[16] and the chapter on Libavius,[17]

[11] Dr. Loth, *Die dem Erfurter Mönch, Altchemisten und Arzt Basilius Valentinus zugeschrieben Handschriften der Kirchenbibliothek zu Neustadt am Ausch,* 1905, and the review thereof by Karl Sudhoff in *Mitteilungen zur Geschichte der Medizin,* V (1906). No initials are given for Dr. Loth of Erfurt.

[12] Goclenius cites *De occulta philosophia,* 1602. The first German edition of *Triumph-Wagen Antimonii* was in 1604; the first Latin edition, *Currus triumphalis antimonii,* in 1646. The German text on the microcosm was printed at Leipzig in 1602; at Marburg, in 1609. The first German edition of *De occulta philosophia oder von der heimlichen Wundergeburt der sieben Planeten und Metallen* was in 1602 or 1603; the second, in 1611.

[13] See previous note. *Haliographia de praeparatione usu ac virtutibus omnium salium . . . ex manuscriptis Basilii Valentini,* Bologna, 1644. German editions at Eisleben, 1603; Leipzig(?), 1612.

[14] Goclenius, *Mirabilium naturae liber,* 1625, pp. 201-4-6, with a Latin translation. The work is posthumous, Goclenius having died in 1621.

[15] Brendel, *Chimia in artis formam redacta,* 1630, p. 209.

[16] T V, 649-51.

[17] T VI, 238-53.

who lived until 1616, involved discussion of the controversy between the Paracelsists and the medical faculty of Paris which began in 1603, of Duchesne or Quercetanus, Le Paulmier or Palmarius, Nicolas Guibert, Hoghelande, and Israel Harvet.

Going back to the first year of the seventeenth century, we encounter several works for the first time. Nicolas Barnaud put forth at Leyden in 1601 a *Tractatulus chemicus* as well as the *De occulta philosophia* which has been already mentioned. The New Alchemy of Giovanni Battista Birelli also appeared in the first year of the century in Italian at Florence,[18] and was printed in German translation at Frankfurt in 1603,[19] while Copenhagen saw the publication of a Latin version in 1654.[20] Yet another alchemical work was issued in 1601 at Magdeburg: Martin Copus, *Apotelesmata philosophica Mercurii triumphantis.* It was in the nature of a key to a poem entitled *Mercurius triumphans* which precedes it and which had been dedicated to Rudolf II in 1599.[21] At Magdeburg, too, in the year following, John of Padua published his *Philosophia sacra sive praxis de lapide minerali,* while at London in 1602 Thomas Russel published a tract on a powder, *Diacatholicon aureum,* and at Antwerp in 1604 Willem Mennens (1525–1608) printed *The Golden Fleece* (*Aureum Vellus*). Thus aside from the Paracelsist controversy which was breaking out at Paris, and the publications of Guibert, Hoghelande and Libavius at Strasburg, Cologne and Ursel, recorded in our previous volume, there was printing of alchemical works in the Netherlands, Italy, England, and elsewhere in Germany.

Jacques Bongars (1554–1612), noted as a historian, a collector of manuscripts, and a correspondent of Tycho Brahe, was also

[18] See Ferguson and BM 1033.i.5. Zedler speaks of a Bologna edition of 1600 and dates the Florentine edition in 1602.

[19] BM 8905.c.15.

[20] Ferguson cites two editions for that year.

[21] In BM 837.g.20 (1.) the title page is missing for the poem, which is in five books and covers 79 pp. BM 837.g.20 (2.) is *Apotelesmata philosophica Mercurii triumphantis . . . in quibus elucidatio et clavis totius operis,* Magdeburg, 1601. It contains in prose 61 *Apotelesmata* for Book I of the poem, 55 for II, 91 for III, 71 for IV, and 105 for V. These are followed by another poem entitled *Eutopia* in one book, and then another poem in honor of Cope.

something of a Paracelsist. In a manuscript preserved at Berne[22] he states that Paracelsus took refuge in loftier and more secret philosophers than Hippocrates and Galen. Bongars especially recommends "our balsam," which is beyond the elements, and declares that one must know the universal harmony of all creation. He says that Paracelsus recognized four elements divided between two globes, the upper containing the heavens or fire and air; the lower, water and earth.[23] Plato set forth three *principia*—God, exemplar and matter; Aristotle's three were matter, form and privation; while Paracelsus substituted salt, sulphur and mercury. Presently Bongars gets to "human astronomy" or man the microcosm, and associates certain diseases with the five planets. Discussing the generation and transplantation (or magic transfer) of diseases, he affirms that from the beginning pure and perfect seeds of things received the power of generation in the Word, but after the fall of man new tinctures came in, by whose mixture "is transplanted the beauty of the universe."[24]

The *Novum Lumen Chymicum* of Sendivogius was first published at Prague in 1604, and subsequently at Paris, 1608; Frankfurt, 1611; Cologne, 1614; Geneva, 1628; and Venice, 1644.[25] A French translation, entitled *Cosmopolite, ou nouvelle lumière de la physique naturelle . . .*, appeared in 1609, 1618, 1639, 1669, 1691 and 1723. Sendivogius was supposed to have achieved transmutation at the Polish court, at Prague in 1604 before the emperor Rudolf, who himself made the projection, and at Stuttgart in 1605 before the duke of Wurtemberg.[26] He was kidnapped by a Moravian count and also by an alchemist of Wurtemberg who later was caught and hanged for it in a robe of tinsel on a gilded gallows.[27] In 1624 Orthelius composed in German a commentary on the *Novum lumen* of Sendivogius which, as later translated into Latin, fills over one

[22] Berne 492, pages unnumbered. There are 14 chapters.

[23] This view is attributed to Severinus in his *Idea of Philosophic Medicine,* 1573, by Barchusen, *Historia medicinae,* 1710, pp. 442-43.

[24] Berne 492, cap. 12.

[25] Carl C. Schmieder, *Geschichte der Alchemie,* Halle, 1832, lists editions of Frankfurt, 1606 and Cologne, 1610.

[26] Ferguson, II, 368.

[27] Christoph Gottlieb von Murr, *Litt. Nachrichten d. Gesch. des sogenannten Goldmachers,* 1805, pp. 54-79.

hundred and fifty pages in Zetzner's *Theatrum Chemicum* of 1661.[28] Sendivogius' Dialogue between Mercury, an Alchemist and Nature also ran through several Latin editions and was translated into French and English.[29] There is not a little alchemical writing by him in manuscripts of the seventeenth century in the Sloane collection of the British Museum.

Martin Ruland the Younger had issued in 1607 a work in three parts consisting of chemical problems and the true way of making the philosophers' stone.[30] The work has sometimes been erroneously ascribed to Martin Ruland the Elder. The dedication, dated from Ratisbon or Regensburg on May 10, 1606, is signed, "Martinus Rulandt." Of the three parts, the first contains 64 problems, followed by chemical remedies.[31] Part Two is an Appendix of Chemical Questions, containing the remainder of the problems, numbered from 65 to 91. Part Three contains two treatises on the philosophers' stone of twenty and twelve chapters respectively. The former was reprinted by Manget in 1702 as the work of Marsilio Ficino, to whom we have already heard it ascribed by Albineus in 1653, but its closing chapter of questions, which a philosopher named Ylerdus in Catalonia put to a spirit, dates back to manuscripts of the fifteenth century in which Elardus or Hilardus or Hylardus, as his name is there spelled, is more frankly termed a necromancer.[32] as indeed he was by Albineus, who called him Illardus.

Among the questions asked earlier are: whether alchemy is a part of philosophy, an art, or a figment of the imagination; whether salt, mercury and sulphur are perfect principles of mixed bodies or inventions of chemists; whether *medicus* and *chemicus* are the same, which is denied. Whether the function of a physician should

[28] VI, 397-458.

[29] *Dialogus Mercurii Alchymistae et Naturae,* Paris, 1608, 12mo; Cologne, 1612, 1614; Wittenberg, 1614, 1623; Venice, 1644.

[30] Martin Ruland fils, *Progymnasmata alchemiae sive Problemata chymica nonaginta et una quaestionibus dilucidate cum lapidis philosophici vera conficiendi ratione,* Francofurti, 1607, 3 parts of 254, 138 and 165 pp. in one vol. in-8: BN R.12460-12461. The third part is dated 1606 on its separate title page.

[31] At p. 226, after the 64th Problem is printed "Finis", but in p. 227 begin "spagyric remedies thus far in common use," and at p. 237 an Appendix of Other Medicaments.

[32] Manget, *Bibliotheca Chemica,* II, 172-83; DWS II, 713, item 1067; T IV, 573.

be distinct and separate from that of a pharmacist, which is likewise answered in the negative. After a group of questions concerned with salts, it is asked whether chemical remedies are safe and whether they are to be repudiated because of their supposed dissimilarity with the human body; whether purgatives lose their force with distillation; whether the natural temperament of a thing is lost thus; whether chemical essences lose the qualities of the elements? That quicksilver is a poison is denied, and that a mercurial girdle can be worn without harm is affirmed. The fifty-first problem is whether the philosophers' stor e is a catholic and universal medicine, and the next whether remedies should be like or contrary to the disease. Can potable gold be made, and is it of use for the conservation and prolongation of human life? Is all fire natural and none artificial? Should oil or spirit of vitriol be rejected from medicine as erosive? Is the making of the philosophers' stone naturally possible? Is it to be sought from imperfect metals and from minerals, or from gold and silver, or from the elements, to which the answer is yes, or from vulgar mercury? Are chemical essences to be shunned as too hot; can medicine do without chemistry and chemical remedies; are the old remedies, because long tested, to be preferred to new unexplored chemical remedies? Whether chemical remedies, because they are sold at a high price, should be abolished as harmful to the state?

It may be noted further that Martin Ruland the Younger issued an alchemical lexicon in 1612.[33]

One might expect from its title that The Chemical Treasury of Most Certain Experiments, collected and proved in use, by Fidejustus Reinneccerus,[34] printed in 1609, would prove to be interesting and even exciting reading and mark a stage in the development of experimental method. But it is merely a collection of medical or pharmaceutical prescriptions, arranged in six books for diseases of the head, thorax, abdomen, diseases of women, various other kinds of diseases, and fevers. Reinneccer had been an apothecary at

[33] *Lexicon alchemiae*, 1612, 471 pp.: BN R.8478.

[34] *Thesaurus chymicus experimentorum certissimorum collectorum usuque probatorum, cum praefatione Joachimi Tanckii,* M.D., Lipsiae, 1609, 191 pp.

Saalfeld, and Janus Baccerus, who succeeded him, published the work at the suggestion of Joachim Tanckius, M.D.,[35] who contributes a preface. In it he discusses how the material of a universal medicine is to be found and prepared, how it must harmonize with macrocosmic and microcosmic stars, and how it works on man the microcosm, adducing past alchemical authorities. Thus in the main body of the book the word, experiment, is used in the sense of a medical recipe, while the preface is based on authority and savors of magic and astrology.

Tanckius had already contributed a preface to the first German edition in 1604 of the *Triumph-Wagen Antimonii* and in 1605 had published a brief manual of chemical instruction,[36] an *Alchimistisch Waitzenbäumlein,* and a German translation of works of Bernard of Treves or Trevisan, besides editing chemical *Opuscula,* as we have seen.

The theme of a universal medicine, which we have just heard discussed by Tanckius in his preface of 1609, was in the following year made the subject of a distinct, although brief, book by Johann Wolfgang Dienheim, professor of medicine at the University of Freiburg-im-Breisgau.[37] A German translation appeared at Nürnberg in 1674.[38] The original Latin edition of 1610 was dedicated to Maximilian, archduke of Austria and supreme master of the Teutonic Knights, while a preface to the reader states that the object of the volume is to demonstrate that a universal medicine can be found, fit to cure all diseases. Dienheim does not deny that diverse medicines may be applied in diverse diseases according to the quality of each disease, but he contends that they may be sharpened and their effects rendered more certain in one universal medicine.[39] It is the quintessence of the four elements, and its maker must

[35] The preface by Tanckius opens, "Janus Baccerus, vir doctissimus, amicus meus singularis, lector benevole," and continues, "e penu chymico sui antecessoris depromptas opes meo instinctu et suasu . . . in publicum emisit."

[36] *Succincta et brevis artis chemiae instructio,* Leipzig, J. Rose, 1605, 106 pp. It will be discussed in Chapter 27.

[37] *Medicina universalis seu de generali morborum omnium remedio liber,* apud Lazarum Zetznerum, Argentorati, 1610, in-8. I have used a copy in NYAM. 78 pp.

[38] *Taeda trifida chimica,* copy at BM 1033.c.12 (4.).

[39] *Op. cit.,* p. 34.

follow the instructions of the Emerald Tablet of Hermes. It is wonderful by how many names the philosophers have designated the material of this medicine, whether in order to conceal or to explain it, Dienheim does not know. He also discusses the philosophers' stone but admits that he does not know how to make it. But the *Medicina Catholica* he has made once and will do so again. But the time has not yet come to publish it to all.[40]

Some say that the elements are discordant and cannot be combined into one inseparable and homogeneous quintessence and universal medicine, but Dienheim asserts that experimentation has proved that they can be.[41] Arguing from Greek mythology and Egyptian hieroglyphs, he identifies Saturn, Mars and Venus (lead, iron and copper) with earth; the moon and Mercury (silver and quicksilver), with water; and the sun and Jupiter (gold and tin), with the other two elements, fire and air.[42]

Incidentally Dienheim alludes to the ability of the salamander to withstand fire, and to spontaneous generation as established facts.[43] He similarly accepts occult qualities, saying:

> I am silent as to those which have a secret in themselves and an occult power of healing, which they work not by force of hot or cold, wet or dry, but from specific virtue which has its place among the hidden causes of things, of which I could bring forward a thousand examples, if need be.[44]

The treatise of Petrus Arlensis de Scudalupis on the Sympathy of the Seven Metals and Seven Selected Stones to the Planets, which appeared at Paris in 1610, and the commentary on it by Albinius, printed in 1611, have already been treated in our sixth volume.[45]

Michel Potier was primarily, if not exclusively, an alchemist[46] who called himself the first Hermetic philosopher of the age, and traversed all Europe asserting possession of the greatest secrets,

[40] *Op. cit.*, pp. 5, 11, 13, 64, 72.

[41] *Op. cit.*, p. 38.

[42] *Op. cit.*, p. 16.

[43] *Op. cit.*, pp. 38-39.

[44] *Op. cit.*, p. 35.

[45] T VI, 301-2, 324.

[46] He is, however, listed in LR, and Zedler, 28, 1878, calls him "ein Medicus und Philosoph."

but who died in poverty and neglect at Dortmund.[47] Numerous alchemical treatises by him appeared at various dates between 1610 and 1648.[48]

The dedicatory epistle by Philipp Müller (1581–1659) of his Chemical Miracles and Medical Mysteries to Maximilian III, archduke of Austria, is dated from Freiburg-im-Breisgau on August 4, 1610. The book first appeared in print at Leipzig in 1611.[49] A second edition soon appeared at Wittenberg together with Sendivogius and the *Tyrocinium chymicum* of Jean Beguin in 1614,[50] and a third there in 1616, and a fourth in 1623.[51] Other editions followed at Paris, 1644; Rouen, 1651; Wittenberg, 1656; Amsterdam, 1656, 1659 and 1668; Geneva, 1660.[52]

Of the five books into which the work is divided, the first has chapters on instruments, with ten figures, on the material of the philosophers' stone in general, on mercury, its preparation and purification, sun and moon (i.e., gold and silver), and the work itself in the usual seven stages. Book Two is on particular transmutations, with a figure of a furnace.[53] Book Three deals with rarer preparations, especially from minerals, and has chapters on those from mercury, sulphur, vitriol, tartar and arsenic. Book Four treats of more secret ways of making, from vegetable simples, extracts, distilled waters, balsams, essences and salts philosophic. But its fifth and last chapter is on extraction of essences and tinctures from all sorts of stones. The fifth and last book, which occupies nearly half of the text, is devoted to various rarer and more secret remedies for all diseases of the human body from head to heel.

[47] Ferguson, II, 221; Hoefer, II, 331, where more titles of works by him are listed than in LR, although not with such full titles. His works appeared at various dates between 1610 and 1648, but there seems to have been no collected edition of them.

[48] They are listed by Borel 189-90; Hoefer, II (1843), 331; Ferguson, II, 221.

[49] *Miracula chymica et misteria medica libris V enucle⟨a⟩ta*, in-18: BM 1036.a.12.

[50] The form of title was now altered to *Miracula et Mysteria chymico-medica*, etc., as found also in subsequent editions. For Beguin see Chapter 27.

[51] BN R.44676.

[52] LR, which lists these editions, also ascribes editions of 1614 to Leipzig and Regensburg. The editions of Paris, 1644, and Amsterdam, 1659, (said to be a reprint of the first edition) went back to the original form of title.

[53] Ed. of 1623, p. 47.

In the Prooemium Müller says that transmutation is not difficult in itself but has been rendered so by the obscurity of writers thereon. Moderns are not merely obscure like the ancients but follow a preposterous order and mix in irrelevant matters. He will try to avoid these faults. He does not, however, pretend to be experienced "in this divine and rare work," but merely to set forth plainly and in proper order what has been said obscurely and in disorder. He has used manuscripts in Germany and Italy as well as printed books.

One of his modes of transmutation takes three months, and he warns not to trust those who say that they can transmute in nine or ten days, for it requires time to remove the impurities from metals. The recipe is in part as follows. Take one portion of aerial earth well washed, of watery fire three parts, of our mother half a part. Calcine what require calcination by *aqua fortis* or in any other way. Sweeten as said above. Mix all together on marble. In the process some blackness will appear as a sign of good mixing. Transfer all this matter into an oval phial so that the bottom third is filled, the other two-thirds empty. In the first month let the fire be of the first degree, in the second of the second, in the third, of the third. The fourth step of fixation is a matter of some weeks. A sign of fixation is the ashen color of the powder which forms about the matter. Now cast the coagulated matter and you will have it.[54]

The pills for headache of Eustachius Rudius, today first professor of the practice of medicine at Padua, which he employs as a great arcanum not to be revealed to others and entrusts to only one apothecary, are really not his but come down from Paracelsus through Guinther of Andernach and Wecker. But Italians know little or nothing of Paracelsus.[55]

Müller says that the action of all the remedies which he lists cannot be explained by their manifest qualities. For sore throat a dried viper's head worn about the neck as an amulet is commended. Also a plaster of swallows' nest, or scrapings of the tooth of a wild boar with sweet almonds and urine. For quartan fever, suspension from the neck of a big spider in a little ring. For all mental alienation, a powder of burnt tortoise; for the frenzied,

[54] Ed. of 1611, pp. 45-46. [55] *Ibid.*, pp. 100-102.

juice of the herb *consolida* mixed with honey and poured into the nostrils; for melancholics, a powder of the herb *fumaria.* Essence of *chelidonia minor* is wonderful for hypochondriac melancholy. Cinnamon taken in any way aids the memory, as do various other simples.[56]

Müller was to write later on the comet of 1618[57] and in 1624 on great conjunctions.[58] In 1622 he examined the questions whether images and numbers of the stars and things celestial portended their fate to church and state, and whether is was the part of an astronomer to interpret mystic numbers, together with an Appendix against Oswald Croll, whether the force of human imagination could exert itself outside its own body and move external objects. Licentiate in medicine at Leipzig and then professor of mathematics there, he was further interested in botany, as his work *De plantis in genere,* Leipzig, 1607, showed.[59] It briefly lists 223 captions for public disputation.

Billich in 1631, under the caption, "rancid Butter of Philipp Müller," said that Müller esteemed his May Dew above all others and as an arcanum which he showed only concealed by a cloud, but that "he would have lessened the admiration, if he had said butter. . . . You, whoever you are, who commit oils to butter, fat and grease,

> . . . da qui custodiat ipsos
> Custodes . . ."[60]

A Poitevin, Arnaud de la Chevalerie, concocted a work on hieroglyphics which were supposed to conceal the secret of the

[56] *Ibid.*, pp. 99, 133, 150, 114-16.

[57] *Hypotyposis cometae nuperrime visi una cum brevi repetitione doctrinae cometicae,* Lipsiae, typis et sumptibus Henningi Grosii senior., 1619, 67 pp. with 219 theses: Col 523.6 Z, vol. 2, No. 19.

Ph. Mullerus, *De cometa anni 1618. Accessit Rockenbachii tractatus de cometis cum enumeratione cometarum in hunc diem,* Lipsiae, 1619, seems a different work or edition, since no treatise of Rockenbach is found with the *Hypotyposis.*

[58] *De comitiis secularibus politiae coelestis s. de conjunctionibus magnis superiorum planetarum,* Leipzig,1624.

[59] See Zedler. Copy in BN S. 3780.

[60] A. G. Billich, *Observationum ac paradoxorum chymiatricorum libri duo,* Leyden, 1631, p. 105, citing *Miracul. et Myster. Chymic.,* lib. 4, cap. 3. For Billich's criticism of Beguin, see our chapter 27 on Chemical Courses and Manuals.

transmutation of metals under the name of Nicolas Flamel,[61] a bookseller of the late fourteenth and early fifteenth century.[62]

Under the title, Secrets of the Arts Discovered, both of pharmacy and distillation, commonly called alchemy, Godefroy Roussel in 1613 printed a dialogue between a master and an aspirant for the mastership in pharmacy,[63] in which the master asks the aspirant to state the rules for making potable gold in the appropriate mystic sense followed by the philosophers in order that so worthy an art may not be profaned, as it would be, if it were intelligible.[64] The work is dedicated to the king of France.

Georg Molther, a medical student who was respondent at Marburg in several disputations on obstruction of the liver, palpitation of the heart, pneumonia and pleurisy, which were printed in 1614[65] and 1615,[66] in the year following published an account of a Rosicrucian pilgrim who had passed through Wetzlar the year before and was admirable for his multiple science, words and deeds.[67] The account appeared in 1617 in German translation.[68] Numerous works on the Rosicrucians appeared at this time.[69] The

[61] *Le livre des figures hiéroglyphiques de Nicolas Flamel, escrivain, ainsi qu'elles sont en la quatrième arche du cimitière des Innocents, entrant par la porte rue Sainct-Denys, devers la main droicte, avec l'explication d'icelles per le dict Flamel, traictant de la transmutation métallique, non jamais imprimé.* Paris, Guillaume Marette, 1612.

[62] L'Abbé Villain, *Histoire critique de Nicolas Flamel et de Pernelle sa femme, recueillie d'actes anciens,* Paris, 1761.

[63] *Les secrets decouverts des arts, tant de pharmatie que de celuy de distiller, vulgairement nommé Alchemie ou Spargarie, par le moyen desquels l'on parvient à la perfection tant par theoricque que practique à rendre l'or potable, succinctement déduicts en forme de Dialogue,* Paris, 1613, in-8, BM 1038.c.13; Paris, 1618, BM 1035.a.7.

[64] *Ibid.,* p. 85; and the same attitude shown at p. 136.

[65] *De obstructione hepatis,* in J. Hartmann, *Disputationes chymicomedicae,* Part II, 1614, in-4. BM 1185. c.l. (16).

[66] The other three in H. Petraeus, *Nosologia Harmonica,* vol. I, 1615, in-4. BM 1179.a.3.

[67] *De quodam peregrino qui anno superiore MDCXV imperialem Wetzflariam transiens non modo se fratrem R.C. confessus fuit verum etiam multiplici rerum scientia verbis et factis admirabilem se praestitit,* Frankfurt, 1616, in-12. BM 1036.a.1.(1.).

[68] *Von einer frembden Mannperson welche inn . . . 1615 Jahr durch . . . Wetzlar gerisst und sich . . . für ein Bruder dess Ordens Rosen Creutzes aussgegeben . . . hat . . .,* 1617, in-9. BM 1033.c.5.(4.).

[69] Borel (1654), 263, lists eight between 1617 and 1619, and six more in the 1620's.

first seems to have been the Common and General Reformation of the whole world and the fame of the fraternal Order of the Rosy Cross.[70]

Pegasus of the Firmament, or a brief Introduction to the Wisdom of the Ancients which was once called Magic by the Egyptians and Persians but today by the venerable fraternity of the Rosy Cross is rightly called *Pansophia,* written for the sake of pious and studious youth, appeared in 1619 under the name of Josephus Stellatus, the pseudonym of Christoph Hirsch. In the preface to the Brotherhood, which is dated 1618, he says that the book was written because almost all our German schools are infected with paganism from reading ethnic and classical authors. Of the three kinds of philosophers in modern schools he prefers the Paracelsan to the Peripatetic or Ramean, and recommends especially the reading of Michael Maier (Meijerus). The true fount of knowledge is the book of Scripture and the book of Nature. Of the latter the first interpreter was Hermes Trismegistus; the second, Paracelsus; and the third, Basil Valentine. Of seven columns of wisdom which students of *Pansophia* must know, the fifth is magnetic, the sixth, crystalline, the seventh adamantine. The light of grace peculiar to those who have been born again comprises three degrees: an instinctive feeling for God, predicting dreams, and prophetic visions. The light of nature manifested to worthy geniuses comes from the stars to the microcosm.[71]

Angelo Sala (1576–1637) of Vicenza left Italy for religious reasons in 1602, was at The Hague from 1612 to 1617, and then in Oldenburg (1617–1620) and Hamburg (1620–1625), and spent the rest of

[70] Borel (1654), 240, lists editions of Cassel, 1614, and Francof., 1615; but Ferguson, II, 290, dates the first sure edition of the *Fama fraternitatis* in 1616. Borel also assigns to Henricus Neuhusius (not in Ferguson) an *Admonitio de fratibus Roseae crucis nempe an sint quales sint etc.*, Francof., 1611, in-8, as well as Danzig, 1628, in-8.

BM and BN have rather: *Pia et utilissima admonitio de fratribus Rosaecrucis nimirum, An sint? Quales sint? Unde nomen illud sibi asciverint et quo fine eiusmodi famam sparserint?* (Frankfurt?), 1618, in-8.

[71] *Pegasus firmamenti sive Introductio brevis in veterum sapientiam quae olim ab Aegyptis et Persis magia, hodie vero a venerabili fraternitate Roseae crucis Pansophia recte vocatur, in piae ac studiosae iuventutis gratiam conscripta,* 1619: BM 1033.b.30.

his life as physician to the dukes of Mecklenburg-Güstrow.[72] Sala was the author of various medical and chemical works. We may notice a few specimens chiefly of the latter. In 1608 appeared a Latin translation from the Italian of his two treatises concerning various errors of both chemists and Galenists in the preparation of medicines.[73] In the same year is dated the dedication to his Anatomy of Vitriol, of which the Latin translation from the Italian was printed at Geneva in 1613.[74] It was in two parts, the first treating of spirit of vitriol, oil of vitriol, salt of vitriol, sulphur of vitriol, earth of vitriol, vitriol rectified, vitriol regenerated, anodyne extract of vitriol, diaphoretic liquor of vitriol, and cordial liquor of vitriol. The second treatise dealt with compounds in which vitriol was one of the ingredients. In 1614 there was a Latin edition at Amsterdam of his treatise on the seven terrestrial planets or metals and their analogy with the microcosm.[75] In it he declared that a water or oil could not be got from gold by any art, but he was later to write on potable gold. At Leyden in 1616 in French appeared his *Ternarius bezoarticorum, ou trois souverains médicaments bézoardiques.*[76]

Sala's Anatomy of Antimony was issued in Latin in 1617[77] almost

[72] A. Cossa, *Angelo Sala medico e chimico Vicentino del secolo XVII*, Vicenza, 1894; Robert Capobus, *Angelus Sala, Leibarzt des Johann Albrecht II . . ., seine wissenschaftliche Bedeutung als Chemiker im XVII. Jahrhundert*, Berlin, 1933, 67 pp. Karl von Buchka, "Angelus Sala," *Archiv für Gesch. d. Naturwiss. u. d. Technik*, VI (1913), 20-26, is a compilation of estimates of Sala from past histories of chemistry and of bibliographical information.

[73] *Angeli Salae Vincentini medici Spagyrici Tractatus duo de variis tum chymicorum tum Galenistarum erroribus in praeparatione medicinali commissis. Opus italice primum ab auctore conscriptum, iam vero . . . in latinam linguam translatum labore et conatu M.A.R.*, Hanoviae, haer. J. Aubri, 1608, in-12. Not included in the *Opera* of 1647.

[74] *Angeli Salae Vincentini Veneti medici Spagyrici Anatomia Vitrioli in duos tractatus divisa . . . accedit arcanorum complurium . . . sylva*, Aureliae Allobrogum, 1613, 75 pp.

[75] *Septem planetarum terrestrium spagirica recensio qua perspicue declaratur ratio nominis Hermetici, analogia metallorum cum microcosmo, eorum praeparatio vera et unica, proprietates et usus medicinales*, Amsterdam, 1614, 98 pp.

[76] It is a different work from his *Ternarius Bezoardicorum et hemetologia seu triumphus vomitoriorum*, published at Erfurt in 1618 with a chymiatric exegesis by Andreas Tentzelius.

[77] *Anatomia antimonii*, Leyden, 1617, 145 pp. in-16. NYAM.

thirty years before Basil Valentine's Triumphal Chariot of Antimony was translated into that language, and illustrates the attention that alchemists and iatrochemists were giving to that metallic substance. Sala notes its virtues and use, first according to ancient medical authors, then according to modern writers. He points out how it can injure the human body and how to guard against this. He questions whether there are gold, silver and copper in it, as some chemists think, whether quicksilver can be separated from it, and whether any liquor or oil or tincture of it can be produced. He tells how to separate from it a more metallic and fixed substance known as *Regulus antimonii,* how to derive vomitories and purgatives from it, how to prepare *Crocus metallorum* and *Flos antimonii,* and a preparation of antimony which purges by means of perspiration and insensible transpiration.

Billich had been a student under Sala and in 1622 published a defense of his master's Chymiatric Aphorisms.[78] Sala's continued influence was to be shown by collected editions of his works at Frankfurt in 1647 by Doctor Hartmann Beyer, and at Rouen in 1650 (*Editio auctior et emendatior*).[79]

The merits or the inertness, medical and chemical, of potable gold were to be argued repeatedly throughout the course of the seventeenth century.[80] In England, as that century opened, the medical practice of Francis Anthony, in the words of the *Dictionary of National Biography,* "consisted chiefly, if not entirely, in the prescription and sale of a secret remedy, called *aurum potabile,* from which he derived a considerable fortune." But it is doubtful if his potable gold contained any gold at all. He is said to have first published his *Panacea aurea* in 1598 at Hamburg, but the 1618 edition of that title there represents it as "now first printed in Germany from the London original." Anthony was in 1600 examined by the London College of Physicians, found ignorant of the medical art, and forbidden to practice it. He disregarded this prohibition,

[78] *Ad Animadversiones quas Anonymus quidam in Angeli Salae Aphorismos chymiatricos conscripsit responsio,* Leyden, Apud Godefridum Basson, 1622, in-8.

[79] I have used the edition of 1647: Col 610.8 Sa 3, but not that of 1650: BN 4° Te131.102.A.

[80] Ernst Darmstädter, "Zur Geschichte des Aurum Potabile," *Chemiker-Zeitung,* 48 (1924), No. 115, on the general subject.

was fined five pounds and imprisoned, but released by a warrant from the Lord Chief Justice. When the College had him recommitted, he submitted temporarily. But soon he was again prosecuted for the same offense and fined heavily. Rather than pay this, he spent eight months in jail, when in 1602 he was released at the petition of his wife and on the ground of poverty. He resumed practice, relying on the support of powerful friends at court; was again in 1609 brought before the College of Physicians, but in 1610 published at Cambridge an Assertion of Chemical Medicine and true potable gold.[81]

Angelo Sala, in a treatise of his own on potable gold, perhaps referred to Anthony as "a Spagyrite of great name in England," since later in the same work he spoke of "that Englishman, Francis Anthony," as if referring back to a previous mention of him.[82] His recipes for making potable gold also were drawn from Anthony as well as Quercetanus, Basil Valentine, Osiander, and others. Sala's treatise was published in Latin in 1631 with the title, *Descriptio auri potabilis,* and was in large measure a commentary upon Oswald Croll, with some expression of dissent. There had been an earlier work in French by G. de Castaigne, almoner to Louis XIII.[83]

Anthony's book of 1610 was answered the next year by Matthew Gwinne (c. 1558–1627), first professor of "Physics" at Gresham College, London, who quoted Anthony's text bit by bit and replied *seriatim.*[84] Anthony had held that the most potent force in medicine resided in metals, that among metals gold took first place in medicinal preparations, and that potable gold deserved to be called the universal medicine. Gwinne dedicated his reply, as Anthony had his Assertion, to James I. He denied gold medical importance,

[81] *Medicinae chymicae et veri potabilis auri assertio,* Cambridge, 1610.

[82] Angelo Sala, *Opera,* 1647, p. 271, "quidam magni nominis Spagyrus in Anglia"; p. 288, "Angli illius Francisci Antonii."

[83] *L'or potable qui guérit tous les maux,* Paris, 1611, in-8. Listed by Hoefer, II, 331; not in LR. BM 1033. f.46 is the 1613 edition.

[84] *In assertorem chymicae sed verae medicinae desertorem, Fr. Anthonium . . .,* London, 1611. "Aurum non aurum" does not appear on the title page, but is the running head for the pages of the text. In the edition of Antwerp, 1613, however, the title is: *Aurum non aurum, sive, In assertorem chymicae sed verae medicinae desertorem, Fr. Antonium, Adversaria.* I have used the edition of 1611 at the New York Academy of Medicine.

and that there was any such thing as a universal medicine. He furthermore condemned Anthony's preparation as not potable gold. In addition he smothered his opponent under a tremendous mass of oratorical verbiage and literary quotation, especially from classical authors, both Greek and Latin, but also Giovanni Francesco Pico *De auro*, Raymond Lull, and other alchemical authorities.

Anthony retorted in both Latin and English editions in 1616. His reply contained a dedicatory poem by another English alchemist, Timothy Willis, who in that same year issued *The Search of Causes, containing a theophysical investigation of the possibilitie of transmutatorie Alchemie,*[85] and who the year before had published a work in Latin on the chemical elements.[86] Another friend of Anthony was Michael Maier, a German alchemist who visited London at about this time.

One of Willis's propositions was that all matter had been created from nothing. As Burggrav spoke of a difference of lives in his Vital Philosophy, so Willis distinguished a triple clasification of lives: (1) simple, of the individual; (2) relative to the species, whence sympathy and antipathy; (3) respective to other individuals both of one's own and other species. The difference between metals was specific from the proper form of each, but each metal could be resolved into a matter that would cook and coagulate. And since gold was the soul and animator of all metallic matter, if it were in the composition or mixture, it would conquer and convert the whole mass into gold "more or less, according to the proportion and regimen of coction and active causes."[87]

In *The Search of Causes* Timothy Willis is called "Apprentise in Phisicke," but writes in a religious tone and vein. The text opens: "The knowledge of truth revealed unto the first friends of God and by succession from them continued unto us their children, is more

[85] Printed by J. Legatt, London, 1616, 87 pp.: BM 1036.a.13 (2.).

[86] *Propositiones tentationum sive Propaedeumata de vitis et faecunditate compositorum naturalium quae sunt elementa chymica.* It was reprinted in 1618, with the alternative title, *Elementa chymica.*

[87] *Propositiones tentationum,* 1615, pp. 3, 7-8, 20, 27, 33 (propositions 7, 16, 46, 56, 73). The work terminates with the 89th proposition on p. 40. Copy used: BM 1033.d.5.(1.). Also BM 1036.a.13(1.), as *Elementa chymica.*

perfect than the wisdom of any philosopher." Much is said about creation and chaos. The ninth chapter opens:

Now let us enquire whether it be possible in nature to produce such a compounded substance in which, after exact digestion, the predominancy of the spiritual causes shall be manifested in true figure of regeneration.

The fourteenth and last chapter begins: "For our better understanding herein let us consider the Historie of Creation." The Testament of Raymond Lull is cited; a heptagonal figure is given, and a ladder of natural magic (*Scala magica naturalis*).

Michael Maier[88] (1568–1622) received his M.D. at Rostock in 1597 and was in the service of Emperor Rudolf II and of the Landgrave of Hesse. In the years 1616–1619 he published a number of alchemical tracts with quaint titles: Squaring the Physical Circle, i.e., Of Gold and its Medicinal Virtue; The Dross of Pseudo-Chemists Examined; Play of Mercury; Symbols of the Golden Table; A Serious Joke; Of the Tree Bird without Father or Mother; and Atalanta Fleeing, i.e., New Chemical Emblems of Nature's Secrets. What I take to have been this last-named work, was reprinted as late as 1687 under another title.[89] There are fifty copper engravings representing alchemical allegories, each accompanied by an epigram of a few lines and a longer *discursus*.

Maier's *Viatorium*, that is, Concerning the Mountains of the Seven Planets or Metals,[90] an Ariadne's thread through the ocean of chemical errors, considered each planet or metal in turn, beginning with Mercury, and stated three things as to each: its use in making gold, in tincture, and in medicine. Each was further accompanied by a picture. That for Mercury showed Thebes in Boeotia with the problem which of its seven gates to enter, while Saturn was accompanied by the combat of dragon and elephant. With the moon went the circumnavigation of the globe by Magellan, and with the sun a representation of the grateful lion. The *Viatorium* was reprinted at Rouen in 1651.

[88] Sometimes spelled Mejerus.

[89] *Secretioris naturae secretorum scrutinium chymicum per oculis et intellectui accurate accomodata figuris cupro appositissime incisa ingeniosissime emblemata . . .*, Francof., 1687, in-4, 150 pp.

[90] *Viatorium, hoc est de montibus planetarum septem seu metallorum . . .*, Oppenheim, 1618, 136 pp. BN R. 7947.

Maier was a Rosicrucian[91] and, when in London, gained Robert Fludd for that fraternity. Two defenses of it by Fludd were printed at Leyden in 1616 and 1617.[92] The chief German inventions listed by Maier in his book on that theme were the Holy Roman Empire, gunpowder and artillery, printing, the religious Reformation, and, in the fields of medicine and chemistry, Paracelsus and the Rosicrucians.[93]

In his Philosophic Week, an imaginary dialogue between Solomon, the Queen of Sheba, and King Hiram of Tyre concerning the world of nature,[94] beginning with simple bodies and then running through meteors, fossils, vegetation, animals and man, rather more attention is given to alchemy than any other single topic.[95] When the queen asks Solomon what he thinks of the new or reformed medicine, which has reduced all the contents of the human body to salt, sulphur and mercury (*cremosum*, sublimate and precipitate) and derives all diseases and remedies from the same, he replies that these are like dreams in which you are the richest king only on awakening to find yourself the vilest of the people. They go on, however, to dicuss these *principia* further, quoting Rosarius, Hermes, Morienus, Avicenna, and "Arnoldus," then pass on to the metals under their planetary names with quotation of other medieval Arabic authors such as Geber and Rasis. Thomas Aquinas is said to call the matter of the philosophers' stone "gross water."

The Cave Philosophic is a book of slight importance by I. B. Besard, a lawyer and physician of Besançon, where he was born about 1576. It was printed at his expense at Augsburg in 1617.[96] It is merely a collection of recipes in six books and 248 pages. At

[91] See his *Themis aurea, hoc est de legibus fraternitatis R. C. . .*, Francof., 1618, in-8, 192 pp.

[92] Mersenne, *Correspondance*, I (1932), 37-38.

[93] *Verum inventum, hoc est Munera Germaniae, ab ipsa primitus reperta . . . et reliquo orbi communicata*, Francof., 1619, in-8, 249 pp. BN M. 29376.

[94] *Septimana philosophica qua aenigmata aureola de omni naturae genere a Salomone Israelitarum sapientissimo rege et Arabiae regina saba . . .*, Francof., 1620, in-4.

[95] *Ibid.*, pp. 76-86, and beyond.

[96] I. B. Besardus Vesontinus, *Antrum philosophicum . . .* (and ten more lines of title), Augustae Vindelicorum imprimebat David Franck impensis authoris, 1617, in-4. I examined the volume at the National Library, Munich, in August, 1935.

first the arrangement is by diseases; then the fifth book is devoted to chemical remedies, and the last book to experiments, of which the final one aims at perpetual motion. Authorities are usually not named. An exception, under Experiments, is, "Albertus tells many ways of winning the love of men."[97] Besard advised, in chemical and medical operations, to note whether the stars were favorable.

Henning Scheunemann, who in 1608 had written on the pest as a mercurial disease,[98] in 1610 on fever as a sulphuric disease,[99] and in 1613 on Paracelsan hydromancy,[100] in 1617 published a Reformed Medicine or Hermetic Decade, in which he traced all diseases to ten roots, four mercurial, three sulphuric, and three saline, also touching on transplantation of disease and astral influences.[101]

In 1618, the year immediately following the publication of Sala's Anatomy of Vitriol, Hamerus Poppius, whoever he may have been,[102] imitating the title of Croll's *Basilica chimica,* produced a *Basilica antimonii*[103] in twelve chapters[104] with a dedication to

[97] *Ibid.,* p. 247.

[98] *Paracelsia de morbo mercuriali contagioso, quem pestem vulgus nominat, ex quintuplici ente, dei nimirum, astrorum, pagoyi, veneni et naturae prognato,* Bamberg, 1608, in-4.

[99] *Paracelsia de morbo sulphureo et quintuplici ente,* Francof., 1610, in-8.

[100] *Hydromantia Paracelsica, hoc est . . ., de novo fonte . . . olim S. Annae fons dicta,* 1613 (colophon dated 1615): BM 1171.g.21 (5.). This copy was unfortunately at the binder's, when I tried to see it.

[101] Henning Scheunemann, Halberstad. Saxo, Ph. et M.D., *Medicina reformata seu Denarius Hermeticus philosophicus-medico-chymicus, in quo mira brevitate dilucide docetur decem entibus omnium morborum radices, productiones, transplantationes, astra, signa, indicationes et curationes compleri et absolvi,* Francof., 1617, 11 caps., 122 pp. BM 1034.c.38; BN 8° Te131.65.

[102] Ferguson, II, 213, says, "I have not met with any notice of this author."

[103] Francof., 1618, 50 pp.: BN 8° Te151.78; BM 1033.h.5.(6.).

[104] The captions of the successive chapters are: i, De natura antimonii; ii (p. 17-), De minera antimonii eiusque fusione et purgatione sive regulis; iii (21), De calcinatione antimonii per ignem coelestem seu radios solares; iv (22-), De calcinatione et reverberatione antimonii; v (25-), De croci et fixi antimonii praeparatione; vi (28), De caementatione antimonii; vii (29-), De praecipitatione antimonii; viii (31-), De praecipitatis ex liquore antimonii per destillationem parato tam simplicibus quam compositis; ix (37-), De fusione vitri antimonii; x (41-), De florum antimonii praeparatione; xi (44-), De liquoribus ex antimoniis per destillationem prolectis; xii (47-), De tinctura sale et liquoribus antimonii per deliquium.

five of his students. It also appeared in later editions.[105]

Meanwhile a Johann Popp, Poppe or Poppius had published *Chymische Medicin* at Frankfurt in 1617, where it was printed again in 1627, in which year his *Hodogeticus Chymicus* came out at Leipzig,[106] followed in 1628 by *Thesaurus medicinae,* also written in German.[107] Johann Agricola, born in 1589, commented on the first of these works in 1638 and added a hundred new processes to it.[108]

Johann Bernhard Hildebrandt published a long poem in German on the philosophers' stone, with the running head, "Das Buch Magnesia," at "Hall in Sachsen" in 1618.[109]

The Dogmatic-Hermetic Handbook of Johann Vincenz Finck[110] dedicated on February 1, 1618, to the margrave of Brandenburg, is largely a collection of chemical remedies from Paracelsus and such recent writers as Croll, Quercetanus, Beguin, Ruland, Philipp Müller, Rhenanus, de Boodt, and Duncan Burnet. Of the *laudanum opiatum* of Paracelsus he found so many different accounts that he scarcely knew which to follow.[111] Despite such frequent repetition, sometimes at second-hand,[112] Finck gravely argues whether to publish such secrets openly or in cryptic form. Except for an initial chapter on universal digestives, the remedies are grouped in 38 chapters under particular diseases such as melancholy, epilepsy, apoplexy, catarrh, ophthalmia, diarrhoea, dysentery, colic, worms and haemorrhoids.

Roderic à Castro tells of a septuagenarian physician who often had nose-bleed and always had ready for it in a capsule ass manure

[105] See BM catalogue: Ferguson, I, 365-66.

[106] It is the only one of the four in Borel (1654), 189, and in BN: 8° Te131.79, 405 pp.; but BM has all but the 1627 edition and a number of other medical works by him.

[107] Ferguson, II, 213-14.

[108] *Ibid.*, I, 11; *Comm. et Observ. in d. Chymische Artzeney Joh. Poppii . . . mit etlich hundert newen Processen,* Leipzig, 1638.

[109] *De lapide philosophico,* with the rest of the title in German, over 100 unnumbered pp. BM 1034.c.5 (2.).

[110] *Encheiridion dogmatico-hermeticum morborum partium corporis humani praecipuorum curationes breves continens,* Lipsiae, L. Cober, 1618, in-12, 224 pp. BN 8° Td30.80.

[111] *Ibid.*, p. 15. Concerning Duncan Burnet see Chapter 27.

[112] He knows of pills of Eustachio Rudio (for whose *De morbis occultis* of 1610, see T V, 43-44) through Müller's description of them. See *supra,* p. 164.

moderately dried. Others prefer that of sows.[113] Finck takes remedies from old-wives[114] as well as from books and physicians. He does not scorn such ancient remedies as river crabs, if washed in rose vinegar and crushed and bound on in place of a plaster.[115] An "experiment known not only to the ancients but also to moderns" is prepared from green frogs. By it Frederick IV, elector Palatine, was completely liberated from a pertinacious epilepsy contracted in his youth.[116] Finck further introduces drugs from the New World. Blackened teeth are made very white by rubbing them hard daily with the ash of Indian tobacco.[117] The front of the human skull is more medicinal than the back, and use should be made of the cranium of a person of the same sex as the patient.[118] Amulets are employed, such as wearing red coral on the hands or about the neck against phantasy, specters, phantasms and melancholy; or a dried powdered toad in muslin in the arm-pits or hands against nosebleed; or the right hind hoof of an elk, worn in a ring so that it touches the skin, or about the neck so that it touches the bare flesh, against epilepsy.[119] Astrological ceremony is to be observed, as in digging a root of peony under a waning moon in March or April—but some prefer dog-days, or gathering mistletoe in a waning moon between the two feasts of Mary [120]

Most elaborate is the preparation of an amulet from an elder-tree growing above a willow. In October before the full moon, the part between two knots should be gathered, cut into nine slips, these bound in linen, and suspended from the neck by a thread until it breaks of itself. Then no one should touch the amulet with bare hands, but it should be picked up with some instrument and thrown into water or some other place where no one will touch it. Finck goes on to explain that the reason for this amulet is not completely occult, since the elder-tree and its seeds are beneficial to this disease (epilepsy). There are those who assert that this elder growing above a willow comes from the putrefied corpse of an epileptic sparrow.[121]

113 *Encheiridion*, p. 69.

114 *Ibid.*, p. 11: "Mulierculárum nostrarum panacea."

115 *Ibid.*, p. 10.

116 *Ibid.*, p. 36.

117 *Ibid.*, p. 75.

118 *Ibid.*, p. 33.

119 *Ibid.*, pp. 25, 69, 32.

120 *Ibid.*, p. 32.

121 *Ibid.*, pp. 42-43.

In view of the opposition of the medical faculty at Paris to chemical remedies, it is a little surprising to hear Finck say that he had seen a *Consilium* of 1612 composed by *medici Parisienses* for a most illustrious and noble boy, in which they prescribed spirit of vitriol against epilepsy.[122] The only quotation of a dissenting voice with regard to such remedies that I noticed was from a letter of Thomas Moffett (Mufetus) disapproving of the common practice of dissolving pearls and coral in vinegar because of the presence of too much acidity and pungency of sal ammoniac.[123]

In 1616, while still a candidate for the medical degree, Johann Daniel Mylius had published the *Iatrochymicus* of Duncan Burnet. Two years later a Medical-Chemical Work of his own appeared in three parts or *Basilicae*,[124] a designation probably suggested by Croll's *Basilica chemica.* The first of these, called *Basilica medica,* was Hippocratic and divided into three books on physiology, pathology and therapeutic. The second or *Basilica chymica* contained seven books, of which three were on metals, the others on gems, minerals, vegetables and animals respectively. The three books of the third *Basilica philosophica* were alchemical, dealing with the philosophers' stone or universal medicine, with vessels and furnaces, and with obscure passages in the "philosophers", i.e., alchemical writers.

In 1622 Mylius published a Reformed Philosophy[125] concerned with the divine science and divine art of alchemy. Its first book or volume was in seven parts, of which the fourth treated of the twelve degrees of the philosophers. It is illustrated by metaphorical pictures in sets of four each, and couched in enigmatic and figurative language, "hiding what is manifest, and revealing what is hid."[126] God created the fifth essence from nothing; then produced from it, first the angels and Empyrean heaven, second the celestial bodies, third our world. Mercurial water is not the mercury of the philosophers; it is the first matter from all metals.[127]

[122] *Ibid.*, p. 44.

[123] *Ibid.*, pp. 18-19.

[124] *Opus medico-chymicum continens tres tractatus sive basilicas,* Francofurti apud Lucam Jennisium, 1618, in-4. LR 567.

[125] *Philosophia reformata,* Francof., 1622. BM 1033.i,7.

[126] *Ibid.*, p. 97, *mg*: "Manifesta occultanda et occulta manifestanda."

[127] *Ibid.*, pp. 171, 179.

What remains at the bottom of the vessel is our salt, that is, our earth, and it is black in color, the dragon devouring its tail. For the dragon is the matter remaining at the bottom after the distillation of water from it. And that water is called the tail of the dragon, and the dragon is its blackness. And the dragon is imbibed by its water and coagulated, and thus it devours its tail. And scorn not the ash which is at the bottom of the vessel, since it is the diadem of your heart.[128]

Also the arcanum of the art of gold is made from male and female, because the female receiving the force of the male rejoices, in that the female is strengthened by the male. So, son, by the faith of the glorious God the complexion is from the complexion between the two luminaries, male and female. Then they embrace and have intercourse, and modern light is born of them, to which no light is similar in the whole world.[129]

The dragon dies not unless slain with both his brother and his sister, sun and moon; and the dragon is quicksilver extracted from bodies having in themselves body, soul and spirit.[130]

Book One closes with an epilogue or Enigma of the Philosophers or Symbol of Saturn in Parable, in the form of a dialogue between a philosopher, a youth and Saturn.[131] Book Two, as we have seen, is a collection of past authorities.[132]

Besides pharmaceutical works,[133] Mylius issued in 1628 an Anatomy of Gold or *Tyrocinium medico-chymicum*[134] borrowing Beguin's title. Its five parts treated of the harmony between the celestial sun and terrestrial gold, and the definition and conflict of opinion concerning the latter; of medicines and recipes, ancient and modern, in which gold was an ingredient; of the preparation of potable gold, both vulgar and philosophic; of their medicinal use; and the idea of the philosophers' stone.

The Hermetic-Dogmatic Encyclopedia of Fabritius Bartolettus or Fabrizio Bartoletti of Bologna was published in that city in

[128] *Ibid.*, p. 195.

[129] *Ibid.*, p. 244. I was led to this and the preceding passage quoted by their representing the sole references in the Index to "Ars notoria" and "Magia."

[130] *Ibid.*, pp. 239-40.

[131] *Ibid.*, p. 312 *et seq.*

[132] *Ibid.*, pp. 365-695.

[133] His *Pharmacopoea Spagyrico-Medica*, Francof., 1628, 1629, two vols. of 989 and 896 pp., is just one prescription after another. BM 1934. b.8.

[134] *Anatomia auri sive Tyrocinium medico-chymicum* . . ., Francofurti apud Lucam Jennisium, 1628, in-4; LR 568.

1619,[135] and dedicated to Ferdinand Gonzaga, duke of Mantua and Montferrat. It was edited by Theodore Bugeus who, in a preface to the readers, states that Bartoletti, who is young, would not have published it, unless many friends had urged him to do so. Although he has composed other larger and more original works,[136] this *Tyrocinium medicum* and Theory for medical students is published now because there is no good book of the sort. Novel features are the exposure of Hermetic impostures as to humors, temperaments, three substances or *principia,* and generation of metals; additional questions, and such matters, hitherto treated by no one, as *aqua benedicta* from the *crocus* of metals, salt of antimony, sweet oil of the same, a method of extracting balsamic mercury from gold and silver, an appendix on oils and waters with a new way of making salts, gemmed liquors of the Grand Duke of Tuscany, *et cetera.* As this preface makes evident, the work combines criticism of alchemy with faith in chemical remedies and processes. It divides into five parts: physiology, hygiene, pathology, *simiotica* and therapeutic, and considers, according to the table of contents, eight Hermetic impostures. But in at least two cases, the germination of metals and the innate spirit or radical balsam, they are not so represented in the text itself. Rather it is stated that the Hermetics have discovered a method by which they make metals germinate as in nature, and prepare medicines that command the highest admiration.[137] Two definitions are given for balsam, and it is held that it is nothing but a substantial aggregate of heat and *humidum radicale.* Negatively, it has no diaphoretic virtue, no caustic faculty, no attractive quality, and no *complexio* from the elements; but affirmatively it is sweet, temperate and consolidative, and Bartoletti refers the reader to the second book of his Surgery for preparations of various balsams.[138]

[135] *Encyclopedia hermetico-dogmatica Sive Orbis doctrinarum medicarum physiologiae hygienae . . . et therapeuticae, ad . . . Ferdinandum Gonzagam Mantuae et Montisferrati ducem,* in-4, 321 double-columned pp. BM 544.g.3.

[136] Bugeus mentions: *Tota logica textualis,* Physics - text, *summae* and questions, *3 libri de anima, tractatus de visu et visibili, de infinito,* works in surgery and anatomy (which I'll not specify), *Iconographie foetus humani, Antidotarium chimico-dogmaticum, de dolore, Compositio medicamentorum secundum Hermeticos.*

[137] *Ibid.,* 129.

[138] *Ibid.,* 136-39.

But other alchemical tenets are severely criticized. The contention that sulphur, mercury and salt differ from both elements and mixed bodies, that no body is made immediately from the elements, and that these three chemical principles are secondary substances, is declared absurd, since they are mixed and do not differ really from mixed bodies.[139] The Hermetics are called malign and tricky in their three arguments against the conception of temperament, one of which is that morals, talents, nature and custom do not come from temperament but from the stars.[140] Their distinguishing seven chief parts of the human body, related to the seven planets, as against the three principal members of the Dogmatics (heart, brain and liver) is also rejected.[141] Also Bartoletti stands by the old doctrine of contraries curing contraries, although he gladly uses some chemical remedies. When they object that a remedy is discovered by its resemblance in form and property to the disease, he replies that these specific cures are also recognized by Galen, but that they are not regular cures.[142]

Marco Cornacchini, ordinary professor of medical practice at Pisa, first published at Florence in 1619[143] a work on a powder which would cure all diseases, especially putrid fevers, as he proceeded to show by "experiments" and prove by reasons. The work was issued at Frankfurt in 1628,[144] and again in 1647 and 1682 with editions of J. Hartmann's *Praxis chymiatrica,* and in 1690 with his *Opera.* Thus, although it claims to employ a method which is both Galenic and chemical, it seems to have been adopted by the iatrochemists.

Rudolph Goclenius, M.D. and ordinary professor at Marburg,

[139] *Ibid.,* 15-16, "sed tantum modaliter ut fusius in Theoria medicina tractabimus."

[140] *Ibid.,* 27-29.

[141] *Ibid.,* 82-83. The four added parts are the lungs for Mercury, spleen for Saturn, kidneys for Venus, and gall for Mars.

[142] *Ibid.,* 316-20.

[143] BM 542.a.26, in-8, 92 pp.

[144] *Methodus qua omnes humani corporis affectiones ab humoribus copia vel qualitate peccantibus genitae tuto cito et iucunde chymice et Galenice curantur,* Francof., 1628. 146 pp. and index. The dedication is dated on the Ides of April, 1620, perhaps according to the Pisan calendar. Preliminary poems run to p. 23 and the preface to the reader to p. 30. The text refers to 1619 as the present year. BM 1034.b.9, where it is bound with Mylius, *Pharmacopoea spagyrica.*

published there in 1620 an Assertion of a Universal Medicine against that commonly pronounced universal.[145] At least his medicine was good for all diseases having a cold cause. It was composed of numerous vegetable simples; the mode of preparation was laborious; the time of collecting the herbs, particular; while the method of using it required practical judgment. The list of ingredients ran to a page and a half, and was repeated in German. Many patients might be adduced who have benefitted by it, but Goclenius contents himself with a letter from one and a noteworthy example of the cure of a serving maid by it in 1610.

Le Sieur de Nuisement, receiver-general of the county of Ligny-en-Barrois, published in 1620 at Paris a philosophical poem on the truth "de la phisique minerale" or alchemy, and the next year a treatise on the harmony and general constitution of the true salt, secret of the philosophers, and on the universal spirit of the world.[146] It was dedicated to le duc de Lorraine. De Nuisement holds that the world is alive and possessed of spirit, soul and body; that all which has essence and life is made by the spirit of the world from first matter. Hermes called the sun the father of these last two, and the moon their mother. Later we are told that air is the root of the spirit of the world, and that earth nourishes it. It in turn may take on body and be converted into earth. Then we hear of the separation of fire from earth and of the subtle from the dense, and of the ascent of the spirit to the sky and its descent to earth. A number of sonnets then pave the way for the philosophical poem which follows.

John Thornborough (1551–1641) is said to have led a gay life at Oxford, "employing Simon Forman," the physician and astrologer whose diary from 1552 to 1602 records his questionable practices and dabblings in alchemy and magic, "as the minister of his pleasures."[147] After he had become bishop of Worcester in 1617, he

[145] *Assertio medicinae universalis adversus universalem vulgo jactatam, authore Rodolpho Gloclenio M.D. et in Academia Marpurgensi professore ordinario.* Marpurgi, 1620, 24 pp. BM 443.d.28 (2.).

[146] *Traittez de l'harmonie et constitution generalle du vray sel, secret de philosophes, et de l'esprit universelle du monde*, Paris, 1621, 333 pp. In BN R.45239-45240, it precedes the *Poeme philosophic de la verité de la phisique minerale*, Paris, 1620, 80 pp.

[147] DNB.

published in 1621 an alchemical treatise in Latin divided into three sections headed Nothing, Something, and All.[148] Repetition of a few of the chapter headings will sufficiently indicate the character of the book. The Nothing of the philosophers arises from corruption, solution, and privation of prior form. The Nothing of the philosophers is a vile thing, which by proper regimen of fire is perfected into a precious thing. Our Nothing is the key to the whole art. The resuscitation of our Nothing from its corruption to a nobler form teaches the resurrection of our flesh in glory. The Something from Nothing of the philosophers, lika a flower in the desert, teaches from the humility of Christ his goodness and glory. Our Something is water extracted from earth and returned to earth to make it fertile. The Something of the philosophers is seen in their water as in a mirror, and in this water lurks their secret, a living fire vivifying dead bodies. Elemental or external fire should excite and nourish the internal, until the birds of Hermes can be caught and kept. Otherwise we shall not obtain Something but sit forever in dark Nothing and the house of death. The dissolution of our individual is made with conservation of the species, and the falling of our dew to earth is the semination of philosophic gold in its earth. Are All in All? What are, and what is the meaning of, All in All? Of the true sun and wine of the philosophers, sanctuary of nature and domicile of gold itself, containing All necessary for gold. All in All taken philosophically leads us to All in All said theologically. Conclusion: of the exaltation of the stone which transforms the imperfect bodies of metals into perfection and has reference to the blessed day of resurrection and the glorious state of the saints in heaven.

Seeing Thornborough's book led Georg Lehmann, professor of theology at Leipzig, to publish towards the close of the century a purely religious work with a similar title.[149]

De signatura rerum of Jacob Boehme, the German mystic, appeared in the same year, 1621, as Thornborough's book, and some-

[148] *ΛiΘoΘεωpikos, sive Nihil, Aliquid, Omnia, antiquorum sapientum vivis coloribus depicta, philosophico-theologice, in gratiam eorum qui artem auriferam . . . profitentur . . .*, Oxoniae, 1621, in-4, 152 pp. and a fantastic chart. BM 1034.h.32.

[149] *Nihil aliquid et omnia theologorum*, Lipsiae, 1693, in-8.

what similarly combined alchemy with religious mysticism. Boehme speaks of the essence of all essences, of sympathy and antipathy, of sulphurean death, of the generation of a water and an oil, and of the wheel of sulphur, mercury and salt.

To Johann Ernst Burggrav are ascribed alchemical works which I have not seen on the Bath of Diana or magnetic key of pristine philosophers and on the magical-physical electrum of the philosophers, printed at Leyden in 1600 and 1611 respectively,[150] and a *Biolychnium* of 1611, of which we treat in a later chapter on Medicine. At present we are concerned with his Introduction to Vital Philosophy, which was published posthumously in 1623. He asserts that the subject has never been treated before but has always lain hidden. The full Latin title suggests the magical character of the book, since it mentions astral diseases, mysteries of cures, and arcana of remedies.[151] Despite the claim to novelty[152] and pretense of revealing great secrets for the first time, the work further professes to expound the ancient medicaments of Hippocrates, Galen and Celsus, and to profit by the experience of such moderns as Paracelsus, Turnheuser (Thurneisser?), and Quercetanus (Duchesne). Fernel is frequently cited, and such late medieval and early modern writers are mentioned as Arnald of Villanova, Zabarella and Scaliger.

[150] *Balneum Dianae seu magnetica priscorum philosophorum clavis,* 1600; *De electro philosophorum magico-physico,* 1611.

[151] Johann Ernst Burggrav, *Introductio in vitalem philosophiam cui cohaeret omnium morborum astralium et materialium seu morborum omnium elementatorum et hereditariorum ex libro naturae codice philosophicae et medicae veritatis additis veterum placitis Hippocratis Galeni Celsi aliorum explicatio atque curatio. In speciali explicatione morborum agitur de curationum mysteriis, indicationum impendiis remediorum arcanis. Et primum Galeni et aliorum veterum medicamenta proferuntur: deinde Paracelsi, Turnheuseri, Quercetani aliorumque Neotericorum philosophorum experientia demonstratur, medicamenta omnium morborum ex anatomia et arte signata tam simplicia quam composita ostendendo.* Francofurti Typis Hartm. Palthenii, Sumptibus Ioh. Th. de Bry et Ioh. Ammonii, Anno MDCXXIII.

A dedicatory letter of Feb. 1, 1623, by the same Ioannes Theodorus de Bry to Hartmann Beyer, M.D., *archiater* of the republic of Frankfurt, alludes to the author as *vir et medicus quondam optimus,* showing that the work is published posthumously, but Burggrav is not named on the first title page, nor on another at p. 45 which repeats most of the first.

[152] *Ibid.,* p. 1.

True philosophy is not within human power, but is the breath of God and the gift of divine inspiration and illumination which is vouchsafed to hardly one person in thousands. Paracelsus, "the true monarch of true philosophy and medicine," spent years as captive among the Egyptians in order to acquire it, and returned bearing rich spoils in the shape of beautiful remedies.[153] "Interior and essential form is the vital principle," the Sphere of the Pythagoreans, "equally diffused through all parts of the world, whose center is everywhere and circumference nowhere." The question is also raised whether first matter may not be identical with essential form.

Scaliger, "subtlest of subtle philosophers," is represented as having said (but no specific passage is cited), that "inferior forms are fostered by superiors, and superiors are not destitute of the benignity of inferiors." The Hebrew Cabala is made responsible for the doctrine that all the virtues of the stars and celestial ideas are received by the moon, and thence passed on to inferior matter. There is a difference of lives and a variety of balsam in the macrocosm. Of all waters antimony is the greatest cordial, in which lies hid so great virtue of balsam that it cures "all deplored and desperate diseases." Burggrav treats of transplantation as well as generation and mixture. Chronic diseases last as long as the course of the planet causing them. Of the four humors, the blood is salt salt; phlegm, sweet salt; bile, bitter salt; melancholy, acid salt. Soon the influence of the stars on disease is again considered, and the relation of macrocosm and microcosm.[154] Evidently all that the vital philosophy amounts to is a thin mixture of alchemy and astrology. The book is primarily medical and concerned with particular diseases and their cure, with inclusion of many chemical remedies. But the astrological factor persists to the end, for the concluding passage of the text, against phlebotomy in certain diseases, is as follows:

It should be observed in certain fevers, especially tertian, and in all other astral diseases and epidemic, mercurial, arsenical, not to open a vein. For it often has been found that those afflicted with diseases of

[153] *Ibid.*, p. 4. Paracelsus of course was never in Egypt.

[154] *Ibid.*, pp. 5-6, 18-19, 20, 29, 38, 43, 50-58, 59.

this sort expire soon after incision of a vein, and this is proved by experience and usage.[155]

The *Restored Physics* of Jean d'Espagnet, with its Paracelsan tinge, is treated in a later chapter. But with it is bound an *Arcanum of Hermetic Philosophy* of the same date, 1623,[156] which may appropriately be considered here. The divine science of alchemy requires fear of God and the whole man. Many alchemists are over subtle in their methods. One should beware of pseudo-philosophers and read few authors but the best, and be suspicious of books which are easy to understand. Espagnet especially recommends Hermes, Morienus, Trevisan and Raymond Lull, whose *Testament* and *Codicil* he advises to pore over perseveringly. Alchemists often express themselves better in enigmatic types and figures than they do in words. The work of transmutation can be performed from sun and moon (i.e., gold and silver) alone, but soon our author is discoursing about mercury and quoting Geber. Presently he treats of material means, operative means and demonstrative means, then of four digestions, then of the duplex wheel and three circles. The fire of the stone is threefold: natural, unnatural, and contrary to nature. We come to vessels, furnaces, and eventually to the end of the treatise.

Espagnet later published a brief popular booklet in French entitled The Mirror of Alchemists[157] in which he called them *lacrimistes* and quoted against them from their own authors. This verbose and fulsome screed seems, however, to be directed only against ignorant alchemists and not against alchemy. It includes some matter addressed to the ladies on beautifying.

In our sixth volume we had occasion to treat of the condemnation by the medical faculty at Paris in 1603 of certain writings of advocates of chemical remedies.[158] Now in 1624 the Sorbonne or theological faculty censured as rash and insolent a number of alchemical theses. These theses denied the three principles of Aristotle, namely, matter, form and privation, or, more specifically,

[155] *Ibid.*, p. 166.

[156] Copy used: BN R. 51612, Paris, 96 pp.

[157] *La miroir des alchimistes,* 1669, 69 tiny pages: copy used, BN 8° Lb35. 856.

[158] T VI, 247 *et seq.*

first matter, substantial form (except for the rational soul in man), and privation. In place of the traditional four elements they held that mixed bodies were composed from ungenerated and incorruptible atoms of five simple bodies, namely, earth, water, salt, oil or sulphur, and mercury. Air was said to be essentially identified with water, and fire with the heavens. Fire was also held to be most humid, and earth to be lighter than water. The theses further denied the existence of virtual qualities, declared that alteration was not merely in accidents but that some of the atoms of the compound must be lost or gained. The five elements or principles, however, could not be transmuted into one another, but Aristotle was wrong to deny that all things are in all things, and that all things are composed of atoms. Except for the rational soul, diversity of genera, species and individuals were the result solely of mixtures of these five simple bodies. The medical faculty supported the Sorbonne in this censure, and the Parlement of Paris confirmed it.[159]

These theses were to have been defended at the house of François de Soucy, Sieur de Gersan or Guerseren—formerly the hôtel of Queen Marguerite—by Jean Bitaud or Bitault of Saintonge, under the direction of Antoine de Villon or Billon, of Plassans en Provence, known as "le soldat philosophe" (*miles philosophus*), and Etienne de Clave, a physician and chemist. Morin says that Villon, wishing to attack Aristotle and all celebrated sects of philosophy, but knowing little chemistry himself, used de Clave as a cat's paw to pull chestnuts out of the fire for him. A crowd estimated at eight or nine hundred came for the discussion. According to Morin, the first president of the Parlement had already forbidden Villon to

[159] I base this account on the contemporary work of Mersenne, *La Verité des sciences contre les sceptiques ou Pyrrhoniens*, Paris, 1625, 1012 pp., in-8, pp. 79-83, (Copy used: BN R. 9668), and the account given by Jean Baptiste Morin de Villefranche, *Réfutation des thèses erronées de Antoine Villon, dit le soldat philosophe, et Etienne de Claves, médecin-chimiste, par eux affichées publiquement à Paris contre la doctrine d'Aristote le 23 Août 1624, à l'encontre desquelles y a eu censure de la Sorbonne et arrêt de la Cour de parlement*, Paris, 1624, in-8, viii, 106 pp.: BN R. 12464; R. 44575. Also upon the deliberations of the Sorbonne and *Arrest du Parlement* printed in d'Argentré, II, ii, 147; III, i, 215-216.

There is a MS of the Theses at the Vatican: Reg. Suev. 952, fol. 47-.

defend the theses, but he distributed them and kept the audience waiting until 3 P.M., before he informed them of the prohibition, whereat he and de Clave were roundly hissed. Later the theses were torn up in the presence of de Clave, who was the only one concerned that the authorities had succeeded in arresting. The three authors of the theses were exiled from the sphere of jurisdiction of the Parlement of Paris, and it was forbidden to teach anything against the ancient and approved authorities under penalty of death.[160]

This vague threat does not seem to have been taken very seriously. No action was taken against Gassendi for his *Exercitationes paradoxicae adversus Aristoteleos,* printed in the same year—at Grenoble, it is true, not Paris. De Clave retired for a time to Brittany, but later both gave chemical courses and published chemical books at Paris. Gaffarel in *Curiositez inouyes,* printed at Paris in 1629, though composed some years before, held that fire was moist rather than dry, as the Theses had, and represented "M. de Claves, one of the excellent chemists of our time," as performing daily the experiment of reproducing an herb or flower from its ashes.[161] It is possible, however, that he refers to activities of de Clave before the condemnation of 1624. It may also be that his own retraction of October 4, 1629, of which we treat in a later chapter,[162] was meant to cover such alchemical doctrine as well as the astrological images with which it has usually been connected. It is further noteworthy that these alchemical theses were refuted in detail by Morin, although he had published an astrological treatise only the year before.[163]

Mersenne, who informs us of this censure, later in the same book advocates the establishment of an academy of alchemists in each kingdom or in the chief city of each province in order better to regulate their activities.[164] Farther on in the same volume he asserts that the works of alchemy attributed to Aristotle and

[160] Mersenne, *Correspondance,* I (1932), 167. The account given there does not quite agree with that by Morin.

[161] *Curiositez,* pp. 139, 209-12; *Correspondance,* I, 168, 326.

[162] See Chapter X.

[163] *Astronomicarum domorum Cabala detecta,* Paris, 1623.

[164] *Correspondance,* I, 105.

Aquinas are not genuine.[165] Strowski has already pointed out that in this same book Mersenne is quite favorable to alchemy based upon experience but does not share the alchemists' profound contempt for Aristotle—yet, as we have just seen, they ascribed alchemical treatises to him—or their admiration for Bacon's *Novum Organum,* and does not approve of their quoting Scripture for their own ends and employing analogies between the alchemical process and the mysteries of religion.[166] When Nicolas de Blegny, decades later in 1694, founded an *Académie chimiatrique,* his action stirred up a scandal.[167]

Returning to the resumption of chemical and alchemical activity at Paris, we may note that late in the year 1628, at a conference attended by Descartes and his friend Villebressieu, by the above-mentioned Morin, by Cardinal de Bérulle, and perhaps by Naudé, a new philosophy was propounded by a chemist named Chandoux. Descartes found it more plausible than scholasticism, but not without difficulties, and inferior to what he himself already had in mind. Chandoux seems to have preached better than he practiced, since within three years he was hanged for counterfeiting.[168] But perhaps his philosophy was counterfeit, too.

In 1629, David de Planis-Campy, who had previously issued works on phlebotomy, musket-wounds, and the pest, put forth at Paris a huge Bouquet, as he called it, of the most beautiful chemical flowers, preparations, experiments and rarest secrets, with pharmaco-chemical remedies drawn from the three kingdoms, mineral, animal, and vegetable.[169] Four years later and also at Paris he published a work on opening a school of transmutatory metallic philosophy.[170] When his collected works appeared in 1646, they

[165] *Ibid.,* p. 167.

[166] Fortunat Strowski, *Histoire du sentiment religieux en France au XVIIe siècle,* 1909, I, 216-18.

[167] Emile Guyénot, *L'évolution de la pensée scientifique, les sciences de la vie aux XVIIe et XVIIIe siècles, l'idée d'évolution,* 1941, p. 154.

[168] *Correspondance du P. Marin Mersenne,* II (1936), 163-64.

[169] *Bouquet composé des plus belles fleurs chimiques, ou Agencement des préparations et expériences ès plus rares secrets, et médicaments pharmaco-chimiques pris des minéraux, animaux et végétaux,* Paris, 1629, in-8, 1007 pp. BN R. 46763.

[170] *Ouverture de l'escole de philosophie transmutatoire métallique . . .,* Paris, 1633, in-8. BN R. 46764.

were said to contain the finest treatises of chemical medicine, "corrected by the author before his death and increased by some hitherto unprinted."[171]

Vincenzo Solombrino, a Jesuit father, wrote at Turin on the marvelous virtue of antimony, with an account of 115 cases in which it had been administered during the three years preceding.[172]

Claude Dieudonné's book on the prolongation of human life to one hundred and twenty years appeared in 1628–1629,[173] and fills three volumes. The full Latin title mentions "new, rare, wonderful mysteries of recondite nature, precious Hippocratic-Hermetic arcana, essences, tinctures, elixirs," and so forth. He recognizes that he lives in an age of evil enchantments,[174] but he believes in the excellence of man and harmonic analogies with the Megacosm, in signatures, sympathies and antipathies, and virtues of animals, vegetables and minerals serving longevity. Nature, including the influence of the heavens, is still as strong as in the first age of the world. Climacteric years do not prevent one from living to be one hundred and twenty, a consideration which leads to further discussion of critical days and the mysteries of the number seven. Dieudonné believes in astrological medicine but not that the stars bring ills to man. He digresses in connection with water to affirm that fountains are derived from the sea. Among foods he mentions caviar, but disapproves of frogs and snails, "those monstrous ban-

[171] *Les oeuvres de David de Planis-Campy . . . contenant les plus beaux traictez de la médecine chymique . . . corrigées par l'autheur avant son deceds et augmentez de plusieurs traictez non imprimez,* Paris, 1646, in-fol., 752 pp. BN Fol. Td[3].13. Hoefer, *Hist. de la chimie,* II, 332, says that there are several MSS by him in the BN.

[172] *L'antimonio, cioe trattato delle maravigliose virtù dell'Antimonio commune et particolarmente dell' Antimonio che con rara preparatione si raffina hoggidi in Turino. Con le annotationi del Signor Filostibio,* Turin, 1628, in-4, 70 pp. Copy used: BN 16° T.417.

[173] Claudius Deodatus, *Pantheum hygiasticum Hippocratico-Hermeticum de hominis vita ad centum et viginti annos salubriter producenda, libris tribus distinctum . . . politico-historica et medico-spagyrica narratione exornatum . .,* Bruntruti (Pruntrut, Porrentruy, Switzerland), Excudebat Wilhelmus Darbellay, 1628, in-4: BM 1088.i.12,13: BN Res. Tc[11]. 114. At I, 180, "Aer Bruntrutanus qualis."

[174] *Prodromos,* "in tam ulcerata nostri seculi aetate tot illecebrarum incantamentis dementata."

quets."[175] Other specimens of his contents are the admirable virtues of the eagle-stone (*aetites*), potable gold, "the balsamic medicine of the ancient Hermetics, sole, unique, true, catholic," a secret to sharpen the sight, and a secret amulet.[176]

Of a symposium instituted in 1634 by Beverwyck whether the term of life is fatal or mobile we treat in a later chapter on Medicine, but we shall presently come to a third discussion by Zaccagnini in 1644 of the question as to prolongation of life.

Two years after issuing the editio princeps of Harvey's epoch-making *De motu cordis,* William Fitzer of Frankfurt-am-Main also published there eight books by another English author, Samuel Norton (1548–1604?). These really belonged to the previous century and now appeared posthumously under the editorship of yet another Englishman, Edmund Deane. Instead of proclaiming a new discovery in science and medicine, they revamped old traditions in the field of alchemy, "considerations of ancient writers in alchemy," and *Mercurius redivivus,* although in one instance they claimed to reveal "a way of making the tincture ... sought by the olden philosophers but as yet transmitted by no one." Even the engravings in these alchemical treatises seem to have been executed by the same man as that in Harvey's volume.

The *Fasciculus chemicus* of Arthur Dee (1579–1651), first-born son of John Dee, "from our Museum at Moscow, March 1, 1629," where he was physician to the Czar, but printed at Paris in 1631,[177] was a selection of extracts from past alchemical authors, chiefly medieval, arranged topically in ten chapters.

Another old work which was printed later in the century, a treatise on the elixir for white and for red and the great philosophic stone, which Jean Saignier of Paris left on his death bed for his son Charles, is represented in the edition of 1664 as having been composed in 1632.

By my hope of heaven I have declared to you what my eyes have seen,

[175] Caps. 43, 45 (pp. 400 and 403 of vol. I). It would seem that book II on foods, should have begun at cap. 20 on p. 235, but at p. 406 we read, "Finis libri primi et corollarium ad lectorem," vol. I ending at p. 408.

[176] Vol. I, 96; III, 79-81, 100, 132, 204.

[177] In-12, 172 pp. BN R.33119. There was an English translation in 1650.

my hands have operated, my fingers have extracted. And I have written this booklet with my own hand and signed it with my name when I was in the last agony, the year 1632, May 7.[178]

Really the date should be May 7, 1432, as manuscripts of the work show.[179] There also appears, in addition to the Latin edition of 1664, to have been a French edition in 1661[180] and perhaps a Low German or Dutch edition in 1600.[181]

Billich, of whose criticism of Beguin and others we treat elsewhere, in 1631 published Chymiatric Observations and Paradoxes in two books, of which the first explained the preparation of chemical remedies, and the second their use.[182] He affirmed the occult hates and friendships of things, which produced new and marvelous effects far beyond the scope of the elements. But he rejected the three principles of the chemists. Mercury, sulphur and salt were neither principles nor prime miscibles. Their *terra mortua* was not sterile, but so fertile and lively that it would resuscitate an herb, if its seed was committed to it. Potable gold was a figment, but the empiric of Verona, Vittorio Algarotta, by his powder of Mercury of Life had not only restored many sick to health, but won a fortune of some thousand gold pieces and an immortal name.[183]

[178] Joh. Saignier Parisinus, *Magni lapidis naturalis philosophia et vera ars in opus deducta et filio suo Carolo relicta in agone mortis propria manu subsignata*, Bremen, 1664, BN R.8486, p. 52, closing words.

[179] Cassel Landesbibliothek Chem. Octavo 15, "Naturalis philosophia et vera ars per Joannem Saignier Lutetianum in opus deducta et filio suo Carolo magni thesauri testamento relicta et in agone mortis propria manu subsignata Parisius, anno domini 1432 die 7 Maii."

Orléans 291 (245), 16th century, fols. 57-74, "Cy commence une doctrine de philosophie laquelle maistre Jehan Saulnier bailla à son fils sur la transmutation des métaux," dated "cccc trente deulx le viie jour de mai." This MS is said to give a fuller text than the printed version mentioned in the next note.

[180] Jehan Saulnier, *Doctrine des philosophes sur la transmutation des métaux*, ed. Gabriel Castaigne, 1661. Not in BM or BN printed catalogues.

[181] Johann Saignier, *Duytsche Alchimie van J. Saignier waer by ghevoecht is Lumen luminum*, Leyden, 1600. But the title suggests another work. BM 1033.c.1.(1.).

[182] *Antonii Guntheri Billichi Frisi archiatri Oldenburgensis Observationum ac Paradoxorum Chymiatricorum libri duo: quorum unus medicamentorum chymicorum praeparationem, alter eorundem usum succincte perspicueque explicat*, Leyden, 1631, in-4, 174 pp. BM 1033.h.19 (4.)

[183] *Ibid.*, pp. 12, 22, 31, 134, 163.

In his *Thessalus in chymicis redivivus* of 1643,[184] Billich was much less favorable to chemical remedies. He asserted that some were stinking, that chemists abused them both in preparing them and in applying them. He also tried to pick out passages unfavorable to chemical remedies from Glückradt's notes on Beguin.

In 1635 de Clave published at Paris a volume in the preface of which he argued in favor of chemical remedies and held that chemistry was the principal and most essential part of medicine.

> Quartan fevers yield within a few days to the excellence of our remedies without vomiting and without violence, as do intestinal hernias or ruptures which we cure in fifty days up to the age of fifty-five by application of plasters only, venereal disease in three weeks.

The ancients had discovered something in the animal and vegetable kingdoms, but very little, and that faulty, in the mineral domain. Since de Clave has delved in it for thirty years past, he feels under obligation to give to the public the fruit of his researches, which he proposes to do in no fewer than forty treatises, of which the present two on stones will be followed by two more; these by four on semi-metals and marcasites, two on gems, two on bitumens, two on salts, and two on sulphurs. He will also treat of the central fire which is the efficient cause of all subterranean generations. There will be two treatises on generation and corruption, four on meteors, ten on vulgar errors in medicine, and two upon Hermetic medicine.[185] Which does not seem to quite total up to forty. In any case, this formidable program of publication seems fortunately not to have progressed very far.

Johannes Pharamundus Rhumelius, the alchemist, should not be confused with his contemporary, Johannes Conradus (or, Janus Cunradus) Rhumelius (1597–1661), an M.D. of Altdorf, 1630, and author of various medical works and two poems entitled *Theolo-*

[184] *Thessalus in chymicis redivivus, id est, de vanitate medicinae chymicae hermeticae seu spagyricae dissertatio*, Francof., 1643, in-8, 235 pp. BM 1033.d.13. The dedication is dated in July, 1639. Its "vii Non. Julii 1639" is probably a mistake for July 2, as July 1 would be the Calends.

[185] Etienne de Clave, *Paradoxes ou Traittez philosophiques des pierres et pierreries contre l'opinion vulgaire*, Paris, Veuve P. Chevalier, 1635, in-8, Au lecteur. Copy used: BN S.20394.

gia Vegetabilis and *Philosophia Animalis.*[186] Rhumelius the alchemist was cited by John Webster in 1671 for his description of *primum ens auri* or primal being of gold. In the fourteenth century the doctrine had been widespread that the philosophers' stone could best be made from mercury alone, rather than a combination of mercury and sulphur.[187] Webster credited Rhumelius with an analogous gold-alone theory and said that he distinguished four sorts or states of gold, namely, the astral, mineral, metal and elemental. The first was primal being of the Sun (i.e., gold) and was a great secret. Potable gold made from it was superior to that from perfect common gold. Elemental gold was any earth, mineral or stone wherein the spirit of gold lay hid.[188]

Webster cited no particular work by Rhumelius for this theory, and the one which I have been able to examine[189] is more miscellaneous in character, taking up such subjects as the most universal medicine, human *mumia,* the influence of the stars in magnetic cures, the mystery of transplantation of disease to an irrational animal, the mineral stone, vegetable stone, and animal stone, potable gold, "the universal menstruum and our pontic water in which gold and all metals are dissolved like ice in hot water," tincture of coral, quintessence of pearls, and the elixir of life. On the title page

[186] LR 559.

[187] T III, 797, Index, Mercury, mercury alone theory.

[188] Webster, *Metallographia,* London, 1671, pp. 39, 119.

[189] *Compendium Hermeticum de macrocosmo et microcosmo totius philosophiae et medicinae cognitionem breviter et compendiose complectens. Additum est Dispensatorium chymicum novum de vera medicamentorum praeparatione,* Francof., 1635, in-12: the sole work by him listed in LR.

In the NYAM copy examined, the *Dispensatorium,* though listed on the title page, is missing, while the *Compendium Hermeticum* is preceded by *Avicula Hermetis catholica* and other tracts—*Elixir vitae, Leo rubeus antipodagricus fixus*—of Salomon Raphael (London, 1639; *Imprimatur* of Dec. 16, 1637, 98 pp.), and by a *Canticum canticorum quod est Schelemonis de medicina universali.* Ferguson, II, 266-67, states that Rhumelius sometimes wrote under the pseudonym, Solomon Raphael.

A German version of the *Compendium Hermeticum* appeared in the same year, 1635, as the Latin text, in a volume entitled, *Opuscula chymico-magico-medica:* BN 8° Te^{131}.93.

Also in German by Rhumelius was *Medicina spagyrica tripartita oder Spagyrische Artzneykunst,* Frankfurt, 1648, of which portions were printed in French translation at Paris in 1932. Ferguson lists the contents of the 1662 edition and of the aforesaid *Opuscula* of 1635.

Rhumelius is called *mathematicus medicus,* and he dedicated the book to August, duke of Anhalt.

Petrus Joannes Faber or Pierre Jean Fabre, an M.D. from Montpellier and a citizen of Castelnaudary, published a number of spagyric works at Toulouse. It may suffice to examine two or three of them. In *Hercules piochymicus* not only the twelve labors of Hercules are interpreted alchemically but also his begetting fifty sons from the fifty daughters of Thespius in a single night and his death from the poisoned robe of Nessus. The conclusion of the whole work is an anathema to those who would have alchemy made perfectly plain to fools in words of one syllable, and a request that believers in transmutation await the appearance of his *Panchymicus.* But then is added a bit from Faber's own medical practice. For a girl whose head was disfigured with ulcers he prescribed an unguent celebrated by Gordon, Guido and Paré. When after three or four months she grew worse rather than better, her family without consulting him covered her whole head with the unguent so that it even filled her ears. She died within twenty-four hours in intense pain, but Faber is shocked that her father blamed his unguent for it.[190]

The *Panchymicus,*[191] to which we heard Faber allude, appeared at Toulouse in 1646 and at Frankfurt in 1651. It treated of such things as birds, fish and insects; vegetation, flowers and fungi; stones, metals and minerals.

In his *Propugnaculum alchymiae*[192] of 1645 Faber inveighed against the opponents of alchemy and indulged in such mutually contradictory and inconsistent clichés as that the quicksilver of

[190] *Hercules piochymicus, in quo penitissima, tum moralis philosophiae, tum chymicae artis arcana, laboribus Herculis apud antiquos tanquam velamine obscuro obruta deteguntur,* Toulouse, Petrus Bosc, 1634, 8vo, 8 fols., 191 pp. Copy used: BN R. 35629.

[191] *Panchymici seu anatomiae totius universi opus.*

[192] *Propugnaculum alchymiae adversus quosdam misochymicos, philosophos umbratiles, naturae humanae larvas, qui se philosophos profiteri audent dum Chymiam stulte rident nec tamen brutorum genia tenent . . . auctore Petro Ioanne Fabro doctore medico Monspeliensi ac Castronovidaurii Tectosagum cive,* Toulouse, 1645, 8vo. 128 pp. Copy used. BN R. 35578.

the philosophers and vulgar quicksilver differ by the whole sky, and that vulgar mercury duly prepared is in every way sufficient to complete the chemical work.[193] Yet gold and silver are also necessary, and the gold of the philosophers differs from ordinary gold.[194] One must have a body which is neither mineral nor vegetable nor animal, but in which nature has joined purest sulphur, mercury and salt, which is always at hand and before our eyes, which you have daily and cannot live without.[195] Yet another chapter states that the philosophers' stone can be made from metals alone.[196] Faber was no better at history than logic, for he asserts that Raymond Lull, who died in 1315, made gold for Edward III of England about 1354.[197] Gui Patin, who was of course prejudiced against alchemy, in a letter of January 27, 1649, referred to Fabre as "un pauvre souffleur."[198]

The problem of the prolongation of human life was again considered by Lelio Zaccagnini in 1644.[199] He held that God from eternity had set an ultimate term of life for every man beyond which neither medicine nor other human effort could prolong it. He accepted the influence of the stars, critical days, and astrology in moderation, but concluded that no sure faith could be placed in astrological prediction of length of life, because the human will was free and future contingents could not be determined. The philosophers' stone might prolong life notably but not perpetuate it as the tree of life could. The alchemists tell wonders of potable gold. For, although it cannot be assimilated by our innate heat as food can, it acts on the human body as a medicament, communicating arcane virtues to the heart and preserving the vital spirits for prolongation of life. They say it receives these properties by sympathy with the sun, but in Zaccagnini's opinion gold most effectively

193 *Ibid.*, pp. 44, 51.

194 *Ibid.*, pp. 62, 71.

195 *Ibid.*, p. 46.

196 *Ibid.*, p. 111, cap. 37, "Quod ex solo metallico genere lapis philosophorum fieri possit."

197 *Ibid.*, p. 117.

198 *Lettres* (1907), p. 642.

199 *Notabilium medicinae libri duo. Primus agit de vitae humanae longitudine ac brevitate etiam quoad astrologos incerta et an arcanis medicinae remediis possit prorogari . . .*, Rome, Bernardino Tani, 1644, in-4, 178 pp. BM 1169.g.11.

gladdens the heart of man, if he possesses it in good quantity.[200]

Ludovico Locatelli, who says that he has already published a pest tract at Venice in 1629, issued his *Theatro d'Arcani* at Milan in 1644.[201] He states that there are few chemists in Milan, and that most medical men there are hostile to the art.[202] The arcana include the preparation of first matter, philosophers' stone, mercury of life, potable gold, transparent vitrified gold, an *Aurum vitae* of Locatelli's own invention as well as three others, a quintessence of silver of his own, and the quintessence of wine of Raymond Lull (really of John of Rupescissa).[203] The arcana are followed by an exposition of Paracelsus on the Aphorisms of Hippocrates, and explanations of the obscure terms of the philosophers and of alchemical characters. Locatelli further promises to publish a *Lucidario Chimico* to explain the obscure passages in his own work. I do not know if this promise was kept, but there were later editions of the *Theatro d'Arcani*.

In 1645 was printed at Frankfurt a brief treatise entitled Chemical Nonentities, or a catalogue of those chemical works and operations, which although they are not in the nature of things and cannot be, yet are everywhere circulated and forced on the world with a great noise by the vulgar herd of chemists.[204] The work was reprinted at Frankfurt in 1670 with a preface by G. W. Wedel,—whom some have in consequence regarded as its author, but this is impossible, as the first edition was in the year of his birth,—and again

[200] *Ibid.*, pp. 1, 10, 15, 27, 31-32. *De termino vitae humanae* by S. A. Fabricius, 1666, is purely medical and free from astrology. So is Giambatista Vertua, *De morte retardanda libri tres*, Milan, 1616, in-4, dedicated to cardinal Federigo Borromeo: BM 1039.d.15, the only work by him in the BM. He was a member of the medical college of Milan and died of the pest in 1630: Corte, *Notizie storiche intorno a' medici scrittori milanesi*, Milano, 1718, p. 170.

[201] In-8, 456 pp. BM 1034.a.4 is the sole work by him in BM.

[202] Castiglione (see Index) had died in 1629.

[203] T IV, 37-38.

[204] *Non-Entia chymica sive catalogus eorum operum operationumque chymicarum quae, cum non sint in rerum natura nec esse possint, magno tamen strepitu a vulgo chymicorum passim circumferuntur et orbi obtruduntur*, Francof., apud Thom. Matthiam Götzium, 1645, 35 small pp. of which 3-13 are "Lectori salutem." BM 1036.a.3 (1.).

Borel (1654), p. 59, is presumably mistaken in dating the work 1606, Francof. "apud Thomam Matthiam Goltzium."

at Berlin in 1674 with the *Destillatoria Curiosa* of Elsholtz. The treatise has also been attributed to Michael Kirsten (1620–1678), who in 1655 became professor of mathematics at Hamburg and in 1660 also professor of physics.[205] It is more likely, in my opinion, to have been by Georg Kirsten (1613–1660), professor of medicine at Stettin and royal physician,[206] who three years later attacked the commentary of Johann Agricola on Johann Poppe for its false and fraudulent use of chemical remedies, potable gold and other panaceas.[207] The 1645 edition contained the statement, "Utis has written in the year from the blinding of Polyphemus 2830 or thereabouts."[208] Utis was amplified to Utis Udenius in later editions.

The text opens with nine *non-entia* from vegetation, of which the first is the resuscitation of a plant from its ashes. Five from animals include the extracting of "live mercury" or true quicksilver from the blood of animals, *zibetha* from human dung or that of cows or any other animal, and the homunculus of Paracelsus. From stones all waters, liquors, spirits, salts, oils, tinctures, essences and extracts which cannot be prepared without destroying their subject and resolving it into its component parts, such as those from pearls or tincture of coral. In the case of minerals, eleven *non-entia* are listed from vitriol or calcanthus, seven from antimony, six from sulphur, and so on. Those from metals include potable gold and silver. Most quintessences are classed as *communia non-entia,* and the final topic is nonenties in the transmutation of metals.

This little booklet seems to have exerted a wholesome influence. We find very similar *non-entia* listed by Rolfinck in his manual of 1661.[209]

Johann Rudolph Glauber was born early in the century and at the age of twenty-one had discovered in a mineral spring at Vienna the salt (sulphate of soda) which has since borne his name. His

[205] Ferguson, II, 489.

[206] *Ibid.*, I, 471.

[207] Georg Kirsten, *Adversaria et Animadversiones in Johannis Agricolae, D. ac physici Breslaviensis, Commentaria in Poppium et Cirurgiam parvam. Darinnen der falsche und betriegliche Gebrauch der chymischen Artzneyen, des Aurum potabile und andere Panaceas beleugend . . . wiederlegt wird,* Alt Stettin, 1648, in-4, 604 pp. BN R.3017.

[208] At the close of the preface to the reader, p. 13, "Vtis scripsit anno ab exoculato Polyphemo MMDCCCXXX aut circiter."

[209] *Chimia in artis formam redacta,* 1661, pp. 419-38.

first published work in 1646 was in German and concerned with potable gold.[210] In 1648 he settled in Amsterdam for the rest of his life and founded a Hermetic Institute. His *Furni novi philosophici* appeared there in five parts.[211] In some respects Glauber was a Paracelsan and he accepted the doctrine of signatures. But whereas Paracelsus had stressed the three principles—mercury, sulphur and salt, and earlier alchemists had regarded all metals as compounds of mercury and sulphur, Glauber made mercury and salt the principles of metals, and salt the source of all things. In other passages, however, he spoke of all minerals and metals as composed of earth and water, which were perhaps equivalents of salt and mercury. He believed in panaceas and marvelous medicines and the transmutation of metals, and his works contain tricks and illusions akin to those of magic. Yet he was perhaps the first to note the existence of chlorine, working with hydrochloric acid and the metallic chlorides. Besides sodium sulphate, he has been credited with the discovery of arsenic trichloride and potassium nitrate.[212] He showed unusual penetration as to the composition and decomposition of bodies, chemical analysis and synthesis, and is said to have been the first to explain a case of double decomposition. Although he was much given to figurative expressions such as iron man for a furnace, and white swan for a stage in the process, his descriptions of experiments and chemical processes are unusually clear for his time. He gave especial attention to practical or technical or applied chemistry. We see in him, then, a combination of alchemical fantasy with detailed chemical progress.

Glauber still indulged in a good deal of mystical and semi-magical patter. He was sure that piety, as well as true knowledge of superior and inferior bodies, was essential for success in alchemy,[213] and that God would never allow pseudo-Christians and the proud and avaricious to discover the secret of the philosophers' stone.[214] He

[210] *De auri tinctura sive auro potabili* . . ., first ed., Amsterdam, 1646; 2nd ed., Frankfurt, 1646, in-4.

[211] *Oder Beschreibung einer new erfundenen Distillirkunst*, J. Fabel, 1646-1649, in-8. BM 1033.b.9 (1.).

[212] John Read, *Humour and Humanism in Chemistry*, 1947, p. 97.

[213] *De vero auro potabili*, Latin edition of 1651, p. 9.

[214] *Furni novi philosophici*, IV (1658), 47.

professed to set forth new inventions, hitherto unknown,[215] and secrets which he reserved for himself or revealed or promised to reveal.[216] He admitted that he himself was ignorant and inexperienced as to the transmutation of metals, on which he had not dared to spend time and money, but it had been believed in for many centuries by the most famous men among Jews, pagans and Christians and proved by most certain reasons.[217] The title, New Furnaces, was something of a misnomer, since in the first part of 65 pages only a page and a half is devoted to the first furnace, and the rest to various oils, spirits, flowers, and other chemicals and chemical remedies. He also believed in a most efficacious and incomparable remedy for all diseases, "for which praise and glory to eternity to God immortal who has revealed to us so great secrets."[218] He asserted that, after years of effort, he had succeeded in obtaining potable gold,[219] and we turn to his treatise on it.

Since it is indisputable that all life proceeds from the heat of the sun, the ancient philosophers sought to unite gold, the terrestrial sun and fixed and perfect body caused by the sun's rays, with man by aid of the spirit of wine. Since alcohol appears not to have been segregated until about the tenth century, these philosophers were hardly as old as Glauber thought. But he goes on to say that gold, "occultly endowed with the virtues and properties of the sun," is reducible by chemical art to what it was before it coagulated, forsooth into a warming and vivifying spirit which will communicate its virtues and gifts to the human body. The ancients employed a water which nature offers spontaneously without need of violent distillation, by which they brought to light that which was hidden in gold and hid what was manifest, separating its soul or tincture

[215] *Furni novi*, I (1658), Praefatio ad lectorem benevolum.

[216] *Idem;* also V (1651), 47, where in closing he says that if the reader likes this work, he will go on to communicate "marvelous secrets, incredible to the world, hitherto hid from envy or ignorance." Then in the Appendix, p. 49, he says that in a new edition he will communicate "many most select secrets omitted for certain reasons in the first edition." There follow seventy-two separately paginated Annotations to the Appendix, "where is treated of various most useful secrets, of the best and unknown, published for the sake of the incredulous and those ignorant of natural secrets."

[217] *Furni novi*, IV (1658), 44.

[218] *Furni novi*, I (1658), 45.

[219] *Ibid.*, IV (1658), 44-45.

from its crass and black superfluous body, choosing for their elixir the subtlest part of gold, which was accomplished by philosophical union of wine and gold, and their volatilization and re-coagulation and inseparable fixation. Gold cannot be rendered volatile without spirit of wine, and spirit of wine cannot be coagulated and fixed without gold.

After this and more by way of preamble, Glauber gives the following instructions for the preparation of true potable gold:

Take one part of live gold and three parts of live mercury, not vulgar but philosophic, obtainable everywhere without expense or labor.

Glauber then remarks parenthetically:

(You can also mix in quicksilver of equal weight with the gold, and better so than gold alone, because of the greater variety of colors resulting from the mixture of male and female. He who is persuaded of the superiority of the tincture resulting from gold alone mixes in gold alone. Not so one skilled in metals who knows the power of radical union of gold and silver when dissolved in one and the same menstruum.)

After this aside, the main instruction proceeds:

Put the mixture in the philosophic vessel to dissolve, and in a quarter hour those mixed metals will be radically dissolved by the mercury and will take on a purple hue. Then increase the fire a little, and they will change to a most vivid green, upon which, removed from the fire, pour water of dew to dissolve it, which can be done in the space of a half hour.

Strain the solution and draw off the water by a glass alembic into B, then pour the water on again, and again draw it off. And do this three times, and in the meanwhile that greenness will be altered to the color of black printers' ink and will stink like a corpse . . .

The water should be drawn off, poured on again, and digested several times, and in the space of forty hours the blackness and stench will vanish and a pure milk white appear. Then all humidity should be drawn off until it is a dry mass, which will still be white. But after a few hours of gentle heat, during which various colors appear, it will turn to a marvelous green, far excelling the previous green.

Upon this should be poured spirit of wine well rectified to the height of two or three fingers' thickness, and that gold which has changed to green will attract that spirit of wine with the utmost friendliness, like a dry sponge absorbing water, and will communicate to it its soul red as blood, by which vivifying process that green is deprived of its tincture and changed to a red color, leaving an ashen superfluity.

The tincted spirit should be decanted, filtered and drawn off into B by a glass alembic from the red tincture attracting the fiery essence of the spirit of wine, so that they are joined most tightly and inseparably, on which account an insipid water falls drop by drop, leaving the virtue of the spirit of wine with the tincture of gold in the form of a fiery red salt, fusible and volatile, of which one grain can tinge a whole ounce of the spirit of wine or any other liquor with the color of blood. For it is soluble in any humidity and hence can be preserved in liquid form as a Panacea of many most desperate diseases.

Glauber goes on to state that this tincture or true potable gold is next to the philosophers' stone the most outstanding of all medicines, and that there is only this difference between them. The soul of gold is volatile and has no ingress into imperfect metals, and so cannot transmute them into pure gold, as the philosophers' stone is said to do. For although the soul of gold is its most potent part, yet it is not fixed in fire but volatile, whereas the philosophers' stone, because of longer digestion, is fixed in fire and permanent.[220]

In his book of 1648 on the old hermetic of the Egyptians and the new medicine of the Paracelsists,[221] Hermann Conring (1606–1681) applied an historical corrective to the alchemy and occult medicine, philosophy and science, of his time. He pointed out that neither Greeks, Phoenicians, nor Egyptians attributed the invention of medicine to Hermes; that chemistry was not very old and that it was not ascribed to Hermes by the first artificers; that the Hermetic writings mentioned by the ancients were lost and in any case were spurious like those now in existence. If there ever was any such

[220] *De auri tinctura sive auro potabili vero, Quid sit et quomodo differat ab auro potabili falso et sophistico, Quomodo spagyrice praeparandum et quomodo in medicina usurpandum,* per Joh. Rudolphum Glauberum, Amsterdam, 1651. Copy used: BN R. 12481. Passage quoted at pp. 12-14.

[221] *De hermetica Aegyptiorum vetere et Paracelsicorum nova medicina,* 1648, 404 pp. Copy used: BN T5.29. *Editio secunda infinitis locis emendatior et auctior: De hermetica medicina libri duo. Primum agit de medicina, de sapientia veterum Aegyptorum; altero non tantum Paracelsi sed etiam chemicorum doctrina examinatur,* Helmstedt, 1669, in-4.

thing as Hermetic medicine, it was full of magic and impiety, like ancient Egyptian medicine and natural philosophy. Paracelsus himself did not claim to be Hermetic, but his school employs magic even more than the Hermetic did, witness their use of weapon ointment. Conring affirmed that Paracelsus gave more attention to metallic remedies, but that the new chemical pharmacy had not yet been reduced to an art, while the Paracelsists had corrupted natural philosophy with their quintessence, three principles, and what-not. In connection with the notion of a fifth essence Conring denied that there was any evidence for the existence of a celestial substance in sublunar bodies.

That alchemy and iatrochemistry stayed about the same, at least in Germany, is suggested by the publication in 1676 and 1701 of works composed by Friedrich Zobell, who had died in 1647. The dedication to his *Chymische medicinische Perle* is dated February 5, 1636, as physician to the prince of Holstein, and the text, which is just a collection of chemical remedies, speaks of a triple kind of astral diseases.[222] Yet the work was printed at Dresden in 1701.[223] Similarly his Spagyric Tartarology was first printed at Jena in 1676 by G. W. Wedel, who states in a preface of September 20, 1675, that he publishes it from a manuscript given him thirteen years before.[224] It was reprinted in 1684 and 1708.

[222] *Op. cit.*, p. 241.

[223] Copy examined: BM 1034.a.13.

[224] *Tartarologia spagyrica*, copy examined: BM 1036.a.3 (2.), 96 pp.

CHAPTER VII

DANIEL SENNERT

Life—Works—Questions debated—Attitude towards nature—Not the founder of the corpuscular theory—Pessimism as to scientific progress—Attitude towards magic—Natural efficacy denied words, characters, images—Occult qualities accepted—Even manifest qualities uncertain—Spiritual qualities—Like attracts like—Signatures—Marvels of nature—Animal sagacity—Power of imagination—Influence of the heavens—Critical days, comets, divination from dreams—Alchemy—Barnerus.

Sennert, our German Galen, a man of the Asclepiadean republic who has earned the esteem of the best men

—E. H. Henckel

Daniel Sennert was born in Breslau in 1572 and died at Wittenberg in 1637 of the seventh recurrence of the plague which had afflicted that town since he first came there as a student in 1593. He became a master of arts in 1597, visited other German universities, returned to Wittenberg for his M.D. in 1601, and from 1602 on was professor there and the first to introduce the study of chemistry at that school. In 1607, he tells us,[1] he published a commentary on Fernel on diseases of the whole substance and reprinted it in 1611 with the first edition of his Medical Institutions. His Epitome of Natural Science appeared in 1618 and again in 1624, 1632 and 1633 during his lifetime. In 1619 came out the first edition of his treatise on the agreement and disagreement between Galenists and Peripatetics and chemists,[2] with a preface to the reader dated on New Years Day. Other editions during his lifetime followed in 1629 and at Paris in 1633. His four books on fevers appeared in 1619 and then in 1627,

[1] *Opera* (Lyons, 1650), I, 851.

[2] *Ibid.*, III, 697 (the title page), "De consensu et dissensu Galenicorum et Peripateticorum cum Chymicis"; but at p. 703, "De Chymicorum cum Aristotelicis & Galenicis consensu ac dissensu." This work will henceforth be briefly cited in the notes as *De Chymicorum*.

and his six long books of medical practice between 1628 and 1635. The *Hypomnemata physica* were published only in 1636, the year before his death. The replies from the theological faculties of eight German universities acquitting him from the charge of heresy and blasphemy, made by Freitag because Sennert had held that the souls of other animals than man were also created by God from nothing, date from December 23, 1635, to May 28, 1637.[3]

Sennert's collected works fill three huge folio volumes comprising some three thousand double-columned pages.[4] Much of this, at least in the case of those portions in which we are especially interested, is direct quotation (sometimes a single quotation is a full page in length)[5] or a setting forth of the views of others from Aristotle and Galen down. The quotations are from historians and poets as well as medical and scientific writers. Later in the century Sennert was accused of plagiarism by John Rhodius and G. C. Schelhammer who charged him with having copied Laurentius (André Du Laurens, d. 1609) *de crisibus* whole cloth and Octavianus Roboretus *de peticulari febre,* word for word. On the subjects which interest us, however, he cites and quotes with such frequency that he often provides what amounts to a valuable review and survey of sixteenth and early seventeenth century literature on the points in question, so that it is regrettable that there are only *Indices rerum et verborum,* and not of authors and persons, to his vast repertories. He not only repeatedly cites such noted authors as Cardan, Fernel, Scaliger and Libavius, but also obscurer persons like Ioannes Wolffgangus Dinheimius or Dienheim. But he was not very up-to-date. Harvey's discovery of the circulation of the blood published in 1628, is unmentioned by Sennert. Another reason for the length of his writings is that he repeats himself a good deal, discussing the same topic in different works, or stating a particular fact over and over again, as that some persons cannot endure the presence of a cat, even if it is concealed in a box.[6] This example of occult antipathy was further to be repeated by many other authors.

An Epitome by Claude Bonnet, professor of medicine at Avignon,

[3] *De origine et natura animarum in brutis: Opera* (1650), I, 851-82.

[4] In the edition of 1650.

[5] See *Opera,* III, 672-73, from Cornelius Gemma.

[6] *Opera,* I, 150; III, 231, 521, 828.

which reduced Sennert's tomes to one volume and expurgated them for use by Roman Catholics, appeared in 1655.[7]

Another reason for Sennert's diffuseness is his digressing to debate controversial questions in scholastic style with presentation of all the conflicting views and rebuttal of those rejected. Of such questions 261 are debated in the first volume, 311 in the second, and 157 in the third. Many of them take us back, in the spirit if not the letter, three centuries to the *Conciliator* and *De venenis* of Peter of Abano. Are women colder than men? Does the species of odor always require vapor as a vehicle? Is taste different from touch? *An res praeter et contra naturam differant?* Whether there are diseases of the whole subtance or of occult qualities? How is the stone, and how are worms generated in the human body? Can the air be made poisonous? Are there poisons which kill at a stipulated time? How can some persons go without food or drink not merely for months but for years? Does ecstacy happen naturally? Does melancholy induce fear and sadness because it is black in color or because it is cold? Whether and why images of dogs sometimes appear in the urine of those bitten by mad dogs? How looking down from a height excites vertigo? Are melancholy temperaments the most ingenious? Should a baby be nourished on its mother's milk or that of a wet nurse? Are contraries cured by contraries? In distinguishing degrees of medicaments should one observe geometric or arithmetical proportion? Is transmutation of metals possible?[8]

Turning to questions concerned exclusively with poisons, we may note the following. Are *bezoardica* and purgatives poisonous? Is all poison inimical to the heart? How does poison reach the heart? May poisons nourish? Do poisons act by manifest qualities? Are external poisons also poisonous taken internally? Can poisons kill by odor, vision, hearing? Is the magnet poisonous? How is the scorpion a remedy for its sting? Is the venom of a mad dog extremely humid? Why is bull's blood harmful? How does the torpedo stupefy?[9]

[7] LR 181-82, for full title.

[8] *Quaestiones controversarum in institutionibus medicinae* 13, 48, 50, 56, 61, 81-82, 88, 90, 98, 105, 111, 115, 130, 136, 196, 215, 249, 250 (*Opera*, vol. I, pp. 270, 304, 305, 310, 318, 358-60, 366, 368, 393, 417, 423, 427, 443, 451, 600, 683, 751, 755).

[9] *Opera*, III, 604, 605, 606, 607, 609, 611, 622, 652, 658, 661, 663.

Such queries and problems remind us of the intellectual milieu in which Sennert read and thought and wrote, and of the direction which his reading and thinking and writing were apt to take. Even Lasswitz recognized that Sennert "made no claim to originality," and that he held fast to the Aristotelian conception of form and matter.[10] Far from any approach to the theory of evolution, he not only held, as we have already seen, that the souls of both men and brutes were created by God from nothing, but also that God gave their matter and form to all natural bodies in the first creation.[11] He believed with Scaliger that "Nature is the power of God in secondary causes, to which He has set certain limits."[12]

Yet Lasswitz, and Gerland and others after Lasswitz, ascribed the corpuscular theory to Sennert and dated its first enunciation and the renewal of physical atomism from the year 1619, when *De chymicorum cum Aristotelicis et Galenicis consensu ac dissensu* was published.[13] Only a small part of a single chapter in that work is devoted to the subject and even in Sennert's later works it occupies relatively little space. Moreover, in the earliest passage Sennert gave credit for the idea to Scaliger, saying:

[10] Kurd Lasswitz, *Geschichte der Atomistik,* I (1890), 447-48.

[11] *Hypomnemata physica,* I, cap. 3: *Opera* I (1650), 142-43.

[12] *Ibid.,* Hypomnema IV, cap. 2; *Opera,* I, 169b: "Causae tamen secundae revera etiam agunt, et Deus agit in omnibus, ut et ipsae operentur, unde eleganter Scaliger, *in Theophrast. de caus. plantar.,* lib. 5, cap. 1, scribit, quod Natura sit Dei potestas in caussis secundis, quibus ipse certas constituit praescriptiones."

[13] Lasswitz, *Gesch. d. Atomistik,* I, 436-54, especially p. 441, "In diesen Ausführungen ist die Korpuskulartheorie so bewusst ausgesprochen, dass wir vom Jahre 1619 ab die Erneuerung der physikalischen Atomistik datieren müssen."

See also Lasswitz's earlier article, "Die Erneuerung der Atomistik in Deutschland durch Daniel Sennert und sein Zusammenhang mit Asklepiades von Bithynien," *Vierteljahrsschrift für wissenschaftliche Philosophie,* III (1879), 408-34. The nationalist interest of Lasswitz appears in a passage at pp. 433-34, where, after admitting the great contributions to science of Galileo in Italy, Bacon in England, Gassendi and Descartes in France, he adds: "Trotzdem kann wenigstens die deutsche Physik sich rühmen, in Bezug auf die Theorie der Materie den Weg der Erneuerung durchaus selbstständig eingeschlagen und die Periode der Corpuscularphilosophie eröffnet zu haben."

On the other hand, the name of Sennert does not appear in the Index of J. C. Gregory, *A Short History of Atomism,* 1931, and Ernst von Meyer, *Geschichte der Chemie,* 4th ed., 1914, although devoting a paragraph to Sennert (pp. 76-77), does not connect his name with the corpuscular theory.

I, to use the words of Scaliger in Exercise 101, if I am forced to speak, confess that I am now won over by the opinion of Scaliger, who defines mixture as 'the motion of very small bodies to mutual contact in order to achieve union' [14]

Sennert cited Scaliger by name no fewer than four times in this same single column of text and repeated the quotation from him in the next column. He then went on to quote passages from Aristotle and Galen, and said of Avicenna:

He defines temperament as a quality arising from the action and passion of contrary qualities in the found elements when the parts are reduced to such extreme smallness that a maximum of each of them comes in contact with a maximum of the other.[15]

In the History of Atomism Lasswitz merely alluded to these quotations from Scaliger and Avicenna,[16] although he noted those from Aristotle and Galen which did not so distinctly militate against his claims for Sennert.

Gerland echoed these claims yet more loudly. Not only was Sennert, despite Paracelsus and his medieval predecessors and the whole Paracelsan revival of the later sixteenth century, proclaimed the first to introduce chemistry into the study of medicine,[17] but it was further asserted without foundation that physics and chemistry should honor him as the new founder of the corpuscular theory.[18] Actually his belief that the heavens were a simple body, his faith in occult qualities and in the existence of spiritual qualities—of which

[14] *Opera*, III, 779a. Scaliger, whose *Exotericae Exercitationes* were published at Paris back in 1557, goes on to say: "Nor do our corpuscles come in contact as the atoms of Epicurus do, but so that one continuous body is formed . . . Erit igitur eorum mistio quorum extrema cum aliorum extremis unum fieri poterunt."

[15] *Opera*, III, 780a.

[16] In the earlier article Lasswitz did quote the passage from Scaliger but rather unobtrusively and without Sennert's acknowledgement of indebtedness to it.

[17] This may be regarded as an elaboration on Zedler, who affirmed that Sennert was the first to borrow the *studium chimicum* from the Paracelsists and introduce it into the universities.

Lest the reader think that Germans have been the only ones to distort the history of science from a nationalist bias, I may note that Sir Clifford Allbutt, *Greek Medicine in Rome . . . with other historical essays*, London, 1921, p. 514, assures us that Robert Boyle "made chemistry for the first time an academic study."

[18] E. Gerland, *Geschichte der Physik*, 1913, pp. 467-68.

we shall presently treat—were all quite inconsistent with the developed mechanistic corpuscular theory, which accounted for everything in nature by such corpuscles alone.

Isaac Beeckman would seem to have a superior claim to that of Sennert to be regarded as founder of the corpuscular theory, since already in 1613–1614 he wrote in his Journal of pores corresponding to asperities and *vice versa,*[19] and in 1614 explained the magnet's attracting iron by subtle spirits from the magnet entering the pores of the iron and so reducing the air pressure on the side of the iron facing the magnet, which was thus forced towards the magnet by the greater air pressure on its opposite face.[20] And before March 16, 1618, he explained the making of quicklime as due to the pores and asperities of the water and lime fitting one another.[21] That which did not have pores could not be broken with the hammer.[22] Atoms for him had but four figures, one for each of the traditional elements.[23] In 1618 he held that the essential difference between things depended upon the position of the atoms.[24] With which he combined such views as that a vacuum was possible, that a candle shot from a gun would penetrate a door, that what was once moved would never stop unless impeded, that air could be condensed, that ice had greater volume than water, and that the movement of light was not instantaneous.[25]

It would, however, be rather to the advantage of our own thesis, if these exaggerated claims of priority and scientific eminence for Sennert were true. For then he could serve as a shining example of a leader in the advance of modern scientific thought who was still immersed in the pseudo-science and the magic of the past and who combined the corpuscular theory with belief in a world-soul. But for our purposes I fear that it will have to suffice that he was one of the leading physicians of his day, that the Italian intelligentsia were said to doff their hats at the mere mention of his name, and that Gui Patin ranked him above all the moderns except Fernel.[26]

[19] *Journal tenu par Isaac Beeckman de 1604 à 1634,* ed. C. de Waard, La Haye, I (1939), 23.

[20] *Ibid.,* p. 36.

[21] *Ibid.,* p. 139.

[22] *Ibid.,* p. 109.

[23] *Ibid.,* pp. 152-53; also III, 138.

[24] *Ibid.,* I, 201.

[25] *Ibid.,* pp. 22-24, 46, 60, 99-100.

[26] Zedler also, however, quotes the Dutch physician, Plempius, as calling Sennert an empty rhapsodist.

Sennert was somewhat pessimistic as to the attainments of human knowledge and alluded more than once to the weakness or blindness of the human mind, the obscurity of things of nature, and the lack of agreement as to them among supposed authorities.[27]

Sennert did not have much sympathy with the use of the expression, natural magic.[28] The magic of Paracelsus, which he made, after chemistry, the second foundation of medicine and divided into six parts—the interpretation of preternatural signs, the transformation of living bodies, the power of words, astrological images, wax images of patients and other persons, and the cabala—Sennert rejected as diabolical and impious.[29] In the existence of such magic or witchcraft, based upon pacts with demons and worked with their aid, he had full faith, and rejected Wier's denial that diseases could be so produced.[30]

Words, characters and incantations had in themselves, according to Sennert, no operative power or natural force.[31] He denied that a seal of a scorpion could cure its sting, no matter with what astrological or magical observances it might be combined.[32] The voice as such cannot injure.[33] Love for a particular person cannot be excited by philters, although they may arouse lust and madness in general.[34] Poisons cannot kill by mere sight or sound, but must be applied internally or externally.[35] Fascination is not from sight, for vision is not by extramission but by intramission, but, if natural, must be the result of effluvia which emanate from the entire body but some-

[27] *Hypomnema IV*, cap. 6; *Opera* I, 184b, "in hac humanae mentis caligine et ignorantia"; *De origine et natura animarum in brutis*, preface; *Opera*, I, 851a, "In hac humanae mentis imbecillitate et naturae rerum obscuritate pauci sententiis eadem de re plane conveniant." *Conclusio* to *Practica medicinae*, lib. VI, *Opera*, III, 693a, "in hac naturae obscuritate ac mentis nostrae imbecillitate." Also III, 232, "explicare in mentis humanae hac caligine impossibile est."

[28] *De Chymicorum*, cap. 10; III, 749, he cites Giovanni Francesco Pico della Mirandola and Erastus against its use. See also III, 671-72 in the sixth book of his Practice.

[29] *De Chymicorum*, cap. 13; III, 781-83.

[30] *Practica*, lib. VI, pars ix, cap. 4; *Opera*, III, 675-76. The entire *Pars Nona* (pp. 666-93) is devoted to the repulsive subject, then worn threadbare, "De morbis a fascino et incantatione ac veneficiis inductis."

[31] *De Chymicorum*, cap. 18; III, 825-27.

[32] *Opera*, III, 653.

[33] *Opera*, III, 232.

[34] *Opera*, I, 424.

[35] *Opera*, III, 611.

times especially from the eyes.[36] To such effluvia Sennert, like Fracastoro,[37] was ready to attribute great effects, and he cites Pliny concerning certain families in Africa who wither trees and kill infants.[38]

For, on the other hand, Sennert accepted many of the ideas and beliefs upon which what was then called natural magic was largely based. He was firmly convinced of the existence and importance of occult qualities. The sixth book of his *Practice of Medicine* was entitled, "Of Occult Diseases," and in its first chapter he emphatically declared that to his mind no more blameworthy and harmful opinion had crept into medicine than that of those who tried to account for every natural phenomenon by manifest qualities and the four elements.[39] He repeated this statement the next year in his *Hypomnemata physica,* of whose five parts the second was on the subject of occult qualities.[40] Such qualities might characterize all members of a species, in which case they were derived from the specific form. Or, in the case of animate beings, they might be possessed by some individuals and not by others of the same species. They might be generated in living beings which had not at first possessed them, or they might be found in things which once were alive but now were not, such as dried toads and herbs.[41]

The potent effects produced by very small quantities, as in the case of poisons, could not be accounted for by heat and cold, dry and moist. Nor could the marvelous instances of sympathy and antipathy be explained in terms of these manifest qualities.[42] Strong purgatives also exercised their force by occult qualities, as did specifics and amulets.[43] For Sennert, while admitting that the advocates of occult virtues were too ready to accept the most idle

[36] *Opera,* III, 231.

[37] T V, 496.

[38] *Opera,* III, 232.

[39] *Opera,* III, 521b. Sennert had treated more briefly of "Diseases of the substance as a whole or of occult qualities" in his *Institutiones,* II, i, 4; *Opera,* I, 318-23. Among his *Disputationes,* too, Zedler lists one "De occultis medicamentorum qualitatibus" as of the year 1630. And see *Opera,* I, 618, "An dentur occultae medicamentorum qualitates et unde ille oriantur"?

[40] *Hypomnema* II, cap. 1; *Opera,* I, 148a.

[41] *Opera,* I, 153-54; III, 528.

[42] *De Chymicorum,* cap. 8; *Opera,* III, 733-34.

[43] *Hypomnema* II, cap. 4; *Opera,* I, 157.

fables and superstitions, would not agree with those who declared that all ligatures and suspensions were fabulous and without truth. He still believed with Galen in the virtue against epilepsy of a peony root hung round the neck.[44] He cites Albertus Magnus for wearing a precious stone next one's skin to ward off poison.[45] He appeals to Pliny for the truth of the belief that the little echeneis stops ships.[46] These and the marvelous properties of elk's hoof and the nephritic stone are proved by experience. Frommann in 1675 stated that Sennert repeated on the solemn assurance of trustworthy persons that animals similar to small puppies were generated from the foam of mad dogs which adhered to clothing.[47]

Even as to manifest qualities there was in Sennert's time much uncertainty. He tells us that it is disputed among medical men whether camphor is of hot or cold temperament. The ancients call it cold; the moderns, hot. For its taste and inflammability so manifestly convince one that it is hot, that it seems strange that there is any doubt about it. However, it has a cooling effect. But he holds this to be accidental and not *per se*. It happens as a result of its extracting from the body and drawing to itself atoms of fire and sulphur.[48] In the fourth book of his Practice Sennert again questions whether camphor is cold or hot and whether it extinguishes lust. Some say that there are two substances in it, as in rhubarb and roses, the one cold and the other hot, but Sennert denies this. Scaliger (*Exercit.* 104, *Sect.* 8) contended that the notion that it was an anti-aphrodisiac had been disproved experimentally. For a youth who held it in his hand *coivit validissime*, and Scaliger had given it to a dog with its bones and drink, and applied it to the animal's nostrils without effect. But Sennert holds these experiments to be insufficient.[49]

Sennert, like Fracastoro,[50] further presumed the existence of spiritual species and qualities. Some poisons act by these rather than by very minute corpuscles. The torpedo fish numbs the hand of the fisher through the length of his trident, which corpuscles could

[44] *Opera*, I, 151.

[45] *Opera*, III, 616.

[46] *Opera*, I, 151.

[47] *Tractatus de fascinatione*, Nürnberg, 1675, p. 242 (wrongly marked 224).

[48] *Opera*, I, 146.

[49] *Opera*, III, 96: "Verum experientia ista non sufficit."

[50] T V, 496.

not penetrate. The action of the magnet is to be similarly explained.[51] Besides such passages in the sixth book of his Practice, in his chemical treatise he adverts to *species spirituales.* Many of them are not perceived by the senses, and they would be the chief basis of natural magic, if there be any such thing.[52] Or in the *Hypomnemata* he speaks of an architectonic spirit or spirits which forms or form metals and gems underground, and is or are varied according to the different species of gems and metals.[53] Averroes in the *Colliget* criticized Galen for ascribing the fears and gloom of melancholy to the black color of the bile. Averroes argued that color was not so efficacious a quality as to affect the actions of the soul, and that the soul could not see without eyes. Ergo it must be because it is cold, that melancholy induces fears and sadness. But Sennert would interpret the passage by Galen as referring to obscurity, impurity and dimness of the animal spirits.[54]

Sennert also held the semi-magical doctrine that like attracts like. This was the reason why camphor drew atoms of fire and sulphur to itself.[55] It was thus that he explained how placing the carcass of a scorpion over the wound it had inflicted would cure the same.[56] But this would not explain the experiment of Galen which Sennert repeats without any expression of scepticism. Galen saw an enchanter kill a scorpion by thrice murmuring an incantation and spitting on the scorpion each time. Afterwards Galen tested it without any incantation and further found that the scorpion died sooner, if the spittle came from a fasting and thirsty man.[57]

Closely related to theories of likeness is the doctrine of signatures, already advanced by Croll and others, that certain plants by their figure and appearance suggested the parts of the body for which they were remedial. This was also carried farther to associate plants with one of the seven planets or four humors. Sennert attributed it to the chemists, but was inclined to think that there was something in it, since the Creator had endowed plants with formative virtue,

[51] *Opera,* III, 611-12, 663.

[52] *De chymicorum,* cap. 10: III, 748.

[53] *Hypomnema* IV, cap. 6: I, 184b.

[54] *Institutiones* II, iii, ii, 4; I, 423.

[55] *Opera,* I, 146.

[56] *Opera,* III, 652.

[57] *Ibid.,* p. 653a-b. Similarly at p. 232b he cites Vallesius, *De sacra philosophia,* that human saliva poisons a viper.

and form determines both internal and external properties, which may well be analogous.[58]

Sennert was still credulous as to marvels of nature, which have a close association with natural magic. He knew that Cardan discredited everything that was said about the basilisk, and doubts himself whether there is any such animal hatched from the egg of a decrepit cock. But then he tells a long story of a boy and girl in Warsaw in 1587 who hid in a cellar. A maid who couldn't wake them by calling went down the steps to arouse them and died too. Then a criminal condemned to death was offered his life, if he would descend. His whole body was clothed in leather, his eyes protected with goggles, in one hand he held a poker, in the other a flaming torch. Finally mirrors were attached to all parts of his person, before and behind, so that the basilisk might stare itself to death in them. At first the man saw nothing, but then the dead basilisk. It had the figure of a fowl, but eyes and extra feet like a toad, and was highly colored and spotted. But Sennert denies that any such animal can kill or be killed by mere sight. Death must be caused by a venomous exhalation, just as the presence of a menstruating woman clouds a mirror.[59]

One *Hypomnema* was devoted to the subject of spontaneous generation, including a long discussion of barnacle or, as Sennert calls them, Scottish geese.[60] Some think that in the Orcades along the banks of streams there is a tree which bears fruit resembling ducks. When ripe, this fruit, if it falls into the water, becomes a bird; if it falls on the land, it putrefies. Others say that such geese come from driftwood or shells or eggs. Sennert next quotes Turner indirectly through Fortunio Liceto, then Hector Boethius, Scaliger, Petrus Pena and Matthias Lobelius for more than a page, and then Michael Meier (or Maier), Clusius and Gerhardus de Vera (Gerrit de Veer) of Amsterdam for two-thirds of a page. He then concludes for himself that these birds are not born from rotting wood but from shells which adhere to it. Meier says that if the shells are opened,

[58] *De chymicorum*, cap. 18; III, 822-24. In the earlier *Institutiones*, V, i, i, 22 (I, 645b), he spoke more briefly and less favorably of signatures.

[59] *Opera*, I, 119-20, III, 612.

[60] *Hypomnema* V, cap. 8, "De anseribus Scoticis": I, 239b-241b. Also, more briefly, *Opera*, III, 775.

one sees tiny foetus like chicks in their eggs, which have a bill, eyes, feet, wings and the beginning of feathers. They may, however, receive nourishment from the driftwood to which their shells are attached. Clusius and de Vera said that they have seen such birds hatching eggs, but this does not upset Sennert who argues that, since many insects which are spontaneously generated afterwards themselves engage in intercourse and generate, why should not birds which are more perfect animals do the same? But then he remarks rather disconcertingly and inconsistently that he does not regard their generation as spontaneous, "but I think that they are generated like other shell-fish."

Animal sagacity is also marveled at by Sennert who asks,

> Who would deduce from the nature of the elements the sagacity of dogs, the prudence so to speak of the elephant, the slyness of the fox, the magnanimity of the lion, the wonderful works of bees, ants and other brutes, so that some have questioned whether they are not rational? [61]

Imagination was not accepted by Sennert as a means of working marvelous external effects and natural magic. He not only denied that it could affect other bodies but even that it could affect one's own body directly. Indirectly, however, by causing fear and terror which stir up noxious humors lurking in the body, it might even bring on the pest, or, on the other hand, cure diseases by inspiring confidence and cheerfulness. Its direct action is limited to the brain. No other reason, however, than the effect of imagination can be given why a pregnant woman, who is startled by a mouse, dog or ape, imprints on the foetus some mark of the animal. But it is hard to tell how this happens. The phantasies formed in the brain of the mother cannot be carried to the foetus, for they cannot mingle with the blood. Although the foetus had its own soul from the moment of conception, and this soul directs the formation of the child's body, Sennert suggests that it is still in close continuity with the soul of the mother, as fruit is with the tree, and so the same powers of the soul that move in the mother, also may move the faculties in the seed. Such mutations are rare at the time of conception and chiefly occur during the period of gestation, while the

[61] *Hypomnema* II, cap. 3: I, 152b.

formative faculty is at work.[62] Sennert's explanation of somnambulism was that the sleeper's imagination excited his motive power.[63]

For Sennert the heavens were immutable and incorruptible themselves but had power of acting upon inferior, sublunar and elementary bodies in various ways. Although this action was exercized most potently by the stars, yet the heavens too had their, or rather its—for they were thought of as a single, most simple body—own form and force of action. Light was conceived of by Sennert as neither a body nor an incorporeal substance, but as an accident and quality. Rays, rather than either the motion of the sky or light, were the cause of heat. The heavens act on inferiors by occult influence as well as by their light and motion, directly upon material things, but only indirectly upon the human soul. Astrologers can predict the weather, diseases, and other natural change with fair probability, but not always with certainty, much less human actions dependent on the will, or contingent and fortuitous affairs. But Sennert holds that the better astrologers agree to this. On the other hand, beyond doubt the causes of many sympathies and antipathies which sublunar things have with the heavens are to be sought in the occult influences of the sky.[64] But the sky is not the cause of spontaneous generation any more than of generation from seed.[65] Forms are not from the sky.[66] Sennert rejects the explanation of the eighth month's child dying, that Saturn rules both the first and the eighth month, but suggests instead an equally astrological explanation, that the sun in the fourth and eighth months returns to a sign of the same triplicity.[67]

Sennert still attributed critical days in disease to the influence of the moon.[68] But he felt that the analogy between macrocosm and microcosm had been pushed too far, especially by the Paracelsists.[69]

[62] *De Chymicorum*, cap. 14; III, 786-90; *Practica medicinae*, IV, ii, iv, 7; III, 118-19.

[63] *Institutiones* II, iii, ii, 4: I, 418.

[64] *Epitome scientiae naturalis*, II, 2; I, 33-36.

[65] I, 170.

[66] III, 527.

[67] III, 143: "Cum enim Sol singulis mensibus signum unum peragret, solo mense quarto & octavo ad signa eiusdem triplicitatis pervenit qui propterea etiam gravidis (*sic*) plerumque gravissimi esse solent.

[68] I, 561 *et seq.*

[69] *De Chymicorum*, cap. 6; III, 725-28.

He agreed with those astronomers who located comets in the ethereal region above the moon. Their long regular movement and the fact that their parallax was less than that of the moon convinced him that they were not meteors. As rare phenomena they certainly must signify something, but just what was known to God alone, not men, although great political changes usually followed their appearance.[70]

On divination from dreams Sennert was also brief and noncommittal, but not unfavorable. Some presage, others not. To explain the figurative ones requires trained interpreters acquainted with the resemblances and relationships of things.[71]

Sennert held that the transmutation of metals had been proved by experience, and that therefore it was a waste of time to dispute about it further.[72] He classed chemistry as an art and not a science, but traced it back to such mythical personages as its inventors as Tubal Cain, Hermes Trismegistus, and Mary, the sister of Moses.[73] The possibility of a single universal medicine he discussed rather sceptically.[74] Concerning such a favorite chemical remedy as potable gold he has little to say in his collected *Opera*, merely telling how to prepare it and referring to Libavius for further details.[75] But Zedler lists a disputation of 1630 by him on the universal medicine of the chemists and potable gold.[76] Sennert was accused by Freitag of founding an new Sennertiano-Paracelsica sect, but he pointed out the errors of Paracelsus and the defects of his personal character very freely.[77] In this he seems to have followed the lead of Erastus,[78] whom he cites freely. An example of his mistakes in chemistry is given by Boyle who says in *The Sceptical Chymist:* "The vulgar chymists are wont to ascribe colours to mercury; Paracelsus in divers places attributes them to salt; and Sennertus, having recited their differing opinions, dissents from both; and refers colours rather unto

70 I, 39b.

71 I, 115b, "qua de re agunt libri oneirocritici."

72 *De Chymicorum*, cap. 2; *Opera*, III, 706b.

73 *Ibid.*, caps, 1 and 3; *Opera*, III, 703-6, 709-713.

74 *Ibid.*, cap. 18; III, 817-21.

75 I, 806, under *De tincturis et extractis*. At least this is the only passage listed in the indices of his three volumes.

76 *De chymicorum medicina ut vocant universali et auro potabili.*

77 *De Chymicorum*, caps, 4, 5; III, 713-25.

78 T V, 657-60.

sulphur."[79] Many subsequent authors followed Sennert in this.

Sennert still remained a name to conjure with in 1674, when Jacobus Barnerus, doctor of philosophy and medicine, published at Augsburg a Prodrome of the New Sennert, or, Delineation of a New System of Medicine. As Sylvius dealt with *methodus medendi* in seven lectures, so Barnerus promises to teach all medicine in six weeks to one already acquainted with anatomy and chemistry. But the program which he outlines seems to call for more time, since he takes into account not only such ancients as Hippocrates, but Paracelsus and his followers, especially Severinus, van Helmont, Sennert and Hofmann, Willis, Sylvius and Barbette, and "such others as are deduced from the Harvaean, Gassendian or Cartesian principles," and the cures of "the famous Dr. Michael," late of Leipzig, and his "select and happy experiments," but also will give his own judgment, formed on anatomical and chemical principles, of the cause of every known disease. He also includes celestial influences.[80]

[79] *Works* (1772), I, 556; quoted by L. T. More *The Life and Works of the Honourable Robert Boyle*, 1944, who, however, misquotes "dissents" as "differs."

[80] *Prodromus Sennerti novi seu Delineatio novi medicinae systematis*, Augsburg, 1674, in-4; reviewed in *Philosophical Transactions*, X, 435-39.

CHAPTER VIII

VAN HELMONT: SPIRITUAL SCIENCE AND MYSTIC MEDICINE

General estimate—Autobiography—Elements: water— Seed and ferment—Confusion of spiritual and physical—Influence of the stars—Astrological medicine—Astrological images rejected: true talismanic power—Horoscopes—Experimentation—Medicine—*Magnum Oportet*—Signatures—Poisons—Butler's stone—Amulets—*Recepta injecta*—The devil and human magic—Magnetic cure of wounds—Alphabet of nature—Some promise of geology and palaeontology—Publication of Helmont's works—Their future influence: Schott, Rattray, Polemann, Conti, Boyle, Schoock, Kerger, Stirk, Helvetius, A. O. Faber, Hoffmann, Du Hamel, Rolfinck, Webster, Willis, Shirley, Wirdig, Ammann, Pantaleon, Simpson, von der Beck, Ettmuller, Horst, Vigani, Henckel, Wolff, Schelhammer, Mercklin, Martius, Baker, Jungken, Garmann, Heer, Barchusen, Stolle.

. . . ne proposant rien par songes, mais par longues recherches des choses et experiences controuvees

—VAN HELMONT

. . . je suis venu à là que je m'abstraicte de tous livres, veu qu'il y a un livre en nous, escrit du doigt de Dieu, duquel nous pouvons lire le tout

—VAN HELMONT

Nunquid aliud novi ratiocinii in medicina post Helmontii discessum e vita obortum?

—BARCHUSEN

Johannes Baptista van Helmont (1577–1644), or Helmont, as we shall henceforth call him for short, was the most original alchemical or iatrochemical writer of the first half of the seventeenth century, in fact the most so since Paracelsus. The Introduction to the 1707 edition of his writings compared him to such other innovators as Francis Bacon and Descartes. His commentator, Ettmuller, placed him next to Celsus, Fernel and Paracelsus. He introduced a new

---

Barchusen, *Historia medicinae*, 1710, p. 490.

terminology of his own, of which the word "gas" came into universal chemical usage, whereas its fellow, "blas," indicating an astral or meteorological efflux or influence, never came into general use. Other words, like Alcahest[1] and Archeus, whose use by Paracelsus Helmont developed further, continued in widespread use for a time. Other terms, such as ferment, alkali and acid, if he did not originate, he helped to popularize. Although he often mentions Paracelsus, and may be regarded as developing further the latter's natural philosophy and chemical medicine, he spoke of himself as having no light from his predecessors, not interpreting the findings of others or disputing with authorities (yet he often attacks the schoolmen and their Aristotle), but as a new author of medicine, setting forth everything new and unheard of. This change came to him as the result of an inner experience, when a light illuminated his soul compared to which the visible light of this world seemed continuous darkness.[2]

In his autobiography Helmont tells us that he finished the course in philosophy at the age of seventeen.[3] He studied the Sphere and Theory of the Planets, Logistic, algebra and Euclid. Then the *Cyclonomica* of Cornelius Gemma called his attention to Copernicus, and he learned that excentrics were vain and all astronomy uncertain. He refused to take the degree of master of arts, because he felt that he was not yet even a disciple. When the Jesuits began to teach philosophy at Louvain and included the subject of geography and an exposition by Martin Delrio of magic arts, Helmont attended both courses with avidity but only collected a harvest of empty stubble and poorest rhapsodies. He turned to the Stoicism of Seneca and Epictetus and the mysticism of Thomas à Kempis and Tauler, but was warned in a dream of the danger from paganism. Reading Mattioli and Dioscorides, he realized that botany had made no progress since the days of Dioscorides[4] and found the doctrine of degrees unsatisfactory. For a time his thoughts turned toward the law, but he soon gave up the idea. He then read intensively in

[1] On Alcahest or Alkahest see J. R. Partington, *Annals of Science*, I (1936), 362.

[2] *Opera* (1707), I, 10, 13.

[3] *Opera* (1707), I, 16 *et seq.*

[4] Had he read *Circa instans*, Rufinus and Albertus Magnus, he might have thought differently.

medicine: the works of Fuchs and Fernel, all of Galen twice, Hippocrates once, almost memorizing the Aphorisms, all Avicenna, and some six hundred Greeks, Arabs and moderns, studying them attentively and taking notes. Finally he realized that the time was wasted. "I had learned, 'tis true, to dispute problematically concerning any disease, but I did not know how to cure even a toothache completely." Meantime he had begun to study herbs directly, and for thirty solid years thereafter he labored at his own expense and peril of life to learn the natures and properties of vegetables and minerals. "Now I am an old man, useless and displeasing to God, to Whom be all honor."

Helmont objected that Aristotle tried to subject nature to mathematics, whereas the rules of mathematics accord ill with nature, "for man does not measure nature; but it, him."[5] Logic was useless;[6] and the physics of Aristotle and Galen, ignorant.[7]

There were three elements: the heavens, earth and water. Fire was neither an element nor matter; it was not for generation but destruction. There was a thousand times more water underground than above. Earth is the matrix, not the mother. It persists unchanged and departs not from its primeval constancy. Therefore it never contributes to natural and seminal generations.[8] The influx of the heavens is most general in its application and has no seminal power in it.[9] But in his treatise on the magnetic cure of wounds it is stated that the seed of *usnea*, the moss that grows on the skulls of thieves or those broken on the wheel, falls from the sky.[10]

Presently, however, Helmont affirms that air and water are primigenial elements and not to be changed into each other by cold or heat.[11] A part of earth may be reduced to water by art but not by nature alone. But generations and mixtures occur in nature only through the impregnation of water, and all visible things are materially from water alone.[12] Helmont believed that he had demonstrated this experimentally by weighing a growing tree and the earth in which it was contained. The weight of the earth remained constant,

[5] *Opera* (1707), I, 37-38.
[6] *Ibid.*, p. 39.
[7] *Ibid.*, p. 44.
[8] *Ibid.*, pp. 50-54.
[9] *Ibid.*, p. 36.
[10] *Ibid.*, signature, F 3 verso.
[11] *Opera* (1707), I, 58, col. b.
[12] *Ibid.*, 66b, 100a.

while that of the tree increased, and this, according to Helmont, was to be attributed to the water it had absorbed.[13]

But there is no generation so long as the water is deprived of seed. All things in nature, even diseases, have invisible seeds.[14] For the mingling of water and seed, ferment is required; hence ferment is the original beginning of things. It is further a formal created ens, which is neither substance nor accident but neuter, by way of light and forms established from the foundation of the world in the places of its monarchy, "to prepare, excite and precede seeds."[15] "The image of the ferment impregnates the mass with the seed."[16] Rays of light are not bodies, or there would be two bodies in the same place, light and air, or light and water.[17] As for the origin of forms, Helmont holds that the Schools incorrectly teach that forms are from the heavens. Rejecting the opinions of Aquinas and Scotus, he maintains that the form of each thing is created from nothing by God.[18] The Archeus consists of the connection of vital aura as material with the seminal image which is the inner spiritual nucleus. Disease is the vital matter in which is born or inserted a seminal character or idea of a badly affected Archeus.[19] Similarly Helmont's doctrine that earth remains unchanged encourages him to explain earthquakes as produced supernaturally by angels. He also believed in apparitions of spirits made immediately in place, color, figure and light, but not in body.[20]

Thus Helmont confused spiritual and physical, natural and supernatural. He even held that the moon had a light of its own, trusting what the Bible said of two great luminaries rather than the astronomers.[21] And he affirms that there are as many species of lights as there are of things in nature. Since angels are included among things, it follows that there are many more species of lights than there are of material things.[22]

[13] *Ibid.*, 100b.
[14] *Ibid.*, 556b-557a.
[15] *Ibid.*, 34-35.
[16] *Ibid.*, 107a.
[17] *Correspondance*, III, 87.
[18] *Opera* (1707), I, 123b-125b.
[19] *Ibid.*, 38, 553a.
[20] *Ibid.*, 97a, 155b. Helmont's assertion that no good angel ever appeared bearded provoked an interchange of letters between Samuel Hartlib and John Worthington in 1661: Harriet Sampson in *Isis*, 34 (1943), 473-74.
[21] *Opera* (1707), 135a.
[22] *Ibid.*, 139b.

We have heard Helmont deny the stars control over seeds and generation. He further declared that they did not signify as to the life, body or fortunes of the individual. To date, he says, the Church allows weather prediction and as to agriculture, perils at sea, the death of primates, pests, floods and whatever does not depend on the direction of our will. The inclination of the individual to this or that calling—medicine, geometry, music—is given to the soul by the Creator himself and does not come from the stars. The inclination to evil from a corrupt nature comes from the seed, which, as we have seen, is not under the stars; the inclination to good from grace, free will and exercise. A third inclination from the weakness or strength of the seed is entirely subject to the directing Archeus, as no one but an astrologer will deny. In short, the stars have no causative power over us except through the *Blas meteorum*, and the present uncertain state of astronomy makes astrology the more difficult, so that it is not surprising that the devil takes a hand in it. In his younger days Helmont ascribed much to the significations of the stars. But his faith waned, when no winner claimed the reward of 600 gold pieces which he had offered to anyone who could argue back to the exact time of nativity from the future indications of a horoscope he had drawn up at London. He came to the somewhat lame conclusion that he would not inquire into celestial secrets, when he knew so little of terrestrial.[23]

Yet after a few pages we find him talking of the cold *Blas* of the moon and stating that wounds inflicted by moonlight are difficult to heal, that there is solar light in bird and quadruped, lunar in fish, and that the two great luminaries correspond to the two primary elements, namely, sun to air and moon to water.[24] Earlier he had said that *Magnale* had not its like among created things. It was not light but a certain form assisting the air and transmitting the *blas* of the stars instantaneously.[25] The key to Helmont's terminology by Michael Bernhard in the edition of 1707, defines *Magnale* as a mean between body and not-body, contagion flying through the air,

[23] *Opera,* I, 111a-123a.

[24] *Ibid.*, 135a, 137a, 142a-b. On the other hand, we have heard Linemann ask why wounds of the head incurred in daylight were far more dangerous than those incurred by moonlight.

[25] *Ibid.,* 83a.

of which Hippocrates spoke as "the divine." In a letter to Mersenne, Helmont said that it penetrated all elements and transmitted the force of the stars to us.[26]

Van Helmont's penchant for astrological medicine is well shown in another letter to Mersenne, in which he holds that "in my poor judgment" the skin disease *herpes mordax* does not come from the liver but from the influence of the planet Mars, which by virtue of its mere aspect converts the balm of the martial places of the physiognomy into a mineral salt of a sort that he sees no hope of recovery from except in a planetary remedy to transplant the adverse radiations. Every recurring ailment, whether connected with the moon or other star, laughs at elementary and qualitative medicines. Van Helmont therefore has little hope in the waters of Spa and less in liver pills. For both macrocosm and microcosm are ruled by the invisible and the astral. However, if Mersenne does not have the right planetary remedy at hand, he may try touching the sore with the hand of one who has died a slow death, until the patient feels a great chill.[27]

Astrological images, however, Helmont rejected. In a letter to Mersenne concerning Gaffarel's book on talismanic sculpture, he declared that Gaffarel had composed a superstitious rhapsody concerning the non-existent. He had attributed everything to the positions of the stars instead of their general influence; did not properly understand signatures, dreams and the power of words; while there was nothing more ridiculous than attempting to read the stars by the Hebrew alphabet.[28] In a later letter, however, Helmont recommended a rod or ring of iron, which was to be carried in the palm of the left hand, against sudden death; on the ring finger of the left hand, against vertigo and epilepsy; on the second finger of the right hand, against gout, unless it was hereditary; and so on. It was to be made of a nail with which one had started to shoe a gelding at least five years old. The nail must not be allowed to touch the fire again or it would lose all its virtue. His explanation was that our imagination impresses equine forces on

[26] Mersenne, *Correspondance,* III (1946), 34, 111.

[27] Letter of June, 1630: *ibid.,* II, 497-98.

[28] Letter of Sept. 26, 1630: *ibid.,* II, 532-36.

the nail, while a gelding was required "inattentive to generation," in order that all its forces might be at our service. This was the true talismanic power, which Gaffarel had overlooked.[29] It certainly is an extreme instance of the power of the human imagination over external objects.

When the ecclesiastical authorities found horoscopes of Richelieu, Helmont's daughter Clémence, and Helmont himself among his effects, he testified on March 21, 1634, that he had not studied astrology and that his daughter's horoscope was false, since she had died at the age of only four, although the astrologer had predicted that she would live to be seventy.[30]

Helmont performed various chemical experiments and believed that he had found an inextinguishable and "insuffocable" fire,[31] but inclined to secrecy in such matters.[32] He assured Mersenne that the latter woud be astonished at the uniformity and natural simplicity of his inventions.[33] At the same time, he could still engage in scholastic explanations. Thus, when asked by Mersenne why a bullet or the kick of a horse had more force farther away than close to, he replied that every finite thing had a beginning, middle and end; that, when a mover impressed motive force upon a mobile body, its motion had an increment at first and a decline towards the end.[34] Despite the publication in 1630 of the book of Jean Rey,[35] he would not admit that calcined tin weighed more than crude tin, and asserted that he could demonstrate the falsity of this hypothesis.[36]

In March, 1631, Helmont was anxious to offer his medical services to end the impotency of Anne of Austria and gave Mersenne a list of similar cases in which he had been successful.[37] Similarly Gaf-

[29] Letter of January 11, 1631; *ibid.*, III, 14-15.

[30] *Ibid.*, III, 130, citing Broeckx, "Notice sur le manuscrit 'Causa J. B. Helmontii,'" *Annales de l'Académie d'Archéologie de Belgique*, IX (1852), 74-75.

[31] *Correspondance*, III, 75.

[32] *Ibid.*, pp. 76, 119.

[33] *Ibid.*, p. 107.

[34] *Ibid.*, III, 78.

[35] *Essays de Iean Rey docteur en medecine sur le recerche de la cause pour laquelle l'estain et le plomb augmentent de poids quand on les calcine.* Bazas, par Guillaume Millanges, 1630. There are said to be only four copies extant of this original edition. Gobet reprinted it in 1777.

[36] *Correspondance*, III, 181, letter to Mersenne of July 21, 1631.

[37] *Ibid.*, III, 153.

farel left an unpublished manuscript on the generation of male offspring.[38]

Helmont made much of the *magnum Oportet,* or flux and reflux of life, of which there are three by the three Monarchies of things.[39] Even vegetables and minerals have their three lives. If you cut an apple in two, rub the pulp on warts, and sew the two halves together again, the warts will disappear as the apple rots. "For at the same time the last life of the apple perishes." But if a pig or mouse eats the apple before it rots, the warts will not disappear, because the animal's stomach conserves the last life of the apple in retrocession to its middle life.[40]

In yet another letter to Mersenne Helmont states that sin has left its mark on the bodies of the sinners and their offspring in various signatures, whence physiognomy had its beginning. Or they come from the inclination of the stars, or from God who, according to Job, did not make one line of the hand in vain.[41]

In the action of poisons, formal and utterly abstract properties emanate from the forms and are luminaries and firelets, as it were, of the form itself. They have the power of penetrating the Archeus through all its light, life and the forms of its parts. Whether the mad dog bears in its saliva some singular fantasy, which converts ours to itself and so produces hydrophobia, or whether our Archeus fabricates spontaneously to itself a virulent image, is a mere matter of words. Properly speaking new poison is not stirred up in the Archeus, but the formal lights of poisons penetrate the vital light, and penetrate our middle life to its roots. Do they transfer our life to theirs? Or madden the Archeus to ruin itself? Or mortify by privation of light? Different poisons act on us differently, some by ferment and not by luminous firelets. In any case, they do not act by contraries, for Nature knows no contraries.[42]

A treatise by Helmont which was much cited by subsequent writers was called Butler after an Irishman of that name who was a prisoner in the castle of Guildford and cured a Franciscan of

[38] *Ibid.*, p. 155, and MS Carpentras 703.

[39] *Opera,* I, 144a.

[40] *Opera,* I, 149a-150a.

[41] *Correspondance,* III (1946), 102. Letter of Feb. 14, 1631. Yet on page 222 above we heard Helmont deny that the stars incline.

[42] *Opera* (1707), I, 154a-155a.

erysipelas in the left arm by dipping a little stone in a spoonful of milk of almonds and giving it to the prison guard to give to the friar to drink, with the promise that he would be cured within an hour, which happened without his knowing what had cured him. Helmont arrived the next day, made Butler's acquaintance, and saw him cure instantly an old washwoman who had suffered for sixteen years from an intolerable headache, this time dipping the stone in olive oil and anointing her head with a single drop. He would have cured a prince of Ghent of gout by having him touch the stone with the tip of his tongue and three weeks later bathe his joints in his own urine, if the prince had not offered him money, but reduced the weight of a fat man by that process. When Helmont's health declined, because an enemy—who afterwards confessed— had poisoned him, anointing with the olive oil failed to relieve him, but his wife, maids and other women cured their external complaints with it, and Butler told him later that he would have advised a different procedure, had he known that his aches and lassitude were due to poisoning. Helmont argues that, if a very little poison kills a man almost instantly, a remedy like Butler's stone should act on the Archeus far more potently and quickly in less quantity.[43]

It is hardly necessary to add that Helmont accepted the action of amulets, believed in the sympathetic remedy or absent treatment of the wound with powder of *chalcanthos*, felt admonished by sacred Scripture to credit great virtue to stones, and held that there were herbs which by mere touch would stop atrocious pains instantly "or at least relieve them." He had seen the bone from a toad's leg cure toothache at first touch.[44] The reason why a toad born within a rock does not putrefy is that it receives its Archeus from the rock, which suggests that it is a remedy for the disease of the stone. When applied in cases of the pest, it does not swell up, yet relieves the pain immediately and is of aid.[45]

Helmont's credulity extended to the belief that large objects could enter the body through the pores without breaking the skin, or be vomited, although twice the size of the throat. A piece of cowhide the size of one's palm was extracted by a surgeon with forceps after

[43] *Ibid.*, I, 554b-555b, 558-59.

[44] *Ibid.*, 557a, 560b.

[45] *Ibid.*, II, 5-6, 264-65.

the sore had maturated, and a witch who was burned at Bruges confessed that she had injected it. Helmont saw orphan boys at Lire vomit a rack equipped with base, four feet, wheel and ropes. At Antwerp in 1622 he saw a girl vomit a mass of pins, hair and filth, and at Mechlin in 1631 in his presence another virgin spewed forth wood shavings with much mucilage—about two fistfulls in all.[46]

In one place Helmont says that he calls *recepta injecta* what are spiritual portents perpetrated by cooperation of Satan. The demon has power to move bodies but not to alter their form, and he possesses *blas* by which he can stir up air and sea. But he does not have ideal power, unless the witch helps him out. As the basilisk by its visual ray sprays its virus on its victim but not on the locality or any body, however close, but only on that at which its glance was first directed, so seminal ideas in suspended or buried filth are invigorated by the idea of kindled desire and exercise it on the object. So too the most potent force of an incantation is from the natural idea of the witch, and Helmont will show that the aid of Satan is not needed to make a solid body pass through a space much smaller than itself, yet without any diminution of itself. For when body passes totally into the domain of spirit and is transumed and as it were informed by it, then bodies penetrate one another naturally, at least where they are porous, for spirit then closes body within itself and deprives it of dimensions. "So far as penetrations of bodies are concerned, our Archeus absorbs bodies in itself so that they are made quasi spirits." There is therefore a quite different power of incantation from the diabolical, and it is natural and free. The devil is responsible for one thing, however, and that is making the objects invisible when they are injected or penetrate.[47]

Helmont in these passages greatly restricts the power of the devil and demons and makes the witch rather than the devil an indispensable factor in magic. He affirms not merely the possibility of natural magic but of human magic. Man as a spirit has a freedom of will and action which the demons do not possess. But the pseudo-

[46] *Ibid.*, I, 563b, "De injectis materialibus."

[47] *Opera* (1707), I, 537-42, "Recepta injecta"; 563-69, "De injectis materialibus"; 569-70, "Injaculatorum modus intrandi." Schott quoted from these tracts almost *verbatim*, while Mercklin reprinted them *in toto*, as we shall see.

science with which he illustrates his point is extremely faulty. For not only does he attribute vision by extramission to the basilisk, but asserts that the torpedo fish, too, "throws the poison of its glance not by chance on anyone who is nearer but rather and solely on the person who is drawing in the ropes (i.e., of the nets) from a distance."[48] Of all the allusions to the numbing effect of the torpedo that I have seen, this is the only one I can remember that ascribes it to its glance or indeed speaks of it as poison. Moreover, Helmont goes on to attribute the injurious effect of the glance of basilisk or torpedo upon a determined object to an act of will, as well as virulent vision, on their part.[49] We shall encounter a like attitude later in the century in Charas, who denied that the bite of the viper was injurious because it secreted a liquid venom, ascribing the effect rather to its anger and irritated spirits.

The book of Van Helmont On the Natural and Legitimate Magnetic Cure of Wounds, was published at Paris in 1621 against the Jesuit Roberti, who had attacked an earlier work of Goclenius on the same theme. The faculty of Reims censured Van Helmont's book, and twenty-four propositions from it were submitted as heretical to the tribunal of Malines-Brussels and then to the Spanish Inquisition. It condemned them for heresy and magic on October 16, 1625, and prohibited the book in an edict published at Madrid on February 23, 1626. Van Helmont appeared before the official of Malines on September 3, 1627, as to the twenty-four propositions and three others of Paracelsus and submitted September 6 to the judgment of the church. His enemies procured further censures from the theological faculties of Louvain and Cologne, and in 1629 from the college of physicians of Lyon.[50] He wrote Mersenne in June, 1630, that he had lost an important law-suit "contrary to all justice by the subornation and unfairness of my adversary."[51] Appearing again before the ecclesiastical court of Malines, he testified on October 24, 1630, that he had only 23 copies of his treatise on the magnetic cure of wounds, given to his wife by a person unknown,

[48] *Ibid.*, 541b.

[49] *Idem*, "tanquam a voluntate Basilici (*sic*) vel Torpedinis exercentque illam in objectum determinatum dumtaxat."

[50] *Correspondance du P. Marin Mersenne*, II, 499.

[51] *Ibid.*, II, 497.

that he believed there were no more in existence, and that he was ready to burn the work. Yet in March, 1634, violent attacks upon him were resumed; he was imprisoned, and seems to have been left in peace only in 1638.[52]

Helmont's arch-enemy appears to have been a Henry van Heer[53] who had published a book in 1614[54] which led to a violent controversy between them.

The writer of the Introduction to the 1707 edition of Helmont's works contended that the treatise on the magnetic cure of wounds must have been a spurious concoction made by Helmont's enemies in order to discredit him.[55] On the other hand, a Scottish doctor, William Maxwell, in a work on magnetic medicine which was published by Georg Franck, dean of the medical faculty at Heidelberg, in 1679, but composed years before, represented Helmont as having silenced "the idle uproar of certain theologians clamoring as to the superstition" of "that famous sympathetic unguent and our magnetic water and the magnetic powder."[56] Actually, as we have seen, it was Helmont who was silenced.

Helmont was author of a treatise on the alphabet of nature[57] in which he held that the very Hebrew characters for letters represented the motions and configurations of the tongue and mouth which were required to produce the sounds of the letters, and that consequently Hebrew was the easiest language to learn for a deaf and mute who also had weak sight. At first sight this may seem a harmless fancy on his part of no linguistic or scientific significance. But there was the danger that it might encourage the belief in the operative and magical force of words, characters and the Cabala.

But let us conclude our summary of Helmont with a passage of supposedly observed facts showing some promise of future sciences of palaeology, palaeontology, and geology rather than with half-

52 *Ibid.*, II, 589.

53 *Ibid.*, II, 487.

54 *Spadacrene, hoc est, fons Spadanus: eius singularia, bibendi modus, medicamina bibentibus necessaria,* Liége, 1614, in-8. BN 8° Te163. 1713.

55 *Opera,* leaf with signature, A 3, recto.

56 William Maxwell, *De medicina magnetica . . . ,* Francof., 1679, p. 34.

57 *Alphabetum naturae.* A correspondent of Robert Boyle (*Works,* 1772, VI, 260) thought that Helmont's authorship "will prove no great commendation to it."

magical or wholly magical doctrines and fancies. A sea-going ship was found under a sandy hill near Maestricht in 1594. In the region of Peele (Peel, Holland?) pines were found in rows underground which do not grow readily except on mountains. In Hingsene near the Scheldt twelve feet underground in a damp meadow was found "the tooth of an elephant with the entire jaw of which I have kept a third part two feet long. And so there once were live elephants in this region." It was only recently that all Groenlandia was covered by the sea. So the center of the earth must have shifted.[58]

Many of Helmont's writings were first published posthumously by his son in the *Ortus medicinae, id est, initia physicae inaudita,* of 1648 at Amsterdam; reprinted at Venice, 1651, with an Index by Tachenius; again at Amsterdam, 1652, and at Lyons in 1655 and 1667. Evidently they had a great vogue. There were also an English translation of 1662 and 1664; a French one in 1670 and 1671; a German version of 1683; and a Flemish text. *Opera omnia* at Frankfurt in 1682 and 1707 professed to add some new treatises.[59]

We turn to some indication, largely through passages encountered at random, of Helmont's influence upon later writers of the century or their estimate of him. His being cited by Meyssonnier was no great compliment, and when Conring suggested to Gui Patin that someone at Paris should write against Helmont, Patin refused to do so on the ground that it would be a waste of time.[60]

Caspar Schott in his *Magia universalis* of 1657–1659 repeated Helmont almost in his own words as to the injection of large solid bodies, but he did so in order to show "how fantastic he is and how prone to invent without any foundation."[61]

Sylvester Rattray in his treatise on the occult causes of sympathy

[58] *Opera* (1707), 54b.

[59] For these editions see J. R. Partington in *Annals of Science,* I (1936), 365-67, who rejects those listed by Ferguson as "mostly imaginary."

[60] Letter of Feb. 20, 1654: *Lettres* (1846), II, 118. On April 16, 1645, Patin had written: "Pour Van Helmont, il n'en fera plus. C'étoit un méchant pendard flamand, qui est mort enragé depuis quelques mois. Il n'a jamais rien fait qui vaille. J'ay vu tout ce qu'il a fait. Cet homme ne méditoit qu'une médecine toute de secrets chimiques et empiriques et pour renverser plus vite, il s'inscrivoit fort contre la saignée, faute de laquelle pourtant il est mort frénétique." *Lettres* (1907), p. 458.

[61] *Thaumaturgus physicus sive magiae universalis . . . Pars IV et ultima,* Würzburg, 1659, p. 531.

and antipathy, first printed in 1658, sometimes opposed and sometimes followed Helmont. Rattray held that the only true remedy was extinction of the exotic ferment, not of a fantastic idea, or calming a perturbed Archeus.[62] According to Garmann, Rattray criticized Helmont's distinguishing essential, vital and substantial form as inept, because all things having form live and feel, and also Helmont's assertion that God kept creating new forms, whereas the Bible says that creation was finished in six days.[63] On the other hand, Rattray, like Helmont, held that all things which used to be thought mixed are made from water alone.[64] Among five explanations offered of action at a distance, as in the case of the sympathetic powder, Rattray first mentioned Helmont's theory of ecstatic virtue or power excited by the phantasy of things, but preferred his own explanation by ferments.[65] Of course Helmont too had emphasized ferments.

Joachim Polemann wrote a New Medical Light in which he explained the excellent teaching of Helmont of the lofty secret of the sulphur of the philosophers.[66] The work appeared at Amsterdam in 1659 and 1660, in Latin translation in the sixth volume of Zetzner's collection in 1661, in English translation at London in 1662, and once more in German at Frankfurt and Leipzig in 1747.[67]

Luigi Conti of Macerata first published at Venice in 1661[68] a treatise distinguishing the liquor Alchaest of Helmont from the philosophers' stone. In the preface to the reader he states that the Alchaest was real and not a mere fancy of Helmont, and that he has finally succeeded in working it out, but knows that the reader would

[62] *Aditus novus ad occultas sympathiae et antipathiae causas inveniendas per principia philosophiae naturalis ex fermentorum artificiosa anatomia hausta patefactus,* Glasgow, 1658, in-8, 135 pp.: BN R.48035.

[63] Garmann, *De miraculis mortuorum,* 1709, Preliminary Diss., p. 65, ¶ 99, citing *Aditus,* p. 51, but p. 51 in the edition I used was concerned with the growth of plants. The Tübingen edition of 1658 has 216 pp.

[64] *Aditus,* Glasgow, 1658, pp. 53, 82.

[65] *Ibid.,* pp. 109, 123.

[66] *Novum lumen medicum in welchem die vortrefliche Lehre des . . . Helmontii von dem hohen Geheimnüs des Sulphuris philosophorum . . . erkläret wirdt.*

[67] BM has copies of all these editions.

[68] Ludovicus de Comitibus, *Clara fidelisque admonitaria disceptatio . . . de duobus artis et naturae miraculis: hoc est de liquore Alchaest necnon lapide philosophico . . . ,* Venice, 1661, in-4.

not wish him to reveal it openly to the vulgar. He suspects that even Helmont thought that things written concerning the stone were meant to be understood of his liquor, and that it was the chief source and base of the philosophers' stone. It will therefore be well, if he shows in what they agree and are discrepant. But for Van Helmont he has only the highest praise. He was a man outstanding in character, genius, eloquence and every kind of learning, professing in physics, medicine and chemistry a new, unheard of, and wonderful doctrine, and who strove to shake the old foundations of the schools and to overthrow the received dogmas of the old masters. Conti's work was reprinted at Frankfurt in 1664 [69] and in French translation at Paris in 1669 and 1678.

Boyle in *The Sceptical Chymist* in 1661 was of the opinion that Helmont, because of his experiments, was to be more highly esteemed than it seemed to many learned men he should be, although Boyle recognized that the falsehoods in his treatise on the magnetic cure of wounds rendered his other statements suspect. Boyle admired the experiment, lasting five years, in which Helmont had planted a willow in two hundred pounds of earth and watered it. At the end of five years, although the weight of the tree had increased from five to 169 pounds, the weight of the earth in which it grew had decreased only about two ounces. Helmont therefore held that water was the material cause of mixed bodies or the sole element of mixed bodies, and even Boyle says nothing of the possibility of the tree's receiving sustenance from the air as well as from water.[70]

Helmont was unfavorably criticized by Marten Schoock in 1662 for holding that the spleen supplied acid ferment to the stomach, aided in concoction of the blood, was the "nest of Venus," seat of the imagination, and source of sleep and dreams. Schoock added that Helmont's writings deserved to serve as a pillow for Endymion, but were esteemed by those who dreamed upon the Helmontian Parnassus.[71]

Martin Kerger, a physician of Liegnitz, in a work on fermentation

[69] This is the edition I have used: BN Te131.118.

[70] Robert Boyle, *Chymista scepticus*, Roterdami, 1662, in-8, pp. 55-59.

[71] *De fermento et fermentatione*, Groningen, 1662, pp. 404-405.

printed at Wittenberg in 1663, cited Helmont more than once[72] and alluded to *vitae gradus* and *vita media et ultima.*[73]

George Stirk,[74] who was born in Bermuda and graduated from Harvard in 1646, went to England, where he died in 1665, and, under the name of George Starkey, besides other alchemical treatises, wrote on the Liquor Alchahest or a Discourse of that Immortal Dissolvent of Paracelsus and Helmont, a work extant in English both in print[75] and manuscript,[76] and of which a French translation appeared at Rouen in 1704.[77] Another alchemical work of his was dedicated to Robert Boyle and spoke of his admiration for Helmont.[78]

In 1667 Helvetius included extracts from Helmont's *Arbor vitae* and *De vita aeterna.*[79]

When Albert Otto Faber came from Germany to reside in England in 1661, he was spoken of by Samuel Hartlib as "an excellent Helmontian physician... called by his Majesty," and one of his remedies resembled Helmont's oil from the stone of Butler. He also, in his work of 1677 on potable gold, was influenced by Helmont's occultism, his conception of the Archeus, of water as the primordial element, and the hypothesis of three lives.[80] Helmont, however, had declared the use of gold and gems in medicine ridiculous.[81] But it would be too much to expect anyone else to duplicate precisely Helmont's extraordinary combination of occasional scepticism and hard common sense with unbridled theorizing and sheer flights of fancy.

The title of the *Opus de methodo medendi* of Friedrich Hoffmann

[72] *De fermentatione liber physico-medicus,* Wittebergae, 1663, pp. 15, 222.

[73] *Ibid.,* pp. 46, 62.

[74] DNB and George Sarton in *Isis,* IX, 163; 442; XI, 333.

[75] London, 1675, in-8, 16 fols., 55 pp.: Duveen 564.

[76] One dated 1675 was offered for sale by Feisenberger & Gurney, Catalogue 17, item 228.

[77] *L'Alcaest ou dissolvent universel de Van Helmont, révélé dans plusieurs traités qui en découvent le secret,* Rouen, 1704, in-8.

[78] *Pyrotechny,* London, 1658 and again in 1696. Dutch translation, Amsterdam, 1687; French, Rouen, 1706.

[79] Johann Friedrich Helvetius, *Vitulus aureus,* Amsterdam, 1667, pp. 7, 21.

[80] Harriet Sampson, "Dr. Faber and his Celebrated Cordial," *Isis,* 34 (1943), 472-96, especially pp. 473, 475, 486, 490-93.

[81] *Ibid.,* p. 490, note 141.

the Elder in 1668 mentioned Helmontian as well as dogmatic and Paracelsic principles,[82] and the text cited Helmont as recommending use of goat's blood in pleurisy,[83] and his followers as eschewing venesection.[84]

Du Hamel in 1670 suggested that what Helmont had written on magnetic cure of wounds was not entirely false nor what he reported concerning Butler's stone.[85] He also noted that Helmont found that dried powdered toad soaked in water and applied as a poultice lessened the pain of buboes and anthrax, and that Butler advised suspending a toad at the hearth in June. After three days it would vomit flies and other insects. These and the dried cadaver of the toad were to be brayed separately, and with the addition of wax and tragacanth made into trochees to avoid or cure the pest. From the eyes and brains of toads, similarly suspended in July, worms exude, which, prepared in the same way, and applied to the infected places as amulets, draw out all the poison.[86]

In the same year Werner Rolfinck cited from *Liber Butler inscriptus* those passages concerning the virtues of herbs, toads and minerals which we have already noted.[87]

John Webster, discussing in 1671 "those authors that have treated of metals and minerals," included van Helmont, "though he left no treatise (that ever came to light) that was purposely written upon this subject." But he "enriched his writing with much deep mineral knowledge," and had been so much read and studied that "now a Helmontian seems to overtop a common Chymist, Paracelsian and

[82] *Opus de methodo medendi iuxta seriem Wallaeianam annexis fundamentis astrologicis ex veterum ac recentiorum scriptis concinnatum, Dogmaticis Paracelsicis Helmontianis principiis et propriis observationibus illustratum, elegantissimis chymicis flosculis adornatum*, Leipzig, 1668, in-4.

[83] *Ibid.*, p. 157.

[84] *Ibid.*, p. 203, "An venaesectio juxta Erasistrataeos et Helmontianos ex familia medica exulare."

[85] J. B. Du Hamel, *De corporum affectionibus cum manifestis tum occultis libri duo seu promotae per experimenta philosophiae specimen*, Paris, 1670, in-12, p. 27.

[86] *Ibid.*, pp. 444-45. Similarly John Locke wrote on February 11, 1679: "In the Ile d'Elva in Italy are toads above a foot broad which the inhabitants will not suffer any one to kill, imagining that they draw to them the venom of that country," *Travels in France, 1675-1679*, edited by John Lough, 1953.

[87] Werner Rolfinck, *De vegetabilibus plantis suffructibus fructicibus arboribus in genere libri duo*, Jena, 1670, in-4, p. 193.

Galenist."[88] In *The Displaying of Supposed Witchcraft* in 1677, Webster repeatedly cited Helmont and once dwelt for seventeen pages on his discussion of *recepta injecta,* which we have dismissed in a single page.[89]

Thomas Willis in his book on fevers says that he has known many who—and they are wont to use the words of Helmont—by fortifying the *Archeus* with wine and confidence and using no other alexipharmaca, have gone among the pestridden without catching the contagion.

> And on the contrary, others, struck by fear, though living far from all contagion, have sucked in the seeds of pestilence as if derived from the stars.[90]

Dr. Thomas Shirley, in a treatise on the generation of stones published in 1672, showed the influence of Van Helmont when he held that stones, like all other sublunar bodies, were made of water, which was condensed by seeds which worked by virtue of their fermentative odors.[91]

Sebastian Wirdig in 1673 credited Helmont with having reduced the employment of talismans in medicine by disassociating them from the macrocosm or influence of the stars and making their force depend upon man the microcosm.[92]

Helmont was quoted once more as to the use of toads as an amulet against pest by Paul Ammann in 1675, who also ascribed to him the view that vipers are not venomous unless enraged, that *usnea* of the cranium receives seed from the heavens, and that the comforting virtue of drugs consists in their odor.[93]

The Alkahest of Helmont was mentioned as one of the four most famous menstruums in the alchemical Tomb of Hermes by an anonymous Pantaleon (the pseudonym of Franz Gassmann) and also

[88] John Webster, *Metallographia,* London, 1671, p. 34.

[89] *Op. cit.,* pp. 250-66. The preface is dated February 23, 1673, but the imprimatur on July 29, 1676.

[90] *De febribus,* cap. 13 de peste: *Opera,* 1676, pp. 147-55, at p. 152.

[91] JS III, 100-101.

[92] As quoted by Paschius, *De novis inventis,* 1700, p. 357.

[93] *Brevis ad materiam medicam . . . manuductio,* bound with his *Supellex botanica,* but paginated separately: p. 152, citing "Helmontius in Tumulo Pestis, p. 160;" p. 185, citing "Helmont. de magnetica vulnerum cura, p. 602;" and p. 93, citing "Helmontius qui p. 477 in Tr. de Buttleri lapide."

in his *Bifolium metallicum,* both printed at Nürnberg,[94] in 1676. "It cannot be denied," he said, "that Helmont is easily supreme among the saner philosophers."[95] And he concluded with Helmont that all mineral and metallic remedies were of no moment without the Alcahest.[96]

William Simpson or Sympson continued Helmont's stress upon seeds and ferment and water in his "confirmation of the Corpuscular Philosophy taking in Seminal Principles" (i.e., acid and sulphur in small particles of matter constituting seeds) "and Ferment to make up the generality of mixt bodies in the World." Furthermore, "these Principles themselves are also material" and "ultimately reducible into water."[97]

David von der Beck followed van Helmont in holding that water or alkali was the matter of all things, while seeds or fire or acid were the formal principle. He was interested in the strange force of imagination and in the causes of monstrosities. He believed that seeds contained specific ideas, and that ideas or characters remained in the bodies of animals after death. He accepted the resuscitation of plants. His book first appeared in 1674;[98] an enlarged edition in 1684,[99] and a third in 1688,[100] indicating that such interests were still potent.

When Michel Ettmuller died on March 9, 1683, and an edition of his works appeared that year in London, the reviewer in *Philosophical Transactions* noted that he "looks upon Helmont to be very faithfull where intelligible, and to be imitated as far as possible, since he suggests the best method of curing *a priori.*"[101] In the

[94] JS VII (1679), 51-53; VI (1678), 237-38.

[95] *Tumulus Hermetis apertus,* Noribergae, 1676, in-8, 51 pp., p. 29. BM 1034.f.27 (3.).

[96] *Bifolium metallicum,* Noribergae, 1676, in-8, 55 pp., p. 47. BM 1034. f.27 (2.).

[97] The quotations are from the review in PT XII (1677), 883-84, of his *Philosophical Dialogues concerning the Principles of Natural Bodies,* London, 1677.

[98] *Experimenta et Meditationes circa naturalium rerum principia,* Hamburg, 1674; reviewed in *Philosophical Transactions,* IX, 60-64; JS VI (1678), 442-46, where the author's name is spelled Becte, and the date of publication given as 1678. LR 240 spelled it Becke.

[99] LR 240.

[100] R. A. T. Lier, *Short Offer No. 9,* Item 17: D. Becke, *Experimenta et meditationes circa naturalium rerum principia,* Hamburg-Ferrara, 1688, in-16, 12, 516 pp.

[101] PT XV, 1141.

Chimia rationalis by Ettmuller, which appeared in 1684, Helmont was cited more often than any other author.

The *Physica Hippocratea* of Johann Daniel Horst in 1682 was "illustrated by the comments of Tachenius, Helmont, Descartes, Espagnet, Boyle and other recent writers."[102] It is noted that Helmont accepted only two elements, water and air, rejecting fire because it was not mentioned in the biblical account of creation.[103]

Vigani cited Helmont in his *Medulla chemiae* of the same year and treated of a "stone" which accorded in many respects with the stone of Butler that Helmont told of. But, according to Vigani, the first investigator of its virtues was John Iarbrough of Newark in Notts.[104]

In 1690 Helmont was cited by Henckel in a work on philters and by J. W. Wolff in A Scrutiny of Amulets. In the former case it is to the effect that if you crush a certain herb in your hand and then hold another person's hand, it will act as a love charm.[105] In the latter instance it is to the effect that amulets have a *blas* by which they compel objects to obey them as the stars do, "and they act only on their own objects and not on an alien one, though it be nearer"[106]

Schelhammer in 1697 regarded the *Archeus furens* of Helmont as the creature of his brain, just as Pallas was said to have sprung from the brow of Zeus. Helmont's genius, potent in powers of the imagination and prone to fanciful ideas, had invented it, when he tried to find a cause for the excessive commotion of the spirits or humors which results in mental disease and delusion.[107]

In 1698 Georg A. Mercklin, in his book on cases commonly ascribed to incantations,[108] not only repeated the long passage from

[102] *Physica Hippocratea Tachenii Helmontii Cartesii Espagnet Boylei... aliorumque recentiorum commentis illustrata*, Francofurti, 1682, 87 pp. BN R.55281 (1.). The dedication is dated 30 Jan. 1662, but I presume this is a misprint for 1682.

[103] *Ibid.*, p. 20.

[104] J. F. Vigani, *Medulla chemiae*, Danzig, 1682, pp. 11-12.

[105] Elias Henricus Henckelius, *De philtris eorumque efficacia ac remediis*, Francofurti, 1690, viii, 85 pp., pp. 56-58.

[106] J. W. Wolff, *Scrutinium amuletorum*, Leipzig and Jena, 1690, in-4, p. 488.

[107] *Guntheri Christophori Schelhammeri Natura sibi et medicis vindicata*, Kiel, 1697, p. 285.

[108] *Georgii Abrahami Mercklini Sylloge physico-medicinalium casuum incantationi vulgo adscribi solitorum*, Nürnberg, 1698, in-4: BN 4°Te30.29. Another edition at Nürnberg, 1715.

Schott, to which we have already adverted,[109] but went on to quote fragments of Helmont on the same theme.[110] J. N. Martius, in a dissertation of 1700 on natural magic, referred to the same point, and further to Helmont's denying diabolical cooperation even in cases of the witch's making an image of her victim and sticking needles into it, where he attributed the effects to the ideal ens or strong imagination of the sorceress.[111] He also, however, quoted Helmont that the virtue of words was more to be admired than applied.[112]

Thomas Baker in his *Reflections upon Learning*, after enumerating recent "discoveries" and fads, added:

> Anwald's Panacea, discussed by Libavius, and Butler's stone, so much magnified by Helmont, were as much talked of in their time . . . and yet they are dead and have been buried with their authors.[113]

But such an expression as *blas lunare* is more evident in the 1702 than in the 1682 edition of J. H. Jungken's *Chymia experimentalis*[114] or *Medicus praesenti seculo accommodandus*.[115]

Garmann published the first book of his Miracles of the Dead in 1670, but the full text in three books of some 1400 pages was edited posthumously by his son in 1709.[116] The citations of Helmont fill nearly a page of the index. In the text he speaks of "Ioh. Bapt. ab Helmont, whom in many respects I admire and follow."[117] Many of his treatises are cited, and such doctrines of his as those of ecstacy and *vita media*. Or particular statements of his are repeated, such as that amber strengthens phantasy because it attracts straws and particles; or that the light of quadrupeds is hot, that of birds solar,

[109] Ed. of 1698, pp. 154-61.

[110] *Ibid.*, pp. 163-88: "Joann. Baptistae van Helmont Fragmenta De receptis injectis. De injectis materialibus. De injeculatorum modo intrandi, ex eius Ortus Medicinae tractat. de morbis eruta."

[111] Joh. N. Martius, *Dissertatio de magia naturali* . . . , Erfurt, 1700, in-4, pp. 39, 41.

[112] *Ibid.*, p. 26.

[113] *Reflections upon Learning* . . . , 2nd ed., London, 1700, p. 181. Anwald is for Georg am *or* an Wald.

[114] Ed. of 1702, pp. 93-96.

[115] This is the title in the 1682 edition.

[116] Chr. Fr. Garmann, *De miraculis mortuorum libri tres quibus praemissa Dissertatio de cadavere et miraculis in genere*, Dresden and Leipzig, 1709, in-4. The preliminary dissertation has a separate pagination.

[117] *Ibid.*, Diss. p. 23, ¶ 32.

and that of fish lunar.[118] In 1633 Helmont saw his own soul in a vision.[119] According to Garmann, who cites passages in six different treatises of Helmont in substantiation, Helmont said that there was no sensitive soul in man until after the fall of Adam, since none was needed. But Martin Heer (1643–1707), whom Garmann calls an outstanding interpreter of Helmont, in his Introduction to the *archivum archei vitale et fermentale* of that magnificent man, Johannes Baptista van Helmont, philosopher by fire,[120] disagreed with Helmont on this point and said that the sensitive soul in man was illumined by a ray of the mind which had not yet been detected in the case of the brutes.[121]

In 1710 Barchusen in his History of Medicine devoted some twenty pages to Helmont.[122] He introduced barbarous words and obscure opinions, and added two principles. water corresponding roughly to matter, and ferment in the lieu of form. Though he accepted the doctrine of creation, water was for him primordial and seeds essential to generation. Ferment was a sort of formal creation which was neither substance nor accident, but, after the manner of light, fire and form, resident in the principles of things to prepare, excite and precede seeds. After noting his *Archeus* and *gas*, Barchusen states that *blas* indicated a twofold movement of the stars, one local, the other of alteration. *Blas humanum* also was twofold, one the cause of natural, the other of voluntary movements. Since beasts were created a day before man, their *blas* takes precedence in weather prediction, augury and inspection of entrails. He distinguished two classes of disease, *recepta* and *retenta.* He would not admit the tartar of Paracelsus as a cause of disease, held that fever was the material part of the *Archeus* disturbed by indignation, and that medicaments operated most potently by their odors.

Gottlieb Stolle in 1731 in his *Anleitung zur Historie der medicinischen Gelahrheit* said that van Helmont ridiculed the tartar of

[118] *Ibid.*, Diss., pp. 35, 64; Text, 611, 684.

[119] *Ibid.*, Text, 756.

[120] Martin Heer, *Introductio in archivum archei vitale et fermentale viri magnifici Johannis Baptistae Van Helmont, philosophi per ignem,* Laubae apud A. Vogelium, 1703, in-4, 356 pp. BN R.7636.

[121] Garmann, *op. cit.*, Diss., p. 65, ¶ 97.

[122] *Historia medicinae,* 1710, pp. 461-85.

Paracelsus as a cause of disease as much as he did the Galenists who attributed it to the humors. But his own *Archeus irritatus* was attacked by Schelhammer (1649–1716) as an asylum for ignorance. And Bartholomaeus de Moor (1649–1724) in his oration on medical hypotheses asked:

What does he mean by his *Archeus faber*? What does he mean by his water gas? What is his *Blas meteoron*? What *Duelech*? What archeal diseases? What *meteoron anomalon*? That is talking nonsense in a serious matter.[123]

[123] Stolle, *op cit.*, p. 530. Bartholomaei de Moor, *Oratio de hypothesibus medicis habita Groningae . . . anno 1706*, Amsterdam, 1706, in-4, 36 pp. BM 1185.i.15.(42); BN 4°T6 378.

CHAPTER IX

MINERALOGY

Plan of chapter—Morales—Silvaticus on the unicorn's horn and the bezoar, emeralds and pearls—Catelan—Other works on the Bezoar—Guybert—Robert Pitt—Imperato—Neri—Canepario—Hornius—Guidi—Chesnecopherus—Collections of gems—Gans on coral—Bernardo Cesi—de Clave—Barba on mines in Peru—Severino—Seals and Gnostic gems—Bausch—Webster—Shirley—Saint-Romain—Acqueville and Candy—König—Jesuit works on the magnet: Cabeo, Kircher, Zucchi, Leotaud.

There is nothing so hidden and occult as the virtues of stones

—SAINT-ROMAIN

In a previous volume on the sixteenth century, the chapter on "The Lore of Gems" included the work of de Boodt in 1611. In the present chapter we note other books on gems and stones, including the bezoar stone and related topics, of the seventeenth century; also works on metals which are not primarily alchemical; and treatises on other minerals and the subject of mineralogy in general. Finally, notice is taken of a series of publications upon the magnet which followed Gilbert's epoch-making *De magnete* of the last year of the previous century. To the modern reader it may well seem that the bezoar stone, supposed to grow inside of animals, and still more the unicorn's horn, belonged to the animal rather than the mineral kingdom. But we shall shortly find them both associated with pearls and emeralds by Silvaticus, and to the thought of the seventeenth century they went with precious stones, like the jewel in the toad's head. Really they did not belong to either the animal or the mineral kingdom, but to "one vast realm" of marvels.

The very title of a book published in Spanish in 1605 by Gaspar de Morales shows it to be concerned with the marvelous virtues and properties of precious stones.[1] The work is primarily a compilation

[1] *Libro de las virtudes y propriedades maravillosas de las pierras preciosas*, Madrid, L. Sanchez, 1605, in-8, 378 fols. Copy used: BN S.21783

and cites ten or a dozen past authorities for each stone. Such powers in gems as that of the topaz to promote hilarity and of the diamond to win love, come not from the primary qualities but from celestial virtue.[2] In connection with the magnet are mentioned the virtues of aquatic animals—the remora and torpedo.[3] Chapters are devoted to the signification of the twelve stones that form the foundation of the city of God, and of those in the pectoral of the High Priest.

In 1605 J. B. Silvaticus of Milan published a detailed treatment in 160 pages of the reputed medical virtues of unicorn's horn, the bezoar stone, emeralds and pearls.[4] In the first three cases he reviewed the previous literature pro and con at considerable length, and himself in the main assumed a sceptical attitude. He listed the so-called unicorn horns which were in the possession of various princes of Europe[5] because of the widespread notion that they would cure the worst poisons very speedily and emit sweat in their presence. But he insists that this is not proven by authorities, reason or experience, and that many of the horns are not genuine. His final conclusion is that the notion is idle, superstitious, untrustworthy and full of impostures.[6]

Silvaticus makes the surprising assertion that, although the bezoar stone was famed among the Arabs, it was not long after forgotten until the present age, when an epidemic of pestilential fevers had led to resumption of its use.[7] This statement hardly accords with a list of Latin authorities on the bezoar from Peter of Abano down to Mattioli which he gives twenty pages later.[8] Some say that the stone today does not measure up to its old reputation among the Arabs;[9] there is much difference of opinion as to what animal and what part thereof it comes from; and it is difficult to detect adulteration.[10] However, Silvaticus discusses ways of detecting adulteration for some pages, and then when, for how long, and in what doses the stone should be administered.

2 *Ibid.*, fol. 66v.

3 *Ibid.*, fol. 66r.

4 Jo. Bapt. Silvaticus, *De unicornu lapide bezaar smaragdo et margaritis eorumque in febribus pestilentialibus usu*, 1605. Copy used: BM 546.i.12. Brief allusions to the work occur in T V, 264, 457.

5 *Ibid.*, pp. 7-9.

6 *Ibid.*, p. 71.

7 *Ibid.*, pp. 72-73.

8 *Ibid.*, p. 92.

9 *Ibid.*, pp. 77, 93.

10 *Ibid.*, pp. 82-83, 98.

In discussing unicorn's horn, Silvaticus had incidentally rejected all that was written as to the properties of sapphires, topazes and granates, and had declared ridiculous such statements as that of Serapion, that the stone hyacinth or a seal made from it would preserve one from lightning.[11] He now turns to the emerald, but almost all that he says is taken from the Medical Letters of Mundella, published in 1543, as we have noted in a previous volume.[12] But whereas Mundella favored use of the emerald as an amulet, Silvaticus doubts if the occult virtue or specific form of the gem would so operate.[13] This does not mean that he denies occult qualities in general, as we shall soon see. He is ready to admit that there are medicaments which are effective, although they pass rapidly and apparently unaltered through the stomach, as Dioscorides teaches in the case of the chameleon because of its maximum abstersive, biting and burning properties. This leads to a digression as to the ability of the ostrich to digest iron and gold, which is ascribed not to the heat of its maw but to an occult property of its specific form.[14] But the emerald has no more effect taken internally than when worn as an amulet. Let physicians employ it as an amulet, if they insist; at least it won't do any harm, as it might, if taken internally.

After this denial of any medical virtue to emeralds, it comes as a surprise to find Silvaticus affirming the medical properties of pearls both as simples and in compounds, and against the putridity of malignant fevers in particular. But this is confirmed by all physicians, he says, and by reason and daily experience.[15] He also asserts that all writers agree that a celestial dew enters into their composition.[16] They vary with the moon's phases;[17] and of their medicinal properties some come from manifest qualities, others from their substantial form or occult quality. Those from the latter source are by far the greater, especially for the heart, between which and pearls there is a close sympathy.[18] No recent writer has doubt-

[11] *Ibid.*, p. 62.

[12] T V, 457.

[13] *De unicornu . . . etc.*, p. 131.

[14] *Ibid.*, pp. 132-33.

[15] *Ibid.*, p. 144. Mundella, however, had opposed the used of powdered pearls as cordials.

[16] *Ibid.*, pp. 149-50.

[17] *Ibid.*, p. 154.

[18] *Ibid.*, pp. 158-59.

ed that pearls are beneficial in cases of pestilential fevers, but Silvaticus has to do some squirming to explain why Galen and Dioscorides failed to mention it. He urges physicians, however, to employ pearls confidently against such fevers, using the same method as in the case of the bezoar stone.[19] His treatise thus from several points of view presents a curious mixture of scepticism and credulity. In another chapter we find him ten years later writing against the notion of climacteric years.

The combination of unicorn and gems in one volume by Silvaticus may have been suggested by reading Andrea Bacci, whose treatise on the monoceros or unicorn had been published in a Latin translation of Wolfgang Gabelchover at Stuttgart in 1598, while his work on gems and precious stones and their virtues and use had appeared at Frankfurt in 1603, likewise in Latin translation by Gabelchover.[20] There was another edition in 1643. Silvaticus was later imitated in his turn by Anton Deusing, who in 1659 at Groningen combined dissertations on the unicorn and bezoar stone in one volume with others on the mandrake and on manna and sugar.

But long before this, Laurens Catelan, an apothecary of Montpellier, had published there in French a treatise on the bezoar in 1623,[21] another on the lycorne or unicorn in 1624,[22] and, in 1638, yet another on the mandrake.[23] The treatise on the lycorne appeared in German translation in 1625, and that on the bezoar in

---

Alfonsus Nuñez (or, Ildefonsus Nuñes), on the contrary, in an *Assertio . . . de margaritis,* which was first published in 1620, then reprinted with dissertations by Ludovico Settala in 1626, held that pearls were by no means to be classed among *alexipharmaca* which operated by their whole substance or occult property. "Yet because they strengthen the heart, they are called cordials."

[19] *Ibid.*, p. 160.

[20] On Bacci see T V, 484-5; VI, 315-6, etc.

[21] *Traicté de l'origine, vertus, proprietez et usage de la pierre bezoar,* Montpellier, 1623, in-8, 56 pp. Copy used: BN Te151.188.

[22] *Histoire de la nature, chasse, vertus, proprietez et usage de la lycorne,* Montpellier, 1624, in-8, 100 pp. Copy used: BN Te151.696.

[23] *Rare et curieux discours de la plante appellée Mandragore: de ses especes, vertus et usage. Et particulierement de celle qui produict une Racine representant de figure le corps d'un homme; qu' aucuns croyent celle que Josephe appelle Baaras et d'autres, les Teraphins de Laban, en l'Escriture Sainte,* Paris, 1638, 52 pp.

1627. Meanwhile Catelan, who in 1623 was a simple apothecary, by 1624 had become apothecary to the Duc de Vendôme, and in 1638 read his discourse on the mandrake publicly in the auditorium of the faculty of medicine.

Between the appearance of the works by Silvaticus and Catelan several other authors had written on the bezoar. One was Edmund Hollyng of York, who had come to Ingolstadt in 1588 and died there in 1612. His treatise on the bezoar appeared there the year before his death.[24] Others soon followed by Caspar Bauhin, the noted botanist,[25] Giovanni Contarini,[26] and Philibertus Sarazenus.[27] Soon after Catelan, Angelo Sala of Vicenza and Venice was to treat the theme again at Erfurt.[28]

It is doubtful if Catelan was directly acquainted with any of these writers except Bauhin,[29] but he probably had used Silvaticus, although he mentions him only once as among sceptics as to the bezoar.[30] He says that he is the first author to treat of the subject in French and that he possesses one of the finest, rarest and most extraordinary Oriental bezoars, as large as a hen's egg and weighing two ounces. He has written this treatise to reassure princes and great lords as to the probable genuineness of such stones in their possession. He has already published works on the confection Alkermes and on theriac.[31] If the present work is well received, he will write others on the lycorne, elk horn, porce-

[24] *Ad epistolam quandam a Martino Rulando medico Caesareo de lapide Bezoar... Responsio*, Ingolstadt, 1611, in-8.

[25] *De lapidis Bezaar orientalis et occidentalis cervini item et Germanici ortu natura differentiis veroque usu ex veterum et recentiorum placitis*, Basel, 1613; with a reprinting in 1625.

[26] Johannes Contarenus Venetus, *De purgandis enixis et recto bezaharticorum usu*, Venice, 1614, in-4.

[27] *De notis Bezaar Epistola*, with his *Observationum Centuriae IV*, Oppenheim, 1619, in-4, p. 66.

[28] Angelus Sala, *Ternarius bezoardicorum...*, Erfurti, apud Johan. Bircknerum, 1628 et 1630, in-8.

[29] *Traicté ... de la pierre bezoar*, 1623, pp. 16, 45.

[30] *Ibid.*, p. 23: "Hierosme Brscianus en son livre de nova medicina, Manlius ad Cratonem, Massarias de febre pestilentiali, Valesius, Silvaticus, Hercules à Saxonia libro de febribus, Rulandus de Hungarica lue, Thomas Jordanus et plusieurs autres misprisent grandement l'usage de ceste pierre..."

[31] Both were printed in 1614; that on Alkermes reprinted in 1620, and that on theriac in 1629.

lain vases, toad stones, birds of paradise, the remora, salamander, mandrake, chameleon, pelican, asbestos, *mumia,* and what-not. His chief novel suggestion is that the bezoar stone is formed in the pouch of marsupials, where they store their food and ruminate it. They cure themselves of disease by sucking snakes out of their holes and eating them and protect themselves against the venom of the snakes by eating certain herbs. The bezoars gradually form by a marvelous sympathy between the snakes, herbs and diseases in question, and the quintessence of the salutary herbs is extracted by the animal heat in the pouch.

Catelan's treatise was soon followed by another work in French but against the bezoar stone, *Les tromperies du bezoar decouvertes* by Philbert Guybert,[32] one of the medical faculty at Paris. In his dedication to Charles Bouvard, first royal physician, he refers to Bouvard's having thrown into the ocean at La Rochelle the horn of a fabulous lycorne, *mumia*—"a true poison," powder of pearls, and other useless drugs. He blames the recent popularity of the bezoar upon Spanish and Portuguese physicians who visited the Indies, and who also had been extensively cited by Catelan. Guybert's work is in two parts, the first affirmative in which he notes accounts favorable to the bezoar, but does not mention Catelan; then a longer negative section in which he criticizes them and lists various experiments and authorities, including Silvaticus, to the contrary. Here again he cites many of the same writers as Catelan had, but of course this was to be expected.

When Johann Michaelis wrote on bezoardic tincture in 1678, he explained that the bezoar stone was not a constituent in it, since it was not usable in a tincture.[33] But the mere name was evidently still impressive.

Robert Pitt (1653–1713), writing early in the next century,[34] said that the bezoar stone "has held its name and reputation almost

[32] I have used the second edition of Paris, 1629 (BN 8° Te151. 189. where the pages are numbered 531-694. The dedication is undated.

[33] *Dissertatio pharmaceutico-therapeutica de natura tincturae Bezoardicae* . . . Hall. Saxon., 1678, p. 68.

[34] *The craft and frauds of physick expos'd,* London, 1702, in-8: BM 1038.e.38. I have used the second edition of 1703. There was a third edition in the same year.

sacred with us," though they were exploded long since in almost all other parts of Europe. He cited Guybert, *Les tromperies du bezoar decouvertes,* and affirmed that it was despised and condemned by Massarias, Severinus, Sanctorius, Minodeus,[35] Ruland, Silvaticus, Minderer, Sennert, Untzer, Pauli, Diemerbroek, Patin, Bontius and others, and had been omitted from the Leeuwarden Dispensatory.[36]

Coming back to Catelan and his treatise on the lycorne, we note that he again claims to be the first to write on the subject in French, and to be the possessor of an entire horn, "recovered at great pains from the depths of Ethyopia," beautiful to see, and corresponding to the descriptions in Pliny, Aelian and other authors. His text opens with the statement that Nature keeps its best treasures secret. Such are stones which quench flames unbelievably; plants, which, crushed in the hand, indicate the day and hour of death; among birds, the Ephemeris of marvelous plumage, born and deceased on the same day; among fish, the brave little *remora* which stops ships; among reptiles, the *draconcalopedes* with the face of a beautiful virgin and of an attractive coloring; among quadrupeds, the renowned lycorne, unicorn or monoceros.

The first kind of monoceros or unicorn is a bird; the second is a marine animal called *vletif* (swordfish?); the third, a sort of snail. Of the fourth variety of quadrupeds some eight varieties have been distinguished, including the rhinoceros, onager, elk and reindeer, and last but not least the lycorne, which is the size of a horse, has the mane of a lion, a head like a deer, and feet like an elephant. Catelan gives eighteen objections to its existence and then answers them at length.[37]

In a work of mineralogy published in 1610, Francesco[38] Imperato of Naples professed to set forth points untouched by others in "marvelous order," and to give new interpretation of hieroglyphs

[35] Minadous Rhodiginus may be meant.

[36] Pitt (1703), 32-34.

[37] Some of them are attributed to Andrea Marini, a Venetian, *De falsa opinione erga Unicornum,* or, more accurately, *Discorso . . . contra la falsa opinione dell'Alicornio,* Aldus, Venice, 1566, in-4: BM 975.d.8.(2.).

[38] Vander Linden incorrectly ascribes it to Ferrante Imperato, his father: LR 274.

by stones.[39] His marvelous order of presentation is as follows: kinds of earths, concreted juices, general description of stones, gems, the varieties of marbles known to antiquity, silicates, rocks, tufas, pumices, stones, those which consist of perfect humid vaporous exhalation, and the metals. Imperato still had faith in the occult virtues of gems. "Small in size, they produce wonderful effects." An amethyst, placed on the navel, frees its bearer from intoxication. Wearing a sard fulfills the function of recreation and dispels fear. The sapphire cleanses the eyes; the ancients thought that wearing it extinguished lust. Crystal by its frigidity restrains poisonous draughts, and as the celestial rainbow announces coming rain and fair weather, so the crystal may be taken for future adversity or felicity. *Nefrites* is employed against gravel in the kidneys and stomach-ache. Wearing a jasper checks haemorrhage and *menses*, and augments the natural virtue of the stomach; but experience shows that the oriental heliotrope is more efficacious. Some cordials and antidotes for poison are compounded from fragments of hyacinth, topaz, sapphire, sardonix and granate. *Ophites*, in Imperato's opinion, has greater potency against headache, if bound on.[40] That his views with regard to the virtues of gems did not alter with time is seen from a passage in his *Discorsi* of 1628, in which he asserts that gems have occult as well as elemental virtues, such as to counter-act poison, make a man victorious or beloved, and many other qualities and virtues.[41] But he did not believe that gems engraved with characters or celestial signs, or astrological and magical images acquired any further powers thereby unless from the demon. Such engraved images were a superstition introduced by Satan and justly condemned by "our holy Catholic religion."[42]

Of the fourteen *Discorsi* of 1628, besides that on the virtues of gems above mentioned, three others dealt with stones such as *pyrites*, crystal, and the fossil bezoar, which, we are told, differs

[39] *De fossilibus opusculum, in quo miro ordine continentur naturalis disciplinae scitu dignissima eisque professoribus omnino necessaria, ab aliis minime excogitata. Multa quae hieroglyphice per fossilia noviter interpretantur nonnullaeque icones fideliter ad vivum delineatae*, Naples, Jo. Dominico Roncaliolo, 1610. Copy used: BN S. 5525.

[40] *Ibid.*, pp. 40-46.

[41] *Discorsi intorno a diverse cose naturali*, Naples, 1628, p. 61.

[42] *De fossilibus*, p. 39.

from the bezoar stone found in goats and other animals. A fossil bezoar was given to Imperato's father, who had composed a Natural History in the previous century, by a physician, who had been preserved by it while caring for the plague-stricken in a pest which killed a third of the population. He wore it over his heart, with an oriental topaz bound on his upper left arm.[43]

The recent author of the catalogue of the *Museo Calceolario* had refused to accept the statement of Imperato's father that the so-called toad-stone (*Rospo*) is not found in the heads of toads, preferring to accept other authors. Some said that the stone should be removed from the toad's head in August; others, under a waning moon; others, on dog day. Others advised that, after the toad had been killed, its carcass should be left near an anthill until the insects had stripped it of all but its bones and the stone. But Imperato's father, who had spent much time in observing toads, had disproved all these assertions and had shown that other stones had been taken for the *Rospo*.[44]

Since the *Historia Naturale* of Imperato's father, although first published in 1599, had further editions in the late seventeenth century,[45] a word more may be said of it. Hallevord[46] and Toppi[47] cite Vincent Placcius[48] that the true author of the work was Nicolaus Antonius Stelliola, to whom Imperato paid 100 *scutati* for the privilege of putting his name on the title page. The book gave more attention to the mineral than to the animal and vegetable kingdoms, devoting its first five books to mining and nine others to alchemy.

A work on coloring glass and making artificial gems by Antonio Neri, first published at Florence in Italian in 1612,[49] both emphasized

[43] *Discorsi*, pp. 57-59.

[44] *Ibid.*, pp. 27-31.

[45] Venice, 1672; Leipzig, 1695; Cologne, 1695.

[46] Joannes Hallevordius, *Bibliotheca curiosa*, Regiomonti et Francofurti, 1676, p. 77.

[47] Nicolo Toppi, *Biblioteca Napoletana*, Napoli, 1678, pp. 77-78.

[48] *De scriptis et scriptoribus anonymis atque pseudonymis syntagma*, Hamburg, 1674, p. 213.

[49] *L'arte vetraria distinta in libri sette del R.P. Antonio Neri Fiorentino nei quali si scoprono maravigliosi effetti e insegnano segreti bellissimi dei vetro nel fuoco ed altre cose curiose*, in-4. Other editions and translations continued throughout the century. Icilio Guareschi, *Sui colori*

experimental method and stressed secrecy and marvelous results.[50]

It is difficult to decide whether to classify the work on inks by Pietro Maria Caneparío,[51] a physician and philosopher of Crema who practised medicine at Venice, under mineralogy, alchemy, technology or natural magic. Its first part or *Descriptio* is on the stone *pyrites*, "stem of inks and metals," in which sulphur and mercury flourish from which metals and inks are generated, and considers such other mineral substances as *cadmia*, magnesia and marcasite. It also discusses whether semen with spirit and soul is in metals and stones; concerning "roots and veins of metals, stony plants and vegetables which flourish without any manifest roots;" and whether, in striking a stone, the spark of fire elicited comes from the stone or air or motion; and in how many ways fire can be produced. Also wax candles and a candle burning in water. Avicenna is cited that, if pyrites is worn on an infant's neck, it defends him from all fear.[52]

The second *Descriptio* is about metallic ink, especially "chalcity," "unknown to almost all men of the present age, for in the composition of theriac they use another ingredient different from the chalcity (*chalcitis*) of the ancients." The relation of chalcity to misy and sory is set forth, and Agricola and his sect are refuted. The third *Descriptio* in twenty-two chapters on shoemaker's blacking is chiefly concerned with vitriol. The *sulphur vitriolatum* obtained from iron is called by moderns *crocus Martis*. Iron and all other metals by the aid of art revert to vitriol. We are told how to segregate sulphur from vitriol, how to extract quicksilver from

degli Antichi; Parte seconda, 1907, says that the third edition appeared at Venice, 1663. Besides the editio princeps of 1612, I have seen the English translation: *The Art of Glass . . . with some observations . . .*, London, 1662; and the Latin edition of Amsterdam, 1686, *Antoni Neri Florentini De arte vitraria libri VII & in eosdem Christophori Merretti M.D. & Societatis Regiae Socii Observationes & Notae* (these had appeared anonymously in the English translation).

[50] These points are emphasized in the original Italian title, in the preface of January 6, 1611 to Don Antonio Medici, where Neri's further experimentation in chemistry and medicine is mentioned (*sperimentato molti effetti utilissimi credibili & mirabili*), and throughout the text.

[51] P. M. Caneparius, *De atramentis cuiuscunque generis. Opus sane novum hactenus a nemine promulgatum, in sex descriptiones digestum,* Venice, 1619, 368 pp.: BN S. 5387.

[52] *Ibid.*, p. 47.

it, and how to make a new kind of vitriol. The quality and nature of vitriol are perceptible in many foods, as in the atrabile superabounding in our bodies.

The fourth *Descriptio*, on inks for books and writing, asks what the ancients used; describes printer's ink as compounded of a pound of varnish, an ounce of soot, and enough linseed oil or nut-oil to mix them; and devotes several chapters to encaustic. It then turns to ways of erasing letters, of secret writing, and of magic writing. Letters and images may be executed on the shell of an egg so that they will appear on the albumen inside after the egg has been hard boiled. Dissolve rock alum in sharp vinegar and write on the egg-shell with this solution. Dry the writing in the sun, then soak the egg in sea water or very salt water or vinegar for three or four days. Then let it dry before hard-boiling it. Or cover the egg with wax, trace the letters in the wax with a stylus, and fill them in with salt water and alum. Soak the egg in vinegar for a day, then clean off the wax and you will have the letters. Another feat of natural magic is to grow peaches, almonds and quinces with letters inscribed on their pits. First plant the peach stones until they begin to open up. Then take out the pits and write on them with cinnabar dissolved in mucilage (*Aqua gummata*), taking care not to mar the pits lest their prolific faculty perish. Let them dry a while and then replace them. Giovanni Battista Birelli of Florence, *XIII de diversis arcanis*, cap. 239,[53] advises to soften almonds in water until the shells begin to separate. Then take out the pits, trace the image or letters lightly on them with a knife, put them back and sew up the shells and plant them. Two chapters then deal with tattooing and how to remove the marks.

After the fifth *Descriptio* on writing inks of different colors, the sixth *Descriptio* (at pp. 247-368) reverts to vitriol in the shape of oil of vitriol and its fifth essence, with argument for chemical remedies and alchemy. Many hot medicaments nevertheless have an occult property from the heavens of evacuating hot humors. But those who claim to segregate the four elements in vitriol at one distillation by means of a vessel with four outlets go too far for Canepario, and he calls them tricksters (*praestigiatores*). Yet

[53] The reference is probably to his *Opera*, Florence, 1601.

he treats not only of the mastery of vitriol but also of the salamandric stone, elixir and universal medicine. He gives a *Praxis* of oil of vitriol from the *Monarchia* of Paracelsus, a secret experiment of an aged man, an oil of vitriol by Bernard Penotus of Aquitaine,[54] and other medicines from Moffett and Philipp Müller. He tells how to soften and sweeten oil of vitriol, howsoever fiery and corrosive it may be. "Innermost experiments of this sort conclude the whole work," with an "elegant figure of the arcanum," a transformation of the terrestrial planet Saturn into effulgent Apollo, and an *oleum benedictum*.

The printed catalogue of the Bibliothèque Nationale, Paris, lists compendia of logic and of theology by Conradus Hornius or Horneius, printed in 1654 and posthumously in 1655. But neither it nor the printed catalogue of books in the British Museum has his Compendium of natural philosophy concerning stones, metallic and mineral mediums, published thirty years earlier in 1624.[55]

That mineralogy had made little advance during the previous hundred years might be inferred from the fact that Giovanni Guidi of Volterra in 1625 dedicated to Ferdinand II, grandduke of Tuscany, and published at Venice a treatise *De mineralibus* by one of his ancestors who lived from 1464 to 1530, and that there was another printing of the book in 1627.[56] Such an inference, however, would not be quite justified, since Guidi was a jurisconsult and the volume is chiefly concerned with legal questions involving alchemists, gems, and so forth.

Johann Chesnecopherus (1581–1635), professor of medicine and anatomy at Upsala, was interested also in astronomy and incidentally in mathematics, and published a work on eclipses in 1624.[57] In the year following was printed a disputation over which he presided concerning congealed juices and precious earths.[58] It

[54] LR, pp. 129-31, for his works from 1594 on.

[55] *Compendium naturalis philosophiae de lapidibus metallicis et mineralibus mediis*, Helmstadii, 1624, in-8.

[56] I have used this edition in BN [* E. 544: J. Guidius, *De mineralibus tractatus absolutissimus*. For the 1625 edition, BM 33.b.25.

[57] David Eugene Smith, "Medicine and Mathematics in the Sixteenth Century," *Annals of Medical History*, 1917, p. 134.

[58] *Disputatio physica decima octava de succis concretis et terris pretiosis*, 1625, in-4: BM B. 236 (3.).

lists four kinds of fossil salts: sal ammoniac, sal gemmae, *sal nitrum*, and *sal Indicum*. Of these the first three are identical with the first three of the medieval seven salts of Hermes,[59] which do not include *sal Indicum*. Amber is held not to be the sperm of whales, since it is not found where they abound. More likely it is bitumen from the depths of the sea. I have not found a dissertation by Chesnecopherus on three *terrae sigillatae:* sun-grease, moon-grease, and sun-soul; also of mercury of iron as a cure for gout.[60]

Petrus Stephanonius of Vicenza published at Rome in 1627 under the title, Gems of Antiquity,[61] a collection of 51 Plates "médiocrement gravées par Valeriano Regnart."[62] The only text was distichs accompanying the Plates. But the book was of sufficient interest to that period to be reprinted twice at Padua, in 1646 and 1653.

An illustrated catalogue of the collection of gems of Ludovicus Chaducius fills 218 leaves of MS Ste. Geneviève 1168 at Paris. It was drawn up in 1628.

Coral at this time was usually grouped with stones. The History of Corals by Johann Ludwig Gans, M.D., of Frankfurt, was largely chemical or alchemical and offered little that was new.[63] In the last chapter on the virtues of coral, which fills a third of the volume,[64] it is said that, although many deride this as idle and fictitious, yet experience is witness that there is some power in coral against incantations. For Gans remembers a physician of great name who dispelled many diseases that had been brought on through witchcraft, by making use of a powder in which coral was a prime ingredient.[65] Ettmuller in 1665 said that Gans believed coral to be superior to all gems in its almost divine virtue, and

[59] *Isis*, XIV (1930), 187-88; XXVII (1937), 53-62.

[60] *De tribus terris sigillatis: axungia solis, axungia lunae atque anima solis. Item de mercurio ferri in quo solo est podagrae topicum dissertatio*, Upsaliae, 1629. I derive the title from Mollerus, *Suecia literata*, 1698, p. 93, where are further listed *Disputationes physicae plures.*

[61] *Gemmae antiquitus sculptae et declarationibus illustratae*, Romae, 1627, in-4: BM 277.k.32. *Bibl. curiosa* (1676), 330.

[62] Graesse, VI, 491.

[63] *Corallorum historia qua mirabilis eorum ortus locus natalis varia genera praeparationes chymicae quamplurimae viresque eximiae proponuntur*, Frankfurt, 1630, 170 pp. in-16. BM 446.a.12.

[64] *Ibid.*, pp. 115-74.

[65] *Ibid.*, p. 171.

further to be preferred to all herbs.[66] The demand for Gans' book was sufficient to induce an enlarged edition thirty-nine years later.[67] A contemporary review of this second edition is silent as to the original edition but says of that of 1669:

His opinion is that coral is form'd out of a glutinous juice which, being turned into stone by a salt abounding in it, riseth up in the form of a shrub, the salt being the cause that maketh plants spread into branches.[68]

Bernardo Cesi (1581–1630), who was born at Modena and died there of the plague, was a Jesuit who taught at Modena and Parma. He composed a very diffuse book of little worth in mineralogy which was published posthumously by his Order at Lyon six years after his death. Its full title, which gives promise of the marvelous and occult, may be translated as,

Mineralogy or Treasures of Natural Philosophy, in which are contained the miracles of metallic concretion and of medicated fossils, the price of earths, collections of colors and pigments, the virtue of concreted juices, the dignity of stones and gems.[69]

Webster in 1671 criticized Cesi as too digressive and as mixing tares with the wheat.[70]

The work opens by listing the evils and benefits of mineralogy. Mining is disappointing and dangerous. There are specters of demons underground, as Agricola and Thyraeus testify. Mining uses up fields and forests, etc. Why lacerate the vitals of kindly mother Earth? Ills from riches, iron weapons, and especially gold are enumerated, with much poetical quotation in the case of gold.[71]

[66] Michael Ettmuller, *Examen coraliorum tincturae*, Lipsiae, 1665, fol. D verso. BM B. 426 (7.).

[67] At Frankfurt, 1669, in-12: BM 987.b.19; 234.a.43 (1.).

[68] PT V, 1202.

[69] R. P. Bernardus Caesius, S.J., *Mineralogia, sive naturalis philosophiae thesauri, in quibus metallicae concretionis medicatorumque fossilium miracula, terrarum pretium, colorum et pigmentorum apparatus, concretorum succorum virtus, lapidum atque gemmarum dignitas continentur*, Lugduni, Jacobi et Petri Prost, 1636, in-fol. 2 col. 626 pp. and Index: copy used, BN S. 1260.

[70] *Metallographia*, p. 29.

[71] *Mineralogia*, 4b-6b. On Petrus Thyraeus, T VI, 533, note 51.

On the other hand, the study of mineralogy helps one to understand the Bible, as Augustine said.[72] It supplies us with medicines and money; with ornaments for religious purposes, while iron and other metals are useful in agriculture, war, industry, painting, the arts, music and alchemy. Cesi then answers the aforesaid objections to mineralogy,[73] and further inquires whether it is respectable or sordid, liberal or mechanical, practical or speculative, and whether it may be considered a part of philosophy.[74]

Cesi's numerous citations give evidence of wide reading. On the single point of gems as religious symbols he cites Anselm of Canterbury, Bonaventura, Pierius, Clement of Alexandria, Petrus Berchorius, Caussin, Paschasius Balduinus's letter to Francis Rueus, Rueus himself, Cornelius a Lapide (Gemma), Alcasarius, the Bible, and numerous church fathers.[75] On the immediate matter of minerals, besides the seven differing views of Democritus, Avicenna, Gilgil, the chemists, Albertus Magnus, Agricola and Aristotle, he cites Seneca, Vatable, Cardan, Pierre Sainctfleur of Montpellier, Theophilus Raynaud, Gregorius Reisch in the *Margarita philosophica,* and Clavius on the *Sentences.*[76]

Cesi still regards it as certain that the heavens and stars act upon and influence this lower world, which they do by their motion, light, and occult influence. But this does not abrogate the contingency and variety of sublunar things.[77] After digressions to discuss whether bodies subject to attrition are necessarily diminished, whether the heavens are fluid, comets, and the vulgar error that metals and gems were all formed at creation as we find them now,[78] he notes that Hermes Trismegistus held that metals are produced by the planets, and gems are caused by the fixed stars.[79] He then turns to astronomical images, which he here tends to oppose,[80] but in a later passage he quotes Albertus Magnus[81] concerning images and seals of gems:

[72] *Mineralogia,* 7a, quoting *De Chris. doctrina,* II, 16.

[73] *Ibid.,* 12b-15a.

[74] *Ibid.,* 15a.

[75] *Ibid.,* 7b.

[76] *Ibid.,* 23.

[77] *Ibid.,* 31a-b, 36.

[78] *Ibid.,* 32a-b, 33b, 36b-37a. Comets are again discussed at 48b-51b.

[79] *Ibid.,* 37b.

[80] *Ibid.,* 39a-b.

[81] *Lib.* 2 *Mineralium,* tract. 3, caps. 1-6.

You ask first to what science belong images and seals of gems. I respond, to that kind of necromancy which is subordinated to judicial astrology . . .[82]

You ask sixth whether gems bound on and suspended have power of influx especially on human bodies. I reply affirmatively, which can easily be proved by the following experiments.[83]

The moons of Jupiter raise the question whether the number of planets is seven or eleven, but we are next assured that we learn by daily experiments that the force of the magnet is dulled by garlic.[84] Another digression, why children resemble their parents, is followed by an account of the operation of mines and their officials, and the use of three shifts working eight hours each.[85] We then pass to the final cause of minerals,[86] the three regions of air, and the altitude of mountains.[87]

These last two topics are digressions from the main theme of the location of minerals.[88] Cesi discusses whether they are generated in fire, air and water, and treats of mineral waters and baths, "other miracles of waters," and "marvels of waters," listing many particular fountains, before he at last comes to "earth, their real fatherland."[89] Even this leads at once to another digression, whether the earth moves or not? We then descend into the caverns of earth and encounter subterranean exhalations, fires and so forth.[90] The divining rod is discussed and Agricola is cited as authority that it is better not to use it[91]

So much for the first book of the Mineralogy of Cesi. From the remaining four books[92] we note only his attitude towards the marvelous properties attributed to gems. Raising the question whether Pliny is to be trusted, since some say that he mixes falsehoods with truth, Cesi decides that his authority in what he has written on gems is entitled to great respect and is vouched for by

[82] Cesi, *Mineralogia*, 543b.
[83] *Ibid.*, 544b.
[84] *Ibid.*, 40a-b.
[85] *Ibid.*, 42b, 43b-45b.
[86] *Ibid.*, 46b.
[87] *Ibid.*, 52a-53a, 53b-55a.
[88] *Ibid.*, 47 *et seq.*, "De locis mineralium."
[89] *Ibid.*, 115, "De tellure propria fossilium patria."
[90] *Ibid.*, 117b *et seq.*
[91] *Ibid.*, 124b, 126a.
[92] *Ibid.*, 173-, Liber II De terris insignibus; 288-, III De succis concretis; 513-, IV De lapidibus et gemmis; 609-, V De metallis.

St. Jerome.[93] In general Cesi simply lists a great many properties of gems, twenty-two for jasper alone, from past writers without accepting or rejecting them. The fifth property of jasper is that it is believed to dispel all apparitions. Jerome says so, and Mylius and Rueus assert it categorically. For the eighth property, that suspended from the neck it soothes the stomach, Galen and these last two are cited. But for its tenth virtue that worn on the person it keeps one safe and powerful and frees from adversities, while Isidore, Albert, Mylius and Mattioli are cited, it is said that they well add that this is magical and superstitious. But Cesi's citations are not now so extensive as in the instances noted above. For example, I failed to see any mention of either Marbod or de Boodt. Similarly, while Porta is quoted a great deal concerning the magnet,[94] Gilbert is unmentioned. The philosophers' stone is discussed inconclusively for only two pages, and in the book on stones and gems rather than that on metals.[95]

Gilbert's work on the magnet was not to pass unnoticed by other Jesuits, however, and we shall presently consider four successive works by members of that Order which in large measure consist of reflection or grotesque refraction of the inspiring scientific light that had streamed forth from his *De magnete.*

In his book of 1635 on stones Etienne de Clave held that they were generated from seed and produced by the action of subterranean fire. He takes up the opinions of Aristotle, Theophrastus, Avicenna, Agricola, Falloppia, and Scaliger against Cardan as to the matter of stones, and of Aristotle, Theophrastus, Avicenna, Albertus Magnus, Agricola, Falloppia, Cardan, and Empedocles as to the efficient cause of stones; and engages in a longwinded discussion which it does not seem worth while to follow further.[96]

A book entitled *Mundi lapis Lydius,* by Antoine de Bourgogne,[97] has nothing to do with gems but is a moralizing dialogue between *Vanitas* and *Veritas,* somewhat after the fashion of Petrarch's *De remediis utriusque fortunae.*

[93] *Ibid.*, 548-49.

[94] *Ibid.*, 532-39.

[95] *Ibid.*, 539-41.

[96] *Paradoxes ou Traittez philosophiques des pierres et pierreries contre l'opinion vulgaire,* Paris, Veuve P. Chevalier, 1635, in-8, pièces liminaires et 492. Copy used: BN S. 20394.

[97] Printed at Antwerp, 1639. Copy used: Col 246.5 B66.

The work on metallurgy by Alvaro Alonso Barba was first printed at Madrid in 1640.[98] Barba was born in Andalusia but when he wrote the book was priest of a parish at Potosi in Peru, the center of an important mining region. The volume was composed at the command of Don Juan de Liçaraçu, and its prime purpose was to give information for the owners of the mines of the provinces under Don Juan's jurisdiction and which Barba had examined in person. Edward Montague, earl of Sandwich, translated only the first two of its five books into English in 1669, the first book being printed in 1670 and both in 1674. This translation was reviewed in both *Philosophical Transactions* and the *Journal des Sçavans*.[99] The work was printed in Spanish again at Cordova in 1675, appeared in German translation in 1676, and continued to appear in Spanish, French, German and English editions through the eighteenth century.[100]

In the opening chapter of the second book Barba states that the mountain and city of Potosi have already produced between 400,000,000 and 500,000,000 pieces of eight, enough to cover the ground for a space sixty leagues square, but that a lack of care and needless waste have accompanied this glut of wealth. There is a crying need for better refiners. In the next chapter he tells of obtaining 900 pieces of eight per quintal, where previous miners had gotten only four or five, and then had abandoned the mine, and sixty to the quintal in another mine which had been given up as containing almost no silver. In succeeding chapters he notes that hardly any ore in one mine resembles that which bears the same metal in another mine, gives instructions for sorting the ore, cleaning it from copperas or vitriol so that this may not consume the quicksilver which is later used to attract the silver, removing other impurities, grinding the ore fine enough so that the silver may be attracted by the quicksilver, burning it, but not too much or the

[98] *Arte de los metales en que se enseña el verdadero beneficio de los de oro y plata por açogue. El modo de fundir los todos y como se han de refinar y apartar unos de otros.* Compuesto por el licenciado Albaro Alonso Barba, natural de la villa de Lepe, en la Andaluzia, Cura en la Imperial de Potosi de la Parroquia de S. Bernardo, Madrid, 1640.

[99] PT IX, 187-91; JS IV (1675), 52-54, 145-47.

[100] Ferguson, I, 70-71.

copperas will increase, and in no case burning any ore with salt. To test an ore one should pulverize it not too finely and throw the bits onto a red-hot iron plate. If a black or white smoke arises, it indicates the presence of bitumen; if yellow, of orpiment; if red, of sandarac; if yellow within and green without, of sulphur.

Earlier, in the first book, Barba said that little use had been made of quicksilver, until it was employed to collect the silver from the ore. Before that, it was only wasted in mercury sublimate, cinnabar, vermilion, "and the powders called Precipitate and used to such mischievous purposes."[101] Yet a few pages further on he tells how to make mercury sublimate.[102] But, between 1574 and 1640, no less than 204,600 quintals of mercury had been received at Potosi. In the three closing books Barba goes on to discuss mineral waters, the process of founding metals, furnaces and other apparatus and instruments employed in metallurgy, the process of refining and separating metals, and such matters as how to assay silver to tell if it contains gold. Speaking earlier of the direction in which veins of metal run, he says that miners in Europe put first those that run east to west in the northern part of a mountain, next veins running north and south in the northern part of a mountain, and last those running north and south on the eastern side. But Barba holds that experience has shown the contrary both in Europe and America. In the case of the Potosi mines, he would esteem most those running north to south on the north side, and second those running north to south on the south side.[103]

Such expression of personal opinion and experience, and such direct practical knowledge of mines and metallurgy, are accompanied however, by theoretical views which one might not expect and which seem backward or occult. Barba still accepts substantial forms[104] and occult qualities. Some of the virtues of minerals work through their occult essential qualities or specific form, others by their elementary or manifest qualities. A sapphire taken internally in drink is beneficial for scorpion bites. And of those gems that cure

101 English edition of 1674, chapter 33, p. 140.

102 *Ibid.*, chapter 34, p. 145.

103 *Ibid.*, I, 25, pp. 102-3.

104 *Arte de los metales*, p. 12, "Tienen las piedras sus formas sustanciales con q̄ se constituyen en sus proprias especies."

by occult qualities, some, as *Haematites,* prevent the blood from flowing to a particular part of the body; others, like jasper, strengthen the stomach if merely suspended from the neck; others, like *aetites* (the eagle stone) restrain abortion, if tied to the left arm. But applied differently, it and jasper produce just the opposite effect. The magnet purges gross humors; other stones make one vomit; some dissolve warts; others corrode the flesh or make it putrefy; some minerals, like chalcitis, misy and alum, heal wounds.[105]

Although Barba grants that the name, alchemist, has been made hateful by ignorant pretenders, he repeats the view that metals are generated by the influence of the heavenly bodies, and defends the doctrine that they are compounds of sulphur and mercury. He also affirms that more difficult transmutations than that of metals are performed both by art and nature. By art, he says, wasps and beetles are made out of the dung of animals, and scorpions from the plant *alvaca* or *alvahaca,* "peresta en el lugar y modo que conviene." In Scotland ducks are engendered from driftwood and the fruit of trees that falls into the sea.[106] There are sympathies and antipathies between metals and minerals, and strong waters which not only separate gold from silver but liquefy gold itself.[107]

Although Barba accepts the heavens as a universal cause, he regards as not free from vain curiosity those who attribute to the stars and planets particular influence and dominion over particular things, associating gems with the fixed stars and metals with the seven planets. Bismuth, which was discovered a few years ago in the mountains of Bohemia, must be reckoned a metal and placed between tin and lead. Maybe there are other metals as yet unknown.[108]

This brings us to the matter of Barba's citations. On the one hand, he is acquainted with the work of Agricola and even with authors, such as Marsilio Fierno,[109] whom I do not find listed in the printed catalogues of the Bibliothèque Nationale and British

105 Bk. I, cap. 36.

106 I, 18-20.

107 III, 2; V, 7 and 14. But at I, 26 he was non-committal as to the claims made for potable gold.

108 I, 20, "los cielos que como causa universal"; 22, "Los que no sin nota de vana curiosidad" etc.

109 III, 3. Fierno may be a misprint for Ficino.

Museum. On the other hand, he still employs such medieval authorities as Albertus Magnus and "the learned Raymond" Lull. Touching on the pictures and engravings which nature makes on stones,[110] he cites Albertus as saying that he saw five hundred snakes on a stone which had been presented to him, upon which a serpent was depicted, and which possessed the occult quality of attracting serpents. In the same chapter, referring to the fossils of shell-fish found on mountains which the sea could never have reached, Barba declares that they are so lifelike that only the Author of Nature could have produced such a piece of workmanship.

Two letters on fungi-bearing stones by Marco Aurelio Severino, professor of surgery and anatomy at the University of Naples, dated in 1642 and 1644, were printed at Padua in 1649, at the close of a new edition of the fifteenth century *Coena* of Battista Fiera.[111] As in his *Vipera Pythia* of 1643, Severino makes much of a fermentatory spirit. In the longer letter there is a deal of quoting authorities, including Imperato on tubers, until we finally come to the opinion of rustics and the author's own observation.[112] And the fungi-bearing stones are shown in plates.

A brief treatise on the astronomical seals of the Arabs and Persians by an anonymous Persian was edited by John Greaves (Gravius), professor of mathematics at Oxford and a Persian scholar, and consists of an Arabic text with Latin translation and notes.[113] A work on Gnostic gems by Jean l'Heureux or Macarius, a canon of Artois who had died early in the century,[114] was

[110] I, 17, "De algunas accidentes de las piedras y sus causas."

[111] *Baptistae Fierae Mantuani medici sua aetate clarissimi Coena notis illustrata a Carolo Avantio Rhodigino, cui novissima hac editione accesserunt Cl. V. Marci Aurelii Severini in regio Neapolitano gymnasio chirurgiae et anatomes publici professori Epistolae duae: altera de lapide fungifero* (at pp. 167-201, dated 1642), *altera de lapide fungimappa* (at pp. 202-208, dated 1644), Patavii, 1649. BN V. 14679.

[112] *Ibid.*, p. 192.

[113] *Anonymi Persae de siglis Arabum et Persarum astronomicis. Tract. cum annot.*, London, 1648, 16 pp. BN X. 1757 (3-4), where it is bound with other works by Greaves such as that on the Elements of the Persian language.

[114] The edition of 1657, pp. 4-5, gives the date of his death as June 11, 1614, whereas *Bibliotheca curiosa*, p. 187, and other works of reference, state the year as 1604.

printed only in 1657,[115] with additions by Jean Chiflet, a canon of Tournai.[116] Evidently in these two works gems and images carved in them were considered primarily as things of the past and of antiquarian interest.

Du Hamel's discussion of the mineral kingdom in 1660, in the second book of his work On Meteors and Fossils, is postponed to a later chapter.[117]

J. L. Bausch (1605–1665), municipal physician of Schweinfurt, connected the Academy of the Curious as to Nature, of which he was president and founder, with a pair of treatises which he issued in 1665 on the bloodstone and the eagle-stone.[118] Although he described both stones and their different species in some detail, his primary interest was in their medical virtues, and nearly half of the treatise on the bloodstone is a prooemium in regard to loss of blood. As to magical virtues, however, he was sceptical, declaring fabulous, ridiculous and fantastic the attribution to the bloodstone of power to win victory, freeze hot water, and preserve crops from hail or locusts.[119] He also states that those who have written of the eagle-stone have promised many vain things which exceed the limits not only of medicine but of nature.[120] He was credulous, however, as to the generation of snakes, frogs and fish in rocks and marbles from the more humid putrescent *substantia* of the stone, as worms are generated. But, because of their solider substance

[115] Joannes Macarius, *Abraxas sive Apistopistus quae est antiquaria de gemmis Basilidianis disquisitio. Accedit Abraxas Proteus seu multiformis gemmae Basilidianae portentosa varietas exhibita et commentario illustrata a Joanne Chifletio canonico Tornacensi,* Antwerp, Plantin, 1657, in-4, Col AK 5523 M 11. The book passed the censors in 1651. For Abraxas and Basilides: T I, 372, 379.

[116] Other works by him of 1634, 1652 and 1653 dealt with *Mater sacrorum* of Germigny, the *ampulla* of Reims and royal unction, and quinine.

[117] Chapter 29, "Other Exponents of Experimentation."

[118] *Schediasmata bina curiosa de lapide haematite et aetite ad mentem Academiae Naturae Curiosorum congesta,* Lipsiae, 1665. The two treatises have separate paginations, and, after that on the bloodstone ends at p. 164, that on the eagle-stone begins with a new title-page: *De lapide aetite schediasma ad modum et mentem Academiae Naturae Curiosorum congestum a Joh. Laurent. Bauschio M.D. et physico reipubl. Suinfurtensis patriae ordin.,* Lipsiae, 1665, 79 pp. BN R. 13095-13096.

[119] *Ibid.,* cap. 6, p. 141-.

[120] *De lapide aetite,* caps. 5-6.

they are livelier and wear away the stone and live and grow on it, as Cardan says.[121]

Although these two treatises seem to be compilations from previous authorities, I did not notice any reference to Lauremberg's work on the eagle-stone[122] and perhaps Bausch did not know of it.

Soon after this publication, Bausch died, leaving unprinted another *Schediasma* on *coeruleum* (which he used as a synonym for *lapis Armeniacus* and lapis lazuli) and *chrysocolla.* It was issued posthumously at Jena in 1668.[123] It too is primarily medical and largely composed of citations. Since I found many of the leaves still uncut in the copy which I examined at the Bibliothèque Nationale, Paris, it would not seem to have been very influential or widely read.

The *Metallographia* or history of metals of John Webster, dating from London in 1671 and dedicated to Prince Rupert,[124] praises Bernard Trevisan, Basil Valentine, Paracelsus and Van Helmont, and belongs as much to the history of alchemy as to that of mineralogy. As its full title shows, it is devoted in large part to "the discussion of the most difficult questions belonging to mystical chemistry, as of the Philosophers' gold, their Mercury, the liquor Alkahest, *aurum potabile,* and such like." The early pages of the volume are devoted to a critical bibliography of those authors that have treated of metals and minerals. It divides them into three groups: the speculative, who have produced more chaff than corn; the mystical or chymists; and experimental observers. Morhof two years later criticized Webster's book as largely taken from German authors with a few observations of his own.[125] Boyle's *Sceptical Chymist* is not mentioned, while the alchemical views of Rhumelius are set forth at some length in the eighth chapter. Eighteen chapters are then devoted to the seven metals; chapter 27, to antimony, bismuth,

[121] *Appendix de lapidibus gravidis . . .*, p. 74, citing Cardan, *De rerum varietate,* VII, 29.

[122] Treated later in our Chapter 24.

[123] *Schediasma posthumum de coeruleo et chrysocolla,* Jena, 1668, 168 pp. BN R. 12578. LR, p, 626, lists another *Schediasma curiosa* by him on fossil unicorn of 1666. For both *caeruleum* and *chrysocolla* see the Index to Pliny's Natural History.

[124] London, 1671, in-4, 388 pp.

[125] *De metallorum transmutatione ad Joelem Langelottum epistola,* 1673, pp. 11-12.

zinc and cobalt; chapter 28, to *cadmia, chrysocolla, caeruleum, aerugo,* talc, the magnet, bloodstone, schist, lapis lazuli, etc.; and the final chapter, to transmutation. A contemporary reviewer was impressed by the apparent inconsistency that, while drinking the scrapings of rusty brass is fatal to many animals, driving a brass nail into meat and especially game keeps it from decay.[126]

Dr. Thomas Shirley set out to write a medicinal work on the causes and cure of the disease, the stone, but got no farther than *A Philosophical Essay declaring the probable causes whence stones are produced in the greater world.*[127] The purpose of this First Essay was to show that not only stones but all bodies "owe their original to seeds and water." He followed Helmont in holding that everything on earth was made of water, "condensed by the power of seeds . . . with the assistance of their fermentive odors," and that stones were no exception to this rule, but had their seeds. The Essay consists largely of quotations, often of a page or more, from Helmont, Sennert, Kircher, Boyle, Willis and others. For instances of petrification he cites over a score of authorities, including the statement of Helmont (*De lithiasi*, cap. i) that there was still to be seen between Turkey and Russia in 64° north latitude a whole army which had been turned to stone in 1320.

This worthless screed was honored by a review in the *Journal des Sçavans*[128] and was promptly translated into Latin and published at Hamburg for the delectation of the German scientific world.[129]

In the same year, 1679, that he issued a longer work, of which we treat elsewhere, on Natural Science Freed from the Chicanes of the Schools, G. B. de Saint-Romain published a brief treatise of only fifty-two small pages on the marvelous effects of the divine stone.[130] This was not, however, as one might have expected, an alchemical tract on the philosophers' stone,[131] but on the employ-

[126] JS VI (1678), 303.

[127] Printed at London, 1672, in-8. Copy used: BN S. 20412. The author's name is spelled Sherley on the title page.

[128] JS III, 100-101.

[129] *Dissertatio philosophica explicans causas probabiles lapidum in macrocosmo . . .*, Hamburgi, 1675, in-8.

[130] *Discours touchant les merveilleux effets de la Pierre Divine*, Paris, 1679, in-12. Copy used: BN 8° Te97. 18.

[131] Gmelin, *Geschichte der Chemie*, II (1798), 19, listed it in his bibliography.

ment of jade as an amulet against the disease of the stone and related complaints. Its action was explained as from effluvia, which the heat of the human body caused the jade to emit and which then entered the pores of the body and dissolved the stone or gravel in the kidneys and bladder. Despite this professedly scientific explanation, the divine character and wonderful effects of this stone are ever emphasized, and one passage speaks of adding marvel to marvel.[132] More than a thousand *écus* had been spent to get such a jade. Saint-Romain would answer for the curative properties of only those which he had proved and approved, since they have to be quarried under a certain aspect of the stars which is not known to all, and he has a way of testing whether this astrological condition has been observed which is not known to the lapidaries or dealers in precious stones. Saint-Romain's treatise was dedicated to Daquin, or d'Aquin, first royal physician, and bore his approbation at its close. It enjoyed greater currency than its author's longer work, further editions appearing in 1681, 1689, 1715 and 1750, whereas *La science naturelle* seems to have had only one other edition, which, however, was in Latin and at London in 1684.[133]

Moreover, the other editions of the *Discours* were not in Saint-Romain's name. That of 1681 was issued in the name "du sieur d'Acqueville, Prieur du dit lieu," who, in an *Avis au lecteur,* speaks of having distributed the said stone since eighteen months, though he admits that the *Discours* had originally been composed by "M.D.S.R., docteur en medecine." The edition of 1689 was by Louis Candy, who, after being associated with Acqueville, had a falling out with him but finally made an agreement by which Acqueville ceded to him all his rights and privileges, and all the stones remaining in his possession. Both Acqueville and Candy reprinted "D.S.R."'s dedication to Daquin and the latter's approval and such portions of his *Discours* as the opening passage in which he said:

There is nothing so hidden and wonderful as the virtues of stones. The

[132] *Discours,* p. 38: "Mais pour ajouter merveille sur merveille . . ."

[133] *Physica sive scientia naturalis scholasticis tricis liberata,* London, 1684, in-12: BM 445.a.6.

hardness of their substance and the firmness of their composition do not allow us to penetrate to the center of their being, which encloses all their best. I have said elsewhere that stones have a life of their own, and, if the coagulating spirit had not formed barriers and closed all avenues to the dissolving spirit, we would see the effects of that life.

But while they profess to enlarge the *Discours,* they omit his last eight pages with his astrological condition. Acqueville calls the divine stone Yiade instead of jade, while Candy calls it the nephritic stone. Both alter the order of the five *Reflexions* into which Saint Romain divided his *Discours* and introduce an appeal to past authorities, where he appealed only to reason and experience.[134] But the chief innovation, aside from the negative omission of astrology, is the further omission of Saint Romain's impressive list of cures or "Experiences," which began with M. de Bourgneuf, "excellent Ingenieur du Roy," and concluded with his most serene Highness, Monseigneur le Prince,[135] and the listing of other cases of their own.[136] But both assert that "one can say that one has never seen cures so surprising."[137]

Whereas, in the edition of 1679, after Daquin's approbation came a very brief permit to print,[138] Acqueville in 1681 printed a long privilege, "with which His Majesty has honored me." In Candy's edition of 1689 the *Privilege du Roy* was still to Acqueville, but dated of March 23, 1684.[139] Candy, however, noted on his title page, "One will find at the close of this *Discours* the address of the aforesaid sieur Candy, who alone sells the said divine stone."

The reviewer in *Journal des Sçavans* of the 1689 edition thought it more likely that the effluvia from the jade dissolved the stone or gravel by setting up some fermentation than by acting on them as a file or as rust would disintegrate iron.[140]

Emanuel König followed his book on the Animal Kingdom, of

[134] Acqueville divides the *Discours* into six *Reflexions;* Candy has eight.

[135] *Discours* (1679), pp. 16-29.

[136] *Discours* (1681), pp. 27-41. "Candy, priest," at p. 30 is the second. In Candy's ed. of 1689, "Reflexion IV, Des experiences certaines de ma pierre," he gave no individual names, but at the close of the volume are a number of signed Attestations.

[137] Ed. 1681, p. 45; ed. 1689, p. 30.

[138] P. 52, "Permis d'imprimer. Fait ce 26 septembre, 1679. De la Reynie."

[139] Ed. of 1689, pp. 85-89.

[140] JS XVII, 231-32.

which we treat in a later chapter on Natural History, by one in 1687 on the Mineral Kingdom.[141] It divided into four sections on minerals in general, metals, stones, and on salts, sulphurs and earths. Some magic virtues were stated, as that the Hindus believed that the gem, cat's eye, would preserve and increase the wealth of its possessor. Coral vied in virtue with almost any stone.[142] The book was reprinted in 1703.

We now turn to works upon the magnet by members of the Company of Jesus. Although published before the *Mineralogia* of his fellow Jesuit, Cesi, of which we have already treated, the Magnetic Philosophy of Niccolò Cabeo, who also, like Cesi, taught at Parma, was probably composed later, since it was printed only in 1629, the year before Cesi's death.[143] Cabeo was born at Ferrara in 1585, and died at Genoa, whither he had gone to teach mathematics, on June 30, 1650.[144]

Cabeo's book contains little that is new but bears witness to the appeal that the subject then exerted. He was not even the first of his Order to treat it, since Leonardo Garzoni, a patrician of Venice who became a Jesuit, had left unfinished at his death in 1592, a work in Italian on magnetic nature, of which Porta had made use. Cabeo liked Gilbert's sure and accurate method of experimentation, but thought that reasons and physical causes for what he said were still desiderata. On the other hand, Garzoni's philosophical method (*philosophandi ratio*) pleased him, but Garzoni was sometimes deceived in his experiments and not so accurate as he should have been. Many of his conclusions had since been proved false and he made many omissions.[145] Cabeo also cited the thirteenth century work of Petrus Peregrinus more than once.[146] He added some further experiments to those of Gilbert, and thought that he had explained the action of the magnet without having recourse to occult virtue, ascribing it to the two faces or surfaces. "In magnetic

[141] *Regnum minerale*, Basel, 1687, in-4. BN S. 5594.

[142] *Acta eruditorum*, VI (1687), 55.

[143] Nicolaus Cabeus, S.J., *Philosophia magnetica in qua magnetis natura penitus explicatur*, Ferrara, 1629, in-fol., xiv, 424 pp.: BN R. 673. There was also a Cologne edition of the same year and number of pages.

[144] Alegambe and Zedler.

[145] *Philosophia magnetica*, preface.

[146] See especially, *ibid.*, II, 3, pp. 109-15, "Expanditur opinio P.P. de magnetica attractione."

bodies only contrary faces join together; similar faces always shun each other.[147]

Cabeo considered first what he regarded as the first effect of the magnet, that is, turning towards the poles; second, its other property of attracting magnetic bodies; third, experiments relating to polar attraction or direction; and fourth, experiments relating to attraction by the magnet. He believed that there was magnetic force in the whole terrestrial globe, but confuted Gilbert's opinion that the earth was a great magnet.[148] He held that from electrum or any body attracting electrically there flows off a most tenuous effluvium which is diffused through the air but then returns to the electric body and brings with it straws and other particles that it encounters.[149]

Cabeo, influenced no doubt by Gilbert, is sceptical as to many marvelous properties attributed to the magnet. Already in the Preface he complains that some authors labor to explain marvelous effects which are non-existent. Like Gilbert and unlike Cesi, he rejects the notion that garlic by its odor or rubbing dulls or destroys the force of the magnet. Other false beliefs which he lists are that a diamond placed between the magnet and iron prevents the former from attracting the latter; that goat's blood frees the magnet from being thus bewitched by the diamond; that married couples can be reconciled with a magnet; that gold may be extracted from the deepest wells; that the magnet acts as a love philter, makes one eloquent, and *persona grata* to princes. Cabeo also doubts if perpetual motion can be achieved by use of the magnet.[150]

With regard to the influence of the heavens Cabeo is somewhat less sceptical. He asks why in so many centuries of revolving about the earth, the heavens may not have so affected it as to produce polarity, and again asserts that the earth must receive influences and virtues from the heavens and stars.[151] Later on, after citing Ficino as to planetary influence, he adds that this opinion may seem to lean towards dreams of astrologers, but, lest he seem to condemn

[147] *Ibid.*, p. 286: "Contrariae facies in magneticis corporibus solae se mutuo coniungunt; similes semper se fugiunt."

[148] *Ibid.*, I, 17-19, pp. 57-, 66-, 72-.

[149] *Ibid.*, pp. 192-93.

[150] *Ibid.*, p. 338.

[151] *Ibid.*, pp. 61, 66.

them, he will give these dependencies from the stars in earthly things, although he does not personally approve of them.[152] We shall find Cabeo more favorable to the influence of the stars, rather than less so, seventeen years later in his commentary on the Meteorology of Aristotle.[153]

Another Jesuit to treat of the magnet was Athanasius Kircher in a brief volume of 1631 at Würzburg,[154] followed a decade later by a long book on the same subject at Rome, with a second edition at Cologne in 1643,[155] and a third and further enlarged edition, which I have used, at Rome in 1654.[156] Its full title goes on to speak of the prodigious effects of magnetic and other hidden motions of nature and to profess to disclose many hitherto unknown arcana of nature by physical, medical, chemical and mathematical experiments of every sort. The preface to the reader further represents the magnet as a priceless treasure which unlocks "all the sacraments of recondite nature," and Proteus-like assumes the form of every science. After a relatively brief first book on the magnet itself, the second book on its application, after magnetic statics, magnetic geometry, and magnetic astronomy, comes in its fourth part to magnetic natural magic, with magnetic hydromancy, onomatomancy, and *steganologia*. But these marvels are greatly overdrawn. What is labelled the revelation of one's secret thoughts to the absent by force of the magnet turns out to be simply a very thin plate inserted in the wall between two adjoining rooms with letters of the alphabet on its rim to which a pointer may be made to move by applying a magnet to the other side.[157] In the third and last book Kircher drifts off to such topics as the magnetism of earth, planets and stars, thermometers and their use, the natural and artificial production of rain, wind, and thunder and lightning, whether gold and silver attract mercury by magnetic force, the luminous Bologna

[152] *Ibid.*, p. 103. But we find in the Index, "Influentiae inaequales astrorum faciunt res sibi invicem subordinatas."

[153] See Chapter XIII, "The Cursus Philosophicus or Physicus Before Descartes."

[154] *Ars magnesia*, Herbipoli, 1631, in-4, viii, 63 pp.

[155] *Magnes sive de arte magnetica opus tripartitum*, Rome, Ludovici Grignani, 1641, in-4, 916 pp. Kircher, *Mundus subterraneus* (1665), I, 346, "Coloniae deinde in 4to eodem anno."

[156] In-fol. 618 pp. Copy used: Col 621.321 K63.

[157] *Ibid.*, p. 285.

stone, the magnetism of sun and moon on the tides, the magnetic force of plants, magnetic miracles of grafting, heliotropic plants and their magnetism, the magnetism of medicines, poisons and antidotes, the attractive force of the imagination, and the magnetism of music and of love. In this work Kircher holds that the four elements may be transmuted into one another.[158]

In 1667 Kircher published another little book on the triple magnet of nature, inanimate, vegetable and sensitive. In it he stressed how potent, wise and provident Nature was in adorning different things with their appropriate virtues; dwelt upon Nature's hidden operations, and insisted that all things in nature acted by attraction and repulsion, or sympathy and antipathy. The marvelous force of superior over inferior bodies, the wonderful concatenation of things, the marvelous virtue of the serpentine stone in extracting poison, the marvelous properties of certain Indian roots, and of such animals as the remora and torpedo, were among the subjects of his consideration.[159]

Yet a third Jesuit, Nicolò Zucchi, who also was of Parma, added to the second edition of his *New Philosophy of Machines,*[160] a Magnetic Dissertation or Promotion of Magnetic Philosophy published at Rome in 1649.[161] He had once been professor of mathematics in the Jesuit College at Rome but now dedicated his book to the duke of Parma. He disclaimed any "divinatory writing," and based his treatise upon reading since 1612 the literature on the subject (he almost never, however, cites any particular author by name) and testing it by repeated independent experimentation.[162] He denied that the magnet attracted by diffusion of magnetic corpuscles, since its force operated if copper, lead or gold was interposed, or if the magnet was immersed in water. He therefore held that magnetic virtue was the true form of the magnet's "qualificative accident."[163] But his chief purpose is to find in the magnetic

[158] *Ibid.*, p. 409.

[159] *Magneticum naturae regnum sive disceptatio physiologica de triplici in natura rerum magnete* . . . Amsterdam, 1667, in-12, 201 pp.

[160] *Nova de machinis philosophia* . . ., Rome, 1649.

[161] *Promotio philosophiae magneticae* . . ., on the title page, but *Dissertatio Magnetica* at p. 145, where the treatise begins.

[162] *Ibid.*, pp. 145, 168, etc.

[163] *Ibid.*, p. 150.

philosophy a new argument against the Copernican system. He holds that the globe of earth possesses magnetic virtue. But among the bodies which constitute the universe, the heaven, by its poles occupying a constant position as to the earth, is the prime propagator of magnetic virtue, and the earth receives it from the heavens as a whole. But the moon has no magnetic correspondence to sun, earth or sea, and there is no magnetic relation between the sun and planets which may be used to argue that they move about it rather than about the earth.[164] Also the gravity of the earth is an argument against its rotation.[165] Zucchi ends his treatise, however, with assertion of modern progress:

> So much concerning magnetic virtue which, shrouded in darkness and error in preceding centuries, our time has brought to such light of truth that it occupies with dignity a place of its own in philosophy.

The attraction of the magnet for members of the Company of Jesus is seen in yet another book by Father Vincent Léotaud (1595–1672) of Dauphiné. Despite the previous works, he professed to set forth a new magnetic philosophy.[166]

That the magnet made a religious appeal to English Protestants, even at the close of the century, is seen from Sir Matthew Hale's *Magnetismus magnus; or Metaphysical and divine contemplations on the magnet or loadstone,*[167] which was chiefly concerned with conversion of the soul to God. It noted evidences of God from "the parts of the universe and particularly of the magnetic parts," likewise the "wonderful wisdom and power of God appearing in the admirable and various motions of the magnet," as well as the "reasonableness of the Christian religion."[168]

[164] *Ibid.*, pp. 191, 201, 221, 223, 226.

[165] *Ibid.*, p. 218.

[166] *Magnetologia, in qua exponitur nova de magneticis philosophia,* Lyons, 1668, in-4. BN V. 6258; BM 717.g.18. Father Léotaud was professor of mathematics at Dôle for fourteen years, then taught at Lyon and Embrun.

[167] London, 1695, in-8, 4 fols., 159 pp.

[168] *Op. cit.*, pp. 37, 52, 120.

CHAPTER X

NATURAL MAGIC[1]

Natural magic defined—Mixed with theosophy by Khunrath and Riviera—Morestel—Gerosa—Hildebrand—Evenius and Artocophinus—Kornmann on miracles of this and that—Zara—Goclenius—Wenckh—Baranzani—Castiglione—Bulenger—Kirchmann—Campanella: all nature sentient—Suarez and the remora—Campanella continued; his medicine—Naudé's defense of great men falsely accused of magic—Theobald, Cassander and Mornius—Gaffarel's book and retraction—*Ars magica sive magia naturalis*—Jonston—Alexander de Vicentinis—Marcus Marci—Kozak—Bellwood—Martini—Borrichius—Madeira—Two manuscripts—Williams.

It is nothing but recondite knowledge of things, by which, if agents are applied to patients, marvels are produced and effects beyond popular comprehension

—PASCHIUS

Natural magic is the working of marvelous effects, which may seem preternatural, by a knowledge of occult forces in nature without resort to supernatural assistance. It was therefore regarded, unless employed for evil purposes, as permissible, whereas diabolical magic, worked by demon aid, was illicit. Natural magic was also distinguished from natural science, as being more mysterious and less explicable in universal, regular and mathematical terms. Indeed, since demons were often thought to work their magic simply by superior insight into the secrets of nature based on long experience, the connection between natural and diabolical magic was somewhat closer than that between natural magic and classified and generally accepted natural science.

As organized and systematic human scientific experimentation and research on a large scale gradually overhauled the superior but

[1] Chapter 43, "Natural Philosophy and Natural Magic," in vol. VI, included three works of the early seventeenth century which will not be reconsidered here: namely, Hippolitus Obicius, *Dialogus Tripartitus*, 1605; Pietro Passi, *Della magic'arte overo della magia naturale*, 1614; and the Encyclopedia of Alsted. See T VI, 429-36.

undisciplined intelligence and the long but empirical experience of the demons, the need and urge to avail oneself of their assistance has kept diminishing. On the other hand, the frontiers of natural science have been gradually extended over that wild borderland, which was once the domain of natural magic. Forests of occult virtues have been cleared; swamps of erroneous notions have been reclaimed; the old savages that inhabited them have been civilized, and the imaginary gnomes, satyrs and specters that once haunted them have ceased to exist.

The Amphitheater of the Only True Eternal Wisdom, Christian Cabalistic, Divine Magic, also Physical Chemical . . .[2] of Henry Khunrath (1560–1605) perhaps deserves to be classed with works of unqualified magic rather than those of natural magic. It is written in a ranting tone of turgid rhetoric with much theosophic pretense and religious patter. At one point the author declares that all sciences not acquired from God or divine magic or the Christian cabala by prayers and tears are furtive waters and not perfect gifts from the Father of lights. The *Amphitheatrum* is not even good experimental or systematic and detailed magic. Khunrath lists as handmaids of true wisdom the Cabala (not, however, that literal and vulgar Jewish variety, but far superior theosophy), magic, physiognomy, metoposcopy, chiromancy, the doctrine of signatures of all natural things, alchemy, astrology and geomancy.[3] But he never advances far beyond the general attitude which is inferable from his title, which he repeats in virtually the same words again and again, along with a few other pet phrases such as *oratorio et laboratorio* or macro- and micro-cosmically. However, he lauds *Physico-Chemia* also,[4] and the very fact that these words were included in the title of his theosophical ecstacies and cabalistic reveries is a rather noteworthy sign that physics and chemistry were coming into their own

[2] Henricus Khunrath, *Amphitheatrum sapientiae aeternae solius verae, Christiano-Kabalisticum, Divino-Magicum necnon Physico-Chymicum, Tertriunum, Catholicon,* Hanoviae excudebat Guilielmus Antonius MDCIX, 222 pp. Copy used: BN R.964.

[3] *Ibid.*, p. 91.

[4] "Mediante enim Physico-Chemia (quae Sapientiae verae in hoc seculo pedissequa est fidelis et virgo quasi cubicularis a secretis naturalibus) filio disciplinae industrio Domine benigne largitur vegetabilium animalium partiumque eorundem mineralium lapidum gemmarum margaritarum et metallorum essentias praetiosas subtilitatesque salutariter efficacissimas."

in the thought of the time—even in the muddiest and most stagnant and most occult thought.

A brief preliminary sketch or draught of the *Amphitheatrum* had been privately printed in 1595.[5] Also Khunrath had published in 1599 a work entitled *Symbolum physico-chymicum* and an alchemical *Magnesia catholica philosophorum.*[6] He was a disciple of Paracelsus whom he often cites. He tells us that Paracelsus recalled the doctrine of signatures from the shades of oblivion most fruitfully into the light, and that Porta expounded it in his *Phytognomonica* of 1588. Before, however, this last work reached Germany, Khunrath, "first of all after Paracelsus," defended twenty-eight theses on the subject for the M.D. degree at Basel on August 24, 1588, and will continue to defend the doctrine of signatures "as long as I shall live."[7] Khunrath died in 1605; the *Amphitheatrum* is dated at its close, 1602, but was published posthumously in 1608–1609. There was another Latin edition in 1654[8] and a French translation for those interested in occult literature as late as 1900.[9]

Khunrath affirms that the book of nature and theosophy are superior to all past authorities. Even if all existing books were lost, the sciences and arts could be restored by this theosophic method.[10] He adjures the reader not to be the slave or ape of another's opinion and declares that "great men commit great errors."[11] Experience alone is the sufficient mistress of all things, whom it would be worse than stupid to resist.[12] Physico-chemical analysis can be of great aid in such problems as determining the virtues of roots.[13] Gesner is cited as to lunar herbs and secret remedies.[14] But an alliance of alchemy and the cabala is also urged, and seeking the secret of the philosophers' stone not from the writings of gentile philosophers but from Holy Writ.[15] An analogy is pointed out between Christ cruci-

[5] *Amphitheatrum sapientiae aeternae solius verae Cabalae Mageiae Alchemiae cabalisticum mageicum physicochemicum tertriunum Catholicon . . .* (Hamburg), 1595, in-fol., 24 pp. Listed by Maggs, *Catalogue of Strange Books and Curious Titles,* 1932, item 91.

[6] *Lindenius Renovatus,* 1686, p. 396.

[7] *Amphitheatrum,* p. 152.

[8] *Lindenius Renovatus,* p. 396.

[9] Copy used: BN 4° Z.1245.

[10] *Amphitheatrum,* p. 154.

[11] *Ibid.,* p. 171.

[12] *Ibid.,* p. 191.

[13] *Ibid.,* p. 152.

[14] *Ibid.,* p. 129.

[15] *Ibid.,* pp. 73, 75, 89.

fied and the philosophers' stone.[16] Khunrath also speaks of good men as those who strive with all their might under the guidance of the wisdom of Jehovah for the simplicity of the Monad.[17]

The title of Khunrath's book would seem to have been copied by Vanini in his *Amphitheatrum aeternae Providentiae divino-magicum christiano-physicum necnon astrologo-catholicum*, published at Lyons in 1615,[18] and perhaps this called the attention of the censorious to the earlier work. At any rate, it was not until February 1, 1625, that the book of Khunrath was condemned by the Sorbonne as "full of impieties, errors and heresies, and a continual sacrilegious profanation of passages of Holy Scripture." To this was added that it abused the holiest mysteries of the Catholic Religion, and led its readers into secret and criminal arts.[19]

Riviera, in a book written in Italian and printed in 1603 with the title, *The Magic World of the Heroes*,[20] quoted with approval Trithemius who, writing to Joachim, marquis of Brandenburg, said that natural magic not only worked visible effects but marvelously illuminated the intellect in knowledge of God and supplied the soul with invisible fruits.[21] With Ficino, too, Riviera accepted the existence of a world soul diffused through the elements and all its corporeal parts.[22] He also speaks of a mechanical magic with a superior sun and moon and another inferior sun and moon. The superior luminaries are nothing else than most lucid water and the spirit of the world soul with Mercury added,[23] while in mystic theology the water denotes the eternal Word.[24] Like Khunrath, Riviera alludes to the Monad. According to him, St. Thomas says, "Monad begets monad and reflects its ardor on itself."[25] He further mentions Hermes, Orpheus and the cabalists repeatedly. "Our magic heaven" or "the heroic heaven" is called the fifth essence.[26] With Riviera, there-

[16] *Ibid.*, p. 213.

[17] *Ibid.*, p. 172.

[18] For Vanini see T VI, 568-73.

[19] Argentré, II, ii, 162. The date is incorrectly given as March 1 in *Correspondance du P. Marin Mersenne*, II (1936), 139.

[20] Cesare della Riviera, *Il magico mondo de gli heroi*, Mantova, Osanna, 1603, in-4, (19), 217 pp. Copy used: Columbia University, 156.4 R526.

[21] *Ibid.*, p. 112.

[22] *Ibid.*, p. 50.

[23] *Ibid.*, pp. 44-45.

[24] *Ibid.*, p. 47.

[25] *Ibid.*, p. 108. For mentions of the monad in the sixteenth century see T VI, 396, 450, 457.

[26] *Il magico mondo*, pp. 48-50.

fore, as with Khunrath, natural magic dissolves into theosophy and alchemical mysticism. His book was reprinted in 1605, with a longer title which expressly mentioned natural magic and the philosophers' stone.[27] Riviera gives an impression of wide reading, if not deep learning, by citation of various past writers in the margins of his book.[28] But as we proceed, the same names are repeated or the margins are left blank. Sometimes groups rather than individual authors are cited, such as Cabalists, Indians, ancient Magi, ancient poets, ancient heroes.

Two little books by Pierre Morestel, which first appeared in 1607, might seem from the opening words of their titles, Secrets of Nature[29] and Occult Philosophy,[30] to be concerned with natural magic. But, as their full titles suggest and as a glance at their content reveals, they are popular and elementary works, giving in dialogue form a natural explanation of Greek myths and some account of the moral philosophy of the predecessors of Plato and Aristotle. Their sole importance for our investigation is to show that these expressions, Secrets of Nature and Occult Philosophy, were still attractive captions to catch the eye of the reading public, and words to conjure with. Later, in 1621, Morestel published *Artes Kabbalisticae*,[31] and, in 1646, a book on the Lullian art.[32]

[27] *Il Mondo Magico de gli Heroi... nel quale... si tratta qual sia la vera Magica Naturale e come si possa fabricare la reale Pietra de' Filosofi... Ristampato e del... autore ricorretto et accresciuto,* Milano, 1605, in-4, 212 pp. Copies: BM 8630.g.26; BN Z.2934. I have not examined this edition but doubt if it was much enlarged, since the format and number of pages are about the same as in the first edition.

[28] For example, at p. 4, "Davide, Hermete"; p. 5, "Alberto magno, Giovanni Pico, Pitagora Samio"; p. 6, "Dionysio Areopagita, Ariosto, Tacito."

[29] *Les secrets de nature ou la pierre de touche des poetes,* Rouen, 1607: BN J.25112. In the form of questions by Courtisan and answers by Orpheus. Reprinted in 1632: BM 8707.aaa.2. Noted by Hoefer, II, 331.

[30] *La philosophie occulte des devanciers d'Aristote et de Platon en forme de dialogue contenant presque tous les préceptes de la philosophie morale extraite des fables anciennes,* Paris, 1607, in-12, 188 fols. BN V.21888. Reprinted at Bourg-en-Bresse, 1629.

[31] Paris, 1621, in-8: BN A.7729; BM 719.c.25.(3.).

[32] Petrus Morestellus, *Encyclopaedia seu artificiosa ratio circularis ad artem magnam Raimundi Lullii,* in Collegio Salicetano, 1646, in-8: BM 717.e.20; BN Z.19006.

Francesco Gerosa, a physician of Milan,[33] in 1608 issued a dialogue in Italian with quotations in Latin on Magic which Transforms Man to a Better State, with further mention in the title of natural magic and medicinal chemistry.[34] Natural magic is necessary for man in order to know things, and medicinal chemistry in order to prepare them.[35] Gerosa asserts the existence of three worlds—intellectual, celestial and elemental, and the rule of superior over inferior bodies. He has no doubt that God created the world in the springtime. Great secrets are mentioned, and among the topics discussed by the participants in the dialogue are making the philosophers' stone, whether life can be prolonged, and natural remedies against demons and witchcraft.[36] Much attention is given to quintessences. That for melancholy has 95 ingredients; a soporific, only twenty; but that for pest and poison fills three or four pages.[37] There are also quintessences for purging and for the memory.[38]

In 1610 appeared the first edition of a work in German by Wolfgang Hildebrand on natural magic.[39] It contains recipes for coloring the hair, improving the memory, making a man merry or melancholy. To see by night one rubs one's eyes with the blood of a bat, a prescription taken from the *De mirabilibus mundi* current under the name of Albertus Magnus. Other secrets are to see marvels in one's dreams, not to get intoxicated quickly, to make men seem headless or with the heads of animals—an old favorite of medieval manuals of marvelous experiments, and to sleep for three days at a stretch. How to detect magical butter is taken from Luther's Table Talk. There are chapters on horses, wolves, bees, flies, crabs; a whole book on vegetables and plants. In the fourth and last book on artificial magic are instructions how to make steel pliant, as to secret writing,

[33] Corte, *Notizie storiche . . .*, 1718, p. 169.

[34] *La Magia trasformatrice dell' Huomo a miglior stato: Dialogo di F. G. fisico de Lecco nel quale si ragiona del trino mondo, della felicità humano, natural magia e medicinal chimia.* In Bergamo per Comino Venture, 1608, in-8, viii, 96 pp. BN R.37167.

[35] *Ibid.*, p. 26.

[36] *Ibid.*, pp. 5, 22, 13, 41, 57, 84.

[37] *Ibid.*, pp. 33-38, 39, 76-79.

[38] *Ibid.*, pp. 56, 73.

[39] *Magia naturalis, das ist, Kunst und Wunderbuch, darinnen begriffen Wunderbare Secreta, Geheimnüsse und Kunststücke . . .*, Leipzig, 1610, in-8, 554 pp. BM 1034.c.5 (1.).

and chapters on such themes as glass, fireworks, waters, artificial gems, and an artificial flying dragon.

There were further editions of Hildebrand's book at Erfurt in 1611–1612 and 1618, and at Jena in 1625.[40] Meantime he published an astrological work at Erfurt in 1613.[41] Later he issued an astrological prediction for the years 1627–1638.[42]

A Physical Dissertation concerning Magic by Sigmund Evenius is dated in 1612,[43] and an Introduction to the Most Mysterious Mysteries of Nature, ascribed to Henry Artocophinus, was printed at Stettin in 1620.[44]

Meanwhile, in the years between 1610 and 1614, Heinrich Kornmann, a German lawyer who died in 1620, published works on the miracles of the elements (1611), the miracles of the dead (1610), the miracles of the living (1614), and on virginity (1610), which resemble one another in several ways and especially in their emphasis upon natural marvels and their inclusion of superstition and magic. That Kornmann's scientific standards are neither high nor recent may be seen from his citing Pliny, Isidore of Seville and Bartholomaeus Anglicus that the salamander is so cold that it extinguishes fire, and Bartholomew alone that a scorpion does not injure other women as quickly as it does a virgin.[45] But it is not necessary for him to go so far back for the unscientific or magical. He cites Scaliger that sitting on a certain root breaks the maidenhead.[46] That casting the chemise of a virgin having her first menstrual flux into the fire puts it out, he calls, however, a superstition.[47] But he repeats that a tree is injured by a virgin's plucking its first fruit, inquires

40 See BM catalogue.

41 *Ein new ausserlesen Planeten-Buch:* BM 8610.ee.3. BM 7954.b.3 is a later edition at Frankfurt, 1690.

42 *Zehen Jährig Prognosticon und astrologische nützliche Practica von . . . 1627 . . . biss . . . 1638,* 1628. BM 8610.bb.49 (3.).

43 *Dissertatio physica de magia.* Not in the BM and BN catalogues, and I do not know where it was printed. The BM has several dissertations at which Evenius was *Praeses.* He died in 1639.

44 Heinrich Brodkorb, *Prodromus mysteriorum naturae mysteriosissimorum, emissus . . . ab Henrico Artocophino . . . ,* Typis Samuelis Kelneri, Stetini, 1620, in-4. BM 1033.h.3 (1.2.); BN R.6896. I have not examined it.

45 *De virginitate,* 1610, in *Opera Curiosa,* edition of Frankfurt, 1694 (Col B 156.4 K843), pp. 139-40, 274.

46 *Ibid.,* cap. 22.

47 *Ibid.,* p. 137.

why magicians require virgins and chaste boys in their feats, and says that to dream of virgins portends all joy and delights.[48]

The problem of incubi and succubi is discussed in both *De virginitate*[49] and *De miraculis vivorum*,[50] and astrological considerations are introduced in both. The one asks what regions and cities and what members of the human body are under the sign Virgo, and what sort of boy or girl is born under it, quoting Pontano.[51] The other takes up the seventh month birth, the rule of the planets over a month each, and John of Legnano, who foresaw from his son's geniture that he would be hanged, failing to prevent it by educating him for the clergy.[52] One cites the unfamiliar name of Landler on fascination and incantation; the other, the equally unfamiliar Leonhardus Nairus.[53] But these are almost certainly misprints for Tandler and Vairus.

Monsters naturally bulk large in the book on living miracles. Beginning with a general discussion of the microcosm and passing on to giants, pygmies, and hermaphrodites, we soon come to headless men, men with dogs' heads, men with one eye, those who see by night, the tribe having eyes in the breast, the tribe living on odor alone, the race without a mouth, bearded ladies, martyrs who spoke without tongues, huge eaters and drinkers, record-breaking fasters. Aristotle, we are told, testifies that he had seen a man who lived on air and sunlight alone.[54] There follow accounts of men with one hand and one foot, individuals with three testicles, quintuplets and septuplets, and the 364 children at one birth of Margaretha, wife of Hermann, count of Henneberg.[55]

Magical bits from the Miracles of the Dead are that the owl is a fatal omen and the peacock a presage of disease, that suffumigation with the tooth of a dead man expels witchcraft and impotency, that the herb betony protects cemeteries, and that if a mother kisses her

[48] *Ibid.*, caps. 69, 91 (p. 198), and 117 (p. 255).

[49] Cap. 35, p. 103 *et seq.*

[50] *Op. cit.*, ed. 1694, p. 168 *et seq.*

[51] *De virginitate*, caps. 106-7-8.

[52] *De miraculis vivorum*, pp. 145 *et seq.*, 161-62.

[53] *De virginitate*, p. 106; *De miraculis vivorum*, p. 68, "in lib.i de fascino." Fascination is discussed at p. 201.

[54] *De miraculis vivorum*, p. 113.

[55] *Ibid.*, pp. 117, 124, 139. In T VI, 401, the story is told of Margarita, countess of Holland.

dead child, the other children will soon die too.[56] Astrology enters again in the question why thousands of persons with different horoscopes die on the same day in the same battle, and divination in the question what dreams about the dead signify, the discussion of presages of death, and the prophecies of those about to die.[57]

The problem is once more argued whether the witch of Endor really resuscitated Samuel. Joan of Arc's heart was unburned at the stake. Cases are listed of the teeth of corpses growing and a dead woman impregnated. A corpse is heavier than the living body because it is without the levitation of the vital spirits and heat. The size and weight of resurrected bodies is discussed, how men who have been eaten and the cannibals who ate them can both be resurrected in the body, whether abortions will rise again, and whether monsters will be resurrected. The corpse bleeding before the murderer is treated,[58] and if inextinguishable and ever-burning sepulchral lamps are not, they are about the only thing connected with funerals and burials which is omitted.

The work on the miracles of the four elements has the alternative title, Historical Temple of Nature, and is largely drawn from antiquated authors. Not only are the elements still four, but comets are still exhalations in the supreme region of air. Tides, however, are attributed to the moon. According to the tradition of the Magi, there are three kinds of spiritual beings: supercelestial who are very close to God, celestial intelligences for the spheres and stars, and demons for each of the four elements. There are also fiery men and aerial men, and four animals who feed on a single element: the mole on earth, the *alec* on water, the chameleon on air, and the salamander nourished on fire. A type of divination is also listed for each element, but in the case of earth there are eleven others besides geomancy. Besides miracles of each element, there are alphabetical treatments of birds, quadrupeds, mountains, bodies of water, forests, gardens, trees, herbs, flowers, fruits, cities, temples, towers, bridges, and so on, passing from the realm of nature to that of art. But they never amount to much and sometimes are very scanty, as when under

[56] *De miraculis mortuorum*, ed. Francof., 1694, pp. 179, 181, 209, 249, 313.

[57] *Ibid.*, pp. 336, 185, 187, 148.

[58] *Ibid.*, pp. 21-23, 87, 92, 98, 227, 376-77, 382, 392, 425.

aquatic animals only five are given: the sea lion, whale, dolphin, fish and frogs. An example or two will suffice to show the mixture of magic with natural history. Meeting a wolf is a good omen, as was shown in the case of Hiero of Sicily. Under the caption of "A Marvel," we are told that in Crete there are no wolves or foxes or other harmful quadrupeds. The dog has a marvelous sense of nature; a tick taken from the left ear of a dog that is entirely black is a potent remedy; a stone bitten by a mad dog, if put in drink, promotes discord. A dog won't bark at a person who has a dog's tongue in his sock, especially if he is also anointed with the herb, *cynoglossa,* or if he carries a dog's head.[59]

The Anatomy of Talents and Sciences by Antonio Zara, bishop of Biben in Istria, is in four sections divided into numerous *membra.*[60] Of eighteen making up the first section on the dignity and preeminence of man, the eleventh is on dreams, with paragraphs on sleep-walking and "talents from intoxication," the twelfth and thirteenth on chiromancy and physiognomy, and the seventeenth on the influence of the stars.[61] In the second section on imaginative sciences, the second *membrum* is on magic arts, the seventh on mystic arithmetic, and the thirteenth on astrology and astronomy.[62] In connection with chiromancy, after listing the lines of the hand, "mountains," triangle, quadrangle and *rascetta,* Zara continues:

> But the remaining insane nonsense of the chiromancers about mountains of the planets, girdle of Venus, Milky Way, solar line, triangle of Mars, are utterly vain, fallacious, superstitious and deservedly condemned.[63]

In other words, he makes a distinction between plain chiromancy and astrological chiromancy. But he accepts the influence of the celestial bodies upon inferiors, except that it is false to say that the heaven of the fixed stars peculiarly prepares matter to receive the vegetative soul; the crystalline heaven, for the sentient soul; and the empyrean, for the rational soul. But influence upon the body is

[59] *Templum naturae historicum in quo de natura et miraculis quatuor elementorum . . . disseritur,* ed. 1694 (Col B 156.4 K8433), pp. 45, 128, 47-50, 168, 115, 182-83.

[60] *Anatomia ingeniorum et scientiarum . . . ,* Venice, 1615, in-4, 592 pp. BM 527.i.15.

[61] *Ibid.,* pp. 92-95, 96-98, 99-103, and 116-22.

[62] *Ibid.,* pp. 154-92, 223-33, 261-76.

[63] *Ibid.,* p. 98.

specified for each planet, variations as they are in apogee or perigee, five states of the superior planets, the terrestrial regions which each planet is supposed to dominate, and the properties of the twelve signs of the zodiac. Later on he inquires whether the empyrean is square, whether the heavens are animated, but ridicules Avicenna's series of celestial intelligences, and rejects those of Trithemius with their successive rule in history as contrary to freedom of the will. He tells us that Alexander Farra in his *Septenarium*[64] made the Muses the souls of the celestial spheres. Zara himself counts eleven spheres, the primum mobile being the eleventh and last.[65]

Both sacred and profane history demonstrate the existence of magic. One variety is natural and true, based on the occult natures and virtues of things, an intimate part of philosophy, free from all superstition and pure from the artifices and malefices of demons. It divides into physics, mathematics and *praestigia*. Under it belong marvelous fountains and the remora, Archimedes and automata, alchemy. The other variety of magic is diabolical, in connection with which Zara ascribes wide powers to demons. They cannot upset the world order nor fill places which are wide apart without occupying the medium at the same time, nor so alter the quantity of bodies that one is in several places or that several occupy the same place. They cannot transform species, generate human beings from man and brute, or produce the souls of the dead. But they can produce winds, storms, and darkness, stop rivers for a time, draw water from rocks, injure crops and enchant animals. They can generate worms, flies, frogs, snakes and monsters. They can assume bodies. They can alter sex, for that can be done naturally. They can rejuvenate, prevent one's feeling pain, produce dreams with drugs, enable one to fast a long time, and induce a state of ecstacy and seeing visions. They make love philters work. Zara lists various 'mancies alphabetically, with cabala, divination, goetia, theurgia, oracles and lots interspersed between them.[66] One of the least familiar is Tephramancy or Spodonomancy (*sic*) from the ashes of sacrifices.[67]

[64] *Settenario dell'humana riduttiore*, Vinegia, 1571, in-8. BN Z.30990.

[65] *Anatomia . . .*, pp. 263-71.

[66] *Ibid.*, pp. 154-92.

[67] *Ibid.*, p. 191.

Rudolf Göckel or Goclenius the Younger (1572–1621), whose astrology and alchemy have been treated in previous chapters, in 1608 published a treatise on the magnetic cure of wounds by weapon ointment.[68] It was stigmatized by a Belgian Jesuit, Jean Roberti (1569–1651), in a work of 1615,[69] as being necromantic and *idolo-magico-goeticum*, as confusing natural magic with the superstitious, necromantic and diabolical variety, and as full of *praestigiae*, idolatry, blasphemy and divination.[70] Goclenius denied these charges, and, in order to show that he knew what he was talking about, listed twenty-four performances of *magi* which cannot be referred to natural causes,[71] and forty-five kinds of evil magic from *necyomantia* through *alectryomantia* to fascination.[72] He maintained, however, that all bodies in the world are connected, that the virtue of the world soul is spread through all things by the spirit of the universe, and that this was the explanation of such phenomena as signatures in plants and the seemingly marvelous action of weapon ointment.[73]

Goclenius also wrote a book on the marvels of nature and concords and repugnances in plants and animals which was published posthumously in 1625[74] and again in 1643.[75] As late as 1700 it was cited by Martius for the suspension of a live scarab sewn up in yellow linen cloth from the neck as an amulet, or a ligature of the

[68] *Tractatus de magnetica curatione vulnerum . . .*, Marburg, 1608, in-8; 1609, in-12; Francof., 1613, in-12. On weapon ointment in the sixteenth century: T VI, 239-40, 416, 420, 525, 534, 536, 602.

[69] *Dissertatio theologica de superstitione . . . Inseritur magici libelli de magnetica vulnerum curatione authore D. Rodolpho Goclenio . . . brevis anatome*, Trier, 1615, in-12, 72 pp.

[70] Goclenius, *Synarthrosis Magnetica opposita infaustae Anatomiae Joh. Roberti D. Theologi et Jesuitae pro defensione tractatus de magnetica vulnerum curatione*, Marburg, 1617. Copy used: BN V.20949. See pp. 20, 129.

Roberti replied in *Goclenius heautontimorumenos, id est curationis magneticae et unguenti armarii ruina . . .*, Luxemburg, 1618; and Goclenius replied to him in turn in *Morosophia Ioannis Roberti . . .*, in 1619, but we shall not pursue their controversy farther.

[71] *Synarthrosis*, pp. 35-50.

[72] *Ibid.*, pp. 55-82.

[73] *Ibid.*, pp. 193-95.

[74] *Mirabilium naturae liber concordias et repugnantias rerum in plantis animalibus animaliumquo morbis ot partibus manifestans, nunc primo in lucem datus*, Francofurti, 1625, 8vo, 303 pp. BM 1034.e.11; BN 8° Te56. 25.

[75] Same place and number of pages: BN 8°Te56.25A.

nails cut off from the claws of a live crab which was to be thrown back into the water.[76] The work is not well arranged, which may be because its author died before having really completed it. Much of it seems to be notes taken from Pliny's Natural History, as was the case in at least four out of six passages which I selected at random.[77] The book begins with the sympathies and antipathies of beasts. The stomach of a ram cooked in water or wine and given as a drink to sheep "is said not undeservedly" to cure many of their diseases. A ram's horn is turned into asparagus, if buried until it rots, "and this is confirmed by the testimony of many." The same horn, buried near a fig-tree, helps the figs to ripen rapidly.[78] Among later topics are: forces of nature which operate occultly through animals and their parts; medicines from women; lion, elephant, lynx, hyena, crocodile, chameleon, the *scincus* or terrestrial crocodile (see Pliny, NH, VIII, 34; XXVIII, 30); complaints of the tonsils and ulcerated artery of the neck; diseases of the chest; medicines from wool, eggs, dogs, etc. The volume closes with a New Defense of the Magnetic Cure of Wounds, based largely on occult virtue, sympathy and antipathy, and a spirit of the world, with quotation of the *locus classicus* from Augustine:

> There are in corporeal things, through all the elements of the world, seed-beds and occult motives by which, when the opportune moment and cause comes, they burst forth in species and effects befitting their ways and ends.[79]

[76] Joh. N. Martius, *Dissertatio de magia naturali eiusque usu medico ad magice et magica curandum*, Erfurt, 1700, p. 31.

[77] *Mirabilium naturae liber*, pp. 9, 18, 27, 36, 45 and 54. The passage at p. 18 on the juice of wild cucumber corresponds in part to Pliny, NH XX,2. At 27, on wild mint and elephantiasis and the time of Pompey, NH XX,52 is copied verbally. Nicander, Chrysermus, Sophocles, Xenocrates and Hippocrates are cited at 36-37, as in NH XXII,32. At 45, on the Vettones discovering Vettonica etc., is from NH XXV,46. The passage at 54 on *phthiriasis* may be from Pliny, who mentions this disease a number of times, but I did not find it.

[78] *Ibid.*, p. 9; the passage referred to in the preceding note. That asparagus grows from ram's horn we shall also hear from Catelan.

[79] *Ibid.*, p. 179: "Hinc praeclare Augustinus in quaest. super Exod., c. 21 ait, 'Corporeis rebus insunt per omnia elementa mundi quaedam seminariae et occultae rationes quibus cum data fuerit opportunitas temporalis atque causalis prorumpunt in species et effectus debitos suis modis et finibus.'"

Goclenius further defends cures by use of characters and natural seals, and gives images of Chael, Thetel and Hermes,[80] although in 1617 he had denied faith in them.

Rather more notable than the foregoing compilation is the dedicatory epistle or preface which Goclenius's son, Theodorus Christophorus, prefixed to it. He quotes the distinguishing of five kinds of sympathy and antipathy from one of Melanchthon's Declamations: (1) between external objects and the humors and members of the human body; (2) between plants; (3) the marvelous sense in animals seeking antidotes; (4) between animals; (5) the medical action of minerals, as coral benefits the heart, and the emerald is good for epilepsy. But then he repeats Scaliger's wish that those who assert that man loses his voice at the sight of a wolf might be castigated with as many ferules as he has seen wolves without loss of voice, and his assertion that he had seen vipers born without injury to their dam. Bodin did not dare to affirm that an ostrich can digest iron, and Theodore Christopher will not confirm it. He is more inclined to accept what some have called into question, that the feathers of other birds are consumed by those of the eagle. He is not yet persuaded that drinking potable gold will lengthen life, because that effect seems to him to surpass the powers of nature, and he finds it difficult to explain why bitter almonds prevent intoxication. He accepts as a fact that the corpse of the victim bleeds afresh at the approach of the murderer, but he ascribes this to divine providence and not to natural causes, such as a trace of the sensitive soul remaining in the corpse or action of the imagination of the slayer. However, despite such doubts and partial scepticism, he publishes the posthumous and credulous work of his father.

The treatise of Goclenius on weapon ointment continued to excite repercussions among the Jesuits after his death. In 1626 the Notes of Gaspar Wenckh (1589–1634) on the magnetic unguent[81] raised such questions as to what philosophy this cure belonged, how that magnetic virtue was propagated to a distance, whether spirit from the stars could induce such sympathy and bring healing power to

[80] *Ibid.*, pp. 228-, 247-, 254-, 258-.

[81] *Notae unguenti magnetici et eiusdem actionis . . . contra R. Goclenium*, Dilingae, 1626, in-8, 89 pp. BM 1033.f.39.

the wounded part without transfer of accident from subject to subject, whether astronomy by phantasy and strong imagination could affect and transmute a foreign body, whether a soul infused under the dominion of a lucky star acquired the operative force of that star, and whether the whole universe was an animal? Wenckh further inquired as to signatures, characters and natural seals. On such matters his attitude was almost invariably unfavorable.

We have seen that the *Uranoscopia* of Father Baranzani of Vercelli, which posed in 1617 as a new work, contained more old astrology than anything else. Similarly his New Opinions in Physics, issued two years later,[82] is chiefly remarkable for its stress upon natural magic. In a *Prooemiolum* the author proposes to diagnose the causes of the many marvelous effects which nature daily produces, and whence so many noble arts, such as physiognomy, metaposcopia (*sic*), and natural magic, pullulate. The first part of the text is an Introduction to Physics, divided into nineteen *Digladiationes*, at least a novel word for a section of a work, and each of these into theorems. We are soon told in one of these that *Physiologia* divides into *auscultatoria*, cosmic, elementic, uranoscopic, meteorologic, animastic, *plantilogia, animalogia, humanilogia, mixtilogia, lapidilogia, gemmilogia*, and finally *secretilogia.*[83] For a while we proceed in the usual Aristotelian order of the *cursus philosophicus:*[84] matter, form and privation, causes, motion, place and vacuum, duration and quantity, but then branch off to creation, the coefficience of God with creatures, and the work of six days.

But the last *Digladiatio*, with no fewer than thirty theorems under it, is concerned with prodigious actions of natural causes and those especially which can be wrought through natural forces by demons and *arioli.*[85] Magicians can produce real marvelous effects, but there are limits to the powers of demons. It is fabulous and ridiculous to think that they generate living beings, and those persons should be excommunicated who believe that they can transform men into beasts. Satan cannot even change sex or infringe the laws

[82] *Novae opiniones physicae seu Tomus primus secundae partis Summae philosophicae annec. et Physica auscultatoria octo Physicorum libris explanandis accomodata*, 2 parts in one vol., Lyon, 1619, in-8. Copy used: BN R.27525-27526.

[83] *Ibid.*, p. 15.

[84] See our chapter 13 on it.

[85] *Novae opiniones*, I, 178-206.

of quantity. But witches are truly transferred from place to place in their nocturnal conventicles, hear sounds, lead dances, eat and have sexual intercourse with incubi demons. And the devil can deceive the eye so that one thing is mistaken for another. Men can live a long time without food, but it is stupid to hold that a man's *complexio* causes certain prodigious actions, and impious to explain miracles in this way. The imagination has incredible virtue on one's own body, but none on another's. Astrological and magic images are *per se* ineffective.

After indices for the first part, the second opens with a new title page[86] and pagination. Again we are promised "many new views" and "the foundations of a reviving *Physiologia*." Disputations and *Dubitationes* now replace the *Digladiationes* of the Introduction and are primarily in the nature of a commentary on the eight books of Aristotle's Physics. But *Dubitatio 4* is whether, assuming the possibility of making gold, the chemical art is licit and can produce such gold.[87] And as the Introduction terminated with a discussion of natural and diabolical magic, so the ninth and last Disputation of the second part devotes seventy-five pages to the subject of occult qualities,[88] though Baranzani remarks that no philosopher ordinarily treats of it, and some ridicule occult qualities as the refuge of ignorance. Some of them are from the heavens, some from the mixture of first qualities—but others precede first qualities, some emanate from the specific form, some from the material temperament. Some are formal, others virtual. Natural magic is the science of so applying occult qualities as to produce effects surpassing the ordinary workings of nature. The force of antipathy and sympathy is greater than is thought, for several effects are due to it which most philosophers attribute to other causes. Soon we hear of the remora, the drums of wolfskin and sheepskin, the corpse bleeding at the approach of the murderer. Fascination may operate by malign humors from the eye, by use of flattery, which opens up the pores for such effluvia to enter, and by touch. But words,

[86] *Auscultatoriae disputationes quibus methodice tota corporis naturalis in genere cognitio comprehenditur ... Authore R. P. Dom. Redempto Baranzano, clerico regulari Congregationis sancti Pauli, Vercellensi.*

[87] *Ibid.*, p. 294.

[88] *Ibid.*, pp. 847-921, followed by indices at 923-56.

writings, characters, names and other similar signs have no virtue to produce natural marvels. Nor is the sudden transmutation of one species into another natural. Some animals, however, are prevented by occult virtue from living in certain regions. Using a candle of human fat to search for hidden treasure is not only impious but quite unlikely to succeed. Suspending a ring over a vessel full of water in order to tell the time is superstitious. No amatory draught or sorcery or poisoning, with howsoever great occult virtue it may be thought endowed, moves the will directly but only affects it indirectly. The secret friendship of certain sublunar things with the stars is responsible for unusual effects. A slightly more favorable attitude towards the possibility of transformation is now shown, it being said that while the oils listed in books of secrets to transform men into brutes are not wholly credible, yet they lack not some probability. It is also stated that fascination by virtue of the eyes was known to the ancients and is physically possible. But it is incredible that the swords, needles and other objects, which most certainly come out of the bodies of persons bewitched, could have been produced naturally in their bodies. What we encounter by chance when walking about can have no efficacy in disposing our actions for good or evil, but the lineaments of the body and lines of the face, hands and feet have an occult affinity with the qualities of mind and soul.

The Natural Wonders of Pietro Maria Castiglione of Milan are supposed to be limited to stones in the kidneys and their cure, but exceed the limitation suggested by the title.[89] The author died at the age of thirty-five of a fever on October 27, 1629.[90] Other works by him were on pearls in 1618, and on salt in 1629.[91]

"The first secret for removing stones in the kidneys, tested by me more than once with happy outcome," was so effective that not a few persons thought there was something divine about it. It consisted of two scruples of nettle seed, one scruple of powdered licorice

[89] *Admiranda naturalia ad renum calculos curandos,* Milan, 1622, in-8. 168 pp. of text. Index at pp. 169-220. At pp. 221-24 a list of *Jurisconsulti* in the College of Milan. BM 1189.d.27.

[90] Corte, *Notizie storiche intorno a' medici scrittori milanesi* (1718), 171-72.

[91] *Responsio ad Ludovici Septalii iudicium de margaritis,* Milan, 1618, in-4; *De sale eiusque viribus,* Milan, 1629, in-8.

root, one ounce of the juice of unripe lemons, and four ounces of hot chicken juice, mixed and taken on an empty stomach. The nettle indeed is diuretic and, besides breaking the stone, has far nobler faculties. Its leaves, which is a marvel, placed on a fallen womb restore it to its proper position. Licorice is abstersive and lenitive and stops burning urine. Lemon juice reduces stones to nothing. Of whatever sort they may be, if they are put for fifteen days in a glass vessel full of lemon juice and buried in horse manure, they are resolved into their first principles. Bathing in sulphurous waters is also beneficial, since they possess some divine and supernatural property.[92] Whereupon Castiglione cites Raymond Lull and Arnald of Villanova, Geber and Avicenna, Croll and Quercetanus.[93]

The root of a blackberry bush, besides other wonderful virtues, has this that, reduced to ashes and cooked in wine, it drives stones from the body. But it has other incomparable properties. Its decoction taken internally checks diarrhoea and menstrua, heals wounds, stops watering eyes. Applied to the abdomen as a plaster it stops vomiting. It also checks bleeding, which to some seems a miracle.[94] *Aqua ardens* and *aqua fortis* have such marvelous force that touching syphilitic ulcers with a single drop cures them and prevents further spread. "And by its specific form and extreme dryness it dispels all putridity."[95] So Castiglione babbles on. Water of lemon and the herb *saxifragia* (stone-breaker) distilled in a bath of Mary, breaks stones in the kidneys by its peculiar and specific form. "These are marvels and yet true."[96]

Who could ever imagine that in the fish called perch there is a white oblong stone which, reduced to powder, benefits sufferers from kidney disease beyond what would seem human capacity for improvement? There are those who contend that this stone operates solely by a recondite property. But in this case Castiglione abandons his previous affirmation of occult qualities and asserts that there are only two causes of the generation of stones in animals, cold and heat. He goes on, however, to quote Albertus that the stone alectory, often found in the crop of an old cock, worn about the neck as an

[92] *Admiranda naturalia*, pp. 45-47.
[93] *Ibid.*, pp. 48-49.
[94] *Ibid.*, pp. 49-50.
[95] *Ibid.*, p. 51.
[96] *Ibid.*, p. 71.

amulet, makes a man invincible in war, "which, whether it be true, I do not here dispute." But he would not deny that there may be some celestial force in that stone. He could say much more of the nature and virtues of these stones, but they are not the subject of the present work.[97]

Among the greatest miracles of nature is that the kidneys of a hare or ass, dried and pulverized, are a very present help in disease of the kidneys, operating, Castiglione thinks, by sympathy. Digressing to epilepsy, he states that it is incurable, if hereditary; otherwise such remedies as the bone of a human skull of the same sex are in order. But the action at a distance of human mould, fat or blood in weapon ointment and the like, he holds contrary to the principles of natural philosophy and due to the demon.[98]

Having once broken away from his subject proper, Castiglione continues to digress, treating of the soothing effect upon wild animals of being bound to a fig tree, of the echeneis or remora, and antipathy. Although a frog or toad delights in the shade of sage, it is killed by its root. The use of wooden pegs in ship-building in the Maldive Islands is, however, because of lack of iron there rather than due to fear that iron nails would be extracted on passing magnetic cliffs or mountains. Soon we are told of a cure by drinking one's own urine.[99]

A book of Aaron is cited and the Occult Philosophy of Agrippa. Of the genus of marvels is the water in which smiths quench their irons. Drinking it breaks stones in kidneys and bladder, but a more marvelous arcanum is that it cures hydrophobia instantly, if the patient can only be got to drink it.[100] After further outstanding miracles of nature,[101] we presently come to the force of numbers. Some smile at it, but Castiglione knows better. Myrepsus advises taking an odd number of river crabs. The herb pentaphilon resists poisons and demons by its quinary virtue; its leaf taken in wine twice a day cures ephemeral fever; thrice, tertian; and four times, quartan. Similarly vervain should be cut from the third joint for

[97] *Ibid.*, pp. 82-84.

[98] *Ibid.*, pp. 103-107.

[99] *Ibid.*, pp. 109, 111-12, 116, 119, 125.

[100] *Ibid.*, pp. 127, 129, 132.

[101] *Ibid.*, p. 134, "Inter insignia etiam naturae miracula . . ."; 135, "In eodem genere miraculorum naturae . . ."

tertian fever, and from the fourth for quartan.[102] Castiglione also makes a few remarks concerning the Cabala, gives a short account of astrological chiromancy, and a list with brief definitions of various other arts of divination.[103] The Cabala has been admired almost through eternal ages. It is very difficult to acquire, and other sciences are prerequisite and a natural ingenuity. But it does not require knowledge of superstitious arts.

The work of the Jesuit, Julius Caesar Bulenger, on licit and forbidden magic, contains very little as to licit magic, which may have been put in the title as a pious fraud to attract readers. The first of its three books is in part historical, treating of the origin and progress of magic, magi and Chaldeans, the gods of the Persians and sacrifices of the magi, fire worship, the magicians of Pharaoh, and the word *magi*. Otherwise the chief topic is demons and angels. The other two books are chiefly concerning things employed in magic, and are useful only for reference.[104]

The book on rings of Johann Kirchmann (1575–1643) of Lübeck, first published in 1623,[105] is mainly a mosaic of quotations from the classics, and, while it treats briefly of magic rings and talismans and omens from rings,[106] represents the carving of astrological images as a superstitious custom,[107] and states that recent *medici* have disproved the old notion that a muscle or vein connects the ring finger with the heart.[108]

Meanwhile *De sensu rerum et magia* by Thomas Campanella[109] had appeared in 1620. Thirty years before, in *Philosophy Demonstrated by the Senses,*[110] Campanella (1568–1639) had shown himself a disciple of Telesio and also was influenced by the writings of Cardan. From 1599 to 1626 he was imprisoned in Naples for hatch-

102 *Ibid.*, pp. 152-53.

103 *Ibid.*, pp. 153-55.

104 *Opuscula*, Lyons, 1621, I, *De magia licita et vetita,* 437-523, 524-629, 630-48.

105 *De annulis,* Lübeck in-8; then Schleswig, 1657, and Frankfurt, 1672 (Col 393 K63) which is the edition I have used.

106 *Ibid.*, caps. xxi, xxiii; pp. 146-, 169-.

107 *Ibid.*, cap. xi, p. 58.

108 *Ibid.*, p. 17.

109 Campanella's attitude towards astrology has already been discussed, T VI, 173-77, in connection with the papal bulls against judicial astrology and other forms of divination.

110 *Philosophia sensibus demonstrata et in viii disputationes distincta,* Naples, 1591.

ing a conspiracy and revolt in Calabria and for heresy, until his release was procured by the intercession of Urban VIII. Gaffarel, who visited him in the prisons of the Spanish Inquisition and of whose own book we shall treat presently, reported that his legs were all bruised and his buttocks almost without flesh, which had been torn off bit by bit in order to drag out of him a confession of the crimes of which he was accused.[111] Meanwhile in 1617 Tobias Adam had published a compendium by Campanella on the nature of things, which he likewise represented as a completion and clarification of the philosophy of Telesio.[112] The *De sensu rerum et magia* of 1620 also sharply criticizes Aristotle, repeats views of Telesio, and further brings to mind Giordano Bruno's *De rerum principiis* and *De magia* of the previous century.[113] The full title of Campanella's work may be translated as follows:

Four Books of the Sense in Things and Magic. Marvelous Part of Occult Philosophy in which it is demonstrated that the World is the living Image of God and conscious too, and that all its particular parts are endowed with sense, some clearly, others more obscurely, but enough to preserve them and the whole in which they harmonize, and the reasons for almost all of Nature's Secrets are made manifest.[114]

After Campanella came to France, the work was republished at Paris in 1636.[115]

As this title implies, Campanella holds that all nature is sentient. Whatever is in effects was also in their causes, and therefore the

[111] Gaffarel, *Curiositez inouyes*, 1629, pp. 267-71; quoted in *Correspondance du P. Marin Mersenne*, II (1936), 170.

[112] *Prodromus philosophiae instaurendae, id est, dissertationis de natura rerum compendium, secundum vera principia ex scriptis Thomae Campanellae praemissum, cum praefatione ad philosophos Germaniae* (signed at its close. "Tobias Adami,") Frankfurt, 1617.

Adam's preface occupies a number of pages and the *Compendium* proper opens at p. 27 (signature D 2 recto), "Duce sensu philosophandum esse existimamus . . ."

[113] T VI, 424-27, for their contents.

[114] F. Thomae Campanellae, *De sensu rerum et magia libri quatuor. Pars mirabilis occultae philosophiae ubi demonstratur mundum esse Dei vivam statuam beneque cognoscentem omnesque illius partes partiumque particulas sensu donatas esse, alias clariori alias obscuriori quantus sufficit ipsarum conservationi ac totius in quo consentiunt, et fere omnium naturae arcanorum rationes aperiuntur*, ed. Tobias Adam, Francofurti, 1620. Copy used: BN Inventaire R.1923.

[115] Apud Ludov. Boullenger, 1636, in-4.

elements and world, which produce beings with senses, have sense themselves.[116] Instinct is an impulse of sentient nature.[117] All beings abhor a vacuum and rejoice in mutual contact, therefore they are sentient and the world as a whole is an animal.[118] The soul is a tenuous, hot and mobile spirit, apt to suffer and so to feel. Spontaneous generation is caused by heat, while a dying body grows cold. But when a man who is intoxicated by strong wine falls asleep, this is not due to cold, since the wine makes him hotter, but is because the wine sends an abundance of vapors to the brain. To overcome and attenuate these vapors, the spirits leave the other parts of the body and hasten to the brain, leaving the other members without sense and motion, "and this is sleep." The same spirits constitute the conscious, irascible, concupiscent and motive soul, Galen to the contrary. Bones, hair, nerves, blood and spirit all are sentient. When reason seems to contend with lust or anger, if it prevails, they must have been persuaded and yielded; while if reason yields to them, they must have given it good grounds for so doing. Therefore they too are conscious, since they can convince reason. Or, if it is reason that persuades itself, anyway it is all one and the same soul. The passion of love which prefers the young and beautiful is reasonable. In place of the old distinction between animal, vital and natural spirits, Campanella believes in a single spirit "living and working all over the body in the various vessels." He admits, however, that man has a mind and soul infused by God, which brute animals do not possess. They have sense, memory, discipline, discourse and general intelligence—an assertion supported by some very tall stories—but not consciousness of the divine such as man possesses. The world too must have a soul and divine mind, created and infused by God, or macrocosm would be inferior to microcosm. Campanella adds that the Holy Inquisition objected that in this case the world soul would inform vermin and other unworthy objects, but he retorted that it would no more do so than the human soul informs black-heads and tape worms.[119]

[116] *Ibid.*, I, 1.

[117] *Ibid.*, I, 7.

[118] *Ibid.*, I, 9. See also the *Compendium* (1617), 30 (signature D 3 verso): "Vacuum non datur quia omnia corpora sentiunt et mutuo tactu gaudent . . ."

[119] *De sensu rerum*, II, 4-5, 7, 9, 13, 11, 19, 23.

Sky and stars are of the nature of fire and are sentient.[120] The sky moves by its own virtue and the stars by theirs, although perhaps angelic minds are assigned to them. Since we cook with fire and are nourished, grow and move by its virtue, it must be that it controls our feelings and consciousness too. This is further suggested by the fact that the hotter animals are, the more sensitive and alert they are; and the colder, the stupider. Air is a sort of common soul which is available to all and a universal means of communication,[121] as Pliny notes.[122] All waters and liquids are sentient and display sympathy or antipathy. Stones and metals also possess sense and have friendship and hate. This aged rock with such long duration should surely know more than I. The same is true of plants, which Plato called inverted and immobile animals.[123] In an appendix devoted to discussion of how the little *remora* stops a ship, Campanella suggests that it stupefies the vessel and renders it repugnant to its natural motion, as the bite of a mad dog makes the victim inhuman and canine.

In this connection we may recall the discussion of the same echeneis or remora by Suarez,[124] repeated as late as 1690 by Henckel.[125] Some say that, as the hand of the thrower gives an impetus to the missile which keeps it going, so the remora imprints a non-impetus upon the ship which keeps it still. Others say that it detains the ship by innate virtue, as a man holds a stone in his hand so that it cannot fall. Yet others say that it so attaches itself to the ship that it cannot be moved, nor can the ship. But Suarez's conclusion is that, however it happens, there is no doubt that it comes from

[120] *Ibid.*, III, 1. In his *Apologia pro Galileo*, Frankfurt, 1622, p. 54, Campanella corrected this statement, saying that he used to think that the heaven was fiery and the source of all fires, and that the stars were made of fire, as Augustine, Basil and all the Fathers held and recently "our Telesio." But after the observations by Tycho and Galileo of a new star and comets in the starry heaven, and clouds about the sun, "I suspect that not all stars are fiery, which suspicion the augment and decrement of the moon and Jupiter, and the spots on the moon and Jupiter the more confirm."

[121] *De sensu rerum*, III, 3, 5, 7.

[122] Pliny, *Nat. hist.*, II, 38.

[123] *De sensu rerum*, III, 12, 13, 14. Similarly, in Campanella's *Realis philosophiae . . .*, 1623, p. 89, "Sunt plantae animalia immobilia."

[124] *Diss. metap.* 18, *sect.* 8, "ubi agit de propinquitate."

[125] *De philtris*, pp. 34-35.

marvelous and occult virtue aided very likely by some special and connatural celestial influence.

With Campanella's animistic description and disposition of nature, the stage is all set for his further discussion of natural divination, natural magic, and occult marvels. These were fields in which he regarded himself as especially proficient, as may be illustrated by some of his letters. At times he becomes apocalyptic, communes with Joachim and Bridget, ruminates on recent earthquakes in Calabria, inundations of the Tiber, and comets, and comes to the conclusion that the last days are at hand. The solstices and equinoxes are changed by twenty-five degrees, the poles have receded, the planets are nearer the earth. For his predicting from such phenomena and other writings they stoned him and made a devil, rebel and heretic of him. He does not claim to be a prophet, "but I am so addicted to my God, from whom I have seen miracles and angels and demons, and have suffered much, that I am wholly turned into spirit."[126]

If this brings to mind the tone in which John of Rupescissa wrote in the fourteenth century, our next paragraph will as vividly suggest the ambitious program of yet a third friar, Roger Bacon, in the thirteenth century.

For Campanella further regarded himself as something of a magician, since magic consists of religion, astrology and medicine, subjects in which he believed that his equipment was above the average.[127] During his long imprisonment he made extravagant promises of marvels which he would work, if released from captivity. He assured Cardinal Odoardo Farnese that, if set free, he would teach natural and moral philosophy, logic, rhetoric, poetic, politics, astrology, medicine—all within a year's time and in admirable fashion, accomplishing more than ten years of ordinary study in the

[126] *Lettere a cura di Vincenzo Spampanato*, Bari, 1927, (*Scrittori d'Italia*, vol. 103), pp. 65-66, 94-95. "Articuli prophetales" by Campanella are preserved in MSS: Naples I.G.13; Paris, Arsenal 1085, 17th century, fols. 1-114.

[127] *Lettere*, p. 98: "Sed magia constat ex religione, astrologia et physica. Hae mihi facultates adsunt credo non vulgares." This passage is no doubt based on Pliny's statement that magic embraced the three subjects which appealed most to the human mind: medicine, religion, and the divining arts, especially astrology.

schools would. He will reform astronomy and the calendar. He will prove the end of the world by fire against Aristotle, Ptolemy and Copernicus in favor of the Evangel. Under pain of losing all credit as a scholar, if he does not succeed, he will fabricate a marvelous city (his "City of the Sun"?) and ships that move without oars or sails. He will open the whole world like a book from his mouth in two months, and, "when you hear me, your books will seem to you mere tricks of jugglers." If he but opens his mouth at Rome, "you will see a new heaven and a new earth, and from north and south a great rush to the Catholic Faith."[128]

To the pope Campanella wrote of revelations made to himself three years ago by the devil in the guise of an angel, and of other apparitions of demons that afflict him.[129] In *De sensu rerum et magia,* too, he expresses his belief in the existence of both angels and demons, and condemns as impious the opinion that no demons exist.[130] On the other hand, as he states in his Six Books on Astrology, neither good angel nor devil has sovereignty over nature, but they employ magic, applying active to passive at the right time and under favoring conditions of the stars and of matter. "For magic is the flower of all the sciences."[131]

From the predictions (Campanella uses the word, prophecy) of animals and of men and the difference between them the divinity of man is made evident. It is true that other animals sense coming weather changes as men do not, but that is because they are more exposed to the air and more exclusively intent on their food and self-preservation.[132] Vultures can smell carrion at a distance of a thousand miles, because the air is moved by winds, and it is a law of physics that the air as a whole constantly is moved from east to west, as is the sea. Nevertheless man is far superior in divination, because he has presentiments of angels, God and a life after death. Moreover, in sleep man senses the future from the air as he does not

[128] Luigi Amabile, *Fra Tommaso Campanelle, la sua congiura, i suoi processi e la sua pazzia,* Napoli, 1882, II, 379; *Lettere* (1927), 27, 136.

[129] Amabile (1882), II, 383.

[130] *De sensu rerum et magia,* 1620, III, 10, p. 232.

[131] *Astrologicorum libri VI . . . ,* Lyon, 1629, p. 21.

[132] The same idea appears in *Realis philosophiae . . . ,* 1623, p. 197.

when awake. The dreams of Joseph and Daniel, however, were of direct divine transmission and straight from God. Nor does Campanella agree with Ficino in *De vita coelitus comparanda* and many other writers, that melancholy is a cause of prevision. Natural prophecy is made in the spirit, supernatural in the mind. Aristotle, Galen, Avicenna and others believe only in natural prophecy. And it is true that in the world all things are closely associated (*consimiles*) and that he who does not distinguish well is often deceived.[133]

Among many experiences Campanella selects for repetition the case of a girl of twelve, who was possessed by a demon, when the planets reached a certain position. After a year she recovered, married a good husband, and lived in great sanctity until the age of thirty-five, when, because of another position of the planets, she fell into a trance, saw visions and marvels of the world to come, and became very learned in theology without study. She made many true predictions after first praying to God. "From which I learned that natural disposition from the weather and the stars confers much toward superior visions."[134] Campanella has further often found true what Ptolemy says of the association of prophecy with certain planets and of possession by demons with others. He believes that the stars only incline and do not necessitate, but they render the spirit lucid and so favorable to divination, and also apt to receive divine inspiration and angelic visions, as Origen has said.[135]

The fourth and last book of *De sensu rerum et magia* is devoted to magic, once the ancient wisdom of Persia but today declined largely into a superstitious cult of demons, just as astrology has been debased by impostors. Porta tried twice to recall that science, but his treatment was purely historical and descriptive, and did not investigate causes. Magic purges the soul to fit it for occult thinking and attune it to the First Cause. It knows the properties and virtues, the sympathies and antipathies, of herbs, stones and so forth. It knows the right time to operate, and the relation between our affairs and the stars. Incidentally Campanella represents the three Magi in The Gospel of Matthew as led by a comet. Pliny recognized only natural magic. Trismegistus said that man was the miracle of the

[133] *De sensu rerum et magia,* III, 8-11, pp. 221-40.

[134] *Ibid.,* III, 11, pp. 240-41.

[135] *Idem,* p. 239.

world. Campanella further recognizes the existence of diabolical magic, and of divine magic such as Moses worked.[136]

As for natural magic, "Whatever wise men do in imitating nature or aiding it by art, not merely unknown to the populace but to the general run of men, we call a magic work." Before an art becomes known, it is always called magic. Gunpowder, printing and the magnet once were magical. Mechanical clocks soon lost their charm because everyone could see how they ran. But medicine, astrology and religion are very rarely divulged; therefore the ancients restricted the term magic to them.[137] "Natural magic to lengthen and shorten life" does not seem to be much more than diet in the broader sense.[138] Otherwise natural magic is largely due to the action of sense in everything and to sympathy and antipathy between objects. Hardening of the spleen is cured by applying the spleen of an animal and then warming it in the fireplace. Sense is found even in corpses and putridity, and remains in things long consumed, as is shown by the reaction between drums of sheepskin and of wolfskin.[139] Campanella spins yarns of the effects of a tarantula's bite recurring annually so long as the spider lives.[140] In the law concerning leprosy Moses teaches that sense and disease are communicated to and multiplied in the most dissimilar and distant things.[141] Contrary repels contrary better, if given internally; like affects like more, if applied externally.[142]

The remaining chapters of the fourth book[143] are devoted to rules for applying animals, plants and minerals for magic purposes, emanations from the eyes with power to work change, optical illusions—with an appendix on witches, magic power in sounds, whether words can affect persons who are absent, generation, that astrology is necessary for the best magician, and an epilogue on the sense of the universe.

We may consider the discussion of the power of words in somewhat more detail. Earlier in the book Campanella had treated of the evolution of language and the effect of climate upon it.[144] He

136 *Ibid.*, IV, 1 *et seq.*
137 *Ibid.*, IV, 6, pp. 282-83.
138 *Ibid.*, IV, 8, pp. 285-95.
139 *Ibid.*, IV, 9-10.
140 *Ibid.*, IV, 11, p. 303 *et seq.*
141 *Ibid.*, IV, 12, p. 315 *et seq.*
142 *Ibid.*, IV, 13, p. 318 *et seq.*
143 *Ibid.*, IV, 14-20.
144 *Ibid.*, II, 11, p. 84.

held that speech and words originated in the effort of the spirit or spirits within the body to signify its sentiments to others by forming on the air breathed a likeness of the thing meant with movements sounding in the mouth and reflected in various figures, "and so the first speakers formed names from impressions received." Different nations are variously affected and have different methods of expressing the air. Germans pronounce many consonants because the impetus of the spirit in speaking is thrust back by their cold climate. The Venetians, living on the water, have many liquids and vowels. The Spaniards, because of the heat and sharpness of their region, have words abounding in vowels and sibilants. The Italians are, like their climate, midway between these last two. In antiquity, before the sun was so near the earth, they pronounced Latin with more consonants. But with the continuous descent of the sun and mixture of nations in the peninsula the language has undergone continuous variation. In another passage Campanella says that, if verbal language had not yet been worked out, men would understand one another according to the emotion or state of mind impressed on the air. He believes that the gift of tongues at Pentecost, when every man seemed to hear the apostles speak in his own language, is to be explained thus.[145]

Coming now to the chapter on the power of words, its heading states that words have force over absent things, but that often demons craftily intervene. The subsequent text is even more hesitant, stating that it is difficult to see how words can affect the absent, although Campanella asserts that he was cured as a child of an ailment of the spleen by words uttered by an old-wife. Ceremonies *per se* are of no avail. Aside from demon activity, the only natural effect must be by the impression made in speaking upon the air, which might convey it to a distance. But Campanella cannot understand how the stars or weather can be changed by incantations alone or the crops dried up. The voice and imagination cannot alter sky and sea. Nor can mere sound move great rocks and plants, as Orpheus is said to have done.[146]

Since we have heard Campanella say that magic consists of religion, astrology and medicine, it will not be amiss to glance at

[145] *Ibid.*, III, 7, p. 220.

[146] *Ibid.*, IV, 18, pp. 340-45.

his Seven Books of Medicine According to his own Principles.[147] Besides dealing with diseases and cures therefor, he has, as before, a good deal to say about the spirits of the human body. For example, the pulse is described as a vital act of the animal spirits, or it is inquired by what things the native light of the spirit is injured and cured.[148] Occult virtues and the relation of terrestrial things to the planets are also considered.[149] Not merely natural but "trans-natural" medicines are included, for there are diseases which call for religious rather than Galenic procedure, as Paracelsus said. Campanella accepts the existence of diabolical witchcraft, and holds that the physician should not regard the books of exorcists as wholly vain, for Paracelsus praises exorcisms as true and fitting remedies. Campanella recognizes, however, that Paracelsus was somewhat given to superstition, and rejects his further assertion that witchcraft is removed by other witchcraft. Rather it is annulled by the virtue of religion.[150] Thus religion and witchcraft obtrude in Campanella's medicine, which is a step backward rather than forward. Natural as well as diabolical magic also characterizes it. Thus, while accepting an astrological cause for pestilence as well as Fracastoro's seeds or germs, he also speaks of magic and of marvelous sympathy and antipathy. Campanella later claimed to have shown in this book that ancient medicine was blind, and that the natures and causes of diseases and medicaments had not hitherto been explored.[151]

This medical work of Campanella was brought from Italy by

[147] *Medicinalium juxta propria principia . . . libri septem,* edited by J. Gaffarel, Lyon, 1635. 690 pp. in-4. Copy used: BM 544.g.6. In the BN printed catalogue the work is dated 1634.

[148] *Ibid.*, pp. 141, 318.

[149] *Ibid.*, pp. 240, 261.

[150] *Ibid.*, pp. 303-4: "Dixit Paracelsus horum remedium esse in religione non in schola Galeni, licet in superstitione ille locet. Nos autem maleficia cognoscimus arte daemonis a strigibus illata. Non ergo arbitretur medicus vana prorsus quae in libris exorcistarum scripta sunt, nam Paracelsus laudat haec ut remedia vera propriaque. Neque altero maleficio maleficium tollet, ut Paracelsus docet, sed religionis virtute."

In *De libris propriis,* edited posthumously by Naudé in 1642, Campanella gives the following estimate of Paracelsus at p. 57: "Paracelsus denique in destillatoriis et medicinis chymicis aliquid promovit; in speculativis vero ineptit plerumque et accipit pro ratione non rationem."

[151] *De libris propriis,* 1642, p. 25.

Gaffarel, who had been purchasing manuscripts for Cardinal Richelieu and of whose own leanings toward magic and occult science more anon, and was printed at Lyon in 1634-35. Gaffarel had edited a brief synopsis of the writings of Campanella covering only nine pages at Venice in May, 1633, in the preface to which he had suggested to Jean Bourdelot that he might care to publish some of Campanella's works in France. And as the reason for issuing the medical work at Lyon he gave the present scarcity of printers in Italy. Meanwhile Campanella, whose astrology had lost him the pope's favor, had come to France in 1634.

But we must turn back again to the intervening years since the appearance of his *De sensu rerum et magia* in 1620. In 1625, in addition to the posthumous publication above noted of the book of Goclenius on marvels of nature, there was issued by Gabriel Naudé, the celebrated bibliographer, librarian and man of letters, his famous Defense of All the Great Men of the Past who had been falsely suspected of magic.[152] The book is written in French but with a great many Latin quotations. Naudé distinguishes four kinds of magic: divine, theurgic, goetia or witchcraft, and natural magic. He also makes a distinction between licit and prohibited magic. It is from the suspicion of engaging in illicit magic that he defends the great men of the past. Divine magic is beyond human control. Theurgy or white magic, under the color of religion, enforces fasts and abstinences, purity and chastity, piety and integrity, in the effort to free the soul from the contamination of the body and communicate with superior powers. To what extent it is permissible is left somewhat in doubt, but goetia, witchcraft or black magic is the clearly illicit variety. Natural magic, on the other hand, which is all founded on nature, is depicted by Naudé as unobjectionable. "It is entirely false that the magic which was universally practiced by all Egypt was other than natural, with perhaps a few vain and useless superstitions mixed in."[153] The reason that Aristotle and other Greek writers never mention magic is that their lofty sciences, rare doctrine and marvelous disciplines were "nothing else than the practice of

[152] *Apologie pour tous les grands personnages qui ont esté faussement soupçonnez de magie*, Paris, 1625, 22 chapters, 649 pp.

[153] *Ibid.*, p. 39.

that fourth and last kind of magic called natural."[154] It includes astronomy and astrology, chemistry and alchemy, physiognomy, chiromancy and metoposcopy, helioscopy and geomancy.[155]

As this last list suggests, Naudé includes some occult arts of questionable standing under natural magic. He also defends such personages as Anselm of Parma, Henry Cornelius Agrippa, Merlin and Nostradamus, as well as ancient philosophers like Pythagoras, Democritus and Empedocles, writers in Arabic like Alkindi, Geber and Thebit, medieval Latins like Michael Scot, Aquinas, Albertus Magnus, Roger Bacon and Peter of Abano, or more recent personages such as Pico della Mirandola and Savonarola, Trithemius and Paracelsus. Zoroaster was not the author of either goetia or theurgy; Orpheus was not a magician; Numa Pompilius and the poet Vergil are defended as well as Popes Sylvester II (Gerbert) and Gregory VII, or Joseph, Solomon and the three Magi. The book opens with a chapter on historical criticism and closes with an exhortation to writers on demons and witchcraft to be more sceptical. But, as has been indicated, Naudé himself might have shown more scepticism as to the natural basis of certain occult arts. Yet a letter of Guy Patin represents himself, Gassendi and Naudé as in matters of philosophy and religion "all three cured of the fear of bugbears and liberated from the disease of scruples, that tyrant of consciences."[156] Patin also quotes Naudé as warning not to be deceived by four things: prophecies, miracles, revelations, and apparitions, and as saying that it was not worth while to keep changing one's religion.[157] That a man so learned and enlightened should still maintain such a position with regard to astrology, and other supposed arts of natural divination, and to natural magic is significant of the hold which such subjects still had. Naudé, however, could smile at a modern writer who, leaving no stone unturned in his effort to be thought a magician, published a Rhetoric in five parts: the art of Trithemius for invention, theurgy for elocution, the art of Armadel for disposition, the Pauline art for pronunciation, and the Lullian art for memory.[158]

[154] *Ibid.*, pp. 40-41.
[155] *Ibid.*, p. 44.
[156] *Lettres* (1907), pp. 616-17.
[157] *Lettres* (1846), pp. 490, 758.
[158] *Apologie*, p. 32.

The same Jesuit, Jean Roberti, who had attacked the treatise of Goclenius on the magnetic cure of wounds, also criticized Naudé's *Apologie*. From the other side of the fence it was both praised and blamed by Gaffarel four years later in his *Curiositez inouies,* of which we shall say more presently. It was defended by Claudius Forgetus Nancejanus, an M.D. of Padua, at the close of a book which he composed against the vanity of the Pythognomic art.[159] Naudé's sceptical bent was shown again a few years later in a treatise questioning the danger from poisons.[160]

A century and a half later the Abbé Claude-Marie Guyon in the eighth volume of his *Bibliothèque ecclésiastique* (Paris, 1771) was much impressed by Naudé's hypothesis of false accusations of magic, like him held that judicial astrology was the foundation of other occult arts, and distinguished natural from superstitious and diabolical magic. Yet, despite Naudé's warning to writers on witchcraft to be more sceptical, the Abbé recounts in proof of diabolical magic an utterly absurd and incredible tale of shepherds accused of bewitching animals at Pacy.[161] They appealed to the Parlement of Paris in 1688, and the last execution for sorcery by the Parlement of Paris was also at Pacy in 1691.

Three books on secrets of nature and miracles of the universe date from the closing years of the third decade of the century, but I am only able to give their authors and titles. Will[162] informs us of a work written in German by Zacharias Theobald (1584–1627) and printed at Nürnberg the year after his death, on Arcana of Nature collected from trustworthy authorities and his own experience,[163] which was hardly patriarchal. The next year, 1629, Fridericus Casander published at Frankfurt his *Nature Talking* in which were set forth the miracles of the whole universe from its chief parts or

[159] Leo Allatius, *Apes Urbanae sive de viris illustribus qui ab anno MDCXXX per totum MDCXXXII Romae adfuerunt ac typis aliquid evulgarunt* (Rome, 1633), Hamburg, 1711, pp. 157-58. I have failed to find the work of Nancejanus.

[160] *Questio iatro-philologica An magnum homini a venenis periculum,* Rome, apud Facciottum, 1632, in-8.

[161] *Op cit.*, pp. 266-82.

[162] Georg Andreas Will, *Nürnbergisches Gelehrten-Lexicon . . . ,* IV (1758), 25.

[163] Zacharias Theobald, *Arcana naturae, dass ist, sonderliche Geheimniss der Natur, so wol aus glaubwürdigen Autoribus, als aus eigener Erfahrung zusammen getragen,* Nürnberg, 1628, in-4.

kingdoms, ethereal, vegetable and mineral, and their properties and virtues.[164] Then in 1630, at Leyden, Petrus Mornius issued Most Secret Arcana of All Nature, never before detected, from Rosicrucian sources.[165]

Jacques Gaffarel, of whom we spoke above in connection with Campanella, was born in Provence in 1601, educated at the universities of Valence and Paris, where he received the degree of doctor of canon law, became a priest and chaplain of Richelieu, and had a wide knowledge of oriental languages—Hebrew, Arabic, Syrian and Persian. In 1625 he published a book on the hidden mysteries of the divine cabala,[166] in which he replied to the attacks of George of Ragusa[167] and Mersenne.[168] He wrote other works on the cabala, but we are now concerned with his Unheard-of Curiosities which first appeared at Paris in 1629,[169] and then was repeatedly reprinted into the early eighteenth century and translated into Latin and English.[170] It divides into three parts, of which

[164] Fridericus Casander, *Natura loquax, qua miracula totius universi ex praecipuis mundi partibus sive regnis, aethereo, vegetabili et minerali, silvarum nempe, hortorum, pratorum, plantarum etc. proprietatibus, affectis et virtutibus deprompta proponuntur.* Francof. apud Lucam Jenisium, 1629, in-8. From *Lindenius Renovatus* (1686), 305.

[165] Petrus Mornius, *Arcana totius naturae secretissima, nec hactenus unquam detecta, e Collegio Roseano in lucem producta,* Lugd. Batav., 1630, in-12. LR (1686), 901.

[166] *Abdita divinae cabalae mysteria contra sophistarum logomachiam defensa,* Paris, 1625, pp. 79, 88. It was addressed to Richelieu.

[167] T VI, 198-202, for his attitude toward astrology.

[168] Consult the index of Mme. Paul Tannery and Cornelis de Waard, *Correspondance du P. Marin Mersenne,* I (1932), under "Gaffarel" for the foregoing facts. Mersenne replied in a letter to Peiresc written under an assumed name: *ibid.,* 303-7.

[169] *Curiositez inouyes sur la sculpture talismanique des Persans, horoscope des patriarches et lecture des estoilles,* Paris, H. Du Mesnil, 1629, in-8, 644 pp.

[170] I took notes on the Latin edition of Hamburg, 1676, which is sometimes called the best, at the BN in July, 1929, and five other editions at the BM in July, 1933, including one of 1678 not listed in the printed catalogue. The English translation appears to antedate the Latin: *Unheard of Curiosities Concerning the Talismanical Sculpture of the Persians . . .* Englished by E. Chilmead, London, 1650, in-8: BM E.1216.(1.). I have yet to see the French *editio princeps* of 1629, but there are two copies of it at the BN. The French editions of 1631, 1637 and 1650 all have 315 pages against the 644 of the 1629 edition.

Mlle. M. Th. d'Alverny, Conservateur aux Manuscrits, has very kindly compared these editions for me and

the first defends orientals, especially Hebrews, from Christian charges, and the third deals with ancient Hebrew and other oriental astrology. The second part, on the talismanic sculpture of the Persians, especially interests us for its close connection with natural magic.

Gaffarel contends that astrological images are natural and not diabolical. He insists that something is to be ascribed to figure; a square piece of wood for example will not spin as well as a round one. Also he believes in the doctrine of signatures. His further argument for operative force in such images is three fold: from astrology, from sympathy or likeness, and from experience. He will not, however, accept those images which violate free will, and agrees that some of these figures are by now destitute of any efficacy. Ficino was wrong in ascribing to the Rabbis the view that the brazen serpent which Moses lifted up in the wilderness was a talisman to avert the evil influence of Mars and Scorpio. It was not a talisman at all. Nor was the golden calf, as some astrologers think, intended to receive the force of Venus and the moon. On the other hand, many images retain today their pristine virtue. The Druids of Gaul employed them most successfully, and learned men of later times rescued them from darkness and oblivion. Paracelsus applied such diligence that some of those he fabricated are the safest amulets against the plague. Gaffarel also cites Roger Bacon, Junctinus and Naudé's *Apologie,* but condemns the images of Thebit ben Chorat, Trithemius and Goclenius, and the characters of Marcellus Empiricus. He regards it as superstitious to think that the use of certain words is necessary with images. He does not believe that images can assure victory in war, and sorcery with wax images is not natural but diabolical. He does not accept the argument against the influence of such images that action must be by contact. Virtual influence is sufficient. One does not need to touch the fire to be warmed by it, and the attractive influence of the magnet is natural,

---

found that they are similar, except that the royal privilege of the edition of 1629, dated 24 Mars 1629, was suppressed in the subsequent editions. The difference in the number of pages is due to the use of much smaller type in the later editions.

not diabolical. Some of the Fathers have inveighed against images, but the Church has never forbidden their legitimate use. Gaffarel incorrectly represents Aquinas as favorable to them like Albertus Magnus and Cajetan, although he correctly cites William of Auvergne and Gerson as opposing them. But the position of Aquinas was that of William of Auvergne rather than that of Albertus Magnus.[171] But the work that Gaffarel cites most frequently on the subject is that of Roger Bacon on the secret works of art and nature.

Gaffarel further contends that the astrology of the ancients was neither idolatry nor the cause of idolatry, and accuses Scaliger and others of having misrepresented the astrology of the ancient Hebrews, Egyptians and Arabs. It makes him smile that the Hebrew word for the firmament, which means nothing else than air or expanse, should have been twisted into a crystalline heaven. He inquires whether the ancient astrologers of the Hebrews used mathematical instruments, what the method of the patriarchs and old Hebrews was in drawing up horoscopes, and whether one can read anything from the clouds as from other meteorological manifestations.

On August 1, 1629, the faculty of theology at Paris condemned Gaffarel's book as "entirely to be disapproved," and called its doctrine false, erroneous, scandalous, opposed to Holy Writ, contumelious towards the Church Fathers, and superstitious besides. However, if before September first, the author would abjure these perverse dogmas and retract, the faculty would withdraw its condemnation.[172]

It was not, however, until October 4, 1629, that Gaffarel signed a retraction which was couched in vague and general terms.[173] In

[171] T II, 610-11, for Aquinas's position. He held that works of human art receive no new virtue from the stars.

[172] Carolus Du Plessis d'Argentré, *Collectio judiciorum de novis erroribus*, editio nova, Paris, 1755, II,ii,285, "Conclusio sacrae Facultatis qua damnatur Liber des Curiositez inouies a Gaffarello editus."

[173] *Ibid.*, 285-86. Also separately: *Censura sacrae facultatis theologiae Parisiensis lata in Petri Picherelli opuscula theologica Lugduni Batavorum 1629 excusa.—Retractatio Jac. Gaffarelli auctoris libri: Des Curiositez inouyes.* Paris, J. Guillemot, 1629, in-8, 6 pp. BN D.29508.

Bibliothèque de Lyon, MSS français, 1218 (in the catalogue of Delandine, III (1812), p. 138), item 19; "Censure des opuscules théologiques

it he stated that he had not meant to teach or to assert as true what he wrote in the book entitled *Curiositez inouyes,* but merely to record various opinions collected from the writings of the Arabs and Hebrews. He had warned the reader in the preface that he had faith in them only insofar as the Roman Catholic church allowed. But inasmuch as he has been admonished by the Paris faculty of theology that many of these opinions are to be rejected and condemned, he now explicitly condemns and rejects them. And since the most holy Faculty further condemned some views which he had stated as his own, he similarly condemns these. But he does not specify what they are, and nothing is said about altering the text of his book.

Morin tells us that Gaffarel seems to have fooled the doctors of the Sorbonne by not specifying just what opinions he retracted, but that, since there was nothing in his book more superstitious and alien from the Faith than his talismanic doctrine, it is to be assumed that the doctors of the Sorbonne especially desired its renunciation.[174] Yet such astrological images are not specified in the condemnation, and they continued to be the main theme of his work in its numerous subsequent editions.

A different account of Gaffarel's retraction was given by the Jesuit, Claude-François Menestrier, in his *La philosophie des images enigmatiques,* printed at Lyons in 1694, pages 435-440. According to this account it was Gaffarel's work on the divining rod entitled, *De la verge de Jacob, ou l'art de trouver les Tresors, les Sources, les Limites, les Metaux, les Mines, les Mineraux, & autres choses cachées par l'usage du baton fourché,* which caused the action of the Sorbonne. What was objected to particularly was his supposition "that the quality, which is peculiarly characteristic of every animate or inanimate body, depends absolutely or draws its nature from what the star which ruled over his generation impressed on him." Menestrier argued that, if this were so, the stars would be, not merely universal causes, as he granted was the case, but

de Pierre Picherel, imprimés à Leyde, en 1629.—Rétractation de Jacques Gaffarel, auteur du livre des Curiosités inouies."

[174] J. B. Morinus, *Astrologia Gallica,* Hagae-Comitis, 1661, p. 494.

specific causes, whereas he held that the specific germs of each species had been implanted by God at Creation.[175] Gaffarel, on the contrary, had contended that the stars made one man a soldier; another, a merchant, another, a builder; one, fond of affairs, another, of repose and quiet; one, a poet; another, an orator; and finally quoted the apostle Paul that "One has the gift of Faith, another of curing maladies, another of working miracles, another of prophecying, another of discerning spirits, another the gift of tongues, another of interpreting them."[176] He added that the stars were secondary causes which God used in showering their various gifts upon men, and that if one knew his particular inclination and applied himself thereto, he would succeed the better, or know how to avoid such ills as it involved. But the wise faculty of theology of Paris pronounced against this *Livre de la verge de Jacob* which attributed to the stars these virtues on the inclinations of men even to the point of supernatural gifts, and in 1629 obliged Gaffarel to retract what he had written on this subject and to publish his retraction which was printed at Paris by Jean Guillemot.[177] Since, however, we have seen that both the Faculty's condemnation and Gaffarel's retraction apply specifically to his *Curiositez inouyes* and do not mention *De la verge de Jacob,* it seems evident that Menestrier has mistaken the title of the book, although he confirms the view that it was its astrology to which the Sorbonne especially objected.

Leo Allatius, who was charged with transporting the Palatine library from Heidelberg to Rome and was librarian to cardinal Barberini and from 1661 to 1669 of the Vatican, dying in the latter year at the age of eighty-three, had something to say of Gaffarel in his book on illustrious men who were in Rome from 1630 to 1632 inclusive.[178] He states that Gaffarel came to Rome twice, in 1626

[175] *Op. cit.*, p. 435. Also p. 432.

[176] First Corinthians, xii, 9-10.

[177] Menestrier continues (p. 441), "L'Apôtre saint Paul a bien fait des Etoiles la figure des dons que le saint Esprit dispense differement, mais il n'en a pas fait comme l'Auteur de la verge de Jacob le principe & la cause naturelle de ces dons, ce qui est une erreur des Manichéens."

[178] *Apes Urbanae sive de viris illustribus qui ab anno MDCXXX per totum MDCXXXII Romae adfuerunt ac typis aliquid evulgarunt, Rome,* 1633. I have used the edition of Hamburg, 1711 by J. A. Fabricius, where Gaffarel is treated at pp. 193-96.

and 1632. The *Curiositez Inouyes* ran through three editions within six months, twice at Paris and once in an unnamed French town.

It is suspected, and correctly, that booksellers of Rouen from hope of gain published it once and again not without great corruption both of the style and the meaning.

However, Allatius incorrectly gives the initial date of the book's publication as 1630 rather than 1629, and that of Gaffarel's *Abdita divinae Cabalae mysteria* as 1623 instead of 1625. But these may be misprints of the later edition which I used. Allatius adds a list of titles of unpublished works by Gaffarel which fills over a page and of which we may note A New Opinion about Falling Stars and Divination by the Moon according to the Hebrews.

An *Ars magica sive magia naturalis*, printed at Frankfurt in 1631, is, as its long title goes on to explain, also devoted to superstitious or diabolical magic, which subject in fact occupies its first 244 pages out of 571 pages in all. As this section is taken from previous authors such as Delrio, so the part on natural magic, although purporting to have never been seen or known before, cites such sixteenth century writers as Porta, Agrippa, Riolan and Mizauld, while it makes use of the *Secreta* and *De mirabilibus mundi* of the pseudo-Albertus Magnus without acknowledgement. Typical recipes are to enable one to see in the dark or to become invisible. Weather presages, marvelous effects of lightning, and planetary astrology are intermingled with the magic recipes. Then follow chapters on magical seals and astrological images, including those of Chael, Hermes, Thetel and Solomon, while the last two chapters are on memory and the art of memory from Gratarolo. The volume then is a repetition of thirteenth century books of secrets, experiments and magic images, with additions from sixteenth century authors. Its only other notable feature is putting natural magic first in its title, presumably as more likely to attract readers than the superstitious and diabolical variety.[179]

[179] *Ars magica sive magia naturalis et artificiosa, effectus virtutes et secreta in elementis gemmis lapidibus herbis et animalibus secundum certas astrorum ac constellationum figuras et sigilla horasque planetarias exhibens, antehac numquam visa cognitave, nunc primum ex vetustissimis veterum*

The cosmopolitan character which marked many men of letters and science in the seventeenth century is well illustrated in the person of John Jonston or Johnstone (1603–1675), a naturalist of English ancestry who was born and buried in Poland, studied at the universities of St. Andrews and Cambridge, lived a while in London, received the M.D. degree and practiced medicine in Leyden, married twice with German women, traveled extensively on the continent, published his numerous works in Latin at Amsterdam, Leyden and Breda in the Netherlands and Frankfurt-am-Main, Leipzig, Jena and Breslau in Germany, retiring for the last score of years of his life to his country place in Silesia.

The first considerable composition by him and the one which concerns us here was *Thaumatographia Naturalis,* in which the emphasis is upon the wonders of nature, distinguished in ten categories of the heavens, including the new stars, the elements, meteors, minerals, plants, birds, quadrupeds, insects and bloodless animals, fish, and men. It was published at Amsterdam in 1632, and again in 1633, 1661 and 1665; and at London in English translation in 1657 as *An History of the Wonderful Things of Nature.* He later developed various sections of this work into more specialized treatments and also published some medical works.[180]

Alexander de Vicentinis in his book of 1634[181] made a considerable attack not only against astrology, as noted in a previous chapter, but also against the conception of occult virtues and qualities which was a basic tenet of natural magic. The magnet did not attract iron by such virtue but by reason of a likeness in temperament which depended on manifest qualities. This was proved by experiment. For, if the magnet was smeared with garlic, it would no longer attract iron, because the garlic prevented the species of the stone from being represented to the iron as it was—rather, quite differently, since garlic is destructive to iron, making it rust and

---

sophorum et artis magicae peritissimis artificibus eruta . . . cui praeit magia superstitiosa de daemonum variis generibus faunis satyris lamiis et spectris eorumque operationibus diversis . . ., Francofurti, 1631, in-12. Copy used: BM 719.a.40.

180 DNB X, 968-69; *Lindenius Renovatus,* 620-21.

181 *De calore per motum excitato et de coeli influxu in sublunaria corpora,* Verona, 1634: BM 549.e.13.(1.).

deteriorate, if it is anointed with garlic juice. The magnet turns towards the poles, because there are mountains of iron and magnet there, as sailors testify. Amber attracts straws for another reason. When rubbed, it grows hot and draws in the adjoining air and the straws with it, not because of occult virtue but because it has plenty of aerial humidity well tempered by a terrestrial portion. This is plain from its viscidity, sheen, rarity and purity.[182]

To occult virtue are attributed various cures by ligatures and suspensions, such as suspending a root of peony from a child's neck to check epilepsy, suspending a viper by a linen thread with which purple-fish have been choked to death as a cure for sore throat and suffocation, binding the root of *pimpinella* to the thigh as a contraceptive, or a certain kind of spider to the arm to cure quartan fever, coral for the heart, and the right foot of a tortoise on the patient's right foot to allay the pains of gout.[183] Alexander says that the suspension of the peony does not work now-a-days, but that some affirm that we do not have the right species, enumerated by Galen. Out of respect, no doubt, for that great authority, Alexander explains that the peony is hot and very dry, for it has burnt earth in great abundance mixed with aerial humidity. Particles from the suspended root are inhaled, carried to the brain, and prevent epilepsy which is caused by obstruction of the ventricles of the brain by crass matter. But why do other hot and dry roots not have the same effect?

> Because, as the peony differs from any other plant in the variety and dissimilitude of its form, so its specific operations will be different.[184]

But this explanation implies that the peony has a specific form of its own, and such specific form had been practically synonymous with occult virtue. So Alexander gives away his case.

The viper suspension has no truth or foundation of reason that Alexander knows of. *Pimpinella* on the thigh will not prevent conception, although taken internally it has the force of moving the menstrua and purging the womb. The remaining "experiments" enumerated are absolutely ridiculous and proved most false by

[182] *Ibid.*, pp. 128-29.
[183] *Ibid.*, pp. 133-34.
[184] *Ibid.*, p. 136.

daily use. If they ever work, it is by accident or to be attributed to the patient's imagination. In support of the latter assertion Alexander tells the story of a mock incantation by which a practical joker at Padua actually cured a woman of quartan fever.[185]

That a rose growing near garlic has a finer odor, is not due to some secret property but to the fact that the garlic absorbs the impure juices from the earth and leaves the purer sustenance to the rose. That the figs on a fig-tree growing near a wild fig-tree ripen quicker is because midges generated in the wild fig-tree perforate the flowers of the other. The real reason why the laurel tree is not struck by lightning may be that its leaves stay green and are not easily congealed by cold or dried out by heat. That less robust trees may stand the cold better than very sturdy ones is not because of the occult property of some star but is a matter of rarity and density. The cold may penetrate trees and plants of rarer texture, but does not stay as long, since it is free to go. Shell-fish are better at the full of the moon, not from celestial influence, as is commonly believed, nor because they are more nourished then, but because, being bloodless, they rejoice in warmth, and the nights are warmer when the moon is full.[186] But the old explanations seem about as plausible and convincing as do Alexander's new ones.

Pomponazzi in 1520 had written that, as the idea in the divine mind brought forth this sensible world without the aid of any instrument, so an idea in our minds may realize itself through the blood and spirits of the body and produce like effects in other bodies.[187] Severinus in 1571 had published his Idea of Philosophic Medicine,[188] which was reprinted in 1616 and 1660. In the present century Helmont had talked of seminal characters or ideas. Now in 1635 Marcus Marci (1595–1667) of Kronland or Landskron in Bohemia, professor of medicine at Prague from 1620 until his death, dedicated to Ferdinand II of Hungary and Bohemia his Idea of Operative Ideas or detection of that occult virtue which

[185] *Ibid.*, pp. 137-38, 140-42, 145-46.

[186] *Ibid.*, pp. 147-51.

[187] T V, 101-2.

[188] *Idea medicinae philosophicae fundamenta continens totius doctrinae Paracelsicae Hippocraticae et Galenicae*, Basel, 1571. The influence of Severinus seems to have grown in the seventeenth century. Barchusen devoted pp. 442-49 of his *Historia medicinae*, 1710, to him.

fertilizes seeds and produces organic bodies from them.[189] This promising title, however, does not lead to as much magic as might be expected but to discussion of monsters, the force of the mother's imagination, pygmies and giants, *androgyni* and various mixtures of human nature with brutes and of these with one another such as satyrs, nymphs, *cynocephali,* sirens, tritons and harpies. Man is said to consist of three principles: an elementary body, a balsamic humor, and innate heat.

The table of contents outlines a second book with chapters on transplantation; "subordinate" or spontaneous generation from corruption; shadowy generation in vapor, smoke, fire, crystal, or magic mirror, of various apparitions and specters; generation of bodies and metempsychosis of souls; metamorphosis and transmutation of bodies, lycanthropy and witches; separation of soul from body; death and whether it can be impeded by natural means, with discussion of the tree of life and the universal medicine of the philosophers. But this program is not fulfilled in those copies of the book which I have seen.[190]

The title of a work on the rainbow by Marcus Marci likewise gives promise of the marvelous.[191] In cites Roger Bacon through Johann Combach's *Specula mathematica* of Frankfurt, 1614, as well as Kepler on Witelo, and digresses to treat of gunpowder and volatile gold and whether a vacuum is possible. It also inquires whether light is the effect of the form of fire and whether all lucid things are ignited.[192]

A book by J. S. Kozak, which appeared at Bremen in 1636,[193] on the properties which man, as a microcosm, shared with the world about him, left a wide opening for human and natural magic.

189 *Idearum operatricium idea sive hypotyposis et detectio illius occultae virtutis quae semina faecundat et ex iisdem corpora organica producit,* Prague, Typis seminarii archiepiscopalis, 1635, in-4. Pages unnumbered. The last signature is (Tt 4).

190 BN 4°Tb71.21 (1) gives the "Capita libri secundi" only; BM 778.f.8 does not have even these, but only the eight chapters of the first book.

191 *Thaumantias, Liber de arcu coelesti deque colorum apparentium natura ortu et causis,* Prague, 1648, in-4, 268 pp. BN 4°Tb71.21(2).

192 *Ibid.,* pp. 70, 75, 41, 43, 54, 159.

193 *Anatomia vitalis microcosmi, in qua naturae humanae proprietates quas homo cum rebus extra se sitis communes habet,* Bremae, 1636, in-4, 269 pp., 29 caps.; BM 648.g.6.

Paracelsus is quoted that the stars possess reason, wisdom, heat, wrath and other emotions no less than man, who indeed receives these from them and has, as an intermediary between the rational soul and blind body, an astral spirit, which is a most subtle substance, source of all political virtues and natural functions sensual and imaginative, and which preserves the body from decay and death to the term destined by God.[194] Even deluded imagination cures many, and the demon teaches the weak witch an art by which internal imagination operates externally.[195] Dreams are spiritual semblances of things, formed by the spirits of the imagination in the supreme region or lunar sphere of the human body. But not all dreams foretell the future.[196] Occult qualities are scorned as "the anchor of asses," but a world soul is accepted and natural spirits in stars, air, water, forests, and underground. Even the demons cannot harm or even deceive man without his aid, while their operation can by impeded by natural means and their art deluded by art. The magician knows how to remove diseases inflicted by magic art by contrary means, and even how to transfer them back whence they came. Ghosts no longer appear, if the corpses are burned and the ashes scattered on water. But if the specters were demons and not ghosts of the dead, their operation could not be impeded by fire. Kozak disapproves of imprisoning spirits in bottles, rings and the like and making slaves of them. They were placed by God in the elements free, and their service by natural means is permissible provided not employed against God or one's neighbor.[197] Diseases of sulphuric origin are customarily impeded or wiped out by magical remedies. But some things work more by the imagination and credulity of the patient than by their own natural efficacy.

> Such are scrolls worn about the neck inscribed with unknown characters, of which innumerable formulae exist, which I think are concocted by medical men and magicians not for their operative power but to excite confidence in the patient.[198]

[194] *Ibid.*, cap. 4, pp. 23-25.

[195] *Ibid.*, cap. 5, pp. 33-34.

[196] *Ibid.*, cap. 7, pp. 45, 50-51.

[197] *Ibid.*, pp. 97, 102, 118, 121, 130-32.

[198] *Ibid.*, p. 195.

Kozak also makes the acute observation that no diseases arise from the blood but from foreign germs[199] lurking in the blood. "It is the blood which suffers, not which afflicts."

A manuscript in the Cambridge University Library (Dd. VI. 10) contains a work of 12 chapters and 141 pages, written in 1636 by Dr. Marcus Bellwood of London, in which he argues that weapon ointment operates by natural and not evil magic, and by magnetic and sympathetic action.

Valerio Martini, who had published medical works at Venice in 1628 and 1636, issued in 1638 another volume which was primarily concerned with occult and specific properties.[200] The full title page professes to reveal three new Wisdoms: the first, of the substance of a thing as a whole and its individual elementary properties, "abstruse hitherto"; the second wisdom, of the entire substance and its superelemental specific properties, eternal unlike the others, but like them recondite; third, a new wisdom composed from these two new ones. By their means things hitherto enveloped in Nature's secret archives are brought into the open light. Martini, who is a very wordy writer, rejects the view of Fernel and also that of Scaliger as to occult virtues. He cites both their views indirectly through Sennert.[201] Their great mistake was ignoring the whole

199 *Ibid.*, p. 245, "ex seminibus peregrinis."

200 The half title page (or, *Le faux titre*) reads: "Subtilitatum veriloquia in quibus proprietatum totius substantiae quae occultae specificaeque sunt patefactio promulgatur. Itidem de colore luce lumine perspicuo transpicuo opaco ac de aliis visioni inservientibus accurate agitur. Ad quae Epistola de monstri generatione accedit. Autore Valerio Martinio Veneto iatrophysico."

As if this was not sufficient, a fuller title page follows on the next leaf, of which I give only a small portion: "Valerii Martinii Veneti iatrophysici libri duo in quibus rerum proprietates totius substantiae huc usque occultae specificaeque nunc patefactae prorsus refulgent...", Venetiis, MDCXXXVIII Ex typographia ducali Pinelliana. 59 pp. of text with 62 lines to the page.

201 *Ibid.*, cap. 3 (p. 7), "Fernelii opinio de totius substantiae rerum occultis proprietatibus a Senerto posita"; cap. 4 (p. 8), "Scaligeri opinio a Senerto ponitur quam ipse amplectitur, Fernelii existimationem et argumenta divellens"; cap. 12 (p. 19), "Fernelii opinio cap. 3 posita ex Senerto altioribus principiis ab autore examinatur ac profligatur. Hinc patet Fernelii opinionem falsam fuisse"; cap. 13 (p. 22), "Scaligeri opinio cap. 4 posita ex Senerti mente undique ab autore evellitur, cum ea falsissima conspiciatur."

For the views of Fernel see T V, 558-60.

substance, without which one cannot know its properties.[202] Yet he elsewhere states that it is easier to know the occult properties of the whole substance than it is to know its formal essences.[203] Martini does not concede generic occult properties from the entire substance but only specific ones.[204] He distinguishes individuals of the first species of the whole substance or contagious diseases; those of the second species or *alexipharmaca;* those of the third species, which are two-fold, both *alexipharmaca* against contagion and solutive remedies, and also include such occult virtues as those of the magnet and remora; and the fourth species or poisons.[205]

By and large Martini does not make good his promise of revealing new wisdoms. His little book is derivative, albeit indirectly, from Fernel's much better known and superior *De abditis rerum causis.* It is accompanied by two other tracts by Martini. One is a letter on a monstrous birth penned back in 1607,[206] in which he defines monsters as sins of nature rather than *res naturales.* The cause of a monstrous birth is either some impediment or ailment in the womb, or miscegenation, or imagination of one or the other parent. In the case under discussion he decides that the cause must be the imagination of the mother. The other treatise is two books on color described as a work of his youth.[207] In it he cites Buccaferreus and Antonius Scarmilionus (*libro primo de coloribus*) and notes the arguments of Averroes, Albertus, Zabarella, and the Thomists and Scotists. Martini holds that light is required for color, although Averroes is represented as arguing that color existed independently without light, which it required only as a medium, or even that color was visible *per se* without the aid of light.[208] Martini discusses the organ of sight, the medium, the real object and its image. He defines light as a real quality characteristic of elements as well as compounds and having in itself the force of illumination.[209]

Finally, in 1639, Martini put forth a volume on natural magic, based on the same notion of the whole substance and the three

[202] Martini (1638), p. 23.

[203] *Ibid.*, cap. 10 (p. 13).

[204] *Ibid.*, II, 1 (p. 47).

[205] *Ibid.*, caps. 16-20 (pp. 29-45).

[206] *De cuiusdam monstri generatione Epistola*, 12 pp.

[207] *De colore libri duo sua aetate iuvenili collecti*, Venice, 1638. The text occupies 96 and 2 pp. of 62 lines.

[208] *Ibid.*, caps. 6 (p. 10) and 42 (p. 90).

[209] *Ibid.*, cap. 20 (p. 40).

Wisdoms, now called newest instead of new, of which he had treated the year before.[210] The work was further lauded in the long title (which in the seventeenth century obviated the necessity of any blurbs on the jacket) as most curious and most useful, stuffed with celestial and divine cult, and offering keys to all the shrines of most recondite Nature. Martini professed to have revealed for the first time the properties of the whole substance, which had hitherto been occult and were known to God and nature alone.[211] But his explanations are none too clear. Celestial influence has much to do with it. In the generation of elements and mixed bodies there are two agents: one absolute, uncreated and metaphysical, as God is; the other, subordinate, created, and physical, as nature is. Occult qualities result from a union of primary and secondary qualities, and there is a fifth form above all condition of the elements. The virtues of the magnet and remora are, he claims, the only ones that he cannot reduce to terms of manifest qualities, and he regards the latter's property as false. The virtue of the whole substance is nothing else than a twin spirit of implanted functional heat, fecund or anti-fecund. Of these spirits one is elementary, individual, blind, generable and corruptible, acting *per se;* the other is eternal, incorruptible, and so on. Both are in the seminaries and seeds of all natural things. Similarly in the case of the fifth form one occult is as the specific of divine forms everywhere declared, while the other occult is occult as the mixed is aggregate and so confused as a whole. Martini approached this solution long since in 1633 in his other works, but now it is completely made known and he thanks God for it.[212]

In Parts Two and Three of the same work, which were printed two years later in 1641, also at Venice,[213] Martini accounts for

[210] *Magia Physica foecunda coelesti divinoque cultu perfusa trium novissimarum totius substantiae sapientiarum simulque claves reconditissimae adytorum naturae omnium proprietatum divinarumque formarum hucusque occultarum. Opus cunctis studiosis curiosissimum utilissimumque in quatuor libris distinctum. In quibus de tota substantia ac de tribus eius novissimis sapientiis accuratissime agitur.* Venetiis apud Marcum Anton. Brogiollum, 1639, in-4. Copy used: BM 535.e.3.(2.).

[211] *Ibid.,* pp. 9, 197.

[212] *Ibid.,* pp. 37, 61, 179, 180, 184-85, 196-97.

[213] They are briefer, comprising 107 and 90 pages respectively.

substantial forms as supra-elementary, produced by order of God and concurrence of the heavens as a separate instrument from the nature of the whole substance. The heavens in actions of natural generation are the separate instrument of *Natura naturans,* an instrument preceding and disposing, metaphysical, equivocal, universal or generic and confused in fecundating, albeit it acts more than Nature herself. But *Natura altissima* as the servant of God, touching the sky in all directions, *supraelementaria physicans, univocans, specificans et foecundans,* is necessary to all generation. After such indulgence in antitheses, Martini affirms the truth of spontaneous generation, and in a final chapter inquires whether consideration of the heavens with reference to generation is physical or metaphysical. There is much criticism of Sennert in the first two parts.

Olaus Borrichius, or Ole or Oluf Borch (1626–1690), held that no force to dispel disease was implanted by nature in characters, words, seals and images.[214] These are not substances, and only substances are the causes of actions. Contact is requisite between the remedy and the cause of the disease, and an idea conceived in the mind does not act upon external objects. Sckegk and Christian Matthias had argued that a saw would not work unless indented, and that therefore figure was important, but Borrichius replies that a paper saw would not work and that therefore mere figure accomplishes nothing. He further argues that speech is not natural to man, or there would not be so many linguistic differences, and that therefore words have no natural significance. Also their meaning changes with time. Music was invented for its psychic effect, not to dispel disease, although it may help some accidentally by taking the patient's mind off his sickness. That is what happens when those bitten by a tarantula dance to music.

Roger Bacon in his treatise on the secret works of art and nature denied that charms and characters were of avail in expelling disease, except those made at elect times according to the constellations. But Borrichius holds that it cannot be proved that the aspects of the stars have any efficacy over nature, much less artificial repro-

[214] *De cabala characterali dissertatio,* Copenhagen, 1649, 44 unnumbered fols. BM 1473.aa.16. (1.).

ductions of them. Others contend that astral spirits have a sympathetic relation with sublunar things and by their balsamic exhalations influence them occultly and endow them with imperscrutable forces. Reason may seem to be against them, but they appeal to innumerable experiences, so Borrichius leaves this in doubt.

Nor does he venture to agree with Fienus and Erastus that amulets are superstitious, since to the contrary militate the experiments of Galen, Dioscorides, Fernel, Valeriola, Droettus, Henry ab Heer, Rhenanus, Simon Paulli, Blochwitius, Paracelsus, Croll, Rhumelius and others. Nor is he ready to deny occult qualities with Henry ab Heer, for he who denies them, denies nature—magnet, echeneis, torpedo, sympathy and antipathy. Why is it that Spanish flies injure the bladder alone of all the viscera unless because of antipathy? But amulets and sympathy and antipathy are not dependent on the varying positions of the stars, and Borrichius attacks Goclenius for attributing extreme powers to astrological images.

That the asp closes its ears to snake-charmers is no proof of the power of words, since the force of such incantations is from the devil. Nor will Borrichius accept the argument that words cure disease not *per se* but through position and articulation, since position and articulation are neither substances nor qualities. As to the combination of prayers with confidence and imagination, he responds that God nowhere in the Bible promises to endow characters with such force because of anyone's prayers, that natural confidence is not the same thing as miracle-working faith, and that imagination cannot affect a foreign body. It is not permissible to use verses of the Bible in medicine, and the papal practices of trying to overcome disease by certain words and exorcisms, to work miracles, and of adjuring water and other irrational creatures are to be condemned as superstitious. Some say that the beauty of figures and images gladdens man, but magic figures are ill-shaped and the words barbarous. Oger Ferrier ascribes their efficacy to the meeting of minds and confidence of physician and patient, but Borrichius rejects this and seems not to know hypnotism. Adam's naming the animals is not an indication of the power or significance of names.

On the other hand, Borrichius cannot agree with those who

attribute all characteral medications to the devil. Let them first explain those occult virtues which they call magnetic, or tell how some men are able to predict the weather from an affected part of their bodies. "Many things are attributed to the devil which are not his, and not without signal injury to nature." But he will not agree that the imagination of the microcosm can affect the macrocosm, as writers like Croll and Burggrav have asserted.

The brochure of Borrichius was in not exactly friendly company in the copy which I consulted at the British Museum, being bound with Kiranides[215] and a chiromancy in German.[216]

Occult qualities, which Borrichius had not been ready to deny, were affirmed the next year in a New Philosophy and Medicine based upon them, which was published at Lisbon by Edoard Medeira of Arras. He added more "unheard-of philosophy" as to the qualities of the Tree of Life, the powers of music, the tarantula, and electric and magnetic qualities.[217]

"An hundred aphorisms conteyning the whole body of naturall magick" are found in two manuscripts at the British Museum, both of the seventeenth century.[218] Also twelve conclusions, of which the eleventh may be quoted as an example: "In the Excrements blood etc. ye spirit is not so deepely drowned as in the Body, and therefore in them it is sooner infected."[219]

[215] Here entitled, *Mysteria Physico-Medica . . .*, Francof., 1681: BM 1473.aa.16 (2.).

[216] Joh. Praetorius, Leipzig, 1661: BM 1473.aa.16 (3.).

[217] *Novae philosophiae et medicinae de qualitatibus occultis a nemine umquam excultae pars prima, philosophis et medicis pernecessaria, theologis vero apprime utilis. Accedit inaudita philosophia de arboris vitae Paradisi qualitatibus, de viribus musicae, de tarantula, ac qualitatibus electricis et magneticis*, Lisbon, 1650, in-8. No BM-BN. LR p. 255.

I have not had access to the following work: Niccolò Serpetro, *Il mercato delle meraviglie della natura ovvero istoria naturale*, Venice, 1653.

[218] Sloane 1321, fols. 14r-19r; 2220, fols. 251v-255v.

[219] The conclusions are discussed in Sloane 1321, fols. 1-13, but occur without discussion in Sloane 2220 at fol. 256. Sloane 1321 also considers such matters as: fol. 20r, "Of things necessary in a Physitian before he undertake this part of magnetical Physick"; 31r, cap. 9, "Of transplantation"; 33v, cap. 11, "Of the magnet necessary in this art"; 37v, cap. 18, "Of ye parings of ye Nayles & Teeth"; 38r, cap. 19, "Of ye spittle & excrem't of ye nose." It ends at fol. 40r. The hundred Aphorisms have been ascribed to William Maxwell, and in Chapter 34 we shall find him

In 1660 William Williams published a book of 159 pages with the following fulsome title:

Occult Physick or The three principles in Nature Atomized by a Philosophical operation taken from experience in three books. The First of Beasts Trees Herbs and their Magical and Physical Vertues. The Second book containeth most Excellent and Rare Medicines for all Diseases happening to the Bodies both of Men and Women which never yet saw light: An Incomparable Piece. The Third and Last Book is a Denarian Tract, shewing how to cure all Disease with ten Medicaments; And the Mystery of the Quaternary and Quinary Number opened; with a Table shewing the Suns Rising, Setting, Hours of the Day, Hours of the Night, and how many Minutes are contained in a Planetary Hour both Day and Night; with a Table of the Signs Continuance on the Ascendent, fitted for Magical Uses; As Gathering of Herbs, Roots, and the like with their Uses. Whereunto Is added a necessary Tract shewing how to Judge of a Disease by the Affliction of the Moon upon the sight of the Patient's Urine with an Example; Also you are taught how to Erect a Figure of Heaven for any time given.[220]

A preparation of *terra sigillata* is effected by repeatedly burning it with *Aqua vitae.*[221] The medicines are given such names as The Flagrant Flower, The Wild Man's Will, Poste without Haste, and "A Generalissimo Medicine made of all the three former General Medicines put all together, and it is called, The Felicity of Nature."[222] Or we hear of Gribbins Comfort, Gilbers Cordial, and, to make men fruitful, "A dozen of Points to truss up his Hose."[223] The Eagle-stone

is white and round like a Tennis-ball, and hath a stone that shaketh within it. Being worn it delivereth women in their extremity, but at any other time it is not to be used by them that are with Child. It is good to be worn for the Stone... Feavers and Plague. It doth also dissolve the knobs of the Kings Evil, being bound to the place grieved.[224]

---

and Samuel Boulton publishing the twelve conclusions.

220 By W. W. Philosophus; Student in the Coelistial Sciences. London, printed by Tho. Leach, and to be sold by W. Palmer at the Palm-Tree near St. Dunstans Church in Fleetstreet, 1660.

221 *Ibid.*, p. 58.

222 *Ibid.*, pp. 38-41.

223 *Ibid.*, pp. 75-76.

224 *Ibid.*, p. 94.

The following mystic utterance appears without further explanation:

> There is a Crocus Albinatus in a quarry in the end of the new walk in Greenwich in a white Starre, somewhat yellowish.[225]

Some diseases

> arise from the weak and doubting mind, which not knowing its proper strength and faculty, yields to all sorts of Evil. Hence all those Evils which proceed from the Demons of the South, under whom are contained Incantations, Magical works, the Influences, and Curses of Demons, and other Evils which do predominate over everyone . . . Many rare Doctors that were before Hippocrates did only make use of the faculty of the Spirit and Mind, without any Corporeal Medicaments to cure all Diseases meerly by the will and Imagination of the Patients.[226]

They even raised the dead, commanded the elements, and solidified water.

[225] *Ibid.*, p. 101.

[226] *Ibid.*, pp. 122-23.

CHAPTER XI

DISCUSSION OF MAGIC IN PORTUGAL AND SPAIN

Chamisso—Valle de Moura—Torreblanca—Juan Eusebio Nieremberg—Hernando Castrillo—Gutierrez on fascination—Caldera de Heredia.

L'esprit humain est capable de toutes illusions et de toutes seductions
—BOURDELOT

In the Iberian peninsula, where Arabic schools of necromancy were supposed to have flourished in the middle ages at Toledo and Salamanca, and where astrology continued to be taught in the modern period at Salamanca later than at any other university, there seems to have been a somewhat more favorable attitude towards occult science than elsewhere in western Europe, and less of an inclination to account for all magic as diabolical. For this reason I have segregated in this separate chapter the discussions of magic which follow and which are first those of two Portuguese and then those by five Spaniards.

Ioaõ Bravo Chamisso studied arts at Evora and medicine at Coimbra, where he taught anatomy and then medicine until 1624.[1] In 1605 he published a medical work[2] in which he held that spoken words could produce effects by their natural force, and that even

[1] Barbosa Machado, *Summario da Bibliotheca Luzitana,* Lisbon, 1786-87, II, 613: "... Cadeira da Anatomia de que tomou posse a 3 de Abril de 1601 e da Vespera a 7 de Feveiro do 1615, onde jubilou a 24 de Julho de 1624." Valle de Moura, *De incantationibus seu ensalmis,* Evora, 1620, p. 22b: "Ioannes a Bravus Chamisius Conimbricensis Doctor anathomicae lectionis olim singularis, nunc vero medicae Cathedrę moderator subtilis." Hoefer, *Nouvelle biographie générale,* VII, 280, is therefore mistaken in speaking of Bravo-Chamizo as "mort en 1615."

[2] *De medendis corporis malis per manualem operationem,* Coimbra, 1605, in-4. I have not seen the book but follow the account of it by Valle de Moura, *op. cit.,* pp. 22-26.

the demon operated solely by the natural force of an incantation, since he operated only by natural application of active to passive. Chamisso further maintained that words varied both in signification and in power, as their imposition by Adam in the case of every beast and every fowl, and the potency of the name of Jesus indicated. He saw no more reason for denying the efficacy of literary elements, when properly ordered as in the cabala, than for denying the marvelous force of the magnet. He therefore reduced practically all incantations of things sensitive and insensitive to the natural force of words. Alluding to the use of the liver or heart of a fish to dispel demons in the Book of Tobias and to the fact that some demons cannot endure music, Chamisso affirms that after their fall demons were subjected by God to sensible things, and that some spoken words put them to rout.

Valle de Moura, in a book published fifteen years later,[3] did not agree with the position of his fellow-countryman of Portugal, Ioaõ Bravo Chamisso, but held that words and even divine names did not have the natural force to produce the effects then ascribed to incantations. Nor could the modulation of the voice be the cause. Words could not possess specific virtue, as other simples did. But Chamisso had written as an anatomist and medical man; Valle de Moura wrote as a theologian and representative of the Inquisition. Against certain medical men he further maintained that it could not be known from the signs of the zodiac whether a disease or wound was lethal. But he recognized that some serious persons defended and that most Christian princes were said to have made use of astrological rings and images.[4] That prayer would work by natural force he stigmatized as an error of Alkindi,[5] and also condemned the assertion of *Albumasar in Sadan* that, if one sought anything from God when Jupiter and the head of the dragon

[3] Emanuel Valle de Moura, *De incantationibus seu ensalmis. Opusculum primum auctore Emanuele do Valle de Moura doctore theologo ac sanctae Inquisitionis deputato Lusitano Patria Calantica. Precipua quae aguntur in hoc opusculo refert Elenchus ad calcem Epistolae ad Lectorem.* Eborae, Typis Laurentii Crasbeeck, Anno 1620, 552 double-columned pp.

[4] *Ibid.*, p. 179-a-b, for the headings of the chapters in question.

[5] *Ibid.*, 3vb. After fol. 11v, the numbering is by pages.

were in mid-sky, one would necessarily obtain it, "in which configuration Conciliator (i.e., Peter of Abano) says that he had sought science."[6]

Besides his opposition to the power of words and incantations *per se,* and to extremes of astrology, Valle de Moura condemned a number of popular superstitious practices such as offering herbs, which had been blest, to beasts to eat; collecting herbs with certain prayers on certain saints' days and the observance of days in general; and ducking an image to bring rain.[7]

With regard to snakes our Portuguese theologian and inquisitor is somewhat credulous. He believes that the asp really closes its ears to incantations, as stated in the Psalms.[8] He quotes Holcot of the fourteenth century that dealers catch dragons by incantations and sell them to the Ethiopians who eat them for their cooling effect in that hot climate.[9] He discusses why serpents are more subject to charms and incantations than are other animals, and why the saliva of a fasting man kills them.[10] On the other hand, certain bedbugs kill by natural force.[11]

Valle de Moura discussed a greater range of topics, some religious and some secular, than the title of his book would lead one to expect. They include sacraments, miracles true and false, ecstacy and rapture, apparitions internal and external, revelations particularly of women, the poverty and Passion of Christ, comedies, faults of philosophers, and occult virtue. He also raised a number of curious, not to say outré, questions, such as whether the torments of hell cease on the night of the Assumption of the Blessed Virgin and of the Resurrection, whether women cannot survive on a certain island of Ireland, why John the Baptist is cherished by the Church as a true martyr when he did not die for the Faith, whether Pontius Pilate was a suicide and a Manichean?[12] He further states that Jews

[6] *Ibid.*, 4ra. See T II, 900.

[7] The question whether such practices are licit is raised at p. 61 *et seq.* and answered in the negative at p. 67 *et seq.*

[8] *Ibid.*, 5rb. Psalm 57(58), 5-6: "sicut aspidis surdae et obturantis aures suas, quae non exaudiet vocem incantantium . . ."

[9] *Ibid.*, 6ra. For a similar statement by Roger Bacon, T II, 657-58. Valle de Moura also quotes an *Ensalmus* or incantation from Anselm of Parma (T IV, 243-47), whose work was not printed, at fol. 1va.

[10] Valle de Moura (1620), p. 249.

[11] *Ibid.*, p. 262.

[12] *Ibid.*, 3rb, 3va, p. 262, p. 362a.

have a hircine odor, from which they are liberated by baptism.[13]

I have consulted the work of Francisco Torreblanca, a jurist of Cordova, on Demonology or Magic in the edition of Mainz, 1623.[14] But this contains a dedication to Paul V of August 12, 1618, as well as the dedication of September 1, 1623, to the bishop of Würzburg, also *Advertencias contra los libros de la Magia de Don Francisco Torreblanca Villalpando,* numbering six points, and his reply in defense of his books which is dated at its close June 24, 1615.[15] The work is divided into four books devoted to divining magic, operative magic, and its punishment in the forum exterior and juridical, and in the interior tribunal of the soul and confessional. We shall be concerned only with the first two books and will postpone to our chapter on divination consideration of most of the first book on divining magic, confining our discussion here to magic in the narrower sense of the word.

Magic is defined after Proclus and Psellus as an exacter knowledge of secret things in which, by observing the course and influence of the stars and the sympathies and antipathies of particular things, they are applied to one another at the proper time and place and in the proper manner, so that marvels are worked. Examples are the remora, torpedo, magnet, asbestos and marvelous fountains, or Pliny's herb *sabina,* a poison to beasts but for man an antidote against snakebite, or aconite which is poison to man but most wholesome for sparrows; or inextinguishable fires and the ever-burning lamps found in ancient sepulchers. Anyone with the least smattering of philosophy knows that there are occult virtues in nature by which marvels might be worked, if they were well known and adapted to practical use. Moses, Laban, Jacob, Daniel, Tobias, Solomon and the Magi of the Bible were all most skilled in this art, but the exorcisms and incantations which are today ascribed to Solomon are spurious.[16]

These two magics, one good and the other bad, led the Man-

[13] *Ibid.*, p. 327.

[14] *Daemonologia sive de magia naturali daemoniaca licita et illicita deque aperta et occulta interventione et invocatione daemonis libri quatuor,* Moguntiae impensis Ioh. Theowaldi Schönwetteri. BM 719.g.85.

[15] *Ibid.*, pp. 4-9 and 9-74. The text proper which follows has a new pagination.

[16] *Ibid.*, pp. 178a-182b.

icheans, Eymeric says, to their belief in two Gods. Bad are the books of magic ascribed to Abel, Abraham, Enoch, Raziel, Tobias, Paul, Honorius, Cyprian, Anselm of Parma, Picatrix, Cecco d'Ascoli, Peter of Abano, Cornelius Agrippa, Paracelsus, Apollonius of Tyana, Godelmann, Ficard, Wier and Melanchthon. By no means immune from this stigma are Robert Perscrutator on ceremonial magic, Archindus (Alkindi?), Roger Bacon and Geber the Arab. Raymond Lull and Arnald of Villanova are full of superstition and are charged with heresy by Eymeric and Mariana. Thomas Bungey (Bungay?) has a book of natural magic and George Ripley one with the same title. Cardan's *De subtilitate* and *De varietate* have long since been expurgated by the Church. Porta under the deceptive title of Natural Magic tried to veil much superstition and illicit magic, so beware of him. The three books of marvelous nonsense and superstition bearing the name of Albertus Magnus are probably supposititious. Mizauld could not distinguish between what is natural and what is superstitious. Wecher should not be read without an antidote, nor the volumes of Ponzetti and Sante Ardoino on poisons.[17]

Torreblanca turns to diabolical magic and the extent of the powers of the devil which make it possible. It is not denied that he can transfer from place to place, and he can transport human bodies. He did not lose his natural powers by the fall, and can apply active to passive, and by local motion stir up winds and storms. That storms cannot be excited by demons is an error defended by heretics like Wier and Godelmann and even among Catholics by Molitor. Torreblanca thinks that Pharoah's magicians changed their rods into real serpents, for the devil could hide their rods and substitute live snakes. He can effect levitation but cannot resuscitate perfect animals, nor even those born of putrefaction in the same number, as Soto teaches *Sent.* 4, *Dist.* 43, *Quaest.* 1, *Artic.* 3, and Suarez on St. Thomas, *Pars* 3, *Quaest.* 53, *Sect.* 2. The devil cannot violate the laws of the universe or perform genuine miracles. He can impel huge machines without fatigue, but he cannot move the stars or celestial or terrestrial orbs or elements. He cannot produce a vacuum, which is contrary to the principles

[17] *Ibid.*, pp. 196b-197a.

of philosophy. He can induce darkness; corrupt air, water and earth in certain places; and produce earthquakes. Of earthquakes Torreblanca distinguishes four types: brasmantic, when islands are raised in the sea; climatic, when mountains are levelled; chasmatic, when land is absorbed in an abyss, as Plato's Atlantis was; and micemantic, when only noises are heard.[18] With demon aid the magus can destroy all the fruits of the earth, kill herds and cattle, down houses and cities, and extract captives from prison.[19] The devil can give the victory to whom he pleases but cannot give men such strength that they can hold up huge weights or increase their natural speed.[20] He can strengthen or weaken memory and impart science, and can disturb, but not change, phantasy and deceive the senses by the art of perspective and the medium of the air. He cannot render a large colored body invisible but he can form objects from the elements as a painter does from colors and, by moving the air, move them. God alone can transform men into beasts but the devil can deceive them into thinking that they are so transformed.[21]

Change of sex occurs in nature, but the devil can affect it only in the case of hermaphrodites. Torreblanca doubts if mares conceive from the wind. But he believes that old men may renew their youth, since nature grants this to the eagle and the snake, and there is a fountain of youth in the New World. It is natural for white hairs sometimes to turn black. Boys of ten have made their nurses pregnant, and girls of nine have conceived, while the phoenix lives to be five hundred years old because it never indulges in sexual intercourse. It is not possible to restore virginity, although both the devil and doctors may counterfeit it.[22]

The so-called separation of the soul from the body, or state of ecstasy, is really only the ceasing of the senses of the body to function. The devil produces it by stopping up the channels by which the sensitive spirits penetrate to the external senses, or by attracting those spirits from the external senses, as Delrio teaches. The devil can enable a man to fast for a long time, for the chameleon

[18] *Ibid.*, caps. x-xii.

[19] *Ibid.*, p. 229.

[20] *Ibid.*, p. 231. It is hard to see why not, with the powers of local motion and levitation that have been conceded him.

[21] *Ibid.*, caps. xiv-xvi.

[22] *Ibid.*, caps, xvii-xix, pp. 245-55.

according to Pliny has such a large lung that it can live on air alone, and about A.D. 1288 a girl subsisted for thirty years on the eucharist alone. The devil can also induce long trances by perturbing the humors or adminstering soporific herbs.[23]

In explanation of the silence and apparent apathy of witches under torture, we are told—again somewhat inconsistently, it would seem--that the devil lifts the weights appended and loosens the ropes of the rack, or prevents blows or drops of water that seem to hit their bodies from actually reaching them, or interposes some dense and solid medium. He can sometimes prevent objects thrown into the fire from burning by local motion, or by the force of unguents or some other natural cause. With regard to corpses not decaying and speaking, it is held that the devil cannot make corpses speak but he can produce sounds which imitate the human voice. Resurrection of the dead is impossible for demon or magician, and ghost stories are idle dreams and delusions. It is true that Samuel appeared to Saul, but that was by special divine order. The devil, however, can imitate specters by assuming an aerial body.[24]

Torreblanca at first expresses uncertainty as to generation from incubi and succubi, but after he has told of the finding of a cadaver thirty feet long in the mountains of Narbonne in the reign of Charles VII, he suggests that since it is natural that giants be procreated from the power of the seed, the demon might select hot and robust men who abounded in seed and act as succubus to them, and then as incubus to women of similar condition and so propagate giants, or pygmies by an analogous process.[25] Going on to centaurs, satyrs, sirens, tritons, nereids, harpies and other monsters, Torreblanca quotes an alleged sermon of Augustine that, when he visited Ethiopia to preach the Gospel, he saw many headless men and women with huge eyes fixed in their breasts, but otherwise like us.[26] The imagination of the parents at the time of conception alters the form of the foetus. Suddenly Torreblanca waxes sceptical and says that he saw an ostrich in the garden of the duke of Lerma which refused to swallow iron.[27] The next chapter deals with the

[23] *Ibid.*, caps. xx-xxii, pp. 256-.
[24] *Ibid.*, caps. xxiii-xxviii.
[25] *Ibid.*, caps. xxx-xxxi.
[26] *Ibid.*, p. 303a.
[27] *Ibid.*, p. 313.

production, transformation and resurrection of insects. Physical or natural fascination, as distinguished from diabolical, is rather infection or contagion.[28]

Thus far Torreblanca has jumbled together natural and diabolical magic almost inextricably, but his remaining chapters are on witchcraft,[29] with the two last on natural and divine remedies against it.[30]

That the sort of topics which have been included in our previous volumes as closely related to, if not part and parcel of, magic, were still so regarded in the seventeenth century, may be seen by a rapid survey of the contents of the *Oculta filosofia* by Father Juan Eusebio Nieremberg, a Jesuit writing in Spanish, which was published at Madrid in 1633.[31] From the full title we see that in addition to occult philosophy it emphasized relations of sympathy and antipathy, artificial and natural magic, and was a continuation of a Curious Philosophy and Treasury of Natural Marvels which he had published three years before.[32] Most of his writings were religious, but he was to add in 1635 a Natural History, this time in Latin, in sixteen books, with two more books on marvelous and miraculous natures in Europe and one on those in the Promised Land of the Hebrews.[33] The *Curiosa filosofia* was reprinted at Barcelona in 1644, and then with the *Oculta filosofia* at Alcalà in 1649. Juan Eusebio was often cited by subsequent writers of the century, so that we may accept his presentation as fairly representative.

Out of the ninety chapters of the first book of the Occult Philosophy and the hundred and eight chapters of its second book, we shall select only enough to corroborate our contention amply. After noting exhalations from bodies as a cause of marvelous effects, Nieremberg turns to individual properties and to relations of sym-

[28] *Ibid.*, caps. xxxv, xxxvii.

[29] *Ibid.*, caps. xxxviii-liii.

[30] *Ibid.*, caps. lii-liii.

[31] *Oculta filosofia. De la sympatia y antipatia de las casas, artificio de la naturaleza y noticia natural del mundo y segunda parte de la curiosa filosofia*, Madrid, 1633, in-8, 218 fols. BN R.12747.

[32] *Curiosa filosofia y tesoro de maravillas de la naturaleza examinandos en varias questions naturales*, Madrid, 1630, in-8, 264 fols. BN R. 12746.

[33] *Historia naturae maxime peregrinae libris xvi distincta . . . Accedunt de miris et miraculosis naturis in Europa libri duo; item de iisdem in Terra Hebraeis promissa liber unus*, Antwerp, 1635, in-fol., 562 pp. BN S. 1324.

pathy by the pores or the shape and position of things. He then passes on to first and second qualities, the elements, local motion, and the impulse of the air. Affections of the mind lead to a discussion of imagination and of the curative effects of music. This involves reference to the tarantula, the poison of whose bite, recurring at certain intervals, is alleviated by the victim dancing to music.

The question whether there is natural fascination is followed by a chapter on notable properties of animals such as the *catoblepas*, which, Pliny says, fortunately keeps its eyes fixed on the ground, for everyone who sees them drops dead. Returning to the subject of fascination for a number of chapters, Eusebio distinguishes three varieties: superstitious, natural and mixed. He gives the opinion of Avicenna and Pomponazzi as to its cause, also that of the astrologers (*planetarios*), and asks if there is a natural fascination of love, whether one can fascinate oneself, and if the basilisk can stare itself to death in a mirror. Then several chapters are devoted to the corpse bleeding in the presence of the murderer, and a single chapter to weapon ointment.

Do some stones exert virtues by their figures and in accordance with the movement of the stars? Are apparitions of armies in the air produced by some sympathy with the stars or some other natural virtue? The monstrosity of the star of Saturn is noted, animals born in stones, and other great marvels. The instinct of animals is treated, their sympathies and antipathies, and gift of natural vaticination. Ten chapters later the theme is long fasts—how some persons have gone for years without food. But in the next chapter Eusebio returns to the theme of natural antipathy, now illustrated by the familiar example of the asp closing its ear to incantations. The last chapter of the first book raises the question, What is the greatest marvel in the world?

The second book notes Solomon's knowledge of natural history and the science of Adam, whereas the Cabala, magic and metoposcopy are listed as vicious sciences. But physiognomy is expounded as true and certain, and those who deny it are called superstitious. The vanity of the cabalistic art is the subject of another chapter. Climacteric years and critical days are coupled with arithmetic, and the sympathy and antipathy of things with the music of the spheres. Differences of sex are found in plants and stones as well as animals, and both plants and stones have proportion with the stars. But fifteen chapters later it is affirmed that love is the bond and cement of the world. Meanwhile the astuteness of beasts has been remarked.

Signs by which occult virtues may be detected are taken up and the difference between astrology and physiognomy. The books on divination from dreams of Nicephorus, Astrampsychus, Achmet and Artemidorus are condemned, as is the superstitious abuse of characters and astrological images. Chapters on natural magic and secrets of nature are followed by others on the artifices of Anaxilaos and Archimedes, stones of extraordinary movement, marvelous effects of the elements, other natural marvels, the prodigious and magical feats of nature and human industry, the rare properties of fountains, natural transformations, and magic effects in nature of occult qualities. The three last chapters deal with a notable experience of intentional species, the union of supernatural Providence with natural, and a closing pious exhortation to a higher philosophy than natural philosophy, which, however, is enforced by a natural example.

The character of the contents of Nieremberg's other two books of 1630 and 1635 is on much the same order as that of his *Oculta filosofia.* The Curious Philosophy and Treasure of Natural Marvels is concerned with such topics as the marvels of imagination, monsters, the stone Iman, sympathy and antipathy, balsam, the phoenix, prognostication from comets, the tree of the knowledge of good and evil and the tree of life, Paradise, whether there are herbs or other corporeal things with virtue against spiritual beings, whether the force of imagination is from the stars, whether demons are tormented more in certain quarters of the moon, and whether the human soul has power over the world of nature. Marvelous volcanoes and new properties of birds of paradise are other topics. The movement of the earth is denied, and such a question is put as, of what animals did Adam and Eve wear the skins?

The History of Nature holds that celestial motions are found and that meteors are generated in animals, and treats of presages and knowledge of medicine from animals. It inquires whether any animals such as the salamander are generated or live in fire. It mentions fish that feed on gold, plants that bear lambs as fruit, spontaneous generation, barnacle geese, the basilisk, monsters and nations of monsters, extraordinary birds, an herb that indicates life or death, and a marine monster whose mouth could hold a horse and the cavity of whose brain could accomodate seven men. Two cadavers were found in its stomach, and perhaps it was of the same species as the fish that swallowed Jonah.[34] But that the vulture conceives from the wind is denied.

Hernando Castrillo (1586–1667), of the Society of Jesus, composed a work in Spanish on natural magic, which seems to have been first printed in 1636 and then again in 1649,[35] although the license to print is dated in 1643.[36] There was yet another edition in 1692.[37] Apparently only the first part of the work was printed in either case. Castrillo notes how the universe reflects the qualities and attributes of its Maker; inquires concerning signatures in inferior creation; and gives some general rules of physiognomy. He asks if the stars are signs of the virtues in inferiors, discusses occult qualities, and other causes of sympathy and antipathy. Also whether there are new natural phenomena which were not present at creation. Natural magic is in part practical, in part speculative. It is not only a science but is superior to other sciences. Adam and Solomon knew it, likewise the three Magi. Albertus Magnus was superior in it. Two chapters are devoted to its history in Spain.

We pass on from the introductory tractate to five others concerned respectively with the earth; the terrestrial paradise—Huet was still discussing the problem of its location in 1691; mountains—with a chapter whether there are springs on the highest mountains; plains, valleys, forests and the vegetation in them—with a chapter on

[34] For this last, *Historia naturae*, XI, 62, p. 265a.

[35] Alegambe (1676) lists both editions under Ferdinandus Castrillus. I have used *Magia natural o Ciencia de Filosofia oculta con nuevas noticias de los mas profundos misterios y secretos del universo visible . . . Primera parte*, Trigueras, 1649, in-4: BM 719.f.14.

[36] Dedication and *Prologo al Letor* are both undated.

[37] *Historia y magia natural . . . (libro primero)*, Madrid, 1692, in-4: BM 8630.g.14.

some special herbs and their occult qualities; and finally metals and stones. Some stones are precious for their divine virtue; the gem Iman is discussed especially, and the relation of stones to the stars—all of which reminds us of Nieremberg's discussion. It is inquired when precious stones began to be esteemed and which were first known; whether some gems are monstrous and of two species jointly; whether there are living stones and ones that move and bear foetus. The eagle-stone or *aetites* is noted, and stones grateful to other senses than the sight. The last chapter deals with *Bezar* (i.e., bezoar stones) and asks whether there are any potable stones or metals.

It would appear that there is more nature than magic in Castrillo's book, at least in that portion of it which was printed. But he might answer that natural magic is more concerned with nature than other sciences are.

Spanish discussion of fascination, of which two previous examples were noticed in our fifth volume, one in a book by Antonio of Cartagena printed at Alcalà in 1530, and the other in a work by Perez Cascales printed at Madrid in 1611,[38] was resumed by Nieremberg in 1633, as we have seen, and in a work devoted to it primarily by Lazarus Gutierrez, whose treatise of 1653, however, was printed outside of Spain at Lyons.[39] He taught first philosophy, then medicine, at Valladolid.

After reviewing and rebutting the views of others at length, Gutierrez concludes that fascination is naturally impossible, for it is supposed to be produced either by infected eyes or words or both, with a malignant cast of mind. He takes up the customary position that force of imagination cannot immediately alter external objects, although it may affect a person's own spirits and humors and so affect another man by contagion.[40] He then proceeds to refute the theory of vision by extramission of rays, which had long since been discarded by all students of optics, although Cartagena had still maintained it. Gutierrez further rejects the notion that the natural temperament of one man is preternatural to another,

[38] T V, 475-6, 486-7.

[39] *De fascino tractatus*, Lugduni, 1653, in-4. BM 784.m.14. BN R. 7611. Cornell G990.

[40] *Ibid.*, pp. 59-60.

or that there are men naturally endowed with the power of healing others, and the theory of Cartagena that a physician or fascinator may be especially endowed by the stars. Even the royal touch he accepts only as a manifestation of divine grace. He grants that man has marvelous occult virtues with respect to other living beings or mixed bodies, but not with respect to another man.[41] He even denies diabolical fascination, for the devil cannot transmute material bodies by his own virtue, and it has been shown that fascination is impossible by natural means, so by what virtue or instrument can he fascinate? Gutierrez further argues that diseases inflicted by demon aid are fewer than is thought, and that it is permissible to remove diabolical tokens.[42]

The bleeding of the corpse in the presence of the murderer occasions Gutierrez some difficulty.[43] Not a few disbelieve it; others say that it often happens by chance. But he has to admit it as a true experiment, because the gravest doctors of every discipline support it as irrefragable. Peter of Abano may seem to accept fascination in a passage of his commentary on the *Problems* of Aristotle, in which he explains the phenomenon of the bleeding corpse thus. The slayer impresses the slain by virtue of his strong imagination and fury with spirits of hostility aroused at the time of the crime. When the murderer reappears, these spirits tend to return to him where they belong, and so stir the corpse and draw blood from the wound with them. But theologians ascribe the flow of the blood to divine justice. Antonius Santorelius or Santorelli, first professor of medicine at the university of Naples, holds that the breath of a beautiful girl wife will rejuvenate her aged husband, while his impure spirits will injure her,[44] but Gutierrez again disagrees.

The first hundred pages of a large double-columned folio volume by Gaspar Caldera de Heredia[45] distinguish between natural and

[41] *Ibid.*, pp. 74, 89, 92, 108, 153, 156, 170.

[42] *Ibid.*, pp. 126, 182, 192.

[43] *Ibid.*, pp. 161-63.

[44] "In sua postpraxi medice, cap. 21." See Antonio Santorelli, *Postpraxis medica seu de medicando defuncto liber unus*, Naples, 1620.

[45] *Tribunal Apollini sacrum medicum magicum et politicum . . .* (the title runs on for a dozen lines), J. E. Elzevir, Leyden, 1658, in-fol. Parte I ends on p. 534 and is followed by an index. Parte II opens with *Tribunal*

diabolical magic and on the whole leave a large field open to the former. After stating that antiquity believed diabolical magic to be natural, listing signs and instruments of diabolical magic, and indulging in frequent quotation, including a half column from the tenth declamation of Quintilian,[46] Heredia represents the devil as making an imitation and pretense of works of nature. But natural magic is knowledge of secret things "through the revolutions of the heavens and the courses of the stars, or by the essences of things or the essential properties of things," and is especially concerned with sympathies and antipathies.[47] There is also artificial magic, prestidigitory magic, chrysopeian magic—which is making gold by alchemy, and diabolical magic, which involves a pact with Satan implicit or explicit. He classes *Ars notoria* under it but seems to regard physical ligatures as a part of natural magic.[48] Discussing the question what the difference is between natural fascination and bewitchment, he asserts that there is always something exceeding the order of nature in the fascination of sorcery.[49] Predicting the future and speaking a foreign language are not necessarily a sign of being possessed by demons, but may be the result of melancholy or of natural divination, such as is possessed by brute animals. Nor is the generation of monstrous objects within the body a sure sign of bewitchment, for nature can do almost anything.[50] This leads to a discussion of the bleeding corpse. Some think that if there is any vegetative virtue left in the corpse, it suffices to make the blood flow through a certain hostile antipathy. Others attribute this to the survival of hostile spirits in the dead body; others, as we have already heard from Gutierrez and Peter of Abano, to spirits of the murderer which penetrated the body of the victim at the time of the murder and which, returning to their own body as the murderer approaches, cause the blood to come out with them. Yet others ascribe it to the effect of imagination

---

magicum quo omnia quae ad magiam spectant accurate tractantur et explanantur, seu Tribunalis medici pars altera, for 93 pp., after which *Tribunal Politicum* begins at p. 95. Copy used: BN Td30.114.

46 *De sepulcro incantato:* see T I, 540.

47 *Tribunal Apollini,* p. 12b.

48 *Ibid.,* pp. 21a, 42a.

49 *Ibid.,* p. 54b.

50 *Ibid.,* pp. 73a, 77a.

upon the spirits. But Heredia holds that the causes are so hidden and occult that they are known to God alone.[51]

Man ever tries to penetrate to the inner nature of things: the heaven of the philosophers, celestial fire, potable gold, abstruse sympathies, unknown antipathies, known to no other age. So it is not strange that so many marvels are daily observed by us that to the unskilled they seem impossible for human artifice and works of the devil. But Heredia is ready to anathematize those who live at ease, know nothing of nature and do nothing for the state, yet dare to damn precious secrets.[52] He stresses the power of celestial influx and strong imagination and the marvelous consensus both in the elements and mixed bodies. "From this very principle the blood of the goat softens adamant, for it penetrates and inserts itself in the other's narrowest apertures" and reaches its smallest particles and inmost parts.[53] Other examples of consensus are given such as the statement of the son of Mesue in the book on animals that, if a foul woman puts on a man's clothing and then the man wears it before it is washed, he will be cured of quartan fever.[54] After other examples from the thirteenth century *Secreta Alberti*, Heredia dwells again on the admirable efficacy of imagination and magic sympathy, with citations from Marcellus Donatus, Avicenna, Algazel and Albertus *De mirabilibus mundi*.[55]

Heredia next broaches the question whether characters and images acquire operative efficacy from the force of imagination and the influence of the heavens. Delrio says not, but critical days attest the celestial influence, and it seems probable that marvelous effects are produced by images and characters, but it should not be held that love or hate can be induced by them. A long list of past authorities is cited in favor of astrological images.[56]

Similar bits of natural magic are not lacking in the remaining medical portion of the book, where we read, for example, of epilepsy *per consensus* (sic) of the left thumb.[57]

[51] *Ibid.*, p. 79a-b.
[52] *Ibid.*, p. 81a.
[53] *Ibid.*, pp. 81b-83a.
[54] *Ibid.*, p. 85b.
[55] *Ibid.*, pp. 88a-90a. Donatus Marcellus of Mantua (c. 1538-1602), wrote *De medica historia mirabili libri sex*, Mantua, 1586; Venice, 1597.
[56] *Ibid.*, pp. 91b-93b.
[57] *Ibid.*, p. 335b.

CHAPTER XII

INTEREST IN THE OCCULT AT GERMAN UNIVERSITIES

Dissertations and disputations: their general character–Dillingen: Frey, Diem, Schmid–Wittenberg: Tandler, Nymann, Elich, Schmilaverus, Georg and Gregory Horst, Vierthaler, Krevet, Nicolai, Baumgarten, Pompeius, Kirchmaier, Mittendorf, Ziegra, Frenzel, Kiniker, Voigt, Wolff and Wantscher, Pohl, Clodius, Rudinger, Freygang–Frommann at Coburg–Hardt at Leipzig– Basel: medical disputations, Soner–Ostermann of Cologne and Jordanaeus of Bonn; Crusius at Cassel–Freiburg-i-B.: Peterman–Erfurt: Hofmann, J. C. Müller–Strasburg: W. A. Fabricius–Leipzig: Stohr–Jena: Ruttörfer, Frischmuth, the Baiers, Prange, Crausius–Altdorf: Soner, Wurffbain, J. C. Sturm, Esenbach–Regensburg–Tübingen–Leipzig: legal cases referred to the medical faculty–Frankfurt am Main: J. D. Horst–Schweling's seminar at Bremen–Rostock: Dorscheus–Frankfurt am Oder: Huldenreich, Placentinus, Ruttörfer, Mentzelius–Ulm: Geuder–Similar topics at Dutch and Danish universities.

Today we suffer that diversity of opinion, where any inept little holder of the doctorate seeks protection for his ignorance in that philosophic liberty which some . . . extol to the skies

–SCHELHAMMER

Interest in the occult and in subjects on the border-line between magic and science may be detected in the dissertations and disputations of German universities in the seventeenth century. A word should first be said concerning the general character of these dissertations. They are apt to be on trite themes, to deal chiefly with topics suggested by the works of Aristotle in natural philosophy, and to be brief disquisitions upon broad fields, or collections of the opinions of past authorities on some particular topic or question. But they do reflect the average interests and mental outlook of both teachers and students. They include such subjects–with their dates in parentheses– as first matter (1624, 24 pp.), form and privation (1624), material and immaterial forms (Wittenberg, 1648), principles (Franeker, 1613), (1644), and the principles of Descartes

(1684), matter and vacuum (Wittenberg, 1645), place (1624, 1648), the world (1648, 1676), the elements (1604, 1610, 1612, 1651), vacuum (1624), efficient cause (1650), final cause (1619), action and passion (1626), time (1625, 27 pp., 1649, 1657), motion (1681, Kiel 1682, 1684), motion and light (1651), physical nature (1648), eclipses (1616), of the starry heaven and stars in general (1651, 12 pp.), new stars (1644), fixed stars and planets (1651), the globe of water and earth (1657), comets (1688), ignited meteors (1696), thunder and lightning (1675, 1694), intrinsic causes of natural phenomena (1644, 12 pp.), internal causes of natural body (1650), the rainbow (1689), nature of air (1667), cold (1680), colors (1669, 1690), spirits and innate heat (*circa* 1650), the matter of the sun (1672, 1673), earthquake (1691, 7 pp.), the use of mathematics in theology (Kiel, 1667), mixture, generation and corruption, with a few words on atoms (1651, 8 pp.), the sea and its causes, affections and species (Giessen, 1608), tides (1696), origin of springs (1669), mineral media, origin of fountains, and tides (1651), winds (1646), illuminated bodies (1679), affections of natural body (1624), generation (Danzig, 1653), principles of generation (1678), and imaginary space (1672).[1]

Turning to interest in themes related to the occult, we may first note three dissertations at the university of Dillingen under the presidency or sponsorship of Jesuit fathers. In 1603 Ioannes Frey held forth on More Secret Philosophy or Natural Magic under Simon Som, S. J. as *Praeses,*[2] discussing imagination and fascination, sympathy and antipathy, and denying that magic was the result of figures, words and local motion. I do not know if he may be identified with Jean Cécile Frey, who became physician to the queen-mother of France and died in 1631, mentioned in our chapter on Physiognomy. In 1611, under Georg Stengel, S.J., as *Praeses,* Nicolaus Diem delivered A Philosophical Castigation of Certain Arts, partly ancient, partly more recent.[3] He admitted the action

[1] These subjects of dissertations are taken from Niels Nielsen, *Matematiken i Danmark, 1528-1800,* Copenhagen, 1912.

[2] Ioan. Frey, *De secretiore philosophia sive de naturali magia in Acad. Dilingae praeside Simoni Som S.J.,* Dilingae, 1603: BN R. 8160.

[3] Nicolaus Diem, *Castigatio philosophica quarumdam artium partim antiquarum partim recentiarum sub praesidio Geo. Stengel* S. *J.,* Dilingae, 1611. Copy used: BN R. 8162.

of the stars, discussed how far astrological predictions are possible, then turned to catoptrics, geometry, music and arithmetic. After describing optical illusions with mirrors, he branded as diabolical the practice of writing on a piece of paper with one's blood and then displaying it to the moon, which was supposed to act as a convex mirror and reflect the message to a distant friend, so that what was written in Italy could be read in Germany. Passing on to medical, military, hunting and like arts, Diem finally came to "black magic and other nonsense of old-wives." The third dissertation, also in 1611 and by Valentin Schmid under Christopher Brandis S.J., was on things incredible to the vulgar crowd.[4]

Five Physical-Medical Dissertations published by Tobias Tandler, a professor at Wittenberg, in 1613,[5] indicate that a lively interest in magic still prevailed in that cradle of Protestantism. These included a reprinting of Martin Biermann's *De magicis actionibus*, first issued at Helmstedt in 1590,[6] and an oration on imagination which Hieronymus Nymann, a colleague and doctor of philosophy and medicine who had since passed away, had delivered ten years before in 1603.[7] Nymann displays commendable scepticism as to occult action. He admits that imagination by the patient aids in effecting cures, but denies that it can affect other bodies and external objects, although he notes that Avicenna and Albertus Magnus answered this question in the affirmative, holding that as Intelligences move the orbs, so our soul can affect the elements. Nymann further holds that there is no force in characters or weapon ointment and that all cures by such means are really worked by diabolical magic. He also denies that the corpse will bleed at the approach of the murderer, unless by divine miraculous action. But he still believes in "an occult faculty" in amulets, which he explains by effluvia imperceptible to us.[8]

[4] *De vulgo incredibilibus*, Dilingae, 1611: BN R. 8161.

[5] *Dissertationes physicae-medicae de spectris fascino et incantatione, melancholia et noctisurgio, quibus accesserunt non minus desiderata Hier. Nymanni de imaginatione oratio et Martini Biermanni de magicis actionibus*, Wittebergae apud Zach. Schurerum, 1613, in-8. BM 526.g.4.

[6] T VI, 534-35.

[7] *Diss. physicae-medicae*, 201-35. See also BM 719.a.31.

[8] *Ibid.*, pp. 227-29, 224-26, 231-32, 226. A MS at Munich, dated 1616, contains something astronomical ascribed to "Hier. Nymansis": CLM 10675, ff. 131-176.

Tandler himself contributed to the volume a dissertation of 1605 on *fascinum* and incantation, an oration on specters delivered on the occasion of conferring the M.D. on Peter Schmilaverus in 1608, and a question as to the divination and other marvelous effects of melancholy persons by a boy, Caspar Magnus Hetlenbach, and the answer of Peter Schmilaverus. In the dissertation of 1605, after distinguishing between natural and diabolical magic, Tandler says that he is not going to treat of the former nor of all kinds of the latter but only as to the nature and causes of *fascinum*, which he uses as a synonym for bewitching and incantations. He believes that witches do not act of their free will but as a result of alienated reason, emotional outbursts, and delusion of phantasy and by the devil. He therefore holds that witchcraft is merely a dream or delirium of the witches or a delusion of the devil. They may think that they have made a pact with him; "but that contract is of no weight." Tandler grants that demons can transport men through the air, but it cannot be done through incantation. Witches do not banquet nor dance nor have sexual intercourse with diabolical specters. Melancholy and phlegmatic women may think so but they cannot conceive or give birth without masculine seed, and the latter cannot be transported by demons without losing its generative force. Men cannot be transformed into beasts; werwolves do not devour children. Man has no power over the heavens, though the demon is an excellent weather prophet. There are no natural love philters. Infants and women are more subject to *fascinum* than others, but *fascinum* can neither bring on disease nor cure it. Like Nymann, Tandler holds that there is no natural force in weapon ointment. He admits occult antipathies, such as rue's dispelling toads, figs ripening rue and maturing the flesh of fowl, and feathers of an eagle consuming the feathers of other birds, but these do not demonstrate the possibility of fascination. For only visual spirits emanate from the eye—Tandler does not seem to be aware that the theory of vision by extramission had been long since abandoned —not venemous ones, which would also be injurious to the emitter. The basilisk does not kill by its glance but by poison from its entire body. Menstruating women do not injure. After discussing the power of words pro and con, Tandler concludes that images and characters have no force of themselves, and that the virtues of gems and herbs do not extend over spiritual beings.

Tandler adds "A little Sponge to Wipe Out the Calumnies of Master Philipp Ludwig Elich," who had attacked him in a recent publication, although he knew him neither by sight nor studies. In reading Elich's book *De daemonomagia* (1607), Tandler found partly trite opinions of others, partly new reveries of Elich's own. He wrote to the university of Marburg and Giessen, inquiring about Elich, and was told in reply that Elich was an evident rascal. He had disputed publicly concerning diabolical magic, but had been forbidden to print his views, and his belongings and papers had been seized. He promised to abandon his vanity, but instead printed it at Frankfurt with a virulent preface attacking the academic senate at Marburg, then saved himself from arrest by flight and went over to the papists. The work in question by Elich seems to be his *De daemonomagia* of 1607. It states that his foes in Marburg accused him of being a magus, demoniac, necromancer and having a familiar spirit, all of which he denies absolutely. On the other hand, he says that no sane person would deny the existence of natural magic. He affirms pacts both open and tacit between witches and the devil, and asserts that witches can corrupt air and water in certain places, kill sheep and cattle, produce imperfect animals, and go truly and corporally to nocturnal conventicles, of which he gives a circumstantial account, including such a detail as that the witches report what crimes they have committed since the last sabbat, and are beaten by the devil, if they have not committed enough. The devil can transport the witches without use of ointment but prefers to employ it for various reasons, one of which is that many witches cannot endure bodily contact with him.[9] But Elich sets some bounds to the powers of magic. The magi cannot stop the motion of the celestial bodies nor change the order of nature; they cannot reduce in size so as to go through a small hole; they cannot truly transform. He hardly takes up any position of his own in discussing the problem of incubi and succubi.[10]

Having disposed of Elich, Tandler introduces the subject of mel-

[9] Elich, *De daemonomagia*, 1607, (BM 719.b.60): pp. 27, 41, 52-60, 85-86, 121, 131-139.

[10] *Ibid.*, pp. 76, 142, 148, 125.

ancholy and its species.[11] He asks whether the internal causes which render the animal spirits gloomy are from their own substance or the admixture of an alien body. Licit and moderate sexual intercourse often helps melancholics. Bleeding is also good for them, as is antimony properly prepared, and the cautery or trepanning, if other remedies fail. Sometimes they are beaten, or chained lest they injure themselves or others. Sometimes they are left to themselves with an emerald or a powder thereof around their necks, and so often return to their senses of their own accord.[12] Thus Tandler, like Nymann, is not without faith in amulets.

In answering the boy's question, Peter Schmilaverus groups *ecstatici* along with *melancholici.* Antonio Guaineri held that the soul, before its infusion into the body, knew all, but needed to study astrology in order to recover this gift. The influence of the planets meets with less resistance in melancholics. Schmilaverus, however, suggests that what appears to be the gift of tongues may be subconscious memory of a few Latin or other foreign words which, mixed with gibberish, seem to an ordinary hearer genuine discourse in another language. He will not accept the stars as the cause of a sudden gift of tongues or acquisition of learning, nor melancholic humor as the cause of these or of knowledge of hidden and future things. But the demon and melancholy together may cause it. Incidentally he says that animals and some fools possess the power of unreasoning divination, while artificial and rational prediction stem from natural causes and the stars. Whereas Tandler had denied that *fascinum* could cause or cure disease, Schmilaverus holds that witches can inflict disease with the aid of the devil.[13]

There follows a discussion of somnambulism with Tandler as presiding officer and Georg Horst as respondent. It is stated that sanguine persons are seldom sleep-walkers; phlegmatic and melancholic, never; but the bilious, yes.[14]

Last we come to Tandler's Oration concerning Specters which appear to those who are awake. He justifies the introduction of such a subject into medicine by citing Hippocrates that the phy-

[11] *Dissertationes* (1613), 104 *et seq.*

[12] *Ibid.*, 116-17, 137-38, 153.

[13] *Ibid.*, 164-65, 168-69, 171, 173, 176, 178.

[14] *Ibid.*, 180, 187.

sician ought to know τὸ Θεῖον. It is certain that there are specters; in fact, there are three kinds of them. The first appears to the healthy by magic illusion by means of mirrors and the like. The second kind appears to the sick because their organs are disordered and are represented by imagined specters, the phantasms of the intoxicated and of the melancholy. The third kind are illusions of Satan. Paracelsus enumerated seven kinds of spirits: good angels, human souls, fiery beings residing in the upper air below the sphere of the moon and known as Pennates and Salamanders, aerial spirits in the middle and lower regions of the air, such as fauns, satyrs and sylvani, aquatic, earthly and subterranean, and finally infernal. But to profane and blasphemous Paracelsus, Tandler prefers the Bible which mentions only two varieties, good and bad. The souls of the dead do not return as specters, and the reputed transformation of men into brutes is a diabolical illusion.[15]

Another member of the Horst family, Gregory (1578–1636), was author of a disputation not in the present volume but also published at Wittenberg in 1606, in which he questioned whether the flowing of the blood of the slain human corpse indicates the presence of the slayer?[16] He quoted various authorities that such bleeding was no proof of guilt. Five years later in a dissertation at Giessen, Gregory Horst contended that philters and amatory cups were not natural but diabolical.[17] Horst, who had received the M.D. degree

[15] *Ibid.*, 2, 6-8, 10, 18-21, 28-29.

[16] Greg. Horstius, *An fluxus sanguinis cadaveris humani occisi praesentiam infectoris indicet*, Wittenberg, 1606. This may be an abbreviated form of title for his *Σκέψις de naturali conservatione et cruentatione cadaverum ubi ex casu quodam admirando et singulari duo problemata deducuntur, addita exercitatione de somno et somniis*, Witebergae, 1606, in-8.

W. G. Aitchison Robertson, "Bier Right," *Fifth International Congress of Medicine* (Geneva, 1925), 1926, pp. 192-98, gives some account of Horst's and other discussions of the problem, and says on the general subject: "Some hours after death the blood begins to coagulate or clot . . . and this is fairly complete some thirty hours after death . . . Where, however, death is from suffocation or burns or certain poisons, the blood remains fluid longer (p. 198) and in some cases hardly changes at all, and, if the body is moved or handled, may flow . . . (Or) when putrefaction has advanced to a certain stage, gas forms in the cavities of the body exerting great pressure, and the laying on of hands or putting fingers in the wounds may loosen tiny clots and let the blood out."

[17] *Dissertatio de natura amoris ad-*

at Basel in 1606, after teaching at Wittenberg and Giessen, became municipal physician at Ulm in 1622.[18]

Nine meteorological dissertations under Tandler in another volume[19] do not seem to border on the occult, although one of them, by Michael Rollenberg, denies that comets are ethereal.[20] But a preceding dissertation in the same volume, by Martin Vierthaler under Aegidius Strauch as *Praeses* at Wittenberg in 1606,[21] is accompanied by six corollaries which inquire whether witches can produce rain; whether Albertus rightly attributed the roundness of raindrops to their revolving as they fall; whether there are prodigious rains; whether generation occurs in the supreme part of the lowest region of the air; whether manna is honey or a sort of sugar; and whether it is the same as that with which God fed the children of Israel? Of these queries only the third is answered affirmatively. There is no answer to the fourth. The others are answered in the negative. The twenty-fourth of the twenty-seven tracts in the volume is on metals by Heinrich Krevet and in *Coronides* at its close refers to the Silesian boy with the natural gold tooth, but it was published at Hamburg in 1615[22] and not at Wittenberg.

Henry Nicolai discussed witchcraft as a student at Wittenberg in 1623[23] and as a teacher at Danzig in 1649.[24] In the former disputation, under J. Martini as *Praeses,* it was denied that the souls of witches could be separated from their bodies and attend sabbats, and affirmed that their bodies could be borne through the air by demons. Demons, however, could not engage in sexual

ditis resolutionibus de cura furoris amatorii, de philtris atque de pulsu amantium, Giessae apud Casp. Chemlinum, 1611, in-4, BN R. 3394 and three other copies.

18 LR 359a-363b, lists many works by him.

19 BM 531.1.1. (7-16.); (7.) is Tandler's announcement of the series.

20 BM 531.1.1. (9.).

21 BM 531.1.1. (5.).

22 BM 531.1.1. (24.).

23 *Resp. Διασκεψις philosophica de magicis actionibus earumque probationibus . . . Praes. J. Martini,* Typis C. Tham, Wittebergae, 1623, in-4: BM 1395.h.22. Editio secunda, 1623, in-4: BM 8630.ee.i. (3).

24 *De magicis actionibus tractatus singularis existentiam definitionem etc. magicarum actionum discutiens, exemplis et historiis illustrans et obstantia breviter resolvens. Exercitationibus quibusdam in Gymnasio Gedenensi percursus,* Dantisci, 1649, in-4: BM 719.g.3.

intercourse and have human offspring. Men could not be transformed into wolves. But the test of floating on water was maintained against Wier and Timpler. With regard to witches' not weeping when tortured, it was said, "And we sometimes see boys, who by nature cry easily, so obstinate that they do not shed a tear though whipped till the blood flows." In the latter volume there are ten exercises with a student respondent in each case.

Under Samuel Baumgarten (Pomarius) as *Praeses,* a disputation on sleep-walkers by Johann Fabiger as respondent,[25] and a dissertation by Jeremias Schultz on the same subject, were held at Wittenberg in 1649 and 1650 respectively. They received more than passing attention. That by Fabiger attained a fourth edition in 1686,[26] while that by Schulz is preserved in a printing of 1750.[27] Baumgarten's own Treatise on the Consent and Dissent of Natural Bodies, "once written by him in the most celebrated university of Wittenberg," and hitherto much in demand but now out of print, was reprinted at Wittenberg in 1669[28] and again in 1682.[29] It was originally a disputation under Johann Sperling, professor at Wittenberg, as *Praeses,* but now are added nine disputations by other respondents on such topics as signatures in plants, macrocosm and microcosm, weapon ointment, and the corpse bleeding in the presence of its murderer.

The second of these disputations alluded to "the most celebrated physiognomist and astrologer of our university," Nicolaus Pompeius, "my teacher and sponsor," and cited a passage from his Anthropological Physiognomy.[30] This work I have not found, but lectures on chiromancy which Pompeius delivered at Wittenberg in 1653 will be treated in our chapter on Physiognomy.

[25] *De noctambulis disputatio prior* . . ., s.l., 1649, in-4: BN R. 6642; BM 1179.d.10 (15.) and 536.f.7 (5.).

[26] "Editio quarta," Wittenberg, 1686: BM 1179.c.10 (22.) and 7306. i.9 (23.).

[27] *De noctambulis dissertatio posterior,* s.l., 1750, in-4: BN R. 6643; BM 1179.d.10 (16.) and 536.f.7 (7.).

[28] *Tractatus de consensu et dissensu corporum naturalium in celeberrima universitate Wittebergensi ab eodem quondam conscriptus, hactenus diu multumque desideratus, nunc ob exemplarium omnium defectum et materiae utilitatem denuo bono publico exhibitus,* Editio altera, Wittebergae, 1669, in-4, 196 pp. BN R. 2289; BM 444.d.26 (2.).

[29] Wittebergae, 1682, in-4: BM 1175.c.10 (4-13).

[30] *Ibid.,* 1669, pp. 34-35.

When Sperling's *Zoologia Physica* was published posthumously at Leipzig in 1661, it was accompanied by six disputations by or under the direction of Georg Caspar Kirchmaier, upon the basilisk, unicorn, phoenix, Behemoth and Leviathan, dragon, and spider. Another disquisition by Kirchmaier appeared at Wittenberg in the same year on the bird of Paradise, ark of Noah, and the flood. The first six and that on the bird of Paradise were published again together in 1669 and 1671.[31]

August Cademann had been respondent in the dissertation on the basilisk on June 1, 1659.[32] It is asserted that the existence of the basilisk is undeniable, since one was seen at Warsaw by more than two thousand men. But it is false to say that it is hatched from an egg laid by a cock or that it kills by aspect alone. At the close of the dissertation is added a zoological decade, in which it is held that a serpent *per se* and *ex se* cannot understand incantations, that the phoenix is not literally true, but that the gryphon exists. "We laugh at those who think that the entire species of unicorns perished in the flood." That the swan sings just before it dies is a figment; that brutes can talk among themselves is nonsense. Man alone by nature laughs, weeps and speaks, not the ape, crocodile or pie—the hyena is not mentioned. It is absurd to say that there is an insect which lives for only a day, and the pelican's feeding its young with its own blood is a dream. The dissertation on the unicorn, with Johann Friderich Hubrigk as respondent, after listing various animals said to have a single horn, asserts that the monoceros exists and is not to be confused with the rhinoceros. "No one denies that the unicorn's horn resists poisons,"—a very questionable assertion. The dissertation on the phoenix by Peter Oheimb occurred on May 23, 1660.[33] To him the phoenix was nothing but a figment and *non ens*. In four zoological supplements he affirmed

[31] *De basilisco unicornu phoenice behemoth leviathan dracone araneo tarantula et ave Paradisi dissertationes aliquot.* Editio altera locupletior correctiorque, Wittenberg, 1669, in-8: BM 987.a.31 (2.) is the edition which I have used.

[32] BM B. 426 (5.), *De basilisci existentia et essentia sub praesidio Geo. Casp. Kirchmaieri . . . in elect. ad Albim Academia Augustus Cademann.*

[33] BM B. 426 (6), *De phoenice sub praeside G. C. Kirchmaieri in electorali ad Albim Academia Petrus Oheimb.*

that moles have eyes, that dormice do not sleep all winter, bears do not lick their cubs into shape, and not all hares are hermaphrodites. Thirty-six years later at Regensburg the phoenix as a fictitious bird was again the theme of a dissertation.[34] Indeed, the phoenix continued to be the subject of academic discussion in Finland as late as 1748.[35] The unicorn had earlier been the object of an exercise by Christian Sagittarius at the University of Leipzig on September 18, 1652.[36]

Returning to Wittenberg, we find Behemoth identified with the elephant, and Leviathan with the whale. The dragon's antipathy to the elephant leads to a listing of antipathies between other animals. Moreover, the existence of flying dragons is affirmed. On the other hand, it is denied that the bird of Paradise subsists on dew or air, keeps in continual flight, and has no feet.

Bernhard Mittendorf, with Christopher Nottnagel as *Praeses*, discussed unusual winds at Wittenberg in 1661, especially that of December 9 which blew over almost all Europe, with an appendix on the recent comet.[37] He said that there was not a week and hardly a day of the past winter that we were not terrified by such winds, and that now a comet had been seen which was not to be ignored as a sign of future ills.[38] He took up the relation of the planets, signs and fixed stars to winds, and instances of past winds, such as that of 1352 after the appearance of a comet and about the time when the Great Schism began—an example of very loose historical dating. Turning to the comet, he contended that its cause was supernatural and that, like others, it was a warning to a particular part of the world. He opposed Erastus[39] repeatedly as to the signification of comets, and said that no one would deny that the comet of 1618 heralded the Thirty Years War (*bellum nostrum Europaeum*). Since Aquila is the nearest star to the current comet, he doubts not that judgment should proceed from this star. He

[34] Frid. Seuberlich, *De phoenice ave fictitia*, Regiomonte, 2 Iunii 1696: BM B. 426 (19.).

[35] *De phoenice ave*, Acad. Aboensi, Aboae, 1748: BM B. 426 (24.).

[36] *Exercitatio de unicornu in alma Philurae inclutae facultate philosophica:* BM B. 426 (4.).

[37] *De ventis insolentibus . . . cum appendice de recenti cometa*, Wittebergae, 1661. BM 536.f.17. (5.).

[38] *Ibid., Anteloquium.*

[39] T V, 656-57.

then reviews past comets from those of antiquity down to that of 1652,[40] making seventy-eight theses in all.

The dissertation, *De astrolatria*, by J. G. Schwab under J. E. Ostermann as *Praeses*, was on ancient worship of sun, moon and stars rather than astrology.[41]

The case system was employed in a work on the sympathy and antipathy of natural things by or under Constantinus Ziegra, printed at Wittenberg in 1663. Sixty-one instances thereof are given in as many paragraphs, of which the fortieth, to give one example, is concerned with the antipathy between the dragon and the elephant. Besides sympathy and antipathy, Ziegra believed in occult virtues and the influence of the stars.[42] Bound with this work in the copy which I used was another Wittenberg product of the year following on specters in human form which indulges in much citation of Caspar Schott and Martin Delrio.[43] Beughem's *Bibliographia mathematica* of 1688 associates with the same Wittenberg professor, Simon Friedrich Frenzel, a disputation on the star seen by the Magi, printed in 1677 at Wittenberg,[44] where occult themes never seemed to lose their interest. The same topic of specters in human form had been the subject of a Wittenberg dissertation by J. Kiniker in 1664.[45]

Gottfried Voigt discussed the bleeding corpse again at Wittenberg in 1665.[46] Three years later, at Güstrow in Mecklenburg, he issued his Physical Curiosities on the resuscitation of brutes, the resurrection of plants, and the swan's song.[47] Under the fitting

40 After the comet of 1500 only six are listed: in 1506, 1527, 1531, 1539, 1618, 1652.

41 *Diss. . . . de astrolatria in illustri ad Albim academia . . .*, 1663, 20 pp.: Col 156.4 Sch 92.

42 Constantinus Ziegra, Praeses, *De sympathia atque antipathia rerum naturalium*, Wittebergae, 1663. Copy seen: BN Rz. 2105.

43 S. F. Frenzelius, Praeses, *De spectris in specie humanis quae asseruit nuper et produxit in medium P. Caspar Schottus*, Wittebergae, 1664, in-4. Copy used: BN Rz. 2106.

44 S. F. Frenzelius, *Disputatio respondente Dorero de stella . . . a Magis visa*, 1677, in-4. BM 531.l.4 (15.).

45 *De spectris in specie humanis*: BM 8630.e.52.

46 *Exercitatio de stillicidio sanguinis ex interemti hominis cadavere praesente occisore, praeside C. Faselto*, Wittebergae, 1665, in-4: BM 1179.c.7 (9.).

47 *Curiositates physicae de resuscitatione brutorum ex mortuis resurrectione plantarum cantione cycnea . . .*, Gustrovii, 1668, in-8. BM has four copies. I treat of it more fully in Chapter 31.

name of Christian Wolff as *Praeses,* a zoological disputation on the wolf and lycanthropy was engaged in by Christopher Wantscher as *respondens* at Wittenberg on October 15, 1666.[48] Martin Pohl discussed the problem whether Esau was a monster at Wittenberg in 1671.[49]

At Wittenberg in 1675 was printed the dissertation of Joannes Clodius (1645–1733) on the magic of the arrows of Nebuchadnezzar after Ezekiel 21,21: "For the king of Babylon stood at the parting of the way, at the head of the two ways, to use divination; he made his arrows bright, he consulted with images, he looked in the liver."[50] In July of the preceding year Clodius had presided at the dissertation of Johann Christoph Rudinger on familiar spirits as they are commonly called, which was printed four years later.[51] It classed familiar spirits as evil demons, the method of acquiring them as blasphemous, idolatrous and full of superstition, and declared that magicians had no imperium over demons. This Rudinger may have been a descendant of the Johann Rudinger whose book on illicit magic was published at Jena in German in 1630 and 1635. The dissertation of Clodius on the arrows of Nebuchadnezzar was reprinted more than once.[52]

In 1676 C. Freygang engaged in a physical disputation under Johann Müller as *Praeses* concerning sorcerers stirring up storms.[53] He listed ceremonies employed to excite storms such as throwing stones behind one's back towards the west, or tossing sand from the bed of a torrent into the air, dipping a broom in water and scattering drops skyward, making a little ditch, filling it with

48 BM B. 426 (8.).

49 Martinus Pohlius Resp., *Disputatio physica de quaestione an Esau fuerit monstrum.* Praes. N. B. Pascha, Wittebergae, 1671. in-4: BM 1014.b.6 (5.).

50 *Diss. de magia sagittarum Nabuchodonosoris ad Ezech. 21, 21,* Wittenberg, 1675, in-4: BN A. 5600 (754).

51 Joh. Chris. Rudinger, *De spiritibus familiaribus vulgo sic dictis, praeside Johanne Clodio ... ad diem xxii julii 1674 ...,* Wittebergae, 1678. BN Rz. 2015. BM 8406.ccc.36 (5.).

52 By G. Menthen, *Thesaurus theol.-philol.,* tom. I, 1701; Blasius Ugolinus, *Thesaurus antiquit. sacr.,* vol. 23, 1744: while Zedler mentions a reprinting of 1708.

53 C. Freygang, *Disputatio physica de magis tempestates cientibus,* Praes. J. Müllero ... Wittenbergae <1676> in-4. BM 8406.ccc.36. (3.). To Müller himself is ascribed *De corporum defunctorum operationibus,* Wittenberg, 1679: not in printed BM and BN catalogues.

urine or water, and stirring this about with one's finger or a stick.

They boil pigs' bristles in a pot; sometimes they place beams or logs crosswise on the bank, and other madness of that sort.[54]

Freygang's conclusion is that only the devil could have persuaded men that the weather would be affected by such inefficacious rites.[55] He also gives various incantations employed to cure fever and toothache or to stop nosebleed. Witches believe that diseases also may be brought on by use of words, figures and characters, and lost or stolen articles recovered. Likewise in cures by astrological images. When they want to be transported to their conventicles, they first make certain inclinations and circles, then anoint themselves and the broomsticks on which they are conveyed or at least believe that they are. Then finally they say: "Oben aus und nirgend an." But all this is the work of the devil based upon pacts either explicit and expressed or implicit and tacit.[56]

True it is that God alone has control over tempests, but He sometimes permits the devil and witches to produce them. Moreover, the devil is a very skilled meteorologist and, when he sees a storm coming, instigates the witches to ply their rites, and thus persuades them that they have caused the storm. But the natural causes of all storms are the stars, which draw now these, now those effluvia from earth and waters and so produce now rain, now winds, now thunder storms, now other weather changes. Freygang will not concede that either the devil or his addicts have any power over the stars, and he does not agree with Bodin and Helmont that the devil is the cause of all thunder storms. Thunder and lightning are the result of a mixture of effluvia of sulphur and nitre similar to that in gunpowder. Thus the physical causes of storms are God acting through the stars and the stars acting through terrestrial effluvia. The devil and witches are not physical but merely moral causes.[57] Yet Freygang presently states that it is not impossible for the devil to cast fire like a thunderbolt from the air to earth by which many perish and also violent winds which sub-

[54] *Ibid.*, Sect. 4.

[55] *Ibid.*, Sect. 6.

[56] *Ibid.*, Sects. 7-8, 16.

[57] *Ibid.*, Sects. 9-15.

vert large buildings and produce storms at sea and earthquakes.[58] As evidence for this he cites the Book of Job, Herodotus, Olaus Magnus (III,16) concerning the Finns and Lapps selling winds to sailors, and quotes three other authorities who say nothing of any participation by the devil.[59]

Bound in the same volume with this dissertation of Freygang is a series by eight *Respondentes* under J. C. Frommann as *Praeses* at the university of Coburg, dated between 1670 and 1674, which were to form the foundation of his *De fascinatione* of 1675, of which we treat in a later chapter on Illicit Magic. Also in the same binding is a dissertation defended at Leipzig on April 3, 1680, in which it is held that many confessions and sentences prove that the transportation of witches through the air is not only possible but a fact.[60]

Seven volumes of medical disputations at the University of Basel, which were printed in the years from 1618 to 1631[61]—some of the disputations are of earlier date—deal primarily with medicine and not with magic. Nevertheless they show that subjects bordering upon the field of magic or likely to involve semi-magical therapeutic, were discussed at Basel during the first thirty years of the century. Each volume constitutes a *Decas* or Decade, that is, contains ten disputations. The first Decade of 1618 is confined to purely medical themes. But of the second set printed in 1619, the first is on the winds of the microcosm or flatulency, the second on occult diseases and their cure, the third on epilepsy—a disease apt to involve superstitious remedies and ceremonial, the seventh on canine appetite, and the eighth *De ephialte,* i.e., incubus or nightmare. The third Decade of 1620 included a disputation on venomous diseases and a discussion of melancholy by Ernst Soner (1573–1612) back in 1601, when he was a student and candidate for the M.D. degree. In it he briefly (¶ 30) raised the question whether melancholy could be brought on by demons. But, after adducing in the affirmative Scripture, Hippocrates' τὸ Θεῖον, and Guaineri's statement that mel-

[58] *Ibid.*, Sect. 17.

[59] *Ibid.*, Sects. 18-19.

[60] Joh. Gottlieb Hardt, *Diss. physico-historica de Strigiportio,* Respondens L. Hilpertus, in-4, 20 pp. BM 8406.ccc.36 (4.).

[61] *Decades disputationum medicarum selectarum,* Basiliae: BN 4°T31. 240.

ancholy sometimes suddenly endowed illiterates with science and ability to write, he left the question open to the judgment of the learned. The fourth Decade, which also appeared in 1620, contained disputations on rabies, epilepsy and the nature of love and cure of lovers. The fifth volume of 1621 comprised a treatment of incubus. In the sixth Decade of 1622, there was a brief treatment of philters, which opposed the notion that menstrual blood is a love charm, and a much longer discussion of melancholy, both idiopathic and sympathetic. Publication then ceased until 1631, when the seventh volume contained a disputation concerning rumination and ruminating men.

In 1629 Peter Ostermann published at Cologne a legal discussion of various kinds of signatures, characters and stigmata, supernatural and natural, but especially those of antichrist and the marks of witches.[62] At Cologne too appeared a refutation of Ostermann by Joannes Jordanaeus of Bonn, who denied that the so-called marks of witches were a legitimate proof.[63] At Cassel Crusius discoursed on other magic than natural.[64]

At the University of Freiburg-im-Breisgau in 1631, under Leonard Bildstein as presiding officer, Adam Peterman disputed on *Geomagus* and Wolfgang Simon on *Daemonomagus*, each for sixteen pages.[65] Peterman stated that the foundation of magic was sympathy and antipathy, and action at a distance. He accepted sympathy and antipathy but not the action at a distance of a universal spirit of the world. He believed that agent and patient, moved and mover, should be kept separate, and that like did not act upon like. Peterman further defined his position by a series of specific instances. He believed that Archimedes could destroy the foe's ships

[62] *Commentarius iuridicus ad legem Stigmata capitulum de Fabricensibus duodecim sectionibus distinctus . . . in quo de variis speciebus signaturarum characterum et stigmatum tam supernaturalium quam naturalium imprimis vero antichristi et de illorum quae sagis iniusta deprehenduntur*, Cologne, 1629, 102 pp.: BN R. 8154. The treatise has nothing concerning signatures of plants.

[63] *Disputatio brevis et categorica de proba stigmatica utrum scilicet ea licita sit necne in qua pars negativa propugnatur una cum refutatione Ostermanni*, Cologne, 60 pp.: BN R. 8155. The dedication is dated from Bonn, 1630.

[64] Christoph. Crusius, *Discursus de magia non naturali*, Cassel, 1648, in-8: no BM-BN.

[65] See BN R.8158 and R.8159.

with a burning glass. But that, at the time his absent father was killed, the son saddened, the domestic cat died, and the household clock stopped, could not be explained naturally but was due to angelic or diabolic action. Magnetic action was stretched too far in the belief that, if one friend moved the needle on his compass to different letters on its rim to spell out a message, the needle on the other friend's compass in a far off land would move to the same letters. Nor was there magnetic action in the case of weapon ointment. The heliotrope's following the sun was natural, as was the part of the water in a kettle on a fire which was farther from the fire being hotter than the water nearer to the fire. But the transfer of disease by binding the patient's nail clippings to a crab's back and throwing the crab back in the water was nonsense, and the finding of veins of metal with a divining rod was unnatural. That fountains in which torches that had been extinguished burst into flame again, could be explained by antiperistasis or action by which a quality becomes more intense because of its contrary surrounding it. But Peterman rejected the corpse bleeding at the approach of the murderer. Simon discussed the powers of angels and demons as to augmentation, alteration and locomotion.

Johann Hofmann of Culmbach, in a treatise published at Erfurt in 1636, treated of the process in witch-trials, refuted the arguments of witches, and argued for the death penalty for them. He added an Appendix against putting faith in new prophets and prophetesses.[66] In 1687 a Joannes Christophorus Mueller was respondent in a disputation at Erfurt on the sympathetic powder.[67]

Wolfgang Ambrose Fabricius discussed lycanthropy at the University of Strasburg under Johann Rudolf Saltzmann as *Praeses* in 1649, and the signatures of plants in theses printed at Nürnberg in 1653.[68] Follcwing Sennert, he distinguished between natural lycanthropy, which was a derangement of the patient's imagination and akin to melancholy, ecstasy, and mania or rabies, and diabolical lycanthropy, in which the devil also deluded the imaginations or senses of by-standers so that they too thought that the man was

[66] *Apologia principum in qua processus in causa sagarum continetur et maleficiarum argumenta refutantur,* Erfurt, 1636: BN R. 8156.

[67] BM 1185.c.12. (25.).

[68] BM 7004.de.1 (7.) and (6.).

transformed into a wolf. The theses on signatures were illustrated by two sets of illustrations showing supposed resemblances between plants and members of the human body. After giving arguments of the chemists for signatures and of the Galenists against these, Fabricius decided in favor of the latter. He listed, however, seven kinds of sympathetic woods. The ancient belief that certain herbs resist fascination and incantations is a superstition, but not all virtue is to be denied them. Also, as astrologers of old named the stars according to their supposed significance, so it is probable that the herbalists followed a similar method in naming plants. He illustrates the virtue of herbs by telling how his father recently cured a five year old girl at Nürnberg who vomited up needles, sand, mud, worms of all sorts and colors, and even excrement, by administering oil of hypericon as described in Schröder's Pharmacopeia. Wolfgang died that same year, 1653, at Lyon, on his way home from Italy, and his father published the pictures of ancient lamps (*Lucernae veterum*) which were to be the subject of his son's doctoral dissertation.[69]

Similarly in 1661 at Leipzig under the presidency of Joachim Feller, Johann Stohr discussed whether the lamps found in ancient tombs were inextinguishable. After giving the opinions of Augustine, Majoli, Bonamici, Voetius, Gutherius, Porta, Aldrovandi, Aresius, Lazius, Fortunio Liceto, Citesius and others, he rejected them all, except for rendering a little lip-service to Augustine (*approbatur quodammodo*), and came to the conclusion that, if they were found burning, it was either an illusion of the demons or a lighting anew by natural antiperistasis.[70] Christian Haenel disputed concerning the phoenix at Leipzig in 1665.[71]

At Jena in 1665 under J. A. Fridericus as *Praeses*, J. J. Ruttörfer presented a dissertation on the topic of incubus. But he treated it as a natural complaint, merely remarking that some ancient writers had added supernatural causes and believed that mortals were so afflicted by a demon or witch.

[69] BM 7004.de.1 (10).

[70] BM 7004.de.1 (9.): Joannes Stohrius, Resp. . . . , *Elucubratio de lucernis antiquorum substerraneis*, Praeses J. Feller, in-4.

[71] BM 7004.de.1 contains a duplicate copy which is not catalogued. Also in BM B.452 (15.).

But, to use the words of our illustrious Rolfinck, leaders in medicine have eliminated these fables, among whom Aetius was not the last.[72] Furthermore, the other twenty-two dissertations in the collection where Ruttörfer's occurs are all purely medical.

But we have not finished with Jena. In 1676 Johannes Frischmuth presided and Gabriel Reuselius responded at a disputation on the madness, blindness and stupidity of the Jews in presuming to indicate the coming of the Messiah from a conjunction of Saturn and Jupiter in Pisces.[73] There are many quotations in Hebrew. Kepler, Gassendi and Trew are cited against the division of the zodiac into signs. The opinion of Ranzovius, Origanus and Tycho Brahe that such a conjunction indicates an alteration in religion is branded as erroneous. Johann Wilhelm Baier the Elder, professor of ecclesiastical history from 1674 to 1695, engaged in disputations on the cognate themes of superstition or vain observances, and of predictions through ignorance and fortuitous prophecies, in the years 1682 and 1691 respectively.[74]

Similarly Johann Wilhelm Baier the Younger (1675–1729) in a brief disputation of 1699 at Jena, held that presentiments of the future could only be explained as coming from God.[75] After giving several examples, one from Cardan and two from his own experience, he cited Buddeus[76] that they had no basis in natural divination or in a power of the human mind to foresee. Marcus Marci had accounted for them by the influence of the stars, but Baier denies this or that they come from sublunar bodies. He further rejects the attribution of them to a world soul, or to angels, whose powers as

[72] BM 1185.d.4 (19.), page of which the signature would be B 4 v, if marked.

[73] Johannes Frischmuth Praeses, *Disp. Resp. Gabr. Reuselio de Judaeorum amentia coecitate et stupore qui tempus adventus Messiae ex conjunctione Saturni et Jovis in sidere Piscium indicare praesumunt*, Jena, 1676, in-4. BN A. 22788.

[74] *Disputatio de superstitione seu vana observantia*, 1682; *Disputatio de vaticiniis per ignorantiam et prophetiis fortuitis*, 1691. Both are listed by Will, I, 49-50, but neither occurs in the printed catalogues of BM and BN. But a copy of the former disputation is contained in Col. 156.4 Z, vol. 5.

[75] *Disputatio de praesagiis animi*, Jena, 1699, 20 pp.: BN Rz. 2289.

[76] Presumably Johann Franz Buddeus (1667-1729), who taught at Wittenberg, Coburg, Halle and Jena. He was *Praeses* at a disputation whether alchemists were to be tolerated in the state, at Halle, 1702: Ferguson I, 130.

created beings they exceed. There remains only the Creator to whom to ascribe them.[77]

Christian Prange was respondent at Jena under J. P. Hebenstreit as *Praeses* in 1693 in a dissertation on the swarms of locusts which filled the air, and what they were thought to portend as to the future.[78]

Dissertations under R. G. Crausius, dean of the medical faculty at Jena, included alchemical subjects, the universal medicine in 1679, the principles and transmutation of metals in 1686; denial of fermentation in the blood, in 1682; mental disease or bordering thereon, such as incubus in 1683, hypochondriacal vomiting in 1692, delirium in general in 1686, phrenitis in 1689, and nymphomania in 1691; signatures in plants, in 1697. These dissertation subjects did not become any less magical in the first decade of the next century, when enchanted persons were discussed in 1701, philters in 1704, and the spirit of the world in 1707.[79]

Dissertations and Theses at the University of Altdorf which were printed together at Nürnberg in 1644 under the title, *Philosophia Altdorphiana* [80] are largely dialectical with some political, ethical and metaphysical ones, and so offer little approach to magic. However, they include two by Soner: one on the pernicious medicine of Paracelsus, the other upon dreams, an oration dated back in 1610. In the latter he states that dreams caused by the condition of the dreamer's body obviously throw light upon that condition, but that it is harder to explain how they inform us concerning distant friends and future events. Yet any number of cases (*exempla infinita*) of this

[77] Will, I, 55-56, also ascribes to Baier a disputation of 1722 on fancy as the mother of enthusiasm and others chiefly on natural subjects, sometimes connection with the Bible, as that on fossils and the flood, or Behemoth and Leviathan as elephant and whale. They all fall in the eighteenth century.

[78] *De locustis immenso agmine aerem nostrum implentibus et quid portendere putentur*, in-4, 65 misnumbered pp. Three copies in BM, none in BN.

[79] These and other dissertations, with their full Latin titles, will be found listed alphabetically according to the names of those responding under the name of Crausius in the BM catalogue where they occupy three and a half folio columns. It does not give, however, the following: *Disputatio inauguralis de signaturis vegetabilium respondente Geor. Henr. Rosenberg*, 1697.

[80] BN Rés. R. 865.

can be given. Moreover, not all such dreams are from God or demons. Peripatetics do not accept the theory of emanation of *simulacra* of Democritus and Synesius, but Soner contends that to represent external objects as acting directly upon our phantasy without the medium of the senses is good Aristotelianism and no more contrary to action by contact than is the action of the torpedo fish, magnet and weapon ointment, or being struck dumb at the sight of a wolf. It is thus that old-wives fascinate infants by their glance without infecting the medium, and those with an antipathy to cats sense their unseen presence. Soner further affirms that the imagination in some cases has power even over other bodies or minds. As past events leave their vestiges, so hints of future events precede these and affect animals in the case of weather changes.[81]

A dissertation of 88 pages on the salamander at Altdorf by J. P. Wurffbain in 1677 under the presidency of M. D. G. Mollerus is preserved in four copies at the British Museum.[82]

In a disputation at Altdorf on September 27, 1679, with Johann Christopher Sturm as *Praeses* and Christopher Wegleiter as respondent, on the influence of the stars, astrology was attacked and it was said in conclusion that there was today hardly any mathematician of distinction and worthy the name who did not reject it.[83]

Will, in his *Nürnbergisches Gelehrten-Lexicon,* ascribed to Benedict Hopfer or Hopffer (1643–1684) disputations concerning the airy food, or rather marvelous fasting, of the chameleon, and concerning the *pyrausta* and salamander, both famed as living in fire.[84]

In 1692 at Altdorf there was a disputation by Andreas Christian Eschenbach (1603–1722) concerning the auguries of the ancients.[85] He made use of sixteenth century writers on the subject like Nifo and Peucer as well as the classical authors. He had already pub-

[81] *Ibid.,* pp. 539-49.

[82] I have seen 7004.de.1 (1.). For his *Salamandrologia* of 1683 see Chapter 24.

[83] *Siderum influentia, hoc est, efficacia in mundum hunc sublunarem, quaenam et quanta sit,* Altdorf, 32 pp. Col. 156.4 St 97.

[84] A copy of *Dissertatio . . . de victu aero . . . etc.,* 1681, in-4, is BM 1179. d.14 (12.).

[85] *Disp. de auguriis veterum,* Altdorf, 1692, 28 pp. BN J. 7675 is an unbound paper copy. The disputation is noted by Will I (1755), 360.

lished notes on Orpheus at Utrecht in 1689, while journeying through Holland, and was also author of a letter on phosphorus in 1698 and a commentary on the Orphic hymns in 1702, thus combining an interest in chemical discovery with interest in divination and the occult.

A dissertation as to the truth of chiromancy by Cr. Schultz and P. C. Engelbrecht was printed at Regensburg in 1691.

At Tübingen in 1662 were printed dissertations on the denunciation of witches under Ericus Mauritius as *Praeses* by Weininger and Spring as respondents.[86] They still maintained that witches made pacts with demons but rejected the test of floating in water and the supposed marks of witches and advocated greater caution in accepting testimony. Two years later there was yet another dissertation on the same theme.[87] It opposed the views of Wier, Scot and Abraham Palingus of Haarlem who had written in the vernacular in 1659. The author of the dissertation complained that the view that witches were merely melancholy and deluded women was now received in Belgium by almost common consent, even of the learned.[88] It was also at Tübingen that J. A. Osiander (1622–1697) published a theological, etymological and historical treatment of magic.[89] It seems to be of little or no independent importance, but still affirms the miraculous virtue of the echeneis or remora.[90]

Actual medical practice receives illustration from the publication of one hundred cases referred by the courts and magistrates to the medical faculty of the University of Leipzig for its opinion. A few cases from the sixteenth century are included but most are from the first two-thirds of the seventeenth. The book first appeared in

[86] *Diss. de denunciatione sagarum iisque quae ad eam recte intelligendam faciunt*, Tubingae, 1662: BM 897.c.2. (27.28.); JS I, 755-58.

[87] Christ. Daurerus, *Dissertatio inauguralis de denunciatione sagarum ... praeside Erico Mauritio ...*, Tubingae, 1664, in-4: no BM or BN: Cornell D 237.

[88] *Ibid.*, p. 3.

[89] *Tractatus theologicus de magia exhibens eiusdem etymologiam, synonymiam homonymiam existentiam et naturam causas et effectus mirabiles interspersis hinc inde rarioribus subjectis et exemplis ac dilucidatis notabilioribus controversiis, cum indice rerum et verborum necessario adornante Joh. Adamo Osiandro SS. Th. D. P. Cancell. et Praepositi Ecclesiae Tubingensis*, Tubingae, 1687, 328 pp. text and pp. 329-58 Index. Copies at BM and Cornell.

[90] *Ibid.*, pp. 320-21.

German with prefaces of 1669, and then was translated into Latin with a preface dated 1677. I have used a later Latin edition of 1693,[91] which shows that the work was in demand to the end of the century. The collection is especially concerned with questions of illegitimate birth, impotency, abortion and infanticide, with illicit medical practitioners, and cases of suspected poisoning or witchcraft, also with what measures to take against the pest, whether torture should be applied, whether a person is a leper, whether wounds inflicted were lethal, whether an apothecary's prices are fair. Various discourses are added by the editor to the responses of the faculty.

The view is expressed that rare and even miraculous events may occur in medicine, as could be proved by many examples.[92] But those wandering chemists who make a pretense to great secrets and arcana are condemned.[93] Astrologers are pronounced vain and false; witches cannot bring down the moon; a certain Eva in 1663 is declared not a demoniac but hysterical; a doctor of both laws is suffering from melancholy, not witchcraft; and two quacks in 1653 are condemned for using characters and engraving the name of Jesus in an operation for hernia. First the operator asked for a bit of lard from the bystanders, gave some of it to his black dog, and with the rest of it rubbed the soles of the feet of the boy who was to be operated upon. When the boy refused to eat three squares of bread on which he had cut crosses, the quack made him drink warm beer, which made him vomit and put him in agony. But the quack beat the boy and pricked his finger tips to see if the blood would flow, and uttered vain words and the name of the devil.[94] There is not a live crab in cancerous breasts, and so superstitious women try in vain to kill it.[95] Indeed, cancer and hereditary mania are pronounced incurable.[96]

Women are by no means to prepare medicines; barbers (*iners*

[91] *D. Pauli Ammanni Medicina critica sive decisoria, Centuria casuum medicinalium in Concilio Facult. Lips. antehac resolutorum comprehensa*. . ., Latin translation by Christian Francis Paulinus of the Academy of the Curious, Lipsiae, 1693, in-4. Copy used: Col. 610 Am 6.

[92] *Ibid.*, p. 288.

[93] *Ibid.*, p. 194.

[94] *Ibid.*, pp. 37, 269-70, 598, 166-68, 389, 385.

[95] *Ibid.*, p. 256.

[96] *Ibid.*, pp. 257-58, 403.

... *grex*) should not give purgatives (blindness and death in a case of malignant fever were not due to administration of bezoartic tincture but to a previous purgative given by a barber); a bathkeeper should not alone cure syphilitics; and a surgeon ought not to distill and sell medicinal waters.[97]

Considerable scepticism is expressed as to the composition and efficacy of medicines. Trochees of viper may as well be left out of theriac, since they are mostly bread with little or no flesh of vipers. Many ingredients often weaken the force of compound medicines.[98] The stone bezoar does not enter into bezoartic tincture, which is composed merely of vegetables.[99] Neither Frankfurt pills nor Margraf's powder have the power to expel the foetus.[100] Roots in question in 1634 were not *mandragora* but *victorialis* and could not make one invulnerable to weapons without demon aid.[101] Powder of cinnamon and antimony is not a panacea; the Alcahest of Paracelsus and Helmont is in liquid form, not a powder; and the horn which a man bought for unicorn's is not genuine.[102]

On the other hand, a credulous attitude is sometimes displayed. Platter is believed that his greatgrandfather bore a son after he was a hundred and lived to see him married.[103] Marcus Marci and Helmont are cited for operative ideas, and it is affirmed that Christ willed to be born in the ninth month as a good example.[104]

The Hippocratic Physics, illustrated by the comments of Tachenius, Helmont, Descartes, Espagnet, Boyle and other recent writers, of Johann Daniel Horst (1617–1685), printed in 1682,[105] consists of twelve dissertations with his students as respondents. The topics are the principles of natural body, the affections of the same, the world, heavens and elements, mixed bodies in general and imperfect ones in especial, perfect inanimate bodies, plants, animals in general

[97] *Ibid.*, pp. 194, 528, 531, 467, 477.

[98] *Ibid.*, p. 359.

[99] *Ibid.*, p. 528.

[100] *Ibid.*, p. 533.

[101] *Ibid.*, p. 271.

[102] *Ibid.*, pp. 409, 447.

[103] *Ibid.*, p. 499.

[104] *Ibid.*, pp. 165, 38.

[105] *Physica Hippocratea, Tackenii Helmontii Cartesii Espagnet Boylei ... aliorumque recentiorum commentis illustrata*, Francofurti, 1682, 87 pp. The dedication to Ernst Ludwig, landgrave of Hesse-Darmstadt (1678-1739), is by a misprint dated 30 January, 1662, instead of 1682.

and reptiles in especial, birds, aquatic animals, insects, man, and the soul. Creation is precisely dated on September 8, 5428 years ago.[106] There are meteors—floods, fires and winds—in the microcosm as well as the macrocosm.[107] Zoophytes are either plants such as the mimosa and Scythian lamb plant, or animals like the barnacle geese which the Scots call Klekgues.[108] Several instances are given of the emission of flames by human beings, and, although Harvey is quoted that all animals are born from an egg, spontaneous generation is also noted.[109] The suggestion is made that the faculty by which the remora halts ships is contrary to that by which the magnet attracts iron.[110] Some bibliography is given for chiromancy and physiognomy.[111]

At the close of each dissertation are apt to be questions and corollaries which are not further discussed but which illustrate the state of science then and the points in which men were interested. Thus it is asked whether the heavens influence inferiors, whether there are waters super-celestial, and whether a Christian can with easy conscience wear an astrological image about his neck as a safeguard against incantations and in order to win over the minds of princes.[112] The last query is answered in the negative, as is the similar question whether amulets adorned with certain figures are licit.[113] It is still asked whether comets are meteors, whether bloody rains can be natural, and whether the earth produces rain, as Agricola says.[114] The cause of the tides is still regarded as an open question.[115] Does like act on like, must agent and patient be in contact, is there fire in flint, can gold be rendered potable?[116] Will no rainbow be seen for forty years before the end of the world?[117] Do some plants naturally counteract incantations, was the cure of blindness of Tobias in any way natural, and could the smoke from the heart of a fish dispel demons naturally?[118] Do the souls of beasts and vital heat differ, is there a double soul in mule

106 *Ibid.*, p. 16.
107 *Ibid.*, p. 27.
108 *Ibid.*, p. 44.
109 *Ibid.*, pp. 25, 54, 62.
110 *Ibid.*, p. 62.
111 *Ibid.*, p. 72.
112 *Ibid.*, pp. 21, 44.
113 *Ibid.*, p. 44.
114 *Ibid.*, pp. 28, 36.
115 *Ibid.*, p. 21.
116 *Ibid.*, pp. 27, 21.
117 *Ibid.*, p. 28.
118 *Ibid.*, pp. 44, 64.

and leopard, are birds made of water, are there such animals as the unicorn and phoenix?[119] Are there mermen?[120] Several queries assume the truth of spontaneous generation, such as whether animals born from putrid matter differ in species from those born of seed, whether those born spontaneously were created in the first creation, and whether insects are generated in man the microcosm from the soul of the old animal deposited there in excrement, as Fortunio Liceto says?[121] Can man be changed into a beast? Such were the traces of fantastic science and of magic left in academic minds in the last quarter of the seventeenth century.

The reviewer of Horst's book in the *Journal des Sçavans* records one delightful detail which had escaped me. When a squirrel wishes to cross a stream, it pushes a flat piece of wood into the water, leaps onto it, and erects its tail as a combined sail and rudder to guide itself across.[122]

Horst was born in Giessen, taught medicine for a time at Marburg, and next at Giessen, where he was at the same time physician to the Landgrave of Hesse-Darmstadt. He then came to Frankfurt and was admitted to the Academy of the Curious as to Nature with the sobriquet of Phoenix.

In the years 1681, 1682 and 1686, Johann Eberhart Schweling published at Bremen examples of the disputations in his physical seminar. The first contained eighteen decades of miscellaneous theses; the second, twelve groups of twelve each; and the third, twenty-one such *dyodecades*.[123] Many of the theses are distinctly Cartesian. Thus in 1681 the pineal gland is said to be the principal seat of the human mind, while in 1686 it is still more emphatically affirmed that today it is held beyond controversy that the pineal

119 *Ibid.*, pp. 50-51, 58, 51.

120 *Ibid.*, p. 64.

121 *Ibid.*, pp. 51, 70.

122 JS XI (1684), 105.

123 *Specimen collegii physici disputatorii privato-publici quo continentur octodecim decades thesium miscellarum* . . . , Bremae, 1681, in-4, 20 pp. BM 536.f.18 (7.). *Collegii physici disputatorii privato-publici Specimen secundum quo duodecim continentur dyodecades* . . ., Bremae, 1682, in-4, 20 pp. BM 536.f.18 (8.). *Collegii physici . . . specimen tertium*, 1686, 34 pp. BM 536.f.3 (27.). In citing these theses I shall use capital Roman numerals for the *Specimen*, small Roman numerals for the decade or dyodecade, and Arabic numerals for the thesis.

gland is "the palace of wisdom."[124] It is asserted more than once that we are never deceived, if our ideas are clear and distinct.[125] "Cogito, ergo sum," is matched by "Ego cogitans existo."[126] Mind and body, thought and extension, are sharply separated.[127] "Some bodies are vehicles of others, and one is the vehicle of vehicles, which we call the matter of the first element."[128] Primeval light was this very subtle matter of the first element and afterwards was collected in the vortices of the sun and stars.[129] Lead is heavy in the earth's vortex, lighter than a feather in the heavens.[130] Half a page is spent in an attempt to explain the inextinguishable lamps in ancient sepulchers in terms of the first and second elements, which are further declared to be the only bodies without pores.[131] There is no planet which does not have a fiery center; gyrating fire fills the marrow of earth like a star and helps to cook its metals; but the sphere of fire "under the concave of the moon" is a figment.[132]

The face of Schweling's seminar is set against scholasticism. "Nothing is more laughable than substantial form." The opinion of the schoolmen that iron is attracted by the magnet is rejected, there being no such thing as attraction without contact. And the search for final causes is idle and frivolous in physics.[133] "No one on earth can explain what difference there is between virtual extension and formal extension."[134] The terms, maximum and minimum, are discarded.[135] It is madcap talk to say that the heavens are moved by intelligences.[136]

Somewhat less unanimous and uncompromising, somewhat more variegated, are the theses bearing upon the relation of science to religion and of astronomy to Scripture. One asserts that philosophy and sacred annals are not antagonistic; others, that they labor in vain who try to draw from the scriptures accurate knowledge of

[124] I, v, 2; III, v. 12.

[125] I, vii, 1; I, xvi, 2.

[126] I, xvi, 1.

[127] I, v, 6; I, viii, 4; III, iii, 11-12; III, iv, 8.

[128] I, xii, 8.

[129] I, xiii, 8.

[130] I, xiii, 7. Other references to vortices in II, ii, 10bis; III, i, 4; III, xix, 1.

[131] III, v, 8-9.

[132] I, iii, 7; I, x, 10; I, vi, 9.

[133] I, vii, 3; I, ix, 10; I, x, 1; I, xiii, 1.

[134] I, v, 10.

[135] I, xiii, 4: "Dari maximum et minimum in generali Physica negamus."

[136] I, xi, 10.

physical phenomena.[137] Others affirm that philosophy is absolutely essential to a full comprehension of theology, that philosophy alone can refute atheism and scepticism, and that philosophy should not be called the handmaid of theology.[138] To interpret the waters above the firmament as referring to clouds is not violating the meaning of the Bible, nor is the movement of the earth contrary to it, and further seems the best hypothesis to explain the phenomena.[139] But an earlier thesis had held that the earth is at rest and the heavens moved.[140] Later, however, the heliocentric hypothesis is maintained, the sun is represented as revolving on its axis, and a sunspot is regarded as the probable cause of the darkness during the Passion.[141] But in 1686 it is said that no man knows that the sun is the center of the universe, although it is estimated that the world began in autumn with the sun in Libra.[142]

Despite their rejection of scholasticism, separation of mind and body, and affirmation of the independence of science and philosophy from Bible and theology, our theses cannot keep off the theme of angels. One declares that it is possible for angels to assume bodies and appear to exercise corporal functions. Others reject hierarchies of angels as idle dreams of Roman Catholics but admit their mutual irradiation. A fourth states that it is clear to all who understand the nature of angelic mind, that *genius* is nowhere, yet can do two things at once, and so be in two places simultaneously.[143]

Specters and demons also engage the attention of the members of Schweling's seminar. Many specters are nothing but the effects of very strong imagination overcoming the evidence of the senses, and an *ignis fatuus* is not a specter. But specters are not always the effects of a very strong imagination but may also come from an evil spirit.[144] In 1682 an entire *dyodecas* is given over to consideration of specters. The opening thesis is that unusual apparitions without natural cause but produced by the precise direction of a diabolical mind are called specters. They must have divine per-

[137] I, iii, 10; I, xiii, 2; III, xxi, "De abusu verbi divini in physica," especially 6.

[138] I, xv, 8-9; I, xviii, 6.

[139] II, viii; I, xvii, 7.

[140] I, xii, 7.

[141] II, ii, 1, 9-10.

[142] III, xii, 6-7.

[143] I, iii, 8; I, vii, 8-9; I, xiv, 9. II, vi, is also devoted to angels.

[144] I, iii, 9; I, viii, 6; I, xvii, 4.

mission, but the devil can form bodies from the air and so determine the course of the spirits of the human body to certain pores of the brain as to form varied phantasms. This, however, may also happen without the action of the evil mind, and so the final thesis is that many specters are merely the effects of a very strong imagination triumphing over the very senses.[145] In 1686 one thesis contends that there is no natural way of averting diabolical specters from oneself and one's house,[146] and the subject of possession by demons is also considered. Here the demon, using the human body as an organ, works wonders which would otherwise be impossible. Those possessed speak several languages, whereas before and after they know only one. The demon suppresses the forces of the soul and so moves the animal spirits of the possessed as to direct their nerves and muscles. The disputant, however, will not deny that spurious *energumeni* can be cured by natural remedies.[147]

The power of imagination is also seen in the case of somnambulists, who abound in foaming blood and fervid spirit, and whom very strong imagination inflames in sleep to walking "and committing I know not what crimes."[148]

The influence of the stars, too, is still considered to be a potent force. Mental traits follow the temperament of the body, and it varies greatly with the influence of the stars.[149] Man is a most compendious mirror of the entire macrocosm.[150] In 1686 a whole *dyodecas* is devoted to the influences of the heavenly bodies. The sun is physically the universal cause of almost everything produced in this sublunar world. The fixed stars, at an immense distance from us, do not exert so much virtue on sublunars. The planets cannot be other than occasional causes of certain operations happening under the sun, but do not cause heat and cold. Similarly the moon, though near the earth, is not the chief efficient cause of the tides but merely procataractic or occasional.[151] A thesis of 1681, on the other hand, had stated that tides depended not so much on the nearness as presence of the moon in the heaven sur-

[145] II, x, 1, 5-6, 8, 10-12.
[146] III, ii, 4.
[147] III, xvi, 2, 4-7.
[148] I, ii, 1-5.
[149] I, x, 6.
[150] I, xi, 8.
[151] III, xix, 1-5.

rounding the earth.[152] Generation of imperfect animals was ascribed to the moon, and another thesis of 1686 makes the light of the moon and other stars contribute greatly to variation of winds.[153] But the moon's phases are observed in vain in bleeding or purging,[154] and our disputant goes on to argue that it is an error to relate the increase and decrease of shell-fish to the moon, for some of them are fat when others are lean. There is no reason to predict from comets, since they are huge bodies coeval with the universe which pass through immense spaces from one vortex to another. Idle, too, is prediction from conjunctions and oppositions of the planets.[155]

A thesis of 1681 had already held that reason does not teach that comets are producers of evil,[156] and in 1682 a *dyodecas* had been concerned with comets. It stated that they were neither terrestrial exhalations nor a new divine creation, but stars enveloped in a dense covering like sunspots which might become planets and which, like the sun and fixed stars, were very likely coeval with the universe. Although God might make use of them as signs, it was difficult to see why astrologers regarded them as evil rather than salutary, and our disputant, on the eve of Halley's prediction of the return of the comet of 1682 in 1758, declares that science does not know how to foretell the advent of comets.[157]

If our theses offer little comfort to astrology, they accept, at least to some extent, relations of sympathy and antipathy. Iron and steel have "a symbolic nature" with the magnet, and no one can give the special reasons which exist in nature for all sympathy and antipathy.[158] Three successive theses of 1686 declare that poison is often a medicine or even a food for man, that there is power in certain words to cure disease, but that there is no true or solid physical explanation of the corpse bleeding at the approach of the murderer.[159] That witches are borne bodily by demons to nocturnal sabbats may by no means be denied.[160]

The scientific calibre of our theses, like their attitude to the

[152] I, vi, 10.
[153] I, xvi, 9; III, vii, 10.
[154] III, ii, 9.
[155] III, xix, 6-7, 9-10, 12.
[156] I, vi, 4.
[157] II, ix, 2-4, 8, 12, 5, 9.
[158] I, viii, 9; III, ii, 6.
[159] III, ii, 10-12.
[160] III, ii, 7.

occult and magical, has its ups and downs. Heat is identified with the vehement motion of subtle particles, and it is stated that no more action is required to impel a body than to stop it when in motion.[161] But a vacuum is still deemed impossible, and "space in which there is no substance involves a contradiction."[162] Sea water excites rather than extinguishes a fire, and springs are still said to come from the sea.[163]

Beasts who do not understand medicine are healthier than men who do.[164] But cogitation is denied them, and stories of intelligent action by parrots, dogs and elephants are discredited. Descartes explained their actions by purely corporal principles, but inasmuch as they are *animalia,* they must have an *anima* or soul of some sort.[165] Other theses, however, state that their soul consists in the blood, animal spirits, and disposition of organs; their life, in vegetation, locomotion and purely bodily sensation; and that they are living hydraulic-pneumatic automata. [166] If so, it would seem that they should be granted only vital and natural spirits, not animal. On the other hand, we are told that the way in which infants laugh and sigh, and learn to talk, is sufficient proof that they employ reason from the moment of birth.[167]

Distinct from the foregoing theses and disputations appears to be a treatise assigned to Schweling himself on the manner and power of operating outside themselves which are possessed by the devil and magicians.[168]

A theological disputation by Johann Georg Dorscheus, who had been a professor at Strasburg before coming to Rostock in 1654, on possession by demons was printed posthumously at Rostock in 1666 and was sufficiently read to be reissued in 1672, 1683 and 1693. He admitted that it was difficult to find sure signs of such possession because of impostors and the deceit of the devil, but distinguished between primary and secondary indications thereof. He questioned through what part of the body the demon entered, and

[161] I, iii, 8; I, vii, 6.
[162] I, vii, 4; I, xv, 5-6.
[163] III, ii, 5; III, xi.
[164] I, xiv, 6.
[165] II, xii, 2, 5-6.
[166] I, xvii, 8; II, xii, 10, 12.
[167] I, xiii, 10.
[168] J. E. Swelingus, *De diaboli magorumque extra se operandi modis ac viribus,* Bremen, 1677, in-4.

whether unborn children were ever possessed. The question whether witch, magician or diviner was possessed, he answered in the negative, and held that most heretics were not, although some were. Believers might not only be possessed but even die in that state. Demons in possession should not be questioned, and the assistance of papal exorcists should not be sought against them, although medical aid might be requisitioned. Hauber complained that Dorscheus chiefly repeated what had already been said by Thyraeus in the late sixteenth century.[169]

J. F. Huldenreich presented a dissertation at Frankfurt on the Oder in June, 1656, under Samuel Kaldenbach as *Praeses,* in which he treated incubus as a mental disease from a purely medical standpoint.[170] It affected the brain and was a variety of melancholy accompanied by stupor and wild phantasmata. Ridiculous ideas and superstitious notions were current concerning it, but he scorned them as old-wives' tales and unworthy of attention. After treating of its natural causes, he does, however, briefly allude to a supernatural cause, which is sometimes an error of fancy but not always.

In 1654–1655, John Placentinus was presenting to the Dutch States General a scheme for finding longitude at sea. In 1657 he became professor of mathematics at Frankfurt on the Oder, where he died in 1683 or 1687. Of two dissertations of which he was *Praeses,* which are all that the British Museum catalogue has under his name, one in 1657 was on longitude and latitude;[171] the other, in 1672, on the natural harmony of astronomy with chiromancy.[172] This is illustrated by comparing the horoscopes of half a dozen persons born twenty years or so ago with figures of their hands.

Ruttörfer, whose dissertation at Jena in 1665 has been mentioned, in 1666 presented another at Frankfurt on the Oder under the presi-

[169] E. D. Hauber, *Bibliotheca . . . magica,* I (1738), 161-73, whose account of Dorscheus' disputation I have followed.

[170] *De incubone . . .,* in-4: BM 1179.k.5 (2.).

[171] The BN catalogue lists only this first dissertation. Zedler lists other works by him: a Syncretism of Descartes and Aristotle, a Dissertation on heat and the movement of the members of the human body, a Disputation on the tides, and a dissertation by a student of his on *Geotomia* or section of earth (1657, in-4).

[172] *Naturalis harmonia astronomiae cum chiromantia externa et interiore lineae vitalis constitutione ac dimensione . . . deducta* (pr J. F. Rhetius): BM 8610.bb.49. (6.).

dency of J. T. Schenck.[173] This time his subject was frenzy. In the eighth chapter on diagnostic signs of that disease, he accepted astrological indications as offering some certitude, if taken from the horoscope of the patient or from the constellations at the time when the disease set in. In the following chapter upon prognostic signs he made a similar suggestion, stating that astrologers especially considered the ascendent and the moon. In both cases he cited Ganivet of the fifteenth century as his authority, and in the first instance added the astrological tract of the pseudo-Hippocrates.

It was also at Frankfurt on the Oder that J. C. Mentzelius discussed a case of hypochondriac melancholy in 1684.[174] A dissertation there on prodigies of blood by Johann Elias Starck under Johann Christoph Becman as *Praeses* in 1676 reached its fourth edition in 1684.[175]

The problem of the bleeding corpse was discussed yet once more by M. F. Geuder at Ulm in 1684, but he merely offered a compilation of previous utterances on the subject.[176]

A word may be added concerning similar academic exercises in Dutch and Danish universities. At Utrecht in 1675 John Regius was author and respondent under Gerard de Vries as *Praeses* of a disputation on the composition of the *continuum*.[177] Among corollaries at its close is the statement that no philosopher has as yet adequately explained the tides. The moon is too far off to act by pressure through the medium of so fluid a body as air upon so vast a body as the sea. Hence the tides must be attributed to immediate divine action coinciding with the phases of the moon. Objection to action at a distance also causes Regius to deny that weapon ointment and sympathetic powder act naturally. Tenuous effluvia will not suffice as an explanation.

[173] *De phrenetide in alma hac Salana sub praeside Joh. Theod. Schenckii*, Feb. 1666. BM 1179.k.5.

[174] *De aegro melancholia hypochondriaca laborante, praeside Bernhardo Albino*, in-4, 48 pp. BM 1179. k.5 (16). For the connection of that complaint with magic, see Chapter 37, Mental Disease and Magic.

[175] *Diss. de prodigiis sanguinis*, BN G. 7683.

[176] *De probatione per cruentationem cadaverum vulgo Baarrecht, praeside Eberh. Rud. Rothio*, Ulm, 1684, in-4: BM 7004.de.1 (13.).

[177] *Disputatio philosophica de compositione continui . . . sub praeside Gerardi de Vries*, Utrecht, 1675: BM 536.f.18 (3*).

Of twenty-six juvenile exercises, disputations and orations connected with Biblical passages by Johannes Marckius at the University of Franeker in Friesland, founded in 1685, five were on the apparition of Samuel.[178]

Among the subjects of doctoral dissertations listed in Niels Nielsen's *Matematiken i Danmark*, the following show curiosity as to the marvelous and occult: of miracles not miracles or of nature's arcana which are wrongly thought to be miracles by the vulgar crowd; of divinations (1604); of astrology (1607–8, 1635, 1678); of the stars and their properties (1610, 1644); of the matter and qualities of the heavens (1612); of the external causes of natural body (1624); of the qualities, manifest and occult, of natural body (1651); of monsters (1624); of the causes of natural body by accident, fortune and chance, also of monsters (Lund, 1670); on instituting a more secret scrutiny of nature (1643); on the arcana of fountains (1693); on Lot's wife (Kiel, 1669) and on the waters above the firmament (1666, 1693); and that the demon is a marvelous magician in nature (1703).

This chapter may fittingly close by noting a disputation in the next to last year of the century on moderating curiosity in the inquiry after truth.[179] This would seem to indicate that interest in the occult was still prevalent and that the disputant argued that it should be restrained.

[178] *Acta eruditorum*, VI (1687), 379.

[179] Andreas Rinder (1677-1733), *Disputatio de moderatione curiositatis in inquirenda veritate*, Helmstedt, 1699: cited by Will III, 334. I have failed to find the dissertation itself.

CHAPTER XIII

THE *CURSUS PHILOSOPHICUS* OR *PHYSICUS* BEFORE DESCARTES[1]

Two courses in manuscript: Isambert and Boucher—Two posthumous publications: Keckermann and Gorlée, a course and a criticism—The occult slant of Goclenius—Theological of Abra de Raconis and Zanardi—Alchemical of d'Espagnet—Jacchaeus, a Scot, at Leyden; Caspar Bartholinus, a Dane; Aversa, an Italian—Jesuit handbooks of Faber and Arriaga—Backward books of Burgersdyck and Duncan—The Botius brothers—Comenius, educational reformer but fantastic physicist—Neufville, Kyper, Sperling and Cabeo continue the pre-Cartesian tradition.

En un mot, laissons le monde comme il est
—Rey to Mersenne

The usual conviction of the ultimate simplicity of nature
—Bridgman

In the early decades of the seventeenth century, before Descartes had advanced his daring theories of a mechanically operating universe, or Torricelli had performed his epoch-making experiment with the tube of mercury, the ordinary university or seminary course or text in philosophy or physical science still followed closely in the footsteps of Aristotle and his medieval scholastic commentators. It was also influenced by such non-scientific books as the Bible and the works of Augustine. This and the general background of ideas at the time may be illustrated by a number of specific examples, the first two from unpublished manuscripts,[2] the others from contemporary printed editions.

[1] Revised and greatly enlarged from a "Communication présentée au VIe Congrès International d'Histoire des Sciences, Amsterdam, 1950," printed in *Archives internationales d'Histoire des Sciences*, 14 (1951), 16-24.

[2] Some idea of the amount of such material preserved in manuscript may be had from the numerous seventeenth century commentaries upon Aristotle and the somewhat fewer manuals of philosophy and physics listed in Mancini's catalogue of MSS in the public library of Lucca: *Studi italiani*, VIII (1900), 115-318.

In the years 1602–1603 an Augustinian, brother Nicolas Solier, took notes on the lectures of Isambertus (or, Ysambertus), a doctor and professor of the Sorbonne, which were delivered at the Lycaeum, or Augustinian convent, at Bourges. These notes are today preserved in a manuscript of the Bibliothèque Nationale, Paris.[3]

On the first two leaves of this manuscript, before the transcription of the lectures themselves begins, are jotted down various problems suggested by the *Metaphysics, Physics, De cœlo et mundo, De ortu et interitu* (or, *De generatione et corruptione*), and *Meteorologica* of Aristotle and by the treatise on the Sphere, presumably that of Sacrobosco of the early thirteenth century. Some of these questions are: whether a vacuum can be produced by angelic virtue? whether by divine virtue a body can be located in many places? whether the same man can be dead in one place and alive in another? whether the world is eternal? whether the heavens are moved by Intelligences or by their own forms? whether celestial and sublunar matter differ? whether human offspring can be produced by the intercourse of a demon and a woman? whether any animal lives in fire? whether there are only 1022 stars? whether or no the solar eclipse at the time of the Passion was universal? whether mountains have been in existence since the world began? Such a confusion of natural with supernatural forces suggests the likelihood of a further infusion of preternatural and magical factors, and mingling of occult with physical science.

In the notes on the lectures themselves, which are commentaries on the entire natural philosophy of Aristotle, here called *Physiologia*, 74 leaves are devoted to the eight books of the *Physics*, 25 to *De cœlo et mundo*, 17 to the *Sphere* of Sacrobosco, while some 30 leaves are divided between *De generatione et corruptione*, the *Meteorologica*, and what seem to be distinct disputations and tractates on alteration, *mixtio*, and the elements. Sixty leaves are then

[3] BN 6538, fol. 3r, "In universam Aristotelis Physiologiam Commentaria a Domino Ysamberto data et a me fratre Nicolao Solier accepta anno domini 1602"; fol. 36v, "Huic ultimam manum imposuimus die 29a mensii Iunii 1602"; fol. 76r, "Huic ultimam manum imposuimus 11a mensis Ianuarii 1603"; fol. 220r, "Sic ultimam manum imposuimus 20a mensis Iunii 1603"; fol. 251v, "die ultima Iulii anno domini 1603, F. Nicolaus Solier Augustinianus Bituricensis."

spent on *De anima,* after which the manuscript concludes with 29 leaves devoted to the *Metaphysics.*

Along with such time-honored themes as form and matter, *continuum* and indivisibles and infinity, place and vacuum, time and motion, elements and mixed, such questions are again put as whether two bodies can occupy one and the same space? whether by divine power one and the same body can be in different places? whether the world could be produced from eternity? and whether the element fire occupies the concave of the sphere of the moon? [4]

Over a score of years later a similar *Cursus philosophicus* was completed by a P. or Père Boucher on December 20, 1625, and is preserved in another manuscript of the Bibliothèque Nationale, Paris.[5] This course, however, is more inclusive, treating of philosophy in general and of logic and moral philosophy before taking up the *Physics* and other Aristotelian books of natural philosophy, followed as in the earlier course by *De anima* and the *Metaphysics.* In the main, however the method, attitude and content of the two courses are very similar. Boucher still describes four elements and four qualities, asks whether two bodies can be in the same place or one body in several places, whether the world is eternal, whether the heavens consist of both form and matter, and whether they are moved by their own form or by Intelligences.[6] Nor does he forget the Biblical waters above the firmament,[7] or to inquire whether like acts on like.[8]

Isambert, in commenting on the Sphere of Sacrobosco with reference to the extent of the habitable world had mentioned the discoveries of the Portuguese and Spaniards.[9] Boucher is even more up-to-date on occasion, alluding to the new star of 1572 and the comet of 1577, and the question of parallax,[10] and citing as recent and anti-Aristotelian an author as Francesco Patrizi (1529–1597).[11]

[4] BN 6538, fols. 50v, 51v, 83v, 99r.

[5] BN 6549 A, 286 fols. On the front fly-leaf is written. "Commentarii in universam Aristotelis philosophiam," but on the last page we read, "Absoluta stat totius philosophiæ ex diversis authoribus collecta synopsis a. d. 1625 Dec. 20."

[6] BN 6549 A, fols. 208v-218v, 167r, 179r, 182r, 184r.

[7] BN 6549 A, fol. 183r.

[8] BN 6549 A, fol. 202v.

[9] BN 6538, fol. 112r.

[10] BN 6549 A, fol. 224v.

[11] BN 6549 A, fol. 136r.

Both courses considered the topic of fate along with that of monsters and both asked pratically the same questions as to monsters: what a monster is? the diversity of monsters; by what cause they are produced? whether their formation is intended or casual.[12] The chief difference between the two treatments is that Isambert considered fate before he took up the subject of monsters, whereas Boucher first discussed chance and fortune, then monsters, and lastly, fate.

Both lecturers included a disputation concerning alteration and discussed *mixtio* and the elements, with the difference here again that Isambert considered *mixtio* before the elements,[13] while Boucher treated the elements before *mixtio.*[14]

Boucher alludes to occult as well as manifest qualities.[15] The earlier course of Isambert put the question whether the heavens exerted influence by other occult forces and qualities than by motion and light, and answered it in the affirmative.[16] Also whether the future could be foreknown from observation of the stars, to which the answer was, Yes, as a matter of probability.[17] Similarly the course of 1625 still debated whether the heavens acted upon these inferiors, whether the heavens acted upon man, and whether, if the motion and influence of the heavens ceased, the action of natural agents would cease also.[18] Thus both of these traditional courses remained credulous as to the preternatural and occult, the astral and astrological.[19] We turn to printed texts.

Bartholomaeus Keckermann (1571–1609) studied in the Gymnasium at Danzig and the universities of Wittenberg and Leipzig, then taught at those of Heidelberg and Danzig. Although he lived to be only thirty-eight, he was a prolific textbook writer and author of various "Systems." Of these we are here concerned with

12 BN 6549 A, fols. 160r-161r.

13 BN 6538, fol. 139r, "Atque hæc de mixtione dicta sufficiant" (written in very large letters); fol. 139v, "Tractatus de elementis."

14 BN 6549 A, fol. 208v, De elementis; fol. 219v, mixtione.

15 BN 6549 A, fol. 215r.

16 BN 6538, fol. 94r. This question also occurs among the preliminary problems at fol. 2r.

17 BN 6538, fol. 97r.

18 BN 6549 A, fols. 187v-189r.

19 A third Cursus philosophicus in BN 6663, delivered by Padet at the Collège d'Harcourt in 1617, is limited to logic (treated somewhat as Boucher does) and to moral philosophy.

his Physical System, a course of lectures delivered at Danzig in 1607 and published posthumously.[20]

The Physical System divides into six or seven books. The first is about natural body in general; the second is concerned with simple natural body, i.e., the heavens and elements; the third treats of mixed natural bodies in general and of animals especially. Man is the subject of the fourth book; brute animals, plants and metals, of the fifth; while the sixth book deals with meteors, including comets. The seventh book of the original edition, on the universe, is wanting in the edition of 1612.[21] The usefulness of the work as a textbook seems impaired—for the modern reader at least—by too many footnotes, theorems and different fonts of type, and by the absence of an index.

The matter of the heavens is a fifth something, distinct from the matter of all subcelestial bodies. Although outstanding authors, old and new, including Zabarella, Scaliger and Piccolomini, hold that the heavens have only external and assisting form, and not internal form like other bodies, Keckermann does not see how this view of theirs can be reconciled with physical principles. The celestial substance is not so fluid and not so readily dissipated as air is, and it is immutable. Scripture says that it is firm. Light is limited to the stars. Celestial motion is exactly circular and the swiftest of all motions. The stars act upon inferiors by their motion and light (later on he speaks of their influx or influence), and different stars act differently upon inferiors. The stars have no elementary qualities themselves but produce heat and the like in inferiors, acting first on the elements, then on mixed bodies, inanimate and animate, including the human body and indirectly the higher faculties of man.

Keckermann has a chapter of some ten pages on Nature, but is brief regarding time and place, not considering vacuum in connection with place. Physicists have hitherto taught about the

[20] *Systema physicum septem libris adornatum et anno Christi MDCVII publice propositum in Gymnasio Dantiscano a Bartholomaeo Keckermanno SS. Theologiae licentiato et philosophiae ibidem professore,* Dantisci apud Andream Hünefeldum, 1610, in-8. I have used the third edition of 1612 in six books: BM 1478.d.19, 828 pp.

[21] It is brief, occupying only pp. 1044-66 in the edition of Danzig, 1610: BN R.39970.

elements, as if the element were a perfect genus and each particular element a complete species of natural body. But they are less complete than mixed bodies and do not attain as perfect matter and form. They therefore are not entirely and completely different, so that it is no wonder if one element can be changed into another in a moment. Nor are they distinguished by accidents of their own, for these are an outcome of specific form. Noting that Bodin, whom he cites frequently, denied the existence of four primary qualities, Keckermann affirms that heat is the chief quality, since no mixture of the elements is possible without it. But cold is also an active quality, while humidity and dryness are passive qualities. Valla, Cardan, Patrizi and Lambert Daneau have contended that fire is not an element, but Keckermann still accepts it as such. Air and fire tend to move away from the center; water and earth towards it. Like is not affected or transmuted by like, but contraries by contraries.

Air is required for the generation and conservation of fire, as the following three experiments show. If there is not a free movement of air about the fire, but it is everywhere enveloped in thick smoke or even shut up in a furnace, the fire is extinguished. Fire does not burn well in turbid, cloudy and rainy air, as chemists and metallurgists know by experience. Two fires next each other in a hearth impede each other by their excrements and lack of ventilation. Scaliger in his *Exercitatio* 73 observed that a fire outdoors was weakened by the rays of the sun. His explanation of this phenomenon was that the air became too rarefied by the heat of the sun. Keckermann adds that the solar rays consume the subtler parts of the excrements of the fire, leaving the terrestrial and crasser portions to blanket the fire.[22]

In the third book are considered such topics as putridity, color, odor, taste, concoction, the human soul and body, life, health and disease, nutrition, the augmentative faculty, generative virtue, animal spirits, the senses internal and external, appetite, respiration, sleep and waking, and dreams.

Hyperphysical dreams are of divine or diabolical origin, but dreams also arise from the influx of the stars, when the brain is

[22] Edition of 1612, pp. 159-60.

affected by moonbeams shining into the bedroom or "by some arcane influence of other celestial bodies," especially a peculiar constellation of Mercury and the moon. Caspar Peucer is authority that for divining dreams are required an equable temperament, pure and subtle animal spirits, the efficacy of the heavens and freedom of the rational soul from bodily functions. As it is fatuous to observe every dream, so it is rash to scorn and disregard them all.[23]

As to gems Keckermann is very brief, although he speaks of their marvelous efficacy and ascribes their formation from subtler matter to "singular celestial influence."[24] He is correspondingly full concerning comets which he discusses for nearly a hundred pages. He still regards them as terrestrial exhalations produced by action of the planets in the supreme region of air. But God uses good angels or permits bad demons to work with the matter of the comet to produce some extraordinary and horrible effect. Sometimes the same comet may have a good effect in one year and bad in the next, depending on its relation to particular planets and fixed stars. But Keckermann dwells mainly upon their bad effects. "We say that the deaths of kings and changes of empires are merely remote and indirect effects of comets, and that these are denoted by comets more supernaturally than naturally." As the rainbow is a token of divine grace, so the comet is a sign of divine wrath, and those who laugh at them will not escape punishment. But soon Keckermann tells how to predict naturally from comets.[25]

Such is the text of Keckermann, in part adhering to the topics, order of treatment, and opinions of Aristotle, at times relying upon experiment or engaging in bold hypothesis.

Beeckman in his Journal on September 6, 1618,[26] noted that Keckermann was wrong in stating in the last book of his *Physics*[27] in the treatise on the vacuum, that frozen water occupied less space than liquid water.[28] For experience showed that, when a cup full of water froze, the ice would rise above the rim of the

[23] *Ibid.*, 456-60.

[24] *Ibid.*, 603.

[25] This last at pp. 726-39.

[26] *Journal tenu par Isaac Beeckman de 1604 à 1634*, ed. C. de Waard, La Haye, I (1939), 215.

[27] As we have seen, this book is not found in the edition of 1612.

[28] The passage occurs at p. 1064 of the *editio princeps* of 1610.

cup. Also Keckermann should have inferred from the fact that ice floats on water that it must occupy more space. We shall find, however, that Keckermann was far from being the last person in the century to hold that water contracted as it froze.

David Gorlaeus or Gorlée died in 1612, but his Philosophical Exercises, "in which almost all theoretical philosophy is discussed and many leading dogmas of the Peripatetics are overthrown," were printed at Leyden only after his death in 1620.[29] These Exercises, eighteen in number, deal with philosophy, the ens, distinctions, universal and singular, accidents, quantity, quality, things related, motion, place, time, composites, atoms, matter and form, generation and corruption, the heavens, elements, and the soul. This round of topics is of course suggested by the works of Aristotle, but the discussion of them is very brief compared to his. All bodies are said to be composed of atoms, but under divine providence and not as a mere play of chance. From homogeneous atoms are made homogeneous bodies; and from heterogeneous atoms, heterogeneous bodies. The heavens are filled with air rather than a quintessence, but air is a mixture and not an element. Fire also is not an element, leaving only earth and water. That water changes into air is denied. Moving Intelligences for the heavens and stars are likewise denied, and the questions of the Peripatetics concerning the heavens are said to be so frivolous that their discussion provokes nausea. Despite such drastic criticism of Peripatetic astronomy, Gorlaeus still held that the earth did not move.[30] It is doubtful if his Exercises were university lectures, but they may be considered a sharp criticism of the usual *cursus philosophicus*. Espagnet is said to have defended the ideas of Gorlée in his *Enchiridion physicae restitutae*, published anonymously in 1623.[31]

The astrology and natural magic of Goclenius the Younger, professor at Marburg, have been discussed in previous chapters. His General Physics, although described as based on the nature

[29] *Exercitationes philosophicae quibus universa fere discutitur philosophia theoretica, et plurima ac praecipua Peripateticorum dogmata evertuntur*, Leyden, 1620, in-8, 352 small pp. Copy used: BN R.25547.

[30] *Ibid.*, pp. 247, 293, 311, 313, 334.

[31] Mersenne, *Correspondance*, I (1932), 148. The *Enchiridion* of d'Espagnet is treated later in this chapter.

of things and rational experience,[32] starts out more like a manual of natural magic or occult science. He defends astrology, discusses monsters and "illustrious examples of things preternatural," *hyperphysica* and miracles, and asserts that specters cannot be the souls or bodies of the dead and must be demons.[33] But then he considers matter and form, quantity[34] and quality, light, colors, and reflection, odors and tastes, sound.[35] But his interest in the occult continues throughout. Light is that divine vehicle of the virtue of sun and stars and an instrument not passive but active in marvelous ways. And of celestial heat there are many degrees and orders. But, in addition to light and heat, there is the occult influence or action of the stars on these inferiors by virtual contact.[36] Later he discusses the spirit or oil of sweet salt and its marvelous virtues in diseases of the microcosm, or the powers of human urine and its balsam.[37] He tells how, if a marriage is sterile, to tell which person is to blame, or asks why many blind men are learned, deaf men never.[38] He lists many occult qualities, such as the bitter almond's resisting intoxication, and the male peony, epilepsy.[39] Then he further discusses antipathy, amulets, spiritual nature in general, and the world soul. Perhaps the claim made in his title to contain many things unobserved by others applies especially to this preternatural side of his physics.

In 1616 Godefroy Chassin submitted his book on nature or the world to the head of the Jesuit college at Lyons, where he had once been a student, and no doubt would have been glad to have it adopted as a text in place of their teaching the philosophy of Aristotle.[40]

Abra de Raconis was a doctor of theology at Paris, a preacher and royal almoner. His summary of all philosophy, in four parts

[32] *Physicae generalis libri II e rerum natura et rationali experientia deprompti, theorematis distincti ac scholiis illustrati, multa continentes ab aliis inobservata, nunc primum in lucem editi*, Francof., 1613, 470 pp. BN R.10435.

[33] *Ibid.*, pp. 4-5, 19-23.

[34] Book II begins with it at p. 88.

[35] Reached at p. 389. He lists and solves 42 problems as to odors.

[36] *Ibid.*, pp. 113, 118, 123.

[37] *Ibid.*, pp. 365, 367.

[38] *Ibid.*, pp. 373, 394.

[39] *Ibid.*, pp. 126, 404 *et seq.*

[40] Chassin's book has been treated in T VI, 382-86, in the chapter, "For and Against Aristotle,"

devoted respectively to logic, ethics, "physics," and metaphysics,[41] was first published in 1617, and by 1637 had run through half a dozen editions, all at Paris. I have examined the third part or *Physica* in the second edition of 1622,[42] where it occupies over 800 small pages. It is primarily a commentary upon the natural philosophy of Aristotle but indulges in "moral" or religious digressions against sin, on contempt of life and desire for beatitude, the mystery of the Incarnation and sacrament of the Eucharist, the dignity of man, and justification of the sinner.[43] Scholastic authorities cited range from Albertus Magnus and Aquinas through Ockham, Gregory of Rimini and Cajetan to Suarez, the fathers of Coimbra, and Fernel.[44] The world is called the physical tree; the heavenly spheres are the topmost branch; the elements, a second branch; and mixed bodies, the third.[45] But whereas Aristotle had regarded the heavens as a fifth essence, distinct from the inferior elements, Abra considers more probable the opposite view that celestial and sublunar matter are "of the same relation and specific nature, and distinct only in certain accidents."[46] Against Averroes and Durand de St. Pourçain, he affirms that the heavens consist of matter as well as form, and holds that their form is not soul.[47]

Such questions are raised as to the time of year when the world was created, whether there is only one substantial form in all the heavens, and whether the stars are essentially different from the heavens and have a distinct substantial form.[48] Abra believes that epicycles are not only real but solid, and that a planet has no free movement of its own but is fastened to its epicycle.[49] For him John of Sacrobosco is still an authority "whom all accept and revere as the head of astronomy."[50] The motion of the heavens is not absolutely necessary for the action of inferior agents,[51] but it is certain that they influence these inferiors by their motion and light

[41] *Totius philosophiae, hoc est logicae moralis physicae et metaphysicae brevis et accurata . . . tractatio.*

[42] Copy used: BN R. 47881.

[43] *Ibid.*, pp. 54, and 130, 83, 97, 129, 168.

[44] *Ibid.*, pp. 229, 255, 293, etc.

[45] *Ibid.*, pp. 338, 362, 418, 470.

[46] *Ibid.*, p. 49.

[47] *Ibid.*, pp. 365-66.

[48] *Ibid.*, pp. 356, 369-70.

[49] *Ibid.*, p. 378.

[50] *Idem.*

[51] *Ibid.*, pp. 402-3: but is "secundum quid et ad melius ad actiones omnes agentium naturalium sublunarium."

producing heat and drought, and, while some question their occult influence, Abra finally decides in favor of it. He agrees with the Church that they do not cause future events of necessity, but they incline. He holds with Albert and Durand that even the empyrean heaven influences inferiors, and further argues that the stars produce all the dispositions requisite for forming imperfect animals and their spontaneous generation.[52]

As usual in commentaries on Aristotle, monsters are considered in connection with fortune and chance; indeed, they are the main interest.[53] However, nine other causes of them are suggested: including excess of material and seed, lack of material—resulting, for example, in only one eye or one foot, and confusion of matter or seed, the influence of the stars, and the imagination of the parents. Incidentally the question is discussed for some two pages whether from the mingled seed of man and beast a human or brute species would result.[54]

The movement of the sea from east to west is not caused by the motion of the heavens, which hardly stirs the supreme region of air, but by occult celestial influence. The tides Abra attributes to the moon, aided by the aspect of the sun.[55] Rivers and springs which are not perennial may be produced by precipitation, or from air and vapors transmuted into water, but those that flow continually draw their waters from the sea.[56] Augustine is cited for the arcane nature of certain fountains, but Abra does not allow his theological proclivities to go too far: he thinks that roses had thorns before Adam sinned.[57]

Intension is not to be explained as the destruction of prior quality and production of a more perfect quality, nor by mere deeper rooting of the quality in the subject, but by the addition of a degree recently produced.[58]

Abra still clings to the Aristotelian doctrine of comets. The Star of Bethlehem was not a comet but a new meteor produced by divine or angelic virtue and moved by an angel rather than by

[52] *Ibid.*, pp. 411-15.

[53] *Ibid.*, p. 199, "De effectibus fortunae et casus potissimum de monstris."

[54] *Ibid.*, pp. 200-204.

[55] *Ibid.*, p. 456.

[56] *Ibid.*, pp. 461-2.

[57] *Ibid.*, pp. 465, 468.

[58] *Ibid.*, pp. 494-95.

its own virtue. Comets presage disease, death of kings and of tender infants, drought and sterility, winds, storms and earthquakes.[59] Abra had earlier debated the question how far art can imitate nature by applying active to passive, as the demons do in feats of magic. Aquinas denied this absolutely in the *Sentences*, but altered his opinion as he grew older and wiser (*Secunda Secundae*, q. 77, art. 2). Abra had concluded that the making of gold by alchemy is possible but very difficult.[60] Now he treats of metals briefly and has barely a page on stones.[61] The rest of the volume is devoted to the *De anima* of Aristotle with an appendix on human anatomy.[62]

Cursus philosophici encyclopaedia was the original title of Alsted's encyclopedia in the first edition of 1620, although the first two words were dropped in the edition of 1630 and altered to *Scientiarum omnium* in that of 1649, a significant change.[63]

Michael Zanardus or Zanardi (1570–1642) of Milan was a Dominican who studied at Bologna and taught philosophy and theology at Milan, Verona, Cremona and Venice. In the last named city he published commentaries on Aristotle's Metaphysics and Logic in 1616, on his Physics and *De anima* in 1617, and in 1619 the Disputations on the Elementary Universe with which we shall be chiefly concerned here.[64]

Among the twenty-eight questions disputed in the first part are such oft-debated ones as whether the universe is perfect, whether there can be more than one universe, whether it is generable and corruptible, whether it is divided into ethereal and elementary regions, and whether there are the four traditional elements, namely, fire, air, water and earth.

Of the ninety-nine questions of the second part we also note only a few. The query whether, if elements and mixed bodies did not encounter resistance from the medium, they would be moved

[59] *Ibid.*, pp. 525-28.

[60] *Ibid.*, pp. 99-104.

[61] *Ibid.*, pp. 563-64.

[62] *Ibid.*, pp. 776-803.

[63] For the contents of Alsted's Encyclopedia see T VI, 433-36.

[64] *Disputationes de universo elementari in tres partes divisae*, Venice, 1619, in-4. Copy used: BM 536.k.5. The division into three parts is not of much significance, and the second and third parts are differently described at pp. 81 and 253, where they begin, than on the title page.

instantaneously, is answered negatively, it being asserted that it is not merely the resistance of the medium which is responsible for the succession of motion. To the query whether there are celestial as well as elementary qualities, the answer is, Yes, light. But the question whether there are virtual qualities is answered in the negative. Other questions are whether milk and semen are alive. The indivisible is not inalterable *per se* but is so accidentally.[65]

Eighteen questions are put concerning fire in the third part. Can it ever kindle itself? Why is air essential to keep it going? Why does it last longer in the shade than in the sun? Why is it less extinguished by sea water than by other water? Why does flame take a pyramidal form? Why does gunpowder explode from the gun with so much noise? Why does much fire produce less sweat than a little? Why a person who is very cold feels pain on approaching a fire? Does any animal live in fire? Aristotle and Galen say not, and the passage in the former on the *Pyraustes* must be regarded as an interpolation. The salamander is of such a very cold and wet nature that it extinguishes a few coals by its touch but is finally consumed unless it retires. It is further inquired whether any lifeless object can be preserved unharmed in fire. Is there any fire which is nourished by its contrary, that is, wet and cold? If so, such a water must be full of pitch or sulphur. Volcanoes are considered, and why they are found especially on islands; then how fire is produced from stones, from wood, and from glass and mirrors. The seventeenth question is whether there are other marvels of fire besides those enumerated. The last is why there are fires from mountains and not from valleys.[66]

There are eight problems concerning air. The first, whether fire and air are moved circularly by themselves or by the heavens, gives the usual answer that it is by the heavens. Why do empty bladders sink in water and those full of air, not? The levity of fire and of air is not of the same sort. Some things exposed to the sun are colder than in the shade. Why is air that has been long underground injurious to breathe? Why does such air sometimes harm only birds, sometimes other animals but not man, sometimes human

[65] *Ibid.*, II, Qys. 6, 35, 41, 50, 51, 66; pp. 88, 131, 141, etc.

[66] *Ibid.*, III, 12; pp. 279-82.

beings as well? Why is it worse to perspire in cold air than in hot? How is it that air is called the author of the four seasons?[67]

It is rather surprising to find that Isidore of Seville of the seventh century casts the deciding vote in the seventeenth as to whether the ocean is the father of all waters. Of twenty-six doubts as to the sea, one concludes that springs and rivers are generated in the bowels of earth from both rainfall and the sea, and perhaps also from air, as Aristotle held. There are twenty-one doubts as to fish. One is whether there are any which have a human face. When a woman bore a son in the form of a fish, the emperor ordered that it be killed, as Theophanes testified in 582 A.D. But Zanardus does not approve of this, one reason being that there might be a human being under that monstrous form, and another reason that it smacks of Aristotelian superstition that monsters should be killed as presages of future ills. Whether there are fish of marvelous virtue is answered affirmatively by adducing the remora and the torpedo. Zanardus agrees with the philosophers that frogs are born of putrid matter by virtue of the sun, and goes on to other marvels of fishes.[68]

In one place Zanardus says that earth and water make one globe, but in another that the sphere of water is above that of earth. In any case the earth is immobile at the center of the universe.[69]

Many hold that comets are ethereal and of celestial nature, but Zanardus agrees with Aristotle and all Peripatetics that they are not, and even asserts that learned astronomers have proved by mathematical instruments that they are below the moon. From them may be predicted winds, earthquakes, storms at sea and shipwrecks, epidemics and many other mortal diseases, battles, sterility, and death of kings. There can be no good results from those which last more than a week. They may be of divine or natural origin. They deceive astrologers, but Zanardus gives six rules to follow in determining their effects. Those moving from east to north bring pest; those moving east to south cause humidities and famine. A hairy one, vari-colored, and with a long tail, presages

[67] *Ibid.*, pp. 346b-348a.

[68] *Ibid.*, III, 28, 32, 39; pp. 349-54, 369-71, 404-13.

[69] *Ibid.*, pp. 417a, 420b, 423b.

winds. A dark one with short tail and without hairs is followed by sterility. They presage sterility, pest, winds and pest, or fires, according as they appear in earthy, watery, aerial or fiery signs of the zodiac. If one first appears in the twelfth or eighth astrological house, it brings pest and destruction of crops. If one appears in the east in the morning, it signifies heat. But further superstitious rules of Cardan and others as to religious change, birth of legislators and the like are better passed over in silence.[70]

Zanardus, however, lists various weather signs, some of which are astrological. Observance of lucky and unlucky days for business and journeys is condemned by the church, but not of the time for blood-letting or purging or taking medicine. He does not intend to go into prediction of wars insofar as free will is concerned, but only to adduce some signs accepted by Hermes and other astrologers which can move the red bile and so indirectly cause wars. He repeats signs of pestilence and other infirmities from the pest tract of his fellow-countryman, Guglielmo Gratarolo, who became a Protestant and religious fugitive from Italy.[71]

The star of the Magi was an exhalation reduced by God to the form of a star and decorated with great splendor, which was moved by the ministry of angels or other special influence of God, and which vanished after it had served its purpose.[72]

Zanardus also composed Disputations concerning the Small Universe, or man, in which he displayed a favoring attitude towards physiognomy, including even moles and spots on the nails.[73]

Jean d'Espagnet's Handbook of Restored Physics (*Enchiridion physicae restitutae*), first published in 1623,[74] is neither as reactionary nor as revolutionary as one might be tempted to infer from its title. It is made up of 244 short chapters devoted to such topics as God, the world, nature, first matter, creation, the creation of man,

[70] *Ibid.*, pp. 271a-75a, 342a.

[71] *Ibid.*, III, 25; pp. 331-46.

[72] *Ibid.*, p. 276b.

[73] *Disputationes de universo parvo mixto homine usque in senium conservando,* Venice, 1619, in-4, 152 pp., closing, "Venetiis in Conventu Sancti Dominici die 27 Octobris 1618." Questions iv and v at pp. 10a-51a are concerned with physiognomy. Copy used: BM 1179.d.5 (2.).

I have not seen his work on the celestial universe which he cites *De universo elementari*, p. 333b.

[74] Copy used: BN R.51611-51612.

the harmony of the universe. Matter, form and privation are still the three principles of things. Such time-honored conceptions as those of *humidum radicale*,[75] man the microcosm,[76] and the three regions of air[77] are retained. But the existence of a sphere of fire next to the sphere of the moon is denied.[78] Light is the true fire of nature, as Genesis shows.[79] In *humidum radicale* d'Espagnet even thought that there was "something immortal which neither disappears with death nor is consumed by the forces of the most violent fire but remains unconquered in corpses and ashes."[80] As Gray was to write later:

> E'en in our ashes live their wonted fires.

Following perhaps in the path of Patrizi of the previous century, d'Espagnet laid great stress upon light as a force in nature, and upon the sun, which he said was not the eye of the universe, as some of the ancients had thought, but of the Creator of the universe, who thereby had sense perception of His sensible creatures and made Himself conspicuous to them by pouring the rays of His caressing love upon them.[81] Earlier he had said that created light contracted into the solar body and had remarked that some philosophers not without probability located the world soul in the sun,[82] and that God had expressed a triple image of His divinity in the sun.[83]

> A second universal agent is that very light, not however flowing immediately from its source, but reflected from dense bodies illuminated thereby, such as the heavenly bodies and the earth itself.[84]

Love was the genius of nature[85] and there was no contrariety in

[75] *Ibid.*, p. 66, "Ignis ille naturę mixtis insitus humidum radicale tanquam sedem propriam elegit; huius autem domicilium praecipuum est in corde."

[76] *Ibid.*, p. 124; but also, p. 125, "Quodlibet mixtum est microcosmus."

[77] *Ibid.*, pp. 59-61.

[78] *Ibid.*, p. 61, "Superior regio lunae vicina tota aerea est, non ignea, ut falso in scholis dudum invaluit."

[79] *Ibid.*, p. 62 .

[80] *Ibid.*, p. 170, "Aliquid immortale in humido radicali licet observare quod neque morte evanescit neque ullis (p. 171) ignis violentissimi viribus consumitur verum in cadaveribus et combustorum cineribus invictum remanet."

[81] *Ibid.*, p. 65.

[82] *Ibid.*, pp. 15, 22.

[83] *Ibid.*, p. 25.

[84] *Ibid.*, p. 68.

[85] *Ibid.*, p. 69.

the elements.[86] Heat and dryness, which are masculine and formal, proceed from the informing light.[87] "Light and darkness are the principles of life and death."[88]

Espagnet repeatedly affirmed the influence of the heavenly bodies upon inferior creation. Inferiors were governed by superiors.[89] Form was not from the mere potency and virtue of the seed and of matter, for celestial virtues influence the genesis of things.[90] Rocks and stones, it is true, are generated not from a true mixture of the elements but from a concourse of earth and water produced by an external force of heat and cold. But precious stones and gems draw their forms from the most limpid springs of sky and sun.[91] Similarly the multiplicative virtue present in seeds is not from elemental matter but from celestial form as its efficient cause.[92] The heavenly bodies mould the natures of the elements like wax.[93] Since, however, the nature of the celestial bodies and their relations to man are largely unknown, their rule is uncertain and deceptive to us.[94] Thus while d'Espagnet asserts the influence of the stars, he does not hold out much hope for successful astrological prediction.

Espagnet accepts as secondary elements the salt, sulphur and mercury of the alchemists, which he represents as mixtures respectively of water and air, earth and air, and earth and water.[95] He also employs such alchemical expressions and concepts as ferment, matrix and menstruum,[96] or sublimation and decoction.[97]

It must be said further that d'Espagnet's positions are not always consistent; sometimes they seem quite conflicting. He talks of the harmony of the universe,[98] and that the machine of the universe is one and united,[99] but also holds that there are many worlds in the

[86] *Ibid.*, p. 71.
[87] *Ibid.*, p. 73.
[88] *Ibid.*, p. 138.
[89] *Ibid.*, pp. 113, 115.
[90] *Ibid.*, p. 129.
[91] *Ibid.*, p. 132.
[92] *Ibid.*, p. 136.
[93] *Ibid.*, p. 187.
[94] *Ibid.*, p. 113.
[95] *Ibid.*, p. 118.
[96] *Ibid.*, p. 32, "Ex quibus constat alterationis et corruptionis fermentum ac tandem fatale mortis venenum non a qualitatum repugnantia sed ex infecta matrice et venenato materiae tenebrosae menstruo." At p. 101 he calls earth "vas generationis et matrix"; then goes on to speak of "aqua menstruum mundi rerum semina et elementa in se continet et fovet." See also p. 187, "ac continuo suo influxu fermentare non desinant."
[97] *Ibid.*, p. 107.
[98] *Ibid.*, p. 35.
[99] *Ibid.*, p. 163.

universe and that this is not contrary to the Bible.[100] He seems to waver wildly between the heliocentric and geocentric views of the universe. Not only does he say that those philosophers, who not without probability placed the world soul in the sun, also put the sun at the center of the universe.[101] He further affirms that the earth is as much a star as the moon is.[102] But in between these two passages he has stated that the whole globe of the earth is of a no less constant nature than the sky. For since it is the center of the universe, it is as necessary that it be constant as it is that the other parts of the universe be so.[103] He approves of the atoms of Democritus[104] as well as the primary and secondary elements already mentioned. And he holds that the forms of animals and vegetation are rational, not indeed as human beings are but in their own way.[105] Composite living beings are composed of body, spirit and soul.[106]

If this seems an odd hodge-podge and an example of irresponsible eclecticism, which sounds a little as if d'Espagnet tried to please everybody, it also seems to have been to the taste of its century, for his second and third editions appeared in 1638,[107] and yet others in 1642,[108] 1647, 1653, 1673 and 1702,[109] and in French translation of 1651.[110]

A century ago Hoefer in a section on alchemists of the seventeenth century in his *Histoire de la Chimie,* after rapidly listing a number in France, selected d'Espagnet for a paragraph of more particular treatment because of the "notions remarquables sur les généralités de la science" in his *Enchiridion physicae restitutae.*[111] But the first of these, that the water, air and earth which we know are not pure elements but compounds, had been a commonplace of medieval science. That air is essential to life, and fire a very subtle material body, that vegetation is nourished by air as well as earth and water, and that substances are more ready to combine

100 *Ibid.*, pp. 191, 194-5.
101 *Ibid.*, p. 22.
102 *Ibid.*, p. 192.
103 *Ibid.*, p. 96.
104 *Ibid.*, p. 119.
105 *Ibid.*, p. 138.
106 *Ibid.*, p. 125.
107 Ferguson, I, 248, BN Vélins 1990; 3rd ed., BN R.12443-12444.
108 Not in Ferguson; Paris, N. de Sercy, 32mo. BN R. 26815.
109 BN has those of 1647, 1653 and 1673. That of 1653 was at Geneva in the *Bibliotheca* of N. Albineus; that of 1702 also at Geneva in Manget, *Bibliotheca chemica curiosa*, pp. 626-49.
110 Ferguson, I, 248.
111 Hoefer, *op. cit.*, II, 333.

when reduced to a state of minute sub-division, were also ideas which were not peculiar to d'Espagnet.

The physical section of 500 pages of the 1624 edition of the *Universa philosophia* of the Jesuit, Hurtado de Mendoza, followed the usual Aristotelian plan and put questions which were to recur in Arriaga and Oviedo and be answered similarly by them.[112]

Gilbert Jacchaeus was a Scot from Aberdeen who became professor of philosophy and medicine at Leyden and died in 1628. His Physical Institutes was first published in 1624, and had subsequent editions.[113] It largely follows the Aristotelian pattern and divides into nine books: 1) introductory, including the three principia of matter, form and privation; 2) on nature; 3) on motion; 4) on time and space, vacuum, finite and infinite; 5) on the heavens; 6) on mixed bodies; 7) on meteors; 8) on the soul and 9) on the rational soul. The heavens act upon inferiors by light, motion and influence. Pico della Mirandola and others deny occult influence, but experience proves it. Only such influence can reach metals, for light stops at the earth's surface, and motion in the second of the three regions of air. It is also this occult influence of the heavens which makes the magnetic needle turn towards the pole. The heavens act upon man but are not the principal cause of any perfect animal, whereas imperfect animals owe their origin to the heavens. Jacchaeus appears never to have heard of the Copernican system. He still accepts four elements and reckons the earth's circumference as 19080 miles.

The sixth book on mixed body includes first qualities and alteration, that like does not suffer from like, reaction, generation and corruption, mixture, the problem of how the elements are in the mixed, temperaments, and putridity. Comets are still classed as meteors or exhalations in the supreme region of air, where with it they follow the circular movement of the heavens. The sea is salt for the sake of the fish in it. It moves from north to south, and east

[112] Bernhard Jansen, S. J., *Die Pflege der Philosophie im Jesuitorden während des 17-18 Jahrhunderts,* Fulda, 1938, pp. 25, 27-28, where a fuller account of its contents is given. I presume that BN R.1931, *Disputationes de universa philosophia,* Lyon, 1617, in-4, is an earlier edition of the same work.

[113] *Institutiones physicae.* I have used that of Amsterdam, 1644, which is described as *editio postrema* and *emendatissima,* BN R.12417. But there was another at Jena in 1646.

to west, and also has the tides. Their cause is much disputed, but Jacchaeus favors the moon. For the origin of fountains he prefers the sea rather than the Aristotelian formation of water in caverns from the dissolution of air and vapors.

His psychology still includes a chapter on the intentional species of the schoolmen, and still debates, without reference to Alhazen, Witelo, or the science of optics, the question whether vision is made by emission of rays. Plato in the *Timaeus* said Yes; Aristotle, *De anima*, II, 7, said No. At any rate, fascination by sight is pronounced nonsense. After the external senses are considered the internal.

The Physical Handbook of Caspar Bartholinus the Dane (1585–1629), printed at Strasburg in 1625, is a chunky volume of 865 small pages.[114] A special feature is the emphasis upon disputed points as to nature with listing of the arguments on both sides. The work is in eight books, of which the first adds the three *principia* of the chemists to the form, matter and privation of Aristotle. But then follows his program of causes and reference to fate, fortune, chance and monsters; of quantity, place, time, the infinite, vacuum; first and second qualities, adding third or occult qualities, natural magic, sympathy and antipathy; the relation, action and passion of natural bodies; motion and rest; generation and corruption, and motion in special. All this, however, is covered in 34 brief pages.

The other seven books are longer and deal with cosmology, uranology, the elements, mixture, meteors, perfectly mixed bodies both inanimate and animate, and the soul. In large part these books had been preceded by separate treatises: the first, by *De principiis rerum naturalium* in 1622 (50 pp. in-12); that on cosmology by *De mundo* of 1617;[115] uranology, by *De astrologia* of 1616—already discussed in our chapter on Astrology to 1650; that on the elements, by two books on waters in 1617,[116] and one on earth, air

[114] *Enchiridion physicum ex priscis et recentioribus philosophis accurate concinnatum et controversiis naturalibus potissimis utilissimisque illustratum*, Argentinae, 1625. BM 537.a.4.

[115] *De mundo quaestiones et controversiae nobiliores ex sacro codice, rationibus atque experientiis formatae et firmatae. Accessit brevis Uranologiae summa ex iisdem fundis fontibusque derivata*, Copenhagen, 1617. BM 538.a.17 (3.). The three chapters on uranology are very elementary, and it is explained that all is set forth more fully *in astrologia nostra*.

[116] *De aquis libri II*, Copenhagen, 1617: BM 538.a.17 (1).

and fire in 1619;[117] that on mixture, by *De mixtione* of 1617.[118]

Aristotle held falsely that the world is eternal. It is discussed at what time of year creation occurred, and all beings are said to have been created for man. The world is perfect, but Caspar goes on to debate whether God could have created and can create a more perfect one, and whether He created the world in a moment or successively. The Talmud says that the world will last 6000 years, and Chaldean estimates of long ages already elapsed are excessive. Sisto da Siena collected about thirty discrepant estimates of the time elapsed from the beginning of the world to the birth of Christ. A fair estimate is that 5563 years have passed from creation to the present year, 1614.

Caspar cites Falloppia, Cesalpino and Kentmann of the sixteenth century, but not Helmont or Galileo of the seventeenth. Waters have many qualities and virtues which are either manifestly occult or very difficult to understand. Marvelous fountains are described. After repeating thirteen explanations that have been given of tides, Caspar concludes that those are nearer the truth who make the moon the main cause and the sun an auxiliary cause. Some attribute them to the motion of the earth. The chief source of springs is from the sea, although rain and snow are an auxiliary cause. Testing mineral waters in the common way without fire is distinguished from the chemical method. Air is described as a hot element, very moist—more so than water, and very light. Three regions of it are distinguished. It will putrefy either from lack of movement and ventilation, or from the admixture of vapors by quality either manifest or occult and poisonous, or from the influence of certain stars. Fire is nothing but some body being burned, and is either carbon or flame. It is above earth or subterranean, the material of the latter being sulphur and bitumen. We see here again the influence of the chemical view-point, and in his adding putridity, combustion and petrification to his discussion of mixture.

First humors and then spirits are discussed in treating of the

[117] *De terra aere et igne institutio physica succincta cum praemissa elementorum theoria generali*, 1619: BM 1135.a.7 (2.), in-8, 44 pp. Also at Rostock, 1619; and Greifswald, 1624.

[118] *De mixtione eamque consequentibus temperamento, coctione, putridine, petrificatione . . .*, Copenhagen, 1617, pp. unnumbered: BM 538.a.17 (2.).

parts of animals. Insects are imperfect animals, lacking blood and not breathing. They are spontaneously generated from slime, rotting vegetation, putrefying water, or from other animals, as beetles from an ass, bees from a bull, and wasps from the carcass of a horse —for which Caspar cites Aristotle's History of Animals, V, 19; then soon turns to zoophytes. In the book on the soul, sleepwalking is spoken of, and dreams are classified as divine, diabolical, animal and natural. Animal are those which repeat works with which man is occupied or on which he is intent. Natural are from affection of bodies, temperament, incursions of humors, and the like. If dreams are causes or signs of the future, it is easy to conjecture something from them. But if they are merely accidents of future things, then they are deceptive.

A *Systema physicum* by Caspar Bartholinus in 1628 is presumably an enlargement or revision of his *Enchiridion physicum* of 1625.[119]

The Philosophy, embracing metaphysics and physics, of Raphael Aversa of San Severino, professor of theology, who died in 1657 in his sixty-eighth year, appeared at Rome in two volumes in 1625 and 1627 and was expressed in the form of the discussion of various questions.[120] The first volume is very abstract and adheres closely to Aristotle. Concerning it we may note further only that the discussion of chance, fortune and fate[121] has nothing about monsters, which word is also not found in the index of either volume. Turning to the second volume and the 31st question as to the *Mundus*, we find Aversa asserting that the world was created by God from nothing, that it could not be from eternity, that God could reduce it to nothing again, but that as a matter of fact it will endure to eternity, although the motion and influence of the heavens will cease, the action of the elements and generation will stop, and men will have another state.[122]

[119] *Systema physicum ex autoris genuinis partim editis partim non antehac editis libris sequenti pagina indicatis coagmentatum*, 1628, in-8, 197, 874 pp. Listed by Niels Nielsen, *Matematiken i Danmark, 1528-1800*, Copenhagen, 1912, p. 14. For the location of copies of the *Systema* in Germany, see *Deutsches Gesammtkatalog*.

[120] *Philosophia metaphysicam physicamque complectens quaestionibus contexta* . . ., Rome, 1625, 1627, 2 vols. Copy used: BM 1136.1.27.

[121] *Ibid.*, I, 397 *et seq.*

[122] *Ibid.*, II, 26-39.

Another question is whether there are other heavens beyond that of the fixed stars. Ptolemy and others added a *primum mobile.* Thebit, Alfonso X, and others added yet another called crystalline. Theologians posit an empyrean heaven. Aversa contends that it is not necessary to suppose other mobile heavens beyond the fixed stars, because their varied movements are the only reason for such an hypothesis and Scripture says nothing of such heavens. But Aversa accepts the empyrean heaven as the seat of the blest and holds that the heavens are solid because the Bible talks of the firmament. This leaves the problem of explaining "how with a solid heaven and without distinction of heavens and orbs and without penetration or scission, all the movements of the planets and appearances can be saved," and Aversa makes a feeble attempt to solve it by substituting zones for spheres of the planets, citing Tycho Brahe repeatedly. Although most of his astronomical variations thus far have been motivated by the Bible, he denies that there are true elementary waters above the heavens and holds that the heavens are incorruptible. The new stars of 1572 and since were made *de novo* by an accidental change in the heavens.[123]

But the question whether comets are celestial phenomena Aversa answers in the negative, holding that the argument from parallax can be turned against those who answer in the affirmative, and accordingly postpones further consideration of them until he comes to meteors.[124]

Heavens and stars are probably composed of matter and form but possibly are simple bodies. In any case they are of a different matter from inferiors and are not animated by a soul as form, and are not moved intrinsically but by Intelligences. This assumption of several heavens moved by Intelligences seems inconsistent with his previous hypothesis of one solid heaven without distinction of heavens and orbs. Now he says that whether the heavens differ from each other in matter as well as form is uncertain. At any rate they do not possess elementary qualities, nor true colors, and they make no sound.[125] Aversa has been citing Kepler and Galileo as well as Tycho, but he refuses to accept Galileo's explanation of sun-

[123] *Ibid.*, 52-89.
[124] *Ibid.*, 91-100.
[125] *Ibid.*, 103-9, 147, 119, 135-37.

spots or his statement that the moon receives some light from the earth, because for the earth to illuminate a celestial body would evidently be to invert the order of the universe. Instead he suggests that it is illuminated by reflexion from adjacent parts of the heavens.[126] The fixed stars do not seem to receive their light from the sun.[127]

Aversa asserts more than once and in very emphatic terms that the heavenly bodies rule and govern these inferiors.[128] "So all the theologians and sacred doctors teach and holy Fathers, and it is expressly stated in many passages of Scripture."[129] He thinks that this action is limited to the planets and fixed stars, and that the heavens which contain them do not operate upon inferiors, but, like Abra de Raconis, he discusses the question whether the empyrean heaven does.[130] The heavenly bodies act upon the earth not only by their light and motion, which latter "is the condition by which the celestial bodies distribute their operations in these inferiors," but also by occult virtues and influences. "Living beings which are generated without propagation ... seem to be made by the sky itself as principal cause."[131] Here again he seems to contradict his previous position that the heavens containing the stars and planets do not act upon inferiors. He goes on to make the usual caution and qualification that the stars act upon man only by way of inclination and not compulsion. But this does not prevent his concluding that from the stars and other causes, if well noted, corporeal effects can be predicted.[132]

Descending to inferior bodies, Aversa accepts the traditional four elements, with fire next to the heavens, and three regions of air. Water forms one globe with earth which is at rest at the center of the world. Indeed, the earth is somewhat higher than the water, and Aversa contends that Holy Scripture does not teach that the water is higher than the earth, nor that it is kept from overflowing the earth by a standing miracle, nor that rivers have their origin from the sea. Although a frequent caption has been, "Certain

[126] *Ibid.*, 154-60, 169-71.

[127] *Ibid.*, 173.

[128] *Ibid.*, 114; 174, "Sed plusquam certum et evidens est absolute loquendo vere et realiter corpora caelestia in haec inferiora operari ..."

[129] *Idem.*

[130] *Ibid.*, 176, 178-81.

[131] *Ibid.*, 181-87, 195-97.

[132] *Ibid.*, 201, 205.

passages in Aristotle are explained," he rejects the view which he attributes to Aristotle that one unit of earth will make ten of water; a hundred units of air; and so on. He holds that there is more earth than water.[133]

From the elements Aversa turns to generation and corruption, quantity and quality, place, motion and rest. Aristotle is represented as saying that natural motion is swifter at its end, violent motion in the beginning, and that of *proiecta* midway; but by *proiecta* he meant the movement of animals.[134] Aversa asserts that the acceleration of falling bodies is not because the medium offers less resistance as the motion progresses but because the medium impels it down more, the more it falls. In the case of violent motion, however, he rejects Aristotle's explanation and adopts the impetus theory, although he grants that the air may aid somewhat.[135]

Coming to comets again, he states that their supposed effects seem utterly groundless naturally, but that they may be divine signs. The principal cause of stones and metals, on the other hand, is the celestial bodies. The transmutation of metals is difficult but not impossible.[136]

Passing on to *De anima*, Aversa argues that the internal senses are really one and not multiplex.[137] Some things can be divined from natural dreams, but this kind of divination is so weak and fallacious that it is to be regarded as almost nil. Physiognomy considers impressions in the body which indicate internal forces. Imagination by commotion of the humors indirectly produces such impressions on its owner's own body, but not on another body except in the case of the foetus.[138]

In 1626, the year between the appearance of Aversa's two volumes, Johannes Rodolphus Faber, a Jesuit of Grenoble, published a *Cursus Physicus*.[139] In the preface he states that in years past he

[133] *Ibid.*, 207-37.

[134] Aristotle was usually represented as holding that the violent motion of projectiles increased in velocity at first.

[135] *Ibid.*, 358-59, 371-76.

[136] *Ibid.*, 423-24, 493, 496, 498.

[137] *Ibid.*, 731-39.

[138] *Ibid.*, 756-57, 768-70.

[139] *Cursus physicus in quo totius philosophiae naturalis corpus . . . explicatur*, Geneva, 1626, in-8, 496 small pp. and Index. Copy used: BM 536. b.6.

had issued a *Cursus Logicus* combining the precepts of Aristotle and Ramus, and that he hopes soon to issue a *Cursus Metaphysicus*. Eventually, however, he turned to law and in 1643 published a commentary on the *Institutes* of Justinian.[140] His physical text is in two parts, one general, the other special. In the former he treats of the nature of physical science, the essence of natural body and of Nature, the principles of natural body in general, matter, form, and the efficient and final principle; then of motion, rest, the infinite, place and time. He maintains the existence of substantial forms.[141] Monsters are mentioned only incidentally.[142] Unlike Aversa, he attributes the increasing velocity of falling bodies both to increasing pressure from the air above, which rushes in to prevent a vacuum, and diminishing resistance from the air underneath. And he attributes the violent motion of projectiles to an impulse given to the air and not to impetus transferred to the projectile. But he held that the velocity of a projectile was greater in the beginning than in the middle or at the end.[143] Siphoning and artillery demonstrate amply for him the impossibility of a vacuum.[144] In the case of the so-called perpetual lamps, the oil will rarefy for a time and so fill the space, but when it has attained the maximum of rarefication, it is converted into air, "and so will neither burn forever nor produce a vacuum." As for the objection—such objections and his replies to them occupy most of Faber's volume—that, if a vessel full of very hot water is hermetically sealed and put in a very cold place, the water will freeze and so condense and leave a vacuum, Faber retorts, not that water expands in freezing, but that either it won't freeze or will break the vase rather than leave a vacuum.[145]

Faber's second and special part divides into ten tractates dealing with the world, heavens, elements, mixed bodies, imperfect mixed bodies, perfect and inanimate mixed bodies, animate bodies in general, vegetation, sentient bodies, and intelligent animate bodies, that is, men. It is highly probable that creation occurred at the

[140] Further described as *Systema juris civilis criminalis canonici et feudalis*, Geneva, 1643, in-fol.

[141] *Cursus physicus*, pp. 46-47.

[142] *Ibid.*, p. 63.

[143] *Ibid.*, pp. 95-97.

[144] *Ibid.*, p. 127, "concursum fortissimum aeris in bombardis."

[145] *Ibid.*, p. 129 *et seq.*

vernal equinox.[146] Faber's treatment of the heavens duplicates that of Aversa on many points, but his *Cursus* was printed before Aversa's second volume. The heavens are neither generable nor corruptible, as the elements are, they cannot be altered, and their matter and form differ from those of the elements. They are not animated by a soul as form, but are moved by external assisting Intelligences. They move about the earth, act on inferiors by motion, light and an immaterial and occult quality, and make no sound, the reputed music of the spheres being a metaphorical expression to indicate their harmony. If the motion of the heavens ceased, actions of inferiors would not all cease at once. The waters above the firmament are only like water, not real water.[147] Faber states the nature and properties of each of the planets but holds that experience is against the signs of the zodiac determining human occupations. Astrologers can predict eclipses, the weather, disease and other natural effects, but not what is subject to divine or human will. The eclipse at the crucifixion was not natural.[148]

Faber retains the four elements, which may be transmuted into one another not wholly but in part, and three regions of air. He ascribes the tides to the diurnal movement of the moon but derives rivers from the sea and considers the earth immobile.[149] He also retains the Aristotelian explanation of comets and believes that they presage high winds, storms at sea, sterility, drought and failure of crops, earthquakes and distempered atmosphere, whence disease, war, commotion and sedition. But it is ridiculous to argue that they presage the death of kings because kings are of a delicate temperament, for many other persons are more so, and many comets are not followed by royal mortality.[150]

Unlike Aversa, Faber holds that the metals have different substantial forms and so cannot be transmuted any more than a man and brute can be. He also attacks the notion of edible or potable gold. It can be liquefied, it is true, or reduced to a fine powder, but has no food value and is indigestible. While hesitating to accept the ascription to gems of moral effects, such as to make a

[146] *Ibid.*, p. 148.
[147] *Ibid.*, pp. 162-91.
[148] *Ibid.*, pp. 194-202.
[149] *Ibid.*, pp. 206-46.
[150] *Ibid.*, pp. 286-91.

man chaste or vigilant, he is certain that they possess rare occult virtues dependent on the influence of the planets and stars, and a better temperament of elements and qualities in the more refined and subtle matter of the gems.[151] He says nothing of the doctrine of signatures in connection with plants. Some animals are born of putrefaction and the heavens are the efficient cause of this, but perfect animals cannot be so produced. Some species of animals, however, are born either from seed or from putridity.[152]

Unlike Aversa, Faber accepts three internal senses; common sense, phantasy and memory, but rejects *aestimatio*. He makes little distinction between animal and vital spirits, affirms the influence of the mother's imagination upon the foetus, and takes up somnambulism and dreams but not divination from the latter.

Rodriguez Arriaga (1592–1667) became a Jesuit in 1606, taught briefly at Valladolid and Salamanca, and spent the rest of his life at Prague. His Course in Philosophy was first published at Antwerp in 1632,[153] then reprinted at Paris in 1637, 1639, 1647 and 1669, and at Lyon in 1644, 1653 and 1669.[154] Beginning with logic, Arriaga reaches Physics at page 240 of the *editio princeps*,[155] and his treatment thereof is primarily a paraphrase of the work of Aristotle as interpreted in the Middle Ages, starting with first matter and form, and concluding with the question whether any creature was or could be from eternity. In his explanation of the rarefaction and condensation of water, however, we encounter an adumbration of the corpuscular theory.

It is to be said then with Occam in the opuscule on the Eucharist; Gabriel on the Canon, Lectio 45; Vallesius IV Physics, text 84, and in the Controversies to Tyros, Question 27; and with many recent writers of our society, that water is rarefied by the introduction of certain corpuscles of air or other substances (*de quibus infra*). Moreover, by reason of these

151 *Ibid.*, pp. 338, 341-42.

152 *Ibid.*, pp. 388-92.

153 *Cursus philosophicus,* Antwerpiae, 1632 (with approbations of 1630 and 1631), in-fol., 891 pp.: BN R.713.

154 De Backer and Sommervogel, *Bibliothèque de la Compagnie de Jésus,* Nouv. ed., I (1890), 578. The edition of 1669, also in folio, has 1017 double-columned pages against the 891 of the 1632 edition, and is described as, "Iam noviter maxima ex parte auctus et illustratus et a variis objectionibus liberatus, necnon a mendis expurgatus."

155 At p. 277 in the edition of 1669.

more space is occupied by the rare body than before, while in condensation corpuscles of this sort are driven out, and so less space is occupied.[156]

Turning to the heavens, although Arriaga admits that many points are customarily discussed here which bear upon the interpretation of Scripture, and others which have more to do with divination than with truth, he contents himself, if not his hearers, with a single disputation on the nature, number and movement of the heavens.[157] Do they differ in species? Probably not, since we see no evidence of diversity. But the stars differ in their varied light and influence, which astrologers and others have noted. Arriaga momentarily grants that recent astronomical investigation with the telescope has shown the heavens to be corruptible or at least fluid and he feels forced to abandon the Aristotelian doctrine that all comets are sublunar. There may be natural exhalations below the moon which burst into flames, but they are not comets properly speaking. Comets are divine miracles rather than natural phenomena, since they portend events which involve acts of free will by human beings which God alone can foresee. In the case of such miraculous occurrences, the heavens need not be either liquid or corruptible, as they would have to be, if comets were natural celestial phenomena, for God by His supernatural power can raise them to any altitude.[158] Arriaga now further argues that the moons of Jupiter and spots on the sun do not prove that the heavens are liquid and corruptible.[159] He devotes much space to the question, which does not seem of great importance, whether the heavens are solid or liquid. Possibly he does this for no better reason than to avoid discussing other questions which might prove to be more embarrassing. He holds, it is true, that the influence of the sky and stars would continue, even if their motion ceased. He supposes that the sun is opaque behind, so that it may transmit light to us the better, while he would concede to the moon and other planets

[156] Ed. of 1632, p. 484a. The edition of 1669, after "many recent writers of our society," adds, at p. 582a, "et novissime Patre Oviedo *controv.* 7, *de gener.* puncto 5." Since, however, such additions and alterations in the 1669 edition reflect a state of mind *after* and not before Descartes, they will not concern us further in this chapter.

[157] Ed. of 1632, p. 497 *et seq.*

[158] *Ibid.*, p. 500.

[159] *Ibid.*, p. 501.

some light of their own.[160] But other problems which might be disputed concerning meteors, comets and tides, "I prudently omit," for they are very dubious matters whose causes are totally unknown and which cannot be discussed without having recourse to their occult qualities and secret influences, and consequently always having to divine or guess. The Fathers of Coimbra treat this more curious than useful field very curiously and very learnedly, but it would be easier to refute their explanations than to offer anything better. But Arriaga, like Galileo, although for different reasons, doubts whether tides are caused by the moon.[161]

Arriaga accepts only two primary qualities, hot and cold, and is aware that some persons regard cold as mere absence of heat.[162] He speaks of the theory of four elements as still generally accepted, but makes fire and earth both hot, water and air both cold.[163] In his view, unlike that of Aversa, some water is higher than any earth. Springs come from the sea and are found on the highest mountains. But he recognizes that the four elements cannot be arranged in concentric spheres, since air is for the most part in immediate contact with earth.[164]

Gravity Arriaga inclines to regard as substantial form.[165] The old notion that a heavier body falls faster must be given up. He has often tested it himself and found that a small crust of bread dropped from a height fell as swiftly as did a rock which he could hardly lift.[166] But the heavy falling body makes a greater impression upon another body resisting it. Arriaga even contends that a falling body does not increase in velocity, but merely makes a greater impression, the greater the height from which it falls.[167] He still adheres to the theory of *impetus (impulsus)* and has no conception of inertia.[168]

From Arriaga's discussion of *De anima* we may note a single point, that he believes the blood to be animated.[169] In this con-

[160] *Ibid.*, pp. 507-508a.

[161] *Ibid.*, p. 508a-b.

[162] *Ibid.*, p. 569a.

[163] *Ibid.*, pp. 568, 575a.

[164] *Ibid.*, p. 577a-b.

[165] *Ibid.*, p. 581a.

[166] *Ibid.*, p. 582a; also in 1669 ed., p. 690b.

[167] Ed. of 1632, p. 582a-b; ed. of 1669, p. 692a.

[168] Ed. of 1632, p. 584; ed. of 1669, p. 695.

[169] Ed. of 1632, p. 627a.

nection he does not mention Harvey's recent discovery of the circulation of the blood, which was presumably unknown to him. As his *Cursus* does not include physiology and anatomy, this is the only place where he would have occasion to allude to it.

On the whole, however, Arriaga seems somewhat more venturesome in his views and somewhat more conscious of recent trends in scientific thought than were most of his predecessors in the Cursus. But he also shares much of their conservative and traditional attitude, and is still favorable to astrological influence. He has abandoned the Aristotelian explanation of comets as terrestrial exhalations. But some of his views are distinctly backward. Arriaga's book was frequently cited by the Franciscan, John Poncius, in his Scotist Cursus,[170] and Poncius, in his turn, was cited in the 1669 edition of Arriaga's work. Otherwise, most of the additions there consist of answers to the criticisms of recent Jesuit writers like Oviedo. Otherwise, the wording of the passages which I have utilized here have undergone no change, so that one would think there had been no advance in science during the thirty years and more since the book first appeared.[171] The chapter, "De vacuo," still repeats such arguments as that the influence of the heavenly bodies could not pass through a vacuum, or that the Ascension of Christ and Assumption of the Virgin would leave an empty space. But no mention is made of the Torricellian experiment.[172]

Franco Petri Burgersdicius, Burgersdijk or Burgersdyck (1590–1635) was born near Delft in Holland, studied at Leyden and Saumur, and then taught in reverse order at Saumur and Leyden, where he gave instruction in logic and ethics from 1620 on, and

[170] *Integer philosophiae cursus ad mentem Scoti*, 1643, 1648, etc. I have used an edition of 1672 which is the only one in BN (R.1124). Zedler's *Cursus philosophicus ad mentem Scoti, übersehen verbessert und mit einigen Zusatzen vermehrt*, Paris, 1639, in-fol, seems dated too early. Perhaps there is some confusion with Commentaries on Scotus which Poncius published in 1639, in-fol.

[171] Arriaga, who had died in 1667, before the work was published in 1669, says in the prologue, "Ante triginta et amplius annos philosophicum cursum in lucem edidi."

[172] Ed. 1632, p. 446a-b, paragraphs 158, 160; ed. 1669, pp. 539b-540a, 548b, paragraphs 231, 236. A few new experiments of his own with falling bodies are added to that with the huge stone and small crust of bread: ed. 1669, p. 691a, "Nuper ex cuppula nostri templi Pragensis quae valde est alta . . ." etc.

in natural philosophy after 1628.[173] So say the official records, but he himself informs us, in dedicating his *Idea Naturalis Philosophiae* to the magistrates of Delft in 1622, that it was composed for certain adolescents to whom he was giving private lessons in physics. Under twenty-six captions, which parallel the books of Aristotle on natural philosophy, he assembled theses concerning which these private pupils of his might dispute, as he states three years later in a second edition, with references to recent fuller works such as the Coimbra commentaries.[174] But we shall be concerned here primarily with the much fuller *Collegium physicum disputationibus xxxii absolutum* which he composed as professor of natural philosophy for his public course. This later work, which I have examined in the second edition (Leyden, Elzevir, 1642), "augmented by the hand of the author," contains a preface of the printer dated in 1637 which states that Burgersdyck had died two years before.[175] The book was published yet again, in 1650 at Amsterdam and Cambridge.[176]

Burgersdyck holds that the heavens are made of the same form and matter as other bodies, and that their form is not soul. He accepts the Copernican rather than the Ptolemaic theory, the decisive factor for him being the tremendous distance and speed that Saturn and the fixed stars would have to travel, if it were true that they revolved daily. Similarly the fact that they have no parallax has convinced him that new stars and comets exist in the heavens. The tail of a comet is the light of the sun or some other star shining through the comet with evident refraction, and therefore the tail is always in the opposite direction from that star. Burgersdyck regards the notion of solid spheres of the planets as a figment—and so it was, for few pre-Copernican astronomers entertained it. But he thinks that the moon has some light of its

[173] *Album scholasticum Academiae Lugduno-Batavae*, 1941, 26 Molhuysen, Brennen, II; *Nieuw Nederl. Biogr. Woordenboek*, VII K, 229.

[174] I have used a later edition of 1657, of which I own a copy: *Franconis Burgersdici Idea philosophiae naturalis sive Methodus definitionum et controversiarum physicarum*, Editio novissima, Amsterdam, Apud Joannem Janssonium, 1657, in-12, 86 pp.

[175] The year of his death has sometimes been given as 1629.

[176] I have examined that of Amsterdam, 1650 and found the text the same as in the 1642 edition.

own, while the fact that its spots always have the same location proves that it is not moving in an epicycle. It is probable that the fixed stars are either absolutely immobile, as Copernicus held, or, if they move according to the Ptolemaic theory—which is the more accepted opinion—that they are carried along by the motion of their spheres.[177]

Burgersdyck retained not only the four elements but the old relationship of the four primary qualities to them. He distinguished three regions of air: the lowest from the earth's surface to where the rays of the sun ceased to be reflected from that surface; the middle region, from that point on to the tops of the highest mountains; the uppermost region, from there to the sphere of fire. The natural place of water is between earth and air, but by singular bounty of God a large part of the earth is raised above the waters, and sea and land constitute one globe and have the same center of magnitude (but not of gravity). The tides go with the moon but cannot be due to its light. Along with other antiquated notions Burgersdyck still believes that streams which flow from mountains are fed by vapors from subterranean caverns,—for water cannot ascend unless first resolved into vapor—and that this is accomplished by subterranean fires as well as by the heat of the sun. He doubts, however, like Aversa, the Peripatetic dictum that one particle of earth makes ten of water, a hundred of air, and a thousand of fire. For he believes that air has a higher ratio of rarity to water than water has to earth or fire to air.[178]

As for the mixture of the elements in compound bodies, Thomas and his followers wish to do away entirely with substantial forms and hold that a new form is introduced into the matter of the four elements by which the compound is what it is. Avicenna preferred to retain the forms of the elements in the compound and have them coalesce into its form. Burgersdyck rejects both these views and agrees with Averroes that the form of the compound is composed of the forms of the four elements in a remiss and

[177] *Collegium physicum* (1642), pp. 97, 101, 113, 110, 115, 112, 108, 112, 112-13, for the passages cited in this paragraph in that order.

[178] *Ibid.*, pp. 118, 122-25, and about 132 for the contents of this paragraph.

altered state. He does not approve of the opinion of Fernel, in chapter 8 on the elements, that the forms of the elements survive in the compound unchanged, and that only the qualities are mixed and equally disseminated through the entire compound, or, in chapter 2 on temperaments, that the qualities of the elements combine in mixture and temperament.[179]

It is not surprising to find Burgersdyck retaining the belief in spontaneous generation and even developing it further and refining it. Those animals are said to be generated spontaneously which are produced by occult causes, as when worms and other animate beings arise in rotting corpses, or in fruit, seeds, tears and excrements, which nevertheless retain vestiges of their own soul and life. "So you may see fleas born from the sweat of dogs, wasps from the carcass of a horse, beetles from that of an ass, bees from that of a calf, and from other animals worms of a determinate kind." Strictly speaking, they are not produced from putridity itself or the humor which exudes from it, but from the parts which have not yet corrupted. Their efficient cause is not God or any finite Intelligence, not the world soul, not the heavens in general or some peculiar aspects of the stars, but an occult nature which lurks in matter. This is why the living being is said to be born of its own accord, because in its origin it does not receive its form from another source, but merely is freed from impediments to its birth.[180]

A monster is so called *a monstrando* (from demonstration), either because men are admonished by them as to the future, as is commonly believed, or because, which Burgersdyck thinks more likely, they are unusual things which are exhibited to be admired for their rarity. But they have no physical force of prediction. Neither pygmies nor giants, if there are such beings, are to be classed as monsters, nor even those dwellers near the Straits of Magellan who have an eye in the breast.[181]

Burgersdyck occasionally implies the existence of animal spirits in the human body, and reckons the internal senses as four in number: common sense, phantasy, estimation and memory. Those

[179] *Ibid.*, pp. 170-72, 179-80.
[180] *Ibid.*, pp. 251, 253, 255, 259.
[181] *Ibid.*, pp. 262, 265.

who add an imaginative, cogitative, and reminiscent faculty lean, in his opinion, on no probable foundation. The organ and seat of the internal senses is undoubtedly the brain and not the heart, as Aristotle thought. They are not four faculties of the soul but four aspects of a single faculty.[182]

Burgersdyck also determines how much confidence should be placed in astrological predictions. Such celestial happenings as conjunctions, oppositions and eclipses may be predicted with certainty, because the movements of the heavens are regular. Meteorological changes, fertility and sterility, pestilence and other epidemic diseases, and the natural gifts and *mores* of individuals, insofar as they depend on bodily temperament, can be predicted but not too confidently. For the virtue of the stars is diversely received by sublunar bodies, and it is most difficult, indeed beyond human power, to know all the forces of the stars exactly. Hence would-be prophets and interpreters of the stars are very often deceived—much more often than they hit the truth in their predictions. Finally, those matters which depend on human free will, such as marriages, treaties, wars, good and adverse fortune, cannot be predicted by men.[183]

Despite his superior views as to comets, tides and the Copernican hypothesis, and his rejection of solid spheres, the outlook of Burgersdyck otherwise still seems sufficiently antiquated, credulous and superstitious. Yet the fact that his work not only was published posthumously, but also was reprinted in Holland and England in 1650 shows that there was still a considerable audience for such a work at those dates.

Indeed, even the earlier and briefer *Idea philosophiae naturalis*[184] was republished as late as 1657,[185] when it still assumed the existence of occult qualities and of sympathy and antipathy, that nature abhorred a vacuum, that the heavens act on inferiors by occult qualities as well as by motion and light. That, if celestial motion ceased, the action of the heavens on inferiors would not

[182] *Ibid.*, pp. 293, 298-99.

[183] *Ibid.*, p. 105.

[184] Paquot, *Mémoires pour servir à l'hist. litt. des Pays-Bas*, Louvain, 1763-1670, II, 243, lists an edition of 1652 at Leyden in-16.

[185] See note 174 above.

cease, nor their motion cease. But if the action of the heavens on inferiors were removed, their action would stop too. That acts dependent on human free will cannot be predicted from observation of the stars; that natural effects below the moon are predicted only probably; but those above the moon, certainly. If comets appear in the heavens beyond the moon, they are not meteors but simple bodies which seem to be composed of condensed celestial substance. Comets are not merely signs but also causes of storms, sterility, pest, war, death of princes and political change.[186]

Another similar example is supplied us in the *Physiologia* of William Duncan, who is described as a veteran professor of philosophy, and which is a very backward book that was published posthumously at Toulouse in 1651,[187] and is still distinctly Aristotelian. Duncan says that the stars move in their spheres not like fish in water or birds in air but like nails fixed in a wheel. Modern astronomers list twelve celestial spheres, but Duncan specifies only eleven: empyrean, primum mobile, crystalline, with vibration or trepidation, the eighth sphere of the fixed stars, and those of the seven planets. He regards comets as exhalations in the supreme region of air, and as portending drought, immoderate heat, sterility, failure of crops, pestilence, war, political change and the death of leading men. Springs and rivers are vapors generated in subterranean caverns which are then condensed into water, but the great part of the water comes from the sea by hidden underground channels. Tides, however, are attributed to the moon, and the question is asked, If springs and rivers derive from the sea, why do they sometimes dry up, while the sea remains unexhausted? Mineral virtue comes from the heavens, and stones differ in their occult virtues. Some heat, others chill; some are astringent, others are laxative; some strengthen the heart and resist poison, others dispel intoxication; some break the stone in kidneys and bladder, others have other medical properties.[188]

[186] *Op. cit.*, pp. 20, 22, 32-33, 53.

[187] Apud Arnaldum Colomerium regis et academiae Tolosanae typographum. 259 pp. in-4, but only 26 lines of large type to a page. Copy used: BN R.2982. I have failed to find any account of the life of William Duncan or date of his death.

[188] *Ibid.*, pp. 91-92, 121-24, 153, 158-59, 167, 169. At p. 191 begins the concluding tractate, *De anima*.

Duncan states that the four humors exist in the mass of the blood and he thinks of the blood in the veins as spread from the liver with the natural spirits to all parts of the body, while the vital spirits are formed in the heart from the purest blood and thence disseminated through the arteries to all parts of the body, just as animal spirits are formed in the brain and distributed through the nerves for sensory and motor purposes. He says that some deny the existence of natural spirits and recognize only the vital and animal spirits.[189] This would appear to have been the position of Burgersdyck, who speaks only of the heart distributing the vital spirits through the arteries to all parts of the body to serve the functions of the vegetative soul, and the brain supplying animal spirits through the nerves for the functions of the sensitive soul.[190] Neither mentions Harvey's discovery of the circulation of the blood, although Burgersdyck gives an interesting descripton of the lesser circulation.[191]

The widespread tendency at this time to criticize Aristotle and turn to some other form of philosophy is illustrated by a manuscript containing a Reformed Natural Philosophy or Verified Condemnation and Solid Confutation of the Peripatetic Physiology and Introduction of a New and Truer, by Girardus (1604–1650) and Arnoldus Botius (1606–1653), brothers, Hollanders and doctors of medicine which they went to London to practice.[192] Actually to the modern reader the work seems to run in the old Aristotelian ruts more than it strikes out a new way. Scaliger and Piccolomini

[189] *Physiologia*, pp. 201-2, 209-10.

[190] *Collegium physicum*, ed. of 1642, p. 227; ed. of Amsterdam, 1650, pp. 227-28.

[191] *Idem:* "Dexter sinus trahit sanguinem ex vena cava eundemque rursus expellit per venam arteriosam in pulmones. Huius sanguinis quantum alendis pulmonibus superest una cum aere sinister sinus retrahit per arteriam venosam eademque calefactum aerem una cum fuliginibus repellit in pulmones. Sanguinem vero exquisitius excoctum et spiritum mutatum immittit in arteriam aortam, unde per caeteras arterias in omnes corporis partes diffunditur . . ."

[192] BN 12975, a large volume of over 400 leaves with many elisions, rewritings, insertions, and changes of arrangement. On the fly leaf is written: "Philosophia Naturalis Reformata sive Physiologiae Peripateticae accurata damnatio et solida confutatio, Et novae et verioris introductio per Girardum ac Arnoldum Botios fratres Hollandos medicinae doctores." The catalogue incorrectly dates our MS as of the 16th century.

are among the recent authors cited. Only the first book of the work was printed at Dublin in 1641,[193] and there were to be four books more: the second and third on matter, the elements and the nature and properties of things; the fourth, on generation and generating causes "and all variety of efficients"; the fifth, *De anima.* Although the Botius brothers had found that they had some predecessors in the character of their first book, they promised that in the remaining books they would "proceed almost alone and introduce a kind of philosophy not only different from the Aristotelian but evidently new." In their preface to the University of Leyden, where their training had been Aristotelian, they give the chemists credit for having first made them suspicious of Aristotle, but add that they were more disappointed in their books, where everything was uncertain, and much was futile and monstrous. For a time they despaired of ever finding a true philosophy, but then set about constructing their own.[194]

[193] Copies at Paris: BN Rés. R.1013, and R.4325; 368 pp.

[194] This preface does not appear in the MS, which opens immediately with the first chapter of the first book, which is in five sections, as in the printed edition. So is the second chapter, and at fol. 140v we read, "Finis libri primi philosophiae reformatae." Fols. 141-147 are left blank but then fols. 148-172 are marked "lib. I, Cap. i." At fol. 173r, "Caput tertium De principio effectivo. Sectio Prima, Causae nomen soli efficiente competere"—a thought already expressed in the edition, p. 351, "Causae nomen soli efficienti damus." In the MS subsequent sections of the third chapter continue as follows:

fol.	Sectio	
184v	2	An qualitates activae sint efficiens principale vel instrumentarium
208r	3	De conditionibus efficientis
233v	4	Divisio efficientium
258v	5	De causa creante et creatione
291r	6	De generatione et causis generantibus
297r	7	De augmento et causa augente itemque de appositione
318r	8	De alteratione et causis alterantibus.

At fol. 338r the 33rd numbered paragraph of this eighth section ends, and the rest of that leaf and fol. 339r-v are blank, while on fol. 340 are scattered jottings in Greek and Latin. Fol. 341r begins without any heading, "De naturalium rerum principiis et affectionibus actinxi necessarium habemus ab ipsa natura auspicari et quidnam illa sit indagare." On fol. 342v is cap. 2, "Dissentientes philosophorum opiniones super essentia naturae generalis," and at fol. 384v, cap. 4, "An deus sit universalis natura," but I could see no heading for cap. 3; perhaps a blank space left on fol. 355r was intended to be so filled in. At fol. 393r, cap. 5, "De

Borrichius, writing in 1649, called the Botius brothers "that new scourge of the Peripatetic," and said that they had greatly excited the Aristotelians, because they held that qualities were the instruments of forms in acting.[195]

The Physical Synopsis of the noted educational reformer, Comenius (1592–1670), seems first to have been printed in 1643.[196] But the long preface was written at Lesna in Poland on September 30, 1632.[197] In it Jacopo Aconzio of the previous century is quoted approvingly to the effect that no one should publish a book, unless it embodied new observations of his own, was conducive to the glory of God and the edification of the church, and was such that one reading it could not employ his time to better advantage.

> For few writers offer anything of their own; the things and words of which they make up their books are stolen goods.

Comenius himself added:

> If you look only at the titles, they are always new and specious. But when you come to the contents, the same thing is recooked a thousand times and is warmed over *ad nauseam*.[198]

Comenius, however, asserted that he was offering something new and different from the received way of philosophizing. He had found Vives' criticism of the state of learning negative. He then read Campanella's *Prodromus realis philosophiae* and *De sensu rerum* with avidity but was not entirely satisfied, and, after he had perused Francis Bacon's *Instauratio magna,* recognized that Campanella was lacking in particulars that solid demonstration which truth requires. Bacon showed the key to nature but did not himself unlock the doors to all her secrets, giving only a few examples

---

natura particulari ex mente Aristotelis et Aristotelicorum"; at fol. 403r, cap. 6, "Quid sit natura vera opinio"—not a substance, but "sola accidentia"; at fol. 416v, cap. 7, "De principiis," after a few lines of text breaks off and the MS ends. Apparently it is a preliminary draft, left incomplete, and perhaps not observing the arrangement that the brothers had in mind for the full text in five books.

195 Olaus Borrichius, *De cabala characterali dissertatio,* Copenhagen, 1649, fol. A 5 v.

196 Johann Amos Comenius, *Physicae ad lumen divinum reformatae synopsis,* Amsterdam, 1643, in-12, 198 pp.

197 *Ibid.,* p. 33, "Scribebam Lesnae Polonorum ult. Septembris anno 1632."

198 *Ibid.,* pp. 1-2.

and leaving the rest to future centuries of patient observation and induction. But Comenius was convinced that the Peripatetic philosophy must be abandoned and philosophy reformed on the basis of the senses, reason and the Bible. Some persons object that Scripture does not apply to natural philosophy but is a path to life eternal. Comenius does not agree with them, and he hopes, as a result of his little book of less than two hundred duodecimo pages, that there will be no place left for doubts and disputes, such as we have seen characterize, indeed were almost the life-blood of, the *Cursus philosophicus* of the early seventeenth century.

This preface of Comenius is followed by twelve chapters, of which the first does little more than repeat the account of creation given in the Book of Genesis. The remaining chapters deal with the principles of the world: matter, spirit or world soul, and light, with motion, qualities, mutations, the elements, vapors, concretes or the mineral kingdom, plants, animals, men and angels. Which sounds a good deal like a thick chunk of Aristotle sandwiched between two thin slices of the Bible. From the meeting of the aforesaid three principles results motion, from motion quality, from quality varied mutation. The motion of the spirit is called agitation, by which the spirit agitates itself in matter, seeking to inform it. The motion of light is called diffusion, by which light and heat diffuse themselves in all parts. The movement of matter is eightfold: expansion, contraction, aggregation, sympathy, continuity, impulsion, liberation and liberty. Of these the first two are immediately from fire, the four following from other body, the two last from themselves but with mediation of the universal spirit. A leaf is next torn from the book of the alchemists, and sulphur, salt and mercury are pronounced substantific qualities, a concept which is of course a marvelous advance over that of substantial forms. We pass on to tangible quality or touch, taste, odor—which is defined as a most tenuous exhalation of taste, sound, color, and quality perceived by two senses, touch and vision, which is figure. Occult quality is still defined as that which is known only by its effect.

The elements—ether or fire, and air, water and earth—are the same matter of the world but distinct in degree of density and rarity. They therefore are transmutable into one another. Aristotle thought that they were in tenfold proportion, but more recent authorities put it nearer a hundredfold. Thus a drop of water heated will occupy as air one hundred times as much space as it did as water. The same hundredfold proportion holds good between colors, where one drop of ink will tinge a hundred drops of water, but not vice versa. Comenius still holds that the four elements constitute regions or spheres of the visible world and that water naturally surrounds the globe of earth in all directions, but that the Creator has established two waters and a twofold fire: one part of waters above the highest ether, and contrariwise a part of fire detracted from the ether and included within the bowels of earth. The tides are accounted for both by vapors generated by subterranean fire which cause the sea to swell up, and by the sun and moon. Comets are not sublunar, as Aristotle thought, but are generated in highest heaven even beyond the sun.

In his brief treament of gems Comenius is silent as to their marvelous virtues. All plants are hot by nature, but in proportion to the heat of our bodies some are called cold. Vital spirits in the heart have blood as their material, the lungs as bellows, the arteries as canals by which they spread all through the body. Animal spirits are generated in the brain from blood and vital spirits, are purified by the ventilation of respiration, and spread through the nerves to all parts of the body. Thus Comenius resembles Burgersdyck rather than Duncan in omitting natural spirits. The excrements of the brain are ejected through the nostrils, ears and eyes in phlegm and tears.[199] Some animals see better in strong light, others in dim light, because the lucidity of their animal spirits is diversely proportioned. Thus spiders and flies see the most minute objects which escape our vision, and much more that of a horse or elephant, because in a subtler body there are subtler spirits. The motive faculty is given to animals: 1) that they may seek food; 2) for the destined actions of each; 3) to preserve vigor of life. The moving principle is animal spirit which carries the vital

[199] The question, "An cerebrum excrementa sua deponat et expurget per nares, palatum, aures et oculos?" had already been discussed by Bickerus, *Hermes redivivus*, 1612, p. 319 *et seq.*

spirit with it. The enunciative faculty and voice are also controlled by animal spirit. So is the defensive faculty; for if the animal spirit senses the approach of anything hostile, it hurries back to the defense of the part threatened. It directs all generation, for formation of the animal does not begin from the heart, as Aristotle thought, but from the head. For some animals such as fish lack a heart, but none is without a head and brain.

In an appendix to the chapter on animals the tenacity with which animal spirit clings to its body is shown by the fact that flies suffocated in water revive in hot ashes. Especially marvelous is the sympathy of the spirits with blood which has been shed: illustrated by the calf's terror of the butcher; the corpse bleeding at the murderers' approach; the story of the nose reconstructed by plastic surgery that putrefied when the rustic whose flesh had been used died; the sucking of a little of a friend's blood before he goes away in order to sense his ill or grief when far distant—if it be true, and it is very plausible; and the celebrated magnetic medicine and weapon ointment. But in all this talk about blood and its occult properties, not a word is said of its circulation, announced by Harvey fifteen years before. And this continues to be the case in the edition of 1647. Animal spirit is also responsible for spontaneous generation, which, we are assured, is demonstrated by experiment. For example, serpents are generated from the flesh of storks, spiders from that of hens, frogs from that of ducks.

In the next chapter on man, mental operations are ascribed to the animal spirits, but the mind of man is immediately from God. Man is a microcosm, and an angel is man incorporeal. Angels are not generated and do not die; their number is well nigh infinite; they can act upon bodies but cannot be affected by bodies; their power is superior to that of any corporeal creature, their agility greater, and their knowledge far more sublime than human science. These statements concerning angels seem about as well substantiated as Comenius's previous observations regarding the world of nature, but he leaves their fall and the consideration of demons to theology.

Thomas Crenius states that Comenius obtained as a patron a rich merchant of Liége, Louis de Geer, who also aided the ravaged churches of the Palatinate and poor scholars, and that Comenius had eight or ten amanuenses assist him with an encyclopedic *Pansophia* which he had planned but towards which he got nowhere.[200]

Marten Schoock, in his *Physica generalis* of 1660 rejected Comenius's division of motion between spirit, light and matter, and his division of qualities into intrinsic or substantific and extrinsic or accidental, and calling the three principles of the alchemists substantific. He further criticized Comenius for deriving occult qualities immediately from a peculiar spirit infused in each creature.[201]

During the remainder of this chapter we consider four authors—Neufville, Kyper, Sperling and Cabeo—who chronologically come after rather than before Descartes, but whose books seem to belong with those that we have been previously considering, forming one *genre* of like origin and tradition.

In the dedication of his *Physiologia seu physica generalis* of 1645[202] to the consuls, syndics and senate of Bremen, Gerard de Neufville (1590–1648) states that he began to teach there thirty-four years ago, was for a while extraordinary professor of mathematics at Heidelberg, then was recalled to Bremen to teach medicine. Back in 1613, soon after he began teaching, he published a Synopsis of Universal Physics. Now, after many years of teaching, he issues this revision of it.

In the Preface to the Reader he says that natural science has not progressed as mathematics and the mechanical arts have, for the reason that it does not adhere sufficiently to sense and experience. Disputations get us nowhere in it. Some have tried to base it on Scripture, as Lambert Daneau in his *Physice Christiana*,[203] Otto Casmann in *Prolegomena Cosmopoeiae et Uranographiae Christianae Praemissa*, or Conrad Aslacus in *Physica*

[200] *Animadversiones philologicae et historicae*, in 19 parts, 1695-1720, IV (1699), 89-94, quoting the *Antirrheticus* (against Comenius) of Maresius (Samuel Desmarets), Groningen, 1669.

[201] Schoock, *Physica generalis*, Groningen, 1660, pp. 232, 267, 270.

[202] Gerhardus de Neufville, *Physiologia seu physica generalis de rerum naturalium atque etiam substantiae corporeae communi natura, primis principiis et causis communissimisque affectionibus aphoristice proposita et perspicue explicata. Cui praeit Isagoge in Elementa physica etc.*, Bremae, 1645, in-8, 426 small pages. Copy used: BM 1135.g.6, with many leaves uncut hitherto.

[203] T VI, 346-49.

Mosaica. Others seek to rear a structure from a few basic principles, as Euclid does in geometry, but this method is not effective in natural science, where is necessary long and difficult research from varied and multiple experiments, which must be analyzed, and results attained inductively. Experiments of one kind or sort only are insufficient. The chemists are criticized for trying to build up principles for all natural science from chemical experiments alone; Gilbert, for magnetic experiments only; Fludd, for limiting himself mainly to experiments in rarefaction and condensation from heat and cold; Telesio and Campanella, for trying to explain everything in terms of hot and cold. Others try to account for all natural phenomena from a few sensible accidents, especially magnitude, figure and motion, as did Democritus, Leucippus and Epicurus, Sebastiano Basso[204] and most recently René Descartes. But all these erect natural philosophy and natural history on too narrow a basis. The story of nature must be founded on experiments of all kinds, as advocated by Francis Bacon, who is copied—as in the aforesaid criticism of the chemists and Gilbert, praised, and his plan set forth at length.

This is all very well, but it does not affect the main body of Neufville's work as much as might be expected. He still has a chapter on secondary causes and the necessity and contingency of natural effects, and also concerning fate. His first material principles of corporeal substance are atoms, but he denies the possibility of a vacuum, and accepts not only qualities but occult qualities. They

> arise by natural emanation and flux from the essence or essential and formal principle of that body to which they appertain first and immediately and indeed according to nature,[205]

and include sympathies and antipathies. After some consideration of motion, time and alteration, the book ends, and an epilogue informs us that the magnitude and figure of the celestial bodies, elements and mixed bodies will be considered in a volume on Special Physics. It seems not to have appeared until 1668, twenty years after Neufville's death. The circulation of the blood is stated

[204] See T VI, 386-88.

[205] *Physiologia seu physica generalis . . .*, 1645, II, 18, p. 377.

in it, but possibly the passage was inserted after Neufville's death.[206] The subject did not fall within the scope of the volume of 1645.

Albert Kyper (1600–1655), who from 1650 to his death was professor of medicine at the University of Leyden, makes an unfavorable impression upon the present-day reader, when, in the preface of his *Institutiones Physicae*[207] to the reader (of his own day) he affirms that many things are done in this world by the force of demons which we in our ignorance attribute to natural causes. Such a remark is certainly unpromising for the development of natural and experimental science. He adds that much is done by divine providence. For why does God bid us attend not to the courses and efficacy of the stars but to His will and care? We ourselves make the heavens favoring or hostile to us by our acts on earth. God is not the servant but the absolute monarch of the universe.

It is now the eleventh year in which he has been teaching natural philosophy in various ways.[208] Thrice he has dictated the subject from memory and for that reason never uniformly. He has often explained the systems of others but never has satisfied himself. He looks back on such lights and columns of this celebrated university as Jacchaeus and Burgersdicius and hopes that he may be close to them, if not equal or superior. Two of his family have died while this first volume was in press, and some wicked men have tried to injure him secretly, asserting that he held new opinions in philosophy which impinged on theology and might disturb the academic peace. He challenges these enemies to point out any such passages in the printed text. He has always cherished Aristotle, Plato and Galen, but never regarded anyone of them as a god. Aristotle held views contrary to Christian theology concerning the nature and providence of God; the eternity of the world, time and

[206] *Cosmologia et anthropologia sive Physicae specialis partes duae principaliores ad modum physicae generalis quam praedictus author anno 1645 edidit aphoristice explicatae* ... In lucem emissae ab H. Harnes, Bremae, 1668, in-8. BM 536.a.19. At pp. 221-22.

[207] 2 vols. of 600 and 724 pp. in-12, Leyden, 1645-1646: BM 531.a.7, 8. I have used a duplicate with a new title page dated 1647 and with indexes added: BM 718.a.24.

[208] According to the *Album Scholasticum* (1941), 94, he was lector in *physica* at Leyden, 1643-1646, then was at Breda before he returned as professor of medicine.

motion; contingency, the function of the Intelligences, human liberty, the *summum bonum*, the virtues, and so on. There was much concerning nature which he did not know; many of his explanations were insufficient; much that he proposed was false. Yet Kyper thinks that his philosophy should be retained in the schools, but its deficiencies should be supplied, its disordered passages put in their proper places, and its errors corrected. "I have rarely cited new authors . . . Everywhere I have followed my own bent, for I have always hated servitude." He advises beginning students, if they do not have a teacher present, to supplement the reading of his Epitome with Sennert's natural philosophy or Magirus's Peripatetic Physiology[209] and Burgersdyck's *Collegium Physicum.*

Kyper's Institutions are in twelve books: three in the first volume, and nine in the second. The first book upon bodily substance devotes 335 pages to its principles, its origin and essence, and its affections or adjuncts. There follow books on the elements and on mixed bodies in general. The order of books in the second volume is on the stars, minerals, living beings in general, vegetation, animals in general, brutes, man, meteors and the world. The treatment gives little evidence of an experimental basis. It is bold and original in a way, but also too conjectural. Idle and inconsequential questions are sometimes raised—especially since Kyper himself does not always seem to know the answers—such as why ships float better in salt than fresh water, why swimmers are more easily submerged in salt water, why nearer the shore a ship sinks deeper in the water, why drowned bodies come to the surface after a few days.[210] And, as a matter of fact, he cites recent authors by name frequently.

Kyper notes that many recent writers argue for the existence of a world soul. His discussion of monsters in the first volume is general and abstract. In the second volume he holds that, despite the stars, monsters can occur and chance dominate, that the form of a monster generated from man and brute is not rational but

[209] Johann Magirus, professor at Marburg, had died in 1596, and his *Physiologiae Peripateticae libri sex* seems to have appeared posthumously with a dedication dated April 1, 1600. By 1608 it had reached its fourth editon.

[210] *Institutiones physicae*, I, 567-68.

material, and that giants and pygmies are monsters,[211] on which point we have seen Burgersdyck hold a contrary view. The movement of falling bodies is slow at first, faster in course, and fastest at its end; violent motion is fastest at first, slower in course, and slowest at its end.[212] He dares not deny a vacuum utterly but, like Arriaga, does not mention Torricelli's experiment, although he quotes Lucretius for nearly four pages on the subject and cites Scaliger.[213] On the other hand, he has recourse to pores for explanations more than once.[214]

Kyper holds to the four elements and thinks that fire is not only an element but also not different from primeval light.[215] Occult qualities are imperceptible to sense and so do not seem deducible from first and secondary qualities, which are perceptible to sense. In one passage he says that their existence may be doubted, but that it is certain that the elements have relations of sympathy and antipathy. But experience favors the existence of occult qualities. They cannot be produced by immaterial spirits, for these cannot impress a material quality upon bodies. Since the stars are themselves endowed with occult qualities, they cannot be the universal or exclusive cause of them. Therefore they are from idiosyncracy and the specific form of each thing.[216] In the second volume Kyper affirms that the occult qualities of the stars cannot be denied, in view of the turning of the magnetic needle to the north and the influence of the moon on tides.[217]

Kyper prefers one fluid heaven, in which stars and comets can move freely. "All motion is of the stars," not spheres or Intelligences. But he regards the empyrean heaven as a natural body and thinks it very likely that there are waters between it and the other heaven, as the Book of Genesis seems to state.[218] He does not believe that the new stars of 1572 and 1600 are coeval with the world. Rothmann and Galileo held that they were made of sublunar matter elevated to the heavens, but Kyper objects that the whole earth or a major portion of it would be consumed in the

[211] *Ibid*, I, 144, 191-98; II, 84, 513, 568.

[212] *Ibid.*, I, 285-86.

[213] *Ibid.*, I, 307-14.

[214] *Ibid.*, I, 303, 365.

[215] *Ibid.*, I, 361, 364.

[216] *Ibid.*, I, 593-94, 451-3.

[217] *Ibid.*, II, 100.

[218] *Ibid.*, II, 9-15, 50.

operation, and prefers the hypothesis that they are made of stellar exhalations, but not from condensation of the ether.[219] He is opposed to the Copernican theory. He holds that inferiors are influenced by the stars and not by the containing heaven.[220] He questions whether innate heat is of celestial or elementary nature.[221] Inquiring if the magnetic cure of wounds is due to the stars, he says that some deny this mode of cure, but experience proves its validity. It operates actively to some extent from the stars but especially from a specific medicament, passively from a special convenience of curing the body and humor to which the medicament is applied. Astrological images, on the contrary, he rejects as diabolical, and likewise the making a wax image of a man.[222] Spontaneous generation is accepted as a fact, the question being whether God is its immediate cause or heaven or stars or fire or heat.[223]

Kyper agrees with Linemann that the tail of a comet is generated by its head intercepting the rays of the sun and this shadow being illuminated by the neighboring rays. He agrees that most comets are celestial phenomena but holds that sublunar comets cannot be entirely denied. He adds nothing special as to presages from comets, since many vain things have been said by some on this point; moreover, their natural effects are clear from general considerations. He has touched briefly on the Star of Bethlehem and held that the eclipse at the time of the Passion was miraculous.[224]

Like is preserved by like and destroyed by contrariety. Innate heat can be weakened and corrupted by elemental cold. Contraries are the cause of contraries. Transmutation of baser metals into gold is not impossible but difficult.[225] Long fasts are explained on the supposition that the nourishment or solid parts to be consumed acquire such a specific property, that they foster the innate heat, yet are not consumed by it. A like quality undoubtedly exists in self-perpetuating candles.[226]

219 *Ibid.*, II, 64-69.

220 *Ibid.*, II, 104, 103.

221 *Ibid.*, II, 180. Fernel had attributed celestial essence to it, but Pompeius Caimus, *De calido innato*, 1616 and again in 1626, held that it was of elemental essence.

222 *Ibid.*, II, 101.

223 *Ibid.*, II, 251-63.

224 *Ibid.*, II, 73, 56, 603-4, 75,89.

225 *Ibid.*, II, 270, 180, 182, 475, 151.

226 *Ibid.*, II, 295.

For Kyper, as for Burgersdyck, there are four internal senses, since he does not reject *aestimatio* as Faber did, but cites Keckermann for it.[227] Interpretation of dreams is declared possible in the case of those from natural causes, which naturally signify their effects and their causes and their adjuncts. Whether natural dreams can come in another way from the stars or some other spiritual force diffused through all things or from the very virtue of the rational soul, he leaves to more learned men to meditate on.[228] The origin of rivers is a serious problem, but the most probable opinion seems to Kyper to be that they come especially from the sea, although evaporation by the sun and precipitation of rain and snow help.[229] He denies the eternity of the world.[230] He is sufficiently up-to-date to discuss the circulation of the blood, but he ascribes muscular movement to the animal spirits.[231]

A few years later in 1650 Kyper published another volume entitled *Anthropologia* and devoted more particularly to man and medicine.[232] In it he still accepts the four elements and occult qualities and celestial influence.[233] But he now, in agreement with Aristotle, reduces the internal senses to two in number, and in discussing sleep and dreams says nothing of divination from the latter.[234] For him the heart is, as for Aristotle, the "member absolutely first in which the soul is first and originally rooted."[235] Recent anatomical research has shown that there are excrements in the ventricles of the brain and that consequently the animal spirits cannot be generated there, so he puts their generation in the medulla oblongata.[236] He continues his practice of asking superfluous questions: such as why men do not menstruate—the answer

[227] *Ibid.*, II, 352-53; Keckermann (1612), Bk. III, caps. 17-19, pp. 330-39.

[228] *Institutiones physicae*, II, 493.

[229] *Ibid.*, II, 591.

[230] *Ibid.*, II, 633.

[231] *Ibid.*, II, 428-33, 501.

[232] *Anthropologia corporis humani contentorum et animae naturam et virtutes secundum circularem sanguinis motum explicans*, Leyden, 665 pp. There is no date of publication on the title page, but the dedication is dated 1650: BN 4o Tb^7.11. The other copy, BN 4o Tb^8.11A, is dated 1660, but Kyper was dead by then. Despite the title *Anthropologia*, the running head throughout is *Universae medicinae contractae liber primus.*

[233] *Ibid.*, pp. 16, 23, 664.

[234] *Ibid.*, p. 620 *et seq.*

[235] *Ibid.*, p. 215.

[236] *Ibid.*, pp. 214, 216.

being that they are hotter and exercise more, what use the lips and various other members are for, why those born deaf are at the same time mute.[237] The circulation of the blood is emphasized yet more than in the other work.

Johann Sperling was a professor at Wittenberg who was favorably and not infrequently cited by his former students. The reprinting of his *Physica anthropologica* also indicates that he exerted considerable influence. I have had access only to the third edition, at Wittenberg in 1668, but the dedication to the work is dated on September 10, 1647.[238] His influence was rather favorable to the occult, whereas his scientific attainments appear to have been meager. Even the edition of 1668 discusses such questions as why, when a vein is cut, blood also is evacuated from the arteries, without distinctly stating the circulation of the blood, only anastomoses being mentioned, although in a later passage the lesser circulation is set forth.[239] I detected no trace of the influence of Descartes.

"It is a perpetual law of nature to hide its work," and Sperling exclaims at the industry of nature and providence of God.[240] In connection with the question whether the analogy of macrocosm and microcosm is fundamental to philosophy and medicine, as the Hermetics affirm, Sperling quotes the Emerald Tablet in full, but finds no philosophy or chemistry in it, and otherwise leaves the question unanswered.[241] He answers in the negative such questions as whether the speech of those absent can make one's ears ring, and whether serpents are born from the human body.[242] But the mere putting of such questions shows the existence of a rather unwholesome and unscientific curiosity. He goes on to tell a story of a king of Poland who killed his uncles and was pursued by enormous rats which were engendered from their corpses. In vain he climbed, swam rivers, and even went through flame in the effort to escape them; they ultimately devoured him and his wife and two sons. This he has learned from most learned and trustworthy men. But it was a miracle, not a work of nature.[243]

[237] *Ibid.*, pp. 458, 250.

[238] Copy used: BN R.12400, 780 pp., but they are very small.

[239] *Ibid.*, 475-76, 703-6.

[240] *Ibid.*, 658-59.

[241] *Ibid.*, 27.

[242] *Ibid.*, 296, 314.

[243] *Ibid.*, 314-17.

What is to be thought of chiromancy? After quoting John ab Indagine[244] at length, Sperling says for himself that a prudent man should distinguish what is natural from what is voluntary, and probability from necessity, and combine chiromancy with physiognomy and astrology.[245] According to Moncaeus,[246] Sperling published a work on magic (*De magia*) in 1646, but I have not found a copy of it.

One of Sperling's former students, Johann Daniel Major, admitted that there were errors in his *Physica*, explaining that the loss of his left hand kept him from specializing in anatomy and botany, that he was very inexperienced in technology, lapsed into mere speculation, and engaged in too many controversies.[247]

The huge commentary of the Jesuit Cabeo on the Meteorology of Aristotle touches on so many matters already considered in this chapter, that we may note some of its views, although it is on the one hand limited professedly to only one department of physics or natural philosophy, and on the other hand is too long and full a treatment for a lecture course, although perhaps an outgrowth from one.[248] But many questions are raised which are not strictly meteorological, such as why men become seasick, why persons with hot stomachs have cold livers, whether vision is by extramission, and how animals are spontaneously generated from putrefaction.[249] Cabeo mentions the contention of some that the fourth book of the Meteorology should come after the second book on Generation and Corruption. This view he opposes and does not recognize that this fourth book is spurious.

Cabeo holds that the Ligurian Sea is higher than the Adriatic, asks whether the velocity of all falling bodies is equal, and treats of pendulums and the three principles of the chemists.[250] He admits that the material cause of apparitions of armies and the like in the

[244] See T VI, 683, Index.

[245] *Physica anthropologica*, 768-72.

[246] *Disquisitio de magia divinatrice et operatrice*, Francof., 1683.

[247] J. D. Major, *Genius errans*, Kiel, 1677, cap. iv, "De erroribus genio condonandis in J. Sperlingii physica."

[248] Nicolaus Cabeus, *Commentarius in Meteorologica Aristotelis*, Rome, 1646, 4 vols. in-fol. BM c.54.f.9. The last two volumes, however, contain little of interest for us.

[249] *Ibid.*, II, 161; II, 176; III, 78; IV, 81.

[250] *Ibid.*, I, 52-56, 97, 98, 113.

sky is vapors and evaporations, but insists that their efficient cause is neither nature nor chance nor the stars, but God through the ministry of Intelligences, to teach us to fight against impiety. Atheists, however, say that such apparitions are reflections in the clouds.[251]

Aristotle said that hot water freezes faster than cold, but Cabeo and others have found by experiment that just the opposite is true. But he still argues that water congeals by condensation and not rarefaction, although he is aware that ice floats on water and bulges out of a cup, when a cupful of water freezes.[252] As for the origin of rivers, he still maintains that some come from the sea by way of subterranean vapors, but that more come from precipitation. Evaporation, however, is greatest from the sea, so that in that sense rivers may be said to come from the sea.[253]

For the height of mountains Cabeo repeats various estimates by others. Maurolycus said that Etna was visible for 200 miles; Fromondus, that Teneriffe was visible four degrees away; Alhazen, that the highest mountains were eight miles in height. Blancanus was certainly wrong in affirming that no mountain was more than a mile and a half high. If the world were eternal, erosion would have reduced all mountains to a plain. To call the heavens a fifth essence Cabeo condemns as a pernicious doctrine.[254]

Towards Aristotle's assertion of the influence of the heavens Cabeo is much more favorable, Who, indeed, can doubt it? The only questions are whether they act only by their motion and light or also by other more occult qualities, and whether their action is universal or has particular effects and dominates the individual acts of man. The heavenly bodies do not act directly by their motion, but their effects are varied by it. Mere heat from light will not account for all their effects. The light of Mars or Mercury is slight, so that their notable effects cannot be accounted for by it. Cabeo therefore inclines to agree with the astrologers that certain points in the heavens have the greatest efficacy, such as the horoscope, *pars fortunae,* and the cusps of houses. But the stars do not act upon our souls or destroy liberty of action; the predictions

[251] *Ibid.,* I, 141-42, 223-24.

[252] *Ibid.,* I, 324, 322-23.

[253] *Ibid.,* I, 368-71.

[254] *Ibid.,* I, 383-88, 414, 418.

of the astrologers often go awry; the whole art is built on very questionable foundations; and there have not been enough experiments to determine the nature of each star, sign and degree.[255] In another passage, however, he states the relation of celestial comets to the planets according to the astrologers and does not deny this. Their colors are for the most part martial and saturnine, and so they are believed to share the evil influence of those planets. If they are fed by exhalations from the stars, these might come from Mars and Saturn. Their effect is also judged by the part of the sky whence they come or where they first appear. Or they may be signs from God. He has previously stated that no physical cause can be assigned for their motion, which must be "from some free cause."[256]

Four "experiments" are adduced in favor of action at a distance. The first is weapon ointment, of which Cabeo himself has had no experience. The second is that, if the excrement of any animal is mixed in a certain manner with a certain herb and put in a certain place, the animal will have diarrhoea until it dies, or will not be able to evacuate, as long as he who plants the mixture pleases. But when it is removed from the place in question, the charm is dissolved. This second "experiment" Cabeo does not believe. The third is that the effects of being bitten by a tarantula cease, as soon as that particular spider dies. The fourth is the phenomenon of sympathetic clocks.[257] Cabeo did not ascribe the spontaneous generation of animals from putridity to the influence of the stars, but to vagrant animal spirits, expired from dying bodies.[258]

Riccioli, in his *Almagestum novum* which appeared five years after Cabeo's commentary on the Meteorology, repeated Cabeo's explanation of the tides. After discussing the opinions of others most diligently, including that of Kepler, Gilbert, Zanardus and the school of Coimbra, which attributed the tides to the magnetic attraction of the moon, and that of Contarenus and Faber, who ascribed them to occult influence of the moon, Cabeo accounted for them by an occult faculty of the moon which excited sulphurous and sal-nitrous spirits from the bottom of the sea, moving them in

[255] *Ibid.*, I, 33-37,
[256] *Ibid.*, I, 213-14, 211.
[257] *Ibid.*, I, 30-33.
[258] *Ibid.*, IV, 81.

much the same way as the moon moves the humors in animals, and exerting this influence even at the time of new moon and when the moon was below the horizon.[259]

[259] *Almagestum novum,* I (1651), 74, citing (at p. 73), "Cabeus, lib. 2, quaest. 5, ad 12 textus 6."

CHAPTER XIV

MERSENNE AND GASSENDI

Why considered together—Their lives—Estimates of them—Mersenne's position in the history of science—His experimentation—Questions on Genesis—Peiresc and Mersenne against astrology—Mersenne's credulity and love of the marvelous—Alchemy—Relations of Mersenne and Gassendi with Fludd—Gassendi and Peiresc—Attitude of Gassendi towards astronomy and astrology—Towards alchemy, divination, fascination—Difficulty of the sceptic in natural history—Stretching the corpuscular theory—An English version of Gassendi's views: the wolf, the bleeding corpse, basilisk, tarantula—A varying view-point.

Reverendo patri domino Marino Mersennio Mimimo sed charitate et doctrina maximo.

—CLAUDE BREDEAU

homme sage, savant et bon, tempéré et habile homme, en un mot un vrai épicurien mitigé.

—PATIN CONCERNING GASSENDI

In this chapter we consider together two Frenchmen who became Parisians and whose life-spans roughly coincided: Marin Mersenne (1588–1648) and Pierre Gassendi (1592–1655). The former was born in Maine; the latter, in Provence. Both began their careers as clergymen—Franciscans, in fact—and teachers; both ended primarily men of science. Their first books appeared almost simultaneously in 1623 and 1624, and dealt with the time-worn themes of Genesis and Aristotelianism. Both broke away from, or developed beyond, these first interests. Both saw the value of observation and experiment. Both were acquainted with the famous patron of the arts, letters and sciences, Nicolas Fabri, seigneur de Peiresc, who lived in Paris from 1616 on and who got Gassendi a canonry in the cathedral of Digne.*

* For recent publications on Peiresc see the review by H. J. Martin of Georges Cahen-Salvador, *Un grand humaniste, Peiresc. 1580-1637*, Paris, 1951, in *Bibliothèque de l'école des chartes*, 110 (1953), 290-91.

Marin Mersenne was educated at the Collège du Mans, the then new Jesuit school of La Flèche, attended a little later by Descartes, and the University of Paris, where he entered the Order of Minimes. From 1614 to 1619 he taught in their convent at Nevers, and afterwards resided at the Parisian convent near la Place Royale except for travel—in 1625 to Rouen, in 1630 in Flanders and Holland, in 1639 in Champagne, in 1644 in Spain, in 1645 in Italy, and in 1646 in the south of France. In 1623 appeared his Questions on Genesis;[1] in 1624, *L'impiété des deistes, athées et libertins de ce temps;* in 1625, *La verité des sciences contre les sceptiques ou pyrrhoniens;* in 1627, the first two books of *Traité de l'harmonie universelle;* in 1630, *Nouvelles pensées de Galilée;* in 1634, the *Questions théologiques, physiques, morales et mathématiques,* and the *Questions inouyes;*[2] in 1636, the full text of *Harmonie universelle;* in 1644, *Cogitata phisico-mathematica.* His last illness was aggravated by a surgeon's severing the artery in his right arm, and Gassendi was said to have died from excessive phlebotomy.

Pierre Gassendi taught rhetoric at Digne, his birthplace, in 1608, and philosophy at Aix in 1611, returning to Digne as lecturer in theology in 1612. In 1615 he made his first visit to Paris, next year became a doctor of theology, and in 1617 was ordained a priest. He taught philosophy again at Aix for a while, and in 1624 published his first book, *Exercitationes paradoxicae adversus Aristotelem.* After the series of works of this sort from Ramus to Patrizi, there was nothing very novel about this approach, and Gassendi presently dropped the negative attitude of assailing Aristotle for a positive exposition of the philosophy of Epicurus and atomism. Meanwhile he visited Paris again, met Mersenne, returned south, determined the latitude of Grenoble and made astronomical observations at Vizelle, and in May, 1628, came to Paris once more with a letter from Peiresc to Mersenne and traveled in the Low Countries. In

[1] F. Marini Mersenni, O.M. Francisci de Paula, *Quaestiones celeberrimae in Genesim cum accurata textus explicatione. In hoc volumine athei et deistae impugnantur et expugnantur.* Lutetiae, Seb. Cramoisy, 1623, in-fol.

[2] *Questions inouyes ou récréations des sçavans qui contiennent beaucoup de choses concernantes la théologie, la philosophie et les mathématiques,* in-8, x, 276 pp.

1629 he wrote to Henricus Renerius on parhelia at Rome in that year. Mersenne persuaded him to write a work on Fludd which appeared in 1630, and the year following he observed the transit of Mercury. His astronomical observations were especially full and frequent in the years 1633 to 1638. In 1641 appeared his Life of Peiresc; in 1642 *Disquisitiones Anticartesianae* and *De motu impresso a motore translato*. He became royal professor of mathematics and astronomy at Paris in 1645, but pulmonary disease soon forced his retirement to Digne. The year 1647 saw the publication of his *Institutio astronomica* and *De vita et moribus Epicuri*. In 1649 his *Syntagma* of the Epicurean philosophy appeared as an appendix to his *Animadversiones* on the tenth book of Diogenes Laertius. He returned to Paris in 1653 and in 1654 published his Lives of Peurbach and Regiomontanus, Copernicus and Tycho Brahe. Most of his works were published posthumously in 1658. They were in Latin and so voluminous that François Bernier published an abridgement of the philosophy of Gassendi in 1674–1675, Doubts on some chapters of it in 1682, and a second edition at Lyon in 1684.

Gui Patin wrote in a letter of January 6, 1649, of "the incomparable Gassendi . . . a great man of small stature, an epitome of moral virtue and all the fair sciences, yet among others of great humility and goodness, and with a knowledge of mathematics which is quite sublime."[3] After Gassendi's death, Patin called him "homme sage, savant et bon, tempéré et habile homme, en un mot un vrai épicurien mitigé."[4]

Mersenne, whom Voltaire was to call "le minime et très minime père," made little positive contribution to science except perhaps in music and mathematics. But he encouraged and stimulated others, and acted as a go-between and clearing-house for the many persons who visited him at the Convent des Minimes near la Place Royale, or with whom he corresponded. Even scholars who would not communicate directly did so indirectly through Mersenne as

[3] *Lettres* (1846), I, 423. Earlier on September 4, 1641, Patin wrote: "Gassendi est un des plus honnêtes et des plus savants hommes qui soient aujourd'hui en France": *Ibid.*, I, 83.

[4] *Ibid.*, III, 67; letter of November 7, 1656 (wrongly marked 1655).

a medium. He was both curious and receptive, asked many questions, and suggested works for others to write. In 1648 he introduced Huygens to logarithms.[5] Beginning in 1623 with the motive of combatting atheism by discussing philosophical and physical problems which had a bearing upon religion and showing that Roman Catholics were neither unscientific nor superstitious, he turned in 1625 to a defense of science and continued condemnation of astrology and magic arts against sceptics such as Pomponazzi and Vanini.

Mersenne's respect for science and his intellectual tolerance kept increasing. Whereas in 1623 he had opposed the Copernican theory, already by 1624 in *L'impiété des deistes* he was saying that Copernicus could not be refuted, and that his hypothesis was very useful, although science was not yet in a state to decide definitively for or against it.[6] He corresponded with former adversaries, assisted the publication of works by those with whom he had once disagreed, or issued translations of them.[7] On February 1, 1629, he wrote to Galileo that he understood that the New System of the Movement of the Earth was completed, but that Galileo could not publish it because of the prohibition of the Inquisition. Mersenne offered to print it, if Galileo would send him the manuscript.[8] In letters to Peiresc in 1635 and Galileo in 1637 he declared Campanella and Galileo the two greatest men in Italy.[9] Yet Campanella was a devotee of the astrology that Mersenne had condemned. Conversely, Mersenne was admired even by Hobbes, and had Descartes as a confidential correspondent.

Mersenne's published works are in large measure compilations. He was said to have "une rare habilité pour se servir des idées des

[5] *Oeuvres complètes de Christiaan Huygens*, XXII (1950), 507, note 41.

[6] P. Boutroux, "Le P. Mersenne et Galilée," *Scientia*, 31 (1922), 285.

[7] *Correspondance du P. Marin Mersenne religieux minime*, publiée par Mme. Paul Tannery, editée et annotée par Cornelis de Waard avec la collaboration de René Pintard, I (1932), xlvi.

[8] *Correspondance du P. Marin Mersenne*, II (1936), 175, lines 42-46: "Praeterea te systema novum de motu terrae perfectum habere prae manibus, quod tamen ob prohibitionem Inquisitionis non possis divulgare; quod certe, si nobis confidere velis, et tuta via illius exemplar ad nos transmittere, illius editionem, prout praescripseris, audemus polliceri."

[9] *Correspondance*, I (1932), xlvi, note 3.

autres"; La Mothe le Vayer called him "le bon larron."[10] He included résumés of the works of others, like Snell and Maurolycus.[11] His books tended to be catch-alls, and consequently their titles may not exactly or completely describe the contents. Thus the Questions on Genesis digress from the apparition of angels to devote forty columns to the topic of optics, while the title, Truth of Science Against Sceptics, fits only the first quarter of the text.[12] The remaining pages are purely didactic and mathematical. Moreover, the first quarter was chiefly occupied with a discussion of alchemy, pro and con, although it also contains condemnations of astrology and chiromancy, and an estimate of Francis Bacon which fills a dozen pages[13] and is well taken. Mersenne also could compose a telling sentence, such as, "Ignorant Columbus discovered the New World; yet Lactantius, learned theologian, and Xenophanes, wise philosopher, had denied it."[14] But his particular scientific views were almost as likely to be wrong as right. Thus he wrote:

> We always have more than fifty thousand leagues of air on our heads, for it extends to the moon and perhaps to the firmament and beyond.[15]

He not merely refused to believe in the acceleration of falling bodies but even held that their speed decreased at the end.[16] He put other erroneous questions to Helmont, such as why iron does not give forth fire when struck with steel.[17]

Mersenne experimented not a little, and held that he had disproved the acceleration of falling bodies experimentally, and had also demonstrated that balls of iron and wood of the same size would fall at the same speed, although the iron sphere weighed eight times more than the wooden ball. Jean Rey disagreed with him on both counts and urged him to repeat his experiments more exactly. He further disagreed with Mersenne's statement that at the instant when he turned a burning glass to the sun, its heat

[10] J. H. Reveillé-Parise, Introduction to *Lettres* of Gui Patin, 1846, I, xxiii.

[11] *Correspondance*, II, 146, 161.

[12] 225 out of 1008 pages.

[13] *La verité des sciences*, pp. 206-18.

[14] *Ibid.*, p. 26.

[15] *Harmonie universelle*, I (1636), 8; *Correspondance*, II, 358.

[16] *Correspondance*, II, 58. Helmont and Jean Rey both told him that this opinion was false: *ibid.*, III, 78, 239.

[17] *Ibid.*, III, 85.

was as great as after a long exposure to the solar rays. Rey said that he had demonstrated the contrary a thousand times. Mersenne, however, in his reply maintained all his previous positions, and asserted that he was truly astonished that Rey should doubt his experiment of the equal velocity of a bullet of iron and a bullet of wood. He assured him that several persons of quality who had witnessed and participated in it would vouch for its authenticity.

As for the mirror, if you ever come here, I hope to show you one only a foot in diameter which sets a green willow branch on fire the moment it is exposed to it, although the hottest furnaces can do so only after some time.[18]

Mersenne also affirmed that he had disproved Rey's assertion that a dead body weighed more than when alive by actual experiment with a dog and a fowl, which he had strangled to prevent loss of blood. But Rey, who accepted as unchanging the law that weight is increased by addition of matter or restriction of volume, held that after death bodies normally shrank in size, and that Mersenne must have weighed the animals immediately after strangling them, which had kept the air in the lungs and the spirits in veins and arteries, so that the bodies had not yet contracted. Even if the dog and fowl had lost blood and other exhalations, if their bodies were left to grow cold over night, they would be found to weigh more, as Rey had proved by experiment since receiving Mersenne's letter.[19]

This exchange of letters well illustrates the uncertainty and insufficiency at that time of the experimental method. Not only might persons with the best of intentions arrive at diametrically opposite conclusions as a consequence of performing identical experiments, but either or both of them might be right in one of his conclusions and wrong in another. We must not therefore be too critical of their contemporaries who were slow to accept experimentation which seemed to them contrary both to authorities and to reason.

Mersenne's views may be further exposed by a rapid consecutive survey of one of his several voluminous works. His first major publication, the Questions on Genesis, is a formidable folio of 1911

[18] *Ibid.*, III, 188-89, 190, 239, 242, 279.

[19] *Ibid.*, III, 190, 242-43.

columns, with 440 more columns of Observations and Emendations to the Problems of Francesco Giorgio or Zorzi of Venice in the first half of the sixteenth century.[20] The first 462 columns aim to demonstrate the existence of God against the atheists. Mersenne goes on to argue that apparitions of angels are not to be denied, that miracles cannot be accounted for by the power of imagination or bodily exhalations, and that the stars are not the causes of miraculous cures. Neither demons nor necromancers can raise the dead; and, despite Paracelsus, heat mingled with putridity cannot be the cause of resurrection. Evidently Pomponazzi, Cardan and Vanini are also being refuted.

A long argument whether the firmament is solid or not involves a discussion of the nature and position of comets, in the course of which Mersenne asserts without citing chapter and verse, that Messahala and Haly on the *Quadripartitum*, Book II, hold that comets are made of celestial material, and that Albumasar admits that a comet was seen above Venus.[21] Mersenne is ready to accept the evidence and arguments of Tycho Brahe and others that comets are in the heavens, partly probably because for him that does not prove the heavens to be liquid, unless the comets are earthly exhalations which have passed through one or more of the heavenly spheres. Mersenne is therefore ready to abandon that doctrine of Aristotle also. He further contends that Scripture is not decisive one way or the other as to the solidity of the heavens, and finally arrives at two conclusions. 1) That all the heavens in which stars are seen to be moved are liquid like air, seems to him not improbable, of which heavens parts immediately coalesce where

[20] On Giorgio, T VI, 450-53.

[21] *Quaest. celeb. in Genesim*, 1623, cols. 827, 820. In his printed works Albumasar displayed only an astrological interest in comets, and his account of their significance in each of the signs of the zodiac was often repeated by Latin authors. But in *Albumasar in Sadan*, a work found only in MSS, Albumasar is quoted as saying: "The philosophers say, and Aristotle himself, that comets are in the sphere of fire, and no part of them is formed in the heavens, because the heavens undergo no alteration. But they are all wrong in this opinion. For I with my own eyes saw a comet beyond Venus. And I knew that it was beyond Venus, because it had not affected its color. And many persons have told me of seeing a comet beyond Jupiter and sometimes beyond Saturn." The passage occurs in Latin MSS BN 7302, fol. 122ra, and BL Laud. Misc. 594, fol. 140ra. See my "Albumasar in Sadan," *Isis*, 45 (1954), 22-32.

stars pass from place to place by their own motion. 2) It seems to be more probable that the eighth heaven in which the (fixed) stars reside is solid, nor is it absurd, if we retain the solidity of the remaining planetary heavens. Apparently the reader is left free to make his choice between these two conclusions.[22]

Presently we come to the question whether anyone, without indication of heresy or danger of error and temerity, can believe and defend that the earth is mobile, the heavens immobile.[23] After giving the decree of the cardinals in 1616 that the *De revolutionibus* of Copernicus be suspended until corrected and entirely prohibiting the recent work of Foscarini,[24] Mersenne concludes, "Therefore it is certain that the earth is immobile."[25]

After the old familiar question whether the earth is animated, Mersenne considers what future events are signified by the stars. He grants that they are signs of things depending upon natural causes such as the weather and health and disease. For example, Saturn in conjunction with the navel of Andromeda is a sign of clouds, rain or snow; with the stars of Cetus it denotes rough weather; with the horn of Capricorn, cold; with the Greater Dog, winds, thunderbolts and rain; with the tail of Aries, disturbance of the air; with the Dolphin, cloudy weather; with Arcturus, winds and rain; with Lyra, clouds; in *praesaepe*,[26] rain or wind, and so with Aselli; with the head of Medusa, cold and long humidity; with the stars of Orion, rain-storm and wind; with the Pleiades, turbid air with snow and rain; with Spica, sudden changes, rain, thunder and lightning.[27] Aquinas concedes something to genethlia-

[22] *Prima conclusio* at col. 843; *Secunda conclusio*, col. 845.

[23] *Ibid.*, col. 902.

[24] Paolo Foscarini, *Sopra l'opinione . . . del Copernico*, 1615: BM 531. e.2(4); BN R.12953. Despite this complete prohibition, Tiraboschi states that it was reprinted with the *Dialogues on the Two Systems* of Galileo: VIII (1824), 346, "Il P. Paolo Antonio Foscarini carmelitano stampo in Napoli nel 1615 una lettera sulla mobilità della terra e sulla stabilità del sole, in cui cercò di conciliare questa opinione co' testi della sacra Scrittura, che ad essa sembrano opporsi; ed essa fu poi aggiunta, tradotto in latino, a' Dialoghi del Galileo sullo stesso argomento."

[25] *Quaest. in Genesim*, 1623, col. 904.

[26] The space between the two stars called *Aselli* in Cancer.

[27] *Op. cit.*, col. 960; and so for the other planets in turn. At col. 961, he lists the weather following conjunctions of the planets.

logy, but here freedom of the will checks the influence of the stars. Furthermore, Mersenne opposes the whole theory of astrological houses.[28] But he speaks of the parts of the third region of the microcosm.[29] Albertus Magnus approved of astronomical images in the *Speculum astronomiae*, but Mersenne rejects that work as either supposititious[30] or erroneous, ignoring the fact that Albert also favored astronomical images in his *De mineralibus* of unquestioned authenticity.[31] He agrees with Aquinas that "even the images which they call astronomical have their efficacy from the operation of demons,"[32] and disagrees with Cajetan who ascribed the virtue of images and characters to co-principle of operation and a sympathy with things celestial.[33] Yet Mersenne could say in 1631 that the center of anything was its noblest part.[34]

Mersenne rather hurries over the question as to the marvelous virtues of gems, giving a long alphabetical list of stones but only a word or two as to the medicinal virtue of each.[35] For example, the topaz cures lunacy, Varach checks all haemorrhage, while Ziazaa excites terrible dreams.

The four rivers of Paradise lead to praise of the number four.[36] Adam possessed all arts and sciences, and is compared with Solomon.[37] There are such questions about brute animals as whether they spoke in the beginning of the world, how irrational animals do such stupendous things, whence the natural hatred of the serpent for man, and more as to sympathy and antipathy.[38]

Mersenne denies that names depend on the stars, argues against what Galeotto Marzio in the fifteenth century ascribed magically to letters and names, and rejects onomancy. Incidentally something is said of the use of the Hebrew alphabet by cabalists and Magi.[39]

With the question whether the blood of Abel flowed from his corpse against Cain is raised the corollary whether the blood of

[28] *Ibid.*, col. 967, 974 *et seq.*

[29] *Ibid.*, col. 1132.

[30] For its genuineness, T II, 692-719; *Speculum*, 30 (1955), 423-27.

[31] T II, 588.

[32] *Quaest. in Genesim*, 1623, col. 1153; on Albertus, col. 1151.

[33] *Ibid.*, col. 1152, Col. 1165, "sculpturae virtus reiicitur."

[34] *Correspondance*, III, 187.

[35] *Quaest. in Genesim*, cols. 1167-70.

[36] *Ibid.*, col. 1173.

[37] *Ibid.*, col. 1214.

[38] *Ibid.*, cols, 1262, 1270, 1360.

[39] *Ibid.*, cols. 1384-92. Concerning Galeotto; T IV, 399-405.

the victim flows again in the presence of the murderer, and other instances are given of the dead retaining properties of the living. Marvelous antipathies are again noted.

For who will tell why a snake killed and put in the shade of an ash-tree keeps squirming until it is taken away? Who can find out why, if someone suffers from a tumor of the spleen and suspends the spleen of another animal in the smoke of a fireplace, the tumor and spleen dry up as the suspended spleen does so? Unless we have recourse to communication through the air by whose medium the spirits act on each other.[40]

After discussion of mechanical and liberal arts, and much on music,[41] we come back again to man the microcosm but to a denial of astrological chiromancy. Robert Fludd, that heretico-magus, seems to Mersenne to be mildly insane, when he affirms that the hand is as it were a table of the geniture and nature, on which in occult wise are carved the mysteries of one's nativity.[42]

That the interest, even among the learned and students of nature, in curious questions suggested by the Book of Genesis, continued through the century, is seen from a review in the *Journal des Sçavans* of a book on that theme which appeared in 1685.[43] The four questions which the reviewer selected for his readers were whether at the resurrection the rib from which Eve was formed would revert to Adam, whether a serpent or a demon tempted Eve, how many children Eve had, and where the terrestrial Paradise was or is still situated.[44]

Peiresc, the patron and friend of Gassendi and Mersenne, was already interested in attacks on astrology and divination. In 1620 Paolo Gualdo of Padua sent him the letters of George of Ragusa against such arts, and, after their author died in 1622 at the age of only forty-three, Peiresc had them printed shortly before he left Paris in August, 1623.[45]

Mersenne in his *Questions on Genesis* of 1623 had devoted some

[40] *Quaest. in Genesim*, col. 1408.

[41] *Ibid.*, cols. 1514-1700.

[42] *Ibid.*, col. 1743.

[43] *Questions curieuses sur la Genese expliquées par les PP. & les plus doctes Interpretes*, Paris, 1685, in-12.

[44] JS XIII, 224-26.

[45] *Epistolae mathematicae seu de divinatione libri II. Non solum astrologia verum etiam chiromantia, geomantia, cabala, nomantia, magia ...*, Paris, 1623. The work has already been dicussed in T VI, 198-202.

forty columns to what was on the whole a condemnation of astrology. In *La Verité des sciences* of 1625 he continued the attack, stating that you will hardly find two astrologers who agree as to the direction of *promissores* and *significatores,* that they have yet to answer the old argument regarding Jacob and Esau, and that they invent or take from Ptolemy "imaginary principles."[46]

Yet later in the same work he affirmed that it was necessary for a medical man to know not only the phases of the moon but the courses of the stars and planets and their effects.[47] In his *Préludes de l'Harmonie universelle* of 1634, however, he included the Sentence of May 22, 1619 which the Sorbonne had pronounced against the practice of judicial astrology.[48] It was perhaps owing in part to the influence of Mersenne that Gassendi developed his attack upon astrology. Mersenne was somewhat more favorable than Gassendi to comets announcing the death of kings. Strowski has called his explanation why kings are affected by comets more than other men, silly,[49] but it was very similar to those which had been offered for centuries past. And Gui Patin, in a letter of March 4, 1661, wrote that the Hugenots interpreted a recent comet "with two horns" as indicating that the pope and Mazarin would die soon.[50]

Despite his expressed opposition to astrology, divination and magic, much of the attraction of modern science for Mersenne lay in its marvelous character. In the preface to *La verité des sciences* he says that statics, hydraulics and pneumatics produce such prodigious effects that it seems that men can imitate the most wonderful works of God.[51] He also believed in natural prodigies. Strowski has already noted the passage in the same book in which, to demonstrate human superiority to brute animals, Mersenne affirms that man can give birth to anything: colts from a woman of Verona in 1254, a half-bird at Ravenna in 1517, a half-calf in Saxony and a child with a frog's head at Boileroy in 1517, a half-dog in 1493, and a dog with human head in 1571[52]—exam-

[46] *Op. cit.,* p. 31.

[47] *Ibid.,* p. 242.

[48] *Correspondance,* I, 42.

[49] Strowski, *Hist. du sentiment religieux . . .,* I (1909), 214.

[50] *Lettres* (1846), III, 334. Mazarin did, on March 7.

[51] The passage occurs near the end of the preface.

[52] Strowski, I (1909), 213.

ples of credulity and deficient historical criticism which might have given the sceptics whom he was attacking cause for mirth. He was fond of raising such questions as whether one could construct a mirror which would burn in any place you wished up to infinity,[53] or what the power of the voice should be in order to carry from the earth to the firmament.[54] His answer to the former query was in the affirmative, provided there were an incombustible material that would not lose its polish from which to make the burning glasses. He delighted in such paradoxes as that it was more difficult to break the least chord of a spinet than to overthrow the whole world.[55]

The books of Mersenne were not free from the recipes and secrets noted so often in our previous volumes. In the Questions on Genesis he included a recipe "to create the macrocosm artificially." This was reproduced in *Recréations mathématiques* of Rouen, 1628, in which a second and third parts were added to the original text of 1624. Mersenne was therein further credited with "an excellent secret" of casting any metal quickly which he was said to have practiced himself. There was also a recipe for an inextinguishable lamp which may have come from Mersenne, since a correspondent of the same year asked him for the secret of it.[56] On February 1, 1629, Mersenne wrote to Galileo that he was at work on a most extraordinary and incomparable invention of telescopes by which objects on the moon and stars would appear of their actual size.[57] But Galileo had several years before informed another correspondent that the idea was impossible of realization.[58] Jean Beaugrand was the author of mathematical works and "mathématicien de Gaston de France,"[59] brother of the king, in which capacity his duties were very possibly largely astrological. But when Mersenne wrote him that he was in possession of the sym-

[53] *Questions inouyes (1634)*, Question 25, pp. 157-59.

[54] *Questions théologiques etc.* (1634), Question 44; but *Correspondance* II, 434-35, notes that some copies have, instead of this question, a paraphrase of the first Dialogo of Galileo.

[55] *Préludes de l'harmonie universelle* (1634), Question 8, pp. 188-203.

[56] *Correspondance,* II (1936), 77, 87.

[57] *Ibid.*, II, 173-76.

[58] *Ibid.*, II, 180, citing Galileo, *Opere,* XIII (1903), 213, 231, 237-38. See Chapter 19 on Descartes for his efforts in the same direction.

[59] *Correspondance,* II, 504.

pathetic unguent, Beaugrand replied that the effects of the magnet and many other natural phenomena for which he could not account prevented his believing anything impossible. "You will permit me, however, to suspend judgment on this subject until the experience that you have with it has made me more certain."[60]

Already, before the long discussion of alchemy in *La Vérité des Sciences,* Mersenne had shown his interest in the subject in *Questions on Genesis,* where he discussed *aurum fulminans,* stating that a powder was made from gold which, when set on fire by the rays of the sun or fire, exploded with a louder noise than gunpowder, but downward, not forward or upward.[61] Mersenne followed Petrus Arlensis de Scudalupis [62] in his account except that he did not agree with him that the powder could not be kindled by natural fire but only by the rays of the sun. This powder was again discussed in the preface to *Traité de l'Harmonie universelle* of 1627, where Mersenne said that it could be made in half an hour without use of fire, "as I have experimented." Another experiment convinced him that the hottest summer sun would not explode it, although some held that mere heat of the body from carrying it in one's pocket would do so.[63] Again in *Questions théologiques, physiques etc.,* of 1634 he asked, Why the powder of gold, called fulminant, made so loud a noise, when it felt heat?[64]

In the meantime Mersenne had been inquiring of a chemist of Rouen named Lefebvre, whom he met there in May, 1625, concerning such matters as changing mercury into silver by means of an oil drawn from the dung of a goose fed on lead filings, the generation of silver by an oil of mercury, how much sulphur there was in each of the metals, the weight of refined mercury compared to gold, and how to render *aqua vitae* as hard as crystal.[65] Lefebvre kept his powder of gold a secret. He informed Mersenne, however, that those who thought that the phlegm of wine intoxicated had never separated its substances as they should, for there was nothing

[60] *Ibid.,* II, 514.

[61] *Op. cit.,* col. 116.

[62] *De sympathia septem metallorum ac septem selectorum lapidum ad planetas,* Paris, 1610, p. 372; cited *Correspondance,* I, 287. Petrus Arlensis de Scudalupis has already been treated in T VI, 301-2, 324.

[63] *Correspondance,* I, 297.

[64] Question V, pp. 20-23.

[65] *Correspondance,* I, 275, 322.

that prevented drunkenness better, except the fixed salt of wine, "drawn off and separated as a good chemist ought to do, which is a secret known to very few persons." He also told him that, if a salt was derived from good vinegar distilled once, then redistilled with turpentine, with pearls dissolved in it, and then the salt was sweetened in liquor in the months of June, July and August in a damp place like a cellar, a few drops of this liquor would curdle well-rectified spirits of wine into a butter that you could cut with a knife. He added a yet longer recipe for reproducing a plant from its salt.[66]

* * *

In 1617 Robert Fludd (1574–1637) published at Oppenheim the first part on the Macrocosm of a work which was to be on Macrocosm and Microcosm, and in the same year issued at Leyden a defense of the Rosicrucians against Libavius and others. In 1619 appeared the second volume on the supernatural, natural, preternatural and contra-natural history of the microcosm, with a section on genethlialogy, physiognomy and chiromancy.[67] In 1623 came out his Triple Anatomy, of which the first part dealt with "bread, easily the chief nutriment," its dissection by fire, its elements and their occult qualities. In 1629 at Frankfurt-am-Main appeared the first volume of Catholic Medicine, devoted to the Celestial and Elementary Mystery of Health and its Preservation, in which he answered attacks by Mersenne, declared his philosophy in accord with the Bible and Christianity, defended the *lapis Lydius* in particular, and praised the sciences of magic, cabala, and true alchemy. Volume two on the Mystery of Disease and the Signs of Morbid Meteors followed in 1631. In it he discussed crises and critical days, arithmetical divination which he regarded as based on Pythagorean superstition as to numbers, and onomancy or prognostication from names. More reliable in his opinion was meteorological prognostication, from which he proceeded to presages of cardinal diseases and signs of recovery or death. He described the cardinal "com-

[66] *Ibid.*, I, 321-23.

[67] *Tomus secundus de supernaturali, naturali, praeternaturali et contranaturali microcosmi historia* . . ., Oppenheim, 1619.

plexions," natural and preternatural, the presence of diseases or morbid meteors, their origin from the four fountains of the winds, their times and courses. He then inquired what could be predicted from the celestial or astrological *figura coeli,* and what from the terrestrial figure by geomancy. Next came contemplation of the patient's members, physiognomy of the face, and chiromancy of the hands. There is an astrological section on *urina non visa,* as well as deductions from inspection of it and other excrements, and observation of the pulse. Thus a small amount of regular medicine was mixed with much occult science.

On October 3, 1619, Father Jacques Saint-Rémy, S. J. (1578–1647), rector of La Flèche, answering an inquiry from Mersenne,[68] wrote that astronomical chiromancy was to be rejected, like judicial astrology, but that natural and physical chiromancy, "of which alone Aristotle speaks and the philosophers," was not to be condemned, provided it did not exceed its limits. The fact that the life line in the palm of the hand did not arise from the heart or other principal part of the body, did not prove that it was not a sign of long life. For experience showed that anyone with a longer life line was long-lived, although many without a prolonged life line were also blest with longevity. It was probable that such a line was produced by more temperate blood and better cooked food, which could indicate a more temperate liver and nobler vegetative faculty.

Mersenne, however, in his *Quaestiones celeberrimae in Genesim* of 1623, where he refuted the treatment of chiromancy by Fludd, took a more uncompromising tone, asserting that there was nothing solid in that art and challenging Fludd to interpret a figure of a pair of hands "designed from nature by an excellent painter," which he reproduced in his volume.[69]

Mersenne had called Fludd an evil magician, an heretical magician, and a doctor of horrendous magic.[70] Franciscus Lanovius

[68] He was too busy to look up the matter himself and passed the question on to Father Brossard, in charge of the classes in philosophy and theology. For the letter, *Correspondance du P. Marin Mersenne,* ed. Mme. Tannery, C. de Waard and R. Pintard, I (1932), 40-41.

[69] *Ibid.*, pp. 42-43. *Quaest. celeb. in Gen.*, cols. 1739-46.

[70] Gassendi, *Opera,* 1658, III, 215.

(de la Noue) in a letter to Mersenne of November 20, 1628, condemned Fludd's use of Scripture for alchemical purposes as blasphemous and sacrilegious and meriting the same censure as the Sorbonne had meted out to Khunrath. He also attacked Fludd's belief in a familiar genius or demon.[71] Fludd had complained that Mersenne borrowed from his writings without acknowledgement.[72] In a work (*Summum bonum*) issued under the name, Ioachimus Frisius, Fludd distinguished between good and bad magic, defending the former and holding, like Naudé, that Roger Bacon, Trithemius, Ficino and Henry Cornelius Agrippa had cherished only it and been wrongly accused of evil magic. He also distinguished between true and spurious alchemy, defending the former and the Rosicrucians, and also the Cabala.[73]

At the request of Mersenne, Gassendi composed an Examination of the Philosophy of Robert Fludd which was published in 1631.[74] It is a bit surprising to learn that Mersenne entrusted the printing of this work to La Mothe le Vayer, who was something of a free-thinker.[75] In it Gassendi agreed with Mersenne and Lanovius that alchemists should not abuse Scripture and the mysteries of religion,[76] but he did not, like Mersenne, consider Fludd to be an evil magician or an utter atheist.[77] Indeed, Gassendi expressed scepticism as to diabolical and evil magic,[78] and took little stock in the Cabala, especially the alchemical variety.[79]

Gassendi summarized Fludd's philosophy as follows.[80] Light emanating through the Sephiroth is the chief agent of all things. Its union with ethereal spirit constitutes the World Soul, of which all individual souls are particles. The empyrean heaven is angelic

[71] *Ibid.*, pp. 267-68, "Ad R. P. Marinum Mersennum Francisci Lanovii Iudicium de Roberto Fluddo." Reprinted, *Correspondance*, II, 132-41.

[72] Gassendi, *Opera*, III, 228.

[73] *Ibid.*, p. 215.

[74] *Epistolica exercitatio in qua principia philosophiae Roberti Fluddi medici reteguntur et ad recentes illius libros adversus R. P. F. Marinum Mersennum . . . respondetur*, Paris, 1630, in-8, 352 pp.: BN R.13424. As printed in Gassendi's *Opera*, vol. III, it has the briefer caption, *Examen philosophiae Roberti Fluddi*. A seventeenth century MS of it is Bassano del Grappa 1482, "Examen philosophiae Roberti Fluddi medici, Petri Gassendi."

[75] *Correspondance*, II, 446.

[76] Gassendi, *Opera*, III, 259.

[77] *Ibid.*, pp. 215, 240.

[78] *Ibid.*, p. 251.

[79] *Ibid.*, p. 254.

[80] *Ibid.*, pp. 221-28.

nature itself, the flower and purer portion of ethereal spirit illuminated by divine light. There are nine orders of good angels and nine classes of bad angels. On the fourth day the sun was formed in the middle of the ether, then Mercury from the sun and the inferior region, then the moon from Mercury and the inferior region, thirdly Venus from reflexion between the sun and Mercury. Jupiter was formed fourth from reflexion between sun and fixed stars; Mars, fifth from reflexion between the sun and Jupiter; Saturn, sixth from reflexion between Jupiter and the fixed stars. As there were nine orders of angels and nine classes of demons, so there are nine elemental regions arranged in groups of three each. Pure earth, minerals and vegetation make up the lower region; fresh water, salt water and the lowest of the three regions of air constitute the middle group; the highest consists of the middle and upper regions of the air, and that of fire. Man the microcosm corresponds to three heavens: the intellect in the head, to the empyrean; vitality and free will, in the heart, to the ether; natural functions in the abdomen, to the elemental spheres. While the superior or formal Diapason is divided harmonically from the sun to the supreme hierarchy of angels, the inferior or material is divided arithmetically from the earth to the sun.

In a letter of the same year Mersenne criticized Fludd for six impieties, of which the first was that all sacred Scripture had an alchemical significance. Secondly, he identified God with light and an ethereal spirit or quintessence which resided especially in the sun and was the cause of the generation of all things. In this way God was the form of all things, and secondary causes did nothing *per se*. Thirdly, the world soul was a composite of God and that ethereal spirit. Its purest part was angelic nature and the empyrean heaven, but demons and all souls of men and brutes were particles of the same world soul. Fourth, he identified the world soul with Christ and the rock on which the Church was founded, and made it the chief part of the philosophers' stone. Fifth, the just man was the alchemist who had found the philosophers' stone and become immortal by its use, and such were the Rosicrucians. Sixth, creation was not from nothing in the vulgar sense but from matter. Moses was an alchemist in describing creation, as were Wavid, Solomon, Jacob, Job and other Biblical worthies. Similarly true cabalists are nothing but alchemists and so are Magi, philosophers and priests.[81] Mersenne said further that Fludd contended that a dead man was heavier than when alive, because the light and ethereal spirit had left the body. Mersenne, as we have already noted from his correspondence with Rey, claimed to have disproved this by most accurate experiment, since a dog weighing seventeen pounds and a hen of fifty-two ounces when alive were found of the same weight or less after death. Yet he had just quoted Fludd as saying that man has a greater abundance of light than other animals, so that experiments with them would seem hardly germane. Also Fludd had made lightness not merely a matter of weight, but of lightness of movement. However, Mersenne goes on to say that he hears that Santorio of Venice had weighed the same man alive and dead at almost the same hour, and found the corpse a little lighter than the living body. He also adds that many animals which are larger and weigh more than man surpass him in mobility, as Fludd might have learned from Gesner and Aldrovandi.[82]

In a Key to the Fluddian Philosophy and Alchemy, published in 1633, Fludd again defended himself against Mersenne and further against Gassendi and Franciscus Lanovius. Finally he set forth his views in a volume entitled Mosaic Philosophy which appeared posthumously first at Gouda in 1638 and then at Amsterdam in 1640.

Fludd has been characterized as "a philosopher, physician, anatomist, physicist, chemist, mathematician and mechanician," and credited with "a rare gift of observation in the exact sciences."[83] But he still thought it possible and advisable to combine with this science and medicine not only a cloak of religion but also much of the occult science that had come down from the past: magic and cabala, astrology and alchemy, physiognomy and chiromancy, geomancy and weather signs. Natural, preternatural, supernatural,

[81] Letter of April 26, 1630, to Nicolas de Baugy, French ambassador to the United Provinces: *Correspondance,* II, 440-42.

[82] *Ibid.,* pp. 442-43. For Gesner and Aldrovandi, sixteenth century naturalists, see T VI, Index.

[83] Hoefer, *Histoire de la chimie,* II (1843), 185-86.

and things contrary to nature, were all closely related and even confused by him. Magic was science, and nature a mystery.

Even such a devotee of the occult as van Helmont had a very poor opinion of Fludd whom, he said, he had known in England as a poor physician and worse alchemist, garrulous, superficially learned, inconsistent. He found nothing but dreams in Fludd's writings and did not think it worth while to waste time and effort in refuting him.[84] Towards the close of the century Garmann repeated Fludd's assertion that demons were composed of the matter of the empyrean heaven, acted and suffered, and had bodies which were crass or refined according to the element to which they were proscribed, fire, water or earth.[85] But Garmann condemned this view as insane.

* * *

Gassendi's Life of Peiresc affords some notable instances of scepticism and credulity on the part of both men with regard to matters of magic and science. Peiresc wrote long letters concerning the case in 1611 of Louis Gaufrid, accused of evil magic, and Magdalena Paludana, supposed to have become possessed of a demon through his sorceries. At first Peiresc believed in marks on the body as evidence of witchcraft but then began to suspect imposture. Nevertheless he always defended the court's sentence ordering the magician to be burned, because he had in other ways lewdly violated the holy mysteries of religion. Peiresc said that sorcerers do not have the commerce with the devil that they think they do, but they ought to be punished for wishing it. Not long afterwards a priest of Marseilles was accused of magic but was acquitted, although there were punctures all over his body. Three years later came a report from Flanders of a canon, similarly marked, who was acquitted. Peiresc therefore inclined to the conclusion that such supposed marks of sorcerers were either a skin disease or self-inflicted.[86]

[84] Letter of December 19, 1630, to Mersenne in *Correspondance*, II, 584.

[85] Garmann, *De miraculis mortuorum*, ed. of 1709, pp. 733-34, citing Fludd, *Historia utriusque cosmi*, tract. I, liber iv, cap. 2, p. 109.

[86] Gassendi, *Opera*, V, 276b-277a.

In connection with the comet of 1618, Gassendi denies that comets are either signs or causes of calamities. Later he represents Peiresc as disserting at length on the characters used for the planets, which he thought had been derived from the majuscules for the Greek vowels.[87]

The theme of human longevity was raised by a report from England of a man who had died at the age of 152, but then Peiresc heard of a Persian who lived to four hundred. The subject of Tritons was brought up by the escape of a merman from fishermen. Peiresc bought a crocodile and counted its teeth, and further manifested his interest in zoology and experimental verification of tradition by keeping chameleons and proving that they did not live on air. As for medicine, when a cat and one of its three kittens died of the same disease, dissection revealed white worms which looked like cucumber seeds in the intestines. Peiresc, not approving of the treatment recommended by the attendant surgeons or physicians, had the happy inspiration of calling for theriac, which, to the amazement of all, liquefied the hard contents of the intestines. Given to the two remaining kittens, it cured one and killed the other, or at least one survived and one didn't.[88] Simply marvelous!

Gassendi was more active in astronomy than in any other science. His observations of the heavens over the years from 1618 to his death in 1655 fill more than 400 pages in his collected works, of which more than 300 are on the years 1633–1638.[89] They began at Aix on November 28, 1618 with observation of a comet, concerning which Gassendi's biographer, Bougerel, states that Gassendi made conjectures which the event verified. He also measured the distance between Jupiter and Venus, and in June, 1619, the distances of other planets and of the moons of Jupiter, as well as a lunar eclipse. According to Bougerel, he was the first to give the name, aurora borealis, to that phenomenon, which he saw in September, 1621. At Aix in 1623 he observed lunar eclipses of April 14 and June 7, and the distance of Mars from Sagittarius.[90] He

[87] *Ibid.*, pp. 286b, 326a, under the year 1636.

[88] *Ibid.*, pp. 326a-b, 329a, 276a-b.

[89] *Opera,* IV, 77-480, and 110-424 respectively.

[90] Bougerel, *Vie de Pierre Gassendi,* 1737, pp. 10, 14-15.

recorded his observations of the transit of Mercury at Paris in 1631, and of the solstitial altitude at Marseilles in 1636.[91] His observation of the transit of Mercury showed that the Danish Tables of Longomontanus erred by 7°13′ from the true place of Mercury—a discrepancy which Longomontanus had called immense in the case of the Alfonsine and Copernican Tables, that the Prutenic Tables erred by 5°, Ptolemy by 4°25′, Lansberg by 1°21′, and the Rudolfine Tables by 14′, so that Bullialdus (Boulliau) justly dedicated to Gassendi the tenth book on Mercury of his *Astronomia Philolaica.*[92]

Gassendi's interest in astronomy was further demonstrated by his Lives of Peurbach and Regiomontanus, Copernicus and Tycho Brahe, by his exposition of the Roman calendar,[93] and by the section of his *Syntagma Philosophicum* which was devoted to celestial phenomena.[94] It was therefore noteworthy that one who had given so much attention to the stars and to astronomy, should have abstained entirely—with the possible early exception suggested above by Bougerel—from any predicting based upon them, and should have totally rejected astrology. Bougerel tells us further that Gassendi, like the astronomer Cassini, had pursued the study of astrology in his youth, but was soon disillusioned, and, as early as 1623–1624, while representing his cathedral chapter in litigation at Grenoble, succeeded with some difficulty in weaning his friend Valois from his interest in that art.[95]

Gassendi began to ventilate his doubts as to astrology in writing as early as 1629 in his letter to Henricus Renerius on the Parhelia or four spurious suns seen at Rome in that year.[96] So far as presages of natural occurrences were concerned, he said that it seemed ridiculous for anyone to undertake to divine from a meteor at Rome as to weather changes in Belgium, or to predict for several days, months and years. As for preternatural happenings, he ventured

[91] These follow his astronomical observations in *Opera*, IV, where is also "Epistola I, Novem stellae circa Iovem visae a rev. patre Rheita Coloniae exeunte anno 1642 et ineunte 1643."

[92] Riccioli, *Almagestum novum*, 1651, I, i, xv.

[93] *Romanum Calendarium compendiose expositum, Opera*, V.

[94] *Opera*, I, 495 *et seq.*

[95] *Vie de Pierre Gassendi*, pp. 10, 15-16.

[96] *Opera*, III, 651 *et seq.*, Epistola ad Henricum Renerium ubi et de parheliorum genesi eorumque praesagiis fuse edisseritur.

to affirm that nothing could be predicted certainly. All the meteorologists might agree that parhelia portended destruction of empires, strife of princes, popular factions, and coalitions such as the Second Triumvirate. Astrologers might predict after the event as to princes and nations. God might employ such apparitions as warning signs. But to Gassendi it seemed not merely childish but utterly stupid to tremble at these signs which men vied in dreaming. What actual connection was there between such vapors and human affairs? A certain analogy between the sun and the prince was insufficient.[97]

In his *Institutio astronomica*, first issued at Paris in 1647 and of which a second edition appeared in 1653 at London, Gassendi devoted a chapter to the aspects of the planets,[98] an astrological matter, but he mentioned it only to by-pass it, although he did not venture to ignore it entirely. "I pass over," he said, "these aspects so celebrated among the astrologers," who ascribe to them the greatest force on the weather and human fortunes. He was not entirely silent on the subject, however, vouchsafing the additional information that the astrologers called aspects of opposition and quadrate bad, trine and sextile aspects good, and conjunction indifferent. He further "passed over" the fact that Kepler had recently introduced additional aspects[99] on the ground that the traditional five just mentioned were insufficient to account for all changes observed in the air. Gassendi also by-passed the astrological theme of great conjunctions, such as those of Saturn and Jupiter every twenty years, or of Saturn, Jupiter and Mars every eighty years. But this noncommittal attitude changed to one of violent opposition to astrology in the *Syntagma Philosophicum*, which was not published until after his death.

In the sixth book of the second part of his *Physica*[100] Gassendi deals at length with the question whether the stars

[97] *Ibid.*, pp. 658-60.

[98] *Astron. Instit.*, II, 15: *Opera* (1658), IV, 40-41.

[99] *Ibid.*, p. 41: "nempe semi-sextum seu duodecilem, decilem, octilem, quintilem etc."

[100] In *Opera*, I (1658), the discussion fills some forty double-columned folio pages. The *Physica* is itself a section of the *Syntagma philosophicum complectens logicam, physicam et ethicam* which fills the first two of the six volumes of *Opera*.

influence inferiors and with the vanity of astrologers. He so to speak turns up his cuffs, rolls up his sleeves, spits on his hands, flexes his biceps, and prepares to deliver the coup de grace and knock-out blow to that inveterate delusion which had come through so many previous fights but is now reeling and groggy in the late rounds of the present combat. He has just finished with comets, denying their influence upon these inferiors,[101] and next turns to the consideration of the supposed influence of the stars.

If qualities and generation and corruption are from the stars, it exceeds human capacity to measure this. The action of the stars seems to be general and indifferent, and in no respect special. That Saturn is cold and Mars, hot, are mere figments. Mercury is called the lord of the air, but since its position and aspect are identical for the entire earth, there should be the same winds everywhere, which is not the case. Gassendi denies that Genesis I, 14, "and let them be for signs and for seasons, and for days and years," justifies astrology, and quotes Isaiah and Jeremiah against the art.[102]

Gassendi next devotes several columns to the history of astrology, tracing its origin to the Chaldeans and inclining to accept the statement of Vitruvius that Berosus introduced it among the Greeks, while he represents Epicurus as speaking of the "slavish technique of astrologers" and "inane astrology."[103]

The last four chapters of this sixth book deal with the vanity of astrologers: as to general principles (chapter 2), special rules for weather prediction (chapter 3), nativities and human events (chapter 4), and last the evasions by which they try to strengthen and defend their tenets (chapter 5).[104] Gassendi charges that they jump from the effects of the sun and moon to those for the other planets without any justification. Their houses, dignities and so on are unfounded, and, even if they fitted Chaldea, would not apply to other regions. The division into twelve rather than eight or ten houses is purely arbitrary. Such features of their technique as *apertiones portarum* are mere dreamings.[105] The innumerable new

[101] *Opera,* I, 711b-712.

[102] For the points in this paragraph, *ibid.,* 713a-716a.

[103] *Ibid.,* 715b-716a.

[104] Chapters 2, 3, 4 and 5 begin at page 719, 727, 733 and 740 respectively.

[105] *Ibid.,* 728a.

stars disclosed by the telescope call for new rules, and Gassendi asks why no special effects are assigned to the Milky Way.[106] He is sceptical whether astrology has a firm basis in long and accurate observation in ancient Chaldea. Thales made his famous olive crop prediction upon a physical, not an astrological, basis, while modern astrologers do not observe the stars but merely use Ephemerides, and so prediction is a matter of luck, and the astrologers might as well throw dice for it as trust to their Aphorisms. Gassendi does not deny that God attributed some virtues to the stars, but the question is whether the astrologers know these. Lucian is quoted as to astrology for half a column.

Like Heurtevyn earlier, Gassendi weakens his case by making the unwarranted assertion that astrologers in general and Stoeffler in particular predicted a general flood for February, 1524, whereas the entire month turned out to be fair and beautiful weather.[107]

In the course of the fourth chapter Gassendi argues that astrological predictions as to individuals may contribute to their own fulfilment by their stimulating or depressing effect upon such individuals. Or by putting too much trust in a prediction that he will gain wealth or office, the person in question may fail to work and strive sufficiently to attain that end. In general Gassendi is opposed to curiosity as to one's future. Returning to criticism of astrological method and technique, he questions if planets can exert influence when below the horizon, and stresses the difficulty of determining the exact moment of birth. He ridicules such refinements as directions, *apheta, promissor* and *significator,* then turns to criticize interrogations and elections, and ends the chapter by declaring that nothing is more inane and inept than seals and astrological images.

In the fifth and last chapter, the appeal of the astrologers to antiquity and experience is again denied. Gassendi holds incorrectly that the *Quadripartitum* is not by Ptolemy, and calls incorrect a prediction by Nostradamus. Other reasons than astrology itself for the predictions often coming true or seeming to come true are fortune or chance, the cunning of the astrologers, and

[106] *Ibid.,* 729b.

[107] *Ibid.,* p. 729a.

the ignorance and stupidity of those who consult them. He presently discusses whether the death of Pico della Mirandola was predicted by the astrologers, and soon after that, his sixth books ends.

These chapters against astrology were translated into English "by a Person of Quality" in the year following the publication of Gassendi's *Opera*.[108] Bernier included them in his abridgement of Gassendi's philosophy published at Paris in 1675.[109] It may be doubted, however, whether they did astrology as much harm as might have been expected from Gassendi's reputation in scientific and astronomical circles. Other persons might discount his attack as emanating from an Epicurean and sceptical near-atheist. And the fact that he himself had died in his sixty-third year or Grand Climacteric might afford the astrologically-minded some small satisfaction.

Martin Hortensius was born at Delft in 1605 and died an early death in 1639. He edited Snell's *Doctrinae triangulorum* at Leyden in 1627; published a Latin translation of Philip Lansberg's *Commentationes in motum terrae* at Middelburg in 1630; replied to Kepler's *Additiuncula* in 1631, became professor of *Mathesis* at the University of Amsterdam, founded in 1632, where in 1634 he printed a translation of Willem Blaeu, *Institutio astronomica* and an Oration on the Dignity and Utility of Mathesis. This last dealt mainly in generalities but stressed the use of Mathesis in theology, medicine and nautical matters. Meanwhile he had been stirred by Gassendi's observation of the transit of Mercury to write a *Dissertatio de Mercurio in sole viso et Venere invisa instituta cum . . . D. Petro Gassendo* (Leyden, 1633). In closing it, he remarked the great uncertainty of judicial astrology because the movements of the heavenly bodies were so difficult to measure exactly. This should be done before men could pass upon effects in inferiors at certain times. Hortensius did not, however, agree with Gassendi that the small size of Mercury militated against its exerting influence. But he did not mean that it influenced the particular actions of men,

[108] *The Vanity of Judiciary Astrology, or divination by the stars.* Lately written in Latin by . . . P. Gassendus. Translated into English by a Person of Quality. London, 1659.

[109] F. Bernier, *Abrégé de la philosophie de Mr Gassendi,* Paris. 1675, in-4, 280 pp.: BN R.3554.

because he agreed with Gassendi that prediction of these rested on idle and frivolous principles.[110]

Jean Baptiste Morin (1583–1656), in the Apologetic Preface to his *Astrologia Gallica,* a work printed posthumously in 1661, spent some seven folio pages in rebuttal of Gassendi's attack upon astrology.[111] Although Gassendi was Morin's colleague in mathematics, he was, Morin avers, of all writers against astrology the most ignorant of that science and borrowed most of his arguments against it from its previous opponents without acknowledgement. Like them again, he indulged in questions, doubts, sarcasms, and exclamations of surprise rather than physical reasons against astrology. He misrepresented statements by Morin, misunderstood astrological technique, and was insufficiently read in the literature of the subject. He incorrectly held that the circles of the equator, ecliptic, horizon and meridian were imaginary and of no virtue, which is true only of the tropics, polar circles and colures.

When Morin died, Gui Patin wrote that it was fortunate that he and Gassendi were buried in different churches, so that they couldn't bite each other.[112]

Upon alchemy Gassendi made no such a general onslaught as his attack against judicial astrology. But he regarded the notion of a universal or catholic medicine and elixir of life as a chimera. Roger Bacon represented Artephius as living a thousand years by it, but didn't reach a hundred himself. Paracelsus, Khunrath and Fludd all failed to live long. The alchemists and Rosicrucians even promised immortality here on earth and regarded the glorified body after the Resurrection as attainable by their art, a shocking pretense to pious ears. The *Medicina Catholica* of incorruptible gold, having received the rays of the sun, readily dissolved into potable gold and bestowed homogeneity with the spirit or soul of the world, so that one could know and do anything and never perish.[113] Gassendi knew of no one who had succeeded in the transmutation of baser metals into silver or gold, but it was not to be utterly rejected, as there might be some truth in some reported cases of it, especially

[110] *Op. cit.,* pp. 66-67.

[111] *Op. cit.,* pp. ix-xvi.

[112] *Lettres* (1846), III, 67. Letter of November 7, 1656 (misdated 1655 in the edition of 1846).

[113] *Opera,* II, 614-16.

the famous nail of the Grand Duke of Tuscany of which that part had turned to gold which had been immersed in a gold-making liquor by someone whose name Gassendi could not recall,[114] or the pound of mercury turned into gold in the house of Thaddaeus Hagecius at Prague by the Englishman Kelley by infusing a single very red liquid drop.[115]

Gassendi decried divination from dreams[116] and said that it was useless to waste time in arguing against other forms of divination, when he had demonstrated that astrology, the chief of them all, was futile.[117] As the first volume of his collected works and the second section, *De rebus caelestibus,* of his *Physica,* had terminated with the long argument against astrology which we have already summarized, so its third section ended with a discussion of climacteric years, which he declared of equal vanity with the elixir of life.[118] He died nonetheless at sixty-three.

We have already seen him sceptical as to diabolical magic. He contended that the demon of Socrates was nothing but his own prudence and sagacity.[119] He believed that tales of long fasts, like those of possession by demons and bewitchment, were often based upon imposture.[120] He denied the power of imagination over other bodies, but was inclined to concede that a woman fascinator, with eyes and imagination intent on the tender body of an infant, might throw off maleficent rays and injurious effluvia and so affect its state of health.[121]

Gassendi laid down the fundamental principles that there is no effect without a cause, no cause without motion, and no action at a distance without contact. He therefore, in place of such explanations of the marvelous as occult virtues and qualities, sympathy and antipathy, signatures,[122] and like loves like, resorted to the action of atoms or corpuscles, imperceptible because of their

[114] The name was Thurneisser and his trick was to be exposed in the *Hippocrates Chymicus* of Otto Tachenius, 1666, where it is explained that the nail was half iron and half gold soldered together and the golden half given the color of iron by a wash which came off in the oil in which he immersed it.

[115] *Opera,* II, 142-43.

[116] *Ibid.* p. 421b.

[117] *Ibid.,* p. 854a.

[118] *Ibid.,* pp. 618-19.

[119] *Ibid.,* p. 857a-b.

[120] *Ibid.,* p. 616.

[121] *Ibid.,* p. 424b.

[122] He speaks slightingly of signatures at *Opera,* II, 166.

tenuity, as the material and physical cause of such phenomena. That shellfish fatten and that marrow increases in the bones of animals at full moon he explains, not by an occult influence of the moon, but by particles of moisture on the moon which are excited by sunlight and then borne by the sun's reflected rays to earth in greater number than at new moon. That sheep shun a wolf which they have never seen before is because of corpuscles shed by the wolf which are offensive to the sheep. But that a man is rendered speechless on sighting a wolf is ascribed by Gassendi to fright. Lucretius says that the reason a cockcrow scares a lion is that the corpuscles emitted by the cock hurt the lion's eyes.[123]

With all due respect to Lucretius and Gassendi, it must be said that there is more than one objection to this explanation. In the first place, what proof is there that the cock emits corpuscles? If so, why should they be any more injurious than those emitted by the hen, especially considering that the female of the species is more deadly than the male? In the third place, why is it that these injurious effluvia are thrown off only when the cock crows? In the fourth place, how and why do they injure the lion's eyes instead of his nose or ears or paws or mane? In the fifth place, why do Lucretius and Gassendi dodge the more obvious explanation that the sound of crowing startles the king of beasts, and adopt the extremely far-fetched theory that the effect of a noise is felt by an organ of vision? We cannot have much respect for Gassendi's attack upon astrology, when we find him swallowing hook, bait, line and sinker, such a feeble example of the corpuscular theory as this tidbit from Lucretius. Anyone could readily think up a dozen more plausible explanations. But just so long as it is corpuscular and atomistic, it is good enough for Gassendi.

He goes on to say that, just as a menstruating woman clouds a mirror, so an old-wife can injure an infant who is present, but not an absent person, by the malign material spirits which she emits. The stupefying effect of the torpedo, when only touched with a long pole, must be due to its emitting corpuscles which enter the pores and dull the spirits of the person affected.[124] The effect of

[123] *Opera,* I, 450a-451a, 456a, 451b, 453b, 454a.

[124] *Opera,* I, 454b, 455a.

poison results from penetration by subtle mobile particles, and its cure by binding on the wound the scorpion or spider or hair of the dog that bit you must be by their absorbing the poison in a sponge-like manner.[125] That victims bitten by a tarantula dance may be because the venom alters the temper of the body and especially affects the organ of hearing so that it acquires commensurateness with the sounds by which the tarantula itself is affected. Indeed, Kircher says that different kinds of tarantulas are excited by different kinds of music.[126] Snake-charming and the apparent effects of incantations may be produced by sound but not by the sense of the words uttered. But incantations may also help by giving the patient hope who has faith in them.[127] Unless the basilisk be a mere fable, it must emit deadly rays or spirits from its mouth as well as eyes.[128]

Gassendi, as the last sentence suggests, was inclined to reject as false some reported instances of occult virtue. He feared that there was little truth in the supposed antipathy between chords or drums of sheepskin and of wolfskin.[129] But he had such respect for the old belief, that the tiny echeneis or remora can stop a ship, that he devotes most of a long column to suggesting other possible explanations, such as the action of an adverse current.[130]

Duhem has pointed out that a theory, current since the thirteenth century, that the acceleration of falling bodies was caused by the air, was still held by Gassendi in his *Epistolae tres de motu impresso a motore translato* of 1640 and was abandoned by him only in 1645 in a letter to P. Casrée, "où la théorie actuelle de la chute accélérée des graves se trouva, pour la première fois, formulée d'une manière complète."[131]

125 *Ibid.*, p. 455b.

126 *Ibid.*, p. 454a.

127 *Ibid.*, p. 454b.

128 *Ibid.*, p. 453b.

129 *Ibid.*, p. 452b.

130 *Ibid.*, p. 455a.

131 Pierre Duhem, *Les origines de la statique*, I (1905), 139. The Jesuit father, Petrus Cazraeus (Pierre de Cazre) had addressed a letter of forty-four pages to Gassendi in which he opposed Galileo as to the acceleration of falling bodies: *Physica demonstratio qua ratio mensura modus ac potentia accelerationis motus in naturali descensu gravium determinatur adversus nuper excogitatam a Galilaeo Galilaei . . . de eodem motu pseudo-scientiam*, Paris apud J. du Brueil, 1645: BN R.3539. Bound with it is BN R.3540: *Petri Gassendi de proportione qua gravia decidentia accelerantur Epistolae tres quibus ad totidem epistolas R.P. Petri Cazraei S.J. respondetur*, Paris, apud

Gassendi does not entirely abandon the notion of occult quality in favor of the corpuscular theory. He grants that there is occult quality in general, as "the conspiracy of the parts of the universe" against the existence of a vacuum (i.e., the continuity of nature), and the influence of the celestial bodies upon these inferiors. But there is not so great a relationship and association between us and the heavens that our particular acts or sufferings are prescribed by them. And Galileo attributed the tides to the movement of the earth rather than to the influence of the moon.[132]

In treating of plants in particular Gassendi again availed himself of the term, occult or specific or fourth qualities, so called because they were not from the first qualities—hot, cold, dry and moist—or even from the secondary or tertiary manifest qualities, but came immediately from the form or whole substance or property of the whole substance.[133]

Gassendi opened his discussion of weapon ointment and sympathetic powder by declaring the effects attributed to them fabulous, and that reported cases of cure by them had really been worked by nature unaided. He repeated his axiom that nothing acts on a distant object. But even if there were a world soul to diffuse the force of the unguent, it would bear it to other wounds and not merely to that from which the blood came. When Helmont asked Gassendi what other explanation he could give, he suggested that as naphtha takes fire from a distant flame to which its vapor has penetrated, so there might be an insensible exhalation between the wound and the ointment, adducing the distance to which odors are diffused.[134]

Gassendi found it difficult to maintain an attitude of Epicurean scepticism in the midst of the marvel-mongering natural history of his day. It seemed ridiculous to him that a tree should grow with an iron pith which made an armor impenetrable to iron. Yet Norimbergius[135] wrote that he had learned from a trustworthy

---

Ludovicum de Heuqueville, 1646.

132 *Opera*, I, 451a.

133 *Opera*, II, 164a.

134 *Opera*, I, 456a-457a.

135 The author meant is probably Joh. Eusebius Nierembergius of Madrid, *Historia naturae maxime peregrinae libris xvi distincta, in quibus rarissima naturae arcana etiam astronomica & ignota Indiarum animalia ... plantae metalla lapides ... describuntur ...*, Antwerp, 1635.

man that in the island of Zeilan (Ceylon?) there was an herb with a marvelous power of drawing iron even at a distance, while Kircher told of an herb in Bengal which attracted and connected wood. Gassendi could only suggest that such phenomena might be accounted for "by attraction, partly magnetic, partly electric." And the fact that an iron knife was entirely consumed, if left for a single day in the Brazilian fruit Ananas (pineapple?), could hardly be accounted for by the action of corpuscles of corroding salts, since the fruit was sweet to the taste and innoxious to the system.[136]

Certainly there were worms in fruit on trees, but did trees in the Hebrides bear the birds known as Bernachiae or Klakies (barnacle geese), or did these grow out of driftwood or in shells? Yet Scaliger was authority that a shell not very large had been brought to king Francis with a little bird inside almost perfectly formed, with the tips of its wings, beak and feet adhering to the outer edge of the shell.[137]

Plants and animals which were generated spontaneously seemed to require only the putrefaction of matter, in which however were contained their seeds, although not manifest, as in other plants and animals.[138] And Gassendi still believed with Ovid that coral hardened as soon as it touched the air, but

mollis fuit herba sub undis.[139]

He did not know whether the statement about the herb *esula* was true, that if pulled up upwards, it purged by vomiting, if downwards, by the stool.[140] There was a plant in the Philippines of which the leaves which faced east were salubrious; those facing westward, poisonous.[141] An alder rod will bend above water underground because it is weighed down by exhaled vapor, but Gassendi

[136] *Opera,* II, 167a.

[137] *Ibid.,* p. 168a. "Scribit Scaliger fuisse concham non admodum magnam ad Franciscum regem optimum allatam cum avicula intus pene perfecta alarum fastigiis rostro pedibus haerente extremis oris ostraci."

[138] *Opera,* II, 114b.

[139] *Ibid.,* p. 118b.

[140] *Ibid.,* p. 165a. This superstitious passage is not found in the discussion of *Esula* in the thirteenth century *Herbal of Rufinus,* edited from the unique manuscript by Lynn Thorndike, assisted by Francis S. Benjamin, Jr., 1945, pp. 127-28.

[141] *Opera,* II, 165b.

doubts if a hazel rod will do the same for those seeking hidden veins of metal.[142]

Some virtues of plants which are called magical may be natural, as when a bull is tamed by the odor of a wreath worn around its neck, or an herb freezes water in summer because it contains salt and nitre. But others are quite fabulous, as Pliny says. The employment of words and ceremonies in plucking herbs either has no virtue or is diabolical magic. Gassendi accordingly "passes over" such impostures of the Magi as that they divined by the herb *theangelis,* evoked demons with the herb *aglaophotis,* and dispelled them with *hypericon,*[143] all of which he derives from Pliny. It will further be noted that he does not here express scepticism as to diabolical magic.

An extreme and mechanistic instance of using the corpuscular theory to explain emotional, aesthetic and moral reactions is found in a passage on why the eye turns away from an ugly or shameful thing.[144] Gassendi suggests that the visible species from such an object consist of corpuscles configured so that they penetrate and puncture the retina and force it to retire.[145] Here Gassendi seems guilty of inconsistency. For, if he excludes all but physical and material forces from the explanation of natural phenomena, he should keep these within their own bounds and not intrude them into psychic phenomena. But "thus is it ever." The patriots who repel the foreign foe from their native soil are seldom able to restrain themselves from pursuing him across the border, and may even annex a little of his territory. Gassendi's assertion in *Philosophiae Epicuri Syntagma* that sense is never deceived and that only opinion is a source of error[146] also is unacceptable. He further felt the necessity of some "lapifidic and seminal" force to account for the formation of stones and metals, and special juices, most pure and limpid, for the generation of gems.[147] But he did not express these in terms of corpuscles.

There is a certain discordance, not to say inconsistency or double

[142] *Ibid.,* p. 167a.

[143] *Opera,* II, 168b-169a.

[144] "rem turpem."

[145] *Opera,* I, 450b.

[146] *Opera,* III, 5-7. In his *Disquisitiones Anticartesianae,* on the other hand, Gassendi says "esse quidem omnem sensum fallacem at non omnem sensionem falsam": *ibid.,* 281.

[147] *Opera,* II, 114, 117.

meaning, in Gassendi's presentation of science against superstition and truth versus error, just as there was in his treatment of Epicurus, Epicureanism and atomism. In the latter case he professed to set forth not his own views but those of Epicurus. Conversely his own attitude toward magic, astrology and occult science is distinctly sceptical. Yet he not merely repeats at great length whatever classical writers have said on the subject, which may be literature but is often not science. He also is apt to give several instances of a belief or a practice only to reject them immediately as fabulous. But more space is given to the error than to its denial. Thus his content is not merely, like that of the naturalist Aldrovandi in the previous century, literary as well as scientific and philosophical; it also sandwiches a sceptical corpuscular philosophy in between slices of superstition and error. This is no doubt partly due to custom, to the difficulty in shaking off the tradition of the past; and partly to consideration, not entirely unselfish, for his audience. But sometimes it seems due to a lingering interest on his own part in what he rejects or ought to reject, but cannot quite part with.

Otherwise, why, treating *de multiplici foetu* and turning from other animals to man, does he put first, "whether you deem it true or fabulous," what is told of a certain countess of Holland named Joanna who delivered as many *foetus* at one birth as there are days in the year?[148] Or why, when not dwelling upon the powers attributed to stones for three reasons; first, because the greatest part of them are fabulous; second, that many others are dubious; third, that they often go beyond human shrewdness and perspicacity; does he not merely give several instances of the first and second groups, but add this long story anent the third? It concerns the power of the *chelidonium* or swallow-stone, "not that which is said to be found in the swallow's head (some say within its crop), but that which the swallows seek out" to restore the sight of their young. He "would not obstinately deny" that the eyes of swallows are of a sort that can be self-restored, but meanwhile he tells this little story.

An industrious young man named Antonius Agarratus Sammaxi-

148 *Opera,* II, 286a.

mitanus, in whose garret swallows nested, at 11 A.M. in the month of July pierced the eyes of the young with a needle so that humor flowed out. He then withdrew to watch what would happen, making a hole in the door so that he could see the window and nest. When the parent birds returned and the young neither peeped nor opened their mouths for food, the older birds finally withdrew from the nest and one sat patiently at the window, while the other flew away and did not return until 5 P.M. Sammaximitanus expected it to bear the herb *Chelidonia* (swallow-wort) in its beak, but it was something else. But he could see the swallow rub the young birds' eyes with it, and they gradually came to life again and after hardly a quarter of an hour were peeping and showing their hunger. He then examined the nest, saw that they had recovered their sight because they winked when he drew his hand over their eyes, although there was a white spot where he had pierced these with the needle. But what was the chief point, he discovered a little stone the size of a bean in the nest. Three days later he repeated the experiment on one bird with the same result except that the stone was only a third as large and more conical in shape, but similar in color.[149]

Or why does Gassendi detail a number of superstitious practices connected with the root of the herb mandrake?[150]

* * *

Although most of Gassendi's works were not printed until 1658, it was in 1654 that Walter Charleton (1619–1707), an Oxford M.D. and one of the first elected fellows of the Royal Society, published at London a work in English entitled, *Physiologia Epicuro-Gassendo-Charltoniana:* or, a Fabrick of science natural, upon the hypothesis of atoms, founded by Epicurus, repaired by Petrus Gassendus, augmented by Walter Charleton. In it he opposed the plurality of worlds. Van Helmont's notion that the rainbow is a supernatural meteor is characterized as delirium,[151] and visible species are called substantial emanations from the objects seen.[152] Light is corporeal

[149] *Ibid.*, pp. 121a-122a.
[150] *Ibid.*, p. 169a-b.
[151] *Op. cit.* p. 58.
[152] *Ibid.*, p. 136.

and is described as flame attenuated.[153] Sound too is corporeal and material.[154] Atoms are very small, but of different shapes, and keep moving.[155] These three characteristics are enough to explain the origination of all qualities.[156] This gives some indication of the general nature of the work, which considers such other usual topics as odors and savors, rarity and density, magnitude and figure, motion and gravity, heat and cold, fluidity, stability and humidity, softness and harshness, generation and corruption.

We turn to consideration of its attitude towards occult virtues and sympathy and antipathy as the feature most germane to our investigation. This section of the work has many passages similar to those already noted from Gassendi's works, as will become evident as we proceed.[157] The conceptions of occult virtue and of sympathy and antipathy are regarded as equally obstructive to the advance of natural science,[158] and Charleton abhors any such idea as that the tides are due to some immaterial influx from the moon.[159] He also rejects the attribution of cock-crowing to the sun and suggests several substitute explanations. It may be natural for the cock to crow as often as his imagination is moved by a copious and fresh afflux of spirits to his brain. Or he may have his set times of sleep and waking. Or the sudden invasion of increased cold soon after midnight may arouse him.[160] Shell-fish, the brains of rabbits, and the marrow in the bones of most land animals increase as the moon waxes, because it raises more mists—as we have already heard Gassendi argue more fully. That selenites or the moon-stone reflects the phases of the moon is probably because it has some thin, fluid and subtle matter, similar to quicksilver, which is altered by the moonbeams falling upon it. "The secret amities" of gold and quicksilver, brass and silver, may well be referred to a close correspondence between the particles which gold and brass emit and "the pores, inequalities and fastnings" in quicksilver and silver. "But what those figures are . . . is above our hopes

153 *Ibid.*, pp. 224, 207.

154 *Ibid.*, p. 208 *et seq.*

155 *Ibid.*, Book II, p. 84 *et seq.*

156 *Ibid.*, Book III, p. 127 *et seq.*

157 I read and selected the passages from Gassendi first, and have not gone back to either author for further parallels.

158 *Ibid.*, p. 343.

159 *Ibid.*, p. 349.

160 *Ibid.*, p. 351.

of determination."[161] So the relationship would seem to still be occult! Charleton, however, goes happily on to account for the discord between lutestring of sheep and wolf gut by their different contexture.[162]

Some traditional antipathies, however, he rejects as mere fables, such as that between drums of sheepskin and wolfskin, which Gassendi had rejected before him, the notion that a wolfskin placed near a sheepskin will soon consume it, and the belief that, if the feathers of an eagle are mingled with those of other birds, they will devour them.[163] If certain plants grow best near together, it is because they require different kinds of nourishment.[164] And the reason why all sheep run from a wolf is essentially that already advanced by Gassendi:

> When the Woolf converts his eyes upon a sheep as a pleasing and inviting object, and that whereupon appetite hath wholly engaged his Imagination; he instantly darts forth from his brain certain streams of subtle effluvias, which being part of those spirits whereof his newly formed idea of dilaniating and devouring the sheep is composed, serve as forerunners or messengers of destruction to the sheep; and being transmitted to his common sensory through his optick nerves most highly misaffect the same and so cause the sheep to fear and endeavour the preservation of his life by flight.[165]

We pass on to the mystery of the corpse bleeding at the reapproach of the murderer.

> The cruentation (and, according to some reports, the opening of the eyes) of the carcass of a murthered man, at the praesence and touch of the homicide, is in truth the noblest of antipathies: and scarce any writer of the secrets or miracles of nature hath omitted the consideration thereof.[166]

Some writers regard it as a miraculous and supernatural intervention of divine providence, but Charleton offers a natural explanation. In every vehement passion there is formed an idea of the object on which the imagination is most intent, and this

[161] *Ibid.*, pp. 352-53.
[162] *Ibid.*, p. 357.
[163] *Ibid.*, p. 358.
[164] *Ibid.*, p. 359.
[165] *Ibid.*, p. 363.
[166] *Ibid.*, p. 364.

idea is "impressed by a kind of inexplicable sigillation upon the spirits, . . . those angels of the mind." They in turn transmit the same idea to the blood and nerves and muscles. When emanations from the approaching murderer enter the pores of the corpse and affect its blood, which still has the idea of hatred for and revenge upon the murderer, it gushes forth—through the nose and mouth, if the victim was strangled or suffocated and there is no wound.[167] This explanation of an "idea . . . impressed by . . . inexplicable sigillation upon . . . angels" seems all too similar to the old doctrine of magic seals and astrological images.

With regard to "disanimation of the blood in living bodies by the mere presence" of the basilisk, Charleton like Gassendi is more sceptical and he wonders, "If natural historians have herein escaped that itch of fiction, to which they are so generally subject when they come to handle rarities." He will not accept that the basilisk destroys a man by merely seeing him first or that the basilisk is identical with the cockatrice, or that it is hatched from the egg of an old, decrepit cock, or that it has wings, legs, a long spiral tail, and a crest or comb like a cock. He also rejects the tradition that the sight of a wolf causes hoarseness and obmutescence, and the antipathies supposed to exist between lions and cocks, and elephants and swine.[168]

As to the effects of the bite of a tarantula, Charleton is much more credulous. It makes a man "dance most violently at the same time every year" as when he was bit, "till he be perfectly cured thereby, being invincible by any other antidote but Musick," which affects the spirits in the brain and so the whole body and attenuates the poison "by a way very like that of fermentation," which sets the patient dancing until the venom is expelled by a profuse sweat. Also different victims require different tunes and musical instruments, according to the type of tarantula that has bitten them and also according to their own temperaments. The melancholy need drums, trumpets and sackbuts; the choleric and sanguine are cured by stringed instruments. The musicians of Taranto seek out a tarantula like the one which bit the patient, find out what tunes the spider will dance to, and employ them with success upon the

[167] *Ibid.*, pp. 364-65. [168] *Ibid.*, pp. 365-67.

patient.[169] Charleton's discussion of the tarantula resembles Gassendi's but is longer and more detailed.

These data concerning the tarantula make it easy to believe that snake-charming with "a wand of the *cornus* or dog-tree" is natural, not diabolical, magic, and that the snake is affected by invisible emanations from the wand and not by the accompanying words and incantations.[170] Fascination of infants by an old crone is explained by effluvia, as it was by Gassendi. The action of the torpedo fish is accepted, but the supposed stopping of a ship by the tiny remora is ascribed to an adverse sea current, as it had been by Gassendi. Also Charleton refuses to believe that the remora or echeneis is ominous of death or disaster to the chief person aboard.[171] And he rejects the "armarie or magnetic unguent and its cousin german the sympathetic powder or Roman vitriol calcined" of the disciples of Paracelsus, Croll, Goclenius and Helmont,[172] in much the same terms as Gassendi.

Charleton had published a book against atheism in 1652.[173] He was also the author of Physical-Anatomical Exercises on the Animal Economy, which had several editions,[174] and of Pathological Exercises, printed in 1661.[175] Of these the former professed to be based on new hypotheses in medicine and explained mechanically; the latter, to proceed from new findings in anatomy. In it he held that the air was affected by unusual configurations of the stars, by the varied movement of the heavens, and by the rising and setting of outstanding stars; that the moon affected shellfish, the bones of other animals, and the tides; that certain effects such as crises in disease and pest years could be attributed only to the influence of

[169] *Ibid.*, pp. 367-70, and on to 372.

[170] *Ibid.*, p. 373.

[171] *Ibid.*, pp. 374-77.

[172] *Ibid.*, p. 380.

[173] Walter Charleton, *The darkness of atheism dispelled by the light of nature, a physico-theological treatise*, London, 1652, in-4, 355 pp. BN R. 7142.

[174] *Exercitationes physico-anatomicae de oeconomia animali novis in medicina hypothesibus superstructa et mechanice explicata*, ed. secunda, Amsterdam, 1659, in-12, xviii, 244 pp. Also Leyden, 1678, and The Hague, 1681. The first edition had the different title: *Oeconomia animalis* . . . etc., London, 1659, in-12.

[175] *Exercitationes pathologicae in quibus morborum pene omnium naturae, generatio et causae ex novis anatomicorum inventis sedulo inquiruntur*, London, 1661, in-4, xx, 268 pp. BN Td9.16.

the stars.[176] While he had accepted the influence of the moon on shellfish and the marrow of bones in the work of 1654, in other respects this appears to be more favorable than before to the influence of the heavenly bodies, and serves to accentuate a point which I have made repeatedly, that the same man will express varying views according to the standpoint from which he writes. Here in one case Charleton wrote as an Epicurean atomist and follower of Gassendi; in the other, as a medical man.

Such were some of the skirmishes, ambuscades, truces and alliances in the ebb and flow of warfare of Epicurean scepticism and atomism against the occult and magical tradition, marked, like most such struggles, by victories and retreats, by desertions or by being taken prisoners. Or perhaps we should shift the metaphor and say that the war was declared rather against the traditional philosophy of the schools, and that the occult and the marvelous were sometimes attacked as its allies, sometimes were courted as possible allies of the new trend.

[176] *Ibid.*, pp. 61-63.

CHAPTER XV

THE SCOTIST REVIVAL

Mastrius and Bellutus—John Poncius—Backward science—Astronomy—Attitude towards alchemy—Creation—Comets—Experiments *re* rarefaction and condensation—Other views—Impetus theory—Land and water—Influence of the stars—Attitude towards divination and astrology—Comets as natural signs—Poncius on the universe and celestial influence—On comets—Power of imagination.

Coelum per formam suam agit
—Duns Scotus

We hear of a revival of Scotism in the schools of Italy in the seventeenth century under the leadership of two young scholars from the south: Bartholomaeus Mastrius of Meldola (1602–1673) and Bonaventura Bellutus of Catania (1599–1676). They met as fellow-students at the College of Saint Bonaventura in Rome and afterwards taught together at Cesena, Perugia and Padua. They were in such remarkable agreement in interpreting the philosophy of Duns Scotus that their students said that they spoke with one tongue, wrote with one pen, and thought with one head.[1] Their

[1] Zedler lists commentaries by Mastrius on the *Physics* at Rome, 1637; *Organon*, Venice, 1639; *De coelo et meteoris*, *De generatione et corruptione*, both at Venice, 1640; *De anima*, Venice, 1643; *XII libri Metaphysicorum*, Venice, 1646. Mongitore, *Bibliotheca Sicula*, 1707, I, 113, article, "Bonauentura Bellutus," omits the *Metaphysics* and ascribes the other commentaries to both men.

The library of Columbia University has: *Disputationes in Aristotelis Stagiritae libros Physicorum quibus ab adversantibus tum veterum tum recentiorum iaculis Scoti philosophia vindicatur a ... Bartholomeo Mastrio de Meldula ... et Bonaventura Belluto de Catana ... In hac secunda editione ... additionibus ... locupletatae*, Venetiis, typis M. Ginammi, 1644, 1028 pp.

reputation as revivers of Scotism has perhaps been somewhat exaggerated. Philip Faber of Faenza of the Order of Minorites had printed a volume on the natural philosophy of Scotus at Venice in 1602 in quarto;[2] on the *formalitates* of Scotus at Paris in 1604;[3] on the philosophy of Scotus in 1616;[4] and theological disputations based on Scotus at Venice in 1613 and 1619, and at Paris in 1620.[5] Gaspar de Fontis (Fueñtes), an Observantine of the same Order, had issued at Lyons in 1631 lectures on the dialectic and physics of Scotus which he had delivered at Rome before he became professor of theology at Alcalà.[6] No doubt the schools of the Franciscan Order maintained chairs of Scotism all along.[7] But the better known works of Mastrius and Bellutus may serve to illustrate such teaching. We shall also take some note of the Complete Course of Philosophy according to Scotus by John Poncius, who was at times inclined to disagree with Mastrius.

John Poncius was an Irish Franciscan who, after studying philosophy and theology at Cologne and Louvain, was transferred

---

Also *RR. PP. Bartholomaei Mastrii de Meldula et Bonaventurae Belluti de Catana Ord. Min. Conuent. Magistrorum Tomus Tertius continens Disputationes ad mentem Scoti in Aristotelis Stagiritae libros De anima, De generatione et corruptione, De coelo et Metheoris. Editio novissima a mendis innumeris quibus priores scatebant repurgata,* Venetiis, 1688 apud Nicolaum Pezzana, in-fol., 598 double-columned pp. with 86 lines to a column. Three pages in long-hand have been substituted for p. 562, and again for p. 571. By an error the page numbers 514 and 515 have been repeated, paragraphs 53-66 falling on the first 514-15, and paragraphs 67-82 on the second two pages thus numbered. The subject of local motion, as is explained in the concluding passage of the Disputations on the *Physics* at p. 1028 of the 1644 edition, was deferred to those on *De coelo.*

BN R. 327-331 (five vols. in 3) is a 1678 edition by the same printer, Pezzana. The BM printed catalogue lists only later editions of Venice, 1708 and 1727.

[2] BN R.1778.

[3] BN R.38088.

[4] Leo Allatius, *Apes Urbanae,* 1633, p. 228.

[5] BM 3837.h; BN D.1593-1594.

[6] *Quaestiones dialecticae et physicae ad mentem Scoti (quas Romae legit),* Lyon, 1631, in-4: Allatius, p. 120. Not listed in printed BM and BN catalogues.

[7] Later in the century come M. de Villaverde, *Tractatus in octo libros Physicae, in quo sententiae Scoti proponuntur . . . ,* 1658, in-4; Rabesanus de Montursio, *Cursus philosophicus ad mentem Doctoris Subtilis Ioannis Duns Scoti . . . ,* 1665, in-8; and *Scotus Academicus,* published by Claude Frassen, Paris, 1672, in-fol, and Paris, 1677, 4 vols. in-fol., also at Venice, 1744, 12 vols. in-8.

by the influence of Luke Wadding to the Collegium Romanum (or, S. Isidori) of the Irish Franciscans. There he covered twice in his lectures the entire course in philosophy, and taught theology for many years.[8] He then went to the convent in Paris, where he died about 1660.[9] His lectures in philosophy were first printed at Rome in 1643,[10] at Paris in 1648[11] and 1656,[12] and, after his death, at Lyons in 1672.[13] The last is the edition here cited and shows, like the 1688 edition of Mastrius and Bellutus, what views still prevailed in the Order at that late date.

Scientifically the views of Mastrius and Bellutus sometimes are quite backward. They still hold with Scotus and Aristotle that the heart, not the brain, is the seat of the soul. They inquire concerning the soul separated from the body: what it loses and what it retains in the way of knowledge and method of knowing, what it knows and how, its place and local motion.[14] They accept the existence of animal spirits, but so did most scientists even to the end of the century. They count the number of simple bodies as five, namely, the heavens and traditional four elements,[15] although noting the objection of the followers of Telesio that fire is moist,[16] the position of Cardan—since adopted by Tycho, Kepler, Patrizi and others, that air extends all the way to the sphere of the moon,[17] and the argument of Galileo in *Il Saggiatore* that the polished surface of the sphere of the moon could not draw fire with it in its circular movement.[18] They also retain the four traditional qualities. Similarly Poncius keeps the fifth essence and four elements and four

[8] Wadding, *Scriptores,* Rome, 1906, p. 149b. BM 3835.f is his *Integer theologiae cursus ad mentem Scoti,* Paris, 1652.

[9] Sbaralea, *Supplementum,* Rome, 1921, II, 118b.

[10] "Apud Lodovicum Grignanum, sumptibus Hermanni Scheus": Wadding, *loc. cit.* See above Chapter 13, note 170, for a misdated edition.

[11] "Tomo uno in folio sumptibus Antonii Bertier . . . auctior et correctior": Wadding, *loc. cit.*

[12] Sbaralea, *loc. cit.*

[13] The word order of the title now changed from *Integer philosophiae cursus ad mentem Scoti* to *Philosophiae ad mentem Scoti cursus integer,* Lyon, 1672, in-fol., 974 pp. Copy used: BN R.1124, pp. 373a-715a, Physica.

[14] *De anima,* Disp. VIII, Qu. II, 12, 17, 24: ed. of 1688, III, 261b-266a.

[15] *De coelo,* Disp. I, Qu. I, Art. ii, 9; ed. of 1688, III, 502.

[16] *De gen. et corr.,* Disp. I, Qu. IV, Art. i, 71; III, 292b.

[17] *De coelo,* IV, II, i, 9; III, 561a.

[18] *Idem,* ii, 23; III, 564a.

primary qualities, but doubts if fire is immediately below the sphere of the moon.[19]

Mastrius and Bellutus place the earth at rest in the center of the universe and will not even admit that it revolves on its axis.[20] Poncius likewise rejects the Copernican theory.[21] They accept, however, the movement of Venus and Mercury about the sun rather than the earth and note Tycho Brahe's observations that Mars is sometimes beyond the sun, sometimes nearer to the earth than the sun is, which was fatal to the belief in fixed spheres for the different planets.[22] They recognize the existence of the four moons of Jupiter, two satellites of Saturn (really portions of the rings), sunspots, and irregularities on the surface of the moon.[23] Indeed, Fontana with his telescope had recently observed similar cavities and eminences on Jupiter and Mars.[24] They cite Scotus (IV, Dist. 14, Quaest. i) that the first matter of the heavens is the same as that of sublunars.[25] Poncius adds that the heavens can be incorruptible, even though they are of the same matter as sublunars.[26]

In living beings immediate resolution into first matter is not possible, but it may be accomplished in other compounds, a conclusion favorable to the transmutation of metals.[27] Galen and Agricola are cited that lead kept in damp places increases in weight, the explanation being that this occurs "by conversion of damp air or aqueous humor" into the lead's "own nature." Scaliger made a like assertion as to rock salt; Ficino and Fernel deduced that this virtue was found in all stones and metals; Cardan extended it to all bodies.[28] Poncius affirms that it is no more difficult to make gold than to generate a worm from horsehair, which last is attested

[19] Poncius (1672), 687a, 614, 691b, 698b.

[20] *De coelo*, I, II, 16 *et seq.*; IV, IV, iii, 112 *et seq.*; III, 503a, 584b.

[21] Poncius (1672), 625a.

[22] *De coelo*, II, I, ii, 27-28; III, 510b.

[23] *Idem*, II, iii, 90-91; III, 516b-517a.

[24] *Idem*, 96; III, 518a. The reference is presumably to Francisco Fontana, *Novae coelestium terrestriumque rerum observationes et fortasse hactenus non vulgatae*, Naples, Gaffaro, 1646. He claimed to have constructed a telescope in 1608 and to have invented subsequent improvements: *ibid.*, p. 20.

[25] *De coelo*, II, II, ii, 57; III, 514b.

[26] Poncius (1672), 617a.

[27] *De generatione et corruptione*, V, III, i, 78 *et seq.*, ii, 89 *et seq.*; III, 376, 379.

[28] *Ibid.*, VII, V, 52: III, 424b.

by daily experience, or to make iron from a stick immersed in certain waters, "which many most serious historians testify occurs in our Ireland." But Poncius doubts whether present alchemists are able to make gold, and cites the decretal, *Spondent* . . .[29]

It was the opinion of Aristotle that the world was without beginning, but the common opinion of Catholics asserts that it could be, and as a matter of fact was, produced in time, although they disagree as to just when and how it began. Augustine and Cajetan say that everything was created instantaneously, but the Bible and almost all of the Church Fathers and Schoolmen state that this occupied the space of six days. As for the question at what time of the year it occurred, Mastrius and Bellutus prefer the vernal equinox to either July or the autumnal equinox.[30]

They still insist that comets, if they are natural phenomena, are sublunar, although those of divine origin and miraculous character may be in the heavens. They are aware that Tycho Brahe and other recent astronomers have placed comets above the moon, but they contend that measurement by parallax is not so trustworthy for comets as for the stars, and assert that other eminent astronomers have shown by mathematical instruments that comets are below the moon. They discuss comets under meteors rather than the heavens in a disputation concerned with heavy and light objects.[31] Poncius too, although giving seven other opinions, adheres to that of Aristotle that comets are terrestrial exhalations.[32]

As we have already seen, our authors are fairly well informed as to recent scientific opinion, although they may not agree with it. Mastrius and Bellutus refer to recent experiments with glass tubes and vessels bearing on rarefaction and condensation,[33] and do not hesitate to cite such authors as William of Ockham and Pomponazzi, although they do not refer to Richard Suiseth's discussion of rarefaction and condensation. Poncius says that rarefaction of a body is produced by the reception of imperceptible

[29] Poncius (1672), 456a. For the decretal of John XXII see T III, 31, 48-49, 515.

[30] *De coelo*, I, V, 36 *et seq.*; *Opera*, III, 505a-6a.

[31] *De coelo et meteoris*, II, VII, iv, 225; IV, II, iv, 34, 39, 47; *Opera*, III, 537b, 566b-567b, 569b.

[32] Poncius (1672), 705b.

[33] *De generatione et corruptione*, IV, II, 9-12; III, 335a-336a.

corpuscles within its pores and their consequent dilation, while condensation comes from withdrawal of those corpuscles and sensible compression of the pores.[34]

An example of the richness of their citations may be given in connection with the doctrine of Scotus that "by intension a new quality is acquired and lost by remission." Again Suiseth is not named, but it is stated that this opinion is followed by all the Nominalists as well as the Scotists, and among more recent writers by

> Fonsec. 8 *Met.* cap. 3, quaest. 2; Vasqu, citat. Suarez, disp. 46 *Meth.* sect. 1; Ruu. 1 *de gen.* tract. 3, quaest. 5; Hurt. disp. 5; Ares. quaest. 25; Morisanus, disp. 2, dub. 4; Murcia, disp. 2, quaest. 4; Arriaga, disp. 5, sect. 1; Aversa, quaest. 23, sect. 6; Coinch. *de actibus supernat.* disp. 22, dub. 3; Malderus, 2, 2, quaest. 24, art. 5; Azor. libr. 3, cap. 22; Tolet. 4 *Phys.* quaest. 12; estque communis.[35]

Despite such evidence of wide reading along certain lines, Mastrius and Bellutus discuss the veins and arteries without mentioning the circulation of the blood.[36]

On the other hand, they are aware, as indeed were men in the days of Scotus, that vision is by intramission rather than extramission of rays.[37] They reject the Platonic doctrine of the *magnus annus* and the transmigration of souls of Pythagoras.[38] They accept spontaneous generation within limits and quote Fortunio Liceto upon that subject.[39] Animals like horses can be generated only from seed and sexual intercourse. Insects such as bees are produced only from putridity without seed; but mice are multiplied in both ways. In recently built ships they are first generated spontaneously, but then breed rapidly.[40] Many theologians hold that the tree of life possessed a natural virtue of preserving human life in perpetuity. But our Scotists deny this, and for the same reason deny that there is such a virtue in the medicines of the alchemists.[41]

[34] Poncius (1672), 685b.

[35] *De gen. et corr.*, III, II, 11; III, 319a. Poncius (1672), 654a, "Quando fit intensio qualitatis?"

[36] *De anima,* V, XII, ii, 357; III, 136b.

[37] *De anima,* V, II, iv, 80-81; III, 93b-94a.

[38] *De gen. et corr.*, V, VII, ii, 167-68: III, 398b-99a.

[39] Fortunius Licetus, *De spontaneo viventium ortu libri iv,* Vicenza, 1618.

[40] *De gen. et corr.*, VIII, III, 36 *et seq.;* III, 435-36.

[41] *Ibid.*, VIII, IX, 135; III, 456b.

Aristotle's explanation of violent motion is modified by the medieval theory of impetus, that the continued movement of projectiles comes chiefly not from the medium through which they move, but from an impulse impressed on them by the thrower. But Mastrius and Bellutus concede that the medium concurs in such motion, since the air is sucked in behind the missile in order to prevent a vacuum. This confirms them in the old error that violent motion attains a greater velocity in mid-course than at the start.[42] Poncius, too, adopts the impetus theory as "the commonest opinion of more recent writers," although he recognizes that Aristotle, Albertus and Aquinas, Scotus and Camerarius, held otherwise. He also believes that the motion of projectiles is swifter in mid-course, after which the impressed quality begins to be remitted.[43]

Mastrius and Bellutus repeat the very common view that mountain peaks transcend the second region of air and occupy the third or uppermost region. Land is higher than the sea, or at least such mountains are, but mountains are also found in the sea. The Bible does not deny that there were other rainbows before the flood.[44] Poncius sets forth three movements of the sea: from north to south, from east to west, and the tides, which Scotus, following Albumasar, attributed to the magnetic influence of the moon. Aristotle recorded the opinion of some ancients that springs and rivers are the result of rainfall, but himself ascribed their origin to new generation of water within the earth from air and vapors. Poncius prefers the sea as their chief source.[45]

It is with such a scientific and unscientific, credulous and critical, ancient, medieval and recent background and equipment, that our commentators approach the question of the influence of the heavens and stars, and the validity and scope of judicial astrology.

In a disputation as to the power of the imagination Mastrius and Bellutus incidentally note the influence of the stars upon grapes and wine, and speak of dreams caused by the stars and which may

[42] *De coelo et meteoris, Disp. III de gravibus et levibus,* Quaest. V, paragraphs 65-70. III, 552a-553a.

[43] Poncius (1672), 592 *et seq.;* Disp. xx, De motu proiectorum.

[44] *Disp. IV de elementis,* III, i, 67, IV, i, 100, III, iii, 92; III, 574b, 582a, 580b. On the history of measuring the height of mountains see my query and Florian Cajori's long article in *Isis,* IX (1927), 425-26; XII (1929), 482-514.

[45] Poncius (1672), 712b-713b.

have future significance as to the weather and bodily health.[46] In commenting on *De generatione et corruptione* they state that Scotus denied that God moved the heavens immediately;[47] in other words, he accepted their being moved by intervening Intelligences. But it is especially in their Disputations on *De coelo et mundo* and *Meteorologica* that they first tacitly assume and then openly declare specifically different influences exerted by the individual planets.

> The planets produce diverse and contrary effects. For Saturn is cold and dry and causes the same temperament in the new-born babe, so that those born under Saturn are melancholic, thoughtful, timid, saying little, serious, solitary, avaricious and toilsome, shunning merriment.[48]

After remarking that the heavens are incorruptible extrinsically, since no natural agent is strong enough to corrupt them, but corruptible intrinsically, because composed of form and matter, and that their movers, whose number is debated, are extrinsic, incorporeal and spiritual Intelligences, operating by knowledge, Mastrius and Bellutus flatly assert that it is clear to everyone that the celestial bodies act upon and influence these inferiors.[49] And this not merely as universal causes concurring with sublunar causes, but also as special and total causes. Not only the planets, but also the fixed stars and the primum mobile itself exert such influence. Moreover, "we say that the stars can produce the bodily form of imperfect living beings totally," in other words, generate without intervening parents and seed.[50]

Foreknowledge of the future which is based upon knowledge of natural phenomena is unobjectionable. Such occult arts as necromancy and geomancy are vain superstition and become effective only through the aid of demons. But physiognomy, metoposcopy and chiromancy are permissible, if it is true that the stars are responsible for the lines in the hand and for other physical features

[46] *De anima,* V, XII, ii, 368; i, 352; III, 138b, 136a.

[47] *De gen. et corr.,* VIII, III, 43; III, 436a: "Scotus ipse negavit Deum immediate movere coelum."

[48] *De coelo et meteoris,* II, II, iv, 100; III, 518a. Similar passages follow concerning the other planets.

[49] *Idem,* III, 107, IV, iii-iv, VII, 184; III, 519a, 523a-4a, 530a, citing Scotus 2, dist. 14, quaest. 3.

[50] *Idem,* VII, 184; i; 185; ii, 201; III, 530a-b, 533a.

of the body. Of these three arts only chiromancy was definitely condemned by Sixtus V in his bull against astrology and divination.[51]

It is admitted that astrology is a difficult subject, and that there is disagreement among both astronomers and astrologers. But this does not mean that it is an impossible science. Some go so far as to attribute the Flood, Incarnation and Passion to the stars, the origin of Judaism to Saturn, that of Islam to Venus, and so on. At the opposite extreme are those who deny that it is possible to predict anything except the movements of the stars themselves and such celestial happenings as eclipses and conjunctions. The middle course, in which theologians and philosophers agree (*ita communis Theologorum sensus et Philosophorum*) concedes that natural effects in this sublunar world may be foreknown, but denies that those dependent on free will can be, except in general events as a matter of inclination (i.e., mob psychology). The time of natural death can be predicted, but the person may die sooner from other causes. A man's inclinations, natural aptitudes, temperament and physical constitution can be forecast and his consequent prospects of happiness or unhappiness, although heredity, food and education may qualify this somewhat. A good or bad outcome in business undertakings (*in negotiis peragendis*) may be predicted fairly well, but only as a moral rather than physical certainty, since the element of free-will enters in here. Neither fortuitous events nor those which depend upon our will can be certainly predicted from the stars, and the bulls of Sixtus V (1586) und Urban VIII (1631) against judicial astrologers forbid them to predict such happenings even as probable. But these bulls allow the exercise of astrology in medicine, agriculture and navigation. Our authors admit, however, that astrologers are the more likely to err, the more they descend to particulars.[52]

After this rather slight recognition and cavalier dismissal of the papal bulls against judicial astrology, our authors turn to Biblical passages concerning the stars as signs. They further note that, in a certain volume entitled The Narration of Joseph, it is reported

[51] *Idem,* iii, 212; III, 535a. For the bull of Sixtus V see T VI, 156-58.

[52] *De coelo et meteoris,* iii, 213-219; III, 535a-536b.

that the patriarch Jacob said to his sons, "Read in the tables of the sky whatever will happen to you and your children, where there are recorded even fortuitous and free effects." But the Church does not accept that volume as canonical. Also Augustine in *Genesis Against the Manicheans* teaches that even our thoughts are not unknown to the celestial bodies. But here by heavenly bodies he means the saints above.[53]

Comets are called natural signs of sterility and disease because they result from an abundance of many exhalations in the air, by which men's humors are altered. Comets are called signs of the death of the prince, because they are sometimes used by divine providence to warn men and announce some future calamity, so that men may repent. And since the prince is of common interest, and the kingdom is easily upset by his death, therefore comets are quite properly said to announce death to princes. They are not natural signs of effects dependent on free will, nor even invariably signs of sterility, pest and death of princes.[54]

"In the opinion of Scotus it is clear that the stars . . . by their own substantial forms produce immediately sublunar substances."[55] A long debate follows of the old problem whether, if the movement of the heavens stopped, all motion and effects would cease also. The final conclusion is then reached that the stars act by occult qualities as well as by their movement and light.[56]

It will hardly do to try to discount the taking up of such positions as behind the times and harking back to Scotus and the fourteenth century, for editions of the huge folios of Mastrius and Bellutus continued to appear into the sceptical eighteenth century and age of reason.

Mastrius and Bellutus terminate their discussion of the heavenly bodies with the statement that if they have asserted anything which does not square with truth, it is not surprising, and that they deserve to be pardoned therefor, quoting the Book of Wisdom (ix, 16), "We esteem difficult those things which are on earth, and we discover with labor what lie before our eyes. So who will investigate

[53] *Idem*, 223-224; III, 537a.
[54] *Idem*, 225; III, 537b.
[55] *Idem*, 226; III, 537b.
[56] *Idem*, 246; III, 540b.

those things which are in the heavens?" and Job (xxxviii, 37), "Who has known the ways of the heavens?"[57]

Poncius holds that the world cannot perish naturally but only by a miracle, and that it will last to eternity. "This is the common opinion of theologians and philosophers." Also the motion of the heavens ought not ever to cease. The heavens are a simple substance distinct from the elements, are not animated, are naturally incorruptible and can be so, even if they are of the same matter as sublunar bodies. He favors solid heavens but grants that the Bible leaves the question unanswered. The empyrean heaven is the only one other than the eight containing the planets and fixed stars. They are not moved of themselves but either by God alone or by angels.[58]

But neither God alone nor separate Intelligence alone nor the celestial bodies alone produce all the corporeal substances which are produced anew, for sublunar corporeal agents also participate in such production.[59] A created agent can act immediately at a distance and not merely mediately. So the celestial bodies bring forth minerals in the bowels of the earth; so fire will kindle tow at some distance away; so the sun heats sublunar things but not the intervening celestial spheres.[60] The celestial bodies have various influences on sublunars and concur in the production of all living beings including man, disposing matter to receive souls, but without affecting immediately the production of any soul, even in imperfect animals. They do not have any direct action upon the rational soul, but affect both intellect and will indirectly.[61]

Poncius does not enter into consideration of astrology and other arts of divination as Mastrius and Bellutus did. His treatment is more general and less detailed. He allows somewhat less direct and sweeping influence to the heavenly bodies. But he still allows them a great deal and enough to support moderate astrological prediction.

[57] *Idem*, 248: "Quis cognovit coelorum rationes?" whereas the Vulgate reads: "Quis enarrabit caelorum rationem?"; III, 540b.

[58] Poncius (1672), 609-10, 614-615b, 617a, 618b, 620-22, 626a. Yet he has said that angels are not an object of physical consideration, p. 374a.

[59] Poncius (1672), 467a.

[60] Poncius (1672), 589b-590a.

[61] Poncius (1672), 630a-633a.

Poncius notes that Scotus listed various effects or rather significations of comets, namely: winds, heat, earthquakes, crop failures, floods, great mortality of beasts, contentions and wars, death of princes, political change and religious change. These had been proved by the observations of the past. Poncius doubted if any one comet would produce all these effects and thought it not easy to see a natural connection between a comet and these effects, although Scotus tried to trace them to dry and hot exhalations.[62]

The old problem of the power of imagination was again discussed by Mastrius and Bellutus. The commentators on Aristotle at the Portuguese university of Coimbra, had followed Avicenna and Algazel in holding that one's imagination could produce physical effects not only in one's own body but on other bodies, as in fascination, impressions produced by parents on the body of their offspring—especially by the mother during the period of gestation, and even rain and storms, though this power was not conceded to every person. Our Scotists agree that the humors in one's own body may be affected by imagination, but deny that it can act directly on another body. Such apparent cases of physical effects are really divine miracles or worked by magic art and demon aid. Fascination is not the result of imaginative virtue but of vapors from the eye infected with some malignant humor which injures the body of another person. Imagination may, however, affect the foetus and child by affecting the formative virtue of the parents either *in semine deciso* or in the womb.[63]

[62] Poncius (1672), 706b.

[63] *De anima*, VI, XII, ii, 367-68; III, 138a-b.

CHAPTER XVI

MORIN'S *ASTROLOGIA GALLICA*

Time of writing—Other works—Conversion to, and practice of astrology—*Praefatio apologetica*—Outline of the text—Natural basis of astrology—Its principles and technique—Astrological images rejected—Manuscripts on weather prediction—Astrological medicine and chemical remedies—Virtues of gems—Probably slight effect of Morin's book.

Futura praedicere proprie divinum est
—MORIN

Well-nigh the last attempt upon a large scale to defend, rehabilitate and reconstruct judicial and genethliacal astrology was made by Jean Baptiste Morin (1583–1656), an M.D. (Avignon, 1613) and royal professor of mathematics at Paris, 1630–). His *Astrologia Gallica* is a monumental work, in the edition of 1661 comprising a *Praefatio Apologetica* of 36 folio pages and 784 more pages of text mostly double columned.[1] In a preliminary preface to the reader by "G.T.D.G.V." in this posthumous publication it is stated that the work was finished in 1648, but was delayed by the Fronde and other circumstances. The book as published, however, refers to events after 1648 and up to 1656, such as a

[1] *Astrologia Gallica principiis & rationibus propriis stabilita atque in xxvi libros distributa . . .*, Hagae-Comitis, Ex typographia Adriani Vlacq, MDCLXI. This appears to be the first and only edition, although Cornelius van Beughem, *Bibliotheca mathematica*, Amsterdam, 1688, p. 96, lists an edition at The Hague, 1656 & 1660 in fol. But he is probably referring to our edition.

The division of the text into sections and chapters seems to have been somewhat altered by the posthumous editor. Thus in chapter 3 of Sectio iii of Book XII (p. 276a) we read, "iam probatum est cap. 27." But the preceding Sectio ii has only 26 chapters, so that apparently what was its 27th has become the first chapter of Sectio iii. Similarly at XXVI, i, 3 (p. 760b) the text refers to the third chapter of XII, iii, of which we have just been speaking, as cap. 29.

ten-page refutation of a book on the Pre-Adamites, of which "the report reached me in January of this year 1656."[2]

The book on the Pre-Adamites was the work of Isaac de La Peyrère (1594–1676), librarian of Condé. It was condemned by the Parlement of Paris, and its author arrested and imprisoned. Condé procured his release and he disavowed the book and abjured Protestantism. Miron wrote of him:

> La Peyrère ici git, ce bon Israelite,
> Huguenot, Catholique, enfin Préadamite.[3]

An earlier instance of Morin's tendency to let slip no opportunity to demonstrate his orthodoxy is seen in his publishing a refutation of the fourteen alchemical theses against Aristotle which aroused the ire of the Sorbonne and Parlement of Paris in 1624.[4]

Earlier astrological compositions by Morin were on astrological houses[5] and restoring astrology,[6] and he had published works on longitudes[7] and the restoration of astronomy,[8] as well as a dissertation on atoms and vacuum against Gassendi.[9] The dedication to Richelieu of *Astronomia restituta* is dated July 26, 1634, while its seventh part with a separate title page is dated 1638, and the eighth and ninth parts in 1639, both at the author's expense and for sale at his house.[10] The book evoked some criticism from

[2] *Ibid.*, II, chap. 35, pp. 58b-68. Morin also published separately *Refutatio compendiosa erronei ac detestandi libri de Praeadamitis*, Paris, 1656, In-12, 71 pp.: BN D.45113.

[3] *Lettres de Gui Patin,* ed. Paul Triaire, (1907), p. 328.

[4] *Réfutation des thèses . . . d'A. Villon dit le soldat philosophe et E. de Claves . . . contre la doctrine d'Aristote,* Paris, 1624, in-8. In his *Nova mundi sublunaris anatomia* of 1619, however, he had spoken at p. 29 of "Paracelso caeterisque verae philosophiae cultoribus."

[5] *Astrologicarum domorum cabala detecta,* Paris, 1623.

[6] *Ad australes et boreales astrologos pro astrologia restituenda epistolae,* Paris, 1628.

[7] *Longitudinum terrestrium necnon coelestium nova et hactenus optata scientia,* Paris, 1634; and later works in 1636, 1637, 1639, 1647.

[8] *Astronomia iam a fundamentis . . restituta,* 1640; and again in 1657.

[9] *Dissertatio de atomis et vacuo contra Petri Gassendi philosophiam Epicuream,* Paris, 1650; followed by a *Defensio suae dissertationis . . .*

[10] In the copy which I used, BM 533.e.13.(1), *Astronomia iam a fundamentis integre et exacte restituta, complectens IX partes hactenus optatae scientię, longitudinum coelestium necnon terrestrium,* there is then pasted over the lower half of the initial title page a printed slip reading, Paris apud Petrum Menard via Veteris Enodationis iuxta terminum Pontis D. Michaelis sub signo Boni Pastoris, MDCLVII.

Longomontanus (1562–1647) who had been Tycho Brahe's assistant, since 1605 professor of mathematics at Copenhagen, and more recently author of *Astronomica Danica.* Morin replied in 1641,[11] was answered by Fromm in 1642,[12] and replied again to Fromm.[13] It should not be assumed that Morin was without scientific ability. The method of determining longitudes at sea by the distance of the moon from a certain star was "brilliantly developed" by him and for some time enjoyed great vogue, concurrently with Galileo's method of observing the satellites of Jupiter.[14]

Morin tells us himself that he was forced to study astrology unwillingly some forty years ago by a bishop whose physician he was. For ten years he pursued it empirically and could make no sense of it, but finally discovered principles which should satisfy every rational inquirer, and serve to distinguish what is true and what is false in the art, as previously taught and practiced. He declares that these true principles of astrology had not been stated by Ptolemy or anyone else until himself.[15] It seems fairly evident that he is attempting to follow in the footsteps of Descartes. He was praised by the aforesaid "G.T.D.G.V." as having demonstrated astrology more surely and evidently than Aristotle demonstrated physics, or Galen demonstrated therapeutic.

Just as Henri IV had summoned the physician and astrologer, Larivière, to the birth of Louis XIII, so, at the birth of Louis XIV, Morin was concealed in the royal apartment to draw up the horoscope of the future Grand Monarque.[16] Later he selected the favorable astrological hour and minute for the trips of M. de Chauvigny, secretary of state during the early years of the reign, and also the times when he would be well received at foreign courts. He is further said to have failed to predict Chauvigny's imprison-

[11] *Coronis astronomiae iam a fundamentis integre et exacte restitutae qua respondetur ad Introd. in theatrum astron. clarissimi viri Christiani Longomontani . . . ,* Paris, 1641. BM 533.e. 13.(2.).

[12] Georgius Frommius, *Dissertatio astronomica de mediis quibusdam ad restituendum astronomiam necessariis,* Hafniae, 1642, in-4, BM 531.k.17.(4.).

[13] J. B. Morin, *Defensio astronomiae. . . restitutae. . . contra G.F. diss. astron.,* 1644, in-4. BM 533.e.13.(3.).

[14] *Correspondance,* III, 381.

[15] *Astrologica Gallica,* Praefatio Apologetica, p. v.

[16] L. F. A. Maury, *La magie et l'astrologie,* Paris, 1860, p. 215.

ment.[17] Vautier, who was physician to Louis XIV, tried to have Morin made royal astrologer, but his proposal did not go through.[18] There were French ordinances of 1493, 1560, and 1570 against astrology, but they seem to have become dead letters. Morin himself tells us that he owed to astrology his appointment in 1630 by Marie de' Medici to a royal professorship in mathematics. The art had further enabled him to support two nieces in the best nunneries and to marry off a third.[19] Mazarin gave him a pension of 2000 livres, and the queen of Poland contributed 2000 Thaler to the printing of *Astrologica Gallica.*[20]

Morin interprets the Council of Trent's Rule 9 of the Index of Prohibited Books and the Bull of Sixtus V of 1586 as condemning only the prediction of fortuitous events and those contingent upon human free will.[21] He then devotes several pages to explaining away passages of the Bible which had been adduced against astrology.

The devil, in order to defame true astrology, has given the impression that the old arts of divination are mixed up with it, and servers of the devil pretend to be astrologers.[22] As for the relation of religion to the stars, no sane person will ascribe to the stars religions which are of diabolical origin. But man-made religions like Islam, Lutheranism and Calvinism may be referred to the stars insofar as these affected the characters of Mohammed, Luther and Calvin. But we find Jews and Christians among the Turks; the Chinese and American Indians are being converted to Christianity; and so it seems a stupid and impious dogma which holds that religions and especially Christianity are caused or ruled by the

[17] Zedler. He also, however, credits Morin with having predicted the imprisonment of a previous patron, the bishop of Boulogne, and with having foretold the death of Gustavus Adolphus within a few days, and that of Richelieu within a few hours. But such stories, which Zedler repeats after Nicéron, Bayle, and a letter of Gui Patin, are very likely of doubtful authenticity.

[18] Johann Friedrich, *Astrologie und Reformation,* 1864, pp. 32-33, quoting Bailly, *Histoire de l'astronomie,* nouv. éd., Paris, 1785, p. 428.

[19] Praef. Apol., p. xxxi.

[20] Zedler. Gui Patin, *Lettres,* 1846, III, 324, says that she gave two thousand crowns on the recommendation of a secretary who loved astrology.

[21] Praef. Apol., pp. xxxii-xxxiii.

[22] *Ibid.,* p. xxx.

heavenly bodies. On the other hand, Morin defends Cardan's horoscope of Christ, who could choose His own time of birth and who as a natural man was like other men subject in the body to the stars. Morin further argues that it is not impious to believe that Christ employed election of hours, and that the date of His birth may be fixed from the horoscope which fits him and which shows Him teaching the doctors in the Temple at the age of twelve years and three months. But the star of Bethlehem was not a new star or comet but an angel in a lucid cloud.[23] Such are some of Morin's contentions in the *Praefatio Apològetica.* We shall have occasion to notice others in connection with the following résumé of his subsequent text.

Much space is devoted to rebuttal of modern opponents of astrology such as Pico della Mirandola, Alexander de Angelis, Sixtus ab Hemminga and Gassendi. Much space is also given to criticizing and rejecting the errors of ancient astrologers, including even Ptolemy, to purifying astrology from the excesses of the Arabs, Chaldeans, Egyptians and Hindus, and to rejection of the ideas of recent advocates and defenders of astrology such as Lucius Bellantius, Cardan, Giuntini and Kepler.[24]

The first of Morin's twenty-six books defends belief in God and Christ against idolaters, atheists, Calvinists and the like. The second treats of creation, man, and the end of the world. The third book divides the universe into three parts: elemental, ethereal and celestial. Morin still maintains that there are four elements and four qualities, although he recognizes that earth and water form one globe. When he visited deep mines in Hungary in 1615, the idea occurred to him (which he proudly affirms had been maintained by no one before) that corresponding to the three regions of air there were three layers of earth but in reverse order: the first very thin, warm in winter when the lower air is cold, and cold in summer when the air is warm; the middle one hot, whereas mid-air is cold;

[23] *Ibid.*, pp. xx-xxiv.

[24] For criticism of Gassendi, pp. ix-xvi; of Bellantius, *Ibid.*, XXI, i, 4 (p. 501a); of Kepler, *Idem* (p. 501b), XXII, iii, 2 (pp. 561-2); XXV, i, 1 (p. 703a), etc. The others are criticized *passim.* In the Praefatio Apologotica, pp. iv-v, he remarks of Ptolemy's *Quadripartitum* and Cardan's commentary on it, "Multa enim optima habent et retinenda sed plura respuenda."

and the lowest cold, whereas the highest region of air is warm.[25] The earth as a whole is immobile at the center of the universe,[26] for Morin rejects the Copernican hypothesis and elsewhere charges the "insane doctrine of the Copernicans" as denying that the heaven and stars and their motions are made for the sake of man dwelling on earth, and as asserting that the planets and fixed stars are inhabited.[27] Kepler is later twitted with having made the plants and animals on the moon fifteen times as large as ours in proportion to the size of the mountains on the moon,[28] compared to those of the earth.

Beyond the elemental world comes the ethereal, in which the planets move through the ether, which Morin follows Kepler in regarding as a very rare, tenuous and fluid substance.[29] Beyond it is the celestial world, composed of the heaven of the fixed stars and the primum mobile. The two additional heavens which were introduced at the time of the Alfonsine Tables to explain the motion of libration of the machine of the universe are pure fictions.[30] The heaven of the fixed stars, with the Milky Way which forms a part of it, and the primum mobile, are on the other hand, *duo coeli solidissimi.*[31] Morin denies that equator, ecliptic, horizon and meridian are imaginary circles and of no virtue, but grants that the tropics, polar circles and colures are imaginary and of no virtue *per se.*[32]

Book four is on the extension of created beings and continuous quantity. In the next book on space, place and vacuum, it is denied that the Torricellian experiment produced a vacuum and proved the existence of a vacuum. After a book on motion and time, it is

[25] *Ibid.*, III, i, 7 (pp. 76-77). He had already developed this notion in his *Nova mundi sublunaris anatomia,* Paris, 1619, in-8, 144 pp.: BN R.12911 et 44568. Gassendi alludes to it in his Life of Peiresc, stating that Peiresc persuaded Morin to publish the account of his journey to the Hungarian mines, and that Morin prefixed to it his *Mundi sublunaris anatomia:* Gassendi, *Opera,* V, 287.

[26] III, i, 9 (pp. 79-87). Previous works on the subject by Morin are: *Famosi et antiqui problematis de telluris motu vel quiete hactenus optata solutio,* Paris, 1631, in-4, 140 pp.: BN V.7748(1); *Alae telluris fractae* . . . etc., Paris, 1643, in-4, 42 pp.: BN Rés. V.1062.

[27] II, 34 (p. 58).

[28] IX, ii, 7 (p. 175b).

[29] III, ii, 2 (p. 93).

[30] III, iii, 2 (pp. 95-97).

[31] III, iii, 1 (pp. 94-95).

[32] Praef. Apol., pp. xii-xiii.

established in book seven on efficient cause and book eight on the alteration of physical bodies, that they may act at a distance by efflux of virtue.[33] Incidentally there is a chapter on the intension and remission of qualities.[34]

In the ninth book on mixed bodies Morin refutes the opinions of Gassendi and Descartes, approves that of Aristotle, but adds the views of Paracelsus, Severinus, and other chemists. Nicholas of Cusa held erroneously[35] that the earth was neither at the center of the universe nor immobile, that the universe was without center or circumference, and that all the planets and even the fixed stars were inhabited. Morin admits, however, that the planets are not simple bodies but each a different compound, as is shown by observations through the telescope. They combine celestial, ethereal and elemental matter. Morin also abandons the Aristotelian explanation of comets as terrestrial exhalations and holds that they are produced in the ethereal region. He is brief as to their significance.[36] He believes that the fixed stars shine by their own light.[37] In a previous book he had accepted elliptical orbits for all the planets, "as Kepler first of all detected."[38]

The tenth book contends that astrology has a basis in experience. In the eleventh book Morin begins to take up the action of the heavenly bodies, first treating of light, while in the twelfth book he deals with their elemental qualities of heat, cold, dryness and humidity, and with their influence, which is a virtue flowing from their substantial forms.[39] He further holds in support of the doctrine of nativities that the native temperament of a man persists all through his life,[40] and that the native propensities of men cannot

[33] VII, 18 (pp. 143-45), An omnis causa efficiens agat extra se virtutis effluxu; VIII, 8 (pp. 152-53), A corporibus activis in distans virtus perpetuo effluit in sphaeram activitatis ipsorum; VIII, 14, (p. 158), Quo probatur dari posse actionem in distans & non in medio.

[34] VIII, 10.

[35] *De docta ignorantia*, II, 11-12.

[36] In IX, ii, 11 there is only a paragraph on it, at pp. 185-86, which closes: "Vide quid de Cometis diximus in Notis nostris Astrologicis adversus commenta D. De Villennes supra Aphorismos 98. & 99. & 100 Centiloquii Ptolemaei."

But later there is a chapter on the general and particular significations of comets: XXV, ii, 15; pp. 755-56a.

[37] IX, iii, 8; p. 191.

[38] VI, 10; p. 133b.

[39] XII, iii, 4 (p. 277a).

[40] XII, ii, 15 *et seq.* (pp. 253-67).

be determined in any other way more certainly than by drawing up the horoscope at the moment of birth.[41] He rejects employment of a figure of the time of conception.

Having thus built up astrology, to his own satisfaction, on a supposedly firm natural basis in the first twelve books, Morin devotes the last fourteen—which, however, fill nearly twice as many pages, to an exposition of the principles and details of the art. Book XIII on the properties of the planets and fixed stars distinguishes between masculine and feminine, diurnal and nocturnal, beneficent and maleficent planets, and fills eleven pages with elaborate Tables of "the universal lordship of the planets." Book XIV is on the signs of the zodiac, which derive their different virtues from the first heaven. God made the division into twelve signs corresponding to the natures of the planets and the twelve houses, and revealed this to Adam, from whom it came down to posterity by way of the Cabala.[42] Other books deal with the dignities of the planets in the signs, the rays and aspects of the stars, the division into astrological houses, which is illustrated by particular nativities, and the fortitudes and weaknesses of the planets. With regard to aspects, Morin has occasion to refute Ptolemy and Cardan, Bianchini and Regiomontanus, and Leowitz.

After a brief book of definitions, axioms and theorems, Morin considers the universal action of the heavenly bodies upon each other and upon our sublunar world. He contends that his doctrine is purely physical, yet unknown to Aristotle and all previous schools of philosophy. For Ptolemy's division of the terrestrial globe among the signs of the zodiac he substitutes one of his own.[43] He refuses to accept the statement of Aquinas that, if the motion of the heavens ceased, they would no longer heat, and, although illumination from the sun would continue in these inferiors, generation would stop. Morin retorts that, if the sun stood still over one spot, it would burn it the more, and that it is to prevent such destructive combustion that the sun is kept moving. "And so the movement of the heavens works nothing *per se* but merely distributes the active virtue."[44]

[41] XII, ii, 25 (p. 272).

[42] Praefatio apologetica, p. iii.

[43] XX, i, 1-2; pp. 443-44.

[44] XX, iii, 8; pp. 467-68.

Aside from such particular departures from past astrological theory, the chief distinctive features of Morin's system may be summarized as follows, repeating some points that have been noted already. In place of the old distinction between superiors and inferiors, heaven and earth, celestial and sublunar, fifth essence and four elements, he adopts a threefold division of elementary, ethereal and celestial.[45] The planets are no longer simple bodies of a fifth essence, but compound bodies with the elemental qualities of hot, cold, dry and moist, as well as ethereal and celestial matter. Morin distinguishes between their elemental action in heating, moistening and the like, and their influential action by virtue of their celestial nature.[46] The first heaven or primum mobile is a simple body and acts as such, pouring its universal force like a world-soul through the whole world. But it also has a second action, as it is divided into *Dodecatemoria* or signs of the immobile zodiac.[47] The other "most solid" celestial heaven of the fixed stars has *per se* and as a whole no sublunar influence so far as we know, but the particular constellations and stars in it exert virtue of their own.[48] Similarly the ether of the ethereal heaven has *per se* no sublunar virtue, but the planets in it exert a great influence, although their formal virtue is ineffable and incomprehensible to us.[49] Great as it is, the signs signify more fully and efficaciously than the planets.[50] In particular, the degree of the sign on the eastern horizon at the moment of birth signifies more efficaciously than the lord of the horoscope or the planet in the first house. Morin held that it was enough to know the degree for the horoscope and that the exact minute of the degree was not essential.[51]

While Morin affirms the influence of the heavens over elements,

45 A somewhat similar division had been made towards the close of the previous century by Kort Aslaksen (1564-1624), *De natura coeli triplicis libelli tres, quorum I de coelo aëreo, II de coelo sidereo, II de coelo perpetuo. E sacrarum litterarum et praestantium philosophorum thesauris concinnati*, Nassau, 1597, in-8, 214 pp. In 1605, 1606 and 1607 he published three short disputations *De mundo*, of which the last bore the quaint title, *De infima aëris regione et potissimum aquis coelestibus:* Niels Nielsen, *Matematiken i Danmark*, 1528-1800, 1912, pp. 10-11.

46 XX, iv, 1; p. 470b.

47 XX, iii, 1-2; pp. 459a, 456-57.

48 XX, iii, 6; p. 464.

49 XX, iii, 3; pp. 458-59.

50 XXI, i, 4; p. 502b.

51 Praefatio apologetica, p. xxvii.

minerals, plants, brutes and man, he will not admit it in the case of works of art such as astrological images and characters.[52] All artefacts lack seed, and "only the spirit of seed from which things are physically generated... is the proper subject of the inhesion of the influx of the celestial bodies."[53]

Incidentally in the same chapter Morin testifies to the popularity of these images by saying that many lords and ladies had tried to seduce him into making them by offering pay. He had especial difficulty in resisting certain nobles who brought to Paris the sword which Gustavus Adolphus wore at Lützen and desired him to explain the images and characters and golden letters carved upon it. He finally did inspect them sufficiently to prove that they bore no relation to Adolphus's nativity or revolution thereof or the hour of his death. Morin believed, however, that Gustavus had engaged in the battle of Lützen on a most unlucky and lethal day, as could be seen from his horoscope and Ephemerides. It was only on images made by art that he denied action of the heavenly bodies.

Morin's last six books are chiefly concerned with details of astrological technique. Book XXI, rejecting such divisions of the signs of the zodiac as *termini, decani* and *facies,* deals with planetary houses and aspects. It closes with a chapter which asserts that God's method of acting externally is imitated by no physical cause more perfectly than by the celestial bodies.

Book XXII treats of directions, *significatores* and *promissores.* Briefly a direction may be defined as an arc extending across the sky from *significator* to *promissor.* Morin calls this book the chief and most divine of all astrology. From it one learns the times of events which will happen to one after birth. It is "the supreme apex of natural prophecy, and the science which, more than all the physical sciences, is participant of divinity." But great confusion and difference of opinion exist regarding it among astrologers.[54] For instance, Ptolemy and Cardan accepted only five *significatores*

[52] XX, iv, 8; pp. 490-95.

[53] XXVI, i, 3; p. 760b. Also XII, ii, 11 and iii, 3, and XX, iv.

[54] Mersenne, *La Verité des sciences contre les Sceptiques ou Pyrrhoniens,* 1625, p. 31, said that one could hardly find two astrologers who would agree as to "la direction des prometteurs ou significateurs."

for all the future events of life: namely, the horoscope for health and journeys; *pars fortunae,* for faculties; the moon, for character and conversation; the sun, for dignity and glory; and the zenith for other acts and the procreation of children. Haly and Schöner, on the contrary, accepted all seven planets and twelve other points as *significatores.*[55] For Morin both *promissores* and *significatores* are parts of the primum mobile, and when they are quiescent or fixed in the heaven, their effects are produced on earth by their concourse in directions. "This certainly seems to be in the nature of a miracle," the mode of which surpasses the human understanding, although no one who is not ignorant of astrology may doubt its truth.[56]

Book XXIII considers revolutions of nativities, or prediction from the return of birthdays, and illustrates them by many figures of horoscopes. John Stadius rejected them, and no wonder, for he used erroneous tables of the sun's movements. Others today who have good Tables still reject it. Others who accept it are ignorant of its fundamental principles, and have written about it in different ways with many errors and lack of completeness. But it is half of genethliacal astrology, and so Morin tries to purify and unify it on its true foundations, and to leave a correct and complete exposition of it to posterity.

Book XXIV deals with progressions and transits, by which the day and hour of a coming event may be forecast.

Book XXV treats of revolutions of years and planetary conjunctions, eclipses and comets. Past astrologers have written so diversely, confusedly and imperfectly on these subjects that Morin has had a very difficult time in reconstructing a science of them.[57] The great solar eclipse of 1652 was followed not only by political changes but by a great mortality throughout France, so that at Paris alone 100,000 persons perished from malignant fevers, although there was no outbreak of the pest.[58]

In Book XXVI and last of his huge *magnum opus*, Morin comes to interrogations and elections. He asserts that no one hitherto has freed interrogations from figments and errors and established

[55] XXII, i, 3; p. 535b.

[56] XXII, iii, 2; p. 561a.

[57] XXV, Preface.

[58] Praefatio apologetica, pp. ii, xii.

it upon its true foundations. He rejects the Arabic doctrine of interrogations, and notes the errors of Cardan. Morin would limit such questions to certain subjects, but he includes such inquiries of the stars as how long the king will live, and which of two kings will triumph in a battle between them.[59] Elections of favorable times for action he regards as a useful part of astrology, and he gives examples from his own practice illustrated with astrological figures.

And herewith we close the Theory of our astrology, to the honor and glory of the Eternal Wisdom of our Lord Jesus Christ, who made heaven and earth, and endowed celestial bodies with wondrous virtues. To Whom be everlasting praise, virtue and glory, Amen.

In a manuscript of the Bibliothèque Nationale are two tracts on weather prediction by Morin which do not seem to have been printed either separately or in the *Astrologia Gallica.*[60] In the course of *Astrologia Gallica* he sometimes refers to a work of his own which appears to be no longer extant, as in the case of a book on the concourse of the First Cause with second causes, against the Jansenists.[61]

Morin was a doctor of medicine and had been physician to a

[59] XXVI, i, 10; p. 769b.

[60] BN 7485, paper, 17th century, 29 neatly written small leaves. The first tract, entitled, Aëreas constitutiones praedicendi succincta accurataque methodus astrologica, opens, "Futura praedicere proprie divinum est aiebat summus Hyppocrates . . ." Its contents may be indicated as follows:

fol. 1v Signorum et planetarum dominium in telluris partes
2r Planetarum excentricitas
2v Magnae coniunctiones
3r Eclypses
3v Annuae mundi revolutiones
5v Transitus planetarum, ortus et occasus fixarum, aliaque ad diurnum prognosticum conducentia
6v Aerearum commotionum duratio
7r Aphorismi notandi. (There are 27 of them, ending at fol. 10v.)

At fol. 11r begins the second tract, "Aerearum constitutionum varia prognostica ex elementis et astris desumpta . . . a Ioanne Baptista Morino philosophiae et medicinae doctore." Its subheads are, Elementa: Terra, Aqua, Aer, Ignis. Meteora: Nebula, Ros, Pruina, Pluvia, Nix, Grando, Nubes, Coruscatio, Tonitrum (citing Mizaldus), Ventus, Iris, Parelii et Parasilinae, Circuli seu coronae, Cometa et id genus meteora. Astra: Sol, Luna, Stella. It concludes with an Observatio dated 21 July 1628.

[61] I, Theorema xxxi (p. 12b).

bishop, an abbot, and the duke of Luxemburg.[62] He complained that the physicians of Paris were ignorant of astrology and could not draw up a horoscope.[63] Conversely, Gui Patin called Morin a fool, "fantasque, présomptueux, brûlé," and declared that he would not buy his book.[64] Morin regarded the moon as the cause of critical days,[65] and held that the innate temperament was principally from celestial rather than sublunar causes.[66] He approved the drawing up of a *figura coeli* at the beginning of a disease, because diseases were generated naturally from their seeds and had their own symptoms, movements and periods. He further asserted that the Roman Church, "which has never rejected astrological medicine," approved of such horoscopes of diseases.[67]

Morin rails at length against the excessive use and abuse of venesection and phlebotomy in his day. Some physicians, rather than adopt the chemical remedies of Paracelsus, devised a new method of treatment alien from that of Hippocratus or even Galen. They abstained from the strong purgative drugs of Hippocrates, especially hellebore, and contented themselves with infusions of senna, cassia, rhubarb and with clysters, but carried bleeding to excess and were responsible thereby for the deaths of many patients. This, continues Morin, is the chief reason why I gave up the practice of medicine twenty-eight years ago.[68]

The chemical remedies of Paracelsus, however, had gradually won acceptance. The Galenists at first opposed Paracelsus, but then some of them began to use chemical remedies—Ruland, Quercetanus (Duchesne), Croll, Hartmann, and especially the Dane, Peter Severinus, in his *Idea medicinae philosophicae*. Now there are chemical remedies in the pharmacies through all Europe but especially in Germany. Paracelsus aimed to overthrow the Galenic *methodus medendi* and to emphasize rather the vital principle, seed or balsam of vegetables, minerals and animals, and the vital powers of salt, sulphur and mercury.[69]

[62] In 1628, while he was physician to the duke of Luxemburg, Peiresc had wanted to see his observation of the lunar eclipse of January 20. *Correspondance*, II (1936), 20.

[63] Praef, Apol., p. xvii.

[64] *Lettres* (1846), II, 460.

[65] XII, ii, 9; p. 244b.

[66] XII, ii, 14; p. 253a.

[67] XXVI, i, 3; p. 761b.

[68] Praef. Apol., pp. ii and vi.

[69] *Ibid.*, pp. i-ii.

Despite his earlier refutation of the theses of Villon and de Clave, Morin was now favorable not only to Paracelsan remedies but to alchemy itself. "Among physical sciences there are two which surpass the rest in excellence . . . , Chymia and Astrologia.[70] He believed that he had seen the purest gold, far superior to any natural gold, made from lead by projection. He also agreed with the *Chymici* in their books on the philosophers' stone that any mixed body could be reduced naturally, but with the supernatural concurrence of human art, to the highest degree of perfection concordant with its nature, and that, when it reached this supreme stage, it had attained a fixed state and could not in any way be altered to another nature.[71]

Morin furthermore accepted extreme virtues of gems, such as that the sapphire counteracts melancholy and the pest, and represses vain fears; that the emerald checks anger and lust, refreshes sight and heart, and cures epilepsy, leprosy, dysentery and cases of poisoning; and that the diamond makes its wearer intrepid. He attributed such virtues not to any hidden virtue of the stars or planets, but to the formal and specific property of the gems themselves, since such powers worked, whatever the position or movement of the heavenly bodies might be.[72]

Such is the book of Morin, a curious collection of old and new, of progressive and backward views. He welcomes the new medicine of Paracelsus, but opposes the new astronomy of Copernicus. Yet he accepts elliptical orbits for the planets, and to straight and circular motion adds elliptical and spiral. But he won't accept a vacuum. He drops the fifth essence, but holds to the four elements. He is against excessive bleeding, but he is for extreme virtues of gems. He abandons the Aristotelian explanation of comets, but holds that the star of Bethlehem was an angel. He approaches the germ theory of disease, but believes in astrological medicine. He spurns the magnificent chimeras of Fludd as to light,[73] but believes in revelation of knowledge to Adam and its transmission through the

[70] It is with this statement that the Praefatio Apologetica opens.

[71] II, 34; p. 58a.

[72] XII, ii, 2; p. 239a.

[73] XI, 2; p. 212b, "Procul igitur absint Roberti Fluddi circa lucem chymerae magnificae."

Cabala. It is surprising that one who still accepted so much unreformed science of the past, should reject so much of past astrology and attempt to reform that art. But in this jumble of diverse opinions, this juxtaposition of views which seem today inconsistent if not contradictory, this strange mixture of credulity and scepticism, this simultaneous acceptance and rejection of different phases of past tradition, and of recent observation and experiment, he reflects the many conflicting scientific and superstitious interests, the strength and the weaknesses, of the seventeenth century.

As for astrology alone, he found so much fault with that of everyone else, that it is doubtful whether he himself would find many followers. It is equally dubious whether his own destructive criticism would not counteract his detailed rebuttal of other critics of the art and even outbalance his own effort to reconstruct it. A house divided against itself must fall.

Yet in the same year, 1661, that the printing of Morin's *Astrologia Gallica,* with its criticisms of both previous opponents and previous defenders of astrology, failed to reform that art, there was published another book, entitled, *The Sceptical Chymist,* whose criticism of the obscure and mystic mode of writing and of the three principles of past and contemporary alchemists exercised a salutary influence and facilitated, if not the reform of alchemy, the emergence therefrom of the science of chemistry. But the experimental and scientific foundation of its author, Robert Boyle, was superior to that of Morin. Moreover, astronomy had already emerged and largely separated itself from astrology before Morin's book appeared.

CHAPTER XVII

DREBBEL AND DIGBY

Cornelis Drebbel—Elements and alchemical process—Letter to James I: perpetual motion—Inventions—Mingled with magic—Sir Kenelm Digby—*Two Treatises:* arcana, material spirits, experimentation, scepticism—Oration on the sympathetic powder: previous discussion, Highmore, *Theatrum Sympatheticum,* Deusing, Strauss, Meyssonnier—Patin on Meyssonnier—Other works by him—Ross.

I won't leave you out, most illustrious Digby, no more eminent by splendor of birth than for distinguished exploits and learning

—DU HAMEL

Cornelis Drebbel (1572-1633) was probably the most pretentious, secretive and magical figure in the scientific and technical world of the early seventeenth century, although that is a very sweeping statement to make. Born at Alkmaar in West Friesland, he came to London in 1604 and died there, in the meantime, however, oscillating between England and Prague. He was unable to write in either Latin or English, and composed his brief tracts on the nature of the elements, the fifth essence, and the letter to James I on perpetual motion in Dutch.[1] These writings won him a con-

[1] I used the Geneva, 1628 edition of the Latin translation of these three treatises by Peter Lauremberg, whose letter to Georg Schumacher is dated from Hamburg in 1621. Copy used: BN R.54686, in-12, 70 pp. See p. 70, "Sed quia neque latine neque Anglice mentem meam satis exprimere possum, haec Belgice scripsi." Lauremberg states that the German translation was very unsatisfactory.

E. Gerland, *Geschichte der Physik,* 1913, p. 342, speaks of "*De elementis,* 1604," but the first edition seems to have been, *Ein kurzer Tractat von der Natur der Elementen . . . durch C.D. in Nederlandisch geschrieben unnd . . . ins Hochteutsch getreulich übergesetzt,* Leyden, 1608, in-8: BM 1033.c. 2(3). DNB lists a Dutch edition at Haarlem, 1621. *De quinta essentia tractatus* appeared separately in 1621 (Hamburg?); BM 1033.d.16 (8); and also with a Latin translation of the work on the elements and the letter to James on perpetual motion, *Tractatus*

siderable reputation. Alsted incorporated the two last mentioned tracts in the 1630 edition of his encyclopedia, and Bertini, writing as late as 1699, described the treatise on the elements as brief but excellent.[2] But Drebbel was more of an empiric and inventor than he was philosopher and writer, and his fame, real or legendary, as a wonder-worker, rested more upon the testimony of others as to feats which he was supposed actually to have performed than upon those which he asserted in his writings that he could perform. However, let us first examine his scanty literary output.

In it Drebbel repeatedly stresses the marvels and secrets of nature[3] as well as the wonderful works of God. The main contention of the tract on the elements is that they can be transmuted into one another. Earth is of a less simple nature than the other three elements, and is impure and as it were an excrement of the others. Yet we are immediately further informed that fire, air and water are servants of earth. The rays of the sun turn air into fire and water into air, but in the cold upper region of air there is condensation back into water again. Air expands in heated water, condenses when the kettle is removed from the fire. Cold clouds descending and hot air ascending produce winds, and thunder and lightning are caused by hot dry air suddenly reducing a cloud to air and in an instant making it six hundred times larger, so that it requires

---

duo: prior de natura elementorum . . . posterior de quinta essentia . . . editi cura J. Morsii. Accedit . . . epistola . . . de perpetui mobilis inventione, Hamburg, 1621, in-8; BM 1033.d.16 (7). The German translation of the work on elements also was printed at Erfurt in 1624: BN R.53125; BM 1033.f.15 (4.). In 1628 there were Latin editions at Frankfort: BM 1033.d.16 (6.), and Lyons: BM 1036.a.1 (3.), as well as at Geneva. The edition of Erfurt, 1624, and another at Rotterdam, 1632, were with a work of Basil Valentine. A French translation appeared at Paris, 1673, in-12, and there was a late edition of the three tracts at Rotterdam in 1702.

[2] Ant. Franc. Bertini, *La medicina difesa dalle calumnie degli uomini volgari e dalle opposizioni de' dotti,* Lucca, 1699, p. 349.

[3] *De natura elementorum,* 1628, p. 35, "Cogitate quam mirifica ratione"; p. 37, "quanta et quam admirabilis sit Natu(p. 38)rae efficacia"; p. 45, the tract concludes with the statement that Nature herself will teach you her secrets and *maxima miracula. De quinta essentia,* p. 47 *et seq.* holds that the union of the four elements in the fifth essence is a wonderful secret. In the letter to James I he again notes the marvelousness of nature (p. 68), but also says that he is not unaware that even to the most acute geniuses it will seem incredible that "these mysteries can be comprehended by our industry."

more room. The sun draws water up into the extremities of plants and attentuates it into air, leaving to the plants the earthy nutriment. Fire is nothing else than subtle air; air is subtle water; water is subtle earth; earth is crass fire. Earth, either by the force of fire or innate efficacy of Nature, is transformed into water and is made salt. Salt, itself dissolved by fire, is changed to water.

Drebbel describes his alchemical process in only a vague and general way, although he purports to give two methods of preparing the fifth essence from metals and minerals, two from vegetables, and one from animals. He asserts that the fifth essence is found in the purest form in gold, but he warns that the philosophers have not always spoken of the same thing when they made mention of the fifth essence.

It is only in the Letter to James that Drebbel makes boastful claims to specific inventions and thaumaturgy. He begins by saying that the pleasure gained from research into the nature of the elements has alone impelled me, most serene king, to write to your Majesty. What is there that stimulates us more to love, know and adore our Creator? We owe the greatest thanks to sacred letters, and they should ever be held in supreme honor. But the investigation of nature runs them a close second. Drebbel first tried to discover the nature of the primum mobile as a key to all the rest of nature, but without success. He then turned to the elements and tried in vain to make water go up-hill, although fountains would send it up twenty feet or more temporarily. Finally fire gave him his clue. He asserts that he can construct a globe which will revolve every twenty-four hours and indicate the divisions of time and the courses of the planets and stars. He has suspended within a closed glass earth in the midst of water in the midst of air—as their natural spheres are in the world—or vice versa with air in the midst of water and water in the midst of earth. And since he has found the cause of winds, he can construct blowing machines. From knowledge of the tides he makes an instrument that ebbs and flows every twenty-four hours showing the months and their days, the course of the moon and the times of the tides *in perpetuum*, "as your Majesty can see from this present instrument," which is an offshoot from the tree of perpetual motion grafted on true knowledge of the elements. "For I call God to witness that in this I have used neither the writings of the ancients nor anyone's aid, but have discovered these things only by assiduous observation and scrutiny of the elements." All other schemes for perpetual motion are mere *nugae*. Archimedes, it is true, is said to have constructed a globe that moved perpetually with the course of the ether, but war wiped out it and him in a single day. Drebbel concludes with a passage prefering peace and condemning war and praising James for giving his subjects the blessings of peace.

The instrument which Drebbel presented to the king has usually been interpreted as a thermometer.[4] Giuliano de' Medici, ambassador of the Grand Duke of Tuscany at Prague, wrote to Galileo on October 18, 1610, of an instrument of perpetual motion which had been presented by a Netherlander to James I and Rudolf II, and in which water went in a lunette-shaped tube from one side to the other.[5] Daniello Antonini also wrote to Galileo from Brussels on February 4 and 11, 1612, stating that the tube was marked with many equidistant diagonal lines. Drebbel was secretive about it, Antonini said; which was not surprising in view of their efforts to disclose it to Galileo.

Before Drebbel went to England, he had been granted a patent in 1598 for a pump and a clock with "a perpetual motion;" had constructed a fountain at Middelburg in 1601, and received another patent in 1602 for a model of a chimney.[6]

Drebbel also invented a new dye for cloth, getting an intense scarlet by using a salt of tin in cochineal dyeing.[7] By the time of Borrichius, writing in 1668, this invention as well as the thermometer had, despite Drebbel's secrecy, come into general use, "since now common artificers imitate both with equal success."[8]

[4] Ernst Gerland, *Geschichte der Physik*, 1913, p. 347: "Es ist uns schwer, den Apparat nicht als Thermometer aufzufassen."

[5] *Ibid.*, 344 *et seq.* For contemporary allusions to it in English literature see DNB, article on Drebbel, VI, 13.

[6] Gerrit Tierie, *Cornelis Drebbel*, 1932, pp. 4, 92 (sources cited).

[7] Gerrit Tierie, *Cornelis Drebbel*, Amsterdam, 1932, p. 76.

[8] Olaus Borrichius, *De ortu et progressu chemiae*, Copenhagen, 1668, p. 10: "Nec nisi a Cornelio Drebbelio chemico profectus est vel vitri calondarii seu thermoscopii usus vel egregii coloris ignei (Scarletto Italis dicti) quo superbiunt panni Belgici inventum cum utrumque nunc pari felicitate plebeii imitentur artifices."

Alsted also saw perpetual motion in a musical organ (clavichord) invented by Drebbel which played whenever exposed to the sunlight.[9] Drebbel "could make a flat sheet of glass, without any ground edge, in which one could see one's face seven times."[10] This was accomplished by grinding circular depressions in the back and covering it with silver. He asserted that his magic lantern would show him not only in different colors,[11] but in the form of a tree or animal, and Constantyn Huygens, father of the famous physicist, wrote on March 17, 1622, that Drebbel's *camera obscura* was certainly one of the chief features of his sorcery, of which Constantyn's parents had been somewhat fearful.[12] Drebbel is also said to have shown likenesses of persons who were not present, and to have devised an incubator.[13] On a summer's day and in the royal presence he was reported to have so reduced the temperature of Westminster Hall that everyone left,[14] an early instance of air-conditioning.

Not the least extraordinary and one of the best-attested of his inventions was a submarine rowed by oarsmen seated inside from Westminster to Greenwich, the air within being freshened by a subtle spirit (oxygen?) which he had extracted from the atmosphere. Christian Huygens heard of this from his father, Constantyn, who affirmed that he himself witnessed it. Boyle was informed of it by a mathematician who had heard it from one of the surviving passengers. But perhaps the mathematician was Christian and the passenger his father. Leibniz was told of it by Boyle but then verified it from Drebbel's daughter.[15] Kuffler, Drebbel's son-in-law,

[9] J. H. Alsted, *Encyclopedia . . .*, Herborn, 1630, p. 1923, "Hactenus Drebbelius qui idem conficit organum musicum quod mobilis vel potius motivi perpetui virtute soli libero expositum egregiam edit harmoniam sola radiorum solis strictura excitante spiritum inclusum." See also Tierie (1932), 5, 92.

[10] Tierie, *Cornelis Drebbel*, 1932, p. 49.

[11] Garmann, *De miraculis mortuorum*, 1709, p. 802, tells of Drebbel's appearing in different colors in rapid succession.

[12] Tierie, pp. 49-51, 108, 27.

[13] DNB, VI, 14.

[14] *Idem.* In the fifth book on machines of Alessandro Capra, *La nuova architettura famigliare*, Bologna, 1678, are devices for freshening and cooling the air in rooms by cool draughts from fans or the surface of water. See pp. 322-324.

[15] Gerland (1913), 342.

was supposed still to have the secret of the preparation in 1663, as he did that of the scarlet dye.[16]

After the death of James I in 1625, Drebbel was set to work upon the construction of water engines by the office of ordnance, had charge of fireships and water petards on the expedition to La Rochelle, and was one of a company formed to drain the fenlands of eastern England.[17] Pepys records in his Diary on March 14, 1662, that Kuffler and Drebbel's son Jacob tried to get the Admiralty to adopt Drebbel's invention for sinking an enemy ship, which they said had been used successfully under Cromwell.[18]

It would appear that Drebbel was a civil and military engineer of ability and ingenuity, and no mean technologist. His experimental researches in chemistry and physics seem sometimes to have resulted in what might have been momentous discoveries of real significance. But he kept them secret and mysterious, mixed them up with parlor tricks and magical illusions, or used them solely for purposes of exhibition and exciting wonder, or could not quite succeed in making them practical. Also some of his reputed original discoveries may have been simply importations from the continent of instruments and devices as yet unfamiliar in England. The rest, except for the thermometer and scarlet dye, did not long survive him and had to be re-discovered much later. If this was because he was too much of a magician and believer in secrets and marvels of nature, and not enough of a practical inventor and sober scientist, it also suggests that other magicians in times past may have experimented and made similar close approaches to much later scientific discoveries. And it certainly shows that magic and experimental science had not yet succeeded in achieving a complete separation and divorce, either in the general public mind or in the intellects of men of genius themselves.

Paschius in 1700 wrote that some attributed the invention of the thermometer to Drebbel and others to Fludd, but that the latter did not claim it for himself, admitting that he had taken it from a manuscript at least fifty or seventy-five years earlier.[19]

16 Tierie (1932), 68.

17 Tierie (1932), 11-13, 95 (references to *Calendar of State Papers*).

18 DNB, VI, 13-14. Tierie (1932), 75.

19 *De novis inventis*, pp. 624-25.

Sir Kenelm Digby (1603-1665) has left his horoscope in his own handwriting,[20] thereby attesting his interest in astrology, and he ascribed his secret marriage in 1625 to the influence of the stars. "Early in 1633 he and Lord Bothwell were present at a spiritualist séance given by the astrologer Evans in Gunpowder Alley."[21] He also engaged in alchemical and chemical experiments both in London and Paris, but Evelyn, after describing his laboratory at Paris, called him "an errant mountebank," while Henry Stubbes termed him "the very Pliny of our age for lying."[22] He delivered two orations in Italian before the Accademia dei Filomati of Siena on secret modes of writing among the ancients,[23] and—if we accept his own statement—another at Montpellier on the sympathetic powder. Such was his occult outlook and background. We turn next to his chief and almost sole—he later read a paper at Gresham College on the Vegetation of Plants—effort at natural philosophy and experimental science. He was, of course, primarily a man of affairs and action. However, Du Hamel included some account of "Principia rerum ex Digbaeo" in his book on the agreement of old and new philosophy.[24]

The work which we are about to consider was first printed in 1644 at Paris but in English under the title, *Two Treatises, in the one of which the nature of bodies, in the other the nature of mans soule is looked into: in way of discovery of the immortality of reasonable soules.*[25] Several editions at London followed from 1645 to 1669.[26] In the first Latin edition of Paris, 1655, and likewise in that at Frankfurt-am-Main in 1664, the title was modified to Demonstration of the Immortality of the Soul or Two Treatises

[20] BL Ashmole 174, fol. 75: cited DNB, V, 965.

[21] DNB, V, 967.

[22] DNB, V, 970.

[23] J. F. Fulton, *Sir Kenelm Digby: Writer, Bibliophile and Protagonist of William Harvey,* 1937, p. 27, states that the Italian MS of them "has recently found its way into the British Museum."

[24] Jean-Baptiste Du Hamel, *De consensu veteris et novae philosophiae libri duo,* Paris, 1663. I have used the Oxford, 1669 edition, 431 pp. On Digby at pp. 270-72.

[25] Paris, printed by G. Blaizot, 1644, in-fol., pièces liminaires, 466 pp.: BN R. 440 and Rés. R. 449. What seems to be the autograph MS and may have been used for this edition is preserved in the library of Ste. Geneviève, Paris, in 2 vols., MSS 3392 and 3393, anno 1644.

[26] Listed by Fulton, p. 67.

etc.[27] We shall be concerned only with the longer first part on the operation of bodies, which is in thirty-eight chapters.

The general plan may be briefly indicated. The first chapter, in the nature of a preamble, deals with errors in the use of conceptions and terminology in a manner reminiscent of Francis Bacon's four idols. Descartes' identification of matter with extension probably suggested beginning with quantity in the second chapter, proceeding to rarity and density in the third, and from these to four first qualities and four elements in the fourth. But Digby claims to be following a new path and tries to explain natural phenomena in terms of rarity, with which he associates heat, and density, with which he associates cold. For him all sensible qualities are true and real bodies arising from varied proportions of rarity and density between mixed bodies.[28] He holds that moist and dry are generated in dense and rare bodies by the action of gravity upon these, and thus arrives at the old first qualities of hot and cold, moist and dry, and the four traditional elements: earth, water, air and fire. Later on he states that they are found in every part of the world in a pure state in small particles, although not in any great mass.[29]

The fifth chapter deals with the action of the elements. Since earth is the densest, and the power of dividing is natural to dense bodies, it might seem that earth should lead as an active agent. But all philosophers have put fire first in this respect, and Digby explains that its force comes from the minuteness and dryness of its parts which, like troops in squadrons, violently assail its fuel and, like sharpest needles, readily penetrate its porous substance. Also velocity is a kind of density, and the nature of density is rendered more perfect in velocity, and so is more potent in fire than in earth. It will be seen that Digby is mingling atomism and a corpuscular theory with the Aristotelian elements. He says that by atoms he

[27] *Demonstratio immortalitatis animae rationalis sive Tractatus duo philosophici in quorum priori natura et operationes corporum, in posteriori vero natura animae rationalis ad evincendam illius immortalitatem explicantur . . . Praemittitur huic Latinae editioni Praefatio metaphysica authore Thoma Anglo . . .* etc. Paris, F. Leonard, 1655, in-fol. BN Rés. R. 450. I have used the edition of 1664, Francofurti, Secundum Exemplar Parisiense, 1664, in-8. Col 128 D562.

[28] *Op. cit.*, ed. of 1664, p. 350.

[29] *Ibid.*, p. 180.

does not designate anything absolutely indivisible but only the smallest corpuscles in nature. In the next three chapters he identifies light with fire, and affirms that its minute atoms are readily absorbed by the humid air and so hidden in the same that they bear no resemblance to fire.[30] The sun by its rays raises from bodies vapors made up of the most minute atoms.[31] Light with the atoms adhering to it, rebounding from earth, throws off two sets of effluvia, of which one ascends, the other descends in a perpendicular line.[32]

Five chapters are then devoted to local motion and seem worthless. To Galileo is ascribed the view that increments of velocity are always (not merely in the case of falling bodies) in a proportion of odd numbers.[33] The sole cause of violent motion continuing is still held, with Aristotle, to be sought from the medium, i.e., air.[34]

Six chapters follow on mixed bodies and their movements: rarefaction and condensation, attraction, filtration, resilience, and electric attraction, after which come three more chapters on the magnet. Next ensue chapters on living beings in general, animals and plants in particular, and motion in living beings, under which are included the motion of the heart, circulation of the blood, nutrition, growth, and finally corruption or death. The remaining chapters deal with the senses, hearing and sound, vision and colors, sensation, memory, voluntary movement and the natural faculties and passions, the material instruments of cognition and passion, actions of animals which seem rational, and such phenomena as sympathy and antipathy and prescience of the future.

In connection with mixed bodies Digby states that earth and water are the basis of all permanent mixtures,[35] but some bodies are resolved by fire into spirits, waters, oil, salts and earth, which suggests the principles of the alchemists rather than the Aristotelian elements.[36]

One is not much impressed by Digby's claim to originality. He often follows, or thinks that he follows, Galileo, whom he calls

[30] *Ibid.*, p. 77.
[31] *Ibid.*, p. 95.
[32] *Ibid.*, p. 96.
[33] *Ibid.*, p. 87.
[34] *Ibid.*, p. 125.
[35] *Ibid.*, p. 150.
[36] *Ibid.*, p. 169.

"that prodigy of our age to whose incredible perspicacity nothing was ever impenetrable or impervious,"[37] and "Gilbertus noster," who with Harvey has brought to England, he says, credit such as its medieval theologians once gained for it.[38] He repeats experiments with the magnet from Gilbert, but denies that the whole globe of earth is magnetic. His explanation of the origin of the magnet is that the intense heat of the sun under the zodiac attracts effluvia from both poles to the equatorial regions, where the stone was generated in the bowels of earth.[39] When he comes to the question why there is more attraction between the magnet and iron than between two magnets, he prefers his own explanation that iron is humid and hence sticky, either to Gilbert's that there is latent magnetic virtue in the iron, or to Galileo's that the surface of iron is smoother than that of the magnet.[40] Elsewhere he states that the humor in asbestos and gold enables them to resist fire.[41]

It appears that Digby is following Gilbert, Galileo and Descartes step by step, although he refutes Descartes' explanation of refraction, criticizes his account of the movement of the heart, and prefers his own opinion as to sensation to the Cartesian view.[42] His method is more argumentative than experimental. Some of his assertions are true, as that the motion of light is not instantaneous. Others are time-worn clichés, such as that all agents are at the same time patients, or that two hard bodies cannot immediately touch. Some are paradoxical, as that no force is so small but that it can move the greatest weight. Others are wrong, as that ice is not water rarefied but condensed.[43]

Digby is greatly impressed by the secrets of nature, and "how impossible it is for human reason to penetrate and unfold those arcana which Nature has placed in hiding and willed to remain removed from human scrutiny."[44]

Digby is like most seventeenth century physicians and philosophers in accounting for much by the action of material spirits within the body. The vital spirits are the immediate instruments of sense,

37 *Ibid.*, p. 87.

38 *Ibid.*, pp. 220, 230.

39 *Ibid.*, p. 222.

40 *Ibid.*, p. 247 *et seq.*

41 *Ibid.*, p. 167.

42 *Ibid.*, pp. 136, 296, 352-56.

43 *Ibid.*, pp. 68, 176, 150, 80, 190.

44 *Ibid.*, p. 222.

conveying sensible qualities to the brain. Sensation is produced by small solid bodies emitted from objects and stored up in some cell of the brain by the animal spirits, whence they may be evoked by memory and pass once more *per phantasiam.* Pleasing objects dilate the spirits; unpleasant things contract them.[45]

The prescience of beasts as to future weather changes and the like is merely the result of the impression made on them by the first changes which occur in external objects—changes which men are too much occupied otherwise to note.[46]

It has already been implied that Digby made at least a pretense of experimental method. Dr. Fulton has praised him for defending Harvey against Descartes and holding that the heart beats of itself. It is instructive to find that he was led to the experiments upon which he based this correct conclusion as a result "of making the great antidote, in which vipers harts is a principal ingredient," an operation still savoring of magic. Finding that, when he cut up the vipers' hearts, the pieces kept on beating, he went on to the more purposive experiment of removing the heart very carefully and laying it entire on a plate in a warm place, where he asserts that it would go on beating for twenty-four hours, or more, if the weather was warm and moist.[47]

Digby could also at times be sceptical as to natural marvels. A rare experiment which "a noble man of most sincere faith" and a close friend of Digby affirmed that he had seen, was that by which two ounces of powder were collected from rays of light in specially constructed glass vessels. But Digby opines that the powder came from corpuscles accompanying the sun's rays rather than from the light itself.[48] He also questioned the truth of the inextinguishable lamps said to be found in ancient tombs. He suggests that those who dug them up saw a glint of light on their surface and exaggerated this, when later questioned by learned men concerning it. He finds the evidence marshalled by Fortunio Liceto on the subject unconvincing.[49]

Digby denied that any body could act at a distance.[50] But he

[45] *Ibid.*, pp. 357-77.
[46] *Ibid.*, p. 438.
[47] Fulton, *op. cit.*, pp. 58-59.
[48] *Demonstratio immort.*, p. 62.
[49] *Ibid.*, pp. 63-64.
[50] *Ibid.*, p. 175.

believed that poison could be attracted and removed by application of the venomous animal itself to the wound or by wearing an amulet about the neck, from which the effluvia or vapors would have a similar action. So unanimous was the agreement of trustworthy persons as to the action of weapon ointment and sympathetic powder, that he could not doubt it, and, although some called it magic, he preferred a natural explanation.[51] He also accepted the effect of imagination by the parents at the time of conception or of gestation,[52] and the existence of sympathies and antipathies between natural creatures.[53]

This brings us back to Digby's oration on the sympathetic powder, pronounced allegedly at Montpellier. We may consider first its background and setting. Digby claimed to have learned it at Florence in 1622 from a Carmelite who had returned from India, Persia and China, and to have been the first to introduce it in Europe, curing James Howell, Buckingham's secretary by dissolving some vitriol in water and plunging into the water a cloth stained with blood from the wound. The pain in Howell's hand immediately ceased, but when Digby, in the presence of James I and the Duke of Buckingham, held the cloth near a fire, a servant came running from Howell's quarters to say that the pain had resumed. Yet we have just heard Digby deny action at a distance. His explanation now was that the action of light causes particles to separate from bodies into the surrounding air; that the spirits of blood joined with those of vitriol were thus drawn off, while from the wound hot spirits kept pouring out which attracted a flow of air, and that the spirits of blood and vitriol, following up this current, rejoined their own blood in the wound, even though it was a great distance away. Other advocates of the powder said as much as six hundred or a thousand miles.

Weapon salve or ointment, an analogous remedy, had already received much attention in the closing decades of the previous century. Porta accepted it in his Natural Magic, but Adam a Lebenwald in 1580 ascribed its efficacy to the devil. Mairhofer rejected it at the University of Ingolstadt. Biermann, Godelmann and Li-

[51] *Ibid.*, pp. 206-8.
[52] *Ibid.*, pp. 425, 433.
[53] *Ibid.*, pp. 427, 431.

bavius opposed it in 1590, 1591 and 1594 respectively.[54] Again in 1603 Godelmann and Nymann held that any cures which seemed to be worked by it were really diabolical. But such opposition was ineffective, and in the seventeenth century it commanded more attention and support than ever. Tidicaeus took note of it in his treatise on theriac of 1607. We have seen that Goclenius the Younger wrote a special work on it in 1608, and that, although this was violently attacked by the Jesuit Roberti in 1614, Goclenius's defense in 1617, and the further prolongation of the controversy, only advertised the ointment more widely. Meanwhile Croll devoted several pages to it in 1609.[55] At Basel in 1618 appeared a work by Johann Pfanner of Vienna on unusual and prodigious cures and weapon ointment.[56]

We have seen that it received the support of Helmont's name. Daniel Beckher gave a detailed account of the unguent in his *Medicus microcosmus,* first printed in 1622 and reprinted in 1633. Piperno opposed it in 1634, but in the same year Henri de Mohy discussed essentially the same thing under the variant name of Sympathetic Powder, which seems to have given a further impetus to its popularity. His treatise was often reprinted, perhaps in 1640 with that of Nicolas Papin on the same subject.[57] Papin's tract was reprinted at Paris in 1647, at both Rouen and Paris in 1650, and in French version in 1650. Isaac Cattier, professor of medicine at Montpellier and royal physician, published a tract on the abuses

[54] T VI, 420, 525, 416-17, 534, 536, 239-40.

[55] *Basilica chimica,* 1609, pp. 278-82.

[56] *De generali morborum curatione et curationibus insolitis prodigiosis ipsaque armorum inunctione,* Basel, 1618, in-4; also 1628, in-8. Robert Amadou, *Un chapitre de la médecine magnétique, la poudre de sympathie,* Paris, 1953, in-16, 170 pp. has appeared since I completed this chapter. An earlier work which I have not seen is Emile H. van Heurck, *L'onguent armaire et le poudre de sympathie dans la science et le folklore,* Anvers, 1915, in-4.

[57] *Pulvis sympatheticus,* s. l., 1634, in-4. *Brevis . . . pulveris sympathetici praeparandi et applicandi methodus,* (1640?), in-12: BM 1036.a. 5(5.). The fact that in later editions there is prefixed a statement by Johannes Veslingus, professor of anatomy and pharmacy at Padua and prefect of the botanical garden there, that he had read the work with great pleasure and deems it worthy of publication, dated August 22, 1646, might seem to indicate that it had not been previously printed.

of the powder of sympathy and a Response to Papin in 1651,[58] and the latter promptly replied in its defense.[59] Highmore issued a Discourse on it the same year.[60] Meanwhile a *Tractätlein de occulta magico-magnetica . . . curatione* had appeared in 1636, and Petrus Servius had published *De unguento armario* at Rome in 1643.

Digby's oration on the sympathetic powder appeared first in English and French editions in 1658, then was translated into Latin by Lorenz Strauss, professor at Giessen, and printed at Nürnberg as the nucleus of a *Theatrum Sympatheticum,* which was reissued the next year at Amsterdam. In these first two Latin editions it was accompanied by a letter of Strauss to Digby, dated from Darmstadt in October, 1659, and by the aforesaid tracts of de Mohy and Papin upon the sympathetic powder. Helmont was also mentioned in the titles of these editions, and there was a brief reference to him in the preface, but no work by him was included. Then in 1662 at Nürnberg came out a greatly enlarged collection of twenty-six instead of four treatises. It comprised not only the work of Helmont on the magnetic cure of wounds and others which we have mentioned, but criticisms of weapon ointment by Sennert and Kircher, and treatments of sympathy in general or some other particular manifestation of it, such as Fracastoro's book on sympathy and antipathy and Thomas Bartholinus on the transplantation of disease. A Flemish version of this *Theatrum Sympatheticum* appeared in 1665, while Fulton lists four editions in English of Digby's oration and three in German.[61] The French *Discours fait en une célèbre assemblée par le chevalier Digby* of 1658 was reprinted in 1660, 1666, 1669, 1678, in 1681 at both Rouen and Utrecht, and in 1749.[62]

In 1662 Anton Deusing, a professor and physician at Groningen, published there an Examination of the Sympathetic Powder in

[58] *Divers traictez à sçavoir: De la nature des bains de Bourbon et des abus . . . de la poudre de sympathie,* Paris, P. David, 1651 (BN Te163.334) includes Réponse à M. Papin, found also in BN R.12977, *Réponse à M. Papin touchant la poudre de sympathie,* Paris, Martin, 1651; also in Army Medical Library.

[59] N. Papin, *La poudre de sympathie défendue contre les objections de M. Cattier, medecin du Roy,* Paris, Piget, 1651.

[60] It and Beckher's *Medicus Microcosmus* are described in our Chapter 34.

[61] J. F. Fulton, *Sir Kenelm Digby,* 1937, p. 68.

[62] Fulton lists only those of 1658, 1669 and 1678, but BN has the others.

which he pronounced it superstitious and involved in frauds of the evil demon, and refuted the works of Digby, Papin and Mohy.[63] His preface opens with the words, "Marvelous is the force of superstition, candid reader, in the minds of mortals who indulge in superstition!" God has granted us thousands of good remedies, but the devil mixes in superstitious ones. There are too many impediments in the way for the powder to act at a distance; odors do not carry that far; in any case, why not apply it directly to the wound? What reason is there why the spirits of the blood and vitriol should keep together? Since the hot exhalations given off from the wound have their origin within the body, why would it not be better to take the powder internally? Why don't the particles of blood from the powder combine with the similar exhalations from the wound and so never reach the wound itself? Also he doubts if any except near-by air would be attracted to the wound and so suggests suspending the powder from the neck of the patient.

As a consequence of Deusing's treatise with its searching questions and criticisms, Strauss wrote a letter to him which was printed in 1664 with Digby's *Demonstratio immortalitatis*.[64] Strauss explained that he had translated Digby's Montpellier oration only at the request of friends and for exercise, that he had never used sympathetic powder or weapon ointment, "because in such matters I am very incredulous." However, neither did he reject it, and he believed in letting everyone have his say, whether Deusing and Helvetius damning the powder or others defending it. And in his letter to Digby he had often repeated opinions of others rather than convictions of his own. Strauss went on to touch upon mystic passages in the *Birds* of Aristophanes and the Orphic Hymns, upon

[63] *Sympathetici pulveris examen quo superstitiosa ac fraudibus cacodaemonis implicita vulnerum et ulcerum curatio in distans per rationis trutinam ad ipsas naturae leges expenditur, subversis curae sympatheticae fundamentis ab illustriss. comite Digbaeo necnon DD. Papinio et Mohyo positis, autore Antonio Deusingio Med. ac Philos. Doct. illiusque in Acad. Gron. et Oml. Prof. Prim., celsissimi principis gubernatoris ac provinciae archiatro,* Groningae, Typis Johannis Cölleni, 1662, in-12, 657 pp. Copy used: Harvard 24226.5.50. Oml. stands for Oomland, and the book is dedicated to the Estates of Groningen and Oomland.

[64] Strictly speaking in the second part, containing the Peripatetic Institutions etc., pp. 230-36, Francofurti, Secundum Exemplar Parisiense, 1664.

Platonic love, that the hours are longer before noon than after, that there is no dawn among the Arabs, of the emetic force of antimony, Deusing's objection to magnetism of plants, filtration of animal spirits, the problem of vacuum, of the diabetic who voided more urine than his whole body, of the force of imagination, and that a kid should not be cooked in its mother's milk. After this rambling presentation of ideas then occupying men's minds, Strauss called Deusing's attention to a treatise in French by Lazare Meyssonnier of Lyon, entitled, *La poudre de sympathie preuves* (sic) *naturelle et exempte de magie diabolique,* and then concluded:

> It is hard, dear Deusing, to deny the thing by negation of the mode. Extremes are often sensible where means are insensible. It is harder to deny manifest action by taking away material physical contact. It is hardest of all to ascribe immediately to the demon things which are not apparent to our eyes. Nature among its works contains many marvels which surpass and exceed our power of comprehension.

The title given by Strauss for Meyssonnier is different from *La pratique des remedes de sympathie* by him which immediately precedes Strauss's reply to Deusing in the volume of 1664.[65] It likewise differs from an anonymous *Question si la poudre de sympathie peut produire absolument et naturellement la guerison d'une playe simple.*[66] The latter argues from the harmony which exists in the heavens and the influence which extends from heavens to earth, that a similar relation of sympathy may well hold between terrestrial objects. Meyssonnier tells how to apply the sympathetic powder for this and that ill, but gives other remedies, such as herbs which, as they dry up, draw out the ailing humor. Astrological conditions are often observed, and many of the recipes are from previous authors, all the way from Petrus Hispanus to Helmont.

Meyssonnier himself proved experimentally in the case of a gentleman of Dauphiné afflicted with pneumonia, that spitting in a basin filled with tepid water of borage dissolved the clots at the entry of the windpipe, where it branches out between the anastomoses of the arterial vein and the venal artery in the lungs. This was accomplished by the sympathy of the water on the first clots

[65] *Ibid.,* pp. 217-28.

[66] *Ibid.,* pp. 207-16.

expectorated by means of which the others dissolved internally in the body of the patient.[67]

The treatment for gout, of binding the feet of a tortoise on the elbows and knees of the sufferer, which came down from *De physicis ligaturis* of Costa ben Luca, and had been recently approved by Solenander and Argenterius, Meyssonnier employed with such success in the case of a curé of a village in Dauphiné, that the ignorant peasants thought that the cure had been effected by magic.[68]

Although he seems unaware of the discovery of the circulation of the blood, Meyssonnier writes at some time after 1645, since he alludes to a cure at Lyon of that date.[69] Gui Patin, writing to Spon on June 13, 1644, concerning another book by Meissonnier, as he spelled the name, said, "Il tesmoignera toujours de son autheur qui vieillit tous les ans sans devenir sage."[70] In another letter, referring to Meissonnier's conversion to Roman Catholicism, Patin remarked:

> but I have no fear that from Papist he may become a fool, for he is that already, and I have so regarded him for a long while; whoever reads his writings, will not fail to divine as much. The holy bigotry of the superstitious age in which we live has cracked the brains of many others; but the madness of M. Meissonnier is not of that nature; it stems only from his good opinion of himself; he might have some day become a savant, if he hadn't thought himself one already.[71]

Six years later Patin politely thanked André Falconet "for the book of M. Meyssonier.[72] It is attractive and on a curious subject; I shall read it through at my first leisure."[73] The next year, speaking of the Almanach of M. Meyssonnier, Patin asks when the big folio on French medicine which he promised will come out.[74] But in 1656 he attributed the death of Meyssonnier's wife to taking emetic wine,[75] and in March, 1658, wished that Meyssonnier would not send him anything more, for his books are worthless.[76]

[67] *Ibid.*, p. 221.

[68] *Ibid.*, pp. 224-25.

[69] *Ibid.*, p. 224.

[70] Patin, *Lettres* (1907), p. 408.

[71] Letter of October 21, 1644: *Lettres* (1907), pp. 526-27.

[72] The name is so spelled in the edition of 1846 of Patin's Letters.

[73] *Lettres* (1846), II, 547; letter of March 18, 1650.

[74] *Ibid.*, II, 597; letter of November 3, 1651.

[75] *Ibid.*, II, 264; letter of December 5, 1656.

[76] *Ibid.*, III, 81; letter of March 1, 1658.

In 1639 had appeared at Lyons a book by Meyssonnier with a long and pretentious title which may be partially translated as "The Pentagon philosophical-medical, or new art of reminiscence with institutes of natural philosophy and sublimer and more secret medicine . . . And a key hitherto wanting to all the natural arcana of macrocosm and microcosm . . . A new work."[77] The five radii of the universal pentagon are archetype, firmament, planets, elements, and man the microcosm. Noting that the ancients have been far surpassed in anatomy by the moderns, Meyssonnier tries to outdistance them in the fields of astrological medicine, of which his work is full, and the fields of physiognomy and chiromancy. He warns that the bases of his chiromancy are utterly dissimilar to those of the vulgar variety, which are superstitious and justly rejected by theologians. His is based on anatomy, physics and astronomy. He would add divination from the number of letters in one's name whether the ailment is on the right or left side, but fears that it is akin to vanity, for there is no evident connection between right and odd numbers, and left and even numbers. Geomancy is closely related to astrology, as his former colleague, now dead, Henry de Pisis, held. After considering the royal touch, art of seals or astrological images, and weapon ointment, Meyssonnier finds support for astrology in Descartes' demonstration that light is the action or motion of luminous bodies of the finest matter reaching from the stars to our eyes. "And we have learned above from experiments that the coldest substance from the lunar globe and the most ardent fire from the solar extend to us."[78] From the examples of trees whose shade inflicts disease and the occult action of the fish called torpedo it is manifest that forces from the stars influencing some men can be communicated from these to others. The human cranium can be used to cure epileptics, but, despite his addiction to astrological medicine, Meyssonnier is unable to approve fully of observing Egyptian days.

Two years later, in a brief treatise on causes of pestilence,[79]

[77] Lazarus Meyssonerius, *Pentagonum philosophico-medicum*, etc., Lyon, 1639, 104 pp. Copy used: BN 4o T19.21.

[78] *Ibid.*, pp. 84-85.

[79] *De abditis epidemion causis . . . secretioribus theologorum politicorum medicorum physicorum astrologorum*

Meyssonnier found them in the stars and offered "a new and marvelous explanation of astral harmony." In conclusion he gave amulets against the plague. In another treatise of the same year on fevers, dedicated to Richelieu, is a seal of Mercury by Ficino against fevers. He advised against changing the bed-clothes in cases of fever, since the effluvia and dirt from the body were beneficial, and recounted how he had cured two patients of pleurisy by administering their own urine.[80] The titles of both treatises were as blatant as ever. That on fevers purported to present a "new and arcane doctrine": that on the pest, though only 36 pages long, was represented as containing the more secret hypotheses of theologians, politicians, medical men, physicists, astrologers and historians.

A work of 1643 in French on new and extraordinary maladies maintains the influence of the stars and the importance of the animal spirits. While music is the great cure for the sting of the tarantula, it is unavailing in the case of those who have drunk wine in which a tarantula had drowned. On erotic passion Meyssonnier promises to present a quantity of things new and unknown to philosophers and medical men until now, but the remedies proposed reduce to producing a distaste for the object of one's passion. However, they include no superstitious remedy, and against the arts of the devil he recommends recourse to God only. Again in the case of the ailment known as *les Soyes,* he regards a Polish treatment of it reported by Dudith as superstitious, while his own seems sensible,[81] except for belief in "a signature marvelously conformed to several of the things which occur in this ailment."

But when we come to Meyssonnier's Philosophy of Angels in 1648,[82] we drop him, as Patin did.

---

et historicorum hypothesibus instructa..., Lyon, 1641, 36 pp. BN 4o Td60.54 (2).

[80] *Nova et arcana doctrina febrium,* Lyon, 1641, 105 pp. BN 4o Td60.54 (1), pp. 86, 81, 89.

[81] *Des maladies extraordinaires et nouvelles,* 1643, 78 pp.: BN Td4.17B. Chapitre 5, "De ceux qui sont picquez de la Tarente," pp. 31-43; 7, "La Passion Erotique ou le mal d'Amour," pp. 53-62; 8, "Du mal appellé les Soyes," pp. 66-71.

[82] *La philosophie des anges contenant l'art de se rendre les bons esprits familiers avec l'histoire du* S. *Raphael,* Lyon, 1648, in-8.

Sympathy and antipathy, however, so long remained favorite words that they were employed in 1696, in a New System of the Percussion of Bodies, by G. F. Sohier, to indicate movement in the same and opposite directions.[83]

The Philosophical Touchstone of Alexander Ross[84] was a refutation of Digby's *Discourses of the nature of Bodies and of the reasonable Soule.* Ross contended that light was a quality, not a body, and that sound was not motion. He distinguished three kinds of magic: natural, mathematical and diabolical. Weapon-salve was an imposture, but there were sympathies and antipathies in nature for which we can give no reason.

A few years later Ross undertook in *Arcana microcosmi*[85] to refute Harvey's *De generatione* and Browne's *Vulgar Errors.* He classed eels "voided by a maid" as a strange generation, dwelt upon monstrous births—which he ascribed not to the imagination of the mother but to a formative faculty in the seed, and said that serpents were generated from brains of the dead. There was no doubt that centaurs had been produced, partly by the influence of the stars, but they were not human beings and lacked rational souls. The existence of both pygmies and giants was proved. Ross defended the belief that a basilisk was generated from an egg laid by an old cock, and that vipers were generated by the death of their dam. Mice and other vermin were bred of putrefaction.

There were few vulgar errors that Ross did not sanction: the eighth month's child, becoming speechless at the sight of a wolf, women going for years without food, one living some years without a brain, another without a spleen, the tarantula and music, old men becoming young, the virtue of the unicorn's horn, bezoar stones, incubus, the chameleon's living on air, the ostrich digesting iron, the remora stopping ships, the lion's fear of the cock, goat's blood softening adamant, the shade of the ash tree being injurious to snakes, and the existence of the phoenix. It was

likely that the bird Semenda in the Indies which burneth herself to

[83] *Nouveau systeme de la percussion des corps:* JS XXIV, 325.

[84] 1645, in-4, 131 pp. BM E.290 (1.).

[85] 1651: BM E.1405. Ross also had written against Harvey on the circulation of the blood, and in 1634 against the Copernican hypothesis.

ashes, out of which springs another bird of the same kind, is the very same with the old Phoenix.[86]

Ross admitted, however, that it was fabulous to say that the phoenix was seen only once in five hundred years, and he explained that bears lick away a thick membrane in which their cubs are wrapped.

Although Ross made the brain the immediate instrument of sensation and motion, and the seat of phantasy and intellect, he put the heart first as the home of the affections and will. Our spirits were not a celestial substance, but the animal spirits were the chief organ of the soul. The same rose has a stiptic quality in some of its parts, a laxative virtue in others. He could testify from personal experience as to presages of the death of a distant friend. Fascination caused disease and was possible by words, but could be cured by placing the foot of a mole on the child's forehead.

[86] *Ibid.*, p. 287.

CHAPTER XVIII

HARVEY AND PATIN

Discovery of the circulation of the blood—Omne vivum ex ovo—Harvey not mechanistic—Importance of the blood exaggerated—Influence of the discovery of the circulation: Betts, Slegel, Kyper, Webster, Willis, Grube, Walaeus, Bartholinus, Juanini, Tardy—Infusory surgery—Blood transfusion—Patin and antimony—Opposition to chemical and superstitious remedies—Attitude towards magic and demons—Intellectual limitations—Penchant for phlebotomy—Attitude of others—Comparison of Harvey and Patin.

It is such works that deserve to be read and not a great number of portly tomes which only waste paper

—DESCARTES ON HARVEY'S *De motu cordis*

Quantum temporis, quantum sudoris, quantum laboris fastidiosi impendit ille in dilucidando generationis animalium negotio! Quot perdidit menses, imo annos, antequam de inventa sanguinis circulatione triumphare valuerit!

—CLAUDER ON HARVEY

In this chapter, actuated by a motive somewhat akin to that of Plutarch in his Parallel Lives, we bring together the Englishman who discovered the circulation of the blood, William Harvey (1578–1657), and the Frenchman, Gui Patin (1601–1672), who was a leading advocate of blood-letting. The merely, or mainly, verbal antithesis accompanies and covers, however, a deeper distinction between the experimental physician and the conservative doctor and man of letters.

Harvey often cited Galen's experiment with the arteries, and his own discovery of the circulation of the blood was confirmed by nine years of experimentation, ocular demonstration, and frequent vivisection.[1] Fabricius of Aquapendente had made wonderful drawings of the valves in the veins, which "for accuracy

[1] *De motu cordis,* caps. viii, xiv, iv, vi.

and beauty of illustration" are hardly equalled "in anatomical literature," so that it is hard to see how he failed to discover the circulation of the blood.[2] Harvey, who had been his pupil at Padua, said on this point:

> The celebrated Hieronymus Fabricius of Aquapendente ... or, as the learned Riolan will have it, Jacobus Silvius, first gave representations of the valves in the veins, which consist of raised or loose portions of the inner membranes of these vessels, of extreme delicacy and a sigmoid or semilunar shape ... The discoverer of these valves did not rightly understand their use, nor have succeeding anatomists added anything to our knowledge.[3]

Harvey showed that these valves opened towards the heart, not away from it, and so permitted the blood in the veins to re-enter the heart. Barchusen in his History of Medicine in 1710 said that Paulus Servita (i.e., Paolo Sarpi, 1552–1623) observed the valves in the veins more accurately than Aquapendente had and "began to think about the circuit of the blood, in which the incomparable Harvey afterwards proved his diligence."[4]

Aside from any particular discovery, Harvey was of great importance as emphasizing the need of comparative anatomy, studying the movement of heart and blood in lesser animals, and man as an animal. Barchusen attributed to the anatomists of the past century, and especially to Harvey, the discovery that conception did not occur in the womb, as the ancients thought, but in the ovary.[5] His famous statement, that all animals, even the viviparous and man himself, are produced from an egg ("Omnia omnino animalia etiam vivipara atque hominem adeo ipsum ex ovo progigni,"[6] commonly shortened to "Omne vivum ex ovo"), was intended to reduce viviparous animals to the same level as oviparous, rather than

[2] Sir William Osler, *The Growth of Truth,* 1906, p. 17.

[3] *De motu,* cap. xiii. After the English translation by Willis.

[4] J. C. Barchusen, *Historia medicinae,* 1710, p. 486. Paschius, *De novis inventis,* 1700, p. 313, quotes Thomas Bartholinus as writing that Vesling told him that father Paul of Venice, from whom Aquapendente learned of the valves in the veins, had discovered the circulation of the blood, and that Cesalpino defended his claim, but that Clark refuted the defenders of Paul in a letter to Oldenburg.

[5] J. C. Barchusen, *Historia medicinae ...,* 1710, pp. 485-86.

[6] *Exercitationes de generatione animalium,* London, 1651, p. 2.

to assert that all animals were the product of sexual intercourse, since another passage in the *Exercitationes* speaks, like that in *De motu,* of the spontaneous generation of worms from putrefaction,[7] and since Harvey regarded the fertilization of the egg as an incorporeal process like the action of the magnet in passing on its own power of attraction to the iron it touches.[8]

Harvey has been excused for not discovering the capillaries between arteries and veins, and for not making greater progress in the work on the generation of animals, on the ground that the microscope was not yet available. But already in *De motu* he writes: "Nay, even in wasps, hornets and flies, I have, with the aid of a magnifying glass, and at the upper part of what is called the tail, both seen the heart pulsating myself, and shown it to many others."[9] He thus at least did not depend upon unaided and naked vision.

Despite his discovery of the circulation of the blood, it has been noted that Harvey was not a mechanistic thinker.[10] In many ways he remained an Aristotelian. He still believed in spontaneous generation.[11] If he no longer distinguished between vital and animal spirits in the human body,[12] he at least spoke of the blood as "impregnated with spirits and, it might be said, with balsam"[13] (a Paracelsan touch). He spoke of the heart as the sun of the microcosm,[14] "even as the sun in his turn might well be designated as the heart of the world."[15] He conceived a circle of generation as well as the circulation of the blood, and was still convinced "of the supremacy of circular motion in the cosmos at large as well as in the 'microcosm' of sublunary bodies."[16]

[7] *Ibid.,* p. 86, "Et quanquam alia animalia sponte oriuntur sive (ut vulgo dicitur) ex putredine . . ."

[8] *Ibid.,* p. 126 *et seq.*

[9] *De motu,* cap. iv.

[10] Walter Pagel in *Neuburger Festschrift,* 1948, p. 362, and in other papers.

[11] *De motu,* cap. xvii, "grubs and earthworms, and those that are engendered of putrefaction and do not preserve their species."

[12] *Ibid.,* Prooemium. However, he wrote in *Exercitationes de generatione animalium* of 1651, Elzevir edition at Amsterdam, pp. 49-470: "Apud medicos tot sunt spiritus quot partes corporis praecipuae aut operationes: nempe animales, vitales, naturales, visivi, auditorii, concoctivi, generativi, implantati, influentes, etc."

[13] Pagel, *ut supra,* p. 359.

[14] *De motu,* Dedication.

[15] Pagel, *ut supra,* p. 359.

[16] *Ibid.,* pp. 360-61, 358.

Harvey spoke of Nature as a purposive entity, unifying and almost personifying it. Quoting "the divine Galen" as to the relation of the arteries to the heart, he repeats "that nature never connected them with this, the most noble *viscus* of the body, unless for some most important end."[17] Yet we heard Francis Bacon reject the argument from design. Or Harvey asserts that Nature always does that which is best.[18] Even in his work of 1651, seven years after the Torricellian experiment, he could still assert that Nature abhorred a vacuum.[19]

Harvey not merely discovered the circulation of the blood and was credited with the discovery by his contemporaries and century, but even exaggerated the importance of the blood in his later work on the generation of animals. For him the blood alone was innate heat and *humidum radicale*.[20] The spirits in the body, of which previous writers had made so much, calling them aerial or ethereal, the soul's immediate instrument, and of celestial origin and nature, were, according to Harvey, never found separate from the blood. Therefore their tenuity, subtlety and mobility rendered them no more potent than the blood which they ever accompanied.[21] No other bodies or spiritual qualities or diviner heats could be conceded, as Cremonini had stoutly contended against Albertus.[22] Citing Aristotle[23] that in all seed there is a spirit whose nature is related to the element of the stars, Harvey held that there was in the blood a similar spirit or force acting beyond the forces of the inferior elements and related to the starry element. He again cited Aristotle that the inferior world was continuous with the movements of the superior bodies, so that all its motion and change seemed to originate ande be governed thence.[24] After also citing Pliny and Plato, Harvey concluded that the blood had the soul in itself first

[17] *De motu cordis,* cap. v.

[18] *Ibid.,* cap. vi.

[19] *Exercitationes de generatione animalium,* Elzevir, p. 471, "vacuum omne Natura effugiat."

[20] *Ibid.,* 1651, pp. 469, 483.

[21] *Ibid.,* pp. 469-72.

[22] *Ibid.,* p. 472: "Nec sane corpora alia aut qualitates spiritales incorporeae caloresve diviniores . . . concedi possunt; uti Caesar Cremoninus (Aristotelicae philosophiae exime peritus) contra Albertum nervose contendit."

[23] *De gen. animal.,* II, 3.

[24] *Exercitationes de generatione animalium,* 1651, pp. 473-77.

and chiefly, and not merely the vegetative, but the sensitive and motive soul.

So that the blood seems to differ in no way from the soul, or at least should be considered as a substance whose action is soul.[25]

However, it does not seem that Harvey's minimizing the importance of material spirits in the body in favor of the blood had much effect upon subsequent writers of the seventeenth century. We shall find the animal spirits especially made much of by numerous subsequent writers, and they usually thought of them as coursing through the nerves rather than the arteries and veins. Even those who like Harvey associated the spirits with the blood were far from minimizing their importance.

This last point may be further demonstrated by considering a book of 1669 on the Origin and Nature of the Blood by John Betts, royal physician in ordinary and fellow of the London medical college. His association with Harvey is evidenced in the same volume by an appendix on the anatomy of Thomas Parr who reached the age of 152 years and nine months, "with the observations of William Harvey and other physicians present."[26] Betts represents the spirits as the fiery part of the blood and was ready to believe that they sometimes burst forth into flame, as in the cases of Servius Tullius and the apostles at Pentecost. Light too is attributed to the spirits.[27] The head is the citadel where the spirits are housed and whence they go forth as occasion demands. Men with very subtle and fiery spirits are given to the loftiest and most abstract speculation, and diseases are from increment or decrement of the spirits.[28]

Some further instances may be given of the speed or slowness with which the circulation of the blood was accepted,[29] and of different ways in which it was accepted.

Paul Marquard Schlegel had been teaching the circulation of the blood for twenty years before 1650, and in travel through Belgium,

[25] *Ibid.*, p. 480.

[26] Ioannes Bettus, M.D., *De ortu et natura sanguinis,* London, 1669, in-8, pp. 317-325. Copy used: BM 783.e.11.

[27] *Ibid.*, pp. 97, 104-6; and see 147, "Ultimo explicandum restat quo pacto spiritus cum igne in communi materia, pinguedine scilicet, consentiant."

[28] *Ibid.*, pp. 144, 156.

[29] Some information on this point has already been given in Chapter XIII.

England, France, Italy and Germany had met most of the leading medical men and anatomists and did not remember any who could advance anything against it, although he repeatedly raised the question at public dissections. At his suggestion, Alexander Fraser, a Scot, under the auspices of the celebrated Lazarus Riverius, introduced the new theory at Montpellier in 1634. Plempius was for it. Riolan seemed favorable at Paris in 1632. Vesling was gradually won over to it. Schlegel's teacher, C. Hofmann, at first opposed it, but began to alter his view shortly before his death. Conring wrote Schlegel in favor of it in 1640. Walaeus, Descartes, Regius and Hogeland (or, Hoghelande) were others who accepted it. But recently Riolan's variant theory, as expressed in his *De sanguine motu circulatorio*, has come to Schlegel's attention and he writes in 1650 defending Harvey's description of the circulation against it.[30] But he heartily approves of Riolan's condemnation of Germans for their neglect of venesection.[31]

Johan van Beverwyck set forth Harvey's theory of the circulation of the blood in a work published in 1638.[32]

If "Descartes was the first foreigner of any distinction to accept Harvey's views even in part,"[33]—an assertion which overlooks Beeckman, who already in 1633 declared the circulation of the blood to be proved by experiment and named Harvey,[34]—Albert Kyper, in his *Institutiones Physicae* of 1645–1646, was one of the first writers of textbooks in natural philosophy to introduce a discussion of the circulation of the blood into the *cursus philosophicus*. He stressed its importance still more in his *Anthropologia* of 1650, in which he professed to explain the nature and virtues of the contents of the human body and of the soul according to the circular movement of the blood.[35]

[30] Paulus Marquardus Slegelius, *De sanguinis motu commentatio in qua praecipue in Joh. Riolani . . . sententiam inquiritur,* Hamburg, 1650, 135 pp. Copy used: BN 4o Tb36.17.

[31] *Ibid.*, p. 123.

[32] *De calculo renum et vesicae,* Leyden, Elzevir.

[33] J. F. Fulton, *Sir Kenelm Digby,* 1937, p. 58, quoting Osler, *Bibliotheca Osleriana,* Nos. 722, 2030, and citing E. Gilson, *Etudes de philosophie médiévale,* 1921, pp. 191-246.

[34] *Journal,* III, 292. This was quick work on Beeckman's part, since he read Gilbert on the magnet for the first time only in 1627, and did not see Kepler on the movement of Mars until 1628.

[35] *Anthropologia corporis humani*

Anton Deusing of Groningen published in 1655 among other dissertations one on the movement of the heart and blood.[36] Friedrich Hoffmann the Elder, in the preface to his *Opus de methodo medendi*, Leipzig, 1668, compared Harvey to Columbus and Riolan to Vespucci.

John Smith, M. D., in a treatise entitled *King Salomon's Pourtraiture of Old Age*, which was noticed in 1665 in *Philosophical Transactions*, held that Solomon was already acquainted with the circulation of the blood.[37]

John Webster in 1677 stated that for eighteen or twenty years after Harvey's book appeared the circulation of the blood was censured by all Galenists and most of the expert anatomists in Europe. It was "bitterly written against" not only by Alexander Ross and Dr. Primrose, but also by Riolan and Zacharias Sylvius, although the last-named confessed his error in his preface to later editions of Harvey's work.

> Neither could this most clear and evidential verity (which falls under ocular demonstration and manifest experiments) find countenance in the world until that Wallaeus, Plempius, and diverse other judicious and accurate anatomists had found the truth of Harvey's opinion by their own trials and ocular inspection: so difficult is it to overthrow an old radicated opinion.

Even the Royal Society opposed the circulation of the blood, according to Webster, and its defenders were condemned and derided as much as those opposing it would be now.[38]

Thomas Willis (1621–1675) in the preface to his treatise on fevers said that Harvey's discovery of the circulation of the blood had laid a new foundation for medicine. Fevers are fermentation of the blood, which contains the five chemical principles of spirit, sulphur, salt, water and earth. In the fermentation of wine there is no wasting of old parts and coming of new, and the times of crudity, maturation and defection are distinct. But in the blood

contentorum et animae naturam et virtutes secundum circularem sanguinis motum explicans, Leyden, 1650, 665 pp.

[36] BN 8o Tb36.20 bis, in-12, 720 pp.

[37] PT I, 254.

[38] Webster, *The Displaying of Supposed Witchcraft*, 1677, pp. 3-4.

some parts are continually being destroyed and others generated in their place, and the aforesaid three operations go on simultaneously. Fever is an inordinate motion of the blood and a too great heat of it, accompanied by such symptoms as burning and thirst.[39]

Hermann Grube, in his work of 1673 on medical arcana, listed three beneficial results of Harvey's discovery: more intelligent bleeding, infusory surgery, and the realization that many diseases have their origin in the hands or feet and may be best dealt with by bleeding those extremities or making various medicinal applications to them.[40] He even argued that the application of amulets of wholesome medicaments to the hands or region of the heart was explained and supported by the circulation, as their effluvia would alter the blood and so affect other parts of the body.[41]

Further illustration of the effects produced by Harvey's discovery of the circulation of the blood is provided in the works of Johannes Walaeus of Leyden. All the writings by him that are noted in the 1662 edition of Van der Linden and in the *Lindenius Renovatus* of 1686 were connected with it. Two letters by him on the movement of the chyle and blood were frequently reprinted between 1641 and 1673. His lectures at the university of Leyden, on *Methodus Medendi* adapted to the circulation of the blood, were printed at Ulm in 1660, and again with added notes by George Jerome Welsch in 1679 at Augsburg. In 1660 was also printed at London another work by him entitled, *Medica omnia . . . ad chyli et sanguinis circulationem eleganter concinnata.*

Justus Cortnummius, on the other hand, in a work on apoplexy which first appeared in 1672[42] and again in 1677 and 1685, adhered to the "Hippocratic period of blood in the human body" and hoped thereby to stir ordinary physicians to accurate investigation

[39] *The remaining medical works of that famous and renowned physician Dr. Thomas Willis . . .*, Englished by S. P. Esq., London, Printed by T. Dring et al., 1681, 3 pts. in 1 vol. Col. Johnson Coll. K 600 W6792 q. Part I, pp. 53, 57, 64, 68.

[40] *De arcanis medicorum non arcanis commentatio ex inventis recentiorum Harvejanis, Bartholianis, Sylvianis, Willisianis et ceteris,* Copenhagen, 1673, p. 150 *et seq.*, "Quaenam arcana ex inventu Harvejanu."

[41] *Ibid.*, pp. 52-53.

[42] *De morbo attonito,* Lipsiae, 1672 and 1677.

of disease and reading of Hippocrates. But the 1685 edition appeared under the new name of Justus Conradus Michaelis as a "New Useful and Curious Method of curing apoplexy."[43]

The *Anatomicae Institutiones* of Caspar Bartholinus the Elder (1585–1630) first appeared in 1611, were reissued in 1626 and 1632, and were enriched with further observations by his son Thomas Bartholinus (1616–1665) in later editions of 1641 and 1645. But it was only with his third revision of 1651 that Thomas altered the work in conformity with the circulation of the blood, after which there were further printings in 1655, 1660, 1663, to which last was added an appendix by Thomas on the lacteals, thoracic and lymphatic vessels. In 1669, after his death, the work was again revised with reference to the circulation of the blood and the lymphatic vessels, and appeared again in 1673 and 1674.

Juanini (1636–1691), who was born in Milan but went to Spain, wrote in 1679 that the circulation of the blood was not received in the schools of Spain, although the rest of Europe had by that time recognized it.[44]

As late as 1695 Bartholomaeus de Moor asserted that no one as yet had rightly explained the circulation of the blood and especially the impetus with which it burst forth from the heart.[45]

The circulation of the blood was interpreted in an astrological and microcosmic—not to say, magical—sense by Claude Tardy, who had received the M.D. degree at Paris in 1645, and became physician to the duke of Orléans. His treatise on the circular movement of the blood and spirits appeared in 1657. In it he asserted that the situation of the parts of the human body conformed to that of the parts of the universe; that the heart

[43] *Nova utilis ac curiosa apoplexiam seu morbum attonitum curandi methodus,* Hildesiae, 1685.

[44] *Dissertation physique,* French translation from the Spanish, Toulouse, 1685, p. 93.

[45] *Cogitationes de instauratione medicinae . . . ,* Amsterdam, 1695: *Acta eruditorum,* XV (1696), 370. I have not examined the following: Olaus Rudbekius, *Diss. de circulatione sanguinis,* Arosiae, 1652; Pierre Guiffart, *Cor vindicatum . . . probatur cor ipsum chylum immediate in sanguinem convertere,* Rouen, 1652, in-4; Nic. Papinius, *Cordis diastole adversus Harveianam innovationem defensa,* Alençon, 1653, in-4; Fran. Ulmus, *In circ. sang. Harv. exercitatio anatomica,* Poitiers, 1659.

was a sun; that human nature was the perfect original of all the arts; that the four seasons of the year produced the four humors, and that circulation produced in the blood the qualities of the four seasons; that the human body divided into three spheres like the sky, above the sun and heart, of the sun and heart, and beneath the sun and below the heart. Finally, that heat was the chief organ of the soul and lodged in the heart.[46]

* * *

To an age as addicted to blood-letting as the seventeenth century, the discovery of the circulation of the blood was certain to suggest sooner or later the administration of medicines—instead of through the mouth or other orifices of the body—directly into the bloodstream by infusion or injection. The bright idea was also pretty sure to occur to someone that the drawing off of bad or superfluous blood, which had been practiced for so many centuries, might be happily supplemented by the drawing in of good or additional blood from some other animal or human being. Stolle in 1731 tells us that infusory and transfusory surgery made a great stir in mid-seventeenth century.[47]

The *Infusory Surgery* of John Daniel Major, first printed in 1659, touched off a series of publications on the subject, such as John Daniel Horst's Judgment as to the Infusory Surgery of John Daniel Major, with added letters of Bartholinus, Tackius and Hornius,[48] the *Medicina infusoria* of Carlo Fracassati of Pisa (1665), and in the same year the *Clysmatica nova* or Infusory Surgery Applied to Human Beings of Elsholtz.[49] In another edition of 1667

[46] *Traité du mouvement circulaire du sang et des esprits*, Paris, 1657, in-4: pp. 27, 33, 37, 55, 66, 71-72. Copy used: BM 783.f.6.

[47] *Anleitung z. Hist. d. medic. Gelahrheit*, Jena, 1731, p. 840.

[48] J. D. Horstius, *Judicium de chirurgia infusoria Joh. Dan. Majoris*, Francof., 1659, in-12 apud Georg. Fickwirt, 1665, in-12. Cui editioni adjectę sunt Epistolae Bartholini Tackii & Hornii.

[49] Joh. Sigis. Elsholtius, *Clysmatica nova seu chirurgia infusoria hominibus adhibita*, 1665. Two years later he published *Clysmatica nova sive ratio qua in venam sectam medicamenta immitti possint ut eodem modo ac si per os assumpta fuissent operentur; addita etiam omnibus seculis inaudita sanguinis transfusione.* Editio secunda variis experimentis per Germaniam Angliam Gallias atque Italiam factis, necnon iconibus aliquot illustrata. Coloniae Brandenburgicae, apud Daniel. Reichelium, 1667, in-8.

of *Chirurgia infusoria,* Major replied to the judgments of fourteen men on his original volume.[50] This involved his touching upon such points as whether length of life was determined by the stars, fascination, the basilisk, whether being looked at by a wolf takes away one's voice, and as to the potency of human saliva.[51] Which shows how difficult it was to eradicate such thoughts from the human mind.

Nor, as we shall see, did knowledge of the circulation of the blood discourage such notions as that drinking the blood of a criminal might turn one into a criminal.

Another new process was blood transfusion. Priority therein was claimed for various nations, the English asserting that Sir Christopher Wren had precedence over anyone on the continent.[52] Francesco Folli said that the idea occurred to him on reading Harvey's *De motu cordis* in 1652, and that he announced it to the Grand Duke in 1654. But with him it remained an idea which he did not put into operation and test experimentally.[53] At the end of the century Paschius stated that, among the English, Timothy Clerck (i.e., Clarke) and D. Henshaw first thought of this new invention in 1657, but failed to carry it to a successful conclusion. In 1666 Lower, Edmund King, Thomas Coxe and others performed it, but their experiments were limited to brute animals. Jean Denis at Paris was the first to apply it to human beings in 1667.[54]

[50] *Chirurgia infusoria,* Kiloni, 1667, in-4, 328 pp. including Index. At p. 35 begin the judgments of the fourteen; at p. 107 they are tabulated; at p. 129 Major begins to reply to them.

[51] *Ibid.*, pp. 135-41, 273-74, 276-77, 279-80.

[52] Priority was claimed for Wren in PT I (1665), 128. T. S. Patterson wrote in *Isis* XV (1931), 69: "Christopher Wren, about the year 1659, was apparently the first to attempt injections into the blood of animals." Injection and transfusion were of course not quite the same thing. Haller, *Elementa physiologiae corporis humani,* 1757-1766, I, 233, stated that the first experiments in blood transfusion were made by Timothy Clarke in 1657. But no mention is made of this in the article on Clarke in DNB. In PT III, 731-32 is noted an Italian tract which cited Libavius in 1615 as to the possibility of blood transfusion.

[53] Raffaello Caverni, *Storia del metodo sperimentale in Italia,* III (1893), 158-59.

[54] Paschius, *De novis inventis,* 1700, p. 302. See further Harcourt Brown, "Jean Denis and Transfusion of Blood, Paris, 1667-1668," *Isis,* 39 (1948), 15-29. Paschius speaks of Denis as professor of mathematics and "Physic." at Paris, perhaps basing this upon extracts from letters from him in JS (see notes 7 and 8 of Harcourt Brown's

Numerous communications on the subject of transfusion appeared in *Philosophical Transactions* and the *Journal des Sçavans*,[55] and were noted by Joachim Georg Elsner in the first volume of *Miscellanea Naturae Curiosorum* in 1670.[56] As a specimen of these communications may be cited a letter from Paris published in *Philosophical Transactions*.[57] It tells of Denis's transfusion of blood from a young to an old dog, "who two hours after did leap and frisk; whereas he was almost blind with age and could hardly stir before." And a boy of fifteen or sixteen, who had been bled twenty times to assuage the heat of fever, and had become very sleepy from the loss of blood, after an infusion of lamb's blood became more nimble and woke at four in the morning. In the same year, 1667, Claude Tardy published a treatise on transfusion of blood between human beings,[58] and the papal surgeon Riva performed at Rome a triple experimentation, transferring blood from the arteries of three animals into the veins of three men afflicted with different diseases. In the year following Paolo Manfredo published a treatise on the new and unheard-of transfusion of blood from individual to individual, first tested in brutes and then in man at Rome.[59] Meanwhile Oldenburg, secretary of the Royal Society, had written on December 24, 1667, to Robert Boyle: "If these Parisians misrelate not, there hath been freshly made in that town an experiment of transfusion on a madman with a surprising success." "Un fol de la

paper) in which he is called "Professeur de Philosophie et de Mathématique." But Brown calls him merely "Doctor of the Faculty of Medicine at Montpellier" and "physician in Ordinary on the large staff attached to the person of Louis XIV," and speaks of his teaching "various branches of science in semi-public *conférences* in his house in Paris from about 1664 to 1673 or 1674." In a later chapter we shall see that in a *Discours* on astrology, pronounced at one of such "*conférences publiques*" which he held every Saturday, and printed in 1668, he described himself as "Conseiller et Médecin ordinaire du Roy."

[55] For example, JS II, 86-94, 123, 154-55, 178-85, 304-22, 359-61. PT I, 128; II, 501-3, 519-20, 557; III, 710-15, 731-32, 840-41.

[56] *Op. cit.*, Annum I, Observ. 149. Elsner reprinted the account of Riva's experiment and added a Scholion about Manfredo.

[57] PT II, 479, 501-3.

[58] *Traité de l'écoulement du sang d'un homme dans les veines d'un autre et de ses utilités*, Paris, 1667, in-4; BM Pam. 39.

[59] *De nova et inaudita medicochirurgica operatione sanguinem transfundente de individuo in individuum; prius in brutis et deinde in homine experta Romae*, Rome, 1668, in-4.

dernière extravagance," who ran the streets naked, after a second transfusion of calf's blood, slept soundly, confessed, and seemed to be so rational that his confessor permitted him to receive the communion.[60]

Georg Abraham Mercklin Jr., editor of *Lindenius Renovatus* in 1686, wrote in 1679, On the Rise and Decline of Blood Transfusion; in which that made from brute to brute is eliminated entirely from the field of medicine; that which is performed from brute to man is refuted; and that which is practiced from man to man is left to the examination of experience.[61] Bartolomeo Santinelli had previously confuted blood transfusion in a volume of 1668.[62]

At the close of the century Thomas Baker wrote in his *Reflections upon Learning*,[63] a work which went through seven editions and was translated into French:[64]

> What noise have we had for some years about Transplantation of Diseases and Transfusion of Blood, the latter of which has taken up so much room in the *Journal des Sçavans* and *Philosophical Transactions;* and the English and French have contended for the discovery; which notwithstanding as far as I can see is like to be of no use or credit to either nation.[65]

Although belief in the tenuity and mobility of the spirits and their incessant diffusion through the body was of long standing, Baker represented "the circulation of the spirits" as

> a third invention which . . . I should think scarce capable of being prov'd; for neither are the spirits themselves visible nor, as far as I know, does any ligature or tumor in the nerve discover their motion.[66]

He thus thought of the spirits as circulating in the nervous system and not, like Harvey, with the blood.

[60] Boyle, *Works*, 1772, VI, 257.

[61] *Tractatio medica curiosa de ortu et occasu transfusionis sanguinis*, etc., Norimbergae, apud Johannem Ziegerum, 1679, in-8.

[62] *Confutio transfusionis sanguinis*, Romae, 1668, in-8.

[63] *Reflections upon Learning, wherein is shewn the insufficiency thereof, in its several particulars, in order to evince the usefulness and necessity of Revelation*, 2nd edition corrected. By a Gentleman, London, 1700. This is the edition I have used.

[64] *Traité de l'incertitude des sciences*, Paris, P. Miquelin, 1714.

[65] *Reflections upon Learning*, 1700, p. 180.

[66] *Idem.*

Unlike Baker, Garmann in his book on miracles of the dead, published posthumously by his son in 1709, called transfusion of the blood a noble operation.[67]

* * *

Gui Patin was a member of the medical faculty of the University of Paris and shared its attachment to the dogmatic classical medicine of Hippocrates and his successors, and its opposition to Arabic medicine and to chemical remedies. He wrote on October 24, 1646 to a young physician, "Above all, shun books of chemistry."[68] On March 2, 1655, he said, "Chemistry is the false money of our profession."[69] His particular bête noire was the use of antimony and of emetic wine of antimony in medicine, against which the Faculty had passed a decree in 1566,[70] and against which he railed repeatedly.[71] The *tetragonum* of Hippocrates was not antimony but some drug which is today unknown.[72] Patin admitted that the Arabs had added to *materia medica,* although he contended that otherwise all of value in their medical writings had been taken

[67] *De miraculis mortuorum,* 1709, p. 105, § 40. He cited an Exercise by Joh. Cydonius *de sangu. prodig.*, which I have failed to find.

[68] *Lettres de Gui Patin,* ed. Paul Triaire, Paris, 1907, p. 515. *Lettres choisis de feu Monsieur Guy Patin,* from 1645 to 1672, were published at The Hague in 1683, and were reviewed in *Acta eruditorum,* III (1684), 248-53.

[69] *Lettres de Gui Patin, nouvelle édition augmenté . . . ,* ed. J. H. Reveillé-Perise, Paris, 1846, III, 47.

[70] *Lettres* (1907), p. 461.

[71] *Ibid.,* pp. 196, 473; and see *Correspondance de Gui Patin, Extraits publiés avec une notice bibliographique par Armand Brette et précédés d'une introduction par Edme Champion,* 1901, pp. 133, 140, for later letters of December 5, 1653 and January 30, 1654; and that despite the fact that this publication aimed to limit its extracts to political history, stating that the reason why Patin is forgotten today is the excessive space given in his letters to medical dissertation and to worthless discussion and information as to books, new or old. Our interest in Patin's correspondence will be exactly the opposite.

Further examples of his opposition to antimony may be found in his later correspondence; see *Antimoine* in the Index to the 1846 edition of his letters in 3 vols. He was much put out when the Faculty of Medicine authorized the use of antimony on April 16, 1666, and wrote: "These gentlemen say that a poison is not a poison in the hands of a good doctor. They speak contrary to their own experience; for most of them have killed their wife, their children and their friends." (*Ibid.,* III, 609, letter of July 30, 1666.) Earlier he had declared: "L'antimoine seul en a plus tué que n'a fait le roi de Suède en Allemagne". (*Ibid.,* II, 563; letter of November 4, 1650.)

[72] *Lettres* (1907), p. 186.

from the Greeks.[73] He even granted that the writers in Arabic lived at a time when better remedies were available than in Hippocrates' day. But they had abused these and gone to extremes of polypharmacy.[74] Patin belittled Avicenna and Mesue in particular.[75]

When the collected works of Pierre Potier were published posthumously,[76] Patin declared that such books made charlatans rather than great doctors, were full of bad remedies, boasting and falsehood, and that today there were too many chemists and wretched empirics. Potier talked too much about his diaphoretic gold and opium or laudanum, and, on the other hand, too often blamed other remedies by which the public was daily relieved.[77]

The following year Patin wrote of *Observationes medicae* of Lazare Rivière (1589–1655), professor of medicine at Montpellier, while it was still in the press, that it was a wretched book, "charlatanesque, empirique, ou arabesque."[78] After it was issued, he further affirmed:

> The book of M. Rivière is the most miserable work that I have seen. He is neither a philologian nor a philosopher nor a physician. His whole book teaches nothing but charlatanism.[79]

Patin continued to call Rivière a charlatan in 1654 and 1656.[80] Yet Rivière's book was also printed at London in 1646, at The Hague in 1656 and 1659, at Lyon in 1656, 1659 and 1664, and with his collected works at Frankfurt, 1669 and 1674. Potier's works, too, were reprinted in 1666 and 1698 at Frankfurt. Evidently many readers of medical works did not share Patin's views.

Jean Riolan the Younger was another member of the Paris medical faculty who opposed "that antimonial idol which our school has always condemned."[81] M.D.'s from Montpellier introduced chemi-

[73] *Ibid.*, p. 606.

[74] *Idem* and pp. 59-60.

[75] *Ibid.*, pp. 370, 607.

[76] *Opera omnia medica ac chimica*, Lyon, 1645, in-8. Triaire, *Lettres* (1907), p. 449 incorrectly ascribes this edition to Michel Potier, the alchemist.

[77] *Idem:* letter of January 20, 1645, to Spon.

[78] *Ibid.*, p. 497. Letter of March 12, 1646, to Belin.

[79] *Ibid.*, p. 500. Letter of May 10, 1646, to Belin.

[80] Lettres (1846), I, 210; II, 260.

[81] *Curieuses recherches sur les escholes en Medecine de Paris et de Montpellier*, Paris, 1651. I have used the edition of 1657, in-8, 291 pp. and unnumbered pp. BM 1168.c.5.

cal remedies at Paris but were forbidden to practice there by a decree of Parlement of March, 1644. But then they obtained letters "de Conseillers Medecins du Roy." Riolan further complains that even some Parisian physicians are now favoring the Hermetic remedies. He also questions whether "these young doctors of Montpellier" ever were at Montpellier.[82] In other works, as we have seen, Riolan had criticized Harvey's account of the circulation of the blood.[83]

When C. Germain, one of Patin's colleagues on the Paris medical faculty, became suddenly very ill, his alarmed family sent for the last sacrament, and, failing to find his usual physicians, called in "un extraordinaire," who promised an immediate cure by his secret remedy, the product of thirty years labor and investigation, and which would make Germain vomit, go to stool, and sweat profusely. But it only made him vomit violently and long what he had eaten. The giver of the dose—which contained antimony—"tried to dissimulate the sinister effect of his admirable secret," pretended that the chyle which Germain had thrown up was pus from an abscess, and withdrew after stating that the patient could not survive. Germain had a violent fever for sixteen days and his confrères bled him seven or eight times "to quench this extraordinary fire," which they finally succeeded in doing. His first malady then returned, but they cured it in twelve days more. But the report spread that his recovery was due to the secret remedy, and many congratulated him "in favor of antimony and the merit of the one who had composed for me so sovereign a medicine." This was the last straw and induced him to write a book of 442 pages against the abuse of antimony, "to undeceive those who give or take the wine or emetic powder" of antimony and to show that these preparations cannot divest antimony of its poisonous qualities.[84]

[82] *Ibid.*, "Au lecteur."

[83] *Anthropographia*, edition of Paris, 1649; *Opuscula anatomica*, London, 1649; *Opuscula anatomica varia et nova*, Paris, 1652.

[84] *Orthodoxe, ou de l'abus de l'antimoine. Dialogue tres-necessaire pour détromper ceux qui donnent ou prennent le vin et pouldre emetique, où il est prouvé par raisons tirées de l'ancienne et nouvelle medecine ou chymie que ces preparations ne peuvent oster à l'antimoine ses qualitez veneneuses . . .*, Paris, Chez Thomas Blaise ruë sainct Iacques à l'enseigne sainct Thomas proche sainct Yves, 1652.

Germain contended that violent vomiting was dangerous in continuous fevers and unnecessary in intermittent fevers, and that the action of *vomitif d'antimoine* was violent. It was not employed by Paracelsus who treated such fevers quite differently. But while Germain agreed with Patin in attacking the use of antimony, he did not agree with his colleague in rejecting occult qualities.

The approbation of Germain's book by Merlet and Moreau, fellow professors and former deans of the medical faculty, says that for a complete recommendation of this book it could be wished that the author had moderated the heat of his pen, and had not recommended other chemical remedies—"le tartre vitriolé, le gilla vitrioli, le mercure precipité"—which had been condemned by a past decree of the faculty. Germain apologized for having done so in an "Advis au lecteur," and would seem to have expunged the passages to which objection had been taken. The sole reference to any of them in his Index is to "Mercure precipité" at p. 264, and it is there condemned as having a corrosive quality contracted "de l'eau forte."

Bound with Germain's work in the volume at the Bibliothèque Nationale, although by a different printer, is a reply to it by Eusèbe Renaudot, son of the famous Théophraste, a member of the medical faculty and consulting physician to the king, entitled, Antimony Justified and Antimony Triumphant.[85] It is dedicated to his colleague Guenaut, who had employed antimony with success for forty years, and sixty-one doctors of the faculty sign in favor of antimony. Another colleague, Cornuty, was the one who had treated Germain.

In a letter of May 3, 1653, Patin expressed his opinion that the work of Germain was "more reasonable" than that which J. Chartier, son of the publisher of the works of Hippocrates and Galen, had written on antimony.[86] Patin said that it was unworthy to be read and had caused Chartier's dismissal from the faculty.[87] He

BN 4° Te^{151}. 84. Advertissement au lecteur.

[85] *L'antimoine iustifié et l'antimoine triomphant . . .*, Paris, n.d. (1653), 396 pp. BN 4o Te^{151}.85.

[86] *La science du plomb sacré des sages ou de l'antimoine, où sont décrites ses rares et particulières vertus, puissances et qualitez*, Paris, 1651, in-4. 56 pp. BN Te^{151}.82; BM 1179. h.19.

[87] *Lettres* (1846), I, 190.

brought suit, however, and six judges out of ten decided the case against Patin and for antimony.[88]

Jacques Perreau, a physician of Paris who died in 1660, published in 1654, *A Wet Blanket on Triumphant Antimony,*[89] which pleased Patin but did not deter Theodore Kerckring some years later from translating into Latin and issuing a commentary upon the famous work of Basil Valentine which Perreau had opposed.[90]

Already in 1637 antimony had been admitted among the purgatives listed in the *Antidotarium* by order of the medical faculty of Paris, and in 1666 they approved its use by a vote of 92 out of 102, and Parlement soon followed suit.[91]

Claude Germain, too, in the course of time underwent a surprising change of attitude, for in 1672 we find him publishing at Paris an Icon of Occult Philosophy or true method of composing the philosophers' stone.[92] According to Kopp he had become physician to Queen Marie Louise of Poland.[93] At any rate the work is dedicated to John Casimir, King of Poland, and in the dedication Germain speaks of having been a rather unwilling and sceptical witness of "that great and truly royal work . . . undertaken at your Sarmatian court by order of the most serene Queen, Ludovica Maria of Mantua, once your dearest wife of pious and happy memory, and under your august auspices happily accomplished in its first and more important part,"[94] under the guidance of Sendivogius, a Polish knight.[95] Claude Germain's work was printed again at Rotterdam in 1678, and included in Manget's collection of 1702.

[88] *Ibid.,* II, 85: 25 Nov. 1653. On April 21, 1655 Patin recorded that the publishing house was ruined and Chartier "as poor as a painter": *Ibid.,* II, 170-71.

[89] *Rabat-joie de l'antimoine triumphant,* 1654, in-4. It is the sole work by Jacques Perreau listed in the Bibliothèque Nationale printed catalogue.

[90] *Currus triumphalis antimonii commentario illustratus et latinitate donatus a Theodoro Kerckringio.* Amsterdam, 1671, in-12.

[91] JS I (1666), 515, 518.

[92] *Icon philosophiae occultae sive vera methodus componendi magnum antiquorum philosophorum lapidem,* Paris, 1672, 26 fols., 98 pp. Copy used: BN R.37163.

[93] H. Kopp, *Die Alchemie,* 1886, II, 344.

[94] "Magnum illud opus et vere regium me pene invito ac repugnante et de illius veritate dubitante in aula tua Sarmatica susceptum iussu serenissimae reginae Ludovicae Mariae Mantuanae . . . et sub tuis augustis auspiciis ad finem prima et potiore sui parte foeliciter deductum."

[95] This can hardly have been the

When someone sent Patin Glauber's tincture of coral, he thanked him politely but added that what Glauber promised was impossible for coral; "the chemists are liars, as well as the botanists, and the Jesuits with their miracles."[96] Elsewhere he had written: "There are two most mendacious animals, a herbalist and a chemist."[97] When Herman Conring (1606–1681), professor of medicine at Helmstedt and a foe of iatrochemistry, wrote to Patin suggesting that someone at Paris write against Helmont, as Erastus had against Paracelsus in the previous century, Patin assured another correspondent that he, at least, would not waste his time in writing against "ces canailles de chimistes."[98] Conring, whose *De hermetica Aegyptiorum vetere et Paracelsicorum nova medicina* first appeared in 1648, dedicated to Colbert in 1669 an enlarged edition in which he also examined *chemicorum doctrina.*

On August 28, 1668, Patin writes of a Du Moulin who came from Amiens to Paris to make his fortune with secrets of chemistry and who promised to work marvels with his "sirop de Mars." But Du Moulin died of apoplexy in two hours and was labeled "grand charlatan."[99]

Patin was a foe of such remedies as the bezoar stone and unicorn horn.[100] In a hearing before the Parlement of Paris against the apothecaries he waxed eloquent against their bezoars, theriac and confections of alkermès.[101] In an earlier letter he had said that all that the doctors of Montpellier had over those of Paris was their theriac and confections of alkermès and hyacinth. Pliny had well called theriac "a compound of luxury"; the two other confections served only to overheat the sick and profit the apothecaries.[102] The bezoar stone was a figment of the apothecaries to deceive the credulous sick; one need not be a Christian or philosopher or phy-

alchemist who wrote in the first years of the century, unless the meaning is that the directions in his books were followed: "facem lucidissimam in tantis et tam opacis philosophorum mihi praeferente nobili illo Sendivogio Equite Polono ingeniosissimo lapidis physice elaboratore."

96 Letter of July 26, 1658: *Lettres* (1846), II, 410.

97 Letter of Nov. 4, 1650: *Lettres* (1846), II, 568.

98 Letter of February 20, 1651: *Lettres* (1846), II, 117-118.

99 *Ibid.*, III, 680-81.

100 *Lettres* (1907), p. 22.

101 *Ibid.*, p. 526.

102 *Ibid.*, p. 443.

sician to put that bagatelle in its place, since it rested "on no authority, no reason, no experiment."[103]

Patin was continually speaking ill of the Jesuits, whom he called "grex Loyoliticus,"[104] hangmen, and fleas and bedbugs.[105] But he was especially offended to hear that the Jesuits of Lyon were selling a purgative confection.[106] He was even suspicious of quinine, sometimes called the Jesuits' powder. "Monks and empirics make too much of this powder, but people love to be fooled."[107] Even mineral waters were repeatedly belittled by Patin.[108]

Patin's opposition to chemistry was echoed by his son, who delighted his father by maintaining, at his first *Quodlibet*, theses to the effect that the *principia* of the chemists were ridiculous, spurious and chimerical.[109]

"Anti-épileptiques," wrote Patin, "and these very

> deceptive kinds of remedies come from the Arabs, who have misunderstood and badly misinterpreted what they never comprehended in the writings of Galen, namely, *proprietatem totius substantiae.*[110]

Patin would not grant that a property of the whole substance was equivalent to occult virtue, a conception which he rejected. The one thing that he had against Fernel, whom in other respects he greatly admired, was that occult virtues were accepted in *De abditis rerum causis.*[111] Of a writer on weapon ointment[112] Patin said, "He is so foolish and so credulous that he believes in these Paracelsic and Galenian bagatelles."[113]

[103] *Ibid.*, p. 360.

[104] *Ibid.*, p. 115 *et al.*

[105] *Ibid.*, pp. 347-48: the editor of the 1907 edition points out that in previous editions these passages had been suppressed.

[106] *Ibid.*, pp. 337, 347.

[107] *Lettres* (1846), III, 666; letter of October 11, 1667. Patin speaks of it as "quinquina," however.

[108] *Ibid.*, II, 583; III, 470, 541.

[109] *Lettres* (1907), pp. 643-4.

[110] *Lettres* (1846), I, 447.

[111] *Lettres* (1907), pp. 19, 26.

[112] Petrus Servius, *Dissertatio de unguento armario sive de naturae artisque miraculis*, Rome, 1642, 179 pp., in-8. The Surgeon General's Library has a German translation of it without date or place of publication. Servius died at Rome in 1648.

[113] *Lettres* (1907), 373. Servius in 1638 had published an elementary work of medicine, of which Patin had a somewhat better opinion, and *Juveniles feriae*, which he called "fort peu de chose." For other works by Servius, including a *Dissertatio de odoribus*, Rome, 1641, see *Bibliotheca Curiosa* (1676), p. 330.

Monarchs, nobles and great ministers of state were as prone to fall into the hands of quacks and charlatans as they had been in the previous century. The last illness of cardinal Richelieu lasted only six days. He was bled twice, but by the fourth day the doctors despaired of his life, so an old-wife was brought in, who had him swallow horse manure in white wine; then, three hours later, a charlatan, who administered a laudanum pill. But all in vain![114]

To le Fèvre, "un empirique de Troyes," Patin refers a number of times. He had learned his empiricism at Rome, and in Paris was called "l'esgorgeur de ratte." To Patin's great disgust, he cured the archbishop of Bordeaux and was called to the death-bed of Richelieu. Later, together with Vautier—who, after being first physician to Marie de' Medici, was imprisoned from 1631 until Richelieu died, but in 1646 became first physician to Louis XIV—le Fèvre attended the eighteen year old son of the Marquis de Villeroy, who, however, died of small pox. This led Patin to exclaim triumphantly:

> M. le Fèvre, duquel on ne parle plus icy; est-il à Troyes? qu'y fait-il? La petite vérole de l'hostel de Villeroy a esté plus fine et plus forte que tous ses secrets.[115]

In a later letter of January 14, 1651, Patin asked whether "this sleeping pill of your surgeon" was not the same as le Fèvre gave Richelieu the night before his death. "Would to God that he had given it to him twenty years sooner!" Finally on April 21, 1655, Patin announced the death of le Fèvre, "soi-disant médecin de Troyes, bailleur de petits grains" of opium "fardé et déguisé," on the fifteenth of that month at Troyes "of two doses of emetic wine the day before."[116]

Magic arts were referred to slightingly by Patin, and he had little faith in the existence of demons. He asserted that he did not believe and would not believe in possession by demons, or in sorcerers and miracles, unless he saw them himself. Everything in the New Testament he accepted as an article of Faith, but

[114] *Lettres* (1907), p. 255.

[115] See Patin's letters to Belin, an M.D. of Troyes, dated October 12, 1641, July 28 and August 25, 1642, and March 12, 1646: *Lettres* (1907), pp. 211, 228, 236, 498.

[116] *Lettres* (1846), I, 173; II, 172.

he refused to extend this authority to the legends of monks and Jesuits.[117] However, he accepted the existence of purgatory, writing on September 22, 1665, to Falconet:

I send you a printed extract from the registers of the Sorbonne, which I beg you to give to M. Spon, from which he will see how our good doctors believe in purgatory, and I pray God that this will serve toward his conversion and the safety of his soul.[118]

But when, in 1668, a priest was accused at Paris of sorcery, Patin's comment was, "I do not believe in these bagatelles."[119] On March 4, 1661, he wrote to Falconet that he had heard nothing concerning the nuns of Auxonne, but that about two months since he had drawn up a paper for a physician of Dijon against a pretended possession by demons there.

I hate imposture of every kind, but in a matter of religion above all. The devil is no more at Auxonne than elsewhere.[120]

The *Démonomanie* of Bodin, in his opinion, was worthless; and the book of Delrio, "tout plein de sottises."[121] He held that Bodin did not really believe in witchcraft, but wrote the *Démonomanie* to escape the charge of free-thinking and atheism. Then he favored the Huguenots; later he became a Leaguer for fear of losing his office; finally, he died of the pest at Laon, where he was Procureur du Roy, a Jew and non-Christian. He thought that anyone who had passed sixty could not die of the pest, but he did so in 1596.[122] The prophecies of Nostradamus were mere reveries and recalled to mind the couplet:

Nostra damus, dum verba damus, nam fallere nostrum est;
Et quum verba damus, nil nisi nostra damus.[123]

Naudé invited Patin and Gassendi to sup and spend the night at his country place at Gentilly

[117] *Lettres* (1907), pp. 361, 151, 350, 225.

[118] *Lettres* (1846), III, 555.

[119] *Ibid.*, III, 679.

[120] *Ibid.*, III, 334.

[121] *Lettres* (1907), pp. 346, 351.

[122] *Ibid.*, p. 346. These charges against Bodin are repeated on July 27, 1668: *Lettres* (1846), III, 679.

[123] *Ibid.*, III, 50-51; letter of August 30, 1655.

on the understanding that there will be just we three, and that we shall have a debauch. But God knows what a debauch! M. Naudé drinks only water and has never tasted wine. M. Gassendy is in such delicate health that he dare not drink any . . .

As for me, I can only shake sand on the writing of these two great men, and I drink very little. However, 'twill be a debauch, but a philosophical one, and perhaps something more. For, being all three cured of bugbears and delivered from the evil of scruples, that tyrant of consciences, we may approach very close to the sanctuary.[124]

So far Patin has appeared as an enemy of credulity and superstition, a sceptical mind following along somewhat the same lines as Erastus in the previous century, whom he praised more than once. But although he was a man of literary taste and imbued with the classics, there were serious limitations to his intellectual outlook. He had neither Gilles de Corbeil nor Gentile da Foligno in his library, and he could not read Arnald of Villanova, because the book was printed in Gothic type.[125] When he tried to read Saumaise *De annis climactericis*, he found it full of astrology and terms as well as things which he did not understand.[126] He read all of Gassendi's Life of Peiresc but did not understand portions concerning money matters and the Copernican system.[127] In his last known letter of January 22, 1672, he averred that Descartes and the ignorant chemists "try to spoil everything, as well in philosophy as in good medicine."[128] He praised the works of Sennert,[129] who nevertheless believed in occult virtues and in alchemy as well as chemistry. He spoke well of Nifo[130] and so had probably never seen his work on demons, to say nothing of those which were astrological. Despite all the magic and superstition in Pliny's Natural History, Patin called it "one of the finest books in the world."[131] These examples suggest that his scepticism and scoffing were either sporadic or biased or both. In another passage, after exclaiming against popular belief in marvels, "quae omnia rideo," he goes on to say that by *dracunculi* he understands the little worms with feet

[124] *Lettres* (1907), pp. 616-17.
[125] *Ibid.*, pp. 204, 214.
[126] *Ibid.*, p. 592.
[127] *Ibid.*, p. 211.
[128] *Lettres* (1846), III, 795.
[129] *Lettres* (1907), p. 33, and other passages.
[130] *Ibid.*, p. 414.
[131] *Ibid.*, p. 472.

which are born in the veins, "as Galen says in a passage which I will indicate elsewhere."[132]

Finally, Patin had a little foible of his own which he shared with the doctors of Paris,[133] but which has long since been abandoned in medical practice as a superstition, namely, bleeding. A chief reason for his opposition to polypharmacy and Arabic medicine was that bleeding had been neglected in consequence. It was a great remedy in cases of small pox, if administered in time.

> There are no remedies in the world that work so many miracles as bleeding. We Parisians ordinarily take little exercise, eat and drink a lot, and become very plethoric. In this condition one almost never is relieved of any ailment that befalls, unless bleeding goes before potently and copiously.

Purgation alone is insufficient.[134]

Jacques Pons (1538-1612) who became dean of the medical faculty at Lyons in 1576, twenty years later printed there a brief treatise on excessive blood-letting of his day.[135] But Gui Patin affirmed that he would have changed his view, "were he with us today."[136] If "with us" meant in Paris, Patin might have been right, for a third writer tells us that the north winds at Paris make it a better place for bleeding than Lyon or Narbonne.[137] In Italy, too, works on the abuse of venesection by Cotugni and Castelli had been printed at Rome in 1604[138] and 1628.[139] Cotugni's treatise was issued posthumously by his son. Cotugni held that major diseases did not always require venesection, that one should not bleed in cases of exuberance of crude juices, and that sometimes

[132] *Ibid.*, p. 361.

[133] Dupré, *De la saignée fréquente et copieuse des médecins de Paris*, c. 1645.

[134] *Lettres* (1907), pp. 607, 456-57.

[135] *De nimis licentiosa sanguinis missione qua hodie plerique abutuntur brevis tractatio*, Lyon, 1596; 2nd ed., 1600.

[136] *Lettres* (1907), 279: "S'il étoit aujourd'hui parmy nous, il changeroit d'avis."

[137] Pietro Castelli, *De abusu phlebotomiae*, Rome, 1628, xxii, 96 pp. (BN Te10.48.), p. 89.

[138] Jacobus Cotugnus, *Liber de abusu venae sectionis et quando et quibus in morbis et qua ratione ea aperiri deceat . . . in cuius fine etiam est adjecta solemnis illa . . . quaestio eiusdem autoris an in diarrhea possit secari vena et medicamentum exhiberi . . . Julii Caesaris Cotugni ejus filii opera et studio in lucem editus*, Rome 1604, in-4, 59 pp. BN Te10.32.

[139] See note 137.

blood-letting, sometimes purgation, sometimes other remedies and aids were called for. Castelli enumerated twenty-five injurious effects from phlebotomy. It was reported that the University of Paris medical faculty made copious employment of venesection, but he refuses to believe that they are prodigal or brutal in this respect. He says that it is proved by the authority of the legitimate school of Paris that phlebotomy is harmful in bilious complaints, and quotes Duretus, who was a professor there and physician to Charles IX and Henri III, to the scholars of Paris not to bleed under certain circumstances. Unlike Gui Patin, Castelli esteemed Avicenna and other Arabic authors, and asserts that the physicians of Paris employ medicaments which were introduced by the Arabs, although they may not admit it.[140]

Many other works by Castelli had appeared at Rome, and were listed by Leo Allatius in 1633,[141] who further informs us that he had attended the lectures at Rome in the previous century of Andrea Bacci and Cesalpino. After himself lecturing in the Roman university, first in philosophy and then on medicine and botany, he passed to Messina, where he taught medicine, anatomy, and chemistry, and was in charge of the university's botanical garden and dean of the medical college,[142] and in 1642 published a treatise on critical days.[143] Of his treatise on the odiferous hyena we shall speak later.[144]

In Germany Hartmann had held in 1611, that, in cases of great haemorrhages of wounds, venesection from the part of the body opposite the wound was unnatural and injurious,[145] and in 1623

[140] Castelli, *De abusu phlebotomiae,* 1628, pp. 43, 77, 85, 6. On Ludovicus Duretus (1527-1586), LR 761-62.

[141] Leo Allatius, *Apes urbanae sive de viris illustribus qui ab anno MDCXXX per totum MDCXXXII Romae adfuerunt ac typis aliquid evulgarunt,* Rome, 1633, in-8, 270 pp. I have used the 1711 edition by J. A. Fabricius.

[142] In the book to be mentioned in the next note he is called, "Romani nobilis Messanensis philosophi et medici in celeberrimo Mamertinorum gymnasio medicine professoris primarii anatomici publici chimiae extraordinarii interpretis atque academici horti simplicium protoplastae, Messanensis medicorum Collegii Decani, olim in patrio archigymnasio philosophiae primum, tum medicinae et simplicium lcotoris."

[143] *De abusu circa dierum criticorum enumerationem,* Messina, 1642, xxx, 167 pp. BN Td19.17.

[144] Chapter 24.

[145] Joh. Hartmann, *Disputationes chymico-medicae,* Marburg, 1611, p. 16.

Burggrav had opposed blood-letting in some fevers and in all astral diseases.[146] G. F. Laurentius had advised venesection near the right big toe in cases of fever.[147]

In a work on fevers which was first printed at Venice in 1615, Pierre Potier, whom we have heard Patin censure, discussed the question, Whether phlebotomy is the unique remedy for fevers?[148] He said that it was such among barbarous peoples, especially in Egypt and generally in all Asia and Africa. "In Europe the greater part of Spain suffers from the same disease." Formerly France was not much addicted to it, but now it is prevalent there in many places. Italy steers a middle course. As for the specific question, his answer is that it is clear that abundance of blood does not generate fevers; consequently diminishing the amount of blood will not cure them.

But it should not be inferred from this that Poterius never bled a patient. A lady of forty-five had acute fever with vomiting, headache, insatiable thiist, restlessness, coated tongue, and so on. He put her on a scant, humid and cold diet, purged her whole body. The second day he twice administered an attenuating and chilling julep. Finally on the third day, having a benevolent aspect of the stars and observing the hour when Pisces ascended and Jupiter was in mid-sky, he opened a vein on the right side and drew off seven ounces of blood, whereupon all the symptoms ceased and she was safe and free from fever.[149]

The controversy which had been waged in the early sixteenth century as to which side of the patient's body should be subjected to venesection in cases of pleurisy was renewed in the early seventeenth century by Francesco Marziano at Rome[150] and by René Moreau, of Angers, who became royal professor of medicine at

[146] *Introductio in vitalem philosophiam*, Frankfurt, 1623, p. 166.

[147] *Defensio venaesectionis in febre . . .*, 1647, in-4.

[148] *De febribus*, II, viii: *Opera*, 1698, pp. 789-90.

[149] *Ibid.*, p. 63 (Cent. I, 39). This was the first case I noted at random, but bleeding and observing the positions of the stars were not frequent in his collection of cases. However, one notes such phrases as "nisi influxui coelesti" (p. 32) and "sic in maiori mundi," (p. 35).

[150] Fran. Martianus, *Antiparalogismus . . . in quo Hippocratis authoritate recentiorum medicorum abusus notantur circa venaesectionem potissimum in pleuritidis curatione*, Rome, 1622, in-4.

Paris after the publication of his book.[151] Antoine Obert of St. Omer published three treatises on the same subject, in 1629, 1631 and 1635 respectively.[152] Baldo Baldi, professor at the papal university in Rome, brought the matter up again in a treatise printed at Paris in 1640,[153] and Moreau published a letter in reply the next year.[154]

Meanwhile a dissertation on the timely use of phlebotomy and purgation against those who were fearful about it[155] had been included in medical *Opuscula* of François Citois and was praised by Patin in a letter of June, 1639 to its author.[156]

Some examples may be given of the employment of phlebotomy, taken chiefly from Patin's own practice. In 1633 M. Cousinot, now first royal physician, had rheumatism, for which he was bled 64 times in eight months, and then purged and cured. But, says Patin, idiots who do not understand our profession, think purging enough.[157] Patin had M. Mantel bled 32 times for a continuous

[151] Renatus Moreau, *De venae sectione et missione sanguinis in pleuritide,* Paris, 1622, in-8.

[152] Listed by Van der Linden, *De scriptis medicis,* 1662, pp. 48-49. LR, p. 79.

[153] B. Baldi, *De loco affecto in pleuritide disceptatio,* Paris, 1640, 121 pp.

[154] R. Moreau, *Epistola . . . de affecto loco in pleuritide ad B. Baldium,* Paris, 1641, 52 pp.

[155] *De tempestivo phlebotomiae ac purgationis usu dissertatio adversus haemophobos,* in *Opuscula medica,* Paris, 1639.

[156] *Lettres* (1907), p. 159: The following are a few more titles of 17th century treatises on bleeding:

Sohner, E., *De sanguinis detractione per venas,* 1606.

Ramirez, *De ratione curandi per sanguinis missionem,* Lisbon, 1608; Antwerp, 1610.

Moxius, J. R., *Methodus medendi per venae sectionem morb. muliebr. acutis . . . ,* 1612, in-8.

Pellicia, Paolo, *De venaesectione et crisibus,* Venice, 1623, in-4.

Moranus, Did., *De venae sectione,* 1626.

Castellanus, J., *Phylactirion phlebotomiae et arteriotomiae,* 1628.

Angelinus, Facondinus, of Rimini, *Methodus pro venaesectione,* Padua, 1641, 1650; Venice, 1642.

Heunius, Joh., *De hirudinum usu et efficacia in medicina,* 1652.

Cortacius, Geo., *Trutina medica de sanguinis missione . . . in febribus,* Padua, 1654.

Francisco, J. F. de, *De venae sectione contra Empiricos,* Naples, 1655; Frankfurt and Leipzig, 1685.

Scheurl, Christophorus Theophilus, *De arteriotomia,* Nürnberg, 1666, in-12.

Perdoux, Bart., *Statera sanguinis,* Tournai, 1668 (a late edition?).

Holmdorf, Eric, *De phlebotomia,* Upsala, 1671.

Theophilus, Christianus, *De sanguine vetito, Disq. uberior pro Thom. Bartholino. Accessit eiusdem Bartholini de sanguinis abusu Disp.,* Frankfurt, 1676, in-8.

Portius, L. A., *Erasistratus sive de sanguinis missione,* Venice, 1683, in-12.

[157] Patin, *Lettres* (1907), p. 457.

fever, and he was "entirely cured, for which I praise God."[158] Patin once treated a boy of seven who had acquired pleurisy from being over-heated playing tennis or hand-ball and receiving a kick in the side. The young gentleman's guardian was opposed to bleeding, but the consulting physicians favored it, and he was bled thirteen times and in fifteen days was cured "as by miracle," and his guardian converted to a belief in phlebotomy.[159] Although Patin had previously praised the works of Sennert, he became very angry at him,[160] when he found that he was opposed to bleeding children and the aged. At Paris, Patin asserts, we bleed octogenarians and infants who have not yet been weaned. Germany is unfortunate in having such physicians who seek for secrets of chemistry in Paracelsus and Croll who were never *médecins.*[161] When he was sumoned to attend Hobbes, who was suffering such pain that he wanted to kill himself, Hobbes at first refused to be bled on the ground that he was sixty-four and too old. But the next day he agreed and was, according to Patin, much better in consequence, and after that they became great pals.[162]

On July 18, 1642, Patin recorded that Belin's brother, suffering from tertian fever, had been bled four times; and on July 30, that he had been bled eight times. Elsewhere he declared venesection the greatest aid in bloody apoplexy and exulted that "saintly and salutary bleeding begins to spread happily all over France more easily and favorably than ever." In 1647 small-pox and dysentery were rife in Paris, and he boldly and successfully employed venesection without resort to bezoars for both. On January 27, 1649, Patin writes that Nicolas Piètre is very ill and has been bled twelve times for rheumatism. "May he finally recover!"[163]

But the death of Patin's friend Gassendi has usually been ascribed to the excessive blood-letting which he underwent in frequent illnesses, especially pulmonary complaints. He lived, however, to be sixty-three. In his last illness, Sorbière tells us in the preface to the edition of Gassendi's works in 1658, after he had been bled

[158] *Ibid.*, p. 157.

[159] *Ibid.*, p. 457.

[160] "J'en suis tout en colère," Letter of August 27, 1658: *Lettres* (1846), II, 419.

[161] *Ibid.*, p. 420.

[162] *Lettres* (1846), II, 593-94.

[163] *Lettres* (1907), pp. 225, 230, 390, 395, 541, 643.

nine times, he wanted it stopped, and the oldest physician present assented. But another doctor persuaded them to continue, and he was bled four times or more before he died.[164] Although Sorbière gave no names, Patin was enraged by this account which was obviously aimed at himself.[165] But he himself had written, when Gassendi first fell ill:

M. Gassendi insisted on keeping Lent and is much the worse for it. I warned him against it, but he preferred to wait until sickness overtook him, as it has done. Yesterday evening he became very ill with a raging colic, so that he had a huge movement and vomiting which cruelly upset him all night long. He sent for me early in the morning; I went immediately. I found him much affected, very upset, the cholera-morbus continuing with a high fever. I had him bled instantly . . .

Later in the letter Patin continues:

I have just left Gassendi, who is much better than this morning. The blood which was drawn from him is horrible from corruption. He has vomited several times more, but his bowels begin to stop moving. He expectorates easily and freely and abundantly, which relieves the lungs from foul matter . . . I have left orders that, if he has a good night, he should content himself with nourishment and tisane . . .

But if not, that he be bled from the other arm.[166]

Patin, on the other hand, recounts with great satisfaction the death of La Brosse, head of le Jardin du Roy. He had contracted dysentery from eating too many melons and drinking too much wine ("as usual," adds Patin). He had his entire body rubbed with oil of yellow amber for four days. Next morning he swallowed on an empty stomach a great glass of brandy with a little astringent oil. When this did no good, he took an emetic but died the following morning, as it was working. "So vomited forth his impure soul that impure wretch, most expert in killing men." He had refused to be bled, calling it the remedy of sanguinary pedants, and said he would rather die. "The devil will bleed him in the other world,

[164] Gassendi, *Opera,* I (1658), Praefatio, signature a 4 recto and verso.

[165] See his letter of July 5, 1658: *Lettres* (1846), II, 405.

[166] *Ibid.,* II, 153-54.

as one deserves who was a knave, an atheist, an impostor, a homicide, and a public executioner."[167]

A few years after Patin's death Johann Georg Sartorius published at Altdorf an account of a Jesuit, who, in 1681, at the age of fifty-nine lost forty pounds of blood in four days by nosebleed through the left nostril. It had been his custom to be bled every year, but he had failed to do so that year. All sorts of remedies, repercussives, refrigeratives, ligatures, even bleeding itself and the sympathetic powder, failed to stay the haemorrhage until putting rue and fresh nettles well pulverized in his nostrils checked the flood. The amount of blood lost was amazing, since Avicenna and Bernard Gordon said that a man could not lose more than twenty or twenty-five pounds of blood and live, while Bartholinus estimated that there was not that much blood in the human body. In other cases of nosebleed Sartorius had used with success ass-dung and nettle juice mashed together, powdered *bursa pastoris* and henbane, or the hairs of a hare dipped in ink and put in the nostrils.[168]

Opposition to antimony and charlatans and chemists, advocacy of bleeding in and out of season:—these were the A, B, C of Patin's medical philosophy. But on occasion he could write quite dispassionately and sensibly, as in the following passage on scurvy.

For my part, it seems to me that this evil is a disease of the entire system *(morbus totius substantiae)*, a malady of the poor and undernourished, a northern and marine leprosy, which comes from a particular corruption of the blood and internal organs which upsets the natural economy. Good bread, a little wine, clean linen, fresh air, and at the beginning of the disease a moderately strong purging, would do much good, the same as never drinking bad water . . . It is common at sea as well as in Holland, Denmark, Sweden and Poland . . . I think that during my life I have seen more than two hundred theses on it; but this disease is not to be cured either by Latin words or chemical secrets. He who should cure the poverty of the people, would indeed cure scurvy.[169]

[167] *Lettres* (1907), pp. 206-7. For a more favorable estimate, Agnes Arber, "The botanical philosophy of Guy de la Brosse," *Isis*, I (1913), 359-69.

[168] *Admiranda narium haemorrhagia nuper observata et percurata a Joh. Georg Sartorio;* reviewed JS XI (1684) 288-90. The use of dried ass manure had already been suggested in Finck's *Encheiridion* of 1618: see our Chapter 6, note 113.

[169] *Lettres* (1846), III, 732-33; letter of March 13, 1670.

But then he goes on to rail in his accustomed manner against specific remedies.

There are those who say that we must find a specific; but that is how charlatans and chemists talk, who boast of having specifics against epilepsy, quartan fever, small pox, leprosy, gout, etc. When I hear these yarns which are worse than Aesop's Fables, it seems to me I see a man who wants to show me how to square the circle, the philosophers' stone, Plato's Republic, or first matter . . .

Patin was not aware that a specific was at hand in lime juice!

As we look back upon the views of Harvey and Patin, the one an experimenter and discoverer, the other a dogmatic conservative, we find that the conservative is the less animistic, astrological and magical of the two, while the progressive is the more so. The scoffer at superstition has his own cherished delusion. The discoverer of hitherto concealed truth still adheres in many respects to Aristotle and shares many of the mistaken notions of his time! Patin could detect folly; Harvey had faith in Nature. Patin's future influence upon medicine is negligible, although his racy letters provide a valuable contemporary chronicle and commentary. Harvey's discovery not only had its own value in physiology, anatomy and medicine, but led others to further experimentation, as we have seen in their pursuit of blood infusion and transfusion.

Not only did phlebotomy persist in medical practice into the nineteenth century, but belief in occult properties of human blood were still associated in the popular mind with experimental physics and scientific instruments, as may be illustrated by the passage in *Martin Chuzzlewit* in which Dickens tells of Montague Tigg's blood running dull and cold, as he entered the wood where Jonas Chuzzlewit was lying in wait to murder him:

If there be fluids, as we know there are, which, conscious of a coming wind, or rain, or frost, will shrink and strive to little themselves in their glass arteries; may not that subtle liquor of the blood perceive by properties within itself, that hands are raised to waste and spill it; and in the veins of men run cold and dull, as his did, in that hour?

CHAPTER XIX

DESCARTES

Estimates of, by Henry More, Huet, Du Hamel—His use of predecessors—Relation to Beeckman—Use of recent discoveries—His dreams and morning thinking—Theory of vortices—Preference for a simple explanation of all from one—Three elements—Terrestrial mixed bodies—Animals automata: animal spirits—Mathematics, mechanics and optics—Experimentation—Old "facts" and beliefs—Revolutionary influence on thought—Did not dispel marvels and superstition: attitude in 1629; mechanical explanation—Cartesian astrology: Placentinus, Gadroys, Magerlinus, Danckwarten—Beasts as automata: Cureau de la Chambre, Pardies, Le Grand and Duncan—Cartesian medicine: Craanen, Froment.

But no one ventured to order all natural science on new principles except René Descartes, who started in philosophizing as if no one had treated physics before

—Paschius

It appears that Descartes wished to decide all questions of physics and metaphysics without much caring if he was right

—Huygens to Leibniz, 11 July 1692

On the first page of the correspondence of Henry More, the Cambridge Platonist, is a letter to Descartes in which More writes: "All the masters of the secrets of nature who have ever existed or now exist seem simply dwarfs or pygmies when compared with your transcendent genius."[1] Pierre Daniel Huet (1630-1721), the erudite bishop of Avranches, records in his Memoirs a similar impression made upon him as a young man by Descartes.

I could not rest till I had procured and thoroughly perused his book; and I cannot easily express the admiration which this new mode of philosophizing excited in my young mind, when from the simplest and plainest

[1] Quoted by L. D. Cohen, *Annals of Science*, I (1936), 49.

principles I saw so many dazzling wonders brought forth, and the whole fabric of the world . . . as it were spontaneously springing into existence.[2]

Thus, despite Descartes' exaltation of natural reason and promise of an easy method by which any man could find in himself the knowledge essential in directing his life and then by further study acquire "the most curious forms of knowledge that the human reason is capable of attaining," his contemporaries thought of him as a magician who controlled Nature's secrets and had evoked a rational and mechanical explanation of the universe and its origin, nay of many worlds, to replace alike the single universe and incorruptible heavens of Aristotle, and the divine creation of one world out of nothing of Christian theology. "I must confess," wrote Du Hamel, secretary of l'Académie des Sciences, "that nothing could be more acutely thought out than those principles from which almost all the miracles of nature are elicited."[3]

But the whole fabric of Cartesian physics was not so spontaneously generated as it seemed to the youthful Huet. "The golden rule of mechanics," that the work needed to raise different weights to different heights remains the same if the product of the weight and the height are the same, which Descartes was once said to have originated,[4] has been traced back to the middle ages.[5] Isaac Voss in 1662 charged Descartes with having derived his law of the refraction of light from Snellius of Leyden (1581-1626).[6] Duhem affirmed that Descartes' presentation of the theory of simple machines in 1637 clearly depends on Galileo and Stevin, although he denied having ever read Galileo in a letter of October 11, 1638, to Mersenne.[7] And Galileo was disgruntled because Descartes made no mention of him in his *Dioptrique.*

We have already said something concerning Beeckman in our

[2] *Memoirs of the Life of Peter Daniel Huet, bishop of Avranches, written by himself and translated . . . with copious notes . . . by John Aikin, M. D.,* London, 1810, 2 vols.: I, 29. In later life Huet abandoned Cartesianism. His *Censura philosophiae Cartesianae* reached its fourth edition in 1694.

[3] Du Hamel, *De consensu veteris et novae philosophiae,* 1669, p. 260.

[4] Gerland, *Gesch. d. Physik,* 1913, p. 446.

[5] GS II (1931), 614.

[6] Gerland (1913), pp. 479-83.

[7] Pierre Duhem, *Les origines de la statique,* Paris, 1905, I, 331-32.

chapter on Francis Bacon but must notice him again in connection with Descartes. Isaac Beeckman (1570-1637) received the M. D. degree at Caen in September, 1618, and was head of a Latin school at Dordrecht. His doctoral dissertation on intermittent tertian fever[8] seems to have been the only work by him which was published during his lifetime,[9] but he kept a Journal, of which the manuscript survived in the provincial library of Middelburg in Zeeland. This shows that he recognized the law of inertia in 1611 or 1612, from 1613 held the atomic theory, and that water did not change to air,[10] and in general was interested in mathematics, physics and astronomy. In November, 1618, he met Descartes at Breda in the Low Countries and they exchanged views on such problems as the indivisibility of a point. Beeckman told Descartes that he had proved experimentally that ice occupies more space than water, and proposed to him the hydrostatic paradox that a bucket of water can weigh as much as all the water in the ocean.[11] Thereafter they corresponded frequently and met again at Dordrecht in October, 1628.[12] They felt that they were the only two men in the world who thought alike.[13]

According to notes in the edition of Mersenne's letters by Cornelis de Waard, who was later to edit Beeckman's Journal, the latter early adopted the hypothesis, which varied somewhat as time went on, of a very subtle matter, ether, air, spirit, fiery particles, or celestial corpuscles and effluvia, which filled the universe and animated the world, and which served to explain problems of the vacuum, falling bodies, and magnetic action. As he wrote in his Journal in 1616, "It does not rest on anything but passes without obstacle." Its action pushed the sun to the center of the universe, nourished the sun, and it was by it that the sun influenced the earth and other planets.[14] Its celestial corpuscles penetrated

[8] *Theses de febre tertiana intermittente,* Cadomi, 1618, in-4: BM 1179. d.9 (3.).

[9] *Mathematico-physicae meditationes* by him were printed posthumously in 1644 at Utrecht.

[10] *Correspondance du P. Marin Mersenne,* II, 122, 118; I, 299.

[11] *Ibid.,* I, 193, 399, 524, 313.

[12] *Ibid.,* I, 415; and consult the Index.

[13] *Journal tenu par Isaac Beeckman de 1604 à 1634,* ed. C. de Waard, La Haye, 4 vols., 1939, 1942, 1945, 1953: I, 244; II, 94-95.

[14] *Correspondance,* II, 118-19.

everything and passed freely through the tiniest pores until they encountered something solid in a heavy body. "They depress everything toward the center, either because the earth is at the center of the universe or because it is moved in a circle," he wrote in 1620, shifting back apparently from a heliocentric to an Aristotelian and Ptolemaic position. Since this spirit, or rather fire, penetrates all things, denser objects and those having more body are touched in more places. And so bodies fall faster or slower and are called heavy or light according to their matter. Thus weight in bodies is produced by bits of fire emitted from the eighth sphere, and effluvia from the eighth sphere are the cause of gravity. Fire is the cause of greater heat and cold; fire is the world soul; fire is why the magnet attracts iron.[15] But the corpuscular nature of these effluvia differs from that of the rays of the sun; otherwise a man could jump higher at night-time.[16]

On turning to the Journal itself, one receives a different impression. It is true that already in 1614 Beeckman explained the fall of heavy bodies by "a fine deflux of subtle bodies from above (*a superioribus partibus*)." It was the force of the stars, not pure light but an energetic substance mixed with light, which he associated with the spirit which passes through the pores without obstacle and so explains the action of the magnet. It was the light of the stars and the concourse of virtues coming from the eighth heaven of the fixed stars which pushed earth or sun, as the case might be, to the center of the universe. It was heat from the same eighth heaven, rather than celestial corpuscles, which depressed everything towards the center (De Waard quoted only, "Omnia deprimuntur ad centrum," instead of, "Omnia autem deprimuntur ab hoc calore ad centrum"). The fire, on the other hand, of which Beeckman spoke, was not from the eighth sphere, but was that within the concavity of the moon. It was the cause of gravity and of the temperature of the air; it was nearer and so stronger than the rays of the sun, and "posset etiam esse loco animae mundi."

[15] *Journal tenu par Isaac Beeckman de 1604 à 1634*, ed. C. de Waard, La Haye, 4 vols., 1939, 1942, 1945, 1953: I, 25, 100-101; II, 119-20, 232; III, 25-26, etc.

[16] *Correspondance*, II, 473: *Journal*, II, 23r-33.

Again, de Waard says that, since 1616, Beeckman attributed the movements of the planets to this subtle matter emanating from the sun. But the Journal says that, if the earth is at the center of the universe, this is due to the concourse of virtues from the stars of the eighth sphere. Or, if you prefer the heliocentric theory, it will be the sun which is similarly affected by those virtues. The sun's exerting its force on the planets is a secondary matter.[17]

In 1620 Beeckman specified corpuscles coming from the eighth heaven to earth and bearing with them the forces of that heaven. Upon reading Gilbert's *De magnete* for the first time on October 8, 1627, he wrote: "I think the stars emit corporeal spirits on earth, the magnet, which going out by the same way they entered compel the earth always to face the same way.[18] After reading Kepler's *De motu Martis* and *Epitome* between October 8, 1628 and February 1, 1629, he was finally of the opinion that "the effluvia of the sun move the others, nor does the sun suffer on that account," and that the space between the planets was filled with effluvia.[19] On the other hand, in December, 1626, he had put the problem to himself, whether all the stars together do not bring us more heat than the sun alone.[20] And back in 1618 he had ascribed the fall of a stone to the earth's drawing it by corporeal spirits[21] rather than to celestial effluvia driving it downward.

There is therefore not much likelihood that Beeckman influenced his friend Descartes in the latter's formation of his conception of the particles of his first element.[22] It is also evident that Beeckman had not freed himself from the notion of astrological influence and celestial virtue. He believed that the planetary aspects exerted force and he would even add a *quintus aspectus* to trine, quartal and sextile.[23] Cold was of two kinds: one was absence of heat, as when the sun sets; but the other cold was positive, as that from Saturn

[17] *Journal*, fols. 44v-45r *mg* (I, 100-101): quoted *Correspondance*, II, 474.

[18] *Journal*, II, 107; III, 17, "... terram eandem semper plagam respicere cogunt."

[19] *Ibid.*, fol. 342r-v (III, 116).

[20] *Ibid.*, fol. 261bis v (II, 376): cited, *Correspondance*, I, 355.

[21] *Journal*, fol. 106r (I, 263); *Correspondance*, II, 473.

[22] Explained later in this chapter.

[23] *Journal*, fols. 43r-v, 44r, 150 bis v; I, 97, 99; II, 139. Also III, 47: "Aspectus qui potentissimi?" III, 140, 207-8, "Planetarum virus in nos."

and other cold stars.[24] From past history he noted a prediction who would become pope,[25] and he suggested that, if astrological images were true, the best way to make them would be by a glass converging the rays of the desired star on one point.[26] He attributed critical days and tides to the moon, and held that at night, even if not shining, it was more potent than the sun.[27] He believed that the moon caused the tides by attraction rather than by its humidity.[28]

Under the caption, "How Incubi tormented me,"[29] Beeckman tells how it seemed as if someone grasped his right arm and tried to pull him out of bed. On another occasion, as he lay awake, someone seemed to seize him by the shoulder four or five times, but there was no one else in the room. He therefore concluded that the cause of these sensations was not external but in himself, although men often attribute them to Satan.

More important than Descartes' association with Beeckman or debt to any particular author or work is the fact that he was *en rapport* with the general scientific movement of his time. He might write to Mersenne that he was now studying chemistry and anatomy all at once, and that he daily learned something which he did not find in books.[30] But he would not have been studying anatomy and chemistry, had he not been aware that others too had been and were making new discoveries of that sort in those fields. He praised Harvey's discovery of the circulation of the blood and took it into account in constructing his own system of nature. He used the astronomical observations of Scheiner and cited the book of Lotharius Sarsius or Horatius Grassius.[31] He knew of the atomism of Gassendi and the pores of the chemists, and that there were machines by which air could be easily dilated or condensed. He might criticize Galileo for merely investigating certain particular phenomena and forces without treating first causes and so building

[24] *Journal,* fol. 64v; I, 155.

[25] *Journal,* fol. 97v; I, 237.

[26] *Journal,* fol. 108r; I, 269.

[27] *Journal,* fols. 50v-51v, 59r-v, 63r-v, 243r, 149bis v; I, 110-11, 137-38, 113, 151; II, 317, 137. Also III, 86, 195, 197, 200.

[28] *Correspondance,* III, 347, citing *Journal,* fols. 156 bis v-157 bis r, 243r, 258r.

[29] "Incubi ut me vexarint": *Journal,* I, 281-82.

[30] Mersenne, *Correspondance,* II, 423.

[31] *Oeuvres,* ed. V. Cousin, vol. III (1824), *Principes,* II, 35; III, 128.

without a foundation.[32] But the discovery of sun-spots by Galileo formed the basis for the theories of Descartes as to the formation of comets and planets and the stratification of the earth's crust.

It was this basing his hypotheses upon or conforming them to recent discoveries, which were a matter of common report and knowledge, that made them so readily and generally acceptable to his time. Sun-spots were hard to explain on an Aristotelian or Ptolemaic basis. Descartes not only explained them but made them explain almost everything else. And that, perhaps, was where his magic came in.

The remote cause of some of Descartes' ideas may have been something that he had read or heard and perhaps forgotten more often than he admitted or even realized. To him their immediate cause was either subconscious cerebration in dreams or conscious thought as he lay in bed after waking in the morning. In this there was something bordering upon magic: divination from dreams or the insight of the mind or soul when free from the distractions of the body.

We may briefly note some of Descartes' principal innovations. He ceased to think in terms of a single system, whether solar or geocentric, surrounded by a single sphere of fixed stars. The sun was not a planet but one of the stars, and the so-called fixed stars, which shone by their own light and not that of the sun, were each near the center of a vortex or tourbillon of its own which whirled about it. These vortices were not of equal size, and there was "incomprehensible variety" in the situation of the so-called fixed stars and their distance from us and from one another. No point in the universe was truly immobile; the vortices were not perfectly spherical and might lose matter to a neighboring vortex or be wholly swallowed up by it; in the movements of the heavens, as in all other natural phenomena, time kept bringing change. There was vast space between our sun and the other stars. Our earth and other planets were once glowing masses like the sun, and each in its own tourbillon; but their "sun-spots" in time enlarged to a solid

[32] Gerland, *Gesch. d. Physik*, 1913, p. 442, citing *Les lettres de René Descartes*, Paris, 1659, II, 391, 394.

crust; their light now comes from the sun, in whose vortex they revolve, although some of them still have moons as relics of their former vortices. As the earth, losing most of the tourbillon of matter which had whirled about it, came closer to the sun, its crust separated into different layers with air outermost, then water, clay, sand, stone and metals. The action of the sun broke up this crust somewhat, raising some land above water, filling subterranean caverns with water, and tipping different strata, while beneath the crust there was still a glowing central core which occasionally expressed itself in volcanoes and earthquakes. Such was Descartes' theory of vortices which has been described as "one of the grandest hypotheses ever imagined to account for the movements of the universe." Some of it was wrong but some of it was more nearly right than anything before. Recent explanations of the structure of the atom have been somewhat suggestive of microcosmic tourbillons.

Long before the belief in one God had affected Hellenic polytheism, the pre-Socratic Ionian philosophers had sought for one world-ground. This sounds somewhat silly, when one contemplates the great variety of scent and color in a single bed of flowers and the diverse seeds from which they grow. But Descartes could not break away from the *materia prima* of the schoolmen.

Similarly Ramus had attacked Aristotle and scholasticism, yet had held that if one began by giving a correct definition of a given science, it would be a simple matter to deduce all its further ramifications of detail. In geometry one started with a few axioms and based the proof of subsequent propositions upon them. So for the philosophy of Descartes the first fundamental fact was self-consciousness; and for his explanation of nature, heavens and earth were of one and the same matter, and its essence was extension. The simpler an explanation, the better. Descartes could not imagine principles simpler or more intelligible or more probable than his own, and was sure that all the phenomena of nature could be explained by their means. This would seem to imply that the Creator too was simple-minded and not given to intricate thinking or complicated activity. One God, one matter, one line of thought! Only when he considered the infinite spaces of the stars, did

Descartes momentarily escape from this simplex and fashion his magnificent hypothesis of tourbillons.

But, as we have seen, everything was in movement. Descartes piously made God the first cause of motion, but since an equal amount of it was always conserved in the universe, His intervention was no longer necessary or, for that matter, possible. No more action was needed for motion than for rest, all moving bodies continued to move until stopped by some other body, and all tended to move in a straight line. Thus in the tourbillons we have centrifugal force, and it explains why the bodies of the sun and stars are round.

Descartes denied the existence of atoms in the sense of indivisible bodies. But as a result of movement and erosion, his one matter had in some mysterious, not to say magical or impossible, way become divided into three elements, which differed only in size, shape and velocity. The smallest and most rapidly moving particles primarily formed the sun and stars and were luminous. The second element consisted of balls, which had been rounded off in the course of time, made up the heavens, and were transparent, transmitting mechanically the light of sun or star. The third element, made of larger fragments than the first and more slow-moving, formed sun-spots, comets, planets and the earth's crust, and was opaque. Obviously there are spaces to be filled or moved through between the balls of the second element, and there are pores in the sun-spots and other formations of the third element. Matter of the first element enters by the poles of a vortex, moves towards its center, and leaves at places farthest removed from the poles. It can pass out through the pores in the sun-spots but not by the same pores through which it came in, and those particles from one pole require different pores than those from the other.

A comet differs from a planet in not revolving always in a single vortex. It describes an orbit which is irregularly curved according to the different movements of the vortices through which it has passed. It gets its light from the sun or star of the tourbillon in which it is for the time being and so may appear and disappear. Its tail always appears away from the sun.

While the third element is the chief constituent of earth and water, its little parts become fire when they are so separated from one another and surrounded by particles of the first element that they have to move with it. Or they become air when similarly encircled by the balls of the second element. All bodies in the universe touch one another continuously without leaving any vacuum, so that even the most distant always exert some action on the others. The different properties and qualities of mixed bodies are entirely produced by the varying size, shape and movement of the particles composing them.

Descartes not merely attempted a purely mechanical explanation of the material universe and of inanimate nature, but held that animals were mere automata and that the sole principles of physiology were motion and heat. "There is nothing in us that we ought to attribute to our soul excepting our thoughts."[33] This meant that he rejected the vegetative soul and the sensitive soul of Aristotle and retained only the rational and immortal soul. For the other two he substituted the animal spirits which we have already often seen such a favorite resort of the sixteenth and seventeenth centuries. He defined them as "nothing but material bodies . . . of extreme minuteness," which were formed in the cavities of the brain from the most animated and subtle parts of the blood which heat had rarified in the heart. They were never at rest but moved with great speed and caused the movement of the muscles. Presently a chapter is entitled, "How all the members (of the body) may be moved by the objects of the senses without the aid of the soul."[34]

Descartes was an accomplished mathematician who made several improvements in algebraic notation and founded analytical geometry by relating curves to equations. But he retained some of the secrecy of sixteenth century mathematicians, writing purposely in an obscure style, although most of his other works are models of French clarity, omitting large portions of the analysis, and not stating all that followed from his conclusion, "so as to leave to posterity the pleasure" of further discovery.[35] And he asserted that

[33] *The Search after Truth*, cap. xvii, in *Philosophical Works*, English translation, by Haldane and Ross, 1911, 2 vols., I, 340.

[34] *Ibid.*, caps. x-xi, xvi.

[35] A. Wolf, *A History of Science, Technology and Philosophy in the Sixteenth and Seventeenth Centuries*, 1935, p. 196 *et seq.*

"infallible reasoning of geometry" proved that the sun was six or seven hundred diameters of the earth distant from us,[36] a figure far too small.

Although Descartes conceived of a mechanical universe, automatic animals, and an almost mechanical man, and although Gerland in 1913 could still affirm that he reached results in mechanics which are still valid through his experimental activity and use of mathematical aids,[37] more recent histories of science hardly mention him in connection with mechanics, and certainly not in the same breath with Galileo and Huygens.

Les Météores offered one of the first correct explanations of the rainbow since Dietrich of Freiberg in the early fourteenth century, although Leibniz and Newton might suggest that it was taken from de Dominis (1566–1624).[38] *La dioptrique* in its day may have seemed a gem of exacting scientific technique. Yet on the one hand Descartes believed that the motion of light was instantaneous, because he observed no delay in eclipses, and, on the other hand, held that light penetrated a dense medium more readily than a rare one.[39] Fermat (1601–1665) soon demonstrated experimentally that the opposite was true.[40]

Descartes had more faith in reason and in mathematics than he did in experimentation. On December 23, 1630 he wrote to Mersenne that, in the case of the more particular experiments, it was impossible not to perform many superfluous and even false ones, unless one knew the truth before making them.[41] Towards the close of his life, however, he wrote to Henry More:

> I am not sure that I will ever bring to light the rest of my philosophy, since it depends on numerous experiments for the accomplishment of which I know not if I shall be granted the opportunity.[42]

[36] *Principes,* II, 5.

[37] *Gesch. d. Physik,* 1913, p. 442.

[38] See Carl B. Boyer, "Kepler's Explanation of the Rainbow," *American Journal of Physics,* XVIII (1950), 366. Recently Boyer has shown that, while Dietrich's explanation of the formation of the rainbow "compares very favorably" with that of Descartes, the latter solved the problem of the shape and size of the bow, which Dietrich had failed to do.

[39] *Oeuvres,* X (1908), 242-43: "Lux . . . facilius penetrat per medium densius quam per rarius."

[40] Henry Crew, *The Rise of Modern Physics,* 1928, pp. 146-47.

[41] Mersenne, *Correspondance,* II, 597-98.

[42] Letter of April 15, 1649, quoted

Indeed, at the time when he wrote the *Discours de la méthode* his conception of experimental science was more like that of Roger Bacon in the thirteenth century than it was like that of Robert Boyle later in the seventeenth century. But we must remember that Galileo's *Two New Sciences* did not appear until the following year. Descartes says that it will be impossible for him to treat in detail of the sciences which are deduced "from rare and well thought out experiments . . ."

for we should first of all have to examine all the herbs and stones brought to us from the Indies: we should have to have beheld the phoenix, and in a word to be ignorant of none of the marvelous secrets of nature.[43]

Thus for Descartes experiments were still associated with, if not synonymous with, marvelous secrets, with far-off wonders of India, and with the fabulous phoenix. He professed to have wiped the slate of his mind clean, to have razed the previous edifice of knowledge to the ground. Yet he retained such beliefs as these, presumably because he classed them not as erroneous opinions or matters of doubt, but as established facts.

Similarly the aim of his new method in philosophy was not to discover new facts but rather a better explanation of the old facts, a more certain ground for accepted beliefs. Gilson has well said that the Cartesian philosophy was in large part a clear explanation of facts which do not exist. But for Descartes they did exist. Nor did he attempt a new classification of knowledge, or to direct the human mind into new channels of inquiry. His purpose was rather to deal with the century-old problems which had long engaged human thought, but to solve them by an appeal to natural reason rather than by scholastic methods and authorities. He still denied the existence of a vacuum. He still tried to answer such questions as why children and old people weep easily. He still believed in animal spirits. He repeated, without acknowledgement Costa ben Luca's tenth century explanation of thought as centering in the movement of the pineal gland.[44] He still endeavored to explain

by L. D. Cohen, *Annals of Science*, I (1936), 55.

[43] *The Search for Truth*, English translation, 1911, p. 309.

[44] T I, 659.

why the sea was not increased by the rivers flowing into it by supposing the return of its water through underground passages to the tops of mountains, although Jacques Besson in the previous century had maintained that evaporation and rainfall sufficed to supply all springs and streams.[45] Descartes still discussed bitumen and minium, earthquakes and volcanoes and comets, and the other natural phenomena which had been the staple topics of scientific treatises for centuries. He suggested a possible explanation for the reputedly inextinguishable lamps which burned hundreds of years without addition of fuel. In general, he was concerned with the same problems, subjects and notions as had occupied the minds of philosophers and scientists for ages past.

But he offered a new explanation of these by his new method, and by his new hypotheses of vortices, of comets, of animals as automata, and of three elements, common to heavens and earth, as against the four terrestrial elements and fifth celestial essence of antiquity and middle age, or the three principles of Paracelsus. Fontenelle said that the Cartesian philosophy shed a new light on the whole thinking world, and that books written since had been better arranged and more precisely expressed.[46] The hypothesis of vortices was for a time generally adopted, and that of three elements was very influential. Most of all, he emboldened others to think for themselves, to forge their own explanations and classifications of natural phenomena, and to base these upon recent mathematical analysis and experimental or observational discovery rather than upon the most ancient authorities.

Barchusen, in his History of Medicine, published in 1710, said that many medical men—Regius, Le Grand, Craanen, Brockhusius, Waldschmid, etc.—had borrowed numerous hypotheses from the physics of Descartes and combined them with those of Galen and the chemists. This had put a new face upon rational medicine.[47]

If, however, we ask ourselves whether the Cartesian attitude of doubt and of discarding authority, and his mechanistic interpretation of the natural universe, were directed against the belief in

[45] T V, 592.

[46] J. Delvaille, *Essai sur l'histoire de l'idée de progrès jusqu' à la fin du XVIIIe siècle*, 1910, p. 211.

[47] *Historia medicinae*, 1710, p. 524.

the influence of the stars and in natural marvels, and whether he was responsible for the abandonment of such views or of superstition in general, the answer will have to be in the negative. Descartes was so confident in his ability to explain anything and everything by his principles, that he was apt to employ them in justifying such dubious beliefs as that in inexhaustible lamps, or in the bleeding of the corpse of the victim in the presence of the murderer, as well as in expounding genuine natural phenomena. He had no doubt that he could discover the art of prolonging human life.

His sharp separation of the spheres of mind and body, and his insistence upon clear and distinct ideas, were noncompatible with the animism of Kepler, the mysticism of the alchemists, the spiritual science and occult medicine of van Helmont. In the long run they would work against the association of magic with experimental science. But for the moment he inclined to preserve many traditional marvels in the persuasiveness of his argument and clarity of his thought, like flies in amber. For the time being he sought an alliance with natural magic in his attack upon scholastic philosophy.

Already in August, 1629 Descartes had written to Mersenne that there was a part of mathematics which he called the science of miracles because by use of air and light it could produce all the illusions that they say magicians cause to appear by the aid of demons. "This science has never been practiced that I know of."[48] If, however, he had in mind optical illusions, use of mirrors and the magic lantern, his science of miracles was of course much older than he imagined.

In October of the same year he wrote that he judged from the title of Gaffarel's recent *Curiositez inouyes sur la sculpture des Persans, horoscope des patriarches et lecture des estoilles* that it would contain only chimeras.[49] Thus he already drew a sharp line between natural or mathematical magic, which could be effected or explained mechanically, and an immaterial magic based on the power of words, pictures and diagrams. But there was one close resemblance between him and Gaffarel. The latter offered un-

[48] Mersenne, *Correspondance,* II, 253.

[49] *Ibid.,* II, 303.

heard-of curiosities; Descartes, a science never before practiced.

Descartes, it is true, felt that he was above being deceived

> by the promises of an alchemist, the predictions of an astrologer, the impostures of a magician, the artifices or the empty boastings of any of those who make a profession of knowing that of which they are ignorant.

Yet he was credulous as to the wounds of a corpse bleeding at the approach of the murderer and as to instant warnings, in dreams or waking, of the afflictions, danger or death of distant friends and kindred. In *The Search after Truth* Epistemon is especially curious concerning

> the secrets of the human arts, apparitions, illusions: in a word, all the wonderful effects attributed to magic. For I believe it to be useful to know all these things, not in order to make use of the knowledge, but in order not to allow one's judgment to be beguiled into admiration of the unknown.[50]

And Eudoxus promises:

> after having struck wonder into you by the sight of machines the most powerful and automata the most rare, visions the most specious, and tricks the most subtle that artifice can invent, I shall reveal to you secrets which are so simple that you will henceforward wonder at nothing in the works of our hands.[51]

Both the phenomena here alluded to and the explanation promised were commonplaces of natural magic.

Again in his *Principles* Descartes affirmed that there were no qualities so occult, no effects of sympathy or of antipathy so marvelous or strange, but that his principles would explain them, provided they proceeded from a purely material cause. His chief suggested explanation was that the long, restless, string-like particles of the first element, which existed in the intervals or interstices of terrestrial bodies, might be the cause, not only of the attraction exerted by the magnet and amber, but of an infinity of other

[50] *The Philosophical Works of Descartes*, rendered into English by Elizabeth S. Haldane and G. R. T. Ross, Cambridge University Press, 1931, I, 86, 310. I have slightly modified the translations.

[51] *Ibid.*, I, 311.

marvelous effects. "For those that form in each body have something particular in their figure that makes them different from all those that form in other bodies,"[52] and they may pass to very distant places before they encounter matter which is disposed to receive their action. According to Barchusen, the Cartesian hypothesis of which medicine had especially made use, was his much-vaunted first element and its passage through pores, "the chief cause of extraordinary effects."[53]

Since Descartes was so confident of his ability to think up a rational and mechanical explanation for all such seemingly occult phenomena, he was likely, for a time at least, to encourage rather than discourage the belief in them. Furthermore, his tendency to advertise the results of his method as marvelous as well as easy of attainment savored more of magic than of science.

Astrology in especial sought support and justification from Cartesianism. Kirchmaier, in a work printed in 1680, says that twenty-one years ago the Cartesian, John Placentinus, professor at Frankfurt-on-the-Oder and *mathematicus* (which probably means astrologer) to the Elector of Brandenburg, held that the principle and origin of human life was a most subtle celestial matter, analogous to the element of the sun and fixed stars, and that the natural motion of the limbs proceeded from that celestial matter. Further Placentinus cited a Jean de Raey, professor at Lyon and author of two dissertations on this subtle matter, as affirming that, after the child is born, it continually fills its lungs with the subtlest particles of this celestial matter. Kirchmaier, however, objects that the nature of celestial matter is as yet unknown, and that the first element of Descartes can be explained far otherwise than as a matter analogous to the sun and fixed stars.[54]

A Discourse on the Influence of the Stars according to the Principles of Descartes, composed by Claude Gadroys, first appeared in 1671[55] and then was reprinted in 1674 without change

[52] *Principles of Philosophy*, IV, 187.

[53] *Historia medicinae*, 1710, p. 524: "decantata illa mundi materia subtilissima . . . causa princeps extraordinariorum effectuum."

[54] Geo. Casp. Kirchmaier, *De phosphoris et natura lucis necnon de igne*, Wittenberg, 1680, BM 1033.h.20, p. 32.

[55] *Discours sur les influences des*

except that to the title was added, "where it is shown that there goes out continually from the stars a matter by means of which may be explained the things which the ancients have attributed to occult influences."[56] This matter is of course Descartes's first element. After chapters on the nature of the stars, of the sun and fixed stars, and of the planets, Gadroys tells how the matter from the stars gets here, and then discusses conjunctions, oppositions and aspects. He not only ascribes to the stars effect on weather and health, but goes on to broach the theme of talismans or astrological images.

Some regard them as vain and superstitious; others consider them useful and natural. They used to be attached to the prows of ships, and Gregory of Tours tells of one against rats and snakes on the bridge entering Paris. When it was somehow lost, the city, which had been free from such pests before, was invaded by them. Gervais of Tilbury in *Otia imperialia* tells of like talismans by Vergil on the gate of Naples. In Gadroys' own time a child relapsed into its former maladies when it lost the image against them. When it was found again, it was restored to its accustomed place only in the presence of several persons assembled for the purpose and who were obliged to recognize its virtue by its instant effect. After all these examples, Gadroys thinks it rash to doubt the efficacy of such images, although we have heard Descartes do so. His explanation is that the matter of the star fills the pores of the metal as it is cast and is conserved there, but the virtue lasts only for a time. But there is no reason why it should not be as effective as carrying about a powdered toad. The sympathetic powder acts at a great distance, the saliva of a mad dog gives rabies; so talismans excite love or hate, and demons can do no more. It only remains then to choose the constellations under which to fabricate the images. In this one should follow the ancients who have discovered the properties of the stars. Some say that they acted capriciously, but Gadroys believes that they gave names to the stars and divided up the heavens as a result of observing their effects. There is much, however, in the art of talismans for which he would not answer; a hundred errors

---

astres selon les principes de M. Descartes, Paris, J. B. Coignard, 1671, in-12, preface and 220 pp. BN V.21795, also V.21824 and 29261.

[56] BN R.13706. Reviewed without naming the author, JS IV (1675), 74-79.

have crept in; and the art is no longer considered other than a superstition.[57]

Gadroys goes on, however, to explain how the stars incline, although they do not compel, us to this or that action. It is through their effect upon the animal spirits, for difference in the animal spirits can produce diversity of inclinations. Descartes says that spirits can differ in four ways as they are more or less abundant, their parts more or less gross, more or less agitated, and more or less equal or equable. Their abundance excites love, goodness and liberality. If they are strong and gross, they make for confidence and boldness; if equally agitated, for tranquillity; if unequally agitated, for desire, promptitude and diligence. If deprived of all these qualities, they engender malignity, fear, inconstancy and disquietude. Sanguine humor is compounded of promptitude and tranquillity of mind, and perfected by goodness and confidence. Melancholy humor is compounded of sloth and disquietude, and augmented by malignity and fear. Choleric humor is a compound of promptitude and disquietude, fortified by malignity and defiance.[58]

It is important to note the positions of the planets at the moment of birth, because immediately thereafter the parts of the brain set themselves and conserve all through the course of life the first impressions which they have received. The force of a star's impressions depends upon its finding in the child dispositions conforming to its quality. Each planet causes certain inclinations in men according to its distance from the sun and consequent solidity. Saturn, being far from the center of its tourbillon, that is to say, from the sun, is very solid and coarse. The planets are solid bodies which are governed by certain laws to which liquid bodies such as the sun and stars are not subject.[59]

But then Gadroys immediately begins to back water again, stating that the Greeks added a hundred superstitions to astrology, and proceeding to criticize astrologers himself, leading to the conclusion that "it is a criminal temerity to pretend to pierce the thick darkness of the future." One should judge only in general con-

[57] *Discours.*, pp. 101-16.
[58] *Ibid.*, pp. 119, 124-25.
[59] *Ibid.*, pp. 144, 146, 149-51, 186.

cerning health or sickness, the fortune or misfortune of the new-born babe, and whether he will be amiable and gentle or otherwise. He holds further, however, that even some of our particular actions can be conjectured from the stars. When they recur in the same disposition as at the first moment of our life, they make us act in a way which we would not do, if we were not extraordinarily agitated. Gadroys then concludes with the statement that he is content to have laid the foundations of a new astrology of which he leaves it to others to rear the edifice. He doubts not that they will succeed in rendering this science very considerable, if they will concentrate upon the nature of each star in particular and upon observing the time of its domination.[60]

In his book of 1675 on The System of the World, of which we shall treat further in Chapter XXIII, only three short pages are devoted to the influence of the stars,[61] and Gadroys' attitude is more non-committal. He says that this question has long been discussed, and that he believes that it will continue to be debated for a long time to come. He would not assert positively the existence of such influence, neither would he deny it, for there are many things of which the causes do not seem terrestrial. Three years ago he discussed the question at length and so will not dwell on it further now.

About 1679,[62] Petrus Magerlinus, who seems to have been a lawyer by profession,[63] composed a Cartesian Astrology or Demonstration of the principles of astrology from the philosophy of René Descartes, with an Epilogue in which certain tenets of Descartes were refuted.[64] This shows that Magerlinus regarded himself as something of a philosopher, and he speaks of thirty-seven

[60] *Ibid.*, pp. 188-214, 214-18.

[61] *Le système du monde selon les trois hypothèses*, Paris, 1675, in-12, pp. 392-94.

[62] Florence, Laurentian library, MS Ashburnham 1530. At p. 203, Kepler's *Harmonice mundi* of 1619 is dated "ante annos sexaginta."

[63] *Ibid.*, p. 209, "negotia juridica quibus sum implicatus," while he closes his work at p. 216 with a quotation from Bartolus.

[64] *Ibid.*, title page: "Petri Magerlini Astrologia Cartesiana autograph," on what was perhaps once a piece of the cover of the MS; then "Astrologia Cartesiana, id est, Demonstratio principiorum astrologiae (here the words "Ptolemaicae sive vulgaris" are inserted) ex philosophia Renati Des Cartes. Cum epilogo apologetico in quo placita quaedam Cartesii refutantur."

years experience in testing the truth of astrology.[65] The underlying idea of Magerlinus' work is that the matter of Descartes' first element, which exists in the sun, planets and fixed stars and also is dispersed between the particles of the second and third elements, furnishes a basis for a mechanical explanation of astrology. The attempts of men like Abdias Trew in Germany and J. B. Morin in France to base astrology upon the philosophy of Aristotle involve occult qualities and "cabalistical relationships" which leave Magerlinus cold. But in Kepler's harmony of the soul of the universe and soul "at the center of the earth," and mutual sympathy of superiors and inferiors he finds a close resemblance to Descartes,[66] although it may seem to the modern reader that the one was animistic and the other mechanistic. Magerlinus is not so insane as to swallow all the *nugae* and superstition of astrologers, but his own experience and observation over many years convince him that many mysteries of nature are contained in the wrappings of astrology, "and have come down to us through so many centuries not without singular divine providence."[67] He still retains astrological houses and the triplicities of the signs, aspects and *antiscii,* exaltations of planets and houses, critical days, directions and revolutions, and prediction of the weather and public events.

Magerlinus criticizes Descartes for setting aside the Word of God and only scrutinizing His works, for holding that the world is infinite, and with regard to the matter of his first element, vortices, comets, clouds and thunder. But it makes no difference to his astrology whether comets are eternal bodies, as Seneca held, or generated from the fixed stars by incrustation and destruction of their vortices, as Descartes held, although Magerlinus personally would attribute their generation rather to mutual adhesion of particles of the first element.[68]

[65] *Ibid.,* pp. 106-7, "Me vero experientia 37 annorum quibus in veritatem astrologiae per quamplurima experimenta diligenter requisivi abunde convincit."

[66] *Ibid.,* pp. 100-102, 108.

[67] *Ibid.,* p. 215.

[68] *Ibid.,* pp. 173, 177 *et seq.* Magerlinus is possibly to be identified with Peter Megerlin (1623-1686) who in 1674 became professor of mathematics at Basel and was much given to astrology and wrote on comets (titles in Poggendorff). BM has his *Systema mundi copernicanum,* Amsterdam, 1682, and *Theses astronomiae* at which

Another who maintained that the Cartesian philosophy was not unfavorable to astrology was Christian Gottfried Danckwarten, M.D., of Hamburg, in 1684, in a work written in German[69] but with so many quotations from authorities in Latin that it almost seems to be in that language.[70]

The biblical account of creation was interpreted in Cartesian terms in a book entitled *Cartesius Mosaisans* and again in 1685 in *Le monde naissant ou la création du monde demontrée par des principes tres-simples et tres-conformes à l'histoire de Moyse.*[71]

After Descartes' characterization of brute animals as automata, Cureau de la Chambre composed a Treatise of the Knowledge of Animals, where all that has been said for and against the reasoning of beasts is examined.[72] First printed in 1647, it had new editions in rapid succession in 1648, 1662 and 1664, and meanwhile appeared in English translation at London in 1657. As the interest which had been aroused by Descartes' pronouncement somewhat subsided, Cureau de la Chambre turned to another theme anent animals, which had long been of perennial interest and also bordered more closely upon magic, namely, the sympathy and antipathy existent between them.[73]

That many did not agree with Descartes is seen from the revival of the sixteenth century work of Rorario (1485–1556), That Brute Animals Use Reason Better than Man. Gabriel Naudé edited a manuscript of it which he had found in Italy at Paris in 1648, and there were other printings at Amsterdam in 1654 and 1666.[74]

---

he was *Praeses,* Basel, 1676. Col 523.6 Z2 is his *Astrologische Muthmassungen von der bedeuttung des jüngst entstandenen Cometen,* Basel (?), 1665.

[69] *Astrosophia coeli terrestris Iatrologia oder dass Gestirnte Erdreich...,* Hamburg, 1684, 79 pp., p. 11. Copy used: BM 1141.c.8 (4).

[70] Especially since the quotations are in Roman type.

[71] Printed at Utrecht, in-12. Reviewed: JS XIV, 109-11.

[72] Marin Cureau de la Chambre, *Traité de la connaissance des animaux où tout ce qui a esté dit pour et contre le raisonnement des bestes est examiné.*

[73] *Discours de l'amitie et de la haine qui se trouvent entre les animaux,* Paris, 1667, 248 pp.

[74] I have used the 1654 edition: Hieronymi Rorarii exlegati pontificii, *Quod animalia bruta ratione utantur melius homine.* Libri duo, 117 pp. The text is dated at the close Aug. 1, 1544.

A MS which seems not to have been printed is Brussels 4013, Alexander Wiltheim, Utrum brutis sit non ratio-natio?

In the later years of the century, however, many writers followed Descartes in his contention that animals other than man were mere automata. Thus the Jesuit mathematician Pardies (1636–1673), in a treatise on the knowledge of beasts,[75] held that they were machines moved by the spirits which formed in the brain and spread to all the muscles. He further contended that Aristotle had often represented beasts as automatic machines, while Augustine mentioned one of which the parts kept moving after it had been cut to pieces. But Pardies did not really believe that brutes were machines for he admitted that they had sensations, although not intellect, whereas true machines have neither senses nor feeling. Similar limitations are apt to hold true of other writers on the subject. Thomas Willis, *De anima brutorum* was published in London in 1671–1672, at Amsterdam in 1674, and appeared in English translation in 1683.

Anthony Le Grand's book on the lack of sense and cognition in brutes was printed at London and Leyden in 1675 and at Nürnberg in 1679.[76] Daniel Duncan's book on The New and Mechanical Explanation of Animal Actions, printed in 1678, will be considered in a later chapter.[77] Theodore Craanen (d. 1688) was a Cartesian in the field of medicine and contended that, to explain most of the bodily functions, it was unnecessary to resort to the soul as mover. He compared the human body to a clock.[78]

As late as 1694, N. Froment published a medical work in which he professed to explain everything by "the principles of the celebrated Descartes and the experience of the best practitioners."[79] He attributed fevers to acid, acrid or salty yeasts in the alimentary canal, and these to sadness or bad sustenance. He recommended

[75] Ignace Gaston Pardies, *Traité de la connaissance des bêtes*, 1672. I have used an edition of 1696, in his *Oeuvres*, BN V.48828. Huygens thought well of a work on refraction by Pardies which was not published and is now lost: *Oeuvres complètes de Christiaan Huygens*, XXII, 677, 692, 904.

[76] *Dissertatio de carentia sensus et cognitione in brutis*, copies of all three editions in both BM and BN.

[77] Chapter 34.

[78] JS XVII, 754.

[79] *Hypothese raisonée, dans laquelle on fait voir que la cause interne de toutes les fievres, et generalement de toutes les autres maladies, vient des levains acides, acres ou salez qui se rencontrent dans les premieres voyes. Le tout expliqué sur les principes du celebre M. Descartes, et confirmé de l'experience des meilleurs Praticiens.* Paris, 1694, in-12. Reviewed: JS XXIII, 27-32.

remedies such as powdered vipers, spirit of hartshorn, spirit of sal ammoniac, and diaphoretic antimony, which increased the amount of spirits in the body. But he explained fermentation in Cartesian terms as matter of the third element surrounded by the sole subtle matter. He rejected bleeding as useless and pernicious, and preferred his favorite chemical remedy, antimonial tartar, even to quinine. His work was severely criticized in *Le Progrès de Medecine* of 1696.

CHAPTER XX

THE UNDERGROUND WORLD OF KIRCHER AND BECHER

Kircher's *Mundus subterraneus*—His credulity—Show of experimental method—Hypothesis of underground reservoirs—Impetus, mountains, fire—Signatures, animal marvels, spirits in mines—Occult action, spontaneous generation—Attitude towards alchemy—Perpetual motion impossible—Influence of the stars—Other topics—Purpose of the work questioned—Kircher's Arithmology—Becher's Subterranean Physics—Three kingdoms—Poisons and medicines—Astrology condemned—Alchemy approved—Other writings—Boyle and Rossetti—Origin of springs: Riccioli, Dobrzensky, Voss, Perrault, Herbinius, Plot, Mariotte—Ramazzini—Fontana.

Where that spirit lurks and what the source is whence it is diffused, is unknown to us in the present blindness and ignorance of the human mind as to the internal and subterranean constitution of the terrestrial globe

—SENNERT

We have heard Morin tell of his visit in 1615 to deep mines of Hungary and his theory of three layers of earth corresponding in reverse of temperature to the three regions of air.[1] In 1641 Giovanni Nardi of Florence had published a treatise on subterranean fire.[2] We come now to two longer and later works on things underground.

The imposing work of Athanasius Kircher on the subterranean world,[3] in two folio volumes and twelve books, was suggested by earthquakes of 1638 in Calabria, at which he was present for fourteen days at great peril to his life, and learned great secrets of nature.[4] The book is illustrated by a number of large plates show-

[1] See Chapter 16.

[2] *De igne subterraneo physica prolusio*, Florence, 1641, 152 pp.

[3] *Athanasii Kircheri e soc. Jesu Mundus Subterraneus in xii libros digestus*, Tomus I ad Alexandrum VII pont. opt. max., Amstelodami, Apud Joannem Janssonium & Elizeum Weyerstraten, 1665, 346 pp. in-fol.; Tomus II, 1664, with dedication of June, 1663, to the emperor, Leopold I, 487 pp. Col. 502 K632.

[4] Praefatio, cap. 2.

ing the interior of Vesuvius and other grandiose, if imaginary, underground scenes. Like most of Kircher's volumes, the work contains a great deal of matter which has only the most remote, if any, connection with the subject stated in the title. The Latin text was republished at Amsterdam in 1668 and 1678, and a Dutch translation appeared there in 1682.

John Webster, writing in 1671, spoke of our author and work in these words:

> Athanasius Kircher, that universal scribbler and rhapsodist, who after a great many huge and barren volumes did promise the world a work by him styled *Mundus Subterraneus* which put all the learned into great expectations of some worthy and solid piece of mineral knowledge. But alas! when it appeared, every reader may soon be satisfied that there is but very little in it except the title that doth answer such conceived expectations or fulfill such great promises.[5]

Webster, who had more faith in the possibility of the transmutation of metals than Kircher, went on to complain that the *Mundus Subterraneus* was "stuffed with scandals and lies against Paracelsus, Arnoldus and Lully."[6] "Historias miras recensuit Athanasius Kircher," said Garmann.[7]

On the other hand, Gabriel Clauder, physician to the Duke of Saxony and member of the Academy of the Curious, writing seven years after Webster, and even more anxious than he to refute Kircher's attack upon alchemy, nevertheless granted that Kircher was more powerful than ten thousand other antagonists. His incomparable brain had produced more works than warriors poured forth from the Trojan horse. So great were the merits of his writings, especially in that part of philosophy related to physics and medicine, that "so long as the world lasts, he will be justly named Athanasius."

> Not our Europe only but the whole world knows how much light he has shed by his laborious dexterity and rare keeness of genius in this current age on many sciences,

[5] John Webster, *Metallographia*, London, 1671, p. 30.

[6] *Ibid.*, p. 31.

[7] *De miraculis mortuorum*, 1709, p. 375.

but especially by his *Mundus Subterraneus.*[8] Similarly, only the year before, Johann Daniel Major had written of "the vast glory radiated" by Kircher's *Mundus Subterraneus.*[9] And Kestler, in the preface to his *Physiologia Kircheriana Experimentalis,* first published at Rome in 1675, called Kircher "the prodigious miracle of our age who has excited the admiration of the whole world by the innumerable experiments on which he has based his universal sciences."

Like Kircher's other works, the *Mundus Subterraneus* habitually calls everything in nature wonderful and marvelous. In the first "centrographic" book, the nature of the center is called mirific and then "admiranda et admirabilis" three or four times more. And the movement of projectiles in parabolas is accompanied by an account of "its marvelous effects."[10] The second book is on the wonderful work of the terrestrial globe.[11] Earth is not homogeneous but of heterogeneous nature with "a marvelous variety of things."[12] In the third "hydrographic" book on the nature of the ocean we hear of the marvels which geographers tell of a certain Minorite at Oxford who was transported by magic to the north pole and saw a vast whirlpool there.[13] Later we have a disquisition on the miracles of waters and the prodigious nature and property of certain fountains.[14] Or we hear of "that vast and inexplicable variety of things which the earth offers us."[15] In its viscera are generated diverse juices which, mingled with waters that they encounter, produce marvelous effects and a marvelous genesis of things according to the earth they are mixed with. Or in the second volume we hear of "the marvelous antipathy of things,"[16] the inexplicable force of mercury and the marvelous properties of mercury.[17]

Kircher was notorious for his credulity, and in the present work

[8] Gabriel Clauder, *De tinctura universali,* Altenburg, 1678, pp. 58-60.

[9] *Genius errans,* Kiel, 1677, fol. Dv: "ille itidem vasta gloria radians Mundus Subterraneus Kircheri."

[10] *Mundus Subterraneus,* I, 30. The subject had been treated by Galileo.

[11] *Ibid.,* I, 55.

[12] *Ibid.,* I, 108.

[13] *Ibid.,* I, 159. Roger Bacon would seem meant.

[14] *Ibid.,* I, 273; and at p. 288, "De reliquis aquarum miraculis."

[15] *Ibid.,* I, 329.

[16] *Ibid.,* II, 110.

[17] *Ibid.,* II, 150-151.

tells more than one tall story. One is of a diver in the time of Frederick, king of Sicily, who spent so much time under water that a web grew between his fingers like that on the foot of a goose or duck, while his lungs became so distended that they contained a supply of air enough to last for an entire day.[18] Another more familiar story is of hibernating bears who sleep in mountain caves for forty days, then suck their paws for the rest of the winter, and grow fat as a result.[19] Or there is the long account of a winged dragon in the island of Rhodes with a poisonous breath which proved to be so invincible that the local king finally forbade anyone to attack it. A certain Deodatus de Gozano from Italy, however, in 1345 constructed an artifical dragon and trained his horse and dogs to attack it, while his servants were provided with drugs to resuscitate him from the venom. Thus prepared, he returned to Rhodes and slew the dragon. The king, however, imprisoned him for disobeying his ordinance, but the people murmured so that the king soon released him and he later became his successor.[20] A man in Kircher's native country drank stagnant water and after two months suffered great pains and felt as if some animal were moving in his stomach. The doctor heated a dish of milk and suspended the patient above it by the feet, whereupon, to the astonishment of all the bystanders, a snake six feet long came out of his mouth to get the milk.[21]

In this connection we may take note of a letter of March 30, 1661, to Robert Boyle from a friend of his at Rome, R. Southwell, who frequently visited Father Athanasius Kircher. Kircher gave this explanation of the origin of "the Soland goose in Scotland." The Dutchmen who visited Nova Zembla saw on the ice near the North Pole enough eggs to feed all Europe. As the ice melts, these eggs fall into the sea, are churned into a caudle, and are washed ashore by the waves on nearby islands and especially the coasts of Scotland. If there are trees near enough to the shore to receive the foam

> of these eggified waves, . . . it may so fall out that, by the specific virtue of the eggs, still inherent in such water, the natural vegetation

[18] *Mundus Subterraneus*, I, 98.
[19] *Ibid.*, II, 88.
[20] *Ibid.*, II, 91-93.
[21] *Ibid.*, II, 126.

of the tree, and the omnipotent influence of the sun, all these combining together may hatch a Soland goose. What you will say to this pedigree, I do not know.

Boyle had requested Southwell to ask Kircher about a cave where mineral vapor rose from a hole in the ground and men were said to sweat and serpents to lick the perspiration off. But a cardinal told Southwell that, if another man stood by with a stick and beat the serpents off, and then wiped the patient's body with a cloth, the same remedial success would follow.[22]

Yet Kircher asserts in the preface that he is so constituted by nature that he does not easily put faith in those statements handed down by authors about the virtues of natural things and prodigies, unless reports by men worthy of all credence have been communicated to him with indubitable attestation, or his own experience and observation have made him sure of his ground.[23]

Kircher makes a great show of experimental method, asserting that in physical matters to philosophize without experiment is the same as if a blind man should presume to be a judge of colors. He praises the Academy of Lynxes for having left the ancients far behind by their investigation of motion, and observation of the mountains in the moon.[24] But his own experiments seem silly and insufficient, as when he adduces to prove the influence of the sun on tides during the interlunium, the fact that, if flowers are put in the outer end of a tube which is carried through a wall or window, when the sun shines on them, their odor will come through the tube and fill the room inside.[25] In discussing movement on an inclined plane, Kircher notes that Mersenne claimed to have disproved Galileo's results experimentally and suggests that the discrepancy between them was due to friction and impediments.[26] On the other hand, he disputes the arguments of Cavalieri and Torricelli as to the motion of projectiles,[27] with many accompanying propositions, diagrams and figures.

[22] *Works,* VI, 299. The cave referred to was presumably the Grotta dei Serpi near Bracciano, of which Bourdelot speaks in our Chapter 24.

[23] *Mundus Subterraneus,* Tom. I, Praef. Secunda, pp. ***1v-***2r.

[24] *Mundus Subterraneus,* II, 168; I, 22, 28, 63.

[25] *Ibid.,* I, 134-5.

[26] *Ibid.,* I, 26.

[27] *Ibid,* I, 30.

The chief distinctive feature of Kircher's book and justification for its title is his hypothesis of vast underground reservoirs in Asia, Switzerland, Africa, and South America beneath the Andes, whence all the chief rivers flow through occult channels. He similarly depicts subterranean canals connecting the Caspian Sea with the Black Sea and with the Persian Gulf, and affirms that the Mediterranean Sea would overflow but for subterranean exits for its waters. He also supposes that a maelstrom on the west coast of Norway communicates by an underground passage with the vortex at the north end of the Gulf of Bothnia, and that a great whirlpool at the North Pole sucks in the waters to a tunnel by which they are finally regurgitated at the South Pole.[28] There likewise are subterranean air reservoirs whence winds rush forth with great violence.[29] Winds may also be produced by the rarefaction or condensation of subterranean waters, and experiment shows how such waters are raised on high by the force of winds.[30] On the other hand, fountains may come from the condensation of air.[31]

Kircher believed that heavy objects seek the center of the earth by innate appetite, and that impetus or impulse is a quality demanding movement of its subject, so that there can be no natural motion, to say nothing of violent motion, without impetus.[32] He held that the circumference of the earth could not be measured exactly and that human genius cannot investigate the solidity or weight of the earth.[33] Mountains are necessary, like bones in the microcosm, to protect dry land from the sea, to ward off winds, grow plants, and produce fountains and mines. He rejected, however, the belief in the great height of mountains, for example, the assertion of Aristotle (*Meteor.* I,4) that the Caucasus was so lofty that the sun was still shining on its peak after sunset for a third of the night and before dawn for another third. Kircher held that no mountain could be seen as much as a hundred miles off, and he inclined to think that the height of mountains did not correspond to the depth of the ocean.[34] He held that the axes of the heavenly

[28] *Mundus Subterraneus,* I, 86, 147-50, 159.

[29] *Ibid.,* I, 115.

[30] *Ibid.,* I, 202, 238.

[31] *Ibid.,* I, 240.

[32] *Ibid.,* I, 21.

[33] *Ibid.,* I, 65, 67.

[34] *Ibid.,* I, 90-91, 96.

bodies were parallel to that of the earth,[35] and denied that there was a sphere of fire next to that of the moon.[36] Fire was neither heavy nor light but indifferent as to position, and his fourth book is on subterranean fire. "What spiritous blood is in the human body, that subterranean fire is in the veins of the earth."[37] He affirmed that there was no meteor which was not generated from the underground world.[38]

The theory of signatures in plants was accepted by Kircher, of which he says that he has already treated in his other works.[39] He included the divining rod among ways to find water underground.[40] He noted the sympathy of waters with this or that earth.[41] He believed that swallows were found in winter beneath the ice in lakes in Poland.[42] After long hesitation he felt obliged to accept the existence of flying dragons, and tells of one with two feet and wings seen in Switzerland in 1619, and of another slain by a Roman hunter in 1660. Its head was brought to Kircher's museum, and he says that it had two feet like those of a goose, but it had putrefied when found, while the hunter had died that night from its poison.[43] Kircher also tells of a monstrous cock with a serpentine tail in the garden of the Grand Duke of Tuscany.[44] He believes in subterranean demons and spirits in mines, but not in pygmies, who, he says, are no longer found today.[45]

Kircher was convinced of the existence of occult and specific qualities as well as of sympathy and antipathy and the power of imagination.[46] But he accounted for fascination and poisonous infection at a distance by insensible vapors emitted from the body.[47] Asps kill by their mere breath. He assumed a close connection between poisons and medicines, and that the flesh of the viper was good for health in general.[48] Application of a sun-dried toad to the buboes cures the plague. Venomous animals seek venomous food. The venom of snakes is by nature hot, not cold, and quick-

[35] *Mundus Subterraneus*, I, 104.
[36] *Ibid.*, I, 171-2.
[37] *Ibid.*, I, 175.
[38] *Ibid.*, I, 219.
[39] *Ibid.*, I, 217.
[40] *Ibid.*, I, 240 *et seq.*
[41] *Ibid.*, I, 329.
[42] *Ibid.*, II, 88.
[43] *Ibid.*, II, 89-90.
[44] *Ibid.*, II, 96-97.
[45] *Ibid.*, II, 101-2.
[46] *Ibid.*, II, 111-12.
[47] *Ibid.*, II, 112.
[48] *Ibid.*, II, 110.

silver is the hottest of all poisons. Some places are pestiferous when shone on by the sun, whereas the shade of some trees is poisonous. Eating the brain of a mad cat infatuates; other poisons affect certain parts of the human body. The bite of a mad dog and of a tarantula do not take effect until a considerable time has elapsed.[49]

In the *Mundus subterraneus* Kircher repeatedly alludes to spontaneous generation. All venomous animals were in the first instance produced from the earth without seed, but then propagated sexually like other animals.[50] He also asserts that animals are truly generated from cadavers and that there are barnacle geese in Scotland.[51] He twice speaks of the ostrich's digesting iron.[52] Coral he classifies as halfway between stone and plant.[53]

Kircher held that all diseases came from sulphur, mercury and salt,[54] but he rejected the transmutation of metals.[55] He argued that alchemists did not agree in defining the philosophers' stone or in the names which they gave it, and that the process of making it was difficult and obscure and went in a circle. He denied that alchemy originated in Egypt and was handed down by Hermes in hieroglyphs. He found no mention of it until Pliny's reference to the emperor Caligula, which involved merely the separation of gold from auripigment. He had not read Zosimus in the manuscripts of the French royal library, and thought that alchemy began with the Arabs who invented the Hermetic writings. Kircher holds that today nothing is viler, nothing more deceitful and fraudulent than an alchemist. He attacks both Paracelsus and the Rosicrucians. He repeats "from the secrets of the Roman College" how the arcanum of vitriol, a great secret, is made, but adds that he has not yet been permitted to witness this prodigy and so classes it as a hyperbole of the chemists. Although he presumes to extract evidence of fraud from alchemical manuscripts, Kircher was none too well acquainted with such literature, representing Lull as contemporary with king Richard of England and asserting that John of Rupescissa plagiarized his work on the fifth essence from Ray-

[49] *Ibid.*, II, 121-32.

[50] *Ibid.*, II, 119.

[51] *Ibid.*, II, 337, 346.

[52] *Ibid.*, II, 112, 220.

[53] *Ibid.*, I, 158, 160.

[54] *Ibid.*, II, 135.

[55] *Ibid.*, II, 231-309, Liber XI Chymiotechnicus.

mond Lull. Salomon de Blawenstein—perhaps a pseudonym[56]—chided Kircher for calling Ripley (Riplaeus) Riphaeus, and making two persons out of Arnald of Villanova, one Arnoldus, the other Villanovus.[57]

Besides the transmutation of metals, Kircher was sceptical as to the possibility of perpetual motion, of which he says ten impractical schemes were sent him in 1661,[58] and as to the fabulous effects of some fountains.[59] In this connection it may be noted that, when in 1678 the *Journal des Sçavans* published a picture of the machine by which a Polish Jesuit, Stanislas Solski, claimed to have solved the problem of purely artificial perpetual motion, the mathematician, De la Hire, sent in a Demonstration of the Impossibility of Perpetual Motion.[60] Despite this, the *Journal* eight years later devoted three pages to a scheme for perpetual motion founded on the equilibrium of liquors and experiments with a vacuum, in which mercury would flow back and forth. Its Italian author held that de la Hire's objections applied only in the case of the employment of solid bodies.[61] L'Abbé de la Roque, after reading the Italian author, came to the conclusion that he had reasoned on false principles and proposed instead a scheme of his own based upon the equally questionable assumption that cold condensed liquids and made them weigh more.[62] One is not surprised to find the same abbé following this up by an account in the same volume of the *Journal* of a cure of haemorrhage by the sympathetic powder.[63] Further objection to the Italian's device was voiced by Denis Papin.[64]

[56] Joh. Kestler, *Physiologia Kircheriana experimentalis*, Amsterdam, 1680, Praefatio, says that the eleventh book of the Subterranean World on alchemy was attacked by Valesianus Bonvicenas, professor of *Physica* at Padua, and by another "fictitio nomine de Blauenstein."

[57] Salomon de Blawenstein, *Interpellatio brevis . . . pro lapide philosophorum contra anti-Chymisticum Mundum subterraneum P. Athanasii Kircheri Jesuitae . . .*, Vienna, 1667, in-4, fols. B 3r, C 2v.

[58] *Mundus Subterraneus*, II, 233.

[59] *Ibid.*, I, 288.

[60] JS VI (1678), 150-54, 315-16. Cornelius van Beughem, *Bibl. Math.*, 1688, p. 126, dated Solski's book at Cracow in 1663, *Tractatus de machina exhibendo motui perpetuo artificiali idonea, mathematicis ad examinandum proposita.*

[61] JS XIV, 14-17.

[62] JS XIV, 49-51.

[63] JS XIV, 156-58.

[64] JS XIV, 172-74; for the Italian's reply, 188-91.

Schemes for perpetual motion are also sprinkled through the early volumes of *Acta eruditorum*. In 1691,[65] for example, it described one which Joannes Bernoulli appended to his dissertation on effervescence and fermentation.[66]

Kircher believed that the stars, by God's will and their own positions in the immense ethereal ocean and mutual influence on and by neighboring stars, acquire specific and peculiar endowments, so that under their perennial influence are produced a great and marvelous variety of terrestrial phenomena,[67] while animals from the Indies adapt themselves with difficulty to the sky of Europe.[68] The rays of the stars do not penetrate the earth, but all of them, fixed stars as well as planets, act on this inferior and elemental world not only by their heat and lucid rays but by specific virtues peculiar to each of the heavenly bodies. Metals are thus generated by the flowing down of astral virtues and pure, subtle, celestial spirits which penetrate all the elements and by magnetic attraction join parts from primordial chaos and produce the whitest sulphur purged from all terrestrial dross, which in the course of years turns into silver or gold.[69] Dr. Collier has noted that

> he even listed among the necessary qualifications of a mine supervisor a knowledge of astronomy in order that from consideration of the celestial regions he might discover extensions of the metallic or mineral veins.[70]

Yet in another passage we find Kircher condemning astrology along with alchemy as "genuine sisters."[71] But in this case he was condemning those who pretended to predict human events together with those who pretended to make gold artificially. On the other hand, he often repeated the concept of macrocosm and microcosm or compared the earth to the human body.

In his *Iter Exstaticum Coeleste,* in which he represents himself

[65] X, 64-66.

[66] *Dissertatio chymico-physica de effervescentia et fermentatione nova hypothesi fundata cum descriptione alicuius perpetui mobilis pure artificialis,* Basel, 1690, in-4.

[67] *Mundus Subterraneus,* I, 57. For similar passages in Kircher's *Iter Exstaticum Coeleste,* see K. B. Collier, *Cosmogonies of Our Fathers,* 1934, pp. 51-52.

[68] *Mundus Subterraneus,* I, 108.

[69] *Ibid.,* II, 165-66, 168.

[70] *Op. cit.,* p. 56.

[71] *Mundus Subterraneus,* II, 304.

as transported by an angel through the heavens, he describes the planets Jupiter and Venus in very favorable terms corresponding to their favoring astrological influence, while a repulsive depiction of Mars and Saturn is given for the same reason.[72]

Other topics touched on by Kircher are how to use a chronometer, whether the sea is equally salt everywhere, volcanoes in such places as Java, Sumatra and Mauritius, a list of eighteen eruptions of Mt. Etna—those since Charlemagne being dated in 812, 1160–1169, 1284, 1329–1333, 1408, 1444–1447, 1536–1537, 1633–1639, and 1650, mineral and medicated waters, hydrometers, baths—of which 45 are listed in Germany alone, the weighing of floating bodies, floating islands, deadly waters, causes of river floods, the incorruptibility of salt, nitre, gunpowder, alum, vitriol, minerals or fossils, medicinal earths such as *terra sigillata* or *Lemnia* and *bolus Armenus*, seven requisites for agriculture, mines and their operation and administration, fungi, insects, grafting, distilling, glass-making, pyrotechnics, and arcana of the mechanical arts.[73]

On the whole, the *Mundus Subterraneus* of Kircher, while paying at least lip service to the new experimental science of Galileo and the Lynxes and supplying a certain amount of useful practical knowledge and applied science, and while condemning current efforts to make gold or achieve perpetual motion, devotes more space to past error and magic, and to fantastic hypotheses of its own, than it does to new scientific truth. And the emphasis is still on the marvelous. The question arises, after reading this and somewhat similar books by other Jesuits of the seventeenth century, whether this is merely a reflection and result of Kircher's own genius, curious and encyclopedic, naive and ostentatious and marveling, or whether such books by members of the same Order represent a concerted effort to offer the reading public in general and Catholics in particular works which profess to cover the physical science and even the occult arts of the day in the hope that they will read these

[72] The work is listed in the bibliography of his works in *Mundus Subterraneus*, 1665, I, 346, as printed at Rome in quarto but without date. There were editions at Rome in 1656 and 1660 and a later edition is at Würzburg, 1671, 689 pp.

[73] *Mundus Subterraneus*, I, 51, 165, 181, 188, 247, 254, 259 *et seq.*, 277, 279, 280, 291, 297, 304, 310, 312, 316, 326, 337, 340; II, 171-224, 339, 353, 382, 390, 450, 467, 481.

rather than, or at least together with, the more radical or more superstitious utterances in such fields? Or whether they aim, by voluminous tomes and disquisitions, enlivened occasionally by some new hypothesis or old superstition, to create a sort of intellectual fog or smoke-screen which may impede and smother too radical departures or innovations and prevent a clear defining of the issue? This would accord with the charges often brought against the Jesuits in the political sphere and with regard to their casuistry and doctrine of probability in the field of morals. But I know of no direct evidence for such concerted action in the intellectual field with respect to science and occult science.

In the same year as the Underground World appeared Kircher's Arithmology concerning the hidden mysteries of numbers.[74] The method adopted was primarily historical and expository, setting forth the superstitions of Gnostics, Arabs and Hebrews, magic amulets, Pythagorean Cabala, and the wheel of life or death, without approving of them. Nearly a third of the book was devoted to magic squares. Its sixth and final part was on the mystic significance of numbers.

* * *

The *Acta* of the Munich chemical laboratory, or Subterranean Physics, of J. J. Becher (1635–1682), first appeared in 1669,[75] four years after the *Mundus Subterraneus* of Kircher. Becher was a doctor of medicine—on the title page of his Chemical Institutes of 1664 he is described as a doctor of mathematics and medicine, taught for a while at the University of Mainz, was physician to the electors of Mainz and of Bavaria, and at Munich had charge

[74] *Arithmologia sive de abditis numerorum mysteriis. Origo antiquitas et fabrica numerorum exponitur; abditae eorundem proprietates demonstrantur; fontes superstitionum in amuletorum fabrica aperiuntur; denique post Cabalistarum Arabum Gnosticorum aliorumque magicas impietates detectas vera et licita numerorum mystica significatio ostenditur,* Rome, 1665, in-4. BM 50.c.23.

[75] I have used the edition of Frankfurt, 1681: *Actorum Laboratorii Chymici Monacensis seu Physicae Subterraneae libri duo.* Imp. Mauritii Georgii Weidmanni, with two supplements. Becher in 1680 listed a third edition: *Physica Subterranea cum duobus supplementis,* Francof. apud Zunnerum.

of the finest chemical laboratory in Europe.[76] He was the author of several industrial inventions and economic projects, and his chequered career took him to various countries including England.[77]

Becher admitted that Kircher's book was the earlier but asserted that his was different. Its opening chapters, however, display considerable resemblance to the work of Kircher. Becher discusses the creation of the earth, the universal production of all bodies from chaos, the general difference between bodies, the movement of water above and below the earth's surface, the center of the earth and the movement of sea water toward it, also the movement of water from the center to the circumference of the earth and springs and fountains, mineral waters, central earthy effluvia, solids and minerals, the perpetual motion of nature, and the definition and nature of subterranean principles. Soft stone or stony earth is improperly called salt; rich or oily earth is improperly called sulphur; and fluid earth is improperly called mercury. Throughout his book Becher tries to explain all physical and chemical phenomena in terms of mixtures of earth and water, with some action of air. To reveal in a few words a great secret of his entire volume, he confides that nitre, common salt and quicklime contain the principles of all things subterranean.[78]

Throughout his work Becher accepted the traditional division of three kingdoms, animal, vegetable and mineral, and treated of their *mixtio* and *solutio* in distinct and separate chapters. Animal dissolution was putrefaction; vegetable dissolution was fermentation; while the dissolution of metals was liquefaction. However, it was most certain that metals, especially imperfect ones, were corrupted in air, fire, water and earth. Metals were also subject to mortification, but not to ordinary putrefaction. The philosophic and natural putrefaction of metals is their regeneration.[79]

Becher declared that his treatment of fermentation was new and hitherto touched on by no one.[80] Sulphuric and saline particles are required in all fermentation, and corrosive solutions are species of

[76] Hoefer, II (1843), 214; *Actorum*, preface.

[77] Zedler.

[78] *Actorum.*, p. 354.

[79] *Ibid.*, pp. 296-97.

[80] *Ibid.*, p. 300 *et seq.*

fermentation. In fermented beverages there is a mean substance hitherto unknown, and the taste of wine resides largely in it.

Like Kircher, Becher made a close connection between poisons and medicine. Indeed, he claimed to know a method by which all venomous animals and their poisons could be converted in a moment into good medicine.[81] He still retained the conception of *humidum radicale*, and believed that in order to comfort it animal remedies were better than vegetable or mineral drugs.[82] Like Kircher again, he still was a believer in spontaneous generation.[83]

Becher condemns astrology even more vigorously than Kircher and does not ascribe such vast influence to the stars as Kircher did. He would relegate far from his physics those "planetists" who assign to each metal or mineral some planet as author and formative cause.[84] He regards as nonsense the conjunctions and constellations of the planets and celestial houses which the astrologers prescribe.[85] He retains, however, the conception of macrocosm and microcosm and believes that both were produced in the same way by angels who arranged particles by rarefaction into "the ideas of various species and bodies," to which the remaining particles flowed and adhered and so various bodies were produced.[86] He further thought that comets poisoned the air, through which they scattered innumerable particles.[87]

But while Kircher had condemned alchemy in far greater detail than he did astrology, Becher's book is primarily chemical and alchemical. Although in one passage he censures chemists for laying too much stress upon making gold,[88] he believed that the most perfect minerals were produced instantaneously, once the requisite principles were present, whereas the generation of animals and vegetation took time.[89] He later states that all metals can be changed into mercury, that in the marvelous sympathy of earth fluidifying with metals the entire secret of the philosophers lies hidden.[90] He also quotes Raymond Lull and various other

[81] *Ibid.*, p. 272.
[82] *Ibid.*, p. 271.
[83] *Ibid.*, p. 288: "unde insectorum et omnium generatio quae ex putridis oriuntur."
[84] *Ibid.*, p. 249.
[85] *Ibid.*, p. 192.
[86] *Ibid.*, p. 212.
[87] *Ibid.*, p. 272.
[88] *Ibid.*, p. 196.
[89] *Ibid.*, p. 250.
[90] *Ibid.*, pp. 394, 398.

alchemical authors with favor,[91] and presently says that he has revealed the *maximum arcanum* in the preceding chapters, if only the reader can pick it out.[92] He asserts that the action of fire makes metals heavier and that, when this occurs, it is certain that they are changed in their whole substance.[93] Soon after he gives a recipe for making copper from half an ounce of lead, a dram of tin, and two or three grains of silver and potash.[94]

Besides such passages in Becher's original work, there are appended to the edition of 1681 three alchemical supplements, each of which covers over a hundred pages.[95] The first of these or "New Chemical Experiment," which had been previously published separately in 1671,[96] shows that from ordinary mud of which bricks are made and any fat, animal or vegetable, like linseed oil, without addition of any other materials, there can be produced within four hours' time true and genuine metal, say iron or another, in notable quantity. The second supplement demonstrates the possibility of transmuting baser metals into gold, and the third deals with Becher's scheme or "New and Curious Experiment" for extracting gold from sea sand.[97] He says that he instituted "two striking and great proofs with best success" at Vienna, and that the Estates of Holland appropriated a hundred marks for a great

[91] *Ibid.*, pp. 299-300: "Plato chymicus, Philippus de Ravilasco, Arda discipulus Aristotelis, Rachaidibi, liber Saturni philosophorum, Bernhardus de Gravia, Melchior cardinalis, author tabulae Senioris, Allegoria Arislei, author de lapide philosophorum in 12 capitulis." For most of these authors consult the Index of Thorndike and Kibre, *A Catalogue of Incipits of Medieval Scientific Writings in Latin,* 1937.

On the other hand, he criticizes Glauber occasionally, and condemns the process of Ludovicus de Comitibus for making *Mercurius lunae* as defective and false.

[92] *Ibid.*, p. 421.

[93] *Ibid.*, p. 445.

[94] *Ibid.*, pp. 453-54.

[95] The first continues the numbering of the pagination from 561 to 678; the second goes on to p. 810; the third has a different date of publication, 1680, and a new pagination of its own.

[96] J. J. Becher, *Experimentum chemicum novum quo artificialis et instantanea metallorum generatio et transmutatio ad oculum demonstratur,* Francof., 1671, in-8.

[97] I have also examined a separate edition: *Minera arenaria perpetua . . . Scriptum hoc inservire potest lectori pro continuatione Trifolii Hollandici et Supplementi Tertii in authoris Physicam Subterraneam,* London, 1680, in-4, 112 pp. BM 1033.h.19 (2.), where it is bound with BM 1033.h.19 (1.), his *Trifolium Hollandicum . . . Drie nieuwe Inventien . . . ,* Amsterdam, 1679, in-4, 20 pp.

test, with a reward of 10,000 florins for him, if it was successful, but that it never came off because of opposition and lack of facilities. From the document itself, of which he appends a copy,[98] it appears that from one hundred and eleven and a half marks he was to make a thirtieth part of a thirtieth part of a million Imperials. To justify himself against some persons who held that he had transgressed against the alchemical tradition of philosophical silence by writing too plainly in the *Minera arenaria perpetua*, Becher added a Synopsis or Catalogue of Secret Passages, some seventy in number, in which the whole truth had been concealed or withheld.[99]

At the close of the third supplement comes a communication of January, 1680, to the Royal Society on measuring time and constructing clocks. Then after it, at the close of the volume, Becher lists his works, some printed, some still in manuscript. Nine are philological, ten concern law and politics, twelve are moral and theological, three are mathematical, and ten fall within the fields of physics, chemistry and medicine.[100] A Concordance of some thousands of chemical processes, which he mentions as still in manuscript form, was printed in 1682 and lists a score of manuscripts and laboratories used as its sources, but only fifteen hundred processes.[101] In 1689 was published posthumously his *Tripus Hermeticus fatidicus, pandens oracula chymica.* Its first part described a portable laboratory; the other two parts were commentaries explaining more fully the aforesaid alchemical first supplement.[102]

The future influence of Becher's Subterranean Physics is attested by further editions of it at Leipzig in 1703 and 1738. In both cases

[98] At pp. 126-27, in the ed. of 1681; p. 107 in the separate ed. of 1680.

[99] Ed. of 1680, Index Conclusio, Synopsis seu Catalogus Secretorum Cuniculorum in hac Minera Arenaria latitantium . . .

[100] Of these last, three or four are in German. *De lapide Trismegisto et salinis philosophicis,* printed in German in 1654 under the pseudonym, Solinus Saltzthal, was republished in Latin in Zetzner's *Theatrum Chemicum,* VI (1661), 675-714. Unfortunately Becher seldom gives the dates of publication for his printed works.

[101] *Concordantia Chymica, or, Chymischer Glücks-Hafen oder Grosse Chymische Concordanz und Collection von funffzehen hundert Chymischen Processen.* BM 1033.i.16; BN R.6941.

[102] *Acta eruditorum,* IX (1690), 83-89.

an addition by G. E. Stahl[103] speaks of Becher's book as "a work without an equal."

* * *

An interest in things underground was also displayed by Robert Boyle, who wrote on "The Temperature of the Subterraneall Regions," as well as on that of the submarine regions and "The Bottom of the Sea."[104]

Applying the notion of macrocosm and microcosm in reverse, Donato Rossetti of Livorno, lecturer on logic in the University of Pisa, in a work published in 1667,[105] held that at the earth's center there was a great heart with two ventricles which dilated and contracted with diastole and systole every twelve hours. This hypothesis explained all natural phenomena such as the tides and winds. Rossetti further maintained that the universe consisted of atoms which attracted or repelled one another by sympathy or antipathy.[106]

As Kircher held that rivers had their sources in underground reservoirs, so others believed that springs and sources of rivers on mountains were supplied with water from the sea which came up through subterranean channels. Indeed, the question as to the origin of springs or fountains, raised by Aristotle in the Meteorology, was a favorite, as we have seen in examining general textbooks of natural philosophy. In 1639 Joachim Burser had published a separate treatise on the subject.[107]

Riccioli in 1651 said that all Catholics should regard not as a popular proverb but an utterance of divine Wisdom, the verse in Ecclesiastes:[108] "All the rivers run into the sea; yet the sea does not overflow; unto the place from whence the rivers come, thither

[103] *Specimen Becherianum fundamentorum documentorum experimentorum subiunxit G. E. Stahl.*

[104] *Tracts about the Cosmicall Qualities of Things*, Oxford, R. Davis, 1671. In the Latin edition of Venice, 1697, *De temperie subterranearum regionum ratione caloris et frigoris*, occurs at I, 523-37.

[105] *Antignome fisico-matematiche con il nuovo orbe e sistema terrestre.*

[106] Maugain, Gabr., *Etude sur l'évolution intellectuelle de l'Italie de 1657 à 1750 environ*, Paris, 1909, p. 130.

[107] *De fontium origine*, 1639, in-8, 136 pp. Copies in BM and BN.

[108] I, 7. Riccioli's discussion occurs in the *Almagestum novum*, II, 13; (1651), I, i, 68-69.

they return again." It had been variously interpreted, however, as indicating merely local motion, or new production of waters from clouds raised from the sea, or in subterranean caverns communicating with the sea and whence vapors were raised. Albertus Magnus said that vaporous air in such caverns drew the water upward, which Riccioli explains by the condensation of the air from cold at night or in winter and the consequent ascent of the water to avoid a vacuum, as in a thermometer. Aquinas and Zanardus attributed the elevation of the water to the occult influence of the stars, which drew it up for the benefit of plants and animals. Scaliger and Geraldinus held that the heavier waters still in the ocean forced up through narrow channels the subterranean waters which had been made lighter by percolation, which would be more likely, were the sea higher than springs. Seneca argued that, as the upper parts of the human body attract the blood, so the upper parts of the earth attract water. Which might seem to show that Riccioli was as ignorant as Seneca of the action of the heart and circulation of the blood. He further suggests that one might resort to a machine or Intelligences which by their impulse cause a perpetual motion of the waters.

Molina attributed springs to external rainfall which penetrated the earth, but Riccioli holds that it does not do so for more than twelve or fifteen feet, and much less the rocky mountains of Peru and Chile, whence there is an abundant waterflow. And Genesis says, "The Lord had not yet rained upon earth," and yet, "A fountain ascended from the earth which watered its entire surface."[109]

The third opinion, that vapors are raised by heat from a subterranean abyss of waters and turned aloft into drops of water which form springs seems to Riccioli too slow a process to account for the great volume of water poured by rivers into the sea.

Dobrzensky, in the opening pages of his New Philosophy of Fountains,[110] discussed the origin of springs and rivers, citing the

[109] Genesis, II, 5-6: "Nondum pluerat Dominus super terram... Fons ascendebat e terra irrigans universam superficiem terrae."

[110] *Nova et amaenior de admirando fontium genio (ex abditis naturae claustris in orbem lucem emanente) philosophia*, Ferrara, 1657, in-fol., 123 pp. BN V.2470; Ferrara, 1659, in-fol., same pagination. BN R.854.

Almagestum novum of Riccioli[111] and repeating some of its argument. Molina (1535–1609) had accounted for them by rainfall, but Dobrzensky repeats Riccioli's objection that it does not penetrate into the ground more than fourteen feet. Also the Bible has fountains over the face of the earth before it rained. The Peripatetics ascribe them to an internal flow or stillicide of waters; the school of Coimbra and those who have recourse to sidereal activity attribute them to subterranean air which is first rarefied by the sun and then condensed by cold at night or in the winter. The followers of Copernicus and Galileo explain them by the earth's rotation (*convulsi orbis vertiginem*); Cardan, by the exhalation and condensation of vapors below and above ground. Some modern philosophers hold that they draw from the sea through subterranean channels and lose its saltiness in passing through sand and clay. But the sea is not higher than the mountains and sources of springs. The conclusion is then reached that the cause is twofold: the one remote, from the tides of the sea; the other proximate, in the condensation of air into water in the caverns of earth. Helmont's *Initia physicae inaudita* is then quoted at length as to "vivid waters," which do not observe the laws of hydraulics but burst forth as streams from a subterranean reservoir containing a thousand times more water than the Ocean.[112]

Isaac Voss, in a volume[113] which was reviewed both in the *Journal des Sçavans*[114] and in *Philosophical Transactions*,[115] declared it impossible for water to rise from the sea through subterranean channels to form the sources of rivers on mountains, and held that all rivers came from rain. This essentially correct position was somewhat spoiled by two supplementary treatises. In one he held that the souls of animals are nothing but fire, that there are no invisible atoms, and no pores, not even in the human skin. In the other he argued that the length of cannon should not exceed thirteen feet

not because the bullet is thrown out of the gun before all the powder is

[111] *Ibid.*, pp. 1-3.

[112] *Nova . . . fontium philosophia*, pp. 5-9.

[113] Isaacus Vossius, *De Nili et aliorum fluminum origine*, The Hague, 1666, in-4. French edition, Paris, 1667.

[114] JS I, 641.

[115] PT I, 304-6 (September, 1666).

fired (as some believe), but because the bullet is beaten back into the gun by the air re-entering into it with impetuosity, when the flame is extinct.[116]

That the souls of animals were of a substance approaching the nature of fire was also maintained by the author of *Philosophia vetus et nova ad usum scholae accommodata*[117] who attributed this opinion to Gassendi, P. Fabry and others.[118]

Petrus Lagerlöf offered a dissertation on the origin of springs and rivers at Upsala in 1675.[119]

Pierre Perrault (1608–1680), brother of Charles and Claude, and previously receiver general of finances at Paris,[120] wrote a work on the origin of springs[121] which was first published anonymously in 1675,[122] and again under his own name, in 1678. After reviewing the opinions of twenty-two previous writers, he rejected the theory that springs and rivers came from the sea and argued that a sixth part of the precipitation would supply all the rivers. For this he has received great credit. Wolf says[123] that his views "were embraced by his contemporaries Mariotte and Halley, and were abundantly confirmed early in the eighteenth century by the work of Antonio Vallisnieri.[124] But Perrault spoiled it all by further advancing the paradoxical thesis that springs were not the source of rivers but rivers the cause of springs. This involved a re-ascent of river water from the bases of mountains to their summits in the form of vapor raised through subterranean channels, and so retained the worst feature of both Aristotle's explanation that air was condensed into water in the bowels of the earth and that which derived springs and rivers from the sea.

[116] *Idem.*

[117] Du Hamel, Tom. 4, Paris, 1678, in-12.

[118] JS VI, 270-71.

[119] Moller, *Svecia literata*, 1698, pp. 261, 438.

[120] Suzanne Delorme, "Pierre Perrault," *Archives Internationales d'Histoire des Sciences*, I (1948), 388-94, especially p. 389.

[121] For trenchant criticism of it by Huygens in 1673 before the book was printed see his *Oeuvres* XXII (1950), 102-3.

[122] *De l'origine des fontaines*, A Paris chez Pierre le Petit, 1675, in-12, 353 pp. It was reviewed in JS IV, 10-13.

[123] *A History of Science, Technology and Philosophy in the Sixteenth and Seventeenth Centuries*, 1935, pp. 361-62.

[124] *Lezione accademica intorno all'origine delle fontane*, Venice, 1715.

It was therefore not very surprising to find Herbinius, in his Dissertations on Cataracts Supra- and Subterranean of 1678,[125] affirming that there was a perpetual circulatory movement of the ocean from the North to the South Pole through the viscera of the earth, and declaring that fountains or springs were produced by the ascent of subterranean waters.[126] He further attributed the tides to the ejection of waters from subpolar abysses, and repeated Kircher's discussion of Scylla and Charybdis.[127] He also accepted the existence of waters above the firmament.[128]

Robert Plot, curator of the Ashmolean Museum at Oxford, writing on the origin of springs in 1685, still held that the volume of water emptied from rivers into the sea was too great to be accounted for by precipitation. Riccioli had calculated that the Volga alone poured into the Caspian Sea every year enough water to inundate the entire surface of the earth. Plot therefore held that there were whirlpools in the ocean which sucked in the water and subterranean channels which carried it back to the springs.[129]

In the same year, the book of a professor at the University of Bourges on the mineral waters of that place, held that springs came from the sea by subterranean channels.[130]

The treatise of Mariotte (1620—1684) on the movement of waters and other fluids, which appeared only posthumously in 1686, held that rain was sufficient to supply fountains and rivers, and that it sinks into the earth until it meets a clayey or rocky soil, and then runs along this until it finds a way out to the air, issuing forth as springs or fountains.[131]

Edmund Halley, in a communication to *Philosophical Trans-*

[125] Johannes Herbinius, *Dissertationes de admirandis mundi cataractis Supra et Subterraneis ...*, Amsterdam, 1678, 267 pp. Copy used: BM 233.i.27. The dedication is dated at Danzig, January 1, 1678. The five books deal with cataracts in general, marine ones, the rivers of Paradise, other rivers (pp. 194-232 are on the Rhine and its falls at Schaffhausen and elsewhere), and artificial ones (at p. 257 a picture of the locks between Padua and Venice).

[126] *Ibid.*, pp. 59, 67.

[127] *Ibid.*, pp. 92, 120-24.

[128] *Ibid.*, p. 9.

[129] Robert Plot, *De origine fontium tentamen philosophicum*, Oxford, 1685, in-8: reviewed in PT XV, 861-65.

[130] Est. Cousturier, *Traité des Eaux Minerales de Bourges*, Bourges, 1685, in-12. Review, JS XIII, 195-96.

[131] Edme Mariotte, *Traité du mouvement des eaux et des autres corps fluides*, Paris, 1686, in-12.

actions which appeared in Latin translation in *Acta eruditorum* the same year, held that water ascended from the sea to mountains by evaporation and not by subterranean channels.[132]

Strata of soil to a depth of sixty-three feet were recorded by Ramazzini in his work on the wells of Modena. It was necessary to go down that distance to reach water. The first fourteen feet were through the remains of an ancient city which did not rise above the present level of the surrounding plain. The next fourteen feet were of marshy soil full of reeds. Then came eleven feet of clay, then more marsh and another stratum of clay but not so thick as before. Last, clay and sand mixed with sea products.[133]

Since we have mentioned Ramazzini, we may notice his discussion of the cause of the rise and fall of the mercury in the Torricellian tube according to the varied state of the air. It would seem that the weight and pressure of the air should be greater when rain is imminent and it is filled with moisture, and so push the mercury up in the tube. But the opposite is the case; the barometer rises with fair weather. Ramazzini suggests that in fair weather the air is weighed down with heavy earthy, saline and nitrous particles. In damp weather these corpuscles are driven out of the pores of the air, and it becomes lighter, and as the weather clears up, they return again. In accompanying letters from two other denizens of Modena other explanations are offered: one that water attenuated by heat into vapor makes a lighter compound with the air than the air by itself; the other that water is suspended in air in fair weather and increases its weight, but leaves the air lighter before and during rain. Ramazzini had further observed that there was no barometric change with new or full moons, or at the time of the solstices and of lunar eclipses, but some change at the time of the equinox and of a visible solar eclipse. This he ascribed to celestial effluvia.[134]

[132] *Acta eruditorum*, XI (1692), 307-12.

[133] Bernardino Ramazzini, *De fontium Mutinensium admiranda scaturigine tractatus physico-hydrostaticus*, Modena, 1691, in-4. English translation by Dr. Robert St. Clair, London, 1697. Review in PT XIX, 734-36. JS XXII, 281-87, varies somewhat as to the lowest strata.

[134] *Ephemerides Barometricae Mutinenses anni 1694, una cum Disquisitione causae ascensus et descensus in Torricelliana fistula juxta diversum aeris statum*, Modena, 1695: *Acta eruditorum*, XV (1696), 41-43.

The work of Carlo Fontana on Running Waters, printed at Rome in 1696, still savors of magic illusion, as the mention of pastimes and tricks in its full title shows.[135] Its second book is devoted to these experiments, pastimes and tricks performed by the element of water by means of the air and fire.[136] The author's science is rather backward for the close of the century. He not only still believes in four elements, but maintains that the sea is higher than the earth, even than the highest mountains. His infallible proof for this is that when one is at sea many miles from land, even the highest mountains are no longer visible—a statement which does not take the curvature of the earth into account.[137] This contention, however, makes it easy for him further to hold that the water on mountains comes from the sea, since it is merely seeking its own level and not running up hill. Fontana is better informed on his specialty of conduits and aqueducts, and his third book describes recent waterworks with accompanying pictures.

[135] Carlo Fontana, *Utilissimo Trattato dell' Acque Correnti diviso in tre libri, nel quale si notificano le Misure, ed Esperienze di Esse, I Giuochi e Scherzi, li quali per mezzo dell' Aria e del Foco vengono operati dall' Acqua*, Rome, 1696, in-fol. 196 pp., 8 fols. Copy used: BN V.2469.

[136] *Ibid.*, pp. 87-176.

[137] *Ibid.*, p. 9.

CHAPTER XXI

ARTIFICIAL MAGIC AND TECHNOLOGY

Books of secrets—Schott's predecessors: de Caus, Leurechon and Ens, Schwenter, Auda, Le Royer—Schott on magic and demons—His limited scepticism—Marvels of optics and acoustics—Mechanics and machinery—Vacuum denied—Medicine and divination—Monsters, portents and marvels of animals—Sympathy and antipathy—Influence of the stars—Other works by Schott: Schwimmer—Lana Terzi—Slow development of inventions and machinery—Veranzio, Zonca, Branca, Böckler, Zucchi—Lipstorp and Deschales—J. J. Becher—Conclusion.

Illa, inquam, post Aristotelis libros ultra solis currum hodie ac equos exaltata triumfalibus laudibus Schotti Physica Curiosa

—J. D. MAJOR, *Genius errans*

Replicatio virtutis potentiae ex vi machinae

—ZUCCHI

In our previous volumes we have seen the books of secrets and recipes, experiments and magic tricks, found in manuscripts of the thirteenth century, more or less duplicated and reproduced in such printed works as the Secrets of Alessio of Piedmont or the *Magia Naturalis* of Giovanni Battista Porta. The same tradition continued in manuscripts of the seventeenth century. In one at Lucca,[1] for example, are varied recipes, partly from the books of secrets of Mapphaeus Spinola, partly obtained from many experimenters, with remedies of the Grand Duke of Tuscany and of Lorenzo Cybo.[2] "And what arcana they are!" It also went on in printed books, as may be illustrated by two works of the Jesuit, Caspar or Gaspar Schott, namely, his Universal Magic of Nature and Art, printed in four fat volumes at Würzburg in 1657—1659,[3] and his Curious Phy-

[1] Lucca 521 (B. 383), chart., mm. 208 × 48, 17th century, 232 fols.

[2] Spinola and Cybo seem to be otherwise unknown.

[3] *Magia universalis naturae et artis,* in 4 vols. of 538, 432, 815 and 670 pages.

sics or Marvels of Nature and Art, of which three editions succeeded one another rapidly in 1662, 1664 and 1667, with one more before the end of the century, in 1697.[4] A collection of three hundred magic tricks under the pseudonym of Aspasius Caramuelius, are also really written by Schott.[5]

It must be admitted that these bulky tomes are more pretentious than the little thirteenth century manuscripts, that they indulge more in generalities, attempt a logical classification and systematic arrangement, add subsequent technological inventions and recent scientific discovery. But under high sounding designations they still include many of the old recipes, secrets, experiments, and magic tricks, although they express scepticism as to some of them. They still emphasize the marvelous, "curious hidden wonders and foreign to the vulgar ken," and "whatever in the universal nature of things is occult, paradoxical, prodigious, and like to a miracle,"[6] "things rare, curious and prodigious, that is, truly magical."[7] Similarly in *Physica Curiosa* we have marvels of demons, marvels of specters, marvels of men, marvels of animals in general, marvels of terrestrial animals, marvels of meteors, miscellaneous marvels, and varied marvels.[8]

Athanasius Kircher himself had planned to write such a work as *Magia Universalis,* but was too busy to do so, and hence put his plan and notes at the disposal of his disciple, Schott. The four tomes deal respectively with optics, acoustics, mathematics, and *Physica* or natural phenomena and marvels. Kircher had already treated of optics in his *Ars magna lucis et umbrae,* of acoustics in *Ars magna consoni et dissoni,* of mathematics here and there in various works, and of *Physica* in *Mundus Subterraneus, Oedipus Aegyptiacus,* and *Obeliscus Pamphilius.*

Books more closely akin to *Magia Universalis* had been composed

[4] *Physica curiosa sive mirabilia naturae et artis libris xii comprehensa quibus pleraque quae de angelis daemonibus hominibus spectris energumenis monstris portentis animalibus meteoris circumferuntur ad veritatis trutinam expenduntur . . . ,* Herbipoli, 1667, 1389 pp., is the edition which I have used.

[5] *Iocoseriorum naturae et artis sive magiae naturalis centuriae tres auctore Aspasio Caramuelio . . . ,* Würzburg, 1665. Ferguson, II, 339-40.

[6] From the unnumbered pages of the *Prooemium* of *Magia universalis.*

[7] From the titulus of Pars III & IV.

[8] *Op. cit.,* I, xvi; and pp. 196-332, 351-519, 678, 776, 1179, 1275, 1365.

in the past by Albertus Magnus, Porta, Cardan, Gerónimo Cortes, Mizauld, and by Giovanni Battista Zapata, whose Marvelous Secrets of Medicine and Surgery, published at Rome in 1586, was reprinted at Venice in 1618,[9] 1641 and 1677, and at Ulm in Latin in 1696.

More recently there had been various treatises along the line of Schott's work. Adriaen van Roomen (1561–1615), the mathematician of Louvain, had published a volume on fireworks in 1611,[10] and François de Malthe, one in French in 1629.[11]

Salomon de Caus (1576–1626), who at that time was engineer and architect to the Elector Palatine, for whom he laid out the castle gardens of Heidelberg, published in 1615 *Les raisons des forces mouvantes avec diverses machines.*[12] He first spoke of machines and their first inventors. He still accepted four elements but described air as cold and dry rather than warm and moist, and denied that water could be changed into air. He was more conservative in denying the existence of a vacuum. He considered ways to raise water, held that the movement of the lever accorded with that of the balance, and that power was increased by wheels and gears. He suggested running mechanical clocks by water power instead of rewinding them. Other devices are mechanical birds which whistle or drink, a saw mill, fire engine, lathe, music boxes run by water power, and organs. All these were usually illustrated by full page pictures. His second book dealt with grottos and fountains

[9] *Li maravigliosi secreti di medicina e chirurgia di nuovo ritrovati per guarire ogni sorte d'infermità.* For other editions see the BM catalogue.

[10] Adrianus Romanus, *Pyrotechnia hoc est de ignibus festivis joculis et artificialibus libri duo,* Francofurti, 1611, in-4: BM 534.f.8 (2.); BN V.9529. For his life and works: Valerius Andreas, *Bibliotheca Belgica,* 1643, pp. 15-16, and Zedler. More recent articles on him are: Ph. Gilbert, "Notice sur le mathématicien Louvaniste Adrianus Romanus," *Revue catholique,* 17 (1859), 277-86, 394-409, 522-27; A. Ruland, "Adrien Romain, premier professeur à la faculté de médecine de Wurzbourg," *Bibliophile Belge,* II (1867), 56-100, 161-87, 256-69; H. Bosmans, "Note sur la trigonométrie d'Adrien Romain," *Bibliotheca mathematica,* V (1904), 342-54.

[11] *Traité des feux artificiels, pour la guerre et pour la récréation avec plusieurs belles observations... de géométrie, fortifications et exemples d'arithmétique.* As the title suggests, only the first of the four parts is on fireworks.

[12] "A Francfort en la boutique de Ian Norton," in-fol.: copy used, BM 535. I, 23.

for gardens; the third, with the construction of organs. Cornier wrote Mersenne that de Caus had a low reputation in musical matters.[13] However, he became royal architect and engineer in 1621, and wrote other books on such themes as perspective and solar clocks. When he died in 1626, he was buried in a Protestant cemetery.[14] He is said to have preceded Huygens in holding that to raise 400 pounds one foot would take the same amount of work as to raise 50 pounds eight feet.

The *Recréation mathématique* of Jean Leurechon, first published in 1624, ran through numerous editions in French,[15] with addition of a second part and a third part on fireworks in 1628. The first English translation was in 1633. In 1636 it appeared in Latin translation, with considerable additions and variations, by Caspar Ens, under the title, *Thaumaturgus mathematicus.*[16] It is believed to be the first book to use the word, thermometer,[17] the chapter upon which is followed, characteristically enough, by one on the proportions of the human body, colossal statues, and monstrous giants. Besides strictly mathematical problems, there are such marvels as to make a door open from both sides;[18] to construct a vessel which will retain the liquid poured into it up to a certain height, when the entire contents will flow out;[19] to build a bridge all round the earth which will not fall when its supports are removed;[20] to keep all the water in the world in the air without a single drop falling to earth.[21] If all the gunpowder in the world were put in a globe of

[13] *Correspondance du Marin Mersenne,* I (1932), 294, 350.

[14] *Ibid.,* 294.

[15] The David Eugene Smith Collection in the Columbia University Library, has the second edition of Paris, 1626; that of Rouen, 1628, in which the title becomes *Recréations mathématiques;* Lyon, 1642; Paris, 1661; and Lyon, 1669. The *Examen du livre des Recréations mathématiques,* by Claude Mydorge, first appears in an edition of 1630. Others in 1639 and 1643.

[16] Cologne, 1636, small octavo. There were further editions in 1651, and at Venice, 1706.

[17] Ed. of 1626, pp. 99-101, "76. Probleme. Du Thermometre, ou Instrument pour mesurer les degrez de chaleur ou de froidure, qui sont en l'air." In the Latin *Thaumaturgus mathematicus* of 1636, pp. 125-28, credit for the invention is given to Drebbel, "Problema LXXXIII. De thermometra, sive instrumento Drebiliano, quo gradus caloris frigorisque aera occupantis explorantur."

[18] Problem 17 in both *Recréation* and *Thaumaturgus.*

[19] *Recréation,* 39; *Thaumaturgus,* 42.

[20] *Rec.* 47; *Thau.* 49.

[21] *Rec.* 48; *Thau.* 50.

glass or paper and set on fire all at once, what would happen? Nothing, since the pressure would be equal in every direction.[22] Similarly, all the men and angels in the world together could not break a spider web, if it formed a perfect circle and their force was the same on all parts of it.[23] Such tricks are included as lifting a bottle with a straw, a self-filling lamp, and pouring out three different liquors from the same orifice of a vessel[24] (which occurs in the medieval *Secretum Philosophorum.*)[25] There are various experiments with mirrors, clocks, artillery, fountains and hydraulic machines. Ens asserts that a ring, in which a diamond or emerald has been inserted, if held pendant by a thread over water in a glass, will strike the glass as many times as there are hours, but gives no explanation of this.[26]

Schott said[27] that the Physico-Mathematical Delights of Daniel Schwenter, a work which first appeared in 1636, after the death of Schwenter,[28] presented many most entertaining machines, hydraulic and pneumatic. It has a description and picture which are said to be the first of a fountain pen.[29] More typical items are to stand an egg on end or to boil it on your head (by putting it in hot bread covered with a napkin),[30] whether there are more stars in heaven than there are children of Israel,[31] how to see the stars in bright sunshine,[32] whether the years now are longer than they were before the Flood,[33] how to regenerate plants, to turn red or white hair black, to restore a frozen finger or nose by putting it in cold water or snow,[34] to make an apple jump on the table,[35] to produce a rain-

[22] *Rec.* 50; *Thau.* 52.

[23] *Thau.* 53.

[24] *Rec.* 55, 71, 81; *Thau.* 58, 74, 87.

[25] T II, 790-91.

[26] *Thaumaturgus,* 111.

[27] *Mechanica hydraulico-pneumatica,* 1657, p. 9.

[28] *Deliciae physico-mathematicae,* Nürnberg, 1636, 574 pp. The Columbia University Library has this edition and also those of 1651 and 1677, with identical pagination. G. P. Harsdoerffer added vols. II and III in 1651 and 1653, of which Columbia has only vol. II in the 1677 edition.

[29] *Ibid.,* pp. 519-20: "Ein schön Secret, ein Feder zu zurichten welche Dinten hält und so viel lässet als man bedürfftig." A quill pen with several inserts is filled by suction, and air admitted and a drop of ink released by pressure of the fingers.

[30] *Ibid.,* pp. 568-69, 439.

[31] *Ibid.,* p. 322.

[32] By looking up from the bottom of a well – a very old trick: *ibid.,* p. 324.

[33] *Ibid.,* p. 325: an old problem.

[34] *Ibid.,* pp. 558, 559, 561.

[35] By placing quicksilver inside of

bow by looking through a prism.[36] The tricks and problems are arranged in sixteen sections under designations from arithmetic to chemistry, and are taken partly from previous authors such as Nicolaus Taurellus and Cardan, or, more recently, Galileo.

Perspective or optics was set forth as a branch of artificial magic by J. F. Nicéron in 1638.[37] His fourth book treated of a marvel of dioptric, "invented in our days," by which, upon a surface where several figures or portraits were shown "in their correct proportions," another different one could be made to appear.

Ens also published a *Thaumaturgi physici prodromus*,[38] which suggests the appeal that such wonder-working titles made to the reading public. It is said to treat of the earthquake of 1640 and whether it portended anything, and such problems as blunting the force of a magnet, striking sparks from sugar, changing air into water, and predicting rain from flying clouds.

The *Breve compendio di maravigliosi secreti* of Domenico Auda was in four books. The first treated of medicinal secrets, some of which we notice in our chapter on Pharmacy. The second book of miscellaneous secrets included face washes, oils and perfumes, dyes, ways to remove spots from cloth and to temper iron, to make letters of gold and varnish. The third book of chemical secrets had quintessence of human blood and masteries of pearls and coral. The fourth part on astrological medicine[39] also included the old familiar superstitious divination according to the day of the week on which the first of January falls.[40] This unoriginal compilation had great currency. The first edition was at Rome in 1655. By 1663 it reached its fifth edition. Many others followed but all were printed in Italy.[41]

---

it, which also goes back to the manuscript age: *ibid.*, p. 563.

[36] *Ibid.*, pp. 258-59.

[37] *Perspective curieuse ou Magie artificielle des effets merveilleux de l'optique*, Paris, 1638. BN V. 1661; Rés. V.171.

[38] *Thaumaturgi physici prodromus, id est problematum physicorum liber singularis, lectu jucundus et utilis*, Cologne, 1649. It does not appear in either the BM or BN catalogues of printed books.

[39] In the edition of Venice, 1668, which I used, it began at p. 209, and pp. 215-36 had been torn out. At pp. 264-301, "Nuova Agiunta."

[40] *Ibid.*, pp. 238-39.

[41] Graesse, I, 250; Jöcher, Suppl. I, 1221; Duveen, 33, 495.

In 1660 Iacques Le Royer, advocat to the Parlement of Rouen, dedicated a very boastful little book to Louis XIV.[42] He claimed to have discovered a perpetual motion machine and to have squared the circle. He would not attempt to find the philosophers' stone, for it would ruin commerce and make us poor. But he is so accustomed to obscure and difficult matters that everything seems easy to him. He will present the king with a triumphal chariot which will go by itself. He will construct a galley that moves without sails or oars. His mechanical eagle will communicate with a besieged town and carry letters two hundred leagues. He will build a house where everything can be heard that goes on. "I dare say the deaf will hear and very easily." Optics, gnomonics and perspective are all at his fingers' ends. He will show how sciences and languages may be learned in a short time, even by children.[43] But the present publication is limited to the causes of the tides, winds and intermittent fevers, which no philosopher nor physician has ever discovered and of which the first passes for incomprehensible, the second is hid in God's treasury, and the third has been unknown till this day.[44] The moon is not the true cause of the tides which are from influences of the sun reflected by the moon. Wind is produced from the same influences reflected by the stars, and these same influences are a contributory cause of intermittent fever.[45]

Despite earlier books, Schott justified his own on the ground that they had mixed with truth what was false, superstitious, obscene or harmful, and had not sufficiently explained the causes of their marvels. He opens his work with a consideration of magic. It was once an honorific term, but legitimate magic was corrupted before the Flood. After the Flood Cham or Zoroaster propagated the later magic which is either natural or artificial or prohibited. But much so-called natural magic is superstitious. Examples of artificial magic are the mechanical eagle of Regiomontanus or the speaking bronze head of Albertus Magnus, astronomical clocks, and

[42] *Les causes du flux et reflux de la mer, des vents, et de la fièvre intermittente*, Paris, 1660. BM 1135.a.7 (3.), while (4.) is a Latin translation of it, of the same date.

[43] *Ibid.*, pp. 15, 17, 33-40.

[44] *Ibid.*, p. 10.

[45] *Ibid.*, pp. 44, 61, 108, 123.

automata. Prohibited or illicit magic is when a pact with a demon is involved, whether explicit or implicit. In the application of active to passive, demons can work many marvels by alteration or mutation of things, of which the causes are unknown to us but perfectly familiar to them. They can dull human sensibility so that a person does not feel torture; they can affect phantasy and through it memory and intellect. Magicians cannot work miracles but they can produce earthquakes and storms, injure animals and crops, burn houses, give victory, free captives, charm serpents, and produce imperfect animals. They cannot contract bodies to a small quantity or penetrate bodies or make replication. They cannot transmute species, nor can they make dumb beasts, statues and trees really speak.[46]

Schott treats at greater length of demons and their relation to magic in *Physica Curiosa.*[47] Angels and demons are incorporeal, but were created with the corporeal world, according to Aquinas in the empyrean heaven, according to Suarez, whose opinion Schott prefers, in the starry heaven. For when Lucifer (Isaiah, XIV, 13) says, "I will ascend into heaven," he means from the starry heaven to the empyrean. Angels and demons can know secret thoughts and have a limited knowledge of the future. They can speak to one another, but as to how this is done there are ten different opinions. They are capable of some spatial extension and can move, contract and dilate themselves *ad lib.* But they cannot move with infinite speed nor instantaneously (whereas it is held in *Magia Universalis* that the motion of light is instantaneous)[48] nor be in two places at once nor pass from extreme to extreme without traversing the mean. Yet several angels can be in the same place at once, the explanation being that they are mutually penetrable. They can move bodies and impart impetus to them; they can move and detain other spirits. Demons can assume bodies formed from sublunar matter or the corpses of the damned and move these, but they are not really united to these bodies as their forms, and cannot

[46] *Magia universalis,* I, 10, 13, 17, 18, 21-24, 27, 35, 38, 39-42.

[47] After Bk. I, devoted to angels and demons, ends at p. 147, an appendix concerning the marvels of demons, including oracles, goes on to p. 195.

[48] *Magia universalis,* I, 72.

exercise such vital operations as seeing, hearing and speaking in them. They have some kind of republic and organization.

Are the marvels which they work through magicians real or only seeming? By control of local motion they can draw fire from the sky, excite winds and earthquakes, render objects invisible, make statues move and dumb animals speak (which he had denied in *Magia universalis*) and modify the animal spirits in the body. They can disturb dreams and, as we were told before, apply active to passive. They also may delude the senses by legerdemain, perspective and the like. They can produce some animals from putrid matter, as Pharaoh's magicians produced true snakes and frogs. After citing many authorities pro and con as to incubi and succubi, Schott finally decides pro. He further holds that demons can transport witches to sabbats. They cannot make bodies penetrate each other or the same body be in two places at one time, or really transform men into beasts. They can bring on sickness and restore health, with divine permission make an old man young but not immortal, enable men to endure a long slumber or fast, alter sex, aid the memory and intellect, and bestow science or excite a state of ecstacy.

After all this, it might well seem that human inventions and machines and experiments and the marvels of nature and art to which Schott's volumes are supposed to be devoted do not amount to much, and that magic is preeminently a diabolical affair and concern. Yet *Magia Universalis* goes on to treat of chromatic magic, magic with mirrors, dioptric magic, telescopic magic,[49] of phonocamptic magic (about echoes), phonotectonic magic, phonurgic magic, phonoiatric magic, musical magic, symphoniurgic magic,[50] of thaumaturgic magic, static magic, hydrostatic magic, hydrotechnic magic, aerotechnic magic, magic arithmetic, geometric magic,[51] cryptographic and cryptologic magic, pyrotechnic magic, magnetic magic, sympathetic and antipathetic magic, medical magic, divinatory magic, physiognomical magic, and chiromantic magic.[52] It is

[49] *Ibid.*, I, 218, 244, 450, 488.

[50] *Ibid.*, II, 76, 135, 176, 218, 251, 380.

[51] *Ibid.*, III, 226, 308, 353, 457, 561, 629, 733.

[52] *Ibid.*, IV, 1, 91, 225, 349, 451, 539, 583, 637.

perhaps just because diabolical magic has been so emphasized in the first book of *Physica curiosa,* that in its remaining eleven books Schott speaks of marvels rather than magic. Or this first book may have been intended as disclaimer of and antidote for the greater emphasis upon the magic of nature and art in the previous volumes. At any rate it constitutes an unwelcome intrusion, as do the subsequent books on specters and demoniacs, which have no place in a work on the marvels of nature and art. But they warn us how close the connection between witchcraft and science still was in the seventeenth century. In the book on specters such familiar questions are argued as whether Samuel really appeared to Saul, why the bodies of specters are cold, how to tell human ghosts from demon specters, and how best to drive them away by faith, prayer, relics of the saints, the sign of the cross, holy water, *agnus Dei,* name of Jesus, and invocation of the Virgin Mary.

We have heard Schott censure previous writers for being too credulous and superstitious with reference to magical recipes and experiments. But he does not always succeed in consistently maintaining this attitude. Thus he repeats and usually agrees with Kircher's censure of the Pseudo-Albertus;[53] and he censures Alexius of Piedmont for his water to make old men young and doubts if there be a medicine which will protect one from snakes for thirty years.[54] However, he not merely cites William of Auvergne, *De universo,* for seeing snakes dancing but asserts that William teaches that, if a banquet is lighted solely by a candle made of the semen of an ass and wax, all the guests will seem to have the heads of asses.[55] Actually William expresses scepticism as to this.[56] Also Schott himself in the *Physica Curiosa* implores the reader not to be so inhuman as to refuse to believe anything unless he sees it with his own eyes, and affirms that many things which antiquity thought fabulous are now proved true by frequent experiment.[57]

The first volume of *Magia Universalis* on optics includes, in addition to projection of images and magic with mirrors, such optical illusions as so disposing the columns in a building that they seem

53 *Magia univ.,* I, 207.
54 *Physica curiosa,* p. 1319.
55 *Magia univ.,* I, 206, 37.
56 T II, 345.
57 *Physica curiosa,* p. 679.

straight upright at a distance but appear irregular and crooked on approaching them, which strikes with horror one ignorant of optical fallacies. This is taken from the last part of the fourth book on Optics of Aguillon.[58]

The second volume on acoustics opens with assertion of the secret nature of sound and voice, their arcane production and propagation, their marvelous and prodigious effects.[59] There is no sound without movement, yet sound is not motion but something consequent to motion. It is borne to the ears by waves of air.[60] Considerable space is devoted to the occult fabrication of instruments which magnify sound, such as loud speakers, speaking statues, speaking tubes for the deaf, and devices to enable the prince in his private chamber to hear everything that goes on in the palace. In this case we are assured that the prodigious sounds, which many authors ascribe to miracle or to the fraud and deceit of demons, actually do not exceed the limits of natural possibility.[61] We pass on to the marvelous effects of music and song, as when the walls of Jericho fell at the sound of the trumpets. Schott discusses sympathetic sound, how music moves men and beasts, the pied piper of Hameln—already treated by Kircher, the virtue of words, whether incantations may be effective naturally, snake-charming, and catching fish by the sound of certain words and song.[62] Kircher argued that, as we call animals by name, so some sounds are proportioned to certain animals. But Schott contends that the fish at Messina have not been accustomed to the words by which they are caught. Therefore, either the words uttered are an incantation involving a pact with a demon, or are a mere imposture and superstition of the fishermen. Discussing the dancing to music of those bitten by a tarantula, Schott holds that the notion, that the urge to dance ceases with the death of the spider which bit one, is contradicted by experience.[63] David's relieving Saul by playing the zither was in part a natural, in part a supernatural, cure, since Saul was not merely given to melancholy but also pos-

58 *Magia univ.*, I, 195.

59 *Ibid.*, II, 1.

60 *Ibid.*, II, 13, 20, 35, "Sonus ad aures propagatur per aëris undationes."

61 *Ibid.*, II, 135, 160, 162, 144, 166-67.

62 *Ibid.*, II, 181, 189, 199, 203, 206, 209, 213.

63 *Ibid.*, II, 238, 249.

sessed by a demon.[64] The hydraulic organ and other mechanical musical instruments are considered, and the vases in theatres for acoustic purposes mentioned by Vitruvius.[65] Symphonic music and counterpoint are also included under the magic of sound.[66]

The first section of the third volume of *Magia Universalis* is entitled *Centrobaryca* and treats of the centers of magnitude and of gravity, and of the place of the earth in the universe. For Schott it is still the center of the universe, but he raises the question whether the globe of earth and water, as a result of continual shifting of its center of gravity, is in a continual movement of trepidation and titubation about the center of the universe.[67] Soon he is discussing rope-walkers and leaning towers, why persons rising from a seat bend their heads forward and their feet backward, and the problem how men would stand and walk, if God removed the upper hemisphere of the earth.[68]

For Schott "mechanical apparatus" includes definitions, axioms, hypotheses and propositions.[69] He believes that the proportion of forces, weights and motions, with regard both to the times in which they move and the spaces through which they move, has been so well treated by Stevin, Guidobaldi, Galileo and Mersenne, that nothing more can be desired. But the physical cause of such effects has hardly been considered by anyone since Aristotle with three exceptions: Honoratus Fabri, Zucchi in his New Philosophy of Machines, and Paul Casati in the manuscript on Mechanics from which he lectured in the schools at Rome.[70] Fabri came to the conclusion—somewhat startling to the modern reader—that by the proportion by which you lessen the motion, by the same you will move a greater weight. Zucchi spoke of increase of power by the use of a machine. Casati made impetus precede motion, but Schott holds that it is acquired through movement. He would not seem acquainted with late medieval writers on impetus. He grants that all machines increase the amount of power and diminish the

[64] *Ibid.*, II, 229.

[65] *Ibid.*, II, 292, 361.

[66] *Ibid.*, II, 380, 389.

[67] *Ibid.*, III, 14, 20, 23.

[68] *Ibid.*, III, 54, 64, 67, 73.

[69] *Ibid.*, III, 90. The word, "apparatus," seems to have been first employed by writers on canon law for their glosses, citations, etc.

[70] *Ibid.*, III, 211-23.

resistance of weight, but how they do it he admits his inability to solve.

The third book of the third volume, entitled Thaumaturgic Magic, is devoted to machinery, with much concerning gears, cyclometers, horseless chariots, astronomical spheres, and the machines by which the Vatican obelisk was erected at the order of Sixtus V. Schott does not know an Italian name for a jack (*Winde* in German; *cric*, in French) and saw only one during twenty-two years spent in Sicily and Italy.[71] Some thought that Kircher knew the art of flying.[72] A machine at Danzig brought a mountain into town.[73] In a later book Schott enumerates the parts of a mill and denies the possibility of perpetual motion by machinery.[74]

Static magic, as interpreted by Schott, does not deal with weighing in balances but with a more occult method of weighing without instruments for measuring weight.[75] Hydrostatic magic is concerned with such problems as why ice floats on water.[76] Under aerotechnic magic, Schott denies at length that a vacuum is produced by the Torricellian experiment, for which he cites Magnanus rather than Torricelli. Spirits from the mercury fill the top of the tube which appears to be left empty by the fall of the mercury. Therefore moderns have without sufficient reason abandoned the Peripatetic doctrine of the impossibility of a vacuum.[77]

Magic squares are included in the eighth book on arithmetical magic.[78] But planetary seals and images, observance of odd and even numbers or the number of a Psalm, and the like are pronounced inefficacious, vain and superstitious.[79]

The fourth volume of *Magia Universalis* is more directly concerned with what is usually regarded as magic or closely akin thereto. Books on cryptography, pyrotechnics and magnetic magic are followed by a discussion of sympathy and antipathy. Schott accepts the action of the magnet upon iron, of the sun on the heliotrope, of the moon on moisture and the brain. After rejecting other views, such as that sympathy and antipathy are the outcome

[71] *Ibid.*, III, 252.
[72] *Ibid.*, III, 270.
[73] *Ibid.*, III, 284.
[74] *Ibid.*, III, 505-8; 521 *et seq.*
[75] *Ibid.*, III, 308.
[76] *Ibid.*, III, 438.
[77] *Ibid.*, III, 568-601.
[78] *Ibid.*, III, 629 *et seq.*
[79] *Ibid.*, III, 727.

of occult qualities implanted by the stars, he attributes them to likeness and unlikeness, if not in the substantial forms of the subject and object concerned, at least in their temperaments and qualities, manifest or occult, expressed by the medium of exhalations.[80] Past authorities, including sixteenth century naturalists, are so uncertain as to the form, size, color and other accidents of the little remora or echeneis, which is supposed to halt ships, that one may doubt if it does. Nor do they agree why it does so. In any case, it is not a matter of sympathy or antipathy. Neither is the stupefying action of the torpedo, nor the feeding on surrounding herbs of the shrub shaped like a lamb and found among the Tartars.[81]

Turning to magical medicine, Schott affirms that, if weapon ointment cures, this is due to diabolical aid. Can a magnetic plaster draw iron objects from the human body? Kircher thought not, because the magnet loses its attractive force when mixed up in a plaster. Mere utterances or glances have no natural power to fascinate, but noxious vapors from the eyes and still more from the nose or mouth may injure another person. Human beings cannot cure by their mere touch or breath or kiss. Schott does not doubt that the corpse will begin to bleed at the approach of the murderer, but as to the reason why there is great disagreement. Demons are able to restore youth to an old man and to inflict diseases either directly or through the agency of magicians. Philters do not compel one to love but do induce a state which renders one more open to subsequent witchcraft and demoniacal action.[82]

Most forms of divination are regarded by Schott as illicit.[83] Dreams can signify only as to the state of mind and body. Such writers as Artemidorus and Cardan are very superstitious in the interpretation of dreams.[84] Physiognomy is founded on probable principles and confirmed by experience; Nicquet shows that it is useful to physicians and educators; but Michael Scot and John ab Indagine carry it too far.[85] Chiromancy is either astrological,

[80] *Ibid.*, IV, 367-69.

[81] *Ibid.*, IV, 410, 414, 437-41 (and in *Physica Curiosa*, p. 1349).

[82] *Magia universalis,* IV, 453-68, 479, 484, 488, 494-95, 498, 505, 512.

[83] *Ibid.*, IV, 541.

[84] *Ibid.*, IV, 564 *et seq.*, especially 580-582.

[85] *Ibid.*, IV, 588-614.

which Schott rejects, or physical, which is less objectionable. But he doubts if much can be correctly inferred from the lines in the hand.[86]

We have already noted the subjects of books I, II and IV of the *Physica Curiosa.* The third and fifth book are devoted to the congenial topic of monsters; the sixth, to portents; and the seventh to marvels of animals in general.[87] We are told that both fish and birds are made from water, although the latter live in the air, while other animals are made of earth.[88] The barnacle geese born from rotting wood are an exception. Such trite topics are aired as whether there are any animals who conceive from wind, as Cappadocian mares were said to do, and whether any live in fire or are generated from fire.[89] Schott thinks that angels may have helped to carry men and other animals to the New World and to islands far removed from the mainland.[90] After the eighth book on the marvels of terrestrial animals, which is in 79 chapters arranged alphabetically with pretty good pictures, comes the ninth on flying animals. Its 68th chapter is on the rhinoceros bird, so called from the horn which it bears on its forehead.[91] The tenth book upon aquatic animals is the last of the zoological section, although Schott recognizes in an epilogue that he might go on to treat of snakes, dragons, worms and innumerable insects and subterranean monsters.[92] He does, however, vouchsafe an appendix on snakes and dragons,[93] especially winged dragons and the biped one of Bologna in 1572.[94]

The eleventh book on the marvels of meteors assumes that there are three regions of air, questions whether laurel, the sea cow, and hyacinth are immune from lightning, gives the opinion of Descartes, and of Aristotle and Schott himself as to thunder, omits comets whether sublunar or superlunar, and is further cut short because of Schott's impending trip to Italy.

The twelfth and last book on miscellaneous marvels is made up

[86] *Ibid.*, IV, 637, 648, 668.
[87] *Physica Curiosa*, p. 678.
[88] Caps. 5-6.
[89] Caps. 7-8.
[90] Cap. 15, pp. 726-31.
[91] *Ibid.*, p. 1047.
[92] *Ibid.*, p. 1151.
[93] *Ibid.*, pp. 1152-78.
[94] T VI, 289-90.

in part of exercises by Schott as a student and roughly corresponds to the fourth volume of the *Magia Universalis,* to which there are occasional references. Many superstitious practices are condemned: divination connected with St. Andrew's eve, the three Thursdays before the Nativity, and Christmas eve; the suspension of an animal's left foot to aid the memory; the use of words; the casting into a cemetery a tooth which had fallen out in order to restore it; the refusal of peasants to allow the corpse of a suicide to be carried to the Rhine through their fields lest the soil be rendered sterile. Schott denies that a secret can be revealed at a distance by using magnets circumscribed with alphabets and which move sympathetically, or that one is immediately affected by the death of a relative afar off.[95]

Sympathy and antipathy are treated in much the same way as before, citing different views as to their cause and various instances of them from previous authors. Of the latter Schott remarks: "I do not approve of all, because I know that some are doubtful, if not false; others superstitious; others perhaps even manifestly false."[96] He opposes use of the divining rod to indicate hidden treasure, and again declares that weapon ointment and sympathetic powder are also due to Satan.[97] Without reason or authority the cabalists suppose some connection of sympathy between letters and syllables and heavens and Intelligences.[98] Schott condemns Cornelius Agrippa but defends Trithemius. Prospero Aldorisio[99] recently propounded at Rome an *Idengraphia* or art of reading natural abilities from handwriting. Schott at first regarded it as on a par with physiognomy, but, after discussing it with Kircher at Rome, decided that it was futile.[100] He is against ascribing any action to figures, characters, images and numbers. He regards as a monument of ancient superstition the belief in critical days, climacteric years, eighth month's birth being fatal, and taking an odd

[95] *Physica Curiosa,* pp. 1277-84.

[96] *Ibid.,* pp. 1285, 1344-45.

[97] *Ibid.,* p. 1286.

[98] *Ibid.,* p. 1287.

[99] The author also of a work on divination from laughter, *Gelotoscopia,* published in the same year (1611) at Naples, which Schott fails to note. *Idengraphicus nuntius,* BN Rés. V.1340, is bound together with it, BN Rés. V.1339.

[100] *Physica Curiosa,* pp. 1288, 1290.

number of pills.[101] He repeats his attitude towards chiromancy and divination from dreams in *Magia Universalis.*

The statement of Albumasar that prayers are surely answered when the moon and Jupiter are in conjunction in the head of the Dragon, is condemned, as is foretelling one's fortune from the letters in one's name, and the assertion in the *Secrets* of Wecker that any year can be judged from those preceding it by twelve, nineteen, eight, four and thirty years.[102] Schott now rejects the notion that the corpse bleeds in the presence of the murderer,[103] which he had accepted in *Magia Universalis,* but is credulous as to the effect of the mother's imagination upon the foetus.

The effects of the stars upon this sublunar world through certain secret forces which are called influences are stoutly affirmed, but it is added that astrologers cannot predict contingent or chance events, and the association of the signs of the zodiac with certain parts of the body or ages is condemned, as is the observance of lucky and unlucky days. In a note Schott adds that he thinks differently about the heavens now than he did when he was young. They are not a fifth essence but, like our pure air, are called ether, i.e., purest air. The planets are not moved by Intelligences by the impression of an impulse, but, like birds in the air or fish in water, are borne by the Intelligences along paths designated by God. The mariner's compass is not attracted to the celestial pole by the influences of the stars but by an ingrained sympathy is turned towards the terrestrial pole now with, and now without declination. Of the influences of the celestial bodies and their marvelous effects on the sublunar world there will perhaps be an opportunity for discussion elsewhere. With accurate observations and hypotheses based thereon astronomers can predict eclipses, the weather, and diseases; but without these, they can do so only probably.[104]

Theurgy is condemned.[105] Delrio showed against Scribonius that the cold water test for witches was illicit.[106] Schott accepts

[101] *Ibid.*, pp. 1294-95.

[102] *Ibid.*, p. 1303.

[103] *Ibid.*, p. 1304.

[104] *Ibid.*, pp. 1305, 1307-8.

[105] *Ibid.*, p. 1310.

[106] *Ibid.*, p. 1313. For others' censures of Scribonius, T VI, 535, 538.

the belief that animal horn is hostile to serpents, but condemns the drinking of the blood of one's beloved in order to cure infatuation.[107] Alchemical remedies lengthen life, if administered in moderation; but shorten it, if taken immoderately.[108] Marvelous trees are next considered for some pages. Botanists say that if rhubarb is plucked upward, it will purge upward by vomit; if plucked downward, will purge downward. But Kircher has well questioned this, since, however it is plucked, rhubarb always purges downward.[109] Kircher and Forerus also rejected as fabulous the statement of Zoroaster in *Geoponica* that, if a man goes with an axe to cut down a tree which has been sterile for many years, and a friend asks him not to do so, promising that the tree will bear next year, and he assents, it will do so. If it does, it must be by demon deceit.[110] But Schott accepts cures by suspension or mere contact of herbs, since by experience and the statement of botanists it is clear that great virtues are latent in herbs. But that a wren will keep turning on a spit of hazel wood until roasted, which Kircher and he had tested experimentally at Rome, is not due to the virtue of the wren but to the twisted fibres of the hazel.[111] The book on the virtues of herbs ascribed to Albertus Magnus is spurious like that on stones and that on the secrets of women.[112] Kircher said, however, that plants could grow again from their ashes.[113] Schott agrees with Deusing, Bartholinus and Kircher that the so-called monoceros horn is really from a fish of Greenland.[114] Bartholinus also gives a number of examples of invulnerability due to a natural cause. Some think that it is obtainable by use of a root, but Bartholinus rejected this as superstitious.[115] Such are a few specimens of the tone and content of the *Physica Curiosa*. While marvels of nature and art are affirmed, little place is left for superstitions, occult arts, and any procedure that is really magical.

Schott further composed a work on secret methods of writing

[107] *Physica Curiosa*, p. 1321.

[108] *Ibid.*, p. 1324.

[109] *Ibid.*, p. 1346, citing *Mundus Subterraneus*, XII, iv, 5.

[110] *Physica Curiosa*, p. 1347. Laurentius Forerus, S. J., *Viridarium philosophicum*, Dilingae, 1624.

[111] *Physica curiosa*, p. 1348.

[112] *Ibid.*, p. 1355.

[113] *Ibid.*, p. 1356.

[114] *Ibid.*, pp. 1376-7.

[115] *Ibid.*, p. 1378.

and communication,[116] in which, however, he declared impossible the supposed sympathetic action of two compasses at a great distance or of friends communicating with each other from afar by having mingled a little of their blood, after which, if one pricked his skin, similar punctures were believed to appear on the body of the other. Schott also cast doubt upon the tale of the nose that putrefied when the man, from whose skin and flesh it had been repaired, died.

A word may be added concerning Schott's Hydraulic-Pneumatic Mechancs.[117] In it he describes and explains a number of machines contained in Kircher's museum, and distinguishes three kinds of machines: tractory, hydraulic and hydraulico-pneumatic. He also outlines experiments, some of which seem scarcely true. But he again declares that it is impossible to achieve perpetual motion artificially.

A book with a title similar to Schott's *Physica Curiosa* is the Curiosities of More Secret Physics which Johann Michael Schwimmer published at Jena in 1672.[118] It has, however, the subtitle of Sympathy and Antipathy, and its fourteen chapters are mainly occupied with these subjects, but in a broad way involving other pseudo-science and magic. The first chapter deals with the nature of sympathy. Chapters two and three take up various causes or explanations which have been suggested for it: a world soul, the influence of the stars on these inferiors (which Schwimmer himself does not accept), influence of the Sephiroth, Platonic ideas, qualities manifest and occult, sense of nature, effluvia and atoms. The fourth chapter digresses to birthmarks, signs of pregnancy, nature of twins, supposititious infants and changelings (*Wechselbälgen, Kielkröppfen*), masticating corpses (*Schmekzenden Todten*

[116] *Schola Steganographica . . .*, Nürnberg, 1665, in-4. Another edition, Nürnberg, 1680.

[117] *P. Gasparis Schotti Regiscuriani e societate Jesu olim in Panormitana Siciliae nunc in Herbipolitano eiusdem societatis academia Matheseos professoris MECHANICA HYDRAULICO-PNEUMATICA . . . Accessit Experimentum novum Magdeburgicum quo vacuum alii stabilire alii evertere conantur.* Sumptu heredum Joannis Godefridi Schönwetteri Bibliop. Francofurtensis Excudebat Henricus Pigrin Typographus Herbipoli, 1657. BM 431. a. 5.

[118] *Ex physica secretiori curiositates . . .*, Jena, 1672. Copy used: Col. 502 Sch 99, 256 pp. with a following Index.

in den Gräbern), infants born with helmets or clothing, philters or *Liebes-Tränncken.* With chapter five come human singularities, thieves' thumbs (*Diebes-Daumen*), and the Abracadabra. The sixth chapter treats of the virtues of gems on man, warm baths, the water of the Nile, and why precocious children are rarely long-lived. Weather-presaging by animals, where the blight of wheat comes from, and from which eggs cocks and hens are hatched, are subjects of the seventh chapter. The divining rod and grafting are taken up in the eighth chapter. It is asked whether fruit will later crack open, if a woman who is menstruating for the first time picks them. There is further discussion of sympathy between metals and the stars, of animals and vegetables, and of characters. The ninth book considers the sympathy between minerals, volcanoes including Etna and Vesuvius, and the problem of Eve's kitchen-fire (*wo sie ihr erst Küchen-Feuer bekommen*). We come next to the tides and the sympathy of the sea with the moon, fountains, and acidulous springs and salt springs. Chapters eleven and twelve treat of the inclination of the magnet to the poles, the mariner's compass, and much more about the magnet. The thirteenth on the enmity or antipathy of things in nature, also shows that man is well called a microcosm, and discusses the bleeding of the corpse in the presence of the murderer. The fourteenth and last chapter continues the topic of discord between natural objects, especially the antipathy of certain men for certain foods, and antipathies between brute beasts. The cause is their specific form, and Schwimmer has no use for those who deny antipathy as Thomas Erastus does.

Schwimmer cites a great many authorities, some of whom are rare and unfamiliar, such as J. H. Engring, *De sex dierum operibus,*[119] Thomas Llamazar, *Disp. V in lib. Arist. de gen. et corr.*, and Stokman *in Hodog. Pestil.*[120] But some are misprints or misspellings, as Daurulcius for d'Averoult.

Bound with Schwimmer's book are Three Centuries of Tricks of

[119] Or Engrinch, *Tractatulus de sex dierum operibus, scilicet Janua per quam patet ingressus Naturae opera et arcana sinceris oculis contemplari et rerum causas cognoscere cupienti . . . etc.*, s.l., 1650, in-4.

[120] I. e., Ernst Stockmann, *Hodegeticum pestilentiale sacrum sive quaestiones quinquaginta . . . de peste,* Lipsiae, 1667, in-12: BM 846. a. 22.

Nature and Art or Natural Magic, in German.[121] This is another edition of the work by Schott mentioned in the opening paragraph of this chapter and which first appeared in 1665. These include many old stand-bys from Albertus, Agrippa, Cardan and Porta, such as making a cat dance, or breaking a stick resting on two glasses of water without breaking the glasses or spilling the water, or carrying water over a mountain, or the Sphere (here Wheel) of Life and Death. More novel possibly is healing a mortal sickness with a draught of beer (this based upon hearsay), and keeping horses and other beasts from eating by greasing their tongues, jaws and mouths, or how to learn Hebrew in a few hours. The work closes with a German translation of Kircher's *Diatribe, oder Beweiss-Schrifft von wunder-seltzamen Creutzen welche so wol auff der Leute Kleider als andern Dingen unlängst nach dem letzten Brand dess Berges Vesuvii zu Neapolis erschienen sind.*[122]

In 1670 a Jesuit father, Francesco Lana Terzi, published at Brescia a work on "some new inventions" as a *Prodromo* to a longer work which he had in preparation.[123] Because of its flying boat, supposed to be raised in air by four large spheres of very thin copper from which the air had been pumped out, the book has been noted in histories of aviation. It begins with ciphers, communication at a distance by cannon, reading signals through a telescope, deaf and dumb language, and proceeds to automatic birds, thermometers, barometers, a perpetual clock, an hour glass that turns itself, other devices for perpetual motion, and the production of a fountain where there is no water supply by distilling the air and converting it into water. There are chapters on agriculture, arithmetic and painting, on chemical transmutation and a panacea, on making flowers and fruit grow in a vase without seed, on a clock run by oil consumed in a lamp, and a cyclometer. Telescopes and microscopes are the chief theme of the last third of the volume.

[121] *Joco-seriorum naturae et artis, sive magiae naturalis Centuriae tres* . . . , Frankfurt-am-Main, 1672, 330 pp. The dedication is by the printer, Johann Arnold Cholin.

[122] *Ibid.*, pp. 2, 8, 199, 214, 54, 59, 93, 278-330.

[123] *Prodromo overo saggio di alcune inventioni nuove premesso all' arte maestra opera che prepara el P. Francesco Lana*, Brescia, per li Rizzardi, 1670, in-fol. iv. 252 pp. and 20 Plates. Copy used: BN Rés. R. 208. See further below, p. 669, note 141.

Fourteen years went by before Lana Terzi began to issue the Mastery of Nature and Art, to which the *Prodromo* was to have been the portico, and its publication in three huge double-columned folio volumes extended over eight years.[124] The long Latin title emphasizes the more occult principles of natural philosophy, experimentation and demonstration, "almost all the inventions of the ancients and many new ones thought out by the author himself." The second and third volumes in twenty-five books discuss as many varieties of motions of natural bodies, manifest and occult: namely, 1, the penetration of corpuscles through pores; 2, the motion of transpiration through the pores of the body or concerning the effluvia of bodies; 3, the motion of the internal parts of any body; 4, liquefaction and concretion or coagulation, where are treated fluidity and consistency; 5, compression and pressure of bodies; 6, elasticity; 7, tension; 8, rarefaction and condensation; 9, tremor of bodies; 10, sound; 11, adherence of parts and resistance to discontinuity; 12, mixture, ready or difficult; 13, configuration or site; 14, assimilation; 15, excitation and fermentation; 16, maturation and crudescence; 17, corruption and putrefaction; 18, coagulation (again?), where, too, of concretion (again?) properly called and incrassation; 19, precipitation; 20, dissolution; 21, fixation and volatilization, alkali and acid; 22, electric attraction (as by amber); 23, magnetic movements; 24, sympathy and antipathy; 25, gravity and levity.

In each of these books there are normally three chapters: the first devoted to observations and experiments; the second, to *Doctrina* in the form of propositions, in order to give the impression that mathematical as well as experimental method is being employed; the third consisting of inventions and *artificia*. Sometimes this third chapter is omitted. Sometimes experiments and inventions which might be made are suggested.

[124] *Magistorium naturae et artis, opus physico-mathematicum P. Francisci Tertii de Lana . . . , in quo occultiora naturalis philosophiae principia manifestantur et multiplici tum experimentorum tum demonstrationum serie comprobantur, ac demum tam antiqua pene omnia artis inventa quam multa nova ab ipso authore excogitata in lucem proferuntur*, Brixiae per J. M. Ricciardum (Parmae, ex typis H. Rosati), 1684-1692, 3 vols, in-fol. of 526 pp. and 24 Plates, 512 pp. and 20 Plates (1686), 571 pp. and 13 Plates. Copy used: BN R. 394-396.

It is with the *artificia* that we are here concerned as mild specimens of artificial magic. Under penetration through the pores we find separation of its quintessence from wine and other subtle liquors, making artificial gems, altering less precious stones to diamonds, changing red coral to white, and sticking a needle into one's arm or leg without feeling any pain. Under transpiration is repeated the distillation of water from air for an artificial fountain already given in the *Prodromo.* We are further instructed how to prepare the sympathetic powder, and the ludicrous spectacle of an egg, inside which has been secreted a live swallow. Among the 37 *artificia* under compression are secret writing and descending to the bottom of the sea in a bell. Among 47 under elasticity is Hero's vase, into which, if you pour water, it will flow out first through one tube, then two, and then three, and the candle which is lighted or extinguished at will. Under facile or difficult mixture are the separation of silver from *aqua fortis* in which it has been dissolved, representing the four elements in a glass vase, rectifying spirits of wine without distillation, separating silver from gold, and solution of ambergris. One egg may be made of many by separating the yolks and whites, and hard boiling them in a pig's bladder, when you will find all the yolks united at the center and surrounded by the albumen. A shell may then be made by painting the exterior repeatedly with a dissolution of powdered egg-shells.

In connection with excitation and fermentation we are told not only how to turn light wine dark and dark light, to make the seeds of gourds germinate at once, and to keep wine from spoiling during a thunder-storm, as well as the ferment of the philosophers' stone, but also how one may swell up without pain. One manuscript says to anoint him with juice of the herb *tarsia;* another, with that of *euforbia.* Lana Terzi does not seem to have put either to the test. Later, under corruption and putrefaction, he states the generation of a serpent from human hair as the narration of a man worthy of faith. Chemical recipes such as "our antimonial panacea" and the preparation of *aurum fulminans* are frequent. Under sympathy and antipathy are "General Precepts or a new art of preparing sympathetic and arcane medicaments," the transplantation of diseases or poisons or noxious qualities, and examples of "magico-physico-medical cures." We are told that insects or animals born of putrefaction have great medical virtue, for "they rise from the most tenuous spirits of the putrefying substance."[125] The reviewer in *Acta eruditorum* justly remarked that many of Lana Terzi's experiments were hardly credible, such as the production of a snake from a human hair and the resuscitation of plants.[126]

Lana's airship of 1670 seems to be imitated in a Physical Exercise on a contrivance to sail through the air, held at the University of Hesse-Schaumburg on March 4, 1676, in which a ship was to be suspended from six or eight globes.[127]

* * *

In the middle of the sixteenth century Cardan stated that a whole book would not suffice to enumerate all the (medieval and modern) inventions unknown to the ancients, such as furnaces in houses, bells in churches, stirrups on saddles, and weights in clocks. He went on to tell of a wheel upon which many threads were spun at once but which was purposely suppressed at Venice out of consideration for the livelihood of poor women and their children.[128] Their lot had been poignantly depicted two centuries before Cardan by the author of *The Vision of Piers the Ploughman:*

> . . . poor folk in cottages,
> Charged with children and the landlord's rent.
> What they save up by spinning they spend in house rent
> And on milk and meal to make porridge with,
> To satisfy their children who cry for food.
> They themselves too suffer much hunger
> And woe in winter time, waking in the night
> To rise up from bed and rock the cradle,

[125] *Magisterium,* III (1692), 478a. K. C. Schmieder, *Geschichte der Alchemie,* 1886, II, 433, was in error in representing the *Magisterium naturae et artis* of 1684, 1692, as a Latin translation of an *Arte maestra,* issued in Italian at Brescia, 1667, in-fol. There probably was a confusion in his mind with the *Prodromo,* which he does not mention as such, and in whose full title (see note 123 above) the words, "arte maestra," occur.

[126] XII (1693), 149.

[127] Franciscus Davis Frescheur, *Exercitatio physica de artificio navigandi per aerem, praeside Philippo Lohmeiero,* printed at Wittenberg, 1679. BM 7004. de. 1 (5.).

[128] *De subtilitate rerum,* lib. XVII; *Opera* (1663), III, 609.

To card and comb, to patch and wash,
To rub and reel, and to peel rushes:
So that it is painful to read or to show in rhyme
The woe of these women who dwell in cottages.[129]

A *filatoio da aqua* and also a spinning-jenny are portrayed in the *Novo teatro di machine* of Vittorio Zonca, first published at Padua in 1607[130] and reprinted there in 1621 and in 1656. But over a century more was to elapse before Hargreaves in 1764 brought it into practical operation and made "a success" of it. Yet even before Zonca's book appeared, a scholastic interest in machines was displayed at Padua by lectures given from 1603 to 1608, "Ad ingeniorum experimentum."

The foregoing single illustration is sufficient to suggest that in the seventeenth century practical labor-saving inventions and "applied science" were neither popularly appreciated nor as yet relentlessly exploited by promoters and capitalists. Drebbel's submarine under the Thames with its possible adjunct of the discovery of oxygen remained an object of mild literary and scientific interest, like the gems "of purest ray serene" which may or may not be found in "the dark unfathomed caves of ocean." Leading scientists like Huygens and Newton themselves ground the lenses for their telescopes. Justus Byrgi or Bürgi (1552–1632), for many years mechanician or *automatopaeus*—as Kepler liked to call him[131]—to the Landgrave of Hesse and then to the Emperor, taught Kepler decimal fractions, was acquainted with logarithms before Napier, and may have invented pendulum clocks.[132] Another pupil of his, Ursus, compared him to Euclid and Archimedes.[133] But he did not know languages and so wrote no books, confining his talents to the construction of celestial globes and other scientific instruments, to astronomical observations and mathematical calculations. On the other hand, the author of a book on machines may turn out to be a Jesuit or a Paulist father, a university professor or even a bishop, or at best an architect.

[129] C version, Passus X, lines 72-82.

[130] *Op. cit.*, pp. 68, 74. For the 1607 edition I used BM 537. m. 8; for the 1656 edition, Col. 621 Y8 Q.

[131] *Opera*, II, 80, 278.

[132] J. L. E. Dreyer, *Tychonis Brahe Dani Opera Omnia*, VI (1919), 346-47.

[133] Kepler, *Opera*, I, 219, letter of May 29, 1597.

Another indication of slow development is that works on the inventors of things were reprinted a century or more after their first appearance. This was partly due to antiquarian interest, the vast majority of the inventors of arts, sciences and religions, if not of mechanical contrivances, being from remote antiquity. But it was also partly owing to a lack of widespread interest in mechanical inventions and to the fact that not many new inventions had been made in the intervening period. Thus the work of Polydore Vergil, first printed in 1499 at Venice, appeared again at Cologne in 1626. And that of Giovanni Matteo of Luni, who lived in the second half of the fifteenth century, and whose unfinished work was edited in 1520 at Paris by Agostino Justiniani—who in the same year also brought out the *Victoria adversus Judaeos* of Porchetus Salvaticensis, written in 1303, and the *More Neuochim* of Maimonides—was reprinted in 1613 at Hamburg.

It should further be said that writers on inventions were apt to be as weak on the historical side as were the critics of astrology. Thus Lorenzo Legati[134] asserted in 1677, and the *Journal des Sçavans* repeated[135] that "Bertault Swart dit le Noir" discovered gunpowder by chance in 1369 and that the Venetians were the first to use cannon in 1380.

Fausto Veranzio, born in Dalmatia in 1551, was a law student at Padua in 1569, married and held various governmental positions and secretaryships. It was only after the death of his wife that he entered the clergy in 1594, became bishop of Czanád in 1598, and a Paulist in 1609. After all this, at some time between June 16, 1615 and July 16, 1616,[136] he published at Venice his work on New Machines,[137] about the last subject for which his previous career would seem to have fitted him. Libri dated the volume too early towards the close of the sixteenth century and gave it excessive credit for remarkable inventions.[138] Besides the suspension bridges

[134] *Museo Cospiano*, Bologna, 1677, in-fol.

[135] JS VI (1678), 311.

[136] This has been demonstrated from documentary evidence by H. T. Horwitz in *Archivio di storia della scienza*, VIII (1927), 168-75.

[137] *Machinae novae Fausti Verantii Siceni cum declaratione Latina, Italica, Hispanica, Gallica et Germanica*, Venetiis cum privilegiis, n.d.: BM 535. l. 16. There are 49 plates and a Latin text of 19 pp., followed by those in other languages.

[138] Libri, *Hist. des sciences*, IV (1841), 47.

and parachute which he noticed,[139] there is a life-preserver in the form of a belt which can be inflated, a wooden dredger like a steam-shovel but worked by a tread mill, and a covered wagon on springs.

Nicolaus Zucchius or Zucchi of Parma (1586–1670) was a Jesuit who taught at and became head of the *Collegium Romanum* and at Ravenna, was papal penitentiary, and holder of other offices. He is said to have cooperated with Scheiner in the observation of sun-spots[140] and to have discovered the spots on Jupiter on May 17, 1630.[141] Alegambe tells of his healing an ulcer of seven years standing in a virgin named Apollonia Caballa by application of a bit of a garment of Francesco Borgia, a "prodigy" which was approved as a miracle of the second order in the subsequent process towards his (Borgia's) canonization. But Zucchi seems to have been a versatile genius, for his New Philosophy of Machines was first printed at Paris in 1646,[142] and then at Rome in 1649, with an answer to criticisms which had been made of it and two additional tracts, one contending that recent experiments demonstrated a plenum and not a vacuum, the other, of which we have already treated in Chapter 9, on magnetic philosophy with a new argument therefrom against the Copernican system.[143] A later work was on optics and the eye.[144]

Three other authors of books on machines were architects, two of them Italians and one German. Zonca, the first in point of time,

[139] *Ibid.*, pp. 48-49: "... les ponts suspendus par des chaines en fer exactement comme on en fait aujourd-hui et le parachute (a square sail) dont la figure est parfaitement dessinée."

[140] *Memorie degli scrittori e letterati Parmigiani raccolte del Padre Ireneo Affò e continuate da Angelo Pezzana*, Tomo settimo ed ultimo, Parma, 1833, p. 668.

[141] H. Bosmans, "Théodore Moretus de la Compagnie de Jésus," *De Gulden Passer*, VI (1928), 57-163.

[142] He so states in the Preface to the Reader of the 1649 edition.

[143] *Nova de machinis philosophia in qua paralogismis antiquis detectis explicantur machinarum vires unico principio singulis immediato ... Accessit exclusio vacui contra nova experimenta, contra vires machinarum. Promotio philosophiae magneticae; ex ea novum argumentum contra systema Pythagoricum, Rome,* 1649: BM 538. i. 2; Col 530 Y82. Third ed., Rome, 1669 or 1670.

[144] *Optica philosophia experimentis et ratione a fundamentis constituta ... Pars prima De visibilibus et eorum repraesentativis*, Lyon, 1652; *Pars altera de naturali oculorum constitutione ...*, Lyon, 1656: BM 537. k. 13.

whose *Novo teatro di machine* of 1607 at Padua has already been mentioned, was architect to the Commune of Padua. Giovanni Branca (1571–1645), whose *Le machine* appeared at Rome in 1629,[145] also composed a manual of architecture[146] and practiced his art at Loreto, Assisi and Rome.[147] Georg Andreas Böckler of Nürnberg wrote on both civil and military architecture in the second half of the century.[148] His *Theatrum machinarum novum* first appeared in German in 1661; then in Latin at Cologne in 1662;[149] again at Nürnberg in German in 1673 and in Latin in 1686.

All five authors who have just been named—Zonca of 1607, Veranzio of 1615–1616, Branca in 1629, Zucchi in 1646 and 1648, and Böckler in 1661 and thereafter—have the word "new" in their titles. Yet all, with the exception of Zucchi, are closely akin not only to one another, but to similar works of the preceding centuries.[150] Thus Zonca shows a machine to prepare gunpowder, an instrument for fulling woolen cloth, and a paper mill.[151] But fulling mills had been known in Christian western Europe since the twelfth century; gunpowder and paper mills, since the thirteenth.

The title of Branca's book of 1629 claims to work marvelous effects *tanto Spiritali quanto di Animali Operatione,* but opens with a batter beater, metal rollers, a pile-driver, powder mill, a mill to press olives, and a device to raise water. Its remaining contents are no more wonderful: water clocks, a wind chariot, water mills, threshing grain, saw mills, a spinning wheel run by water power, and machines to lift weights, drag cannon, and draw boats ashore. A bed on a wagon remains in a horizontal position regardless of the positions of the four wheels. After instruction how to

[145] *Le machine: Volume nuovo et di molto artificio da fare effetti maravigliosi tanto spiritali quanto di animali operatione arichito di bellissime figuro con le dichiarationi a ciascuna di esse in lingua uolgare et latina,* Rome, 1629: copy used, BN V. 7211. Col Egleston D 531 B 733.

[146] *Manuale di architettura,* Ascoli, 1629, in-12; Rome, 1718-19, 1757.

[147] On his life see Zedler and *Enciclopedia italiana.*

[148] *Compendium architecturae civilis,* Frankfurt, 1648; *Compendium architecturae militaris,* Strasburg, 1648; *Hand-Buchelin von der militari-Baukunst,* Frankfurt, 1672; *Arithmetica militaris,* Nürnberg, 1661, Jena, 1671; etc.

[149] This is the edition I have used: BN V. 2455.

[150] See T V, 34-35, 588-89, 593-96; VI, 373.

[151] *Novo teatro,* 1607, pp. 85, 96, 94.

saw through stone, such devices as water wheels, and repetition of some previous objectives, the book is completed by two new series: one of 14 figures of various machines to raise water, the other of 23 figures of *machine spiritali,* i.e., air-pumps and siphoning. These make Boyle's air-pump of some thirty years later seem a bit antiquated. Most such collections of machines include at least one attempt at perpetual motion, but that in Branca's volume seems exceptionally silly. A stream, B, turns a water wheel which works a pump that raises water from the stream to a higher level, C, whence it is further siphoned up to a yet higher level, A, where some of it remains, although most of it falls back into the stream again, in order to maintain the siphoning. The only perpetual motion appears to be that of the stream.[152] There is also a figure for siphoning wine up from a cellar.

Zucchi repeatedly stresses the marvelous character of machines,[153] but he describes and depicts only the fundamental simple types—balance, lever, windlass, pulley, wedge and water screw. His object is to find a single principle which will serve to explain the forces of machines, and he thinks that he has found it in their velocity.

The contents of Böckler's book are less varied than those of Branca's, consisting largely of mills and hydrotechnics, and these seem in large part derived from the earlier work of Jacobus Strada de Rosberg, director of the imperial art gallery at Prague, which had circulated for some time in manuscript copies before it was printed by his grandson, Octavio Strada, at Frankfurt in 1617–18, as *Künstliche Abriss allerhandt Wasser-, Windt-, Ross-, und Handt-Mühlen.*[154] It also reprints Saxon regulations of mills of 1568[155] which Zeisinck had already published in 1612, devices from Agricola's work on mining,[156] and gives a machine for raising water

[152] Number 13 in the 3rd series.

[153] *Nova de machinis philosophia,* 1649; p. 1, "mirabilium quae per illas fiunt"; p. 18, "Huius effectus mirabilitas"; p. 97, "Sed artificium machinae haec habet mirabilia."

[154] Hugo T. Horwitz, in *Archivio di storia della scienza,* VIII (1927), 172, note 8.

[155] *Theatrum machinarum novum,* 1662, in-fol., pp. 50-55. The preceding pages give a very brief text descriptive of the plates which follow. "Zeisinck" is for Heinrich Zeising, *Theatrum machinarum,* Leipzig, 1612; Altenburg, 1614-1621.

[156] Böckler, pp. 28-29, figs. 89, 92.

constructed in Lorraine in 1603.[157] Of its 154 full page Plates the last and most recent, showing a fire engine made by Johann Hautsch of Nürnberg in 1658, by which twenty-four men could raise water to eighty or a hundred feet,[158] had been cut out of the volume which I consulted.[159]

Daniel Lipstorp, who had published a book on the Copernican system at Rostock in 1652,[160] next year in his Specimens of Cartesian Philosophy[161] explained a number of hydraulic-pneumatic machines by Cartesian principles.[162]

A Jesuit, Jean François, wrote on artificial fountains and canals.[163]

Deschales devoted a section of his course in mathematics[164] to hydraulic machines, and in 1675 a royal order directed *l'Académie des Sciences* to prepare descriptions of all existing machinery in France and Europe.[165] The first public exhibition of models followed in 1683.

A treatise which J. J. Becher addressed in 1680 to the Royal Society dealt primarily with the construction of clocks.[166] But he listed a number of other wonder-working and semi-magical machines and contrivances. By order of the king of France, "our Noric Hautsch" had constructed for the dauphin a machine which showed a whole army in conflict. He also reproduced the char-

[157] *Ibid.*, p. 22, fig. 71.

[158] *Ibid.*, pp. 49-50, fig. 154.

[159] BN V. 2455.

[160] *Copernicus redivivus seu de vera mundi systemate*, Rostock, 1652, in-4. His letter, accompanying a presentation copy which he sent to Huygens, is printed in the latter's *Oeuvres*, I (1888), 177-78.

[161] *Specimina philosophiae Cartesianae . . .*, Leyden, Elzevirs, 1653, in-4. The *Copernicus redivivus* was reprinted with it.

[162] Schott, *Mechanica hydraulico-pneumatica* (1657), p. 9.

[163] *L'Art des fontaines . . . avec l'art de niveler et . . . de faire des canaux . . .*, Rennes, 1665, in-4, 120, 32 pp.

[164] Cl. Fr. Milliet Deschales, *Cursus seu Mundus mathematicus*, 1674, 1690.

[165] Huygens, *Oeuvres complètes*, XXII, 694-95, states that Buot with collaborators described the chief ones in use, but this would be impossible after the royal order, if he died in 1673, as recorded by Ernest Maindron, *L'ancienne académie des sciences: Les Académiciens*, 1666-1793, Paris, 1895, p. 24.

[166] Joannes Joachimus Becherus, *Theoria et experientia de nova temporis dimitiendi ratione et accurata horologiorum constructione.* In Col 542 B38, it is bound with his *Physica subterranea* of 1681, following the Index on unnumbered pages, which I indicate by the signatures.

acteristic movements of a great variety of artisans, and a chariot which moved without horses. Instruments have been devised by which one man can weave twelve measures of cloth, and Becher knows of a machine with which three men can produce a hundred ells of cloth a day. He merely mentions mills and the English stocking-frame.[167] But he thinks it impossible to imitate the human gait or voice. The human statues and beasts in clocks revolve on wheels and do not move their feet. The story was current that Albertus Magnus made a walking automaton which saluted and spoke to Thomas Aquinas, when it met him. When Aquinas smashed it, Albertus complained that he had destroyed the work of twenty years. But Becher regards this tale as a fable. He also doubts if artificial animals can be made to fly mechanically, like the wooden dove of Archytas of Tarentum or the eagle which Hautsch made to fly and meet Charles V. However, a few years since, Father Lana exhibited a flying machine, while submarines had been devised by Drebbel, Mersenne, and by a Frenchman recently at Rotterdam.[168] Becher himself claimed to have shown the emperor Ferdinand III his image in the clouds in 1656, and in 1660 to have made a thermostat which would open the furnace door, if the heat began to fail, and close it, when it grew too hot.[169]

The *Journal des Sçavans* occasionally devoted space to proposed inventions. Thus Borelli's machine to breathe under water, proposed in the issue of July 6, 1682, was criticized in that for August 6, 1683,[170] while fourteen pages were given on November 22 to a circulatory statue which would perform all the internal operations of the human body.[171] Another machine was to turn sea water sweet.[172]

There is, however, one pleasing feature of our seventeenth century specimens of technological literature as compared with their predecessors in previous centuries. It is that they are less concerned with engines of war and destruction, and that the word *machina*, employed by Lucretius and Sacrobosco in describing the mechanism

[167] *Ibid.*, K 4 recto.

[168] *Ibid.*, K 4 verso.

[169] *Ibid.*, L 4 r-v.

[170] JS XI (1684), 278-81, with illustration. See also p. 360. See, too, *Acta eruditorum*, II (1683), 73-77, 553-56.

[171] JS XI, 338-51.

[172] *Ibid.*, 380-83.

of the universe—*machina mundi*—but since the invention of gunpowder and artillery used for a time almost exclusively in the sense of an infernal machine, is now reverting to its modern application to machinery in general.

Furthermore, we must admit that machinery and magic do not go together. Magic may employ sleight-of-hand, but not the monotonous regularity of impersonal mechanics, which is the very antithesis of magic. It is true that the magic rite may have become stereotyped by constant repetition but, even though it has lost its original meaning and intellectual content, there is still something emotional and subjective about it. Branca and Zucchi and Becher may boast that machines work marvelous effects, but if so, they do it at magic's expense. Their gain is magic's loss. They are "stealing her stuff." A magician may be needed to invent them in the first place, but once devised, they require only a mechanic to keep them in order. He does not need to know Latin or Hebrew, secrets or ancient lore, or even mathematics, physics and chemistry. All he needs to know is machinery and how it works. In this perhaps there is still some analogy with the unchanging magical rite and time-honored ceremonial.

CHAPTER XXII

HUYGENS

Career and scientific achievement—Relation to Descartes: the laws of percussion—Astronomy and astrology: freedom from superstition and occultism—Terminology—The rainbow and Dietrich of Freiberg—Mathematical and mechanical invention—Late publication and secrecy of method—Use of the experimental method—Conception of nature—Element of marvelousness—Cosmotheoros: the planets inhabited—Sallies into the history of civilization.

I cannot but think of those times with pleasure and of our diverting labor in polishing and preparing such glasses, in inventing new methods and engines, and always pushing forward to still greater and greater things
—Huygens to his brother

Huygens führt zuerst die Mechanik . . . in die Sternkunde ein
—Apelt (1852)

Christiaan Huygens (1629—1695) belonged to an aristocratic and well-to-do family. His father has been called "the most brilliant figure in Dutch literary history."[1] The son wrote chiefly in French or Latin, even in the case of private letters. His brother and he were for a time the leading makers of telescopes in Europe, or, for that matter, the world.[2] Their success was based partly upon correct theory, partly upon long practice, experience and experimentation; partly, Huygens himself said, it was a matter of chance.[3] In 1655, with his superior telescope, he discovered the ring of

[1] Henry Crew, *The Rise of Modern Physics*, 1928, pp. 116-17, quoting Edmund Gosse.

[2] Later they were outdistanced by Italian lens-makers, and Cassini discovered a fourth and fifth satellite of Saturn which Huygens could not see: *Oeuvres complètes de Christiaan Huygens*, publiées par la Société Hollandaise des Sciences, La Haye, XXII (1950), 730-31.

[3] *Ibid.*, XXII, 495-96: "atque ego non experientiae magis quam casui acceptum fero quod eorum perfectissimam fabricam deprehenderim."

Saturn, in 1656 invented the pendulum clock, and in 1657 obtained a patent for it from the Dutch government. He communicated his invention of the watch-spring to the *Journal des Sçavans* in 1675.[4] Already during his lifetime priority in the invention of the pendulum clock had been claimed for Galileo and Justus Borgen. On the other hand, his drawings of Saturn were used in the nineteenth century in determining the variability of its ring; and his drawings of Mars, with reference to the time of its rotation.

Some histories of science speak of a visit to England by Huygens only in 1689, but he was there as early as 1661, when he entertained a group of scientists including Wallis and Wren, and was said by Oldenburg in a letter to Spinoza to have correctly predicted the rebound of two pendulums weighing a pound and a half-pound, if raised to an angle of 48° and then released.[5] In 1663 Huygens was elected to the Royal Society, which in 1666–1668 requested Wallis, Wren and Huygens to set forth their views as to percussion, but then published only the views of the two Britishers. Wren's views were the same as Huygens', but, as the latter complained, not even this fact was noted in publishing Wren's paper. Huygens accordingly published his explanations in the *Journal des Sçavans* of March 18, 1669, and an English translation appeared in *Philosophical Transactions* of April.[6] Huygens had been called to Paris by Colbert, acting for Louis XIV, in 1666 as one of the foundation members of *l'Académie des Sciences* and received a higher stipend than anyone else, six thousand livres.[7]

Serious illness led thrice to his return to Holland: September, 1670 to June, 1671; again, from July, 1676 to June, 1678; finally, in 1681, after which the French government did not encourage him to return. In the first illness, "his fancy was ready enough to suggest the worst"; the second was describd as melancholy, and

[4] JS IV, 67-68: issue of February 25, 1675.

[5] *Oeuvres*, XVI (1929), 173. Huygens referred more briefly to this incident in the Journal of his trip to Paris and London of 1660-1661, 23 Avril, 1661, *Oeuvres*, XXII (1950), 573.

[6] *Oeuvres*, XVI (1929), 173-78. JS II, 531-36.

[7] He had already visited Paris three times: Henri L. Brugmans, *Le séjour de Christian Huygens à Paris et ses relations avec les milieux scientifiques français*, Paris, 1935; J. A. Vollgraff, "Christiaan (ou Christiaen) Huygens," *Archives internationales d'histoire des sciences*, Oct. 1948, 165-79.

he spoke of his "well nigh desperate malady," and said, "I don't believe I'll return to Paris, having found for the second time by too many disagreeable experiences that the life I lead there does not agree with me." When he did return, his family arranged for a housekeeper to look after him. His sister and her family came to take him back to The Hague in 1681. His last illness was diagnosed by the attendant physicians as melancholy and was marked by delirium or insanity.[8]

Huygens' father had been a great admirer of Descartes, and Huygens, as he himself said, was at first carried away by the Cartesian philosophy, as one is from reading a thrilling romance. In 1668 he explained the action of the magnet by a single *tourbillon* instead of by two of opposite directions, as Descartes had done.[9] Hooke in 1690 represented Huygens as making a vortex of ethereal matter the cause of gravitation, with the ether which was about the sun moving forty-nine times faster than that about the earth.[10] Huygens' recent editors speak of him as "croyant toujours aux tourbillons unilatéraux cartesiens," although, after reading the *Principia*, he lost faith entirely in "vortices deferentes."[11] Whereas Descartes had estimated the distance of the sun as six or seven hundred diameters of the earth, Huygens suggested ten or twelve, or ten or eleven thousand.[12] He rejected the Cartesian tenet that the movement of light was instantaneous, and promulgated a wave theory of light.[13]

It has been said that Huygens did not understand the mechanism of percussion, because dynamics were still imperfect, and Newton had not yet founded classical mechanics, with theorems that now seem self-evident but were in Huygens' day still unknown. At least he had absorbed what Galileo had to say on the subject, and expressed correcter views than most of his scientific contemporaries. As early as October 29, 1652, he wrote to van Schooten, under whom he had studied mathematics at Leyden, that, if two bodies collided, one of which was twice as big as the other but moving only half as

[8] *Oeuvres*, XXII, 656, 696-97, 704, 714, 766-68.

[9] *Oeuvres*, XXII, 645-46, 707.

[10] A. R. Hall in *Isis*, 42 (1951), 224, 226.

[11] *Oeuvres*, XXII, 734; XXI, 438.

[12] *Oeuvres complètes*, XXI (1944), 693, 783.

[13] *Traité de la lumiere*, Leyden, 1690, in-12.

fast, they would rebound with the same speed, which shows that he accepted Gailieo's definition of momentum as the product of weight by velocity. Van Schooten, on the other hand, held that only the smaller body would recoil and that it would maintain its original speed, while the larger body would continue its original motion.[14] Two years later van Schooten reproached Huygens for criticizing and refuting Descartes as to motion, and seeming ungrateful to so great a man.[15] Huygens' refutation of the Cartesian propositions was indeed quite crushing, since he said that only the first of the eight rules of Descartes as to the percussion of bodies was correct, namely, that two equal bodies, moving towards each other at equal speed, would rebound at the same speed.

While Huygens esteemed Kepler as a mathematician and astronomer, he regarded his relating the orbits of the planets to the five regular solids in his *Mysterium cosmographicum* as "nothing but an idle dream taken from Pythagoras or Plato's philosophy."[16] Similarly the Ecstatic Celestial Journey of Athanasius Kircher was "nothing but a heap of idle unreasonable stuff,"[17] and full of astrology. Huygens opposed judicial astrology from his early correspondence[18] down to the last work of his declining years, in which he wrote:

> And as for judicial astrology, which pretends to foretell what is to come, it is such a miserable and often mischievous piece of madness, that I do not think it should be so much as named here.[19]

It has been suggested that the person of quality, of whom he speaks in a letter of August, 1666, who believed in horoscopes and wished to draw up Huygens', may have been the astronomer Boulliau or his fellow Academician, Auzout. But astrology seems not to have been discussed publicly at the Académie des Sciences.[20]

[14] *Oeuvres complètes*, I (1888), 185-86.

[15] *Ibid.*, 312.

[16] *The Celestial Worlds Discover'd or Conjectures*, 1722, p. 148.

[17] *Oeuvres complètes*, XXI (1944), 811.

[18] *Ibid.*, I (1888), 307.

[19] *Conjectures*, p. 68; *Oeuvres*, XXI (1944), 737, also p. 541, and the account of his correspondence with Kliner à Löwenthurn about the eclipse of August 11, 1654, in our chapter on Astrology After 1650.

[20] *Oeuvres*, VI, 76; XXII, 629.

Among its members, however, were Frenicle who "owed his celebrity to his knowledge of the properties of numbers and to his studies on magic squares;"[21] Duclos who wrote on mineral waters and the principles of mixed bodies; and Cureau de la Chambre who composed a chiromancy.

In general Huygens was free from superstition and faith in the occult. Such reputed natural phenomena as the Soland or barnacle geese seemed to him simply ridiculous.[22] An idea of examining the motive force of fulminating gold was about as close as he came to alchemy.[23] For the apparent relation of sympathy between two clocks he gave another explanation.[24] But one has a feeling that a subconscious motive for his arguing that the other planets were full of rational beings, and as noble, beautiful, and dignified as the earth, was a desire to restore something of their lost estate as superior celestial bodies by whose influence our inferior terrestrial ball had been governed. Thus he writes:

> Now can any one look upon and compare these systems together without being amazed at the vast magnitude and noble attendance of these two planets (Jupiter and Saturn) in respect of this little earth of ours? Or can they force themselves to think that the wise Creator has disposed of all his animals and plants here, has furnished and adorned this spot only, and has left all those worlds bare and destitute of inhabitants who might adore and worship him; or that all those prodigious bodies were made only to twinkle to, and to be studied by some few perhaps of us poor mortals?[25]

Huygens' recent editors explain his glorification of the stellar number, twelve, in the dedication of his *Systema Saturninum* to Leopold de' Medici as a concession to that prince's mentality.[26] And they account for his brother's calling Saturn *infaustum* by poetical rather than astrological tradition![27] Since Huygens believed that comets move in a straight line, he incorrectly held that a comet seen in November, 1680 was not the same as that of December 26.[28] But

[21] *Ibid.*, XXII, 630.

[22] *Conjectures*, p. 31.

[23] *Oeuvres*, XXII, 680. On Nov. 18, 1660, he bought a copy of the first edition of Le Fevre, *La chimie théorique et pratique: ibid.*, 537.

[24] *Oeuvres*, XVII, 183-86.

[25] *Conjectures*, p. 117.

[26] *Oeuvres*, XXII (1950), 505-6.

[27] *Ibid.*, p. 511.

[28] *Ibid.*, pp. 713-14.

by 1689 he was inclined to accept Newton's opinion in the *Principia* that the orbits of comets were elongated ellipses with the sun at one of the foci.[29]

It has been noted that the terminology of Huygens was not fixed and constant, and that he did not always employ such a word as gravity in the same sense.

Professor Carl B. Boyer, in a paper upon "Kepler's Explanation of the Rainbow," remarks:

It is probably safe to say that more volumes on the rainbow appeared between 1500 and 1700 than during all the years which preceded or succeeded; and most of these were pre-Cartesian, many appearing in Germany.[30]

Huygens' correspondence shows that as early as 1653 he was intent upon the problem of refraction in the rainbow, and that he was not satisfied with Descartes' explanation of the iris.[31] What I want to point out is, not merely that most of these numerous writers of the sixteenth and seventeenth centuries could have saved themselves and their readers trouble, had they known that Dietrich of Freiberg and a writer in Arabic contemporary with him had given essentially correct expositions of the rainbow in the first years of the fourteenth century and in writing still extant, but that the learned editors (from *La Société Hollandaise des Sciences*) of the works of Huygens, who have done so much to correct other misapprehensions in the history of science, were in 1932 (vol. XVII) equally in ignorance of the work of Dietrich, although in the interim it had been printed in part in 1814, in whole in 1914, and discussed repeatedly. You may be able to understand and evaluate the work of Dietrich without having read that of Huygens, but you cannot properly appraise Huygens or his century without knowing of Dietrich and the physics of the fourteenth century.[32]

Huygens spent much time upon mathematical problems and mechanical contrivances[33] as well as astronomical observation and physical experimentation. Pumps and fountains, gears and mills,

[29] *Ibid.*, p. 740.

[30] *American Journal of Physics*, XVIII (1950), 360.

[31] *Oeuvres complètes*, I (1888), 238, 240.

[32] Professor Boyer is of course well aware of the importance of both.

[33] For memoranda and sketches: *Oeuvres*, XXII (1950), 180-324; also 425, 427, 585, 680, 686-87.

flying devices and automata, a motor propelled by the explosion of gunpowder, coaches and music boxes, as well as more scientific instruments, at one time or another claimed his attention. He still thought that squaring the circle was not an insuperable problem.[34] But he believed perpetual motion to be impossible.[35] He seems to have liked to tinker and calculate better than he did to write. He experimented with colors before Newton did, but published nothing on the subject.[36] In 1666 a micrometer, devised by him, was first used in the observation of eclipses by the Académie des Sciences. Although advised in 1665 of the desirability of presenting some mathematical works to Louis XIV, he began work on the final draft of his *Horologium oscillatorium* only in the fall of 1669, and it was published with the dedication to the king only in 1673.[37] Before 1671 he had ample material for another book on Saturn, but put it off until 1672, and then never wrote it.[38]

Some of his writings were printed only after his death, like the *Dioptrica*, or the treatise on the motion of bodies from percussion, although it had been finished between 1652 and 1656. Although he had patented the pendulum clock in 1657, his work upon it, as we have just seen, did not appear until 1673, when, however, it included the results of his intensive investigation of the pendulum and development of the concept of centrifugal force. But *De vi centrifuga* was printed only in 1703, with *Dioptrica* and the treatise on motion from percussion. Of his concept of centrifugal force Apelt wrote a century ago:

> Seine Theorie der Centralkräfte wurde in der That für Newton die Brücke von den Gesetzen Kepler's zu den Gesetzen Galilei's.[39]

A memorandum of 1659, unpublished until 1932,[40] shows that Huygens already had the conception of centrifugal force then, and his modern editors say:

> Cette détermination de la valeur absolue de la force centrifuge est un des grands mérites de Huygens, dont personne, que nous sachions, ne lui a jamais contesté la priorité.[41]

[34] *Oeuvres,* XX, 308-9, 370-74.
[35] *Ibid.,* XXII, 699, 726.
[36] *Ibid.,* p. 618.
[37] *Ibid.,* pp. 654-55.
[38] *Ibid.,* p. 671.
[39] E. F. Apelt, *Die Reformation der Sternkunde,* 1852, p. 245.
[40] *Oeuvres,* XVII, 276-77.
[41] *Ibid.,* XXII, 513.

He presented his treatise on light to l'Académie des Sciences in 1678, but its publication came only in 1690. He kept the secret of the ring of Saturn until March 28, 1658, showing it only to Boulliau in 1657, and begging him in a letter of December of that year "to communicate to no one what you know of the Saturnine world." When he did publish it in 1659, "the ring which the magician Christiaan had discovered in the firmament was not immediately accepted as a fact by everyone."[42]

Huygens wanted to obtain due credit for his own ideas and discoveries. But, had he rushed into print with an explicit account, someone else might have published subsequently a similar treatise with the claim to have composed it long before. Huygens therefore adopted a method which had already been common in the previous century in connection with such matters as the solution of mathematical equations. He first briefly stated the essence of his new idea or invention in a cryptogram or anagram, and then, after sufficient time had elapsed, and no one else had suggested a solution, set the secret forth openly at length.

Thus his discovery of the ring of Saturn was stated in the following mysterious manner: $a^7\ c^5\ d\ e^5\ g\ h\ i^7\ l^4\ m^2\ n^9\ o^4\ p^2\ q\ r^2\ s\ t^5\ u^5$. This was the concealed anagram of a sentence in which the letter n was employed nine times, i seven, c, e, t and u five times each, and so on. The correct solution was, "Annulo cingitur tenui plano nusquam cohaerente ad eclipticam inclinato," which may be translated, "It (that is, Saturn) is belted with a thin flat ring which never touches it and is inclined to the ecliptic." Similarly he wrote to Oldenburg, the secretary of the English Royal Society, on September 4, 1669:

> I send you herewith appended some anagrams which I will be pleased to have you keep in the registers of the Royal Society, which has been so kind as to approve this method of mine for avoiding disputes, and for rendering to each individual that which is rightly his in the invention of new things.

He enclosed fourteen anagrams of which two gave the essential theorems as to the amount of the centripetal force, employing roughly the same method as we have noted above. The solution of these anagrams appeared in his work of 1673 on the pendulum

[42] *Ibid.*, XXII, 519-20, 523.

clock.[43] This secret and occult method is suggestive of magic in general and of magic characters in particular.

But it was an age when one had to be on one's guard against charlatans as well as scientific rivals. The Dutch States-General gave a privilege and two thousand florins to a man whom Huygens correctly regarded as an impostor, and who pretended to be able to find the longitude without observations from the retrogradation of the moon in the firmament.[44]

In the treatise on measuring time and constructing clocks which Becher addressed in 1680 to the Royal Society,[45] it is stated that Huygens claimed the invention of pendulum clocks and received therefor a privilege from the States of Holland and a stipend from the king of France. But count Magalotti, representative of the Grand Duke at the imperial court, and Treffler, who had been clockmaker to the Grand Duke's father, asserted that the first pendulum clock was made at Florence in conformity with Galileo's instructions, and that the original clock or model or a copy had come to Holland.[46] Caspar Doms, who was formerly mathematician to the now defunct elector of Mainz, told Becher that he saw at Prague a pendulum clock which had been made there by the emperor Rudolf's mechanic, Justus Borgen.[47] Becher felt that hitherto there had been no satisfactory water clocks, and that the variety and multiplicity of modern clocks did more harm than good.[48]

Huygens was a staunch advocate of experiment and believed, like Roger Bacon and many another medieval of centuries past, that practical or applied science could only be learned from technicians, clock-makers and artisans.[49] But he was also inclined

[43] *Oeuvres complètes*, VI, 487; XXII, 503; Crew, *Rise of Modern Physics*, pp. 119-20.

[44] *Oeuvres complètes*, IX, 317.

[45] J. J. Becher, *Theoria et experientia de nova temporis dimetiendi ratione et accurata horologiorum constructione*, Ad Societatem Regiam Anglicanam in Collegio Greshamensi, Londini, Jan. 1680.

[46] *Ibid.*, p. (K 6) verso: "cuius exemplar in Hollandiam venit."

[47] *Ibid.*, p. (K 7) recto. Justus Borgen, Bürgi, Byrgi or Buergius (1552-1632), after having served Wilhelm, landgrave of Hesse, since 1579, came to Prague in 1603. See Dreyer, Tycho Brahe, *Opera*, VI (1919), 346-47, who thinks that he invented pendulum clocks.

[48] *Theoria et experientia . . .*, K 3 verso, K 4 recto.

[49] Vollgraff in *Archives internat. d'hist. des sciences*, 28 (1948), 170.

to attempt to prove too much from a single experiment, and that by mere or sheer analogy. For example, he covered the bottom of a cylindrical vessel with bits of sealing wax and partly filled the receptacle with water. It was then whirled about its axis on a revolving table, with the result that the particles of sealing wax went out to the sides of the vessel. When the rotation was suddenly stopped, the water continued to circulate for some time, while the bits of wax returned by spiral paths towards the center of the vessel. This might seem to be simply an example of centripetal force. But Huygens concluded from it that gravity is the "action of the aether which circulates about the center of the earth striving to travel away from the center and to force those bodies which do not share its motion to take its place."[50] Or, as the *Journal des Sçavans,* contemporary with Huygens, put it:

In the thought of Mr. Huygens, it is not the grosser air which produces weight, but a subtle matter which can pass freely through the pores of all bodies and which circulates in the air day and night, just as water goes through a sieve.[51]

Indeed, Huygens himself had spoken of this "matter subtler than air" in the *Journal des Sçavans* nineteen years before, and had expressed the same thought elsewhere since 1669.[52] We are reminded of Beeckman's subtle matter early in the century. Others followed the same line of thought. Oldenburg in 1670 sent Huygens a memorandum by Leibniz, who would explain "all the wonderful and extraordinary effects of nature" by the movement of the ether.[53] In 1687 Huygens spoke of an ethereal matter which served to propagate light and passed easily through glass and all sorts of bodies.[54] Upon hearing that there was a bird in America which sang six notes in order, he jumped to the conclusion, "Whence it follows that the laws of music are unchangeably fixed by nature."[55]

A very interesting instance of experimentation is afforded by a

[50] A. Wolf, *A History of Science, Technology and Philosophy in the Sixteenth and Seventeenth Centuries,* 1935, pp. 164-65.

[51] JS XIX, 351-52, in the review of *Traité de la lumiere . . . avec un discours de la cause de la pesanteur,* Leyden, 1691, in-12.

[52] *Oeuvres,* XXII, 649, 674.

[53] *Ibid.,* p. 663.

[54] *Ibid.,* p. 737.

[55] *Conjectures* (1722), p. 86.

communication of Huygens to the July 25, 1672 issue of the *Journal des Sçavans.*[56] Huygens' series of experiments were suggested by an experiment of Boyle in 1661. In it a glass tube four feet long and filled with water, sealed at one end and open at the other, was erected inside a larger glass vessel from which the air was to be pumped, with the open end of the tube resting in a glass of water. When Boyle had pumped out all the air that he could, the water in the tube fell into the glass until only about a foot of water stood in the tube, leaving its upper three feet empty of air or water. Boyle inferred that the reason why all four feet of water did not drop into the glass was that some air still remained in the larger enclosing vessel. Huygens, however, repeating this experiment, succeeded in bringing all the water in the tube down to the same level as that in the glass in which its open end rested. But when, in December, 1661, he let the water remain there for twenty-four hours, so that it lost all the bubbles of air that were in it when it was fresh, and then filled the tube with it and again exhausted the air from the containing vessel, the water in the tube did not descend at all. But if the tiniest bit of air was let into the tube, the water would fall.

The members of the Royal Society would not believe this until they saw it in 1663 with their own eyes, and Boyle found that the same was true of mercury in the tube, after the mercury had been entirely purged of air during a period of three or four days. Huygens also tried rectified spirit of wine in place of water. In this case, when the air had been almost all pumped out of the container, the spirit of wine would boil, and the bubbles from it would finally occupy the entire tube, taking the place of the spirit of wine. When air was let into the container, the spirit of wine would again ascend into the tube but not fill it entirely, some "air" remaining at its top. But after an hour or two, the bubble would vanish and go back into the spirit of wine.

As for the suspension of the water and mercury, which in Torricelli's experiment was evidently due to the pressure of the air, but in Huygens' experiment occurred without any air pressure, he suggested that it might be due to the stronger pressure exerted

[56] JS III, 111-22.

by some matter more subtle than air and which was able to penetrate glass, water, mercury and other objects which air could not penetrate. Although this solution did not fully satisfy him, two other experiments seemed to support it: namely, the difficulty of separating two polished metal plates in a vacuum, and siphoning in a vacuum. The *Journal des Sçavans* in 1678 spoke of "cette experience fameuse de M. Hugens du vif argent purgé dans le vuide qui demeure suspendu jusqu'à la hauteur de 72 pouces."[57]

For Huygens nature was an entity, a purposive unified system, almost a personality, like its divine maker and originator. Thus in his treatise on light he says that nature in producing so many marvelous effects makes use of an infinite succession of corpuscles of varied magnitude and diverse velocities.[58] Or in the preface to his dissertation on the cause of gravity he speaks of nature's following obscure and intricate paths.[59] He remarked "the artifice of nature" in the construction of the eyes, "organs which nature has destined for the sense of sight." This called for a high degree of geometrical knowledge, more so than in the case of anything else in nature.[60] In New Conjectures concerning the Planetary Worlds he notes "that frugal simplicity nature shows in all her works," that "Nature seems to love variety in all her works," or he admires "the neat and frugal contrivance of nature," and remarks that "Nature might have another great conveniency in her eye."[61]

Yet he wrote in 1679 that it was accepted by almost all philosophers of today that it was only the movement and shape of the corpuscles of which everything was composed "which produce all the admirable effects which we see in nature."[62]

The element of marvelousness which is so constant a factor in magic has not entirely disappeared from the science of Huygens. We have already heard him speak of the many marvelous effects of nature, which he did lest his explanation of light as the subtle

[57] JS VI, 20.

[58] *Opera reliqua*, I (1728), 11.

[59] *Ibid.*, p. 95.

[60] *Oeuvres*, XXII, 634.

[61] English translation, *The Celestial Worlds Discover'd*, London, 1722 pp. 13-14, 22, 44, 75.

[62] *Oeuvres*, XXII, 710. *Ibid.*, XIX, 85, his editors had remarked: "Il y a parfois chez Huygens une légère tendance quelque peu antique, nous semble-t-il, à admettre sans raisons suffisantes la simplicité de la nature."

motion of particles of ether seem absurd or impossible to anyone.[63] Or he remarks the wonderful refraction of Icelandic crystal, which Erasmus Bartholinus (1625–1698), he goes on to say, was the first to describe,[64] and of which his own discussion has been called an "unsurpassed example of the combination of experimental investigation and acute analysis."[65] Or he exclaims at the wonders in generation or at the "wonderful and amazing scheme . . . of the magnificent vastness of the universe . . . And how much must our wonder and admiration be increased when we consider the prodigious distance and multitude of the stars."[66] He still writes that in music "we are compelled to use an occult temperament," and refers to "all the secrets in experimental knowledge."[67] Logarithms were "marvelous numbers" for him.[68] In 1668 he composed an unpublished *De combinationum mirandis.*[69] But he had no interest in numerology or speculative geometry. He wrote to Leibniz in 1691:

> There are certain curved lines which nature often presents to our sight . . . and which I deem worthy of consideration. But to forge new ones, merely as an exercise in geometry, without foreseeing any other utility, seems to me *difficiles agitare nugas,* and I have the same opinion of all problems touching numbers.[70]

Huygens had for some time been planning, and, just before he died, the printing began of a work entitled, *Cosmotheoros* or Conjectures concerning celestial earths and their adornment.[71] It was addressed to his brother, Constantine, but he too had died before the printing was completed in 1698 at the Hague.[72] Whether this com-

[63] *Opera reliqua,* I (1728), 11.

[64] *Ibid., De lumine,* cap, v and p. 40. *Oeuvres,* XXII, 676. E. Bartholinus, *Experimenta chrystallis Islandici disdiaclastici quibus mira et insolita refractio detegitur,* 1669.

[65] Tyler and Bigelow, *A Short History of Science,* 1939, p. 323.

[66] *The Celestial Worlds Discover'd,* London, 1722, pp. 21, 94-95, 151.

[67] *Ibid.,* pp. 91, 41.

[68] *Oeuvres,* XXII, 584.

[69] *Ibid.,* p. 640.

[70] *Oeuvres,* XXII, 769.

[71] *Cosmotheoros sive de terris coelestibus earumque ornatu conjecturae ad Constantinum Hugenium fratrem*; *Oeuvres complètes* (with a French translation on opposite pages) XXI (1944), 680-821, preceded by an Avertissement at pp. 655-75.

[72] *Ibid.,* p. 677 for a facsimile of the title page.

position enhances Huygens' reputation as a man of science for us today may well be questioned, but it was very much to the taste of the time when it appeared and so deserves our consideration. It was published in London the same year in English translation;[73] in 1699, appeared again in Latin and in Dutch translation; in French in 1702, and in German in 1703. Other English editions followed in 1718, 1722, and at Glasgow in 1757; French, in 1718 and 1724 at Amsterdam; German, in 1743, and at Zurich in 1767.[74] Flamsteed recommended it to Dr. Plume, archdeacon of Rochester, and the pleasure which that churchman had in reading it led him to found the Plumian professorship of astronomy and experimental philosophy at the University of Cambridge.

The *Cosmotheoros* is the most human and the least scientific of Huygens' works. "To err is human." Free from mathematical restrictions and guidance, without the mechanic's necessity of making his contrivance work in practice, he substitutes conjectures for experiments, but unfortunately does not supply logic in place of physics. The subjective replaces the objective, but the work is wholly unimaginative. There is a considerable analogy between it and Sir Isaac Newton's *The Chronology of Antient Kingdoms Amended... with three plates of the temple of Solomon.* For one thing, it is frequently religious in tone, with allusion to Divine providence, the Divine architect, contemplation of the works of God, Infinite author of all things, and wise Creator.[75] He should be worshipped, reverenced and adored,

> to the confusion of those who would have the earth and all things formed by the shuffling concourse of atoms, or to be without beginning.[76]

whereas it is

> an absurdity even to think of their being thus happily jumbled together by a chance motion of I don't know what little particles.[77]

[73] *The Celestial Worlds Discover'd or Conjectures concerning the Inhabitants, Plants and Productions of the Worlds in the Planets,* London, Printed for Timothy Childe, 1698. Col 523.13 H98; 160 pp.

[74] *Oeuvres complètes,* XXI, 674-75 for these. I own a copy of the edition of London, Printed for James Knapton, 1722, which differs from that of 1698 in pagination, 162 pp.

[75] *Celestial Worlds Discover'd,* 1722, pp. 11, 73-74, 21, 60, 67, 117, etc.

[76] *Ibid.,* p. 11.

[77] *Ibid.,* p. 21.

Yet he believed that magnetical matter continually passes through the pores of the earth, and that the matter which causes gravity goes through the pores of all bodies at a speed which may seem incredible.[78] Which sounds like Descartes and the corpuscular theory. But the evolution of new forms of animal life found no place in the thought of the author of *Cosmotheoros:*

'tis much more agreeable to the wisdom of God, once for all to create of all sorts of animals, and distribute them all over the earth in such a wonderful and inconceivable way as he has, than to be continually obliged to new productions out of the earth.[79]

In the second place, the line of argument for the other planets being inhabited is woefully weak from the standpoints of both science and logic. Huygens first asserts that it is more than probable that their bodies are solid like that of our earth,[80] whereas actually Saturn has a density only thirteen per cent of the earth's and even less than that of water, while Venus and Jupiter are so enveloped by clouds or vapors that little can be seen of the planet itself. The second step in his argument is not scientific at all but a matter of religion and fitness;

Now, should we allow the planets nothing but vast deserts . . . and deprive them of all those creatures that more plainly bespeak their divine architect, we should sink them below the earth in beauty and dignity, a thing very unreasonable.[81]

The third step is that because there are clouds about Jupiter, there is water there. We can't tell whether Mars and Venus have clouds or not.

But since 'tis certain that the earth and Jupiter have their water and clouds, there is no reason why the other planets should be without them.[82]

And no other reason, it might be added, why they should be with them. Since there is water, there must be vegetation, and further-

[78] *Ibid.*, p. 83; and *De causa gravitatis* in *Opera reliqua*, I (1728), 109.

[79] *Celestial Worlds Discover'd*, p. 31.

[80] *Ibid.*, p. 19.

[81] *Ibid.*, p. 21.

[82] *Ibid.*, p. 27.

more animals to eat it or one another. These must be a great deal like ours, because for Huygens what he is not used to or can't imagine or think of, simply can't exist—even on other planets. Finally, there must be some rational creature like man to enjoy all these things and to adore their and his Creator. He must have sight and the other four senses. And for the sense of hearing there must be sound and for sound, air. Man is not, however, the only rational animal either on earth or on the other planets. Beasts, birds and insects share understanding and reason with him to some extent, and Huygens repudiates Descartes' ranking them with machines and automata. The rational beings on Saturn and Jupiter should study astronomy as well as we, for fear of eclipses gave rise to it here, and should be "of much greater force" there because of the daily eclipses of their moons and frequent solar eclipses. This is another example of Huygens' faulty logic, since no one would fear an eclipse that happened every day. Study of astronomy requires instruments and the art of writing, and so on.

Later Huygens estimates the heat of the sun on the planet Mercury as nine times that on earth.

> And yet there is no doubt but that the animals there are made of such a temper as to be but moderately warm, and the plants such as to be able to endure the heat.[83]

Possibly advancing years and reiterated ill-health had something to do with such feeble ratiocination on Huygens' part, but he had already alluded to the possibility of inhabitants of Saturn in 1659[84] and had again suggested that the planets were inhabited in *Pensées meslées* of 1686, which remained in manuscript.[85]

Huygens correctly states that there are neither rivers, clouds, air nor water on the moon, and so probably no plants or animals. Nor can he agree with Kepler that its craters are human artefacts.[86] These statements, which occur only in the second book of *Cosmotheoros*, might seem to belong in its first book where the question

[83] *Ibid.*, p. 106.

[84] *Oeuvres*, XV, 300, 343; XXII, 510, 512.

[85] *Ibid.*, pp. 733-34.

[86] *Oeuvres complètes*, XXI (1944), 793; *Celestial Worlds* (1722), pp. 130-32.

whether the planets have inhabitants is discussed, and Huygens' recent editors think that his first plan was to begin with conjectures on the moon.[87] It is true that for him the moon is no longer one of seven planets but a satellite, like those of Saturn and Jupiter. Moreover, he would have thrown cold water at the start on his argument for rational dwellers on the planets, if he had admitted to begin with that there were none on the moon. His editors think that he was the first astronomer to deny the moon an atmosphere, whereas Kepler, Maestlin, Galileo, Longomontanus, Giordano Bruno and others had put air or a denser ether about it.[88] Huygens' denial may have been due to the fact that, when on July 2, 1666, he with five other members of the Académie des Sciences gathered at Colbert's house to observe the eclipse of the moon, they observed that "La lune paroit très ronde, egalement noire, sans apparence d'atmosphere."[89] Hooke had said in his *Micrographia* that, if there were animals on the moon, telescopes could be made strong enough to see them, but Auzout in a letter to the *Journal des Sçavans* denied that this was possible.[90]

Huygens' sallies into the history of civilization are no happier or more consistent than his arguments to prove the planets inhabited. In one place he ascribes advance in civilization to war:

> And if men were to lead their whole lives in an undisturbed continual peace, in no fear of poverty, no danger of war . . . they would live little better than brutes.[91]

But in another passage he attributes it to men being born naked and needing to clothe themselves:

> And 'tis this necessity that has been the greatest, if not the only occasion of all the trade and commerce, of all the mechanical inventions and discoveries that we are masters of.[92]

In yet a third passage he questions whether the invention of gun-

[87] *Oeuvres*, XXI, 659: "L'Appendice V qui suit fait voir que Huygens avait d'abord l'intention de faire en premier lieu des conjectures sur la lune."

[88] *Idem*, citing Kircher, *Iter exstaticum*, 1660, p. 65.

[89] *Oeuvres*, XXII, 217.

[90] JS I, 221-25.

[91] *Celestial Worlds Discover'd*, pp. 40-41.

[92] *Ibid.*, p. 75.

powder has done more harm or good and concludes, "I think we had been better without the discovery."[93] A conclusion which *per se* is greatly to his credit but which is not entirely in keeping with his previous justification of war.

The briefer second book, devoted chiefly to such matters as the length of days and years on each of the other planets and how the solar system looks as viewed from each of them, is much more interesting and scientific; and less banal, illogical and faulty than the first book. It also contains two clever contrivances to show how much light Jupiter receives from the sun and to measure roughly the distance of the dog-star, which are good illustrations of Huygens' mechanical ingenuity.[94]

[93] *Ibid.*, pp. 95-96.

[94] *Ibid.*, pp. 119, 153-54.

CHAPTER XXIII

PHYSICS AND ASTRONOMY AFTER DESCARTES

Introduction—Two French Jesuits: Natalis and Gautruche—Two professors at Bologna and Turin: Mazzotta and Torrino—Cartesianism at Leyden and Utrecht: Heereboord, Raei, Clauberg, Caerman, Reyher—Tübingen and Wittenberg: Geilfusius and Scharff—Back to the Netherlands: Voet, Schoock, Voss—A Cartesian and other Italians: Cornelio, Guarini, Bonfioli, Cavina, Honoratus Fabri—A New Physical Hypothesis by Leibniz—Van Sichen's course at Louvain—Natural philosophy at Paris: Rohault, Gadroys, Du Hamel—The Free Philosophy of Cardoso—Schweling and Senguerd—Lana Terzi and Hartsoeker.

Nothing seems more opposed to philosophical pursuits than reducing them to a system

—Huygens

In the second half of the seventeenth century, teaching of natural philosophy less often took the form of a virtual commentary upon Aristotle, following the topical plan of his *Physics, De coelo et mundo,* and other works in that field. Greater stress was now laid in the curriculum upon physics than logic and metaphysics, and the subject might be approached from new angles suggested by Descartes, mathematical method, and chemical and physical experimentation. Set lecture courses decline somewhat in favor of dissertations, disputations and exercises. There are possibly more works of general natural philosophy addressed to a general reading public rather than to an academic audience. Conservative and reactionary treatments still exist. But recent scientific discoveries and new ideas are increasingly recognized, and a Franciscan will contend that there is nothing unfavorable to them in the philosophy of Scotus. In this chapter, then, we combine the study and teaching of natural philosophy in the universities with some consideration of general works in physics and astronomy not intended especially for academic halls. We first note two works by French Jesuits, then two by Italian professors, one a Benedictine, then several from the Netherlands, Germany and so on, roughly in chronological order.

Etienne Natalis (1581–1659), a French Jesuit from near Tulle, taught philosophy for eight years and theology for fifteen and was the head of several colleges. We shall here consider two works published by him in 1648.[1] In the preface to his Physics Old and New[2] he announced his intention of adhering to the Aristotelian principles: matter, form and privation, and to use Aristotle's definitions. At the same time he made a good deal of the conception of corpuscles and of pores, and of that of material spirits. Every body in the world had its accidents, whether they were modes of the body or corpuscles contained in pores or in bodies. Such corpuscles were called spirits when so minute that taken *per se* they escaped all sense, but they were accidents insofar as they were in the body and could leave it without its corrupting.[3] The simple material spirits had been created by God in varied proportion of rare and dense. Mixed spirits were made by the action of natural agents and divided into celestial, solar and natural. Which seems to be ringing changes on Descartes' three elements. Fire contained seven particles of rare for one of dense; earth, one particle of rare for every seven of dense; air, six rare particles for every two dense; and water, two rare for six dense. Sulphur was from the union of solar spirit with air; mercury, from its union with water; and salt, from its union with earth.[4]

The influence, motion and actions of one body on another were nothing but the transmission of these spirits from one body to another. That they were not material qualities nor accidents educed from the potentiality of matter seemed evident from the influence of the planets and stars upon the earth, as the production of lead in such a spot by the influence of Saturn. The spirits contained in the *humidum radicale* divided into material and formal parts, one elementary, the other celestial. The celestial spirit with the solar purged the elemental by sublimation, that is, by vaporization.[5]

1 Other works of physics by him are listed by Alegambe.

2 *Physica vetus ac nova,* Paris, 1648, in-8, 265 pp. BM 534.c.38 (1.).

3 *Ibid.*, p. 80.

4 *Ibid.*, pp. 42, 32, 51.

5 *Ibid.*, pp. 45, 52.

The other treatise, denying that a vacuum had been produced in recent experiments, had previously appeared in a faulty French edition while Natalis was very sick and had not yet seen the experiments in question, and so now was thoroughly revised in the Latin edition.[6] He contends that there are pores in bodies and that ether entered the tube through them as the mercury was falling in the Torricellian experiment. And that fire is separated from air by the compressing of air is proved by the speed of projectiles which are moved against the air by fiery spirits.

Pierre Gautruche, a Jesuit of Orléans, published an *Institutio* of all philosophy and mathematics for the use of studious youth in 1656 in four volumes on universal physics, particular physics, mathematics and metaphysics.[7] We shall be concerned here only with the first two of these volumes. They still consider such topics as first matter, substantial form, quantity, the infinite, place and vacuum. The sphere of activity of natural agents and their action, direct, reflex and refracted, is twofold: by local motion only, and by virtue or active quality diffused from them through the medium, as heat is diffused by fire, and light from the sun. It is more intense near the agent. It may be uniform, difform and uniformly difform—a distinction coming from fourteenth century scholasticism. Under this category much is commonly supposed to be accomplished by sympathy and antipathy, also by some transmission of spirits, or by occult and specific qualities. Magnetic virtue easily takes first place among specific and occult qualities, and is more fully dealt with in the volume on mathematics. Action at a distance is possible by such diffusion of virtue, occult quality, and sympathy or antipathy, but belief in the sympathetic powder is stigmatized as stupid or superstitious, while the bleeding of the corpse in the

[6] *Plenum experimentis novis confirmatum*, Paris, 1648, in-8, 138 pp. BM 534.c.38 (2.). Some works by others for and against a vacuum were: Pascal: *Nouvelles experiences touchant la vuide*, 1647; Guiffart, P., *Discours du vuide sur les experiences de M. Paschal et le traité de M. Pierius*, Rouen, 1647; Paolo Casati, *Vacuum proscriptum*, Genoa, 1649; Zucchi, *Exclusio vacui contra nova experimenta*, 1649; Ant. Deusing, *Disquisitio de vacuo*, 1661; Guericke, *Experimenta nova Magdeburgica de vacuo spatio*, 1672; Vallerius, *De vacuo*, 1678.

[7] Petrus Galtruchius, S. J., *Philosophiae ac mathematicae totius institutio ad usum studiosae iuventutis*, Caen, 1656, 4 vols. of from 300 to 400 pages each. BM 526.a.28-31.

presence of the murderer is ascribed to God. But the simple reason why the water in a kettle on the fire is hotter at the top is that it ascends as it is heated.[8] A vacuum is supernaturally possible but cannot be achieved by man or machines. Aristotle's explanation of the motion of projectiles is rejected, and it is held that it comes from some kind of quality impressed on the projectile by the hurler.[9]

Despite sunspots and comets, the heavenly bodies are still regarded as incorruptible. Comets are formed from the matter of the heavens, either condensed or by the union of several smaller bodies, but by the agency of angels and beyond the ordinary course of nature to serve as announcers of future calamities. Since the heavens occupy the place of supreme dignity and perfection among all natural bodies, they should have the greatest and most universal virtue of action, to which other inferior bodies are subordinated. They may be a particular as well as a universal cause, as in the case of those imperfect animals which are generated from putrid matter, but cannot be the particular cause of such perfect animals as a dog, horse or lion, nor act directly upon human free will. But they act by occult quality as well as by light and heat. Critical days, however, may not be referred to occult action of the moon, since they occur on any day of the moon. Gautruche also rejects astrological houses and most astrology. Physiognomy, chiromancy and talismans are forbidden by papal decree and state legislation and are contrary to philosophy. But then he makes an exception for purely natural—and not astrological—physiognomy and chiromancy.[10]

The principles of the chemists are set forth; potable gold is said to have various uses in medicine; and the transmutation of metals is declared possible but difficult, very rare and undesirable, and the decretal of John XXII is cited against it.[11] The circulation of the blood is accepted,[12] but the distinction between vegetative, sensitive and rational soul is retained.[13] Nothing seems to be said concerning dreams one way or another.

Benedetto Mazzotta was a Benedictine who taught theology at

[8] *Ibid.*, I, 127-50.
[9] *Ibid.*, I, 292-93, 314-16.
[10] *Ibid.*, II, 19-22, 28-38.
[11] *Ibid.*, II, 114-18, 150-55.
[12] *Ibid.*, II, 206.
[13] *Ibid.*, II, 313.

Bologna. His *Triple Philosophy, Natural, Astrological and Mineral,* published at Bologna in 1653, was a conservative and even backward book. He defended fire as an element against Chassin, Vallesius, Aresius, Tasso in a work written in Italian, Telesio, Arriaga and other recent writers.[14] And although in this connection he held that comets could well pass through the sphere of fire and be elevated above the moon,[15] he evidently still regarded them as terrestrial exhalations, and held elsewhere that many of them remained below the moon.[16] He still believed that the earth remained immobile at the center of the universe "against what Copernicus said and the church condemned."[17] He also opposed those who contended that heat and cold were not distinct qualities.[18] He admitted that earth and water formed a single globe,[19] but affirmed that "no fixed truth" had yet been reached with regard to the tides, although he inclined to ascribe them to the moon with the concurrence of the sun and other stars.[20] Earlier, however, he had said that water had its movement of flux and reflux from the sun, moon and other stars, also caused by angelic movers of the waters.[21] He still refused to account for the origin of rivers and fountains by precipitation alone, and, to illustrate the contention that air might be changed to water in the caverns of mountains, told a story of a man who did not weigh over one hundred and fifty pounds or take more than seven pounds of food and drink per day, yet passed thirty-six pounds of urine daily for two months, the air in his arteries turning to water.[22] Mazzotta was, indeed, rather given to the marvelous. He told of an inextinguishable lamp with a wick of asbestos linen and an oil made by repeated distillation of human excrement. This oil possessed the further property that a hook or net which had been smeared with it would catch many fish.[23] He was also prone to attribute phenomena to supernatural causes. Armies seen in mid-air some ascribed to cloud formations and

[14] Mazzotta. *De triplici philosophia, naturali, astrologica et minerali,* Bononiae, I (1653), 5.

[15] *Ibid.,* 12.

[16] *Ibid.,* II, 13.

[17] *Ibid.,* I, 17, "contra id quod ecclesia damnante dixit Copernicus."

[18] *Ibid.,* I, 18-23.

[19] *Ibid.,* II, 134.

[20] *Ibid.,* II, 137.

[21] *Ibid.,* I, 17.

[22] *Ibid.,* II, 140-41.

[23] *Ibid.,* II, 47.

thunder, some to the influence of the stars, and some to God or angels or demons.[24] When five or six suns were seen, it was a sign of some great future event, since it was hard to find a natural cause for such a phenomenon.[25]

Although a theologian, Mazzotta continued the favorable attitude towards astrology which had so long prevailed at Bologna. He urged his readers to consider carefully the annual predictions which had been put forth there by Placido Titi, Antonius Carnevale, Roffeni, Ovidius Montalbanus (whose approval of this work of Mazzotta for the Inquisitor of Bologna appears on its second title page), Artensius, Thebanus, Gessius, Polentanus, Grimaldi and others. They would find that they had come true, and so could arrive at a favorable judgment as to the truth "of this renowned science."[26] Mazzotta believed that through centuries past the stars had produced memorable changes in the air, in kingdoms and cities, and in the whole world. He advised the reader to study history and he would find what eclipses and great conjunctions had occasioned in times past.[27] The stars act mediately upon the will of man, as Aquinas and all the holy fathers and the philosophers agree.[28] Goclenius is quoted to the effect that God impressed signs in the sky of great public events and calamities. What else are eclipses, conjunctions and comets than divine oracles?[29] Mazzotta grants that there are popular impostors in astrology,[30] but he goes on to treat at length of astrological philosophy, with chapters on points to be considered before making predictions; the nature, properties and effects, good and evil, of the planets; great conjunctions; eclipses; the entry of the sun into Aries; significations of the planets according to their various configurations and aspects to one another, conjunction with fixed stars, situation in the celestial houses; and concerning signs pertaining to the weather, medicine, navigation, and so forth.[31] Then, after dealing with risings and settings of the fixed stars with reference to the sun and other

[24] *Ibid.*, II, 55-56.
[25] *Ibid.*, II, 65.
[26] *Ibid.*, II, 211.
[27] *Ibid.*, II, 209.
[28] *Ibid.*, II, 210.
[29] *Ibid.*, II, 211.
[30] *Ibid.*, II, 212.
[31] *Ibid.*, II, 212, 224, 236, 245, 248 and 253.

planets,[32] he gives forty general astrological aphorisms.[33] A brief specimen of his astrology is that planets while stationary increase their effects and significations. For though they may be weaker in their stations, yet, since they operate while fixed and stable in one place, they exert more influence than while wandering about.[34]

As for comets, their elucidation requires more study than that of other fiery phenomena, "since we see by daily experience that they are signifiers of great events." One had recently appeared from December 17, 1652 to January 12, 1653, when it could scarcely be longer discerned. Mazzotta holds that comets do not always portend evil. He names eight of the familiar nine varieties: Veru, Ascone, Aurora, Miles, Rosa, Niger, Pertica and Tenaculum. After telling what they signify in each sign of the zodiac and with each planet, he makes a prediction from that of 1652–53.[35]

More venturesome than the book of his fellow Italian, Mazzotta, was the *Parnassus triceps* of Bartolomeo Torrino, professor at Turin.[36] This work is quaintly divided into three parts or Vertexes, respectively devoted to physics, medicine and mathematics. Then each Vertex in turn is divided between three muses. Clio has natural body; Polyhymnia, inanimate mixed bodies; Thalia, animate body. Melpomene, Terpsichore and Calliope share medicine among them. Euterpe has mathematics; Erato, Perspective and music; Urania, astronomy and astrology. Torrino believes that a vacuum is not only possible but necessary.[37] His three principles are light or spirit, earth or body, heavens or bond, reminding one somewhat of the light, matter and world soul of Comenius. Sulphur contains most of the first, less of the heavens, and least of earth. Salt has most of the first, less of earth, and least of the celestial. Mercury has most celestial, less light, and least earth. Phlegm has most celestial, less earth, and least light. Tartar has most

[32] *Ibid.*, II, 263-65.

[33] *Ibid.*, II, 267.

[34] *Ibid.*, II, 256.

[35] Mazzotta, II (1653), 11-21, for the contents of this paragraph.

[36] *Parnassus triceps seu musarum afflatus Physiatro-Mathematici . . . , Opusculum in quo dum summa naturae et artis mysteria recluduntur et causae deliberantur secretiorum, congessit author physiologiae medicinae et mathematicae Enchiridion*, Turin, 1657, in-fol. BM 545.h.16.

[37] *Ibid.*, p. 5; at p. 63 he gives the experiment with mercury, not mentioning Torricelli.

earth, less light, and least of the heavens. Glass has most earth, less heavens, and least light. Fire is light alone condensed with a little heavens. Water is much heavens condensed with a little earth. Air is celestial with terraqueous vapors.[38]

There is no such property as lightness but only more or less heavy. Heavier bodies not only do not move faster than less heavy ones; they often descend less quickly because they meet with more resistance.[39] Comets are generated all the way from the region of air to the firmament. Sometimes their parallax is four times the lunar, often less than the solar, still oftener non-existent. Tides are explained by the moon's drawing the sea towards itself, not by force of its light or heat, or of salnitrous spirit or other occult quality, but from mere lack of humor absorbed by the rays of the sun. Galileo less plausibly ascribed it to the movement of the earth. The tides rise higher than the mountain tops and supply perennial fountains from the sea, although some springs may come from rain, snow, or from vapors sublimated within the viscera of the earth. Of stones Torrino treats only the magnet and Bologna stone, omitting gems.[40]

Fascination is accepted to a limited extent as effected by contagious breath or fright, and as menstruating women affect mirrors, pearls and tender shoots. But Torrino does not believe in double pupils or images of wild beasts in the eye of the fascinator, unless it be that witches are so marked by the devil.[41] Human imagination has such power that it may alter the body and personality, producing ecstasy, disease, bloody sweat, growth of horns, and bestiality, as no one will deny who has once met with a lycanthrope. The cause is atrabile which affects the spirits and through them produces *idées fixes* in the mind. Pregnant women easily impress their fancies on the foetus.[42] Physiognomy is a real science provided it does not try to predict free actions and fortuitous events. Varieties of it are ophthalmoscopia, metoposcopia, chiromancy, podomancy etc.[43] Critical days and all humor follow the circuits of the moon.[44]

[38] *Ibid.*, pp. 15, 18.
[39] *Ibid.*, pp. 22, 24.
[40] *Ibid.*, pp. 34, 52-53, 58.
[41] *Ibid.*, pp. 78-79. But why is one frightened?
[42] *Ibid.*, p. 82.
[43] *Ibid.*, p. 99.
[44] *Ibid.*, p. 155.

Torrino reduces the number of heavens to three: empyrean, sidereal and aerial. With the empyrean heaven the mathematician has no concern, the physicist little, the theologian much. It is a habitation befitting the glorious bodies of the blest. Its name indicates that its light is fiery, and it is probably nourished by the waters above the firmament. It seems more appropriate for it to be at rest than in motion. Copernicus did not allow the particular properties of terrestrial regions to depend on its influence. But if ecclesiastical authority did not decree the opposite, Torrino would openly pronounce that no other physical or mathematical deduction is as valid as that of Copernicus. Of Kepler's work on the motion of Mars he says nothing.[45]

Torrino believed that the planets affected things on earth by their properties and even admitted that they and also the fixed stars exerted specific influences. A particular planet might be lord and have dignity in any part of the zodiac.[46] He went into astrological technique at length, describing directions, *significatores, promissores* and *pars fortunae*. Here he mentioned Kepler, prefering his method of measuring directions.[47] But wheras Kepler had increased the number of aspects, Torrino would reduce them to three—conjunction, opposition and quadrate, omitting trine and sextile. He suggested that the efficacy of aspects more probably depended on junction with some fixed star, which intensified or weakened the force of the planet. But such aspects as translation, restitution, prohibition, obsession and frustration he rejected as reveries of the astrologers. *Facies* and decans were inane, but triplicities and *termini* could receive some support from the firmament. Retrograde and slow planets were weaker. All were stronger in their perigee than in apogee. Combustion by the sun weakened their force. The division of the zodiac into twelve astrological houses rather than ten had no natural basis. In such matters as religion, hatred and friendship, strife, treachery, imprisonment and honors, the planets could indicate only propensity and inclination. *Pars fortunae* merely indicated that the moon was in such and such a phase of the sun. The lunar nodes were completely inefficacious, as were *chronocra-*

[45] *Ibid.*, pp. 315, 341.
[46] *Ibid.*, p. 368.
[47] *Ibid.*, p. 326.

tores, fridariae and other such dreams. *Animodar* and *Trutina Hermetis* were puerile, as was judging from the time of conception. But Torrino did not reject climacteric years, and both reason and frequent experience showed that revolutions were efficacious to the extent of exciting latent and dormant forces of the stars. A man of note every year for forty years had an intense attack of ephemeral fever upon July 22nd, which was his natal day, and only on that day. The transits of all the planets over the places of others or the ascendent or the zenith were also efficacious. Torrino terminated his discussion with some "astrogeological problems," such as why New England, although in the same latitude as Rome (*sic*), was cold in June, and the horoscope of Augustus Caesar.[48]

Evidently Torrino was more critical of astrology than Mazzotta had been. And, although he still accepted the influence of the stars, he disagreed with Mazzotta on particular points. For instance, whereas Mazzotta held that a planet exerted more influence while stationary, Torrino argued that a planet "delays longer above earth", when direct and swift.[49] Such disagreement shows the weakness of the position of astrology.

At Leyden Cartesianism began to affect the teaching of Aristotle. When Adrian Heereboord came to Leyden—his inaugural oration is dated February 9, 1641, the teaching of philosophy was in a sad state there. A conflict developed whether he should follow the texts of Aristotle in teaching logic, but finally he was allowed to have his way, and to Aristotle as ancient philosophy added various humanists and modern philosophers: Valla, Agricola, Vives, Ramus, Pico, Telesio, Patrizi, Campanella, Francis Bacon and Descartes. But a younger colleague who wished to stick by Aristotle made more trouble for him, and in 1645, when one of his students defended the thesis, "Doubt is the beginning of undoubted philosophy," it was objected to as leading to scepticism. But later he showed that Aristotle himself had maintained this thesis. The recital of these facts in a twenty-page letter to the Curators of the University, prefixed to his *Meletemata philosophica* of 1654,[50]

[48] *Ibid.*, pp. 367-74.

[49] *Ibid.*, p. 369, "Rectus enim et velox diutius supra terram moratur."

[50] *Meletemata philosophica. Maximam partem Metaphysice.* Lugd. Batav., Ex officina Francisci Morgardi. 1654. BN R.2398-2399.

is followed by praise of Francis Bacon and Descartes and disparagement of a decree of May 20 last which forbade professors to make the least mention of Descartes or his opinions.

The *Meletemata* consist chiefly of disputations, one of 1643, a *Repetitio* of 1647, and others which are not dated. These fill two volumes of 362 pages. Then, with a new title page and pagination, follows a *Philosophia naturalis moralis rationalis,* with the same date, place and printer. The first sixty pages are devoted to a Physical College of Physical Theses according to the Peripatetics, but some of them seem quite radical. Thus, in a disputation of 1644–1645, the theses maintained are that all material forms are merely modes of matter, that the heavens are corruptible and of aerial nature, that the stars move by their own motion through a liquid heaven like birds in the air, and that the elements are not transmuted into one another. These are followed by paradoxes. The moon seems to be least illuminated when it is most illuminated, and is never less illuminated than when it is full. Twins who were born at the same time and died at the same can have lived an unequal number of days. There is no theoretical philosophy that is not also practical. Privation is not a physical principle any more than union, but merely a condition *sine qua non.* Bodies when rarefied do not have more quantity than when condensed. Fire is humid. Water is most subtle earth; air, subtle water; fire, subtle air; earth, thick fire. The moon is illuminated by the earth and sun. In man there is only one sense, not five. Logic which fails to contribute to the invention of sciences is useless. All syllogisms are useless to attain truth except induction. Final cause is neither cause nor matter.[51]

Heereboord was cited by Schwimmer in 1672 on the subject of sympathy. His *Philosophia naturalis* was said to have been reprinted at Leyden in 1663 and Oxford in 1668.

Johannes de Raei received the doctorate at Utrecht in 1641 with *Theses Cartesianae.* He lectured on Aristotle at Leyden in 1651, became professor of philosophy there in the following year, and in 1654 published there a Key to Natural Philosophy, or Aristotelian-

[51] *Ibid., Collegium physicum,* pp. 10-11.

Cartesian introduction to the contemplation of nature.[52] It comprised six dissertations on vulgar and philosophical knowledge, general philosophical principles, the nature of body, the origin of motion, the communication of motion and the action of bodies on one another, and the subtle ethereal matter of Descartes. In the second edition of 1677[53] the book was, in the words of its reviewer in *Philosophical Transactions,*[54] "enriched with seventeen discourses," of which the first dealt with the genuine doctrine of Aristotle and the great difference between it and the pretended Aristotelianism of the schools. The next two had to do with man and mind; the fourth was on the origin of error; the next two were concerned with knowledge; and the seventh, with the idea of God. After setting forth the substantial form and soul of man out of Aristotle against the Aristotelians, Raei turned to the system of the world and distinguished three elements, of which the first emitted light and constituted the lucid stars, the second transmitted light as the heavens or ether do, while the third is opaque and reflects light, corresponding to comets, planets and the earth. These remind us of the three principles of Torrino and Comenius. The vital spirits in man and other animals are not only oleaginous but also sharp and aqueous. Heat is identified with motion and cold with the lack thereof. The remaining six disquisitions were on hardness and fluidity, humidity and dryness, place, four rules of logic, physiology or the explanation of phenomena by intelligible causes, and the wisdom of the ancients.

In 1664 Johann Clauberg dedicated his *Opera Physica*[55] to Raei, ten years after the first edition of Raei's Key to Natural Philosophy.

[52] *Clavis philosophiae naturalis seu introductio ad naturae contemplationem Aristotelico-Cartesiana,* Lugduni Batavorum, 1654, in-4, 219 pp. BN Rés. R.1015, where it is bound with Descartes, *Musicae compendium,* Utrecht, 1650, and Huygens, *De circuli magnitudine,* Leyden, 1654.

[53] Amsterdam, 1677, in-4.

[54] PT XI, 790-92.

[55] *Johannis Claubergii Opera Physica, id est, Physica Contracta, Disputationes Physicae, Theoria Viventium, et conjunctionis animae cum corpore Descriptio. Accedunt eiusdem Metaphysica de Ente. Johannis Claubergii Physica quibus rerum corporearum vis et natura, mentis ad corpus relatae proprietates, denique corporis ac mentis arcta et admirabilis in homine conjunctio explicantur.* Amstelodami, apud Danielem Elzevirium, 1664. Col 500 C57.

He called him after Descartes the most celebrated of Batavian philosophers and spoke of spreading to Upper Germany the new philosophy into which Raei had initiated him. Fifteen years ago Clauberg returned from England to teach philosophy at Herborn in Nassau. Three years before that his treatise on the Ens[56] appeared at Groningen while he was travelling in France. Now for the past twelve years he has been teaching at Duisburg in the duchy of Cleves.

Like other writers of the time, Clauberg was fond of giving his works a pseudo-mathematical appearance by chopping them up into numerous brief Propositions or sentences. The *Physica contracta* of 88 pages and 30 chapters has 1210 propositions. The Theory of Living Bodies of 78 pages and 44 chapters runs to 1076 propositions. Clauberg adopts the three elements and vortices of Descartes, and speaks of comets as passing from one vortex to another.[57] Like Raei, he identifies heat with motion and cold with rest—or at least slower movement. He also denies cognition and appetite to plants and inanimate bodies.[58] God is the primary cause of motion, and the same amount of motion is conserved in the universe. Mind and body have no natural relationship but are connected in man solely by the will of God. As the macrocosm moves at His nod, so the microcosm obeys the human will, but there is no immediate connection between the human mind and other bodies.[59] Human life is shorter since the flood, because the earth is colder and drier, and there is less nutriment for animals and plants, but human genius is sharper because the air is purer and more subtle.[60]

Clauberg retains the belief in occult qualities in the case of certain human beings who have a strange aversion to some particular food or drink, or who cannot endure the presence of a cat or the odor of roses. These occult qualities may be inborn or acquired

[56] In the *Opera physica* of 1664 it has a separate pagination of 111 pp. and is described as *Editio tertia*. It is there stated, however, that twenty-two years have elapsed since he wrote the work. The first edition seems to have been: *Elementa philosophiae sive Ontosophia*, Groningen, 1647, BN R. 11212. It is all on method.

[57] *Opera physica*, 1664, pp. 33-38, 49.

[58] *Ibid.*, pp. 246, 252, 274.

[59] *Ibid.*, pp. 167, 374, 377, 393.

[60] *Ibid.*, p. 462.

and may alter with time. They originate from the connection between soul and body, but when once we have connected a certain bodily act with a certain thought, neither can recur without the other. Such aversions may often be traced back to early incidents which have been since forgotten. An infantile headache from the odor of roses or fright from a cat may not be remembered, but one retains the aversion for both. Such sympathies and antipathies may have even been acquired in the mother's womb. Sometimes when a pregnant woman breaks her arm, the child's arm breaks too. A three year old girl, who had never seen Clauberg before, preferred him to the rest of the company then present, because he had often visited and been kind to her widowed mother while she was pregnant with this child.[61]

The first part of Clauberg's book, or *Physica contracta,* was reprinted in 1686,[62] but did not include these passages suggestive on the one hand of modern psychiatry and on the other of olden magic.

What a young person who was devoted to recent scientific development might support as new or progressive ideas in mid-seventeenth century is seen in the corollaries which Antonius Caerman added to the disputation concerning air, over which Jan de Bruyn (1610–1675) presided at the University of Utrecht in 1654.[63] He affirmed that all things are composed of atoms and that "we do not recognize substantial forms in things which are endowed with no cognition." Local motion is the only kind of motion, and there is the same quantity of motion now as at creation. A vacuum is accepted, and the formal *ratio* of a body is said to consist in impenetrability. The moon has no light of its own; black and cold are nothing positive but mere privations; the fixed stars are of the same magnitude as the sun; and no external sense is recognized except that of touch. Among the works ascribed to de Bruyn is a Defense of the Cartesian Doctrine of Doubt, and we shall presently find him defending the Cartesian theory of light.

[61] *Ibid.*, pp. 356-58.

[62] *Dictata physica privata, id est, Physica contracta seu theses physicae commentario perpetuo explicatae,* Francof. ad Moen., 1686, in-4.

[63] *Disputatio physica de aere,* Utrecht, 1654 (the date, 1554, has now been corrected in the catalogue): BM 536.f.17.(1.).

In the same collection of dissertations on the air, winds and other meteorological phenomena mentioned in the previous paragraph is one of 1657 by Samuel Reyher and Johann Laudenbach, published at Leipzig, at whose close are twenty-five very miscellaneous assertions, some of which are political. Of those concerned with nature the third affirms that experience testifies that there is a subtle matter which can penetrate gold, glass and other solid bodies, and therefore it is readily apparent that the vacuum of Valerianus Magnus is really a body not merely mathematical but physical. The sixth asserts that it is not an article of faith that the earth is at rest at the center of the universe. The seventh maintains that the telescope shows that Mercury is much smaller than Tycho held. The eighth holds that *calendariographi* or writers of annual predictions make many vain and impious forecasts, but that nonetheless no one, unless totally unskilled, will deny that this discipline of astrology, if properly treated, is most pleasing and most useful. The tenth opines that it is not likely that the Caspian Sea was once a gulf of the ocean.[64]

Two other dissertations on air in the same collection, although slightly later in date, display no fondness for revolutionary notions and both still adhere to the old conception of three regions of the air.[65] But yet another by Reyher in 1670 includes discussion of the Magdeburg experiment and those of Boyle.[66]

Johann Geilfusius, in Emended Physics published at Tübingen in 1653,[67] and reprinted ten years later,[68] covered all nature briefly but claimed to have often followed recent instead of ancient authorities. But he still adhered to the Aristotelian topics and

[64] *De ventis,* Lipsiae, 1657: BM 536.f.17.(3.).

[65] *De aere G. C. Kirchmaier praeses, respondens Adam Ellinger,* 1659, in electorali ad Albim academia: BM 536.f.17.(4.). *De aere Elias Conradus praeses, respondens Joh. Chris. Laurentius,* Wittenberg, 1662: BM 536.f. 17.(6.).

[66] *Diss. de aere,* Kiliae, Imprimebat Joach. Reuman, Acad. Typog., 1670: BM 536.f.17.(7.).

[67] *Physica emendata in qua universa naturae scientia breviter per theoremata et subjectos commentarios traditur, quaestiones variae resolvuntur et multae difficultates tolluntur,* Tubingae apud Greg. Kernerum, 1653, 151 pp. and index. I have used this edition: BM 536.d.5.(1.).

[68] Only the set-up of the title page is different. Editio secunda, Tubingae, Impensis Johannis Georgi Cottae, 1663. BM 536.d.5.(2.).

order rather closely. In successive "Contemplations" are considered matter and form; causes, chance, fortune and monsters; quantity and quality, including occult; motion and quiet, place and vacuum, time. When we come to the ether and its parts, the matter of the heavens is regarded as different from other bodies and closer akin to form. The heavens are marked by vast quantity, extreme subtlety, perspicuity and incorruptibility. But they are regarded as a continuous body without any real division into distinct orbs. Comets are extraordinary stars situated in the ether, and not the terrestrial exhalations of Aristotle. But the four elements and four qualities are retained. Salt, sulphur and mercury, however, are recognized as the principles of mixed bodies. Salt gives consistency; sulphur, inflammability; mercury, inconstancy. Taste comes from salt; odor and color, from sulphur. Contemplation IX is of action, passion, mixture and mutations, including generation and corruption. The tenth is on vapor, smoke and meteors; the next, on perfect mixed bodies. Minerals, plants and animals follow, and finally the soul, with the three Aristotelian divisions thereof.

It is marvelous that there are plants for different parts of the human body and for varied ailments of those members. Plants retain their virtues when roasted, macerated, pulverized, distilled in oils and waters. In animals spirit is the most subtle corporeal substance, hot and mobile, instrument of action.[69] The earth is immobile at the center of the universe. Springs, rivers, lakes and swamps are from the sea but are augmented by rain and snow. Tides are ascribed to spirits which expand and subside. In finer type it is added that external causes such as the sun and other stars, especially the moon, do not suffice to explain the tides. "For natural motion requires an internal principle."[70]

Geilfusius was also the author of a volume on Pneumatics or the science of spirits, God, angels and the separate soul of man.[71]

Johann Scharff (1595–1660), professor at Wittenberg, in the enlarged edition of his Physical Manual, printed at Leipzig in 1657,

[69] *Ibid.*, pp. 106-7, 111.

[70] *Ibid.*, pp. 48-49.

[71] *Pneumatica seu scientia de spiritibus in qua de Deo angelis et anima hominis separata*, Tubingae, 1652, in-8; ed. secunda, Tubingae, 1662, in-8, 153 pp. BN R.10536.

still adhered to the accustomed Aristotelian order and described his textbook as drawn from the most ancient writers.[72] Beginning with matter, form, nature, and natural causes, he comes to fortune, chance and monsters, and then to the general conditions of bodies. He accepts occult qualities and sympathy and antipathy, and relates natural magic to them. The heavens are for him still incorruptible, the stars are weather signs, but he leaves comets to astronomers and declares the star of the Magi to have been a supernatural miracle. He still holds that animals are generated, live and are conserved in fire, citing Aristotle and Pliny. For him there are still three regions of air, and the tides are a preternatural movement of the sea.[73]

The effluvium of a physical body is either corporeal or intentional. The former appears in smoke, vapors and material transpirations, sweat and the like. But intentional actions are made without bodily contact by spiritual rays and species, and also by occult influences and qualities or hidden virtues. Many experiments force us to admit this, and magnetic action is an instance of it.[74] Dogs can detect footprints; a calf is perturbed when led to the slaughter house; the shade of the yew tree, as Plutarch and Pliny tell us, is very injurious to man. The inevitable torpedo is also adduced, and the popular notion that mistreatment of the afterbirth is very harmful to the mother. Casting excrements into the fire or sprinkling them with scorched wine and pepper pains the buttocks of their depositor, as if these had been touched by hot coals.[75] Like attracts like, as the laying on of the scorpion cures its own bite. "This is true magic. Therefore works of magic are works of nature," says Ficino.[76]

The end of meteors is fourfold. First, the perfection of the whole universe and mutual equalization of the elements. Second, purifying the air. Third, foreknowledge and presage. Fourth, recognition of God, who marvelously orders all these.[77] Scharff then

[72] Joh. Scharffius, *Manuale physicum ordine consueto Aristotelico conscriptum et ex antiquissimis scriptoribus constructum.* Editio altera auctior, Lipsiae, 1657 (Preface dated 1 July 1656), 424 pp. BN R.10509.

[73] *Ibid.*, pp. 62-63, 65, 107, 114, 122-23, 141, 145, 151.

[74] *Ibid.*, pp. 179, 226.

[75] *Ibid.*, p. 181.

[76] *Ibid.*, p. 183, citing Ficino in *Conviv. Plat. de amore.*

[77] *Ibid.*, p. 240.

decides to discuss comets after all. What a comet is, is very obscure. For Aristotelians it is a meteor; for others, an extraordinary star. The more learned mathematicians not without cause teach that comets are extraordinary stars, and that they are generated in the ether or heavens and so have a celestial and ethereal nature and a movement befitting stars.[78] Scharff also departs from Aristotle far enough to speak of atoms as the smallest natural bodies and of salt, sulphur and mercury as the first mixed bodies. The efficient cause of stones is salt and lapidific spirit.[79]

Even at Utrecht there were still conservatives in physics. The *Physiologia* of Daniel Voet (1629–1660), doctor of medicine and philosophy, and professor at the University of Utrecht, who died when only thirty-one, was published posthumously by his brother Paul in 1661.[80] The treatment is quite conservative, retaining *materia prima et secunda* and distinguishing form as substantial or accidental, material or immaterial—the last being the form of spiritual substance. The traditional four elements are also kept, as are both manifest and occult qualities. To the four primary qualities is added impetus or a quality in a body by which it is apt to be transferred from place to place. Examples of occult qualities are seen in the tarantula, the bite of a mad dog or viper, and the pest. Velocity and slowness of motion come solely from interjected delays or pauses, not from adding degrees of motion or parts of motion. Gravity is the effort of parts to rejoin their whole, and is not to be ascribed to the pressure of ethereal globules, or effluvia, or little chords with hooks and barbs which issue forth from mundane globes. Such explanations multiply *entia* and figments unnecessarily and expose "Physiology" to ridicule.

Voet holds to Creation in six days, asserts that the universe is

[78] *Ibid.*, pp. 251-53.

[79] *Ibid.*, pp. 225-34, 274, 292. The existence of lapidific spirit was denied by Wilh. H. Waldschmiedt, in the *Corollaria respondentis* to his dissertation *De auro*, Marburg, 1685.

[80] *Danielis Voet medicinae et philosophiae doctoris. Huius in Academia Ultrajectina professoris Physiologia. Adjectis aliquot eiusdem argumenti disputationibus publice antehac propositis.* Opusculum posthumum, ab eius fratre nunc editum. Amstelodami, ex officina Johannis à Waesberge, Anno 1661. The disputations, dating from March 15, 1654 to June 6, 1660, begin at p. 99. At pp. 198-264, "Sequuntur aliquot disquisitiones auctore fratre Paulo Voet."

not composed of vortices, and denies the world soul of Plato, for holding which Deusing was condemned by judgment of the theological faculties of the entire Belgian Federation. Voet is further opposed to those who contend that the world might be more perfect and convenient. He would appear to have been a most conventional and proper young man.

Mixed bodies are either celestial or sublunar. Five planets go round the sun as their center, but sun and moon move about the immobile earth. Venus and Mercury have phases like the moon. Some fixed stars appear and disappear, as that of 1572 did. Fixed stars and planets are not revolved by their spheres but move like birds through the air or fish in water in virtue of an impetus given them at creation. Sunspots and comets are discussed together, and Voet says that the latter come from the sun, but speaks of sublunar comets in the next book on meteors. *Ignes fatui* are produced by the heat of the sun kindling greasy exhalations in cemeteries.

An unsatisfactory fifth book on minerals devotes much of its space to marvelous effects of the magnet. The first efficient cause of minerals is God; the second, mercurial particles. Experience shows that metals are found more abundant and nearer the surface on the east and south sides of mountains than on the north and west sides.

The sixth and last book on animate bodies keeps the conception of *calidum innatum,* and both animal and vital spirits, but describes and accepts the circulation of the blood. The animal spirits are thought of as passing through both sensory and motor nerves. Three internal senses are recognized: common sense, phantasy, and memory. Some add estimative to these, but it is found only in man and is rather an act of the intellect.

Voet cites no authorities and gives some experiments and pictures, but neither are very illuminating.

There were several subsequent editions of Voet's book at Utrecht in 1678, 1688 and 1694. That of 1688 was accompanied by notes of a later professor of philosophy at Utrecht, Gerard de Vries. He denied the existence of inhabitants in the moon, because the Bible did not mention them, and because the long days there, fifteen times as long as ours, would produce insupportable weather conditions. The review of this edition in the *Journal des Sçavans* asserted that Voet was more influenced by Gassendi than by Descartes,[81] but I would not say that he was carried away very far by either of them.

Marten Schoock first published a General Physics at Groningen in 1660, and then a Celestial Physics at Amsterdam in 1663. The former, dedicated to Hermann Conring, consists of fifteen disputations in which the opinions both of ancient and recent philosophers are discussed with philosophical liberty and so that more regard is had for truth than human authority.[82] Schoock, however, has great respect for divine authority and the Bible. In the preface Schoock lists five kinds of men of whom he disapproves: those who resemble the ancient sophists, recent pretended Peripatetics, those who both oppose Aristotle and scorn recent philosophy, those capable of collating ancient and modern philosophy but who prefer to launch new paradoxes, and those who know only one field yet criticize anyone who disagrees with them.

The fifteen disputations follow roughly the order of the *Physics* of Aristotle, considering the nature of physics, the principles constituting natural body, substantial form, nature and natural body and art, the causes of natural body with fortune, chance and fate and the complex principles of natural body, quantity, space, vacuum, motion and rest, local motion in general, the motion of heavy and light, violent motion, time, and qualities. Recent experiments to demonstrate the existence of a vacuum by Valerianus Magnus, a Capuchin of Milan, and by Mersenne are mentioned, but Torricelli is not named. Schoock suggests that a spirit from the falling mercury fills the apparent vacuum and notes that the Cartesians say that his celestial matter gets in.[83] Descartes and Galileo are cited as to motion, and Galileo is said to contend that all motion is either circular or verging towards the circular.[84] Schoock accepts the existence of occult qualities and says that three varieties are commonly distinguished: idiocratic, which have a singular effi-

[81] JS XVII, 136-39.

[82] *Physica generalis qua sic discutiuntur secundum libertatem philosophicam non minus antiquorum quam recentiorum philosophorum placita ut simul major ratio habeatur veritatis quam authoritatis humanae*, Groningae, 1660. Copy used: BN R.10654.

[83] *Ibid.*, pp. 195-97.

[84] *Ibid.*, pp. 220, 230.

cacy of action, antipathy, and sympathy. Natural magic is connected with occult qualities, but all this will be discussed more fully in his treatise on magic.[85]

The Celestial Physics, is described in the full title as "not only according to the views of the ancient philosophers but also the more accurate observations of recent astronomers."[86] It comprises fourteen disputations which are adorned with many Greek synonyms and lists of Biblical references, and keeps quoting the second book of Pliny's Natural History, whose astronomy is not even up to the Ptolemaic level. And astronomical observations are eked out by the truth of Scripture.[87] Kepler too, however, is quoted, and it is noted that he holds that the planets keep moving about the sun as center of the universe without stationary positions or retrograde movement, but at unequal speed, slower as they recede from the sun and faster as they approach it, and in one part of their circuit move to the north in the ecliptic, in the other towards the south.

> But as this view presupposes that the sun is at rest as the center of the universe, while the earth is moved and at the same time is a planet, we cannot admit it physically.

He admits that the phenomena are best explained by Kepler's hypothesis but does not believe in accepting astronomical hypotheses too hastily as nature's laws.[88] Later, in speaking of the motion of Mars, Kepler is quoted so as to emphasize the difficulty of investigating it rather than his ultimate success, and Gassendi and Hortensius are said to have observed that Mars was covered by the moon on February 5, 1632, contrary to the Rudolfine Tables.[89]

Schoock distinguished the new stars of 1572, 1600 and 1604 from

[85] *Ibid.*, pp. 270-71. I have not found his treatise *de magia*. It is not included in his *Exercitationes variae de diversis materiis*, Utrecht, 1663, in-4, 602 pp., which are all religious and ecclesiastical.

[86] *Physica caelestis . . . non modo juxta antiquorum philosophorum placita verum etiam recentiorum astronomorum accuratiores observationes*, Amsterdam, 1663, 412 pp. Copy used: BN R.10655.

[87] *Ibid.*, p. 4, "per observationes sidereas . . . maxime vero succurrente Scriptura quae varia ex rei veritate de caelo et corporibus caelestibus hinc inde proponit."

[88] *Ibid.*, pp. 101, 107.

[89] *Ibid.*, pp. 148-49.

comets in six ways,[90] but his prolonging the discussion as he does[91] seems rather antiquated for the time at which he writes. He still speaks of three regions of the air.[92] Of the discussion of astrology in his Celestial Physics we treat in a later chapter on Astrology After 1650.

Isaac Voss, in a brief treatise on the nature of light, made a number of bold assertions. The cause and subject of light was fire which was not a body. Much less was light corporeal. There were no pores in glass, water and very pellucid bodies. A vacuum was possible and existed above the air. Colors were not light. Their material was from the quality of sulphur, and flame always followed the color of sulphur and took on all colors except black and pure white. A comet was not some phantastic specter or imaginary illusion but a real body and star which was on fire on all sides.[93]

Three replies to or comments on the treatise of Voss are bound with the copy of it which I examined. Jan de Bruyn, ordinary professor of "physics" and mathematics in the University of Utrecht, defended Cartesian doctrine against it,[94] while Pierre Petit, a physician of Paris, came to the defense of Aristotle. He maintained that the sun was not the seat of fire and that its heat differed from fire. But although it did not have heat in itself *actu et formaliter*, yet it heated our inferior world. Petit denied that there could be vacant space in nature. Since Pliny, Hippocrates and Galen agreed that the Ethiopians were black because of the heat there, he concluded that white results from lack of heat, black from its force and burning, so that we can call cold whatever is white, and hot whatever is black.[95] Incidentally Petit denied the truth of stories of ever-burning lamps being found in ancient sepulchers.[96]

Voss answered the objections of de Bruyn and Petit,[97] after which

[90] *Ibid.*, pp. 232-33.

[91] *Ibid.*, pp. 227-57.

[92] *Ibid.*, p. 231, "in media aeris regione."

[93] *De lucis natura et proprietate*, Amsterdam, 1662, in-4, 85 pp. BM 537.f.30.

[94] *Epistola ad . . . Isaacum Vossium . . .*, Amsterdam, 1663, in-4, 68 pp. BM 537.f.30.(2.).

[95] *De ignis et lucis natura exercitationes ad Io. Vossium*, Paris, 1663, in-4. BM 537.f.30.(3.).

[96] *Ibid.*, p. 54.

[97] *Responsum ad objectiones Joh. de Bruyn professoris Trajectini et Petri Petiti medici Parisiensis*, The Hague, 1663, 104 pp. BM 537.f.30.(4.).

Graindorge concluded the discussion. The position of Gassendi on light seemed preferable to him to those of Aristotle and Descartes, but he was ready to follow a method based on experiments and reason and which investigated not what Aristotle or Descartes or Gassendi thought but what nature itself dictated, and Voss seemed to him to approach close to the truth. But he had not overthrown the arguments of Gassendi that light was a corporeal emission. He agreed with Voss that a vacuum existed, and that there was vacant space beyond the air, but he held that water as well as earth and air had pores, and that diaphanous bodies were a mixture of vacuum and solid parts. Color was a little fire; the properties of flame and color were common; and Graindorge agreed with Voss that the material of colors was sulphur. But he disagreed as to black and white. White was maximum color, while black was darkness and privation of light. Light alone devoid of color was not visible. Voss's theory of comets was very probable.[98]

Some years later, in dissertations on the nature of cold and heat, Petit granted that there was no sphere of fire and that fire was not an element, and that ice was lighter than water, but held that cold was positive, not mere privation of heat, and that the air was naturally cold and was warmed adventitously by the rays of the sun. On the other hand, he thought that all salts, including nitre, were hot.[99]

Italy did not for long remain unresponsive to the new trend in philosophy. Tommaso Cornelio (1612–1688), a Cartesian of Cosenza, issued *Progymnasmata physica* at Venice in 1663, republished at Frankfurt in 1665, Venice, 1681, Leipzig, 1683, Jena, 1685, and with his complete works at Naples in 1688. He divided them into seven exercises which dealt respectively with 1) method, recommending the mathematical study of nature, and chemical and mechanical principles; 2) with the beginnings of natural phenomena where he found the Cartesian explanation the best; 3) with the universe, "where," says the reviewer, "he seems to be in a maze"; 4) with the sun, holding that light is in the sentient, just as pain

[98] Andreas Grandorgaeus, *De natura ignis lucis et colorum dissertatio*, Caen, 1664, in-4, 122 pp. BM 537.f. 30.(5.).

[99] PT VI, 3043-45.

is in the wounded and not in the sword; 5) with human generation; 6) nutrication; and 7) life.[100] When the work was republished in 1683, although the contents remained the same, the title was not only changed to *Physiologia,* but went on to lay claim to "new and hitherto unheard-of . . . weights of reasoning." Yet Cornelio went back to Plato for an explanation of motion which would avoid a vacuum. He noted that the Cartesian system of innumerable vortices was liable to run into the same difficulties as Giordano Bruno's many worlds, but added that Descartes had guarded against this by not making the vortices equal or wholly similar, and had introduced a new refraction of light by which the same star might appear in many places. As for the three systems of Ptolemy, Copernicus and Tycho, Cornelio asked whether any one of them would not do.[101]

The *Placita philosophica* of Guarinus Guarini, of the Order of Theatines, a work published at Paris in folio in 1665,[102] comprised logic and metaphysics but was especially concerned with physics. The author opposed many commonly received opinions. He held that substantial form was a pure power and did not exist by itself. He substituted spirals for epicycles and eccentrics, and denied that the middle region of the air was cold. He also denied that the air was corruptible, and that corruption necessarily preceded generation. He contended that iron attracted the magnet; not the magnet, iron. He even rejected vital and animal spirits, and held "many other extraordinary opinions regarding light, the rainbow, and the tides."[103]

Less radical was a treatise on the immobility of the earth which

[100] PT II, 576-79. These exercises are followed by three letters: *de Platonica circumpulsione, de cognatione aeris et aquae,* and *Epistola M. Aurelii Severini nomine conscripta,* which in the edition of 1663 begin at pp. 113, 142 and 151.

[101] *Ibid.,* pp. 47-48. Corte, *Notizie storiche intorno a' medici scrittori milanesi . . .,* 1718, p. 165, wrote of Cornelio: "nell' osservare le propagazioni della vena porta nel fegato servire come d'arterie, e sì strettamente unirsi al poro biliario, che ciaschedun ramo viene per lo più chiuso in un medesimo invoglio." But F. Bouillier, *Histoire de la philosophie cartésiene,* 1854, II, 510, was obviously mistaken in asserting that Cornelio, credited with introducing Cartesianism into Italy, restricted himself to teaching medicine according to Descartes' principles.

[102] BN R.312.

[103] JS I (1666), 734-36.

a priest of the Oratory at Fano, Orazio Maria Bonfioli, addressed to Carolo Caraffa, cardinal legate. As in the book of Christopher Borri, written back in 1631, three heavens were distinguished—empyrean, stellar and aerial, and the earth was not only said to be in the middle of the universe, but hell to be at the center of the earth.[104]

Much bolder were some of the conjectures of Pietro Cavina as to the nature of the universe.[105] Yet his work was approved by a Jesuit and Inquisitor. However, although he abandoned the doctrine that the heavens are ungenerated and incorruptible, he refused to admit that the earth had either a diurnal or an annual movement. In fact, his attributing variations in the fixed stars to the action of the sun, from which they were therefore not far removed, was in opposition to another Copernican tenet, that the sphere of the fixed stars was at a great distance.[106] He quoted Tycho Brahe that Copernicus had not hesitated to assert that stars of only the third magnitude equalled the sun in size.[107] Other conjectures by Cavina were that the stars were composed of matter which was easily dissipated; that the fixed stars were like torches which went out from lack of fuel or grew large by access of aliment; that the heaven of the fixed stars was composed of a substance similar to oil; and that all the fixed stars were located in the concave surface of their heaven.[108]

Honoratus Fabri (1607–1688) entered the Society of Jesus at Avignon in 1626, taught at Lyons for fourteen years, and then was called to Rome as papal penitentiary. He was a voluminous writer in many fields and has been called the advocate of lost causes. In the domain of science he was familiar with ancient writers such as Aristotle and Pliny,[109] but also was acquainted with contemporary developments. Unlike many members of his Order who limited their writings to laborious compilations from past authorities and seem to have had no ideas of their own, he professed to have anticipated Harvey's discovery of the circulation of the blood, vied with

[104] *De immobilitate terrae,* Bologna, 1667, 110 pp. BM 536.d.22.(2.).

[105] *Congetture physico-astronomiche della natura dell' Universo.* Faenza, 1669, in-4, 44 pp. BN V.7863.

[106] PT V, 2012-13.

[107] *Congetture,* p. 30.

[108] *Congetture,* pp. 37-38, 40.

[109] Fabri left manuscript notes on Pliny's Natural History and keeps citing him in the *Dialogi physici.*

Descartes in thinking things out for himself,[110] conducted experiments on capillarity, engaged in astronomical observation,[111] and wrote against Huygens with regard to the moons of Jupiter. In other words, he attempted to meet developing modern science on its own ground, to fight against it with its own weapons, or, to change the figure, to accost it with diplomatic courtesy and seeming friendliness, to yield a few minor points, and to try to outwit it on more important issues.

Our survey of Fabri's attitude will be based upon two of his works: the *Dialogi physici*[112] of 1665, and the *Physica*[113] in four volumes from 1669 to 1671. Both were printed at Lyons and not in Italy. Petrus Mousnerius, M.D., had already published at Lyons i 1646 a treatise on local motion based upon Fabri's lectures.[114]

The *Dialogi* were the outcome of Cardinal Facchinetto's[115] inviting Fabri to dinner to discuss with other learned guests physical

110 In the fourth volume of his *Physica*, 1671, (after p. 309) he claims to have thought out all his book himself. In opening the third volume he represented himself as shaking the dust of scholasticism off his feet, and was sure that his was the right way to explain the meaning of Aristotle.

111 At p. 208 of *Dialogi physici* are appended two letters of Fabri to Claude Bosset telling of his recent observations through the telescope of the libration or mutation of the ring of Saturn, which, he holds, disproves the movement of the earth. Also as a result of sixty days observation of the moons of Jupiter with a longer telescope, he contends that they do not move in a circle about Jupiter, confirming his opusculum against Huygens. In the preceding text of the *Dialogi*, p. 90, he had held that, if they revolved about Jupiter, the nearest moon at least should cast a shadow upon that planet. He now at last sees shadows on it, but argues that they are mountains. Meanwhile Cassini from observation of the shadows of the satellites on the surface of Jupiter had reckoned that that planet rotated upon its axis in nine hours and fifty six minutes.

112 *Dialogi physici, in quibus de motu terrae disputatur, marini aestus nova causa proponitur, necnon aquarum et Mercurii supra libellam elevatio examinatur.* Lugduni, 1665, 4to. Copy used: BN R.6706.

113 *Physica id est scientia rerum corporearum.* Lugduni sumpt. Laur. Anisson, 1669-1671. 4to. Copy used: BN R.2970-2973.

114 *Tractatus physicus de motu locali in quo effectus omnes qui ad impetum motum naturalem violentum et mixtum pertinent explicantur et ex principiis physicis demonstrantur.* Auctore Petro Mousnerio doctore medico cuncta excerpta ex praelectionibus R. P. Honorati Fabry, S. J., Lugduni apud Ioannem Champion in foro Cambii, 1646.

115 Cesare Facchinetto was born in 1608 and was cardinal from 1643 to his death in 1683.

experiments but especially the question whether the earth moves, the cause of the tides, and the Torricellian experiment. Some months later the cardinal asked him to write out their conversation and the present book was the result. Despite the ecclesiastical censure in 1616 of the doctrine that the earth moves and is not at the center of the universe, and the subsequent submission of Galileo, the cardinal, who appears in the dialogues as Augustinus, is represented as arguing for the Copernican theory against Antimus or Fabri, although Augustinus grants that ecclesiastics especially ought to support the pontifical decrees against that theory. But he calls Galileo "never praised enough." Even Antimus denies that he hates the Copernican hypothesis. He regards Copernicus as the chief astronomer of his time and Galileo as second to none in genius. "But he so weakened the arguments against the Copernican hypothesis that he seemed to confirm it," although Fabri feels sure that Galileo would not deny that so far it has not been demonstrated. Furthermore Fabri believes that he has new arguments to show that the earth is immobile at the center of the universe. "You must be joking," retorts the cardinal. Surely the Copernican hypothesis is supported by such a mass of reasons that, although they by no means equal geometrical demonstration, yet they approach closely to it, and that theory is rightly judged by common agreement of all the learned to be far more probable than any other. When Fabri asks to be enlightened as to those reasons, the cardinal or Augustinus replies, "Haven't you read them in Galileo so simply and clearly explained that nothing in my judgment can be read which is clearer and simpler?" Fabri responds that he has read and reread everything that Galileo ever published.[116] One of Fabri's new arguments against the Copernican theory is that the ring of Saturn is always parallel to the plane of the equator and never to that of the ecliptic.[117]

In the *Physica* Fabri still maintained that the earth did not move,[118] although he admitted that the Copernican theory was most ingenious, worthy of the greatest astronomers, easy and simple for

[116] *Dialogi* (1665), pp. 2-3.

[117] *Ibid.*, p. 91.

[118] *Physica,* III (1670), 151; IV (1671), 415, "Digressio astronomophysica in qua terrae quies adstruitur."

astronomical calculations, and as a pure hypothesis to be preferred to all others hitherto known.[119] But Fabri had his own hypothesis of a spiral movement of the sun to suggest.[120] Another Jesuit, Andreas Tacquet (1611–1660), whose works were published at this time, refused to discuss the Copernican-Ptolemaic controversy because either system seemed probable, but acquiesced in the decision of the ecclesiastical authorities.[121]

Fabri is said to have been hailed before the Inquisition in 1671 for his too favorable attitude towards the Copernican theory in the *Physica,* and to have escaped with only fifty days imprisonment through the influence of Cardinal Leopold de'Medici.

Galileo did not receive such unqualified eulogy in the later volumes of Fabri's *Physica.* It was now said that he was more to be praised for his discovery of the moons of Jupiter than for questioning that the earth is at rest.[122] Furthermore, he was charged with a figment and *nugae* for having inferred from observing water gather on the surface of leaves in globules that there was an occult antipathy between air and water, when there was a most manifest cause of that phenomenon.[123]

Kepler fares much worse than Galileo in both works, partly probably because he was an unregenerate Lutheran. In the *Dialogi* he is belittled by both interlocutors. The cardinal speaks of his having obtained great glory among all for sheer nonsense and fables on which he founded the physical causes of celestial phenomena,[124] while Fabri, when asked why Mars sometimes emerges twice and is hidden once in the morning, says that, so far as he knows, this is asserted only by Kepler, "in whom I think little faith is to be placed."[125] In the *Physica* Kepler is adversely criticized not merely for his treatise on the movement of Mars, but for resorting to the unequal motion of a planet covering unequal segments of

[119] *Ibid.,* IV (1671), 426.

[120] *Ibid.,* IV, 333 *et seq.,* 431. In the appendix to his *Synopsis Optica* Fabri had refuted a spiral hypothesis devised to support the Ptolemaic system and had suggested a new hypothesis of his own to save the Ptolemaic system and keep the earth immobile: PT II, 626.

[121] *Opera mathematica,* Antwerp, 1669; PT III, 869 *et seq.*

[122] *Physica,* IV, 281a.

[123] *Ibid.,* III (1670), 197a.

[124] *Dialogi,* 84.

[125] *Ibid.,* 14.

its orbit in equal times—an allusion to one of his famous three planetary laws.[126]

The names of other seventeenth century scientists appear in the *Dialogi.* The cardinal agreed with Gilbert that the earth was a great magnet, but Fabri said that many denied this.[127] In the *Physica* he himself denied it more definitely, stating that the earth was not a magnet, although it contained magnetic particles.[128] No one should think that the moon is moved by the earth, since there is no contact and no application of motive force. Nor may one recur to magnetic virtue, since in that case the earth would draw the moon to itself.[129] Descartes is criticized in the *Dialogi* for explaining the tides by pressure of the moon on the air and ocean's surface.[130] Mersenne is corrected as to the distance covered in a second by a falling body.[131] Gassendi [132] and Grandami [133] are also mentioned. The experiments of Boyle are cited;[134] Hevelius is called a faithful and accurate observer of lunar phenomena;[135] Huygens, although Fabri has written against him, is "a man indeed most learned."[136] Less familiar names which are cited approvingly are Iavellus and Ruvius on the heavens,[137] Bovius on philosophy, Crescentius and Furnerius on tides.[138]

The tides are discussed with much accompanying geographical detail in the *Dialogi,* but the moon is held to be not the cause but

[126] *Physica,* IV, 415.
[127] *Dialogi,* 33.
[128] *Physica,* III, 150.
[129] *Ibid.,* IV, 420b.
[130] *Dialogi,* 35.
[131] *Ibid.,* 42.
[132] *Ibid.,* 149.
[133] The Jesuit astronomer, if we may so call him, Jacques Grandami or Jacobus Grandamicus (1588-1672), in 1645 published a work arguing that the earth as a magnetic body could not rotate and so was immobile. In 1661 he printed a discussion of the day of the birth of Christ, a work which he later expanded into three volumes on Christian chronology (Paris, 1668). In the interim he composed treatises on the course of the comet of 1664 and 1665, on the lunar eclipse noted by Pachymeres and the solar eclipse of December 30, 1655. In the same year 1668 he published a method of measuring solar eclipses.
[134] *Dialogi,* 185.
[135] *Ibid.,* 77.
[136] *Ibid.,* 89.
[137] Javelli was a sixteenth century commentator on Aristotle; Rubio, a Jesuit commentator on Aristotle of the early seventeenth century.
[138] Theophrastus Bovius, *Corpus totius philosophiae* (in French), Paris, 1614. Bartolomeo Crescentio, *Nautica Mediterranea,* 1601. Georges Fournier, S.J., *Hydrographie,* 1643; *Geographica orbis notitia,* 1648.

only the occasion for them. Gassendi found that the tide is felt for many yards below the surface of the sea.[139] The experiment of Torricelli with the tube of mercury is called "very beautiful and the most celebrated" of the century, while the claim of Valerianus to its authorship is emphatically rejected.[140]

Whereas Lana Terzi had suggested an aerial ship which would be lifted by pumping the air out of four large spheres of very thin copper,[141] Fabri made the ridiculous suggestion of four big tubes filled with a great deal of compressed air, apparently on the theory that more air in the same space would weigh less.[142]

Fabri retained the four elements of old, rejecting the three principia of the chemists and the three kinds of particles proposed by Descartes. But he no longer believed in a sphere of fire in the concave of the sphere of the moon, nor in an ether distinct from pure air, nor in mountains which rose above the middle region of air.[143] Springs for the most part came from rain and snow.[144] The sun was true fire, but mixed, not pure.[145] The moon was a compound body from the four elements, but uninhabited, without animals or vegetation, clouds or precipitation.[146] Its spots were bodies of water.[147] In discussing comets, Fabri based no argument on parallax, which he regarded as still an uncertain matter. He had seen no comet since that of 1618, when he was a small boy. Yet Cassini had observed comets of December 20, 1652 to January 7, 1653; December 20, 1664–March 11, 1665; another in April, 1665; a fourth in March, 1669.[148] Fabri believed that comets were made of matter from the ethereal region, the same matter as occurred in sunspots. Grassi noted that for a month while the comet of 1618 appeared

[139] *Dialogi*, 149.

[140] *Ibid.*, 182-83.

[141] F. Lana, *Prodromo overo saggio di alcune inuentioni nuoue premesso all'arte maestra* . . ., Brescia, 1670. Paschius, *De novis inventis*, 1700, quotes a Latin translation of over ten pages from J. C. Sturm, *Tentamines Collegii Curiosi* . . ., p. 97 *et seq*. See also Sturm, *Collegium experimentale* . . . Norimbergae, 1676, *Tentamen* X, pp. 56-66, with a full page picture of the airship at p. 64.

[142] Paschius, *De novis inventis*, 1700, p. 636, citing Hon. Fabri, *Tract.* I, *Physica, liber* ii, *Propos.* 246, p. 153.

[143] *Physica*, III, 135, 194-5.

[144] *Ibid.*, III, 393.

[145] *Ibid.*, IV, 222.

[146] *Ibid.*, IV, 243, 258.

[147] *Ibid.*, IV, 255.

[148] Rohault, *Tractatus Physicus*, London, 1682, II, 81, *Animadversio* XX by Antoine le Grand.

there was no spot on the sun, the matter having gone into the comet.[149] Fabri had no doubt that new stars appear and vanish like those of 1572, 1600 and 1604. However, they were not new creations but rather coalesced from the collecting of other parts. They were closely analogous to comets. But the star of the Magi was produced by angels and not a physical phenomenon.[150]

Fabri listed fourteen different effects of the moon upon the sublunar world,[151] and held that solar action on the other heavenly bodies produced effluvia from these.[152] But he denied any influence of the stars on the earth aside from the little light which they shed on us.[153] On the other hand, he could be easily induced to believe that the new stars were omens and signs from God, who might also sometimes employ comets as portents. But since they were mere effects of nature, he did not believe that ordinarily they were presages.[154] Thus Fabri left hardly any loophole for a belief in astrology, and the same may be said with regard to divination and magic in general.

In 1671 Leibniz published both at Mainz and at London a New Physical Hypothesis,[155] of which the first part or Theory of Concrete Motion was dedicated to the Royal Society, and the second part or Theory of Abstract Motion to the French *Académie des Sciences.* It sounds almost like a *reductio ad absurdum* of the prevalent mechanistic or corpuscular philosophy, or a feeble effort to outdo Descartes. The reaction of vacuum and plenum is declared to be the origin of all fermentation, acid and alkali, sympathy and antipathy. It was foreshadowed by the red and white of the old chemists, their masculine and feminine, three *principia,* and by Helmont's gas, blas and archeus. Both meteors and fountains come from the earth's interior.

The sea, as Becher ingeniously suggests, perpetually distills its more

[149] *Physica,* IV, 300-307.

[150] On new stars: *Ibid.,* IV, 296-99.

[151] *Ibid.,* IV, 267.

[152] *Ibid.,* IV, 298. In the *Dialogi,* 178, he posited effluvia from bodies in general.

[153] *Physica,* IV, 294.

[154] *Ibid.,* IV, 298, 308.

[155] *Hypothesis Physica Nova qua phaenomenorum naturae plerorumque causae ab unico quodam universali motu in globo nostro supposito neque Tychonicis neque Copernicanis adspernando repetuntur.* In the 1768 edition of his *Opera* at II, ii, 3-34.

bituminous and heavier parts through its spongy bottom to the center of the earth.

That many springs rise from supermontane and submontane cisterns of collected snows and rains, Leibniz doubts not, with Hobbes, Voss and others. But some are produced from subterranean vapors. The chemists' nucleus is their three principles; their cortex, *terra mortua* and phlegm. There is much talk throughout of *bullae* or globules. When the air in them is exhausted (and contrariwise they are distended with ether) we have alkali, what the old chemists called feminine, mercury. When they are distended with air (and contrariwise exhausted of ether), we have acid, masculine, sulphur. "For what is full merely of ether, is a vacuum to the senses." The bigger exhausted globules are alkali or fixed salt; the smaller are volatile alkali. The bigger distended globules are acid or fixed sulphur; the smaller are volatile acid. Whether there is a mean between the two extremes, to match the three kingdoms of nature, can be told only after long and comparative experimentation. Leibniz does not wish to indulge in preposterous divination, but he believes that the admirable wisdom of the Creator has so arranged things that many are produced from few. He would guess that there are thrice three varieties of those two instruments of nature, distension and exhaustion. Both would be of a minimum, maximum and mediocre exhaustion and distention, and of these again each would be subtle, medium and crass. "Therefore there are four larger masses or elements." Acids would be cured by alkalis of similar degree. Hobbes correctly agrees with Descartes that the same mass cannot fill more or less space. Aristotle almost never said what the schoolmen attribute to him. Honoratus Fabri and John Raeus, men of genius beyond mere erudition, agree with the followers of Descartes and Gassendi that all variety in bodies is to be explained in terms of magnitude, figure and motion.

Guillaume Van Sichen (1632–1691) was a Franciscan who taught first philosophy and then theology at Louvain. His Complete Course in Philosophy[156] was written for use in the Order and to

[156] Guilelmus van Sichen, *Integer cursus philosophicus brevi clara et ad docendum discendumque facili methodo digestus*, Antwerp, 1666, in-fol., 2 col., 2 vols. BM 8464.h.6.

obviate the necessity of so much note-taking by the students. It was printed at Antwerp in 1666 in two folio volumes, of which the first was devoted to logic and metaphysics, and the second, which concerns us, to natural philosophy. He is regarded as anti-Cartesian in general but as adopting the views of Descartes on some particular points in astronomy, physics and psychology.[157] More evident is his bold effort to show that the views of Scotus may be harmonized with recent scientific discovery. In this respect his work is in marked contrast to those of Mastrius and Bellutus considered in an earlier chapter.

Van Sichen opens in Peripatetic style with a first tractate on natural body, and disputations on first matter and substantial form. His second tractate on causes, discussing action at a distance, rejects that of the torpedo and tarantula, fascination, the notion that the fire heats the top of the kettle first, weapon ointment, and the corpse bleeding at the approach of the murderer. He also rejects the Cartesian explanation of magnetic action and engages in an animadversion against the foes of qualities and the inventors of corpuscles.[158] Disputations follow on the First Cause and the final cause. In connection with the movement of projectiles, he sets forth the late medieval theory of impetus but fails to mention Galileo. "That nothing which remains unmoved can impress impetus upon another body is contrary to Suarez, Oviedo and others, but the conclusion seems to be made evident in daily experiments."[159] Later, discussing the old problem of the stone dropped into a hole extending through the center of the earth to the Antipodes, he states that the impetus impressed on it would carry it past the center, but after varied undulations it would finally come to rest at the earth's center, when the impetus by which it was moved was exhausted.[160] Now he continues in the usual Aristotelian order to discuss space, vacuum, time, composition, continuum and the infinite.

In the fourth tractate on the heavens and elementary world, comets are not included but left to the seventh tractate on meteors.

[157] *Dictionnaire des Ecrivains Belges*, II (1931), 1969-70.

[158] *Integer cursus*, II, 56a-59b.

[159] *Ibid.*, pp. 87-89a, in Tract. III de affectionibus corporis naturalis.

[160] *Ibid.*, p. 167b.

The question is raised whether the influence of the stars is occult or by the effluvia of very tiny corpuscles, as many modern philosophers think? The stars have no immediate or direct effect on the intellect and will , but it is possible to predict the weather and so forth from them. It is possible that a body move in a circle of itself, but in point of fact the heavens are moved by Intelligences. If the motion of the heavens ceased, not all sublunar motion would stop, but some would.[161]

Van Sichen retains the traditional four elements and four qualities. Fire is hot and dry, water cold and wet, earth dry and cold. But he is uncertain whether air is hot or cold as well as humid. But the immediate transmutation of one element into another can be proved by no reason or experience.[162] Scotus and Poncius [163] suppose a sphere of fire which is located under the concave side of the heaven of the moon and which revolves with that heaven, but it is a question—which neither experience nor reason nor authorities determine—whether fire alone fills that sphere or whether the whole sphere is filled with a fluid body in whose pores there is such an abundance of fiery substance that fire alone seems to dominate there. Van Sichen's hypothesis is that the natural place of fire is in the sun, whence it is diffused and, meeting non-pellucid matter in the heavens and being absorbed by it, constitutes six planets and on our earth fills the pores of the air and water and enters the organ of our sense of sight. He holds that this is not repugnant to the view of Scotus but is rather asserted by him; makes possible the rarefaction and condensation of air, water, earth and other bodies; and agrees with what he has held as to the influence of the stars. It avoids ascribing to the substance of the sun action at a distance in producing heat in inferiors, for according to his hypothesis the sun is like a volcano continually erupting fire to these inferiors and thence absorbing it again in a continuous circle like the sea regaining the waters which have evaporated from it. Van Sichen contends that the authority of the Fathers supports his hypothesis, but the passages

[161] *Ibid.*, pp. 146a, 147a-b, 152a-b, 153.

[162] *Ibid.*, pp. 155, 160, 165-66, 159, 154.

[163] Actually, as we have seen, Poncius doubted it.

which he cites from Basil and Ambrose assert nothing more than the mixture of the elements. Augustine,[164] however, said that fire penetrates all things and Anselm[165] that fire, which is called the fourth element . . . stretches from the moon to the firmament. Van Sichen further asserts that chemical experiments have shown that fire passes intact through water.[166]

Arriaga was wrong in representing the sea as higher than the land, and rivers originate from precipitation and melting snow rather than from the sea, although it is possible that water from the sea in subterranean lakes may be elevated in vapor by subterranean fires. It is dubious whether water or earth has weight in its natural place or not, but the negative opinion is to be preferred on grounds of authority. Some attribute the tides to rivers flowing into the sea; some, to waters which gush out of a profound abyss of the sea and are sucked in again (but why does this happen?); some, to subterranean fires; some, to an angel; some to the moon's moving the sea as it does the humors in an animal body. Cartesians ascribe it to pressure of the moon on the air and through it on the waters. Others hold that the moon attracts the earth away from its center, but Van Sichen objects that it is more likely that the moon would be moved by the much heavier earth. He himself ascribes the tides to the commotion made by spirits bursting forth from the bottom of the sea or an abyss or whirlpools, although the moon may play a part in arousing these spirits, and there are greater tides at full moon and during the interlunium.[167]

The tides had been more correctly discussed just the year before by Théodore Moretus, a Jesuit of Antwerp, who made the moon move them, not by fermentative virtue but by luminous and magnetic virtue.[168] Moretus was the author of various other treatises on physical themes.[169]

[164] *De Genesi ad litteram,* III, 4.

[165] *De imagine mundi,* I, 24.

[166] *Integer cursus,* II, 156-57.

[167] *Ibid.,* pp. 160-61, 162b, 163-65.

[168] *Tractatus physico-mathematicus de aestu maris,* Antwerp, 1665, in-4, 127 pp. BM 537.g.41.

[169] See also H. Bosmans, "Théodore Moretus de la Compagnie de Jésus, mathématicien (1602-1667), d'après sa correspondance et ses MSS," *De Gulden Passer,* new series, VI (1928), 57-163. He points out (*ibid.,* 141-42), that De Gottignies (1630-1689), Moretus and Zucchi differed from each other and were contradicted by experience

In connection with the problem of the tides may also be noticed a dissertation at Giessen by Meno Reiche on the motion of the sea.[170] The first chapter is on motion in general, the second on the latitudinarian movement of the sea, i.e., north and south, the third on its longitudinal movement to the west. Here the candidate, proceeding "according to the celebrated hypotheses of Copernicus and Tycho," holds that the pressure of the moon supposed by Descartes is not the cause of this movement, nor the sympathy of moon with sea which Varenius imagined, but that this movement follows the sun, as the Praeses had suggested in an earlier disputation at Leipzig.[171] Finally we come in the fourth chapter to the altitudinarian movement of the sea, i.e., the tides. Lord Bacon attributed them to the daily movement of the heavens; Eustachius a S. Paulo thought that God meant to keep the cause of the tides secret;[172] J. G. Voss ascribed them to subterranean fire;[173] others have recourse to different depths or cavities of the sea and the mutual tendency of waters to conjunction. Galileo tried to prove from the Copernican theory that it was impossible to give a natural explanation of the tides, if the earth stood still. Descartes explained them by the movement of the moon in an elliptical orbit. They are not from nostrils in mid-ocean nor from innate spirits of the sea. The most commonly received opinion assigns this motion to the moon, but Reiche concludes that moon and sun (or moon and earth in the Copernican system) are co-causes. Incidentally he has remarked that seas differ greatly in their qualities. Not only is one sea more nitrous, sulphurous, salt, and apportioned to

---

as to a problem concerning the equilibrium of a body on an inclined plane, stated by Pappus, and that they also admitted *a priori* that Pappus was right, whereas he has been shown to have been wrong by Duhem, *Les origines de la Statique*, I, 184-87.

170 *De motu maris* (Praeses Frid. Nitschius), 1671, Giessae-Cettorum: BN R.2291.

171 In a list of Nitschius's writings Zedler includes *De triplici maris motu*, etc., Leipzig, 1667, It is not in BM or BN printed catalogues.

172 Eustachius de S. Paulo, *Summa philosophiae quadripartita de rebus dialecticis moralibus physicis et metaphysicis*, Paris, 1609; with other printings in 1611, 1614, 1623, 1626, Cologne, 1616, and Lyon, 1629.

173 If Gerard Voss (1577-1649) is meant, Reiche would seem to have misrepresented his views as to the tides.

the nature of the moon than another, but they also differ in situation and figure.

Returning to Van Sichen, we come to comets. Some are seen between the moon and earth, Fromondus says,[174] and the supralunar or celestial ones can also consist of very tenuous vapors and exhalations set on fire or lighted up. Some say that such vapors ascend from the earth beyond the moon. But these supralunar comets are often larger than the entire earth and so cannot be composed of terrestrial exhalations. So Fromondus agrees[175] with others that the celestial comets consist of effluvia from the celestial bodies and resemble sun spots. Van Sichen further cites Cabeo[176] that they are moved by some Intelligence. They usually presage ill and have power of signifying the future from divine institution. Moreover, as the planets cause many changes in nature and the human body, so noxious and infective exhalations from comets may have some connection with natural and physical ills, such as storms, wars, mortality and sterility, and may induce disease and death, especially in those of tender constitution or who are exhausted by great responsibilities and have little time for rest and recreation, such as princes and magnates.[177]

On stones and metals Van Sichen has less than a column and it is devoted entirely to their generation with no mention of their marvelous virtues.[178]

Géraud de Cordemoy, in a work first published at Paris in 1666[179] and reprinted in 1671 and 1679, accepted the current corpuscular philosophy to the extent of regarding matter as an aggregate of indivisible bodies or atoms and holding that local motion would explain changes of forms, quantity and quality. He traced the cause of motion to subtle matter, and believed with Descartes that the body of an animal moved automatically like a watch.

[174] *Meteor.*, III, i, 5. Libertus Fromondus or Froidmont had been professor of theology at Louvain. His *Meteorologicorum libri* V were printed at Antwerp in 1627.

[175] *Meteor.*, III, ii, 7.

[176] *Lib.* I, *text.* 37, *quaest.* 7.

[177] *Integer cursus*, II, 298b, 300a, 301a-b.

[178] *Ibid.*, p. 318.

[179] *Le discernement du corps et de l'ame*, Paris, 1666, in-12, 230 pp. Copy used: BN R. 13654. PT I (1666), 306-10, is an excellent summary of the book, to which I have nothing to add.

But he not only held that the existence of the soul was surer than that of the body, and that a spiritual First Mover had first set bodies in motion and continued to move them. He further contended that it was as easy to conceive action of spirit on body, or of body on spirit, as it was to conceive action of body on body. Similarly, in a *Discours physique de la parole,* published first at Paris and then in English translation at London, his main argument was that in speech two things are inseparably joined: the voice which comes from the body and the idea or meaning which comes from the soul.[180] In thus opposing a purely mechanistic view of nature, he would seem to have left an opening for occult, magical and supernatural, or at least superphysical, forces. Cordemoy also wrote history and was a member of the Academy.

What might then be envisaged as the scope and task of natural philosophy, is indicated in 1667 by de Launay, who planned a treatment in three parts, of which the first would deal with natural philosophy in general; the second, with the celestial bodies; and the third, with terrestrial phenomena. At that time he published only an introductory dissertation on philosophy in general, and the first two of six books which were to constitute the first part. Of these two books the first was on the universe in general and discussed such long-mooted questions as whether the world was one or many, animated or not, whether from eternity, and how it would end. The three dissertations of the second book treat of place, time and eternity, the exterior vacuum and that scattered through the universe. The other four books were to have considered the material principles of all bodies, their efficient cause, natural qualities, and the themes of motion, generation and corruption. The chief interest of these Physical Essays is their adherence to an Aristotelian order of presentation, while giving the views of Descartes as well as those of the ancients.[181]

Some notion and perhaps a reasonably fair one of how far university instruction had been affected by recent progress in physics

[180] PT III, 736-38.

[181] Gilles de Launay, *Les essais physiques,* Paris, 1667, 44 and 108 pp.; BN R.4288. PT II (1667), 579-80, gives the impression that only the first of the six books of Part I was published.

and by Descartes is afforded by the contest for a vacant chair in philosophy at the University of Paris on October 5, 1669. The panel of thirteen judges had already weeded out three of the seven candidates, and the four who remained each talked for an hour on one of the four following topics: the immortality of the soul, motion, the superiority of the Peripatetic philosophy, and against "la pretendue nouvelle philosophie de M. Descartes which is said to be more addicted to novelty than to truth."[182]

Mathematical theses at the Jesuit Collège de Clermont, Paris, on June 24, 1672, as to varied systems of the world granted that either the Copernican or Tychonic or Cartesian hypothesis could explain all the phenomena, held that the motion of projectiles and falling bodies would be the same, whether the earth moves or stands still, and entertained the supposition, if any planet was immovable, how the others would appear from it. But the hypothesis of an immobile earth was preferred as more conformable to common sense, the experience of the senses, the authority of the wise, and the dignity of man. Affirming that only God knows whether earth or sky moves, it was argued that it was not irrelevant to consult Scripture on such points and to accept its dictum that the earth is immobile, while the sun rises and sets.[183]

Jacques Rohault (1620–1675), after teaching some ten or a dozen years at Paris,[184] was finally persuaded by his friends to publish in 1671 as a *Traité de physique* his course of lectures,[185] which were a combination of the teachings of Aristotle and Descartes in that field. A Latin translation of it by Théophile Bonet with notes by Anton le Grand was printed in London, 1682,[186] and at Amsterdam, 1700; another Latin version with amplifications "from the philosophy of Isaac Newton" by Samuel Clarke appeared at London, in 1710 and 1718;[187] while an English translation by John

[182] Gui Patin, *Lettres* (1846), III, 710.

[183] *Theses mathematicae de vario mundi systemate*, Paris, Collegio Claromontano, S.J., 24 Junii 1672: BN Vp.3541.

[184] Lambert, *Histoire littéraire du règne de Louis XIV*, II (1751), 46.

[185] See his preface.

[186] Copy in the Columbia University Library: 530 R632.

[187] These are called "Editio tertia" and "Editio quarta" on their title pages. Copies in the Columbia University Library: 530 R63 and 530 R633.

Clarke was published in London in 1723. The work was in four books dealing with natural bodies in general, celestial bodies, inanimate terrestrial bodies—earth, air, fire, water, metals, minerals and meteors, and animate beings, especially man. It impressed the *Journal des Sçavans* as including cosmography, anatomy, optics, machines and new experiments, as examining the secrets of various arts such as chemistry, the goldsmith's art, dyeing, and refining, and as treating of the pressure of the air, chemical change, and *les larmes* (Prince Rupert's drops).[188]

The tendency of the seventeenth century to magnify its own importance and to minimize that of the past is well illustrated by the book of Rohault. He not merely speaks of the vast progress made by the great geniuses of his own time, but takes the view that the entire period between Aristotle and Descartes has been barren so far as natural philosophy and science are concerned, so much so that twenty centuries have passed without any new discovery. He eulogizes Descartes and borrows much from him, but he also claims to have overlooked nothing that is good in the ancients and to have taken all the general notions from Aristotle. There are not many things in his book, he says, which are contrary to Aristotle; there are more which are contrary to most commentators on Aristotle; and there are still more which neither Aristotle nor his commentators have treated of at all. The intervening stagnation of two thousand years he attributes to a feeling of inferiority towards the ancients and excessive trust in authority and excessive commenting on Aristotle; to treating physics too metaphysically; to trusting either in reason alone or in experiments alone; and to neglect of the mathematical disciplines.

This too facile and generalized, too slap-dash and cavalier discussion and dismissal of the history of science and problem of scientific method in Rohault's preface, is scarcely justified by his subsequent text. He adopts the Copernican hypothesis and holds that the fixed stars are so many suns.[189] He professes to recognize nothing in bodies except magnitude, figure and motion.[190] But

[188] JS II, 624-29 (June 22, 1671).

[189] Pars II, caps. 24-25.

[190] Edition of 1682, Pars I, p. 170, "Sed cum in corporibus nihil praeter magnitudinem figuras et motus agnoscam."

he rejects the existence of a vacuum and of atoms.[191] He denies the forces of attraction, sympathy and antipathy,[192] but clings to the conception of animal spirits, holding that, besides the sensible parts observed in human bodies, there is an insensible substance like air to which medical men apply the name, animal spirit. It is formed of very rarified and subtle particles which pass from the blood to the brain.[193] In place of the four traditional elements or those of the chemists he adopts the three of Descartes.[194]

Rohault was unfavorable to judicial astrology, arguing that the stars exerted influence only by their light and that consequently the influence of the sun greatly exceeded that of all others.[195] But in a later chapter he gave a mechanistic and Cartesian explanation of tides as being due to a greater pressure upon a part of the earth's surface facing the moon by the matter surrounding the earth, so that the moon at least in this case exerted influence by mechanical pressure other than light.[196] He tried to account for the origin of astrology by the ancient Egyptians marking the course of the sun by the stars which rose as it set. This custom perhaps led to the belief that these risings rather than the sun's course were the cause of rain, drought and other weather changes. Or variations in the weather in successive years might be attributed to the other planets which alone had changed their positions. Rohault denied that the moon corroded stones, or that the marrow in the bones of animals and the size of lobsters increased with the waxing of the moon. For twenty-five years or more he had been observing fish and aquatic animals with this problem in view.[197]

Claude Gadroys (1642–1678) believed that Descartes had dis-

[191] Preface.

[192] Pars II, cap. xi, p. 52 in ed. of 1682.

[193] Pars IV, cap. 17; ed. of 1682, p. 259.

[194] Pars I, caps. xix-xxi.

[195] Pars II, cap. 27; ed. of 1682, vol. II, pp. 82-86.

[196] Pars II, cap. 29; ed. of 1682, II, 95.

[197] Pars II, cap. 27. When Anton le Grand in the dedicatory epistle of the 1682 Latin edition to Thomas Short, M.D. and member of the London medical college, says that everyone knows "how carefully you discern true from false, . . . how accurately you locate places, how exactly you measure the periods . . . of diseases, quam denique in eruendis astrorum influxibus peritiam praeferas," I presume that the last clause should be translated, "finally what skill you display in uprooting (rather than *eliciting*) the influences of the stars."

covered a new world as truly as Columbus and others had discovered America. Although he professed not to accept the explanations of Descartes on every point, he did adopt his theory of *tourbillons* or vortices, his three elements or kinds of particles of matter, and his three laws that everything remains in the state it is, so long as nothing changes it, that a body in movement tends to continue to move in a straight line, and that bodies moving in circles try to break away from the center of their movement. We have spoken of his Discourse on the Influence of the Stars in Chapter 19 on Descartes. We come now to his System of the World according to the Three Hypotheses,[198] in which he maintained that Ptolemy and Tycho Brahe had advanced mere hypotheses, but Copernicus, the simple truth, if his theory was understood in the light of the Cartesian vortices. Copernicus had explained two things which the followers of Ptolemy and Tycho had failed to elucidate: namely, gravity and levity, and the tides.[199] Gadroys, however, regarded the moon as the cause of tides and rejected Galileo's explanation of them from the movement of the earth.[200]

Gadroys accepted as a consequence of the Copernican theory that the earth differed in no respect from the other planets. But he called "an extraordinary opinion", the view which he attributed to many moderns, including Galileo and Kepler, as well as such ancients as Democritus, Heraclides and Pythagoras, that the moon is a world like that which we inhabit, with plains and mountains, seas and forests. But he adds that it is not to be rejected just because it is extraordinary. He ascribed no movement to moon or earth but held that each was at rest in its *tourbillon.*[201] In the grand *tourbillon* of the sun, the particles of the first element were pressed to the center but moved away from it in the case of the *tourbillon* of the earth. Fire existed at the center of each planet.

[198] *Le système du monde selon les trois hypothèses, où conformement aux lois de la mechanique l'on explique dans la supposition du mouvement de la terre, les apparences des astres, la fabrique du monde, la formation des planètes, la lumière, la pesanteur, etc. Et cela par de nouvelles demonstrations.* Paris, 1675, in-12, 18 fols., 457 pp. Copy used: BN R.13664.

[199] *Ibid.*, p. 130.

[200] *Ibid.*, pp. 378-79, 391-92.

[201] *Ibid.*, pp. 131, 309, 306, 132-33.

Detached sunspots made new stars. The occasional appearance of comets was accounted for by their passing from one vortex to another, and "there is nothing to prevent these bodies from re-entering our *tourbillon* many times." Gadroys believed that the matter of one tourbillon could enter another at almost any point and not merely at the poles of these vortices, as Descartes seemed to say.[202] He concluded his book with a consideration of things which he said did not belong to or depend upon any particular system, such as the phases of the moon, the size and distance of the stars,[203] the apparent size of sun and moon on horizon and meridian, eclipses, the four seasons, crepuscles and shadows, and the division of time.[204]

The triumph of natural science over the more scholastic subjects of the curriculum was evident in the enlarged 1681 edition of Du Hamel's Philosophy Old and New for Use in the Schools. Although the work was by an illustrious Abbé who had become one of the great prelates of France, it devoted only one volume each to logic, metaphysics and ethics, and three to physical science. Boyle was cited for a grain of copper coloring blue two hundred thousand times as much water; the sense of taste was explained according to Malpighi, that of smell according to Vernay, of hearing from Perrault. Roemer was used in connection with vision, de la Hire for the telescope and Mariotte for the microscope, Boyle and Papin for experiments with the air pump and respiration.[205] There were a great number of curious experiments and most secret qualities,[206] but occult qualities were said to be only material exhalations.

Yet only the year before the Polish Jesuit Adalbert Tylkowsky still held that natural phenomena could be explained by Aristotelian principles.[207] Several years later a Jesuit who had formerly taught

[202] *Ibid.*, pp. 320-21, 344, 280-81, 284-305, 304, 393.

[203] The Copernican theory has often, however, been represented as requiring a much greater distance of the fixed stars than did the Ptolemaic.

[204] *Ibid.*, pp. 396, 399, 401, 407. 414, 430, 432, 435.

[205] *Philosophia vetus et nova ad usum scholae accommodata. Editio altera multo auctior et emendatior,* Paris, 1681, 6 vols. in-12. JS IX (1681), 285-89, 302-5. For other works of Du Hamel see Chapter 29 on Other Exponents of Experimentation.

[206] JS IX, 288, "des qualités les plus cachées."

[207] *Philosophia curiosa seu universa*

at Lisbon and was now preacher to the queen of Great Britain, published a *Cursus philosophicus* on logic, physics and metaphysics according to Aristotelian and scholastic principles.[208]

The Free Philosophy[209] which Isac Cardoso, a Jewish physician and philosopher, dedicated to the doge and senate of Venice in 1673,[210] was not as progressive as that title might suggest, being free neither from many ancient notions nor from a magical point of view. It is a long double columned folio of 758 pages with 77 lines of fine type to the page, "in which everything that has to do with natural philosophy is methodically collected and accurately disputed." As a matter of fact, the sixth of its seven books deals with man, and the last with God.[211]

Cardoso still accepted four elements, despite the rejection of fire by Cardan and Tasso, and four qualities, although Cardan had held that cold was mere privation of heat. The four elements were true *principia*, simple bodies, not composed of matter and form, and not transmutable. But Cardoso further believed that the elements were composed of atoms or solid, indivisible, insensible and invisible corpuscles. Those of fire came closest to a spherical shape. He also talked of fiery spirits. The reputed first inventor of atoms was Moschus, a Phoenician who lived before the Trojan War, but he borrowed the idea from the ancient Hebrews.[212]

In connection with the element earth Cardoso considers the Copernican theory, which he represents as a resuscitation of the opinion of Philolaus and Aristarchus, first by Nicholas of Cusa and then by Nicholas Copernicus. Coelius Calcagninus embraced it, as did Gilbert, Stunica (Zuñiga), Origanus, Longomontanus, Ar-

---

Aristotelis philosophia juxta communes sententias exposita, Cracow, 1680, in-8. *Acta eruditorum* I (1682), 148-51.

208 Aug. Laurentius, *De triplici ente cursus philosophicus,* 3 vols. in-fol. *Acta eruditorum,* VI (1687), 527.

209 *Philosophia libera in septem libros distributa in quibus omnia quae ad philosophum naturalem spectant methodice colliguntur et accurate disputantur,* Venetiis, 1673, in-fol. BM 526.n.9.(2.). Col B 500 C 17.

210 "Serenissimo principi amplissimis sapientissimis reipublicae Venetae senatoribus."

211 The headings of the first five books are: I. De principiis rerum naturalium; II. De affectionibus rerum naturalium; III. De coelo et mundo; IV. De mixtis; V. De anima et viventibus.

212 *Philosophia libera,* pp. 5-14b, for the statements in this paragraph.

golus, Rheticus, Maestlin, Kepler, Galileo, Lansberg, Boulliau, Gassendi, Cesalpino, Basso, while Descartes inclined toward it. Cardoso incidentally repeats from Origanus the questionable statement that Virgilius was unfrocked by Boniface and pope Zacharias for affirming the existence of the Antipodes.[213] Despite the list of supporters of the Copernican theory, and twelve arguments for it which he repeats, Cardoso holds that the immobility of the earth is proved by sense, Scripture and reason. He gives thirteen arguments for this and answers the twelve of the Copernicans.[214]

Cardoso will not even admit that earth and water form a single globe, but holds that water is higher than earth, and that rivers come from the sea. The tides are "the greatest secret of nature." The fathers of Coimbra, Zanard, Lemnius, Scaliger, Gilbert and Kepler attribute them to magnetic force exerted by the moon; others, to its moist humors; yet others, to its occult influence. Cardoso favors the view that spirits are the immediate cause; and their Mover, the remoter cause. He further inclines to agree with those who say that men die only at the time of ebb tide, a circumstance which he has often observed. He passes on to marvels of waters, marvelous fountains, and an arboreal fountain in the Canaries.[215] As for air, it is not moist but dry. Cardoso, however, still discusses the question whether mountains reach the third region of the air. He quotes Pliny and Aristotle concerning the salamander extinguishing fire and the *pyraustae* born in the midst of fire, and Augustine as to perpetual sepulchral lamps. But he does not believe that any mixed or living substance can long survive or remain unconsumed in fire. The salamander may for a short time extinguish the near-by coals by its sticky humidity, but finally dies and is consumed. The ever-burning lamps are to be explained either as works of the devil, or as exhalations which take fire only when the tomb is opened and afterwards disappear.[216]

Cardoso also discusses the problem of two bodies in one place and one body in two places. On the subject of vacuum, he makes

[213] *Isis*, VI (1924), 369-70.

[214] *Philosophia libera*, 20b-28a.

[215] *Ibid.*, 39 and 54, 45b-51b, 65a-66b.

[216] *Ibid.*, 71b, 72b, 80a.

no reference to Torricelli's experiment but gives experiments against the existence of a vacuum including one shown to the kings of Poland in 1647. Another is that a vessel full of water and hermetically sealed, if the water freezes, which would leave a vacuum(!), breaks because nature abhors a vacuum. Cardoso adds, however, that all the books say this, but no one has tested it.[217]

With regard to falling bodies Cardoso is better informed, noting that both heavy and light increase in velocity—which he ascribes to the air—and that Galileo, followed by Mersenne, Gassendi and Riccioli, made the velocity of falling bodies increase in the proportion of odd numbers, 1, 3, 5, 7, 9.[218] Some ascribe the fall of bodies to magnetic force of the earth; Digby accounted for it by a multitude of descending atoms. If it is because of effluvia and mutual consent or sympathy between the falling bodies and the earth, Cardoso suggests that a stone dropped in a very deep shaft ought to adhere to the sides of the shaft rather than fall to the bottom.[219] Coming to the question whether all weights descend equally, he gives Arriaga's experiments favoring this and those of Riccioli to the contrary,[220] which were performed from the Asinelli tower at Bologna, which is 312 feet high. Of two clay balls of the same size and shape but different weights which were dropped from the tower at an altitude of 280 feet, the one weighing only ten ounces was always at least fifteen feet from the pavement when the twenty-ounce ball hit it and was smashed to bits. Riccioli performed the experiment twelve times, and some professors of philosophy, who were present and who had thought, like Galileo, Cabeo and Arriaga, that the balls would hit the pavement simultaneously, immediately abandoned this opinion.[221] Despite his professed indifference to Aristotle, Cardoso defends the Stagirite's explanation of the motion of projectiles as produced by the air against the impulsus or impetus theory, but says nothing as to their velocity increasing.[222] And after he has finished the discussion of motion, he continues

217 *Ibid.*, *lib.* II, *quaest.* 2 & 3.

218 *Ibid.*, 91a-92b.

219 *Ibid.*, 88a-89a.

220 Citing *Almagestum novum*, IX, iv, 16.

221 *Philosophia libera*, 93b.

222 *Ibid.*, 95a-b.

a Peripatetic order of treatment with chapters on time, the infinite, the composition of a continuum, whether quantity is to be distinguished from substance, alteration, the action of natural bodies, contact of agent with patient, resistance and reaction, and rarefaction.[223]

That the crowing of a cock terrifies a lion, and the grunting of a sow, an elephant, is because of the discrepancy and incommensurability of the corpuscles of sound[224] with the contexture of the organ of hearing, which is so irritated by them that great apprehension of disaster is aroused.[225]

In his third book, which bears the Aristotelian title, *De coelo et mundo,* Cardoso considers at what time of year creation occurred, declares that the world is incorruptible, and inclines to the opinion of the ancient Hebrew sages that the heavens are made of water. The comets of 1577 and 1618 showed that the heavens were fluid, not solid. For their number he prefers ten, in agreement with the curtains of the Tabernacle. The stars are not moved by Intelligences but move themselves, and the heavens are not animated. The celestial phenomenon of 1572 was a new star and not a comet.[226]

Ancient philosophy was content with the light and motion of the heavens and did not use the word, influence. "Now we take frequent refuge in these occult virtues as to a sacred anchorage." The majority of philosophers defend influences from the heavens and stars beyond their light and motion, and astronomers support this by numerous and varied experiments. "From the varied influence of the heavens come the varied condition of men and regions."[227] Yet Cardoso follows this with a chapter on the vanity of astrologers, in which he repeats the usual clichés against their art, all taken at second or third hand.[228]

Comets are not considered in connection with the heavens but in the fourth book on mixed bodies. Their nature is pronounced most obscure, but the Aristotelian explanation of them as terrestrial

[223] *Ibid., lib.* II, *quaest.* 12-20.

[224] This reminds one of the views of Oresme in the fourteenth century: T III, 427, 431-2.

[225] *Philosophia libera,* p. 114a.

[226] *Ibid., lib.* III, *quaest.* 4, 5, 7, 8, 10, 12, 18.

[227] *Ibid.,* II, 20, pp. 173b-174a.

[228] *Ibid.,* II, 21, pp. 176-189b.

exhalations is disproved by the argument from parallax. From all antiquity it has been accepted that comets are heralds of calamities and wars, but many learned men think that they are neither signs nor causes. Cardoso himself would prefer to regard them as divine signs.[229] Later, in discussing prodigious apparitions and strange rumblings and birds fighting in air, he classifies them as arcana of nature, exceeding the forces of nature, and produced by some directing Intelligence. For no one of sane mind would think them the work of chance or fortuitous events, since they always announce slaughter and war, and belong with presages.[230] Yet he attacked astrologers and denied that Intelligences moved the spheres!

Cardoso comes to the conclusion that gold cannot be made by the chemical art, although it works many other wonders. The philosophers' stone is ever sought and never found. But medical virtues are attributed to gems at considerable length. For example, fumigation with gagates removes epilepsy and virginity, expels snakes. It has the virtue of softening and dispersing. It is inflamed by fire, kindled by water, extinguished by oil. Drinking wine in which it has been quenched is beneficial for those with a heart condition.[231]

Marvels of plants and marvelous plants are both noted,[232] and the doctrine of signatures is accepted.[233] Spontaneous generation of plants and animals is admitted, but they are all from seed, whether conspicuous or latent, for seminal virtues are hidden in the earth itself and water.[234] Consideration of melancholy involves the common question whether it can attain a degree at which unknown languages are spoken and the future is predicted. The former point is answered negatively on the ground that no language comes naturally to man. In the other case, the answer is also, No, unless a demon does the predicting.[235] One can foretell from natural dreams only conjecturally and probably.[236] Monsters are

[229] *Ibid.*, IV, 10; pp. 210b, 214b-215a.

[230] *Ibid.*, IV, 15; p. 233a.

[231] *Ibid.*, pp. 262a, 264b, 265a-268, 263a.

[232] *Ibid.*, V, 7, De mirabilibus plantarum; p. 279a, "Sed percurremus aliquas plantas mirabiles."

[233] *Ibid.*, p. 278b.

[234] *Ibid.*, p. 281a.

[235] *Ibid.*, V, 17; pp. 299a-300b.

[236] *Ibid.*, V, 37; p. 357b.

discussed but with the apology that "our stupidity admires not the order but the error of nature."[237] Whether a human being can be generated otherwise than by the union of man and woman is argued at length but decided in the negative—neither chemically, nor by the stars, or demons, or woman alone.[238] The analogy of man to the universe or macrocosm is remarked. Indeed, there are said to be three worlds in man; angelic in the head, celestial in the heart, elemental in the belly. Meteors in man are mentioned without citing the work of Roderic de Castro, *De meteoris microcosmi* (1621).[239]

Physiognomy of the human body is recognized as a part of philosophy, and prediction from physiognomy is accepted, while that from melancholy is again rejected.[240] Natural divination, weather signs, and medical prognostication are distinguished from diabolical and superstitious divination and from prophecy.[241] The number of the month of birth is dwelt upon at some length.[242] It is the order of nature and not any virtue in the number that causes the birth of the child in the seventh month. But Cardoso goes on to discuss seven as a sacred number and to state that males are more often born in the seventh month than females. He rejects the old notion that the eighth month's child dies and its astrological explanation.[243] Passing to the ages of man,[244] he cites Montanus for the sixty-third and seventieth years as terms of life but without saying anything of climacteric years. And in treating the number of the ages of man he gives other estimates as well as seven ages.

J. E. Schweling, whose inaugural oration as ordinary professor of "physics" at Bremen is dated November 3, 1670, six years later published a text of Physical Principles,[245] which developed from aphorisms which he had dictated privately.[246] The work is a somewhat incongruous combination of the old and the new. Beginning

[237] *Ibid.*, VI, 18; p. 473a.

[238] *Ibid.*, pp. 482-489a.

[239] *Ibid.*, VI, 71; pp. 576a-578a.

[240] *Ibid.*, pp. 587b-595b, 626a-b.

[241] *Ibid.*, 622b-625b, 627a-, 636a-.

[242] *Ibid.*, VI, 95-99.

[243] *Ibid.*, pp. 679a, 681b, 685a-687a.

[244] *Ibid.*, VI, 102-103.

[245] J. E. Schweling, *Principiorum physicorum libri tres eiusdemque oratio inauguralis et dissertiuncula theologico-physica de bismortuis*, Bremae, 1676. BM 536.d.22 (1.). The three works indicated cover pp. 1-134, 135-76, and 177-89 respectively.

[246] So he says in the preface.

with a discussion of method and occasions of error in Baconian style, we revert to Aristotle with matter and form, motion and quiet, figure, site and magnitude. But in the chapter on efficient causes are given the same three universal laws of motion as we heard Gadroys take from Descartes. He adds that in the case of percussion between two bodies, the one with less force continues to move but alters the direction of its movement; the one with greater force moves the other with it and loses only as much of its motion as it gives to the other. The movement of falling bodies is slower at the start, faster at the end; the opposite is true in ascent.

After chapters on hardness and fluidity, rarity or porosity and density or solidity, asperity and smoothness, on space and vacuum, place and time, and on fate, chance and fortune, we come to the four Medicean stars or moons of Jupiter and the two satellites of Saturn. Orbs, spheres and epicycles have been abandoned for Cartesian vortices, three elements, and inference from sunspots as to the apparition and disappearance of stars. The earth was once a star which fell to the place which it still occupies today.

There are three kinds of sublunar particles distinguished by their shape or figure. A great part of the waters is under the earth's crust which has hardened but which, it is probable, has undergone various subsidences both when the world began and since. Springs come from the sea through vapors rising from subterranean reservoirs. The pores through which the vapors rise are too small for the water into which they turn to descend again, so that it has to issue outside in springs. The moon is the cause of tides. The three chemical principles are given. Fire on earth is a congeries of terrestrial particles and is generated in three ways: by motion, light, and the mixture of two cold bodies. Mineral baths and earthquakes receive consideration, as do tears of glass (Prince Rupert's drops).

Then come the nature of iron, steel, the magnet, and their properties, and electric power and the causes of stupendous effects, which are attributed, as they had been by Descartes, to particles of the first element in the pores of bodies. The ignorant crowd ascribes them to magic incantations, specters, or the delusions of demons. Here he puts sympathies and antipathies and what they say of the sympathetic powder. Schweling then passes on from the external world of nature to man's external and internal senses.

The instruction given in natural philosophy at the University of Leyden in the academic year 1679–1680 is shown by a compendium based upon his lectures then which Wolferd Senguerd (1646–1724), Ordinary Professor of Philosophy in that institution, dedicated on August 1, 1680, to the curators of the same, and which was printed in 1681.[247] He says that he had been aware that his method of philosophy was regarded by some as scholastic and quite remote from reason and experience. In the present volume, however, he cites no authorities and includes numerous experiments, with about fifty diagrams and pictures, chiefly astronomical and physical. For the sake of conciseness he has not included all the reasonings and experiments which were adduced in his public lectures. His teaching in some respects is reminiscent of Descartes, but especially avails itself of the notion of ferments of contemporary chemists and the subtle particles or corpuscles of contemporary physicists. Zedler represents Senguerd as also librarian of the university. He had earlier published a treatise on the tarantula.[248] Later he printed experimental investigations concerned with air and accompanied by a record of the weather for 1697–98.[249] But in later life he turned from natural philosophy to the law.[250]

Ferment is defined by Senguerd as an irregular and intestine motion of the parts of a body, more rapid than usual, produced by the impeded transit of most subtle particles within that body. This is facilitated by heat and by the pores being impervious, so that the very fine particles cannot escape but keep colliding. Wine

[247] *Wolferdi Senguerdii . . . Philosophia naturalis . . .*, Lugd. Batav. 1681, in-4, 302 pp. and Index: copy used, Col. 194 D45 I3, where it is bound between a 1687 edition of *Opera* of Descartes and a Basel, 1682 edition of Emanuel König, *Regnum animale.*

[248] BN S.8721 is the Leyden, 1667 edition in-4; BN S.13122 is Leyden, 1668 in-12. I consider it in Chapter 24 on Natural History.

[249] The edition of Leyden, 1699, is described as "Editio secunda, priore plusquam altera parte auctior." BN R.4651. Zedler dates the work in 1690. Zedler lists a *Rationis & experientiae connubium*, Rotterdam, 1715 in-8.

[250] He is given as *Praeses* of several legal dissertations in the Columbia University Law Library.

and beer ferment more readily in hot than cold weather; plants grow, and fevers and pestilences flourish in summer rather than winter. When the very subtle particles make their way through the pores, fermentation ceases. But while they are in motion and colliding, they tend to divide bodies and are aided in this by crasser particles.[251] The operations and effects of fermentation are most varied, yet depend entirely on the movement of exceedingly subtle particles and the diversity of the bodies upon which they act.[252] Very subtle particles are also the cause of the movement of fluids.[253] Light is explained, as by Descartes, in terms of round particles.[254] The nature of colors consists in the disposition of the parts of bodies; black bodies are more porous and so reflect light less than white ones do.[255] Newton had communicated his discovery and experimental verification of the composite nature of sunlight to the Royal Society eight years before in 1672. But Senguerd remains unaffected by, and probably ignorant of it.

After treating of matter and form, Senguerd leaves material substantial forms aside as unknown in physics, but admits the existence of accidental forms. God is the cause of motion, and it is probable that the same quantity of motion is conserved continually in the universe. Senguerd rejects all attempts to explain gravity in terms of subtle matter, particles or effluvia, or of magnetic force, and holds that all bodies are moved because of an *impetus* inherent in and impressed on them by divine action.[256] In this first part of his book he contends that the weight of the mercury or other liquid and not the pressure of the air is the cause of its ascent and descent in the curved tube,[257] but in the third part and first chapter on the weight of the air he states that the cylinder of mercury "is of the same weight as the cylinder or cone of air which supports it."[258] Already earlier he has admitted that a vacuum is possible.[259]

In the briefer second part, devoted primarily to the heavenly bodies, the system of Tycho Brahe is still preferred to either the

[251] *Philos. nat.*, 1681, pp. 68-73 (I, 10).

[252] *Ibid.*, p. 269.

[253] *Ibid.*, p. 67.

[254] *Ibid.*, p. 81, "Tales rotundae & exiles particulae ad lucem & lumen constituendum aptissimae sunt."

[255] *Ibid.*, p. 87 *et seq.* (I, 16).

[256] *Ibid.*, pp. 20-21, 31, 56-59.

[257] *Ibid.*, pp. 63-64.

[258] *Ibid.*, p. 175.

[259] *Ibid.*, p. 109.

Ptolemaic or the Copernican. New stars are explained as having been before invisible because covered with spots like those on the sun. The spots on the moon are not shadows cast by mountains there but are spongy and non-reflecting parts. Comets are discussed at some length and pronounced supralunar.

In part three on lifeless bodies the elasticity of the air is recognized, and land and water are spoken of as forming one globe.[260] It is still believed that water is brought up from the sea to springs far above sea level by capillary attraction or by being converted into vapor by subterranean fires.[261] Sea water is represented as tasteless *ex se* and its saltiness due to rigid oblong saline particles with which it is impregnated and which can be seen after it has been allowed to evaporate.[262] The moon is rejected as the cause of tides, which Senguerd is inclined to think may be excited by rivers emptying into the sea.[263] He tries to find an explanation of inextinguishable lamps found in sepulchers in subtlety and purity of fuel and lack of agitation of the air.[264] The action of the magnet is accounted for by pores and particles.[265] Nothing is said of the virtues of gems. The usual seven metals are enumerated, and their transmutation is declared possible, but alchemists disagree as to the method to follow.[266]

The fourth and concluding part on living beings is the briefest of all. Vegetation requires, in addition to the varied motion of subtle particles, larger aerial particles which, impelled by the subtle matter, tear off the larger terrestrial corpuscles and with themselves bear them to the plant and its sprout, lodge themselves in its pores and produce a ferment, just as for the conservation of fire, besides subtle particles, thicker air is required bearing to the fire the matter which serves to keep it going.[267] Put in other words, "It is evident that for the production of plants, corpuscles of a special sort and arrangement, or a ferment, is required which by controlling the motion of the subtle and other particles, brings it about that plants acquire this or that specific nature."[268] Some-

[260] Cap. 2, De vi elastica aeris; 8, De globo terra-aqueo.

[261] *Ibid.*, p. 224.

[262] *Ibid.*, p. 229.

[263] *Ibid.*, cap. xi, pp. 231-39.

[264] *Ibid.*, p. 246.

[265] Cap. xiv.

[266] Cap. xv.

[267] *Ibid.*, p. 270.

[268] *Ibid.*, p. 271.

times a ferment of this kind exists in the soil or parts of other bodies, and then plants of varied sorts are unexpectedly produced from earth, much as animals are spontaneously generated. In grafting, the pores of the tree into which the shoot is inserted must not be of a different shape.[269]

Senguerd still holds that breathing serves to cool the blood. On the other hand, he states that aquatic animals breathe as well as those in the air and on land. Conversely, his acceptation of the corpuscular theory has not led him to abandon the belief in material spirits within the body. When the heart is contracted by the flow of spirits and distributes the blood through the body, the more subtle and spirituous portion of the blood seeks the brain or cerebellum and there constitutes new spirits. These serve the motion and sensation of animals by penetrating the nerves. Sleep occurs when there is an insufficient influx of spirits into the nerves; we wake again as soon as enough new spirits have been generated in the brain. Senguerd distinguishes only three internal senses: common sense, memory and phantasy.[270] König in 1682 distinguished the same three, and also spoke of the spirits as refreshed by sleep.[271]

The text is followed by an Index occupying a score or so of unnumbered pages from which such subjects are absent as alchemy (but *Transmutatio metallorum* appears), astrology, divination, fascination and magic (but Occult Qualities are treated at p. 79).

The book was reprinted in 1685 and 1687, but obviously would not long survive the appearance of Newton's *Principia* in the latter year.

Despite the four volume *Physica* which his fellow Jesuit, Honoratus Fabri, had published in 1669–1671, Franciscus Tertius de Lanis, or Francesco Lana Terzi, of the Society of Jesus, in his three volume *Magisterium naturae et artis* of 1684–1692[272] asserted that,

[269] *Ibid.*, p. 278.

[270] *Ibid.*, pp. 289, 294, 297, 302 and 209, in that order for the statements in this paragraph.

[271] Em. König, *Regnum animale*, 1682, Artic. ix, p. 30; Artic. xi, pp. 33-34.

[272] *Magisterium naturae et artis, opus physico-mathematicum, in quo occultiora naturalis philosophiae principia manifestantur et multiplici tum experimentorum tum demonstrationum serie comprobantur, ac demum tam antiqua paene omnia artis inventa quam multa nova ab ipso auctore excogitata in lucem proferuntur.* Brixiae, per Johannem Mariam Ricciardum, 1684-1692, 3 vols. in-fol.

while there had been many treatises in physics and mathematics recently, there had been no complete treatment of all philosophy. He hardly attained this goal himself, since his three volumes were limited to the subjects of body, local motion, and various motions of natural bodies. In both the preface to the reader and the full title of the work he claims to have thought out much by himself and to follow no one authority. He followed the geometric method which Descartes had made prevalent and further professed to be guided by experiment and observation. He charged Aristotle with having held that the heavens were solid as well as that comets were terrestrial exhalations, but he gladly admitted that forms were latent in matter and the existence of the four elements of the Peripatetics. He also, however, accepted the three Hermetic principles, and such categories as fixed and volatile, alkali and acid. He did not deny magnetism or sympathy and antipathy in nature. In our chapter 20 on Schott and Artificial Magic we have already noted the "Inventions and Artifices" which constitute a prominent feature of his work.

At the close of the seventeenth century, Nicolas Hartsoeker, who back in 1678 had shown Huygens spermatozoids through a microscope,[273] in his *Essay de dioptrique*,[274] *Principes de physique*,[275] and *Conjectures physiques*[276] tried to outsimplify Descartes by reducing everything in the physical universe to only two elements: one liquid and ever in movement, the other composed of hard solid corpuscles of different shapes and sizes which swam about in the first element and never touched one another.[277] Affirming that nature produced that infinite variety of effects which we admire by a few simple and uniform laws, Hartsoeker attemped to deduce the majority of these effects from a single fire burning at the center of a vast atmosphere.[278] Since it was impossible to determine the shape, size and arrangement of the corpuscles, it was

[273] Huygens, *Oeuvres*, XXII, 702.

[274] Paris, 1694, 233 pp. in-4: BN R.4315.

[275] A Paris chez Jean Anisson, 1696.

[276] A Amsterdam chez Henri Desbordes, 1698. In Col 530 H25, it is bound with the *Principes* and the subsequent *Suite des conjectures physiques*, 1708, and *Eclaircissements sur les conjectures physiques*, 1710.

[277] *Principes*, cap. i.

[278] *Conjectures*, Avertissement, pages unnumbered.

necessay to divine (*sic*) them from their effects, just as it was the existence of demons, although Hartsoeker of course does not say this. For example, the corpuscles of acids must be shaped like needles pointed at both ends, while alkalis consist of hollow cylinders in which the needles may lodge.[279] Thus, although Hartsoeker's subject-matter is largely restricted to physical phenomena, his method and his explanation of these, if not magical and divinatory, is almost entirely guess-work. Yet he criticized Descartes for professing to start with only matter and motion, and then boldly assuming "une infinité des choses."—"corpuscles soft as paste or hard as steel, stiff as needles or flexible as eels."[280] In a sense the physics of both these men crowded magic off the stage by its own ingenious, if far-fetched, explanations of the marvels of nature, its own pulling rabbits out of hats. Hartsoeker was made a member of l'Académie des Sciences in 1699.[281]

The *Essay de dioptrique* by no means confined itself to that subject, opening with two chapters on the aforesaid two elements and closing with a discussion of comets, new stars, microscopes and the generation of animals. Although in it Hartsoeker speaks in terms of *tourbillons* and vortices, and explains the fluidity of mercury from its being composed of heavy and well polished little balls which cannot be stopped, he rejects the Cartesian as well as Aristotelian explanation of comets, and declares that a comet is nothing but a sphere which comes forth burning and smoking from the sun. If rays of light are the sole cause of the movement of the planets, we can discover the density and weight of the matter in which each planet revolves, supposing that we know their size and distance from the sun.[282] Rays of light differ in thickness and in velocity. Red color comes from the thickest and swiftest, yellow next, then white, blue and violet.[283]

[279] *Conjectures*, p. 102.

[280] *Conjectures*, Avertissement.

[281] Huygens, *Oeuvres*, XXII, 706, n .14.

[282] *Essay de dioptrique*, pp. 9, 15, 200-202, 14.

[283] *Ibid.*, pp. 32-34.

Printed in the USA
CPSIA information can be obtained
at www.ICGtesting.com
JSHW021320221024
72173JS00011B/1619